Nanoclusters

Nanoclusters

A Bridge across Disciplines

Purusottam Jena
Department of Physics
Virginia Commonwealth University
Richmond, Virginia, USA

A. Welford Castleman, Jr.
Department of Chemistry
Pennsylvania State University
University Park, Pennsylvania, USA

ELSEVIER
AMSTERDAM • BOSTON • HEIDELBERG • LONDON
NEW YORK • OXFORD • PARIS • SAN DIEGO
SAN FRANCISCO • SINGAPORE • SYDNEY • TOKYO

Elsevier
Radarweg 29, PO Box 211, 1000 AE Amsterdam, The Netherlands
Linacre House, Jordan Hill, Oxford OX2 8DP, UK

Library of Congress Cataloging-in-Publication Data
A catalog record for this book is available from the Library of Congress

British Library Cataloguing-in-Publication Data
A catalogue record for this book is available from the British Library

ISBN: 978-0-444-53440-8
ISSN: 1875-4023

Printed and bound in Great Britain
10 11 12   10 9 8 7 6 5 4 3 2 1

Contents

Preface

The field of cluster science has gone through a number of reincarnations, evolving from the fields of fine particles, atmospheric and aerosol research. Subsequently, with the advent of mass spectrometry and interest in molecular beams, clusters were first considered an annoyance, often contaminating mass peaks, making it difficult to resolve and identify species. Soon it was recognized that their study might shed light on phenomena such as nucleation, thus providing an entry into the study of solvation complexes, interactions, and the evolution of condensed matter. It was soon realized that clusters are "condensed matter with a surface." While interesting comparisons could also be made to liquids, they revealed properties in their own right.

In recent years, many more topics are being investigated. Among these are connections between nanoscale science and cluster science. This aspect has been particularly revealing as it has formed the basis for fundamental investigation of nanosystems since clusters are the ultimate nanoparticles whose size and composition can be controlled with atomic precision. The connection between cluster science and nanoscience first came through the discovery of the C_{60} fullerene. The jellium concept devised following the observation of magic numbers in metallic systems led the way to connections between clusters and nuclear physics as well as between clusters and materials. The former related the nuclear shell closure, which gave rise to the stability of magic nuclei, to that of the electronic shell closure which gave rise to enhanced stability of clusters composed of simple metals. This aspect of the field dramatically expanded in recent years when it was realized that the size, composition, and charge state of simple metal clusters can be tailored to design and synthesize other metal clusters with closed electronic shells that are not only stable but also chemically inert. Clusters can thus be formed which mimic the properties of elements in the periodic table. These clusters, commonly referred to as superatoms, can not only form the basis for a new three-dimensional periodic table with superatoms forming the third dimension, but can also serve as building blocks of an entirely new class of materials. These cluster-assembled materials can open a new frontier in materials science, further linking cluster science to materials science. As the field evolves, it grows in scope and encompasses many other topics, including catalysis, new issues in atmospheric chemistry, electronic materials, biomedicine, and even surface chemistry and reaction dynamics.

Expanding interest in the general subject matter of cluster science and how it relates to problems in many disciplines, such as physics, chemistry, materials science, atmospheric science, biology, and medicine, prompted the editors of this book to organize a special feature issue of the Proceedings of the National Academy of Sciences (PNAS) devoted to "Cluster Chemistry and Dynamics" and provide an overview of many aspects of the subject with a title "Clusters: A Bridge Across Disciplines." The special PNAS volume featured three perspectives devoted to biology, physics, chemistry, materials science, and the environment. Due to space limitations, only limited aspects of cluster science were presented in six other chapters. Realization of the significance of the field and its bearing on so many areas of research led to a meeting at Jekyll Island, Georgia in 2006 which brought together many of the experts in the field, and the idea for this book was born.

The collection of authoritative chapters in this book provides insight into topics that are currently at the cutting edge of cluster science, mainly focusing on metal and metal-compound systems that are of particular interest in materials science, and also on aspects related to biology and medicine. A major theme is demonstrating the fact that in the world of clusters, every atom "counts" and plays a role in governing behavior.

While there are numerous books on clusters, the focus on these subjects via the inclusion of 15 in-depth overviews sets this one apart from others. Here, leading researchers discuss the role clusters play in understanding properties of their bulk, their novel size-specific properties, and the bridge they have built across disciplines. In Chapter 1, Jena and Castleman provide an introduction to clusters, with brief discussions of their history, unique structure, and properties. In Chapter 2, Mahanti and coworkers focus on the use of clusters as models of bulk materials and provide a critical examination of how the properties of clusters characterized by metallic, covalent, and ionic bonding evolve. Schreicher and Das describe the use of clusters as models of biological systems in Chapter 3. In Chapter 4, Salahub and coworkers review the techniques and methods for the structure determination of clusters, with an emphasis on how the synergy between experiment and theory can lead to a reliable and accurate description of the electronic and geometrical structure of clusters. In Chapter 5, Zubarev and Boldyrev provide a description of novel electronic structure of clusters that exhibit of σ-, π-, and δ-aromaticity, antiaromaticity, and conflicting aromaticity in small- and medium-sized planar clusters composed of carbon, boron, borocarbon, and aluminum clusters. In Chapter 6, Armentrout discusses the reactions of cluster cations of several transition metals (V, Cr, Fe, Co, and Ni) with D_2, O_2, N_2, CO_2, CD_4, and ND_3. The work illustrates the relationship between the thermochemistry with that for adsorbates on bulk phase metal surfaces. Concentrating on hydrogen, Alonso and coworkers describe, in Chapter 7, how clusters can be relevant for exploring nuclear fusion, clean energy, and catalysis. In Chapter 8, Compton and coworkers deal with the formation of crystals in saturated solutions as a result of intense sound, or compression waves, generated in the solution by intense pulsed lasers. In Chapter 9, Reber, Khanna, and

Castleman describe the design of clusters as superatoms and properties of materials where superatoms form the building blocks. In Chapter 10, using silica (SiO_2) molecules as building blocks, Bromely discusses the design of a variety of distinct topological forms (e.g., nanopores, nanospheres, nanotubes) leading to promising new applications of silica in photonics and biotechnology. In Chapter 11, Knickelbein describes Stern-Gerlach molecular beam deflection studies of bare and adsorbate-covered transition metal clusters to study the emergence of novel magnetic behavior atom-by-atom, in the size range that bridges atomic and bulk properties. The influence of atomic and molecular adsorbates on the magnetism of small ferromagnetic clusters is also discussed. Manninen and Reimann discuss in Chapter 12 the properties of metal clusters, quantum dots, and trapped atoms. The electronic structure of quantal systems confined to two- and three-dimensional systems as well as the effect of the magnetic field is also covered. Chapters 13–15 describe how clusters are helping to bridge our understanding between physical and biological sciences. Koutecky and coworkers present in Chapter 13 the results of a theoretical investigation into the structural, optical, reactivity, and dynamical properties of free clusters and clusters interacting with different environments such as surfaces and isolated biomolecules. The role of the interface on the optical properties and reactivity is highlighted. In Chapter 14, Leung and Palmer discuss the immobilization of different proteins by size-selected gold clusters. The relevance of these studies in several other modern fields of biology and biotechnology is highlighted. Finally, Degtyarenko and Nieminen review the origin of the complexity in the structure and chemical reactivity of biomolecular systems such as L-alanine amino acid zwitterion in aqueous solution and the computational techniques based on explicit cluster models.

The editors gratefully acknowledge the various authors who have contributed chapters to this book as well as Dr. Anil Kandalam and Dr. Qiang Sun for their help during the preparation of this book.

Purusottam Jena
A. Wilford Castleman, Jr.

Chapter 1

Introduction to Atomic Clusters

P. Jena* and A. W. Castleman, Jr.[†]
*Department of Physics, Virginia Commonwealth University, Richmond, Virginia, USA
[†]Department of Chemistry, Pennsylvania State University, University Park, Pennsylvania, USA

Chapter Outline Head

A *cluster* is defined by the American heritage dictionary as "a group of same or similar elements gathered together." Consequently, clusters have different meanings depending on the "elements" of which they are composed. A few common examples include cluster cereals, cluster bombs, cluster headache, computer clusters, musical clusters, and clusters of stars and galaxies. However, in the physics and chemistry communities, the term "clusters" is typically used to describe an aggregate of atoms or molecules. Clusters can be formed when a hot plume of atoms or molecules in a gas are cooled by collision with rare-gas atoms much as droplets of water are formed when hot steam cools and condenses. Clusters composed of a finite number of atoms and molecules are an embryonic form of matter and have become a robust field of research in the last four decades.

Nanoclusters. DOI: 10.1016/S1875-4023(10)01001-6

Molecules and nanoparticles also represent an aggregate of atoms as do clusters. For example, molecules can consist of as few as two atoms, that is, H_2, to as many as a few thousand atoms, for example, proteins. In contrast, nanoparticles may consist of hundreds of thousands of atoms. In the early stage of development of these fields, nanoparticles were large, typically of the order of 10–100 nm, and clusters were small, typically less than 1 nm. With the progress in synthesis techniques, these size differences have now narrowed: clusters as large as a few thousand atoms or molecules and nanoparticles as small as 1–2 nm can now be produced. What then differentiates a cluster from a molecule or a nanoparticle?

To distinguish clusters from molecules, we provide in Table 1 a summary of some of their properties. As pointed out before, both clusters and molecules are aggregates of atoms and may contain as few as two atoms to as many as thousands of atoms. However, molecules such as H_2, O_2, and N_2 exist in nature under ambient pressures and temperatures, while clusters are made in the laboratory under vacuum or cold flow conditions. Unlike molecules that interact weakly with each other, clusters, in general, interact more strongly and often coalesce to form larger clusters. The size and composition of clusters can be varied easily whereas the composition of molecules is fixed by nature. A given cluster can exhibit numerous isomers where the atoms are arranged in different geometric patterns. The atomic structures of molecules, on the other hand, have specific geometries and only rarely exhibit isomeric forms. The electronic bond between atoms in a molecule is primarily covalent where atoms forming the bonds share their electrons. Clusters, on the other hand, show a variety of bonding schemes starting with weak van der Waals to metallic and strong covalent or ionic bonds. Molecules are abundant in nature whereas clusters need to be formed under special experimental

TABLE 1 Clusters Versus Molecules

Molecules	Clusters
Available in nature	Synthesized in a laboratory
Stable in ambient environment	Atomic clusters are stable only in vacuum or in inert environment
Weakly interact with each other	Interaction varies from weak to strong
Size and composition are fixed	Size and composition can be varied
Typically very few isomers	Numerous isomers
Primarily covalent or ionic bonding	Bonding can be weak van der Waals, metallic, ionic, or covalent
Stable against coalescence	Metastable and coalesce

conditions and their stability varies widely depending upon their size and composition. Thus, molecules are different from clusters. One exception may be C_{60}, which, although discovered as a cluster, has most of the properties of a molecule.

To distinguish between clusters and nanoparticles, we note that the size and composition of clusters can be controlled one atom at a time while in general the number of atoms in a nanoparticle cannot be determined with the same precision. Thus, clusters are the ultimate nanoparticles where the size and composition are known with atomic precision and the evolution of their properties can be studied one atom at a time. In Figure 1, we show a schematic plot of how a given property, be it the interatomic distance or electronic, magnetic, and optical property, varies as a function of size [1]. In clusters consisting of a few atoms, the properties change nonmonotonically, often varying widely with the addition of a single atom. As the cluster size reaches a few hundred to a few thousand atoms, the variations of properties with size become less drastic, and eventually the properties smoothly approach the bulk value. The fields of clusters and nanoparticles have been developing over the years in a parallel way. As clusters became large and nanoparticles became small, the distinctions between the two fields have narrowed and consequently clusters are often referred to as nanoclusters.

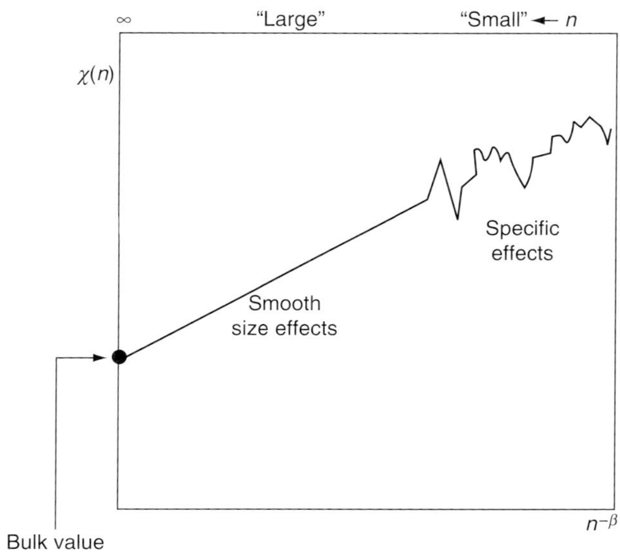

FIGURE 1 The cluster size dependence of a cluster property $\chi(n)$ on the number n of the cluster constituents. The data are plotted versus $n^{-\beta}$ where $0 \leq \beta \leq 1$. "Small" clusters reveal specific size effects, while "large" clusters are expected to exhibit for many properties a "smooth" dependence of $\chi(n)$ which converges for $n \to \infty$ to the bulk value $\chi(\infty)$ (see Ref. [1]).

Thus, nanoclusters can provide complimentary understanding of properties in nanoparticles and in some bulk materials.

I. A BRIEF HISTORY

The history of atomic and molecular clusters dates back to very early times. For example, it has been suggested that in the creation of the universe, very stable clusters such as C_{60} may have been formed [2]. Some of the unidentified infrared bands in interstellar matter are attributed to metal–organic clusters [3]. Similar examples of clusters in nature may be found in biology; ferritin is a shell of proteins that surround an Fe core of up to 4500 atoms [4]. Reference to the formation of aggregates and related nucleation phenomena in smoke and aerosols can be found in the literature [5] dating from the 1930s and earlier. Clusters were also used as models to study properties of extended systems [6–10] such as crystals and proteins by replacing these systems with a few atoms confined to the geometry of their bulk counterpart. This is particularly helpful in studying defects in crystals since carrying out band structure calculations without periodic boundary conditions was not possible due to limited computing power. Here, one assumed that the properties of defects are governed primarily by their interaction with a few neighboring atoms and a finite cluster where the atoms occupied the positions given by their parent crystal structure serves as a good model. In semiconducting or ionic systems, the dangling bonds of the atoms were saturated by hydrogen while in metals this was not necessary due to delocalized nature of the conduction electrons. How large a cluster has to be to account for the defect properties in the bulk remained as a nagging question which could only be solved by increasing the cluster size until the properties converged.

However, the origin of clusters as we know it today can be traced to the first set of experiments [11] in mass spectrometer ion sources in the 1950s and 1960s when intense molecular beams at low temperatures were used to produce clusters by supersonic expansion. Most of early work on clusters involved molecular clusters, clusters of inert gas atoms, and of low-melting-point metals. With the advent of laser vaporization techniques [12], clusters of a vast majority of the elements in the periodic table can now be produced. Since the 1980s, we have witnessed work on clusters of transition and refractory metals as well as semiconductor elements and compound clusters consisting of binary and ternary elements. The early theoretical works were mostly phenomenological in nature [13,14] and first-principles calculations dealt with very small number of atoms or molecules [15–17]. With advancement in computer technology and development of efficient computer codes based on density functional theory, one is now able to model clusters containing as many as a thousand atoms.

The ability to synthesize and characterize clusters consisting of up to a few thousand atoms has given birth to a new field that forms a bridge not only between atoms, molecules, nanoparticles, and bulk matter but also between the disciplines of physics, chemistry, materials science, biology, medicine, and environmental science. The limited size and tunable composition allow clusters to have unusual combinations of physical and chemical properties. Metallic elements can become insulating, semiconductor elements can become metallic, nonmagnetic materials can become magnetic, opaque materials can become transparent, and inert materials can become reactive. It has also been suggested that clusters can be designed and synthesized by varying their size and composition such that they mimic the electronic properties of atoms [18]. These clusters, originally termed unified atoms [19a], are now commonly referred to as superatoms [18,19b]; they can form the basis of a new three-dimensional periodic table with superatoms constituting the third dimension [18,19b,c]. A new class of cluster-assembled materials where clusters instead of atoms form the building blocks can usher an exciting era in materials science with unlimited possibilities for new materials.

A first step in realizing this lofty goal is to understand how the properties of clusters evolve with size and composition and when they mimic properties of their corresponding bulk matter. How large does a cluster have to be before it can resemble a crystal? When does a metal become a metal? Is the evolution of the properties monotonic or does it vary widely with size? It was expected that by systematically studying the structure and properties of this new phase of matter as a function of size, one atom at a time, one can finally answer these fundamental questions. While much work has been done to achieve this understanding [20–23], studies of clusters have raised more questions than answers. Different properties evolve differently, and in most cases, the limiting value is not reached even for the largest clusters studied thus far. The field of atomic and molecular clusters has become a new and growing field of research in its own right. This book describes some of the unique properties of clusters and how they have helped to bridge our understanding in many disciplines.

II. ATOMIC STRUCTURE OF CLUSTERS

Crystals exhibit 14 different lattice symmetries. Among these, body-centered cubic (bcc), face-centered cubic (fcc), and hexagonal close-packed (hcp) structures are among the most prevalent ones [24]. The alkali metals such as Na, for example, form the bcc structure, while alkaline-earth elements such as Mg form the hcp structure. The coinage metals such as Cu, on the other hand, form the fcc structure. An understanding of how these structures evolve and how many atoms are needed for the crystal structure to emerge has been a fundamental question. Studies of the geometries of clusters as a function of size are expected to illustrate this point. However, with existing experimental techniques it is difficult to unambiguously determine cluster structure. Many clusters are too large for precise study by most spectroscopic techniques and often too small for diffraction techniques. Determination of cluster geometries is now possible through a synergy between theory and experiment [25–29]. First-principles theory and well-developed computer codes allow researchers to determine the geometry of the clusters, their isomers, and relative stability up to a hundred atoms. Calculated electronic and vibrational properties of these clusters can be compared with experiments and a good agreement can provide a level of confidence on the theoretically determined structures. The experimental techniques that are frequently used for this comparison are photoelectron spectroscopy (PES) [25–30], trapped ion electron diffraction (TIED) [31,32], ion mobility [33,34], and infrared spectroscopy [35]. We should note that there are many isomers of a given cluster and often the energy differences between low lying isomers are within the accuracy of theoretical methods [27]. Thus, it is again difficult to predict with absolute certainty the ground-state geometry of a cluster. To make things more complicated, it is not always true that experimentally one observes the ground-state structure. Higher energy isomers with a large catchment area in the potential energy surface or having a spin multiplicity that differs from its ground-state spin may be present [36]. In spite of these difficulties, considerable progress has been made and one has a reasonable understanding how the structures evolve.

To demonstrate this evolution, we plot in Figures 2–7 geometries of clusters of nearly free-electron metals such as Na and Be, noble metal Au, transition metal Ni, and semiconductor C and Si. We also discuss structures of compound clusters.

A. Alkali Metal Clusters

We see from Figure 2 that small Na clusters [37] are planar; they assume three-dimensional structures when containing more than five atoms, and exhibit fivefold symmetry in clusters containing as few as six atoms. Note that in crystals fivefold symmetry is forbidden due to space-filling requirements, but this requirement does not hold for clusters. We should recall that quasi-crystals which are mostly made of metals exhibit fivefold symmetry, and in this context, metal clusters and quasi-crystals have something in common. Even for the largest Na cluster studied, the structure does not mimic the bcc crystal structure. However, the average interatomic distance in a cluster rapidly converges and approaches the bulk nearest neighbor distance to within 10% when clusters contain as few as 10 atoms. Although most of these atoms are surface atoms, their arrangements seldom bear any resemblance to crystalline surfaces.

B. Alkaline-Earth Metal Clusters

Alkaline-earth metals such as Be, Mg, Zn, and Cd have a closed ns^2 outer shell configuration. Hence these atoms interact weakly with each other until cluster size increases and the s and p orbitals begin to overlap. The structures

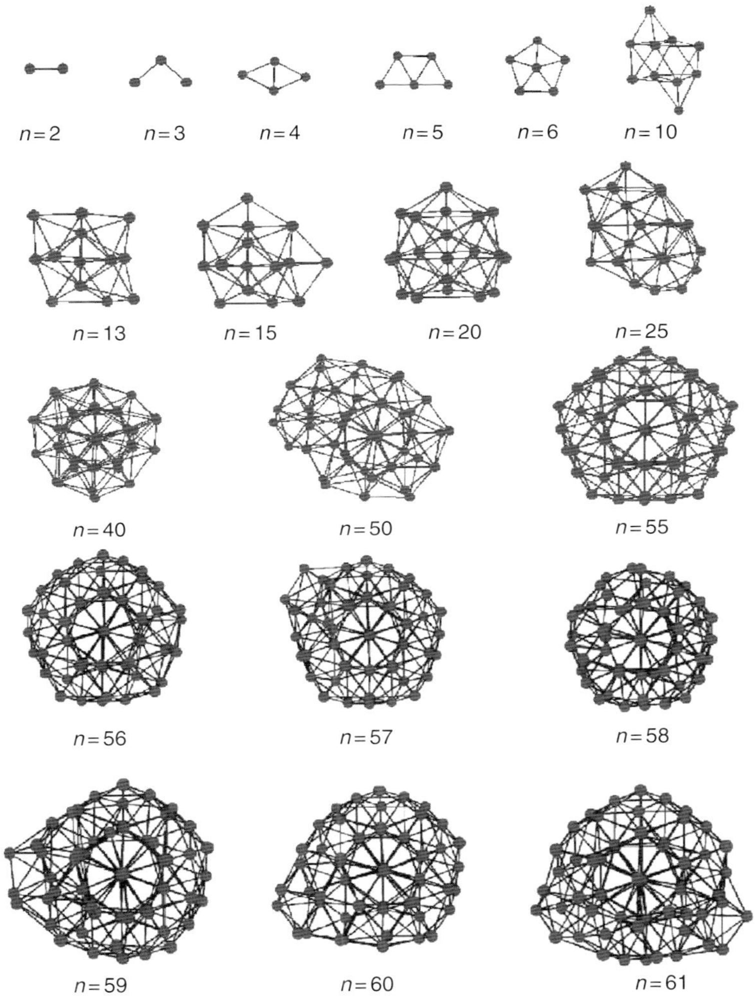

FIGURE 2 Geometries of Na clusters (see Ref. [37], courtesy of Dr M.S. Lee).

of these clusters, therefore, are expected to assume three-dimensional closed packed geometries with as few as four atoms. This is indeed the case. In Figure 3 we show the structures of Be clusters [38]. Note that for very small clusters the structures are compact and follow hard-sphere packing rules. Icosahedric structures with fivefold symmetry do not appear until the clusters contain about 13 atoms. In contrast, alkali clusters exhibit fivefold symmetry in clusters containing as few as six atoms.

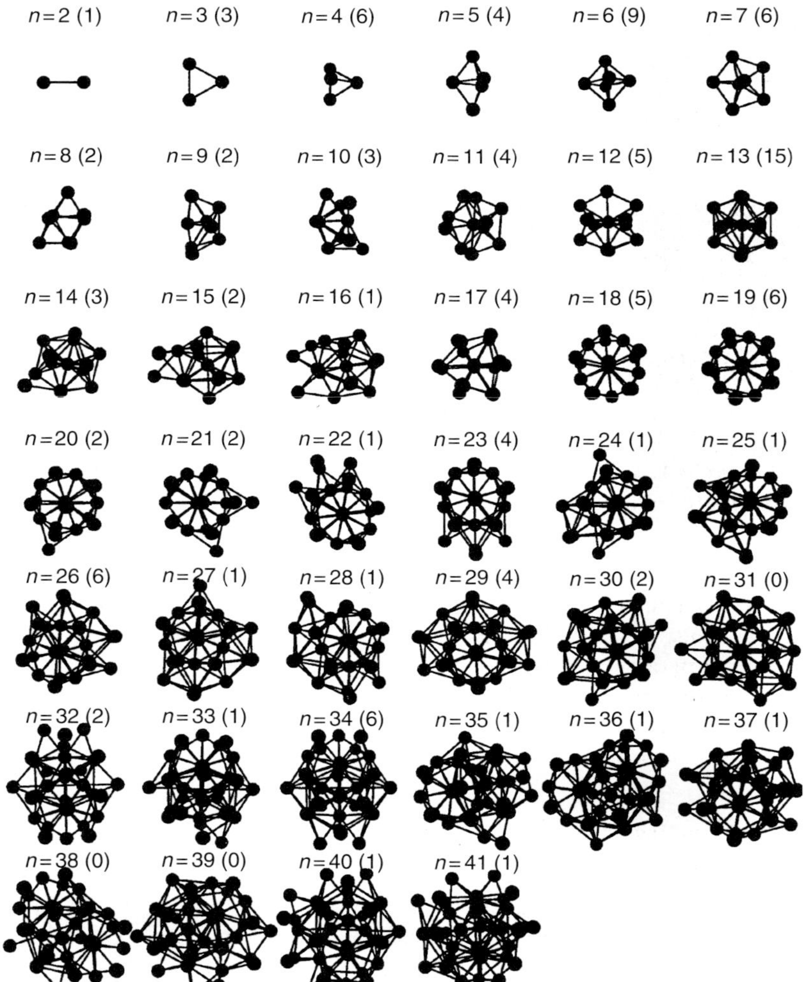

FIGURE 3 Geometries corresponding to the ground state of clusters obtained in the genetic algorithm simulation. The number of reflection planes for each cluster is given in parentheses. Clusters are plotted such that the symmetry as well as "atomic shell" structures is clearly visible (see Ref. [38]).

C. Coinage Metal Clusters

The coinage group metals Cu, Ag, and Au are monovalent like the alkali metals and possess nearly free-electron structure. Although the relative stability of these clusters follows the same pattern as the alkali metals, their structures are different. Among these, gold clusters are unique. While bulk gold is chemically inert, nano-gold can be very reactive [39]. In particular,

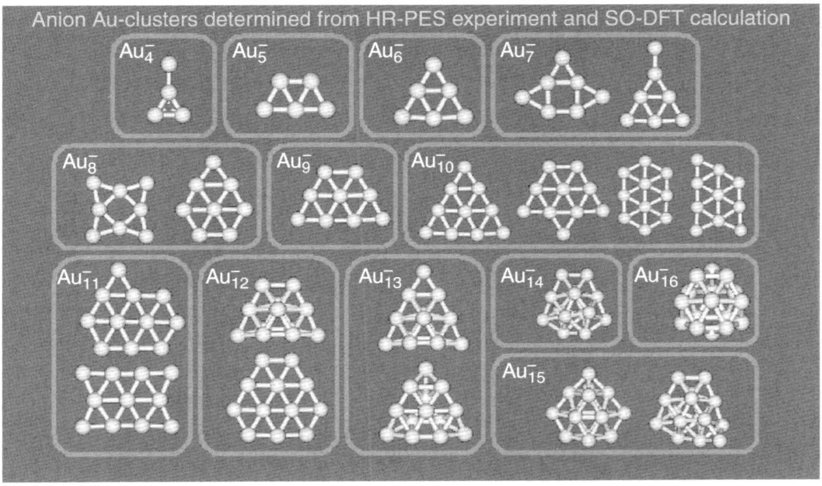

FIGURE 4 Geometries of Au cluster anions (see Ref. [42], courtesy of Prof. X.C. Zheng).

geometries of Au clusters show a very different evolutionary pattern. They form planar structures for clusters containing as many as 11 atoms [40,41] assume a cage structure for clusters containing 14–18 atoms [29], and form a compact pyramidal structure mimicking the (111) Au surface for Au_{20} [25]. The structures of anionic Au clusters are shown in Figure 4 [42]. Sharp departure of Au cluster geometries from those of other monovalent simple metal atoms is attributed to relativistic effects on its electronic structure [43].

D. Transition-Metal Clusters

The properties of transition metals are governed primarily by their unfilled d orbitals which are more localized than the s and p electrons of simple metals. Consequently, the geometry, electronic structure, and magnetic properties of transition-metal clusters are very different from those of the simple metals. The relative stability of transition-metal clusters does not follow any particular rule and their structures evolve differently from those of the simple metals such as Na and Al. For example, most of them assume three-dimensional structures when containing only four atoms. However, as with simple metals, transition-metal clusters also possess icosahedral geometries and bear no resemblance to their crystal structure. As an example, we show in Figure 5 the geometries of Ni clusters [44].

E. Semiconductor Clusters

Diamond and silicon are semiconductors where the sp^3 bonding gives rise to tetrahedral coordination. However, unlike Si, carbon forms both diamond and

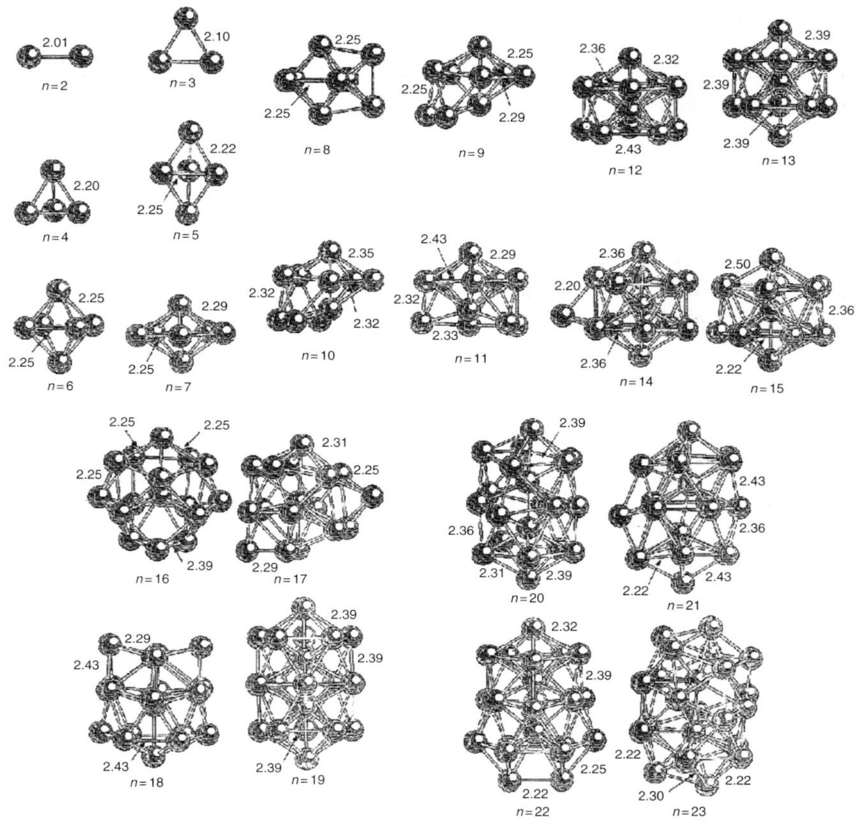

FIGURE 5 Equilibrium geometries of Ni_n ($2 \leq n \leq 23$) clusters (Ref. [44]).

graphitic structures, the later being governed by the sp^2 bonding. Graphite is planar and conducting. Thus, clusters of carbon and silicon exhibit very different structures. Carbon clusters exhibit the famous fullerene structures whereas Si does not. In Figure 6A we show examples of carbon cluster geometries [45] consisting of 10 atoms or less; odd-numbered clusters form linear chains whereas even-numbered clusters form ring structures. Cage structures of carbon clusters emerge with 20 atoms, and 60 atoms of carbon form the well-known fullerene structure which is comprised of 20 hexagons and 12 pentagons with a cage diameter of 6.5 Å (see Figure 6B). Higher fullerenes also exist, although the structures no longer look spherical as C_{60} does. The pronounced stability of C_{60} [2] as well as its synthesis in bulk quantities [46] has been one of the most exciting developments in cluster science and it is the only elemental cluster that has been assembled to form a solid. It is also one of the very few clusters that can be classified both as a molecule and as a cluster.

Si clusters are one of the most studied systems [47–54] and the geometries of these clusters are very different from those of either carbon or metal atoms. In spite of considerable efforts, Si clusters have not been found to form cage structures, and attempts to make Si_{60} in the form of a fullerene by using C_{60} as an endohedral core have not succeeded [55,56]. However, large abundances of the $Si_{12}W$ cluster have been reported which has a metal-encapsulated hexagonal prism structure [57]. Similar structure has also been shown to exist for $Si_{12}Cr$ [58]. Theoretical studies have further shown that it is possible to synthesize Si_{16} and Si_{20} cage structures with an endohedral metal atom [59,60]. In Figure 7, we show the structures of pure Si clusters [61] as a function of size. Note again that the structures do not mimic the Si crystal structure until they reach the size of about 70 atoms.

F. Heteroatomic Clusters

Heteroatomic clusters consisting of more than one kind of atom are similar to alloys with one major exception. Not all elements can be alloyed in the bulk form, and among those that do there are immiscible gaps beyond which alloying is not possible. A classic example is Al and K which do not mix with each other (only 1 in 1000 K atoms can mix with Al in the molten state). However, a single K atom can bind strongly to an Al_{13} cluster, allowing the K concentration to reach 8 at.%. In general, heteroatomic clusters of any

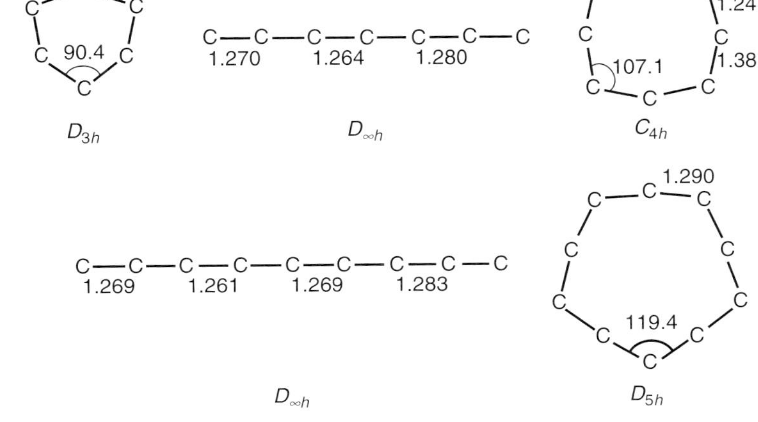

FIGURE 6 (Continued)

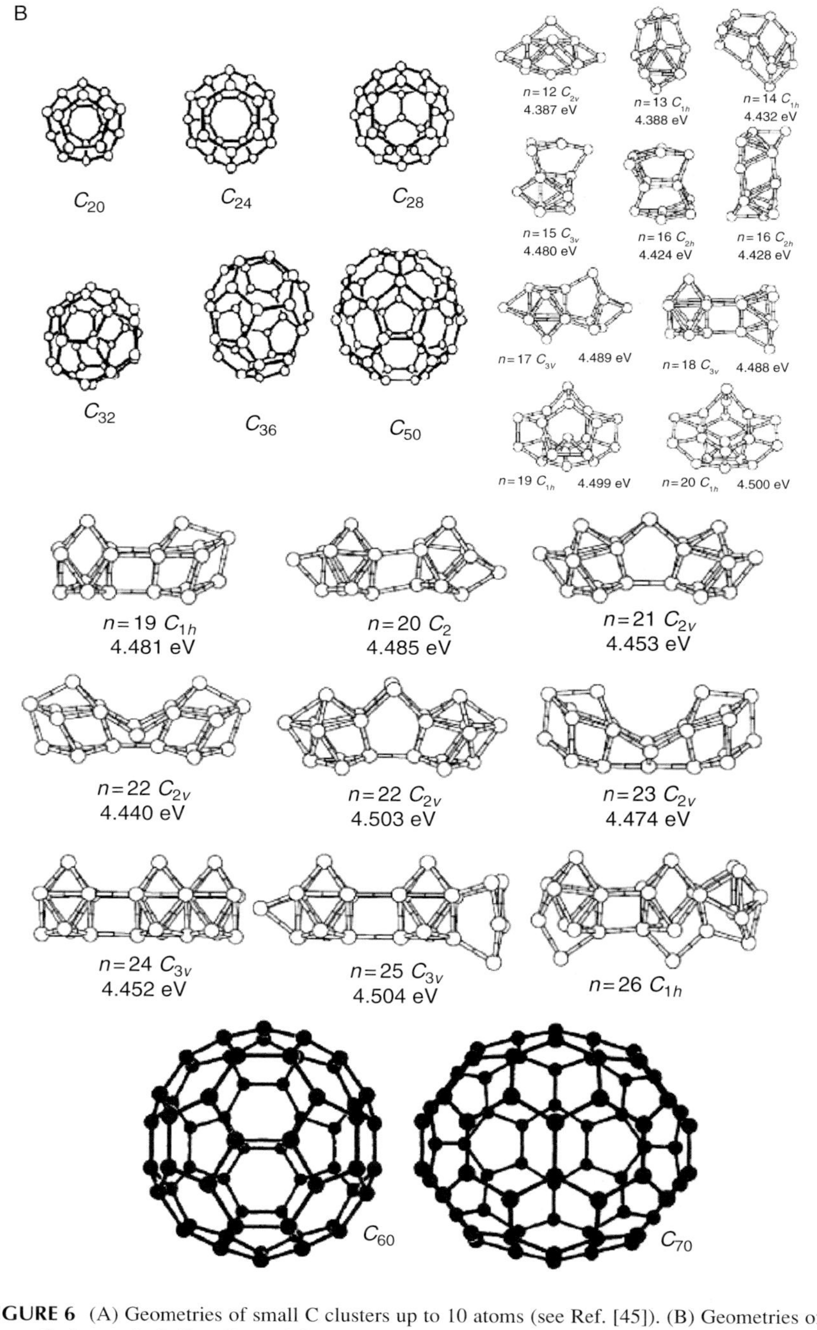

FIGURE 6 (A) Geometries of small C clusters up to 10 atoms (see Ref. [45]). (B) Geometries of larger C clusters including C_{60} and C_{70} (see Ref. [45]).

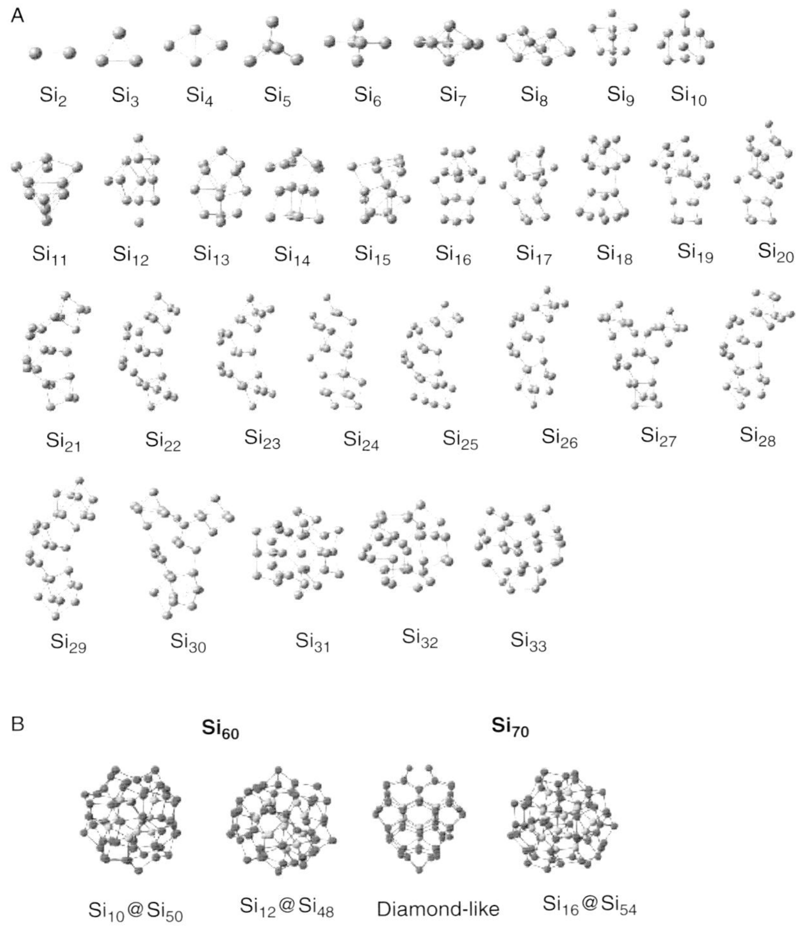

FIGURE 7 Geometries of Si clusters (A) Si_{15-20}, see Ref. [61]; Si_{21-24}, see Ref. [125]; and Si_{25-33}, see Ref. [126,127]. (B) Si_{60} cages; see Ref. 128, and two competitive Si_{70} isomers (Zhao LZ, Su WS, Lu WC, Wang CZ, Ho KM, unpublished) (bulk-like and cage), respectively. Endohedral atoms are colored yellow. (Courtesy of Prof. Ho, KM). (For interpretation of the references to color in this figure legend, the reader is referred to the Web version of this chapter.)

composition are possible even though the corresponding elements do not form an alloy in the bulk phase. Thus, compound clusters consisting of two or more elements provide a fertile ground for understanding how different atoms interact. Numerous studies have been undertaken to study geometries, electronic structure, and properties of compound clusters by changing their composition. These include compound clusters consisting entirely of metal atoms, semiconductor atoms, or semiconductor–metal atoms. Compound metal clusters may be composed of entirely simple metals or transition metals, or a mixture of simple metal atoms interacting with transition-metal atoms. In these clusters, the bonding is reminiscent of metallic character. Heteroatomic clusters also exhibit covalent and or ionic bonding such as those seen in metallo-carbohedrenes or Met-Cars and NaCl clusters. The structures of heteroatomic clusters composed of only metal atoms do not mimic their bulk alloy structure, whereas those bonded covalently or ionically do. The scope of this chapter or the book does not permit a full discussion of heteroatomic clusters. We, therefore, concentrate on a few examples.

The first example is that of heteroatomic clusters consisting of Au and Al. While both Au and Al behave as simple metals, the properties of Al clusters differ very much from those of Au clusters. As has been discussed before, Al clusters form compact three-dimensional geometries with as few as four atoms, whereas Au clusters form planar structure until they reach a size of 12 atoms. Au clusters containing 14–18 atoms form hollow cage structures, whereas Au_{20} forms a pyramidal structure. Al clusters, on the other hand, show icosahedric growth. Which of these characteristics dominate when Al and Au clusters are alloyed together? As an example, we show in Figure 8 the geometry of $Al_{12}Au_{20}$ [62]. Note that the structure is a compact one dominated with Al–Au bonds and the structure is neither that of an Au_{20} pyramid decorated with 12 Al atoms nor that an Al_{12} icosahedron decorated with 20 Au atoms.

The second example we have chosen is that of a heteroatomic cluster where one element is a metal while other is a nonmetal. In particular, we consider Ti_8C_{12}, known commonly as Met-Car [63,64]. The discovery of a strong peak in the mass spectra of this cluster and the original hypothesis that it has a cage structure similar to that of fullerene created a great deal of excitement. However, this cluster has not yet been synthesized in bulk quantities and they tend to interact strongly [65], eventually coalescing as the clusters are brought into the vicinity of each other. Although the preferred geometry of Ti_8C_{12} was originally thought to be a dodecahedron [66], later calculations showed it to be Jahn–Teller distorted [67a,b] (see Figure 9). The structure with the lowest energy has C_{3v} symmetry [67b]. Later experiments [63,68] revealed that prominent peaks in the mass spectra of Ti_nC_m clusters also exist for n and m that are characteristic of the cubic geometric pattern, which is reminiscent of the crystal structure of TiC.

Similar observations of heteroatomic clusters mimicking the structures of their bulk have been seen in ionically bonded systems such as alkali halides and metal nitrides, and oxides [69–72]. Here, unlike the metal clusters discussed earlier, even very small clusters bear the hallmarks of their crystalline structure. Crystal-like growth patterns are often inferred from cluster distributions, as seen in the titanium nitride system in Figure 10 [72]. How small a cluster, characterized by covalent or anionic bonding, has to be for it to bear strong resemblance to the bulk structure? It has been shown that in metal oxide clusters such W_nO_m, this can occur at $n=2$ and $m=3$ [71], which is the stoichiometry of bulk W_2O_3.

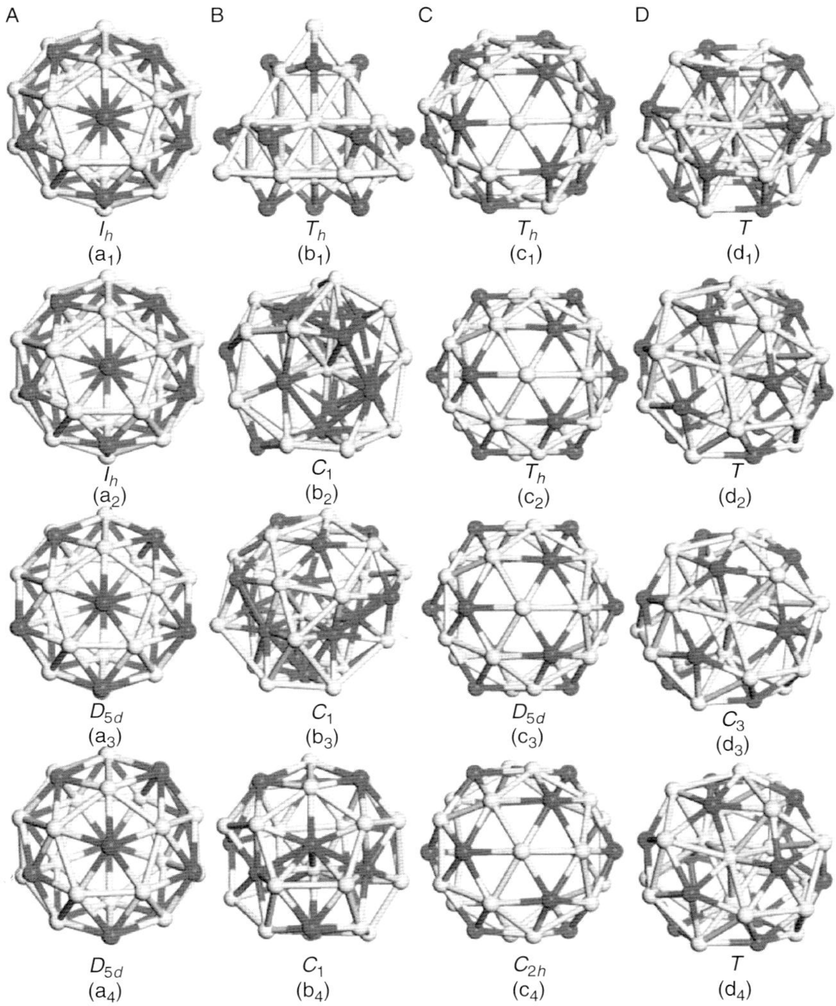

FIGURE 8 (A–D) Starting structures of four isomers of $Al_{12}Au_{20}$. Optimized geometries and their corresponding symmetries in dianionic, anionic, and neutral states are given in (a$_2$)–(d$_2$), (a$_3$)–(d$_3$), and (a$_4$)–(d$_4$), respectively (see Ref. [62]).

III. ELECTRONIC STRUCTURE OF CLUSTERS

In atoms, the electron energy levels are quantized and discrete and the energy gap between the highest occupied and lowest unoccupied orbital (HOMO–LUMO) determines to a large degree their stability, reactivity, and electronic properties. As atoms aggregate to form clusters or molecules, these energy levels overlap, the gap between the highest occupied and lowest

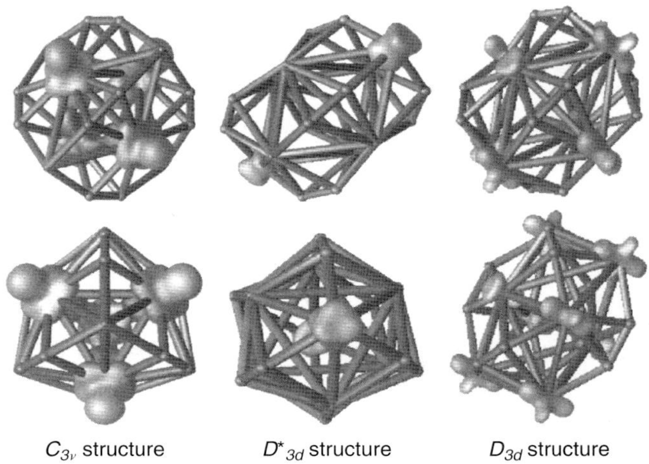

FIGURE 9 Depiction of the spin-polarization isosurfaces of the C_{3v}, D_{3d}*, and D_{3d} structures in two different perspectives (see Ref. [67b]).

unoccupied energy levels change, and electrons are primarily distributed along the bonds formed by the atoms. In crystals, the overlap between the discrete energy levels becomes high and energy bonds give way to energy bands. When the energy bands exhibit a gap at the Fermi energy, the crystals are either semiconducting or insulating and the bonding can range from weak van der Waals to strong covalent, or ionic. However, metallic bonding arises when electrons are delocalized, and the energy gaps at the Fermi level disappear. In small clusters, however, there is always an energy gap between the highest occupied and lowest unoccupied orbital (HOMO–LUMO gap) irrespective of whether the clusters are composed of metallic or nonmetallic elements. Thus, small metal clusters are not expected to exhibit metallic bonding. This is why the demonstration by Knight et al. [73] that the stability of Na clusters can be understood by considering a nearly free-electron model, the jellium model, was at first surprising.

The jellium model for a cluster assumes that the positive charge of the ions are distributed uniformly in a sphere of radius R and the valence electrons are distributed in electronic shells that respond to this uniform charge. The wave functions of electrons corresponding to this spherical potential well are Bessel functions, and zeros of these functions determine the orbital angular momentum and occupancy of each of the shells. The electronic shells arranged in order of increasing energy are $1s^2$, $1p^6$, $1d^{10}$, $2s^2$, $1f^{14}$, $2p^6$, $1g^{18}$, $2d^{10}$, $3s^2$, and so on. As R tends to infinity, the cluster becomes a crystal and the uniform positive charge extends over the entire crystal. Electron wave functions then become plane waves and discrete energy levels become energy bands. The above electronic shells in clusters can be successively closed with

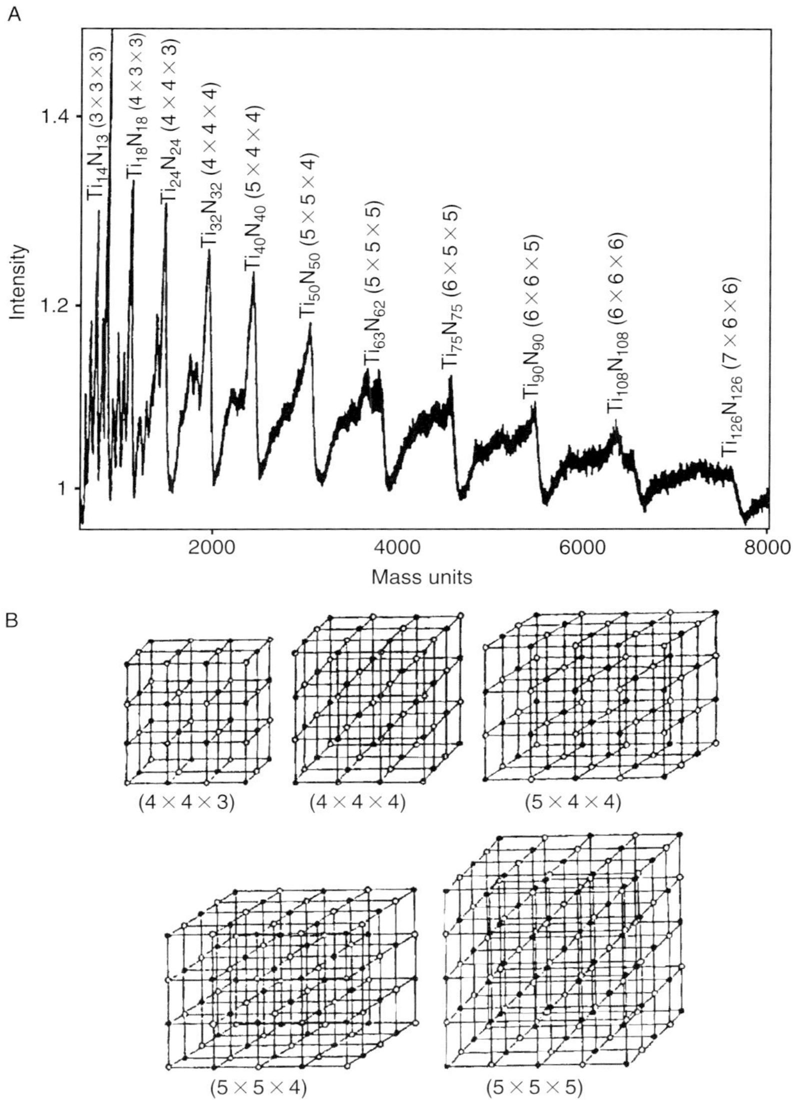

FIGURE 10 Growth patterns of $(TiN)_n$. (A) Time-of-flight mass spectrum of $(TiN)_n$ clusters. Abundance patterns indicate the clusters have cubic structures resembling pieces of the fcc lattice of solid TiN. (B) Proposed structures of $(TiN)_n$ clusters based on magic numbers observed in the mass spectrum (see Ref. [72]).

2 ($1s^2$), 8 ($1s^2$, $1p^6$), 20 ($1s^2$, $1p^6$, $1d^{10}$, $2s^2$), 34 ($1s^2$, $1p^6$, $1d^{10}$, $2s^2$, $1f^{14}$), 40 ($1s^2$, $1p^6$, $1d^{10}$, $2s^2$, $1f^{14}$, $2p^6$), 58 ($1s^2$, $1p^6$, $1d^{10}$, $2s^2$, $1f^{14}$, $2p^6$, $1g^{18}$), 70 ($1s^2$, $1p^6$, $1d^{10}$, $2s^2$, $1f^{14}$, $2p^6$, $1g^{18}$, $2d^{10}$, $3s^2$), and so on, electrons. Since Na is

monovalent, these shell closings correspond to 2, 8, 20, 34, 40, 70, and so on, atoms. These precisely corresponded to the conspicuous peaks in the mass spectra of Na clusters shown in Figure 11 [73], which suggested that a cluster is more stable than its neighbors if it has just enough electrons to close electronic shells. Since the number of electrons in Na_n^+ clusters is one less than those in neutral clusters, Na_n^+ with $n=3, 9, 21$, and so on, will be more stable than those of their neighbors [74]. This is indeed the case experimentally [75].

Since this demonstration, the jellium model has become a very popular model to illustrate the electronic structure of simple metal clusters. Since the stability of magic nuclei was also explained to be due to the nuclear shell

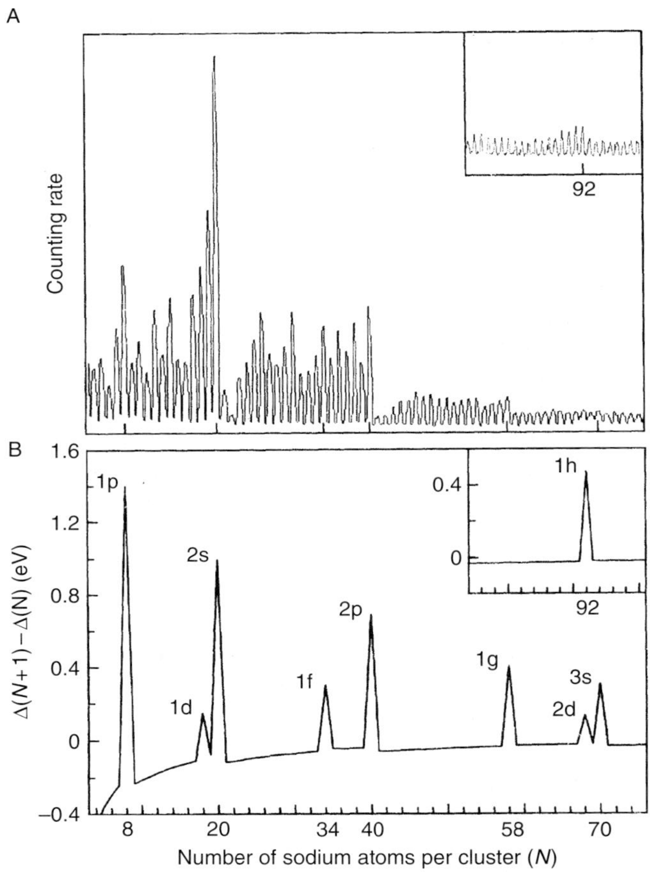

FIGURE 11 (A) Mass spectrum of sodium clusters, $N=4$–75. The full scale-intensity in the main figure is approximately 20,000 counts/s. Source conditions: $P_{Ar}=759$ kPa, $P_{Na\ v}=24$ kPa. The inset corresponds to $N=75$–100. (B) The calculated change in the electronic energy difference, $\Delta(N+1)-\Delta(N)$ versus N. The labels of the peaks correspond to the closed-shell orbitals (see Ref. [73]).

closure, clusters bridged a gap between atomic, molecular, and nuclear physics. This finding opened a new line of research, in which phenomena known to exist in nuclear physics were sought in clusters, and vice versa [76–79]. This included giant dipole resonance and fission [80,81]. For example, it was found that when clusters of simple metals fragment, the most dominant product always involves a magic cluster [74,82]. This is consistent with nuclear fission. It has also been shown that table-top nuclear fusion can be driven by energetic deuterons produced by Coulomb explosion of multi-charged homo- and hetero-nuclear molecular clusters [83].

Clearly, a Na_2 cluster cannot be a metal, and clusters of nonmetallic elements cannot be described by the jellium model. Thus, a proper understanding of the evolution of the electronic structure requires that one consider the actual atomic structure of a cluster and study its bonding and electronic structure accordingly without resorting to a simplified model. One of the ways for studying this evolution from bond to band is to monitor the variation of the HOMO–LUMO gap with cluster size. Two of the experiments where this is measured are PES and velocity map imaging. In the former, one mass isolates a negative ion cluster; photo detaches the electron with a fixed frequency laser, and measures the kinetic energy of the ejected electron. The ensuing PES spectra carry the signature of the electron energy levels of the corresponding neutral cluster. Although there is some discussion regarding whether the measured PES spectra reflects the energy levels of the neutral cluster or the anion, there is no controversy that it provides a measure of the electronic structure and the HOMO–LUMO gap. This is certainly true when the geometries of the anion and neutral clusters are identical. The measured HOMO–LUMO gap, which is the energy gap between the first two peaks in the PES data, can be plotted as a function of size. When the gap reduces to zero for a cluster of metal atoms, one can say that the cluster is a metal. To demonstrate this, we show in Figure 12 the HOMO–LUMO gaps of Mg clusters [84]. We note that the HOMO–LUMO gaps decrease with increasing cluster size, eventually vanishing when a cluster contains 15–18 atoms. While it may be tempting to say that a Mg cluster containing 15–18 atoms is metallic, we note that the HOMO–LUMO gaps appear again as clusters grow further. Thus, one has to explain what it means to say that a smaller cluster has become metallic when a larger cluster has not. The electronic structure of transition-metal clusters is more complex and is dominated by their d electrons. The ground-state spin structure plays an important role, and simple models as that described above do not account for the electronic structure of transition-metal clusters. For clusters of semiconductor elements, the evolution of the HOMO–LUMO gap is different; smaller clusters have smaller HOMO–LUMO gaps and these increase as clusters grow.

The velocity map imaging method has been shown to enable determination of the asymmetry of photoejected electron, and hence, the orbitals involved. Recent studies have revealed that the concept of superatoms can

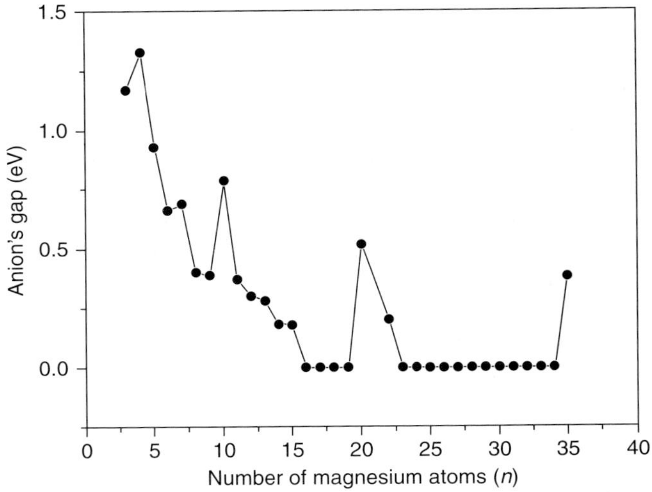

FIGURE 12 Plot of Mg_n gap values versus their sizes n (see Ref. [84]).

be quantified employing the study of isoelectronic systems, for example, TiO^-, ZrO^-, and WC^-, which serve to mimic the elements Ni^-, Pd^-, and Pt^- [85,86].

IV. STABILITY OF CLUSTERS AND MAGIC NUMBERS

Not all clusters are equally stable and their relative stability varies nonmonotonically with size. This is reflected in the intensity distribution of their mass spectra. The mass spectra of simple metal clusters are also very different from those of transition-metal atoms and clusters bound by covalent (e.g., C, Si), ionic (e.g., NaCl), or van der Waals (e.g., Xe) interaction. As mentioned earlier, the intensity distribution in the mass spectra of Na clusters can be understood by a simple jellium model where the valence electrons are free-electron-like. The jellium model for free-electron clusters can also be extended to charged as well as heteroatomic clusters. Consider, for example, positively charge Na clusters. Na_n^+ clusters with $n=3$, 9, 21, 41, and so on, can be magic [74] just as their neutral counterparts with one less atom (or electron). This has been shown to be the case experimentally [75]. Similarly, Al_{13}^- cluster containing 40 electrons can also be a magic cluster due to shell-closure argument. Magic clusters, due to electronic shell closure, also have large HOMO–LUMO gaps and are less reactive than those with open electronic shells. This aspect was demonstrated experimentally where Al_{13}^- was found to be less reactive toward oxygen than its neighboring clusters [19a]. Compound clusters can be designed to fulfill the electronic shell-closure requirement. For example, KAl_{13} containing 40 electrons and Na_2Al_6 containing 20 electrons are also magic clusters [87].

Stability of transition-metal clusters or clusters of semiconductor elements do not follow the above electronic shell-closure rule. This is reflected from the respective mass spectra of Ni and Co clusters [88] (see Figure 13) as well as those of Si clusters [89] (see Figure 14). However, the 18-electron rule provides a good measure to study the stability of clusters based on transition-metal elements. Consider, for example, $Cr(C_6H_6)_2$ or $Fe(C_5H_5)_2$. Cr and Fe are transition-metal elements with $3d^5\ 4s^1$ and $3d^6\ 4s^2$ valence electron configurations. With C_6H_6 and C_5H_5 contributing, respectively, six and five electrons to the valence pool, the total number of electrons involved in bonding are 18 in both cases. These two metal–organic complexes are known to be very stable. Similarly, $Au_{12}W$ cluster also has 18 electrons for bonding, and stability of this cluster has been recently established [90]. The 18-electron rule has been used to design stable clusters that are useful for hydrogen storage [91]. The same rule has also been found to give rise to the stability of $WSi_{12}$where Si was assumed to contribute one electron to the bonding network [57].

Another rule that has been used to describe the stability of boranes (boron–hydrogen complexes) is known as the Wade–Mingos rule [92,93]. This rule states that the stability of boranes requires $(n+1)$ pairs of electrons, where n is the number of vortices of a boron polyhedron. Thus, $B_{12}H_{12}^{2-}$ with 26 electrons for skeletal bonding is a very stable cluster where B atoms

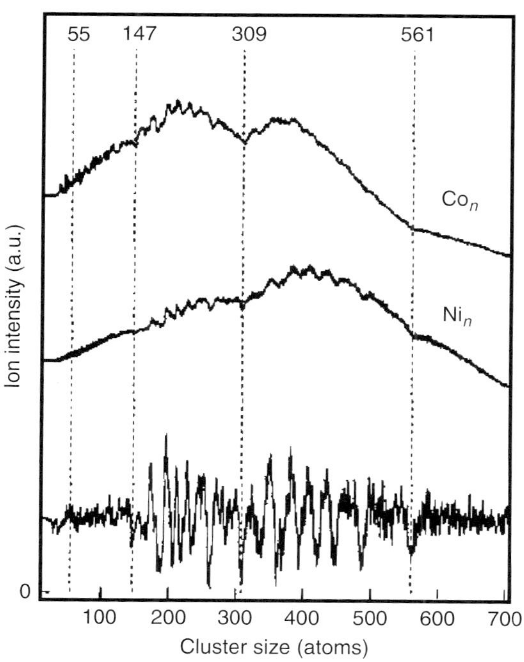

FIGURE 13 Mass spectra of cobalt and nickel clusters. Characteristic sizes corresponding to the major shell effects are reported. The lower spectrum is the same as that of the nickel cluster with suitable mathematical treatment to make structures more apparent. These structures are consistent with icosahedral atomic shell filling (see Ref. [88]).

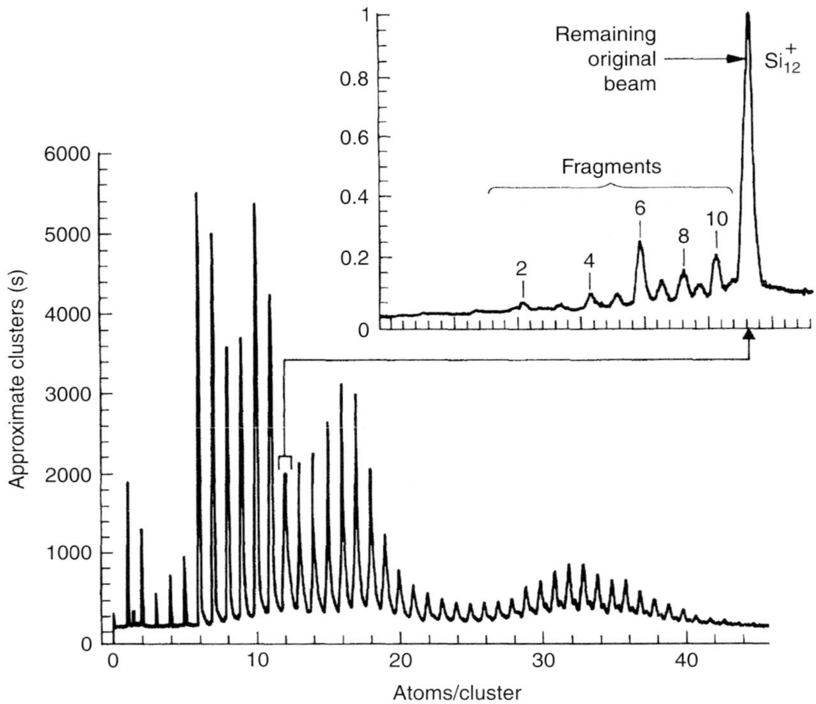

FIGURE 14 Spectrum of small to medium cluster ions. An example of the reselection of cluster size is shown on the upper right where a single ion mass Si_{12}^{+} is isolated and fragmented with 266-nm radiation to produce the fragmentation spectrum shown on the upper right (see Ref. [89]).

occupy the vortices of an icosahedon and H atoms bond radially to each of the B atoms (see Figure 15). This rule has recently been proved to apply to Al_nH_m clusters as well, and numerous hitherto-unknown aluminum–hydrogen clusters have been discovered [94].

Rules of aromaticity and anti-aromaticity have also been applied to study structures and properties of compound clusters, and a separate chapter in this book by Boldyrev and coworkers is devoted to this [95].

For rare-gas clusters, due to filled electron shells of the atoms, the electronic interaction is weak and the stability of the clusters is determined primarily by hard-sphere packing. Thus, clusters with closed atomic shells exhibit icosahedric packing. This can be seen from the mass spectra of Xe clusters in Figure 16 [96].

Atomic shell closure also has been seen to drive the relative stability of large Na clusters. This can be seen in the mass spectra in Figure 17 [97]. A cluster with just enough atoms to close a geometric shell is more stable than its neighbors.

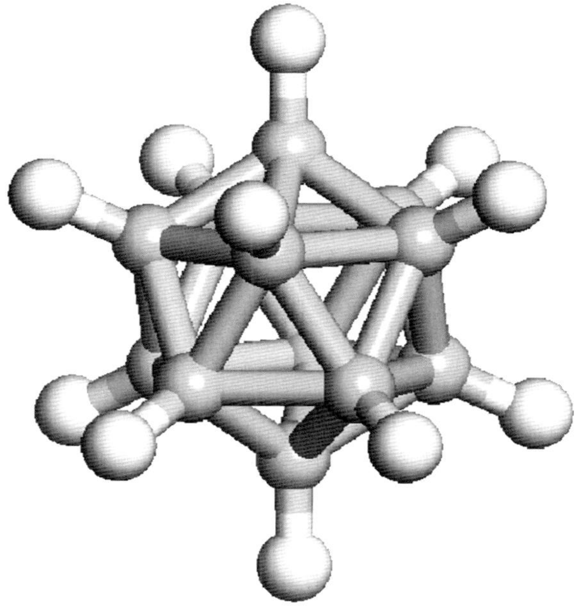

FIGURE 15 Geometry of the $B_{12}H_{12}^{2-}$ borane cluster.

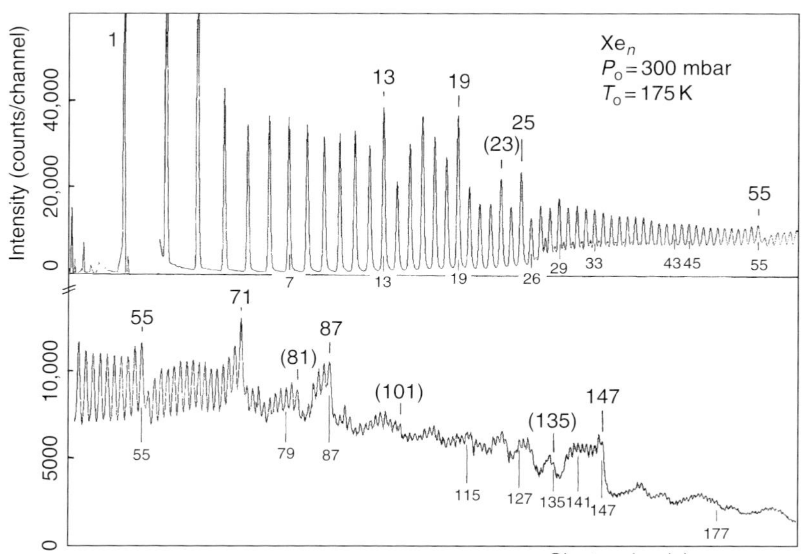

FIGURE 16 Mass spectrum of Xe clusters. Observed magic numbers are marked in boldface; brackets are used for numbers with less pronounced effects. Numbers below the curve indicate predictions or distinguished sphere pickings (see Ref. [96]).

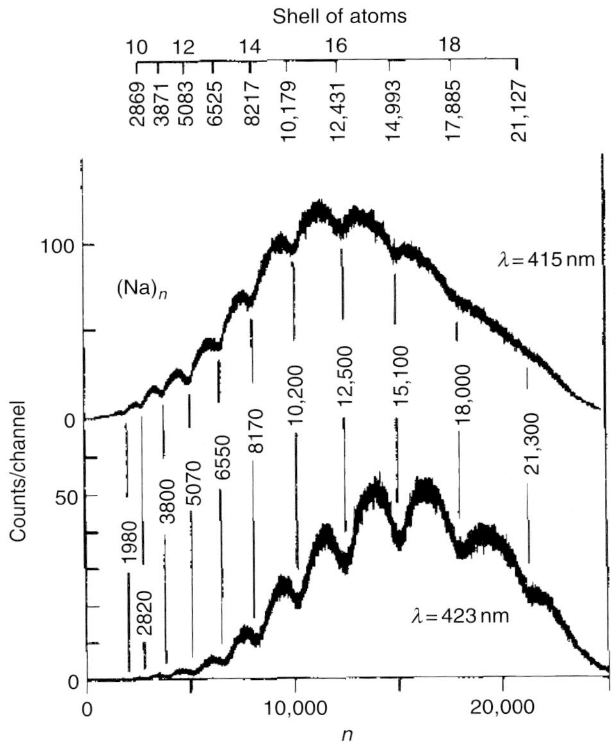

FIGURE 17 Averaged mass spectra of Na clusters photoionized with 415- and 423-nm light. Well-defined minima occur at values of *n* corresponding to the total number of atoms in close-packed cuboctahedra and nearly close-packed icosahedra (listed at top) (see Ref. [97]).

V. PROPERTIES OF CLUSTERS

From the above discussion, it is quite apparent that atomic and electronic structure of clusters evolve differently depending upon their composition, and that clusters containing as many as 100 atoms do not show much sign of bulk behavior. Thus, clusters have yet to bridge the gap in our understanding between atoms and bulk and it has not been possible to answer the fundamental question: How many atoms does it take in a cluster to mimic bulk behavior? However, clusters have displayed unusual properties as a function of size and composition [98] and their structure–property relationships have given the hope that a new class of materials can be synthesized with clusters as building blocks. Some of these properties are outlined in the following.

A. Magnetic Properties

Magnetism of materials has played an important role in technology and its fundamental understanding has been one of the most important fields of research. For a material to be magnetic, each atom has to carry a magnetic moment, and understanding of the origin of this moment, its magnitude, and coupling is key to developing new magnetic materials. The magnetic moments in solids are due to unpaired electron spins as orbital moment is quenched. Note that half of the elements in the periodic table carry a spin magnetic moment of at least 1 μ_B since they have an odd number of electrons. Yet there are only five elements (Fe, Co, Ni, Gd, Dy) that couple ferromagnetically. Some other elements exhibit antiferromagnetism, ferrimagnetism, spin-glass behavior, or paramagnetic behavior. Equally important is the fact that the magnetic moment per atom in the solid is often less than that of the individual atoms. Understanding of the origin of these properties can help in improving magnetic properties of materials. This is where clusters have played an important role.

Magnetic moment and coupling have been known to be affected by local coordination, dimensionality, and interatomic distance [99]. As the overlap between electrons spins at neighboring sites increases, the magnetic moments decrease. Thus, lower coordination, lower dimensionality, and increasing interatomic distance in metal clusters contribute to enhancement of the magnetic moment [100]. Consequently, atoms in linear chains (one dimension) are more magnetic than those on the surfaces (two dimensions), which in turn are more magnetic than those in the bulk (three dimensions). Clusters are often regarded as zero-dimensional units. Since most atoms in clusters are surface atoms, it is expected that the magnetic moment of an atom in a cluster will be larger than that in the bulk. This has been verified by Stern–Gerlach experiments [101,102]. The magnetic moments per atom in clusters are intermediate between the free-atom and bulk value and vary rapidly and nonmonotonically with size as can be seen from Figure 18 [101]. The magnetic moments do not converge to the bulk value even when clusters contain as many as 1000 atoms. Clusters of traditionally nonmagnetic elements also show some unusual properties. For example, V and Rh, which are paramagnetic in bulk, become ferromagnetic when they form small clusters [103–105]. Mn, which is antiferromagnetic in the bulk, becomes ferromagnetic in clusters containing five or fewer atoms and shows ferrimagnetic behavior for larger ones [106–109].

There is also a close link between the topology of a cluster and its preferred spin state. One of the early discussions of the interplay between topology and magnetism was brought into focus in the study of Li_4 cluster [110]. With four electrons, Li_4 can either assume a nonmagnetic state (spin singlet) where the spins are paired, or a magnetic state where two spins remain unpaired (spin triplet). It was found that, when Li_4 assumes a planar structure, spin singlet state is lower in energy. However, when Li_4 forms a tetrahedron, the preferred spin is a triplet, that is, the magnetic moment of Li_4 is 2 μ_B. That a cluster with the same size and composition can either be magnetic or nonmagnetic clearly demonstrates that magnetic transition can be driven by

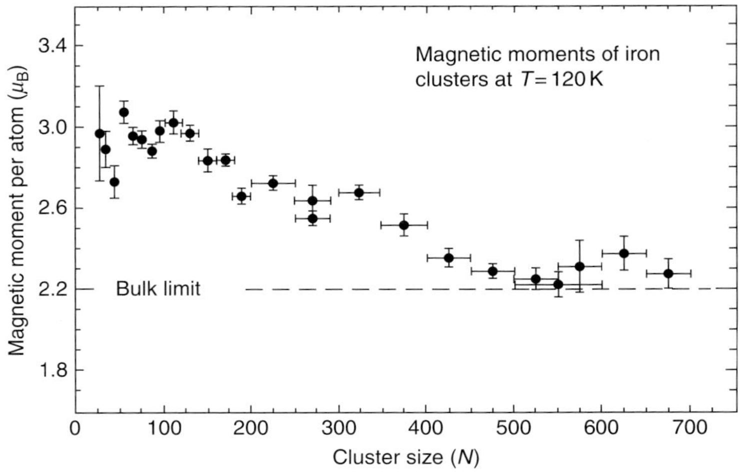

FIGURE 18 Iron cluster magnetic moments per atom at $T = 120$ K. Horizontal bars indicate cluster size ranges (see Ref. [101]).

topology. Thus, one can conceive of a small cluster as a nano-magnet which can be made nonmagnetic simply by changing its structure. This can form the basis of a nano-magnetic switch if a mechanism can be found to manipulate cluster geometry easily. Spins of clusters have also been seen to play an important role in reactivity and photoelectron spectra [111]. In particular, a high-energy cluster isomer can be protected by its spin and exist along with its lower energy isomer with different spin multiplicity [36].

B. Reactive Properties

Clusters also display very unique reactivity. This is attributed to their large surface-to-volume ratio and low coordination of surface atoms. It is, therefore, possible to find clusters of metallic elements that are more abundant and inexpensive that can serve as good catalysts. In addition, clusters also provide a good framework to gain a fundamental understanding of heterogeneous catalysis as one can control both the size of clusters and number of molecules they interact with. One of the classic examples of a metal cluster that exhibits unique reactive properties is gold clusters. While bulk gold is chemically inert and hence earned the reputation as a noble element, it becomes reactive at nanometer length scale and nano-gold can be a good catalyst [39]. Similarly, while Fe does not form a stable hydride, Fe clusters can bind to hydrogen and their reactivity can change by orders of magnitude by simply changing their size over a very narrow range [112] (see Figure 19).

One of the properties that govern cluster reactivity is their electron affinity. Chlorine has the highest electron affinity (3.61 eV) of any element in the periodic table. However, clusters can have electron affinities as high as

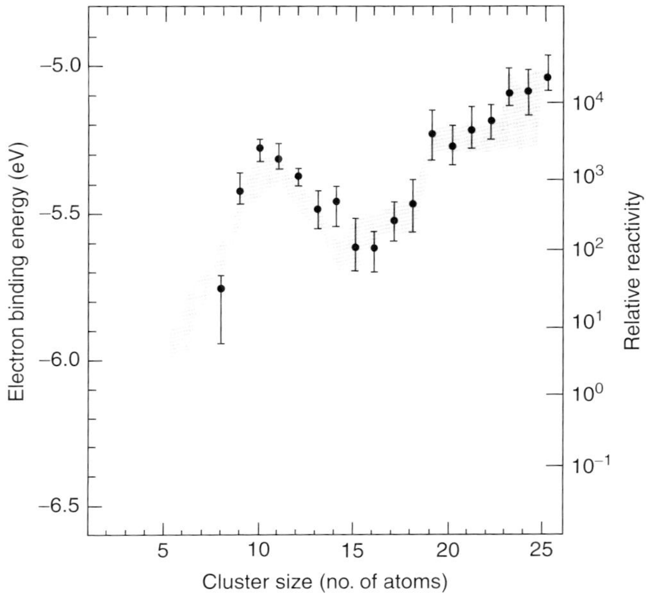

FIGURE 19 Comparison of measured ionization thresholds (left-hand scale) with intrinsic relative reactivities of Fe clusters (right-hand scale). The gray band reflects the uncertainty in ionization threshold measurements, while vertical lines indicate uncertainties in reactivity results, taken from measurements of Fe_x depletion by reaction with D_2 and H_2 (Ref. [112]).

10 eV [113] and these are classified as superhalogens [114]. These consist of a metal atom at the core surrounded by halogen atoms. A good example of a superhalogen is PtF_6 whose oxidative property is dramatically shown in its ability to draw an electron from Xe and form the $Xe^+[PtF_6]^-$ salt [115,116]. Recently, a new class of highly electronegative species has been discovered that has electron affinities even higher than those of superhalogen moieties. These molecules termed as hyperhalogens [117] and consist of a metal atom at the core surrounded by superhalogen moieties.

Interaction of molecules such as N_2 with transition metals has been used to illustrate the geometry of clusters. As mentioned earlier, there are few experimental methods that can provide the structure of clusters. It was shown that, when a metal cluster is exposed to gas molecules under varying pressures, it can adsorb varying numbers of molecules. As an example, we show in Figure 20 the interaction of Ni_7 cluster with N_2 [118]. We see from the plateau in Figure 20 that Ni_7 first adsorbs one N_2 molecule followed by six more. This suggests that a likely structure of a Ni_7 cluster is that of a capped octahedron where the lone Ni atom first binds to one N_2 molecule which is followed by the remaining six Ni atoms in the octahedral binding to six more N_2 molecules.

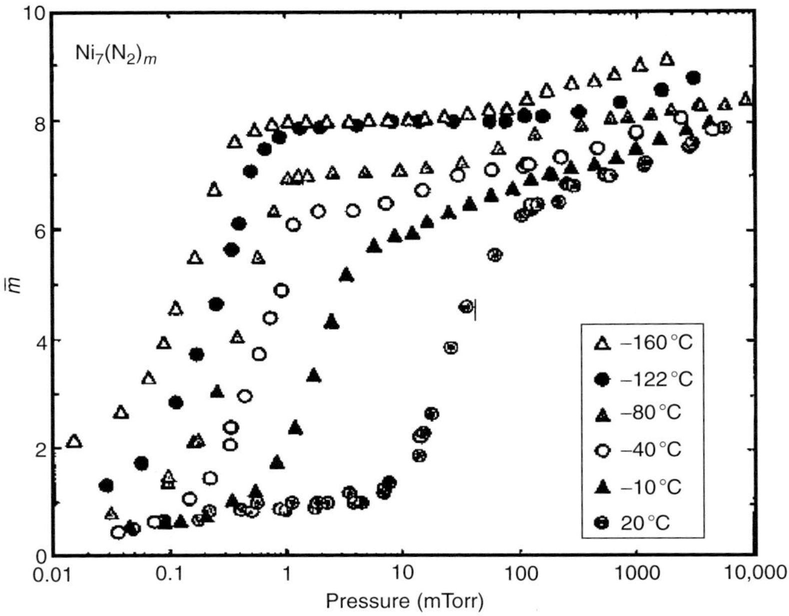

FIGURE 20 Nitrogen uptake plots for Ni_7 for various temperatures of the flow-tube reactor (see Ref. [118]).

C. Optical Properties

The optical properties of materials are determined by their electronic structure and band gap, and these are fixed for crystals. However, in the corresponding clusters, the energy gaps between the highest occupied molecular orbital and lowest unoccupied molecular orbital (HOMO–LUMO) vary with their size and composition. In addition, the HOMO–LUMO gaps can also be modified by coating the clusters with different ligands or surfactants. It is, therefore, possible to design a cluster with a tailored bandgap, and hence tailored optical response, simply by tuning their size, composition, or coating layer. This allows, for example, silica particles coated with gold to absorb infrared radiation and be useful in noninvasive treatment of tumor [119].

D. Melting Properties

Melting of clusters has also revealed some unusual properties. It has been known that the melting point of nanoparticles is smaller than their bulk as these particles are dominated with surface atoms [120]. The surface atoms, due to their low coordination, melt before the bulk atoms and hence the melting point of nanoparticles is less than that of their bulk. However, the melting behavior of clusters can be different from one element to another. While Na clusters have lower meting point than crystalline Na [121] (see Figure 21), the melting point of small Ga clusters exceed that of their bulk [122].

In the above, we illustrated only a few of the many unique properties of clusters. This has led to the suggestion that a novel class of materials can be synthesized if clusters instead of atoms can form the building blocks [18]. One can then tailor the size and composition of a cluster with a specific property, coat them with surfactants [123] or ligands to prevent coalescence, and assemble them into bulk materials. Such cluster-assembled materials can have properties very different from those of the bulk materials. This provides a novel route for the synthesis of atomically engineered materials. Fulleride, a crystal of C_{60} fullerenes, is a classic example of this cluster-assembled material. Currently, considerable effort is devoted to identify stable clusters that can be used to assemble bulk materials.

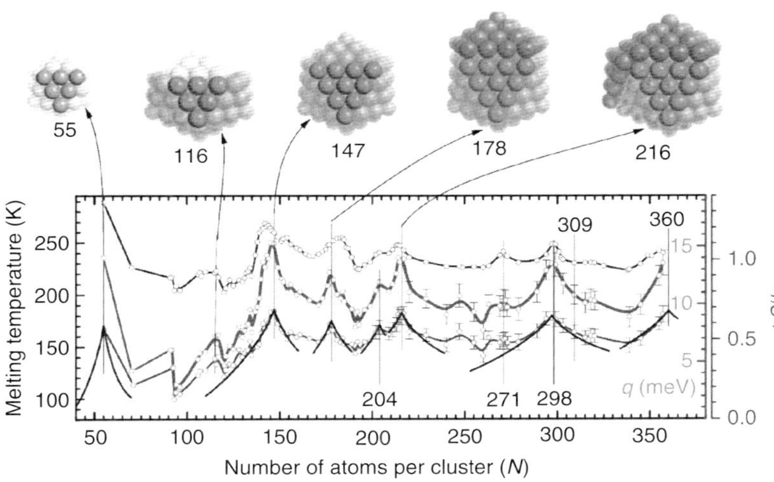

FIGURE 21 Upper panel: clusters of icosahedral growth pattern. The second, third, and fourth layers are given in yellow, green, and red, respectively. One of the 20 triangular faces is colored in a deeper shade. Only the structure of the closed-shell icosahedra at $N=55$ and 147 has been confirmed so far, and for the other sizes different outer layers are possible. Lower panel: size dependence of the melting temperature (T_{melt}, black), the latent heat of fusion per atom (q, red), and the entropy change upon melting per atom (Δs, blue). The data, given by the open circles, are joined (for $N>92$) by splines. Error bars are given only for N above 200, in order to avoid cluttering the figure. The error bars for T_{melt} have about the size of the symbol used. The data for Δs and q have their maxima at the same N, while those for T_{melt} can be shifted. The cluster sizes are indicated for some peaks. The N values are given above the data if the cluster geometry is known or can be guessed. The black solid lines, overlapping partially with the blue line, give the calculated entropy change upon melting. The simple hard-sphere model gives a surprisingly good fit of the peak shapes (see Ref. [121]). (For interpretation of the references to color in this figure legend, the reader is referred to the Web version of this chapter.)

VI. SCOPE OF THE BOOK

In the 14 review articles to follow, this book demonstrates how clusters have been used as models to understand properties of bulk matter and biological systems, the evolution of the electronic structure and geometry of clusters, reactions of clusters and how it has led to an understanding of catalysis at the atomic level, electronic and magnetic properties of clusters and quantum dots, properties of cluster-assembled materials, and role clusters have played in understanding biomolecular processes. While the book attempts to cover some of the important developments in the field, it is by no means comprehensive. The motivation for this book arose from the special volume the authors put together for the Proceedings of the National Academy of Sciences in 2006 [124], which illustrates the role clusters have played in our understanding of many phenomena in physics, chemistry, materials science, atmospheric science, and life sciences. Clusters constitute a field on its own right and provide a bridge across disciplines. The field is robust and expanding and we expect many novel phenomena to emerge.

REFERENCES

[1] Jortner J. Clusters as a key to the understanding of properties as a function of size and dimensionality. In: Jena P, Khanna SN, Rao BK, editors. Physics and chemistry of finite systems: from clusters to crystals. Dordrecht: Kluwer Academic Publishers; 1992. p. 1.

[2] (a) Curl Jr. RF, Smalley RE. Fullerenes. Sci Am 1991;265:321. (b) Kroto HW, Heath JR, O'Brien SC, Curl RF, Smalley RE. C60: Buckminsterfullerene. Nature (London) 1985;318:162.

[3] Rossi MJ. Heterogeneous reactions on salts. Chem Rev 2003;103:4823.

[4] Gorun SM. Oxo—Iron(III) aggregates. In: Que Jr. L, editor. Metal clusters in proteins. ACS symposium series, vol. 372. 1988. p. 196, Washington, DC.

[5] Castleman Jr AW. The formation, growth and nucleation of water cluster ions. In: Borisenkov E, Stepanenko V, editors. Proceedings of the 9th international conference in atmospheric electricity. St. Petersburg: USSR; 1992. p. 67.

[6] Baerends EJ, Ellis DE, Ros P. Self-consistent molecular Hartree—Fock—Slater calculations I. The computational procedure. Chem Phys 1973;2:41.

[7] Averill FW, Ellis DE. An efficient numerical multicenter basis set for molecular orbital calculations: Application to $FeCl_4$. J Chem Phys 1973;59:6412.

[8] Salahub DR, Messmer RP. Molecular-orbital study of aluminum clusters containing up to 43 atoms. Phys Rev B 1977;16:2526.

[9] Jena P, Fradin FY, Ellis DE. Models of electronic structure of hydrogen in metals: Pd-H. Phys Rev B 1979;20:3543.

[10] Raghavachari K, Hallas MD. Quantum chemical studies of semiconductor surface chemistry using cluster models. Mol Phys 2004;102:381.

[11] Becker EW, Brier K, Henkes W. Strahlen aus kondensierten atomen und molekeln im hochvakuum. Z Phys 1956;146:333.

[12] Dietz TG, Duncan MA, Powers DE, Smalley RE. Laser production of supersonic metal cluster beams. J Chem Phys 1981;74:6511.

[13] Kubo R. Electronic properties of metallic fine particles. I. J Phys Jpn 1962;17:975.

[14] Sugano S. Microcluster physics. Berlin: Springer-Verlag; 1991.

[15] Bonacic-Koutecky V, Fantucci P, Koutecky J. Quantum chemistry of small clusters of elements of groups Ia, Ib, and IIa: fundamental concepts, predictions, and interpretation of experiments. Chem Rev 1991;91:1035.

[16] Rao BK, Jena P. Physics of small metal clusters: Topology, magnetism, and electronic structure. Phys Rev B 1985;32:2058.

[17] Martins L, Buttet J, Car R. Equilibrium geometries and electronic structures of small sodium clusters. Phys Rev Lett 1984;53:655.

[18] Khanna SN, Jena P. Assembling crystals from clusters. Phys Rev Lett 1992;69:1664.

[19] (a) Leuchtner RE, Harms AC, Castleman Jr. AW. Thermal metal cluster anion reactions: Behavior of aluminum clusters with oxygen. J Chem Phys 1989;91:2753. (b) Khanna SN, Jena P. Atomic clusters: Building blocks for a class of solids. Phys Rev B 1995;51:13705. (c) Castleman Jr. AW, Khanna SN. Clusters, superatoms, and building blocks of new materials. J Phys Chem 2009;113:2664.

[20] (a) Jena P, Khanna SN, Rao BK. Physics and chemistry of small clusters. New York: Plenum; 1987. (b) Knight WD, Chou MY, Cohen ML. Electronic shell structure and metal clusters. In: Ehrenreich H, Turnbull D, editors. Solid state physics. New York: Academic Press; 1987.

[21] Chapon C, Giller MF, Henri CR, editors. Proceedings of the fourth international meeting on small particles and inorganic clusters. Berlin: Springer-Verlag; 1989.

[22] Scoles G, editor. The chemical physics of atomic and molecular clusters. Amsterdam: North Holland; 1990.

[23] Jena P, Khanna SN, Rao BK, editors. Physics and chemistry of finite systems: from clusters to crystals. Dordrecht: Kluwer; 1992.

[24] Kittel C. Introduction to solid state physics. New York: John Wiley & Sons, Inc.; 1986.

[25] Li J, Li X, Zhai HJ, Wang LS. Au_{20}: A tetrahedral cluster. Science 2003;299:864.

[26] Li X, Grubisic A, Stokes ST, Gantefor GF, Bowen KH, Boggavarapu K, et al. Unexpected stability of Al_4H_6: A borane analog? Science 2007;315:356.

[27] Chen G, Wang Q, Sun Q, Kawazoe Y, Jena P. Structures of neutral and anionic Au_{16} clusters revisited. J Chem Phys 2010;132:194306.

[28] Ji M, Gu X, Li X, Gong XG, Li J, Wang LS. Experimental and theoretical investigation of the electronic and geometrical structures of the Au_{32} cluster. Angew Chem Int Ed 2005;44:7119.

[29] Bulusu S, Li X, Wang LS, Zheng XC. Evidence of hollow golden cages. Proc Natl Acad Sci USA 2006;103:8326.

[30] Huang W, Wang LS. Probing the 2D to 3D structural transition in gold cluster anions using argon tagging. Phys Rev Lett 2009;102:153401.

[31] Xing X, Yoon B, Landman U, Parks JH. Structural evolution of Au nanoclusters: From planar to cage to tubular motifs. Phys Rev B 2006;74:165423.

[32] Johansson MP, Lechtken A, Schooss D, Kappes MM, Furche F. 2D-3D transition of gold cluster anions resolved. Phys Rev A 2008;77:053202.

[33] Furche F, Ahlrichs R, Weis P, Jacob C, Gib S, Bierweiler T, et al. The structures of small gold cluster anions as determined by a combination of ion mobility measurements and density functional calculations. J Chem Phys 2002;117:6982.

[34] Gib S, Weis P, Furche F, Ahlrichs R, Kappes M. Structures of small gold cluster cations (Au_n^+, n<14): Ion mobility measurements versus density functional calculations. J Chem Phys 2002;116:4094.
[35] Gruene P, Rayner DM, Redlich B, Van der Meer AFG, Lyon JT, Meijer G, et al. Structures of neutral Au_7, Au_{19}, and Au_{20} clusters in the gas phase. Science 2008;321:674.
[36] Nayak SK, Rao BK, Jena P, Wang LS. Observation of a spin-protected high-energy isomer of Al_4N^- cluster. Chem Phys Lett 1999;301:379.
[37] (a) Lee MS, Chacko S, Kanhere DG. First-principles investigation of finite-temperature behavior in small sodium clusters. J Chem Phys 2005;123:164310. (b) Lee MS, Kanhere DG. Effects of geometric and electronic structure on the finite temperature behavior of Na_{58}, Na_{57}, and Na_{55} cluster. Phys Rev B 2007;75:125427.
[38] Cerowski V, Rao BK, Khannam SN, Jena P, Ishii S, Ohno K, et al. Evolution of the electronic structure of Be clusters. J Chem Phys 2005;123:074329.
[39] Haruta M. Size- and support-dependency in the catalysis of gold. Catal Today 1997;36:153.
[40] Walter M, Hakkinen H. Photoelectron spectra from first principles: from the many-body to the single-particle picture. New J Phys 2008;10:043018.
[41] Yoon B, Hakkinen H, Landman U, Worz AS, Antonelli JM, Abet S, et al. Charging effects on bonding and catalyzed oxidation of CO on Au_8 clusters on MgO. Science 2005;307:403.
[42] Huang W, Pal R, Wang L-M, Zeng XC, Wang LS. Isomer identification and resolution in small gold clusters. J Chem Phys 2010;132:054305.
[43] Pyykko P. Theoretical chemistry of gold. Angew Chem Int Ed 2004;43:4412.
[44] Nayak SK, Khanna SN, Rao BK, Jena P. Physics of nickel clusters: energetics and equilibrium geometries. J Phys Chem A 1997;101:1072.
[45] (a) Raghavachari K, Binkley JS. Structure, stability, and fragmentation of small carbon clusters. J Chem Phys 1987;87:2191. (b) Ho KM, Shvartsburg AA, Pan B, Lu ZY, Wang CZ, Wacker JG, et al. Structures of medium-sized silicon clusters. Nature 1998;392:582. (c) Kroto HW, Walton DRM. Stable derivatives of small fullerenes. Chem Phys Lett 1993;214:353. (d) Kratschmer W. Fullerenes and fullerites: new forms of carbon. Nanostruct Mater 1995;6:65.
[46] Kratschmer W, Lam LD, Fostopoules K, Hoffman DR. Solid C_{60}: a new form of carbon. Nature 1990;347:354.
[47] Jarrold MF. Nanosurface chemistry on size-selected silicon clusters. Science 1991;252:1085.
[48] Jarrold MF, Constant VA. Silicon cluster ions: evidence for a structural transition. Phys Rev Lett 1991;67:2994.
[49] Kaxiras E, Jackson K. Shape of small silicon clusters. Phys Rev Lett 1993;71:727.
[50] Ho KM, Shvartsburg AA, Pan B, Lu ZY, Wang CZ, Wacker JG, et al. Structures of medium-sized silicon clusters. Nature 1998;392:582.
[51] Shvartsburg AA, Liu B, Lu Z, Wang CZ, Jarrold MF, Ho KM. Structures of germanium clusters: where the growth patterns of silicon and germanium clusters diverge. Phys Rev Lett 1999;83:2167.
[52] Miller J, Lu B, Shvartsburg AA, Ogut S, Chelikowsky JR, Siu KW, et al. Spectroscopic evidence for the tricapped trigonal prism structure of semiconductor clusters. Phys Rev Lett 2000;85:1666.
[53] Wang J, Zhao J, Ding F, Shen W, Lee H, Wang GH. Thermal properties of medium-sized Ge clusters. Solid State Commun 2001;117:593.
[54] Mittas L, Grossman JH, Stich I, Tobik J. Silicon clusters of intermediate size: Energetics, dynamics, and thermal effects. Phys Rev Lett 2000;84:1479.

[55] Menon M, Subbaswamy KR. Structure of Si60. Cage versus network structures. Chem Phys Lett 1994;219:219.
[56] Sun Q, Wang Q, Jena P, Rao BK, Kawazoe Y. Stabilization of Si_{60} cage structure. Phys Rev Lett 2003;90:135503.
[57] Hiura H, Miyazaki T, Kanayama T. Formation of metal-encapsulating Si cage clusters. Phys Rev Lett 2001;86:1733.
[58] Khanna SN, Rao BK, Jena P. Magic numbers in metallo-inorganic clusters: Chromium encapsulated in silicon cages. Phys Rev Lett 2002;89:016803.
[59] Kumar V, Kawazoe Y. Metal-encapsulated fullerene like and cubic caged clusters of silicon. Phys Rev Lett 2001;87:045503.
[60] Sun Q, Wang Q, Jena P, Waterman S, Kawazoe Y. First-principles studies of the geometry and energetics of the Si_{36} cluster. Phys Rev A 2003;67:063201.
[61] (a) Yoo S, Zheng XC. Structures and stability of medium-sized silicon clusters. III. Reexamination of motif transition in growth pattern from Si_{15} to Si_{20}. J Chem Phys 2005;123:164303. (b) Hellmann W, Henning RG, Goedecker S, Umrigar CJ, Delley B, Lenosky T. Questioning the existence of a unique ground-state structure for Si clusters. Phys Rev B 2007;75:085411. (c) Lu WC, Wang CZ, Zhao LZ, Zhang W, Qin W, Ho KM. Appearance of bulk-like motifs in Si, Ge, and Al clusters. Phys Chem Chem Phys 2010;12:8551.
[62] Wang Q, Sun Q, Jena P. Stabilizing a 22 karat nanogolden cage. J Chem Phys 2009;131:204501.
[63] Guo BC, Kerns KP, Castleman Jr. AW. $Ti_8C_{12}^+$-Metallo-carbohedrenes: A new class of molecular clusters? Science 1992;255:1411.
[64] Wei S, Guo BC, Purnell J, Buzza SA, Castleman Jr AW. Metallo-carbohedrenes [$M_8C_{12}^+$ (M = V, Zr, Hf, and Ti)]: A class of stable molecular cluster ions. Science 1992;256:515.
[65] Baruah T, Pederson MR. Stability, electronic structure, and vibrational modes of the Ti_8C_{12} dimer. Phys Rev B 2002;66:241404(R).
[66] Reddy BV, Khanna SN, Jena P. Electronic, magnetic, and geometric structure of Metallo-carbohedrenes. Science 1992;258:1640.
[67] (a) Chen H, Feyereisen M, Long XP, Fitzerald G. Stability, bonding, and geometric structure of Ti_8C_{12}, Ti_8N_{12}, V_8C_{12}, and Zr_8C_{12}. Phys Rev Lett 1993;71:1732. (b) Sobhy MA, Castleman Jr. AW, Sofo JO. A density-functional study of the structural, electronic, magnetic, and vibrational properties of Ti_8C_{12} metallocarbohedrynes. J Chem Phys 2005;123:154106.
[68] Pilgrim JS, Duncan MA. Metallo-carbohedrenes: chromium, iron, and molybdenum analogs. J Am Chem Soc 1993;115:6958.
[69] Yalovega GE, Soldatov AV, Novak K, Riedler M, Lofken O, Kolmakov A, et al. Local geometry and electronic structure of free NaCl clusters. Phys Solid State 2000;42:1942.
[70] Xia P, Bloomfield LA. Accommodation of two excess electrons in sodium chloride cluster anions. Phys Rev Lett 1993;70:1779.
[71] Sun Q, Rao BK, Jena P, Stolcic D, Gantefor G, Castleman Jr. AW. Appearance of bulk properties in small tungsten oxide clusters. J Chem Phys 2004;121:9417.
[72] Chen ZY, Castleman Jr. AW. Growth of titanium nitride: From clusters to microcrystals. J Chem Phys 1993;98:231.
[73] Knight WD, Clemenger K, de Heer WA, Saunders WA, Chou MY, Cohen ML. Electronic shell structure and abundances of sodium clusters. Phys Rev Lett 1984;52:2141.
[74] Rao BK, Jena P, Mannien M, Nieminen R. Spontaneous fragmentation of multiply charged metal clusters. Phys Rev Lett 1987;58:1188.

[75] Bhaskar ND, Frueholz KCM, Cook RA. Evidence of electronic shell structure in Rb_N^+ (N=1–100) produced in a liquid-metal ion source. Phys Rev B 1987;36:4418.
[76] Broglia RA, Pacheo JM, Yannouleas C. Optical response of metal microclusters: Atomic analog of the giant dipole resonance in nuclei. Phys Rev B 1991;44:5901.
[77] Yannouleas C, Jena P, Khanna SN. Optical resonances in bimetallic clusters and their relation to the electronic structure. Phys Rev B 1992;46:9751.
[78] Yannouleas C, Broglia R. Collective and single-particle aspects in the optical response of metal microclusters. Phys Rev A 1991;44:5793.
[79] Poenaru DN, Gherghescu RA, Greiner W. Individual and collective properties of fermions in nuclear and atomic cluster systems. J Phys G Nucl Part Phys 2010;37:085101.
[80] de Heer WA, Selby K, Kresin V, Masui J, Vollmer M, Chatelain A, et al. Collective dipole oscillations in small sodium clusters. Phys Rev Lett 1987;59:1805.
[81] Jena P, Rao BK, Nieminen R. Fragmentation channels and their relationship to magic number studies of microclusters. Solid State Commun 1986;59:509.
[82] Brechignac C, Cahuzac Ph, Carlier F, de Frutos M. Asymmetric fission of Na_n^{++} around the critical size of stability. Phys Rev Lett 1990;64:2893.
[83] Heidenreich A, Jortner J, Last I. Cluster dynamics transcending chemical dynamics toward nuclear fusion. Proc Natl Acad Sci USA 2003;103:10589.
[84] Thomas OC, Zheng WJ, Xu S, Bowen KH. Onset of metallic behavior in magnesium clusters. Phys Rev Lett 2002;89:213403.
[85] Peppernick SJ, Gunarante KD, Castleman Jr. AW. Superatom spectroscopy and the electronic state correlation between elements and isoelectronic molecular counterparts. Proc Natl Acad Sci USA 2010;107:3975.
[86] Gunarante KD, Castleman Jr. AW. Towards comprehending the superatomic state of matter. Chem Phys Lett 2010;489:1.
[87] Zheng WJ, Thomas OC, Lippa TP, Xu SJ, Bowen KH. The ionic KAl13 molecule: A stepping stone to cluster-assembled materials. J Chem Phys 2006;124:144304.
[88] Pellarin M, Baguenard B, Vialle JL, Lerme J, Broyer M, Miller J, et al. Evidence for icosahedral atomic shell structure in nickel and cobalt clusters. Comparison with iron clusters. Chem Phys Lett 1994;217:349.
[89] Bloomfield LA, Freeman RR, Brown WL. Photofragmentation of Mass-resolved $Si_{2\text{-}12}^+$ clusters. Phys Rev Lett 1985;54:2246.
[90] (a) Pyykko P, Runeberg N. Icosahedral WAu_{12}: A predicted closed-shell species, stabilized by aurophilic attraction and relativity and in accord with the 18-electron rule. Angew Chem Int Ed 2002;41:2174. (b) Li X, Kiran B, Li J, Zhai H, Wang LS. Experimental observation and confirmation of icosahedral $W@Au_{12}$ and $Mo@Au_{12}$ molecules. Angew Chem Int Ed 2002;41:4786.
[91] Kiran B, Kandalam AK, Jena P. Hydrogen storage and the 18-electron rule. J Chem Phys 2006;124:224703.
[92] (a) Wade KJ. Structural and bonding patterns in cluster chemistry. Adv Inorg Chem Radiochem 1976;18:1. (b) Wade KJ. The structural significance of the number of skeletal bonding electron-pairs in carboranes, the higher boranes and borane anions, and various transition-metal carbonyl cluster compounds. Chem Commun 1971;792.
[93] Mingos DMP. A General Theory for Cluster and Ring Compounds of the Main Group and Transition Elements. Nat Phys Sci 1971;236:99.

[94] (a) Li X, Grubisic A, Stokes ST, Cordes J, Gantefor GF, Bowen KH, et al. Unexpected stability of Al_4H_6: A borane analog? Science 2007;315:356. (b) Roach PJ, Weber AC, Woodward WH, Khanna SN, Castleman Jr. AW. $Al_4H_7^-$is a resilient building block for aluminum hydrogen cluster materials. Proc Natl Acad Sci USA 2007;104:14565.
[95] Yu Zubarev D, Boldyrev AI, Multiple Aromaticity, Multiple Antiaromaticity, and Conflicting Aromaticity in Planar Clusters. pp. 215–264.
[96] Echt O, Sattler K, Recknagel E. Magic numbers for sphere packings: Experimental verification in free xenon clusters. Phys Rev Lett 1981;47:1121.
[97] Martin TP, Bergmann T, Gohlich H, Lange T. Observation of electronic shells and shells of atoms in large Na clusters. Chem Phys Lett 1990;172:209.
[98] Jena P, Castleman Jr AW. Clusters: A bridge across the disciplines of physics and chemistry. Proc Natl Acad Sci USA 2006;103:10560.
[99] Liu F, Press MR, Khanna SN, Jena P. Magnetism and local order: Ab initio tight-binding theory. Phys Rev B 1989;39:6914.
[100] Reddy BV, Nayak SK, Khanna SN, Rao BK, Jena P. Physics of nickel clusters. 2. Electronic structure and magnetic properties. Phys Rev B 1998;102:1748.
[101] Billas IML, Becker JA, Chatlain A, de Heer WA. Magnetic moments of iron clusters with 25 to 700 atoms and their dependence on temperature. Phys Rev Lett 1993;71:4067.
[102] Billas IML, Chatlain A, De Heer WA. Magnetism from the atom to the bulk in iron, cobalt, and nickel clusters. Science 1994;265:1682.
[103] Liu F, Khanna SN, Jena P. Magnetism in small vanadium clusters. Phys Rev B 1991;43:8179.
[104] Reddy BV, Khanna SN, Dunlap BI. Giant magnetic moments in 4d clusters. Phys Rev Lett 1993;70:3323.
[105] Bucher JP, Douglas DC, Bloomfield LA. Magnetic properties of free cobalt clusters. Phys Rev Lett 1991;66:3053.
[106] Nayak SK, Jena P. Anomalous magnetism in small Mn clusters. Chem Phys Lett 1998;289:473.
[107] Nayak SK, Nooijen M, Jena P. Isomerism and novel magnetic order in Mn_{13} cluster. J Phys Chem 1999;103:9853.
[108] Knickelbein M. Experimental observation of superparamagnetism in manganese clusters. Phys Rev Lett 2001;86:5255.
[109] Khanna SN, Rao BK, Jena P, Knickelbein M. Ferrimagnetism in Mn_7 cluster. Chem Phys Lett 2003;378:374.
[110] Rao BK, Jena P, Manninen M. Relationship between topological and magnetic order in small metal clusters. Phys Rev B 1985;32:477.
[111] Nayak SK, Weber SE, Jena P, Wildberger K, Zeller R, Dederichs PH, et al. Relationship between magnetism, topology, and reactivity of Rh clusters. Phys Rev B 1997;56:8849.
[112] Whetten RL, Cox DM, Trevor DJ, Kaldor A. Correspondence between electron binding energy and chemisorption reactivity of iron clusters. Phys Rev Lett 1985;54:1494.
[113] Scheller MK, Compton RN, Ceederbaum LS. Gas-phase multiply charged anions. Science 1995;270:1160.
[114] Gutsev GL, Bolydrev AI. DVM-Xα calculations on the ionization potentials of MX_{k+1}^- complex anions and the electron affinities of MX_{k+1} superhalogens. Chem Phys 1981;56:277.
[115] Bartlett N, Lohmann DH. Dioxygenyl hexafluoroplatinate(v) O_2^+ $[PtF_6]^-$. Proc Chem Soc 1962;115.
[116] Bartlett N. Xenon hexafluoroplatinate(v) $Xe^+[PtF_6]^-$. Proc Chem Soc 1962;218.

[117] Willis M, Gortz M, Kandelam AK, Gantefor GF, Jena P. "Hyperhalogens: Discovery of a new Class of Highly Electronegative Species", Angew Chem Int Ed 2010;49:8966.

[118] (a) Parks EK, Zhu L, Ho J, Riley SJ. The structure of small nickel clusters. I. Ni_3–Ni_{15}. J Chem Phys 1994;100:7206. (b) Parks EK, Zhu L, Ho J, Riley SJ. The structure of small nickel clusters. II. Ni_{16}–Ni_{28}. J. Chem. Phys. 1995;102:7377.

[119] Sun Q, Wang Q, Rao BK, Jena P. Electronic structure and bonding of Au on a SiO_2 cluster: A nanobullet for tumors. Phys Rev Lett 2004;93:186803.

[120] Buffat P, Borel J. Size effect on the melting temperature of gold particles. Phys Rev A 1976;13:2287.

[121] Haberland H, Hippler Th, Donges J, Kostko O, Schmidt M, Issendorff Bv. Melting of sodium clusters: Where do the magic numbers come from? Phys Rev Lett 2005;94:035701.

[122] Breaux GA, Benirschke RC, Sugai T, Kinnear BS, Jarrold M. Hot and solid gallium clusters: Too small to melt. Phys Rev Lett 2003;91:215508.

[123] Chen SW, Ingram RS, Hostetler MJ, Pietron JJ, Murray RW, Schaaff TG, et al. Gold nanoelectrodes of varied size: Transition to molecule-like charging. Science 1998;280:2098.

[124] See Ref. [98 and other articles on clusters in the same volume].

[125] Yoo S, Zeng XC. J Chem Phys 2006;124:054304.

[126] Zhao LZ, Lu WC, Qin W, Wang CZ, Ho KM. J Phys Chem A 2008;112:5815.

[127] Yoo S, Shao N, Koehler C, Fraunhaum T, Zeng XC. J Chem Phys 2006;124:164311.

[128] Yoo S, Shao N, Zheng XC. J. Chem. Phys 2008;128:104316.

Chapter 2

Clusters: An Embryonic Form of Crystals and Nanostructures

Khang Hoang*, **Mal-Soon Lee**[†], **Subhendra D. Mahanti**[†] **and Puru Jena**[‡]
**Materials Department, University of California, Santa Barbara, California, USA*
[†]Department of Physics and Astronomy, Michigan State University, East Lansing, Michigan, USA
[‡]Physics Department, Virginia Commonwealth University, Richmond, Virginia, USA

Chapter Outline Head

I. INTRODUCTION

A quantitative understanding of the lattice structure, stability, and electronic properties of perfect crystals has been possible due to a number of important developments in solid-state physics in the last century. First, the long-range periodicity of a perfect crystal allows one to use the Bloch's theorem and, in most cases, a unit cell contains one to a few atoms. Second, the development of density functional theory (DFT) and functionals for exchange and

Nanoclusters. DOI: 10.1016/S1875-4023(10)01002-8

correlation potential have permitted accurate determination of lattice structure and cohesive energies, and hence the relative stability, of various crystal phases. However, a quantitative understanding of imperfect systems that contain point defects (e.g., impurities and mono-vacancies) and extended defects (e.g., voids, impurity clusters, surfaces, interfaces, and grain boundaries) have been difficult to achieve because the loss of long-range periodicity invalidates the use of the Bloch theorem. This, however, can be avoided by considering a supercell that contains a large number of host atoms surrounding the defects, thus preventing the defects from interacting with each other. An artificial crystal is then formed by periodically repeating these supercells, and conventional band structure methods can be used to calculate defect properties. Limitations on computing power often do not allow the use of very large supercells, and hence the quantitative accuracy of the computed results may depend on the supercell size.

A parallel method was developed in the 1960s and 1970s where, instead of using Bloch's theorem and long-range periodicity, one models the defect with its surrounding host atoms as a cluster [1–3] and makes use of the real-space techniques based on the linear combination of atomic orbitals (LCAO) formalism. Here, the cluster is assumed to have the same structure as the host crystal. Local distortions due to the defect are accounted for by allowing the host atoms in the immediate vicinity of the defect to relax and minimizing the total energy. Just as the size of the supercell in the conventional band structure technique, the size of the cluster to model defect properties also became a nagging issue. In metallic systems, one could argue that the screening of the defect charge by the nearly free conduction electrons of the host limits the distance over which the defect influenced the host crystal structure. Hence, a "small" cluster may be adequate to model imperfect systems. On the other hand, in systems where the bonding is strongly covalent, one could argue that a large cluster is not necessary to treat the defects, as the interactions are short-ranged. Systems with strong ionic bonds may also be well represented by small clusters because of long-range screening effects. The dangling bonds associated with surface atoms in the cluster can be terminated with hydrogen [4]. The only way one can convince oneself that the size of the supercell or the cluster size is adequate is to calculate electronic and lattice properties as a function of the cluster size until convergence is achieved and compare them with available experiments. The question of how many atoms does it take to model a solid, however, could not be answered in an unambiguous way.

This scenario began to change over the last two to three decades. First, phenomenal growth in computing power could enable scientists to consider large supercells and/or cluster sizes and carry out calculations until the results converged with respect to size. Second, it has been possible to form clusters of varying size and composition in the gas phase using the laser ablation technique, mass isolate these using quadruple mass analyzer or time-of-flight, and study their electronic properties one at a time. One can study the size evolution of these properties and see at what point their properties agree with those of the corresponding crystals. One can finally hope to answer some of the fundamental questions: How many atoms are needed before a cluster looks and behaves like a crystal? When does a metal become a metal?

This chapter addresses these questions. Through examples, we will show that the answer to these questions depend upon the nature of the bonding in clusters and crystals and the property under investigation. We will then discuss our study first by taking a model system where exact answers are possible. We will study clusters composed of metallic elements, oxides, nitrides, halides, and tellurides. We will show that, while as few as a five-atom metal-oxide cluster may mimic the properties of the corresponding crystals, there are systems where even hundreds of atoms may still have nothing to do with the properties of their crystal counterpart.

The outline of the chapter is as follows: We will begin with a discussion of the structure and stability of metallic clusters. First, the clusters of metallic elements will be treated by a simple jellium model where calculations can describe how the electronic structure evolves with size until convergence is achieved. Second, these calculations will be repeated for clusters using actual atomic arrangements and first-principles theory. Models of covalently and ionically bonded clusters will include metal oxides, halides, and metal tellurides. Finally, we will demonstrate how clusters have bridged our understanding of diverse topics such as spintronics and hydrogen storage.

II. CLUSTERS AS MODELS OF THE BULK

A. Metallic Bonding

(i) Jellium Model

Metals are characterized by itinerant electrons, and in simple metals, these electrons are nearly free. When a point defect is introduced to a perfect metallic crystal, the itinerant electrons screen the defect within a short distance so that host atoms lying three to four nearest neighbors away do not feel the presence of the defect. Atoms within the nearest neighbor shell relax the most and this relaxation quickly disappears as one moves farther away. Thus, to understand the properties of defects in metals, one can model it as a cluster with the defect at the center surrounded by three to four nearest neighbor host atoms. Due to limitations in computing resources, the size of the cluster has to be limited. This leaves a nagging question: What should be the minimum size of a cluster that can adequately model bulk behavior? While a convincing answer to this question cannot be provided until one carries out systematic study as a function of cluster size until convergence is reached, it is easy to answer this question by focusing on the jellium model. In this model, the positive charges of metal ions are smeared to provide a uniform positive charge background which is compensated by a uniform distribution of electrons. The electron density of this jellium model is given by:

$$n_0 = Z/\Omega_0, \tag{1}$$

where Z is the valence of the metal atom and Ω_0 is the volume of the Wigner–Seitz cell. It is conventional to represent this density by the electron-gas density parameter r_s, which is defined as:

$$r_s = (4\pi n_0/3)^{-1/3}. \tag{2}$$

An impurity in this jellium model can be represented by embedding it in the uniform density with a charge equal to the nuclear charge of the impurity. For a hydrogen impurity, this is unity. The impurity will then perturb the electron density distribution around it whose behavior in the long range can be calculated exactly. This is known as the Friedel oscillation.

Hintermann and Manninen [5] set out to examine how the long-range behavior of the perturbed electron density around a hydrogen atom in the jellium model would look like as a function of cluster size. Here, the cluster represents a small sphere with a uniform charge distribution. Assuming R as the radius of this sphere, a cluster modeling the finite jellium can be given by:

$$n_+(\mathbf{r}) = n_0\Theta(\mathbf{R} - \mathbf{r}), \tag{3}$$

where Θ is the unit step function. In the limit $R \rightarrow \infty$, the cluster becomes the exact infinite jellium background. A hydrogen atom embedded in the jellium can be given by:

$$n_+(\mathbf{r}) = n_0\Theta(\mathbf{R} - \mathbf{r}) + \delta(\mathbf{r}). \tag{4}$$

Using the external charge in Eq. (4) as a perturbation, Hintermann and Manninen calculated the perturbed electron density around the hydrogen atom as a function of cluster radius R corresponding to the electron density of lithium and compared the result with its exact asymptotic limit, which is the Friedel formula. Their results are given in Figure 1. We see that the calculated radial electron density agrees with the exact limit when the cluster radius is 14.7 a.u. which corresponds to 92 Li atoms. Based on these results, one can conclude that about 100 metal atoms are needed for a simple free-electron metallic cluster to mimic bulk behavior.

Hintermann and Manninen also computed the energy levels as a function of the cluster size. These are discrete with energy gaps between quantum states decreasing as the cluster size increases. In Figure 2, we show their calculated integrated density of states (DOS) and how they compare with the exact limit. We see that the DOS of clusters exhibits a clear structure with a wide plateau below the Fermi level. These correspond to large gaps between the energy levels for a specific cluster size. Even for a cluster containing 125 atoms, the DOS does not converge to the infinite limit. The results in Figures 1

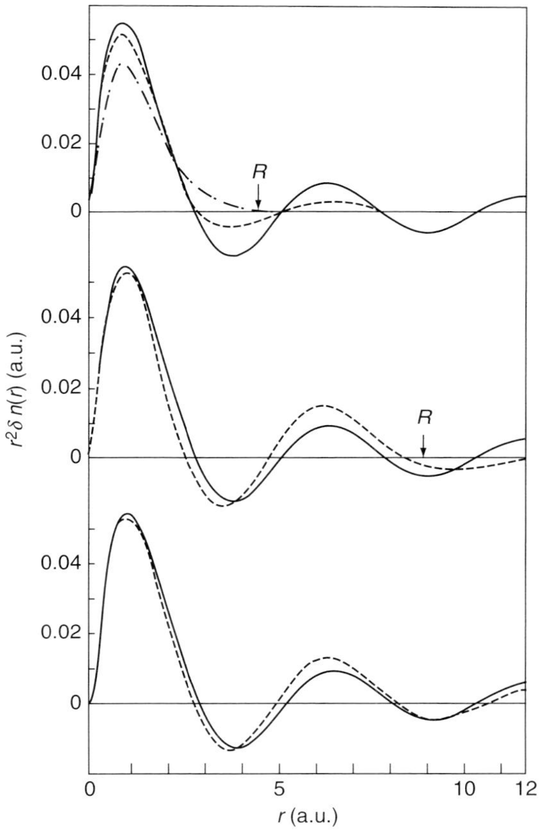

FIGURE 1 Radial-electron density around hydrogen for 3-, 22-, and 92-electron clusters (dashed lines). The solid line in each figure indicates the density for an infinite system. The dot-dashed line in the uppermost figure shows the density of a free hydrogen atom. The cluster radius is denoted by R (for the 92-electron cluster $R = 14.7$ a.u.). Taken from Ref. [5].

and 2 show that different properties evolve differently with the cluster size and there is no unique answer to how large a cluster has to be before it can approach the bulk behavior. It depends upon the property under consideration. This will be further illustrated in the next section where we discuss the evolution of cluster properties where the cluster is not considered within a jellium model but rather within an atomistic framework.

While the above jellium model of the cluster was used to demonstrate the size of a metal cluster that can mimic bulk properties, a pioneering work by Knight et al. [6] on Na clusters demonstrated another fascinating application of the jellium model that linked cluster science to nuclear physics. These authors observed conspicuous peaks in the mass spectra of Na clusters to occur at 2, 8, 20, 40, and so on, atoms. Using the same model as described above, the authors showed that these peaks corresponded to electronic shell closures similar to nuclear shell closures observed a long time ago. Nuclei containing 2, 8, 20, 40, and so on, nucleons were found to be very stable

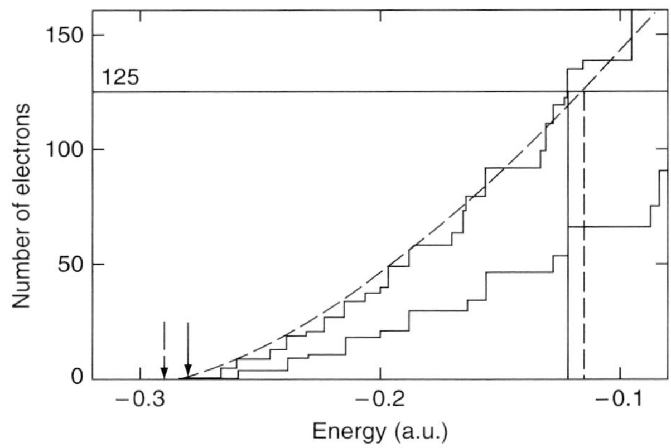

FIGURE 2 Integrated density of states of the 125-electron jellium cluster compared to that of the infinite jellium (dashed line). The lower solid line gives the integrated density of states for the spin-down electrons. The Fermi surface is denoted by a vertical line and the bottom of the band with an arrow. Taken from Ref. [5].

and these were referred to as magic numbers. It was shown that these corresponded to nuclear energy level closings of 1s, 1s 1p, 1s 1p 1d 2s, 1s 1p 1d 2s 1f 2p, and so on. Using the jellium model, Knight et al. [6] showed that similar electronic shell closings could give rise to magic numbers in Na clusters. This correlation not only bridged cluster science with nuclear physics, but also showed that even in small metal clusters electrons can be nearly free. The magic clusters possess large energy gaps between their highest occupied molecular orbital (HOMO) and the lowest unoccupied molecular orbital (LUMO). In addition, the magic clusters are less reactive than their neighbors. This work prompted many investigations where similar shell closings were explored in other nearly free-electron metals such as Be, Mg, Al, Cu, and so on. A unique example of the magic clusters is Al_{13}^-. Castleman and coworkers [7] showed that this cluster not only possesses a large peak in the mass spectra but is also less reactive toward oxygen than their neighbors. This demonstrated that one can manipulate not only the cluster size but also its charge to arrive at magic numbers. Later studies [8] showed that compound clusters can be designed so as to produce magic clusters. $Al_{12}X$ with X=C, Si, and Sn are examples of such clusters. Recent work by Hakkinen and coworkers [9] shows that one can also create magic clusters by choosing the right ligands to coat a cluster. Considerable efforts have been made to use magic clusters as building blocks of new materials [10].

(ii) Atomistic View

In spite of the success of the jellium model in accounting for the relative stability and occurrence of magic numbers in nearly free-electron metal clusters such as alkali metals, one has to realize that real atomic clusters consist of atoms that occupy precise sites and the clusters have precise shapes. An understanding of how clusters evolve with size can be achieved only by studying their geometries and properties one atom at a time. As pointed out in the introduction, the literature is rich with geometries of clusters of simple metals, transition metals, and semiconductors. There is also extensive literature on compound clusters composed of bimetallic species as well metal and nonmetal atoms.

To illustrate the atomistic view of metal clusters, we provide only one example by concentrating on Al. The geometries, average interatomic distances, relative energies, and electronic properties of Al clusters in neutral, cationic, and anionic states up to 15 atoms have been systematically investigated [11] as a function of size. In Figure 3, we show these geometries. Note that neither the geometries mimic their bulk behavior nor do they assume spherical shape. In most cases, the geometries undergo Jahn–Teller distortion. For clusters up to five atoms, the ground-state structures are planar as is the case with alkali-metal clusters. For neutral clusters containing 6–10 atoms, the geometries become three dimensional. Al_9 is the smallest cluster that develops a pentagonal arrangement of atoms, which is the precursor to icosahedric growth. In clusters containing 11–15 atoms, an interior atom with a bulk-like coordination emerges. For the 13-atom cluster, there are two nearly degenerate structures which can be candidates for the ground-state geometry: (1) a strongly Jahn–Teller distorted decahedron where the two pentagons join to form square faces and (2) a weakly Jahn–Teller distorted icosahedron. Their relative energies depend on the selected basis set, energy functional, and so on, as discussed by Rao et al. [12]. More accurate calculations using all electrons and the Perdew, Burke, and Ernzerhof model for exchange correlation show that the distorted icosahdron gives lower energy [13]. However, the distortions are quite small. The geometry of a negatively charged Al_{13}^- cluster with 40 electrons has a closed-shell structure and becomes a perfect icosahedron.

In Figure 4, we show the evolution of the average nearest neighbor distance, the coordination number, and the binding energy (BE) per atom of different Al clusters. With the exception of Al_2 and Al_6, the average nearest neighbor distances generally increase with cluster size and reach the bulk limit of 2.86 Å by the time the cluster contains 14 atoms. However, neither the coordination number nor the average BE approaches their corresponding bulk limit of 12 and 3.39 eV, respectively. Thus, different properties evolve differently.

The evolution of the electronic structure can be seen by examining the content of the s and p electrons of the HOMO level as a function of cluster size. Since the Al atom has a $3s^2$ $3p^1$ configuration and the s and p states are separated by an energy gap of 4.99 eV, it has been suggested [14] that in small clusters Al may behave as a monovalent atom. With increasing cluster size, the orbitals will overlap and Al will resume its traditional trivalent

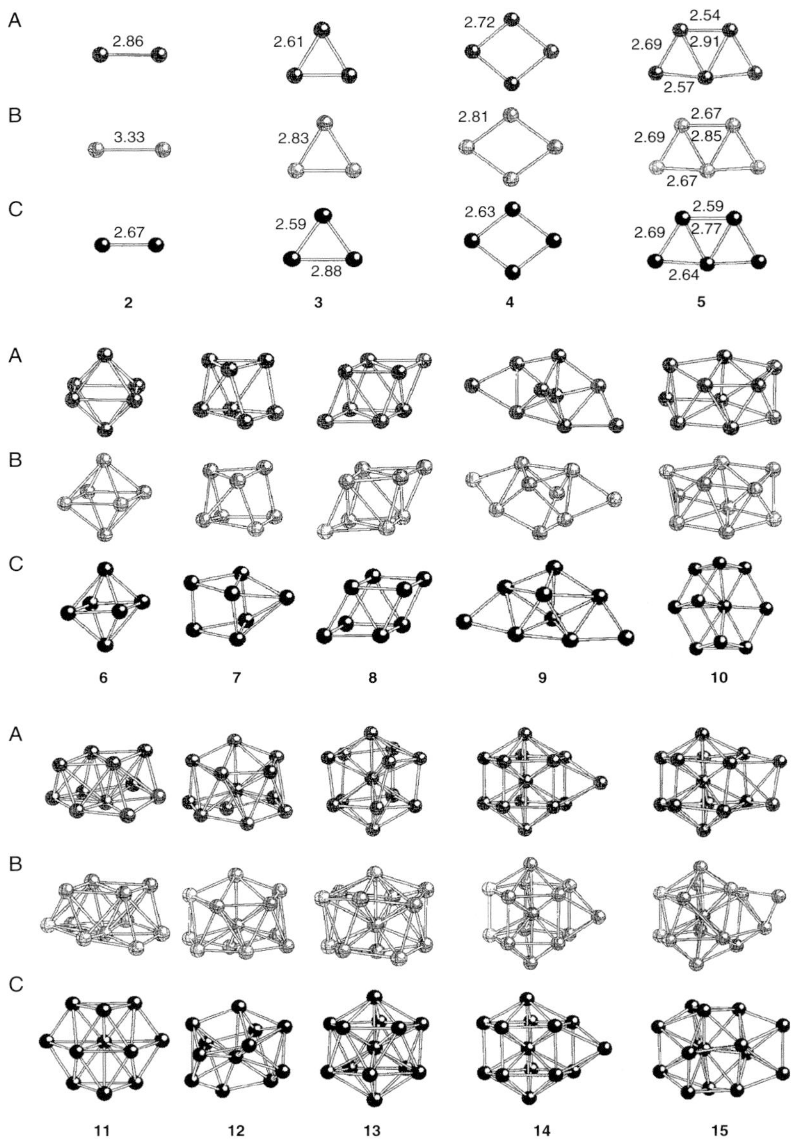

FIGURE 3 Ground-state geometries of (A) neutral, (B) cationic, and (C) anionic aluminum clusters containing 2–15 atoms (see Ref. [11]).

character. How large the Al cluster has to be before Al transitions from monovalent to trivalent behavior? This can be seen from Figure 5. Note that, in the atom, the valence electrons are composed of 66% s electrons and 33%

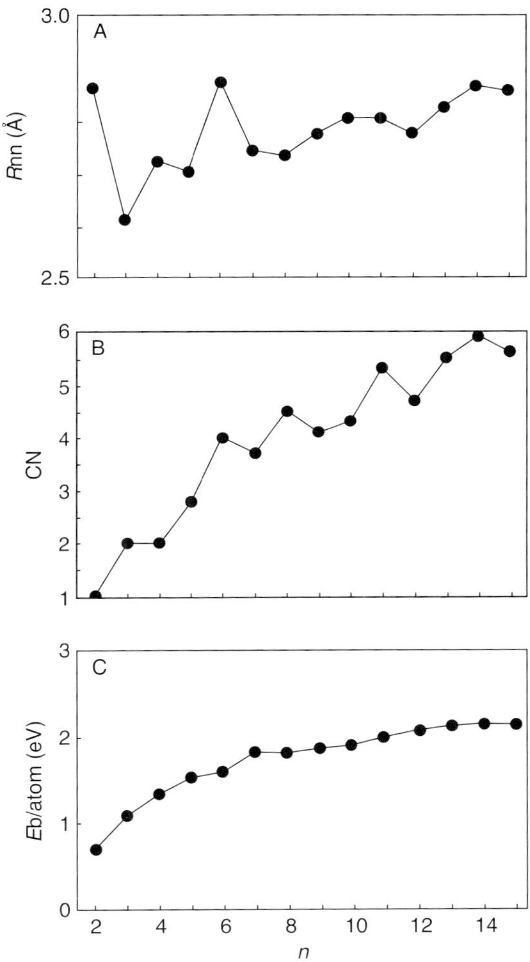

FIGURE 4 (A) Average nearest neighbor distance (B) coordination number, and (C) binding energy/atom as a function of size for neutral Al_n ($n=2$–15) clusters (see Ref. [11]).

p electrons ($3s^2$ $3p^1$). We see that with increasing cluster size the s content begins to decrease with a corresponding increase in the p content. For clusters with less than five atoms, the HOMO is clearly s-like while for clusters containing more than seven atoms it is p-like. Clusters with $5<n<7$ represent a region of transition. It is in this size range that the geometry becomes three dimensional and is accompanied by a sudden increase in the number of bonds and coordination number. Thus, if we consider aluminum to be trivalent in a cluster size where the s and p states overlap, this would make the critical size to be $n\approx 7$.

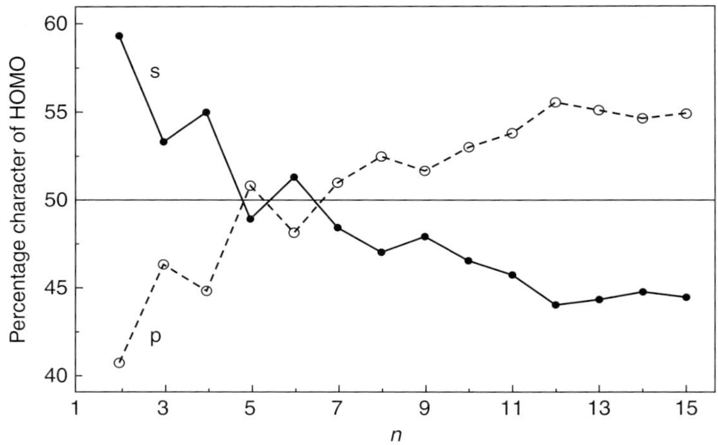

FIGURE 5 Concentration of s- and p-type electrons in the highest occupied molecular orbital (HOMO) of aluminum clusters (see Ref. [11]).

B. Covalently/Ionically Bonded Clusters

(i) Metal Oxides (WO_3) Clusters

As discussed above, small clusters do not mimic the structure and properties of their bulk. For example, metallic elements exhibit fivefold symmetry and icosahedric growth [15], which is nonexistent in crystals. The building block of silicon clusters is a tricapped trigonal prism (TTP) [16,17], which is completely different from the bulk diamond structure. Some of the most stable carbon clusters have the famous fullerene geometry [18], whereas bulk carbon exists in either planar or tetrahedral structures. For coinage metals such as Au, Ag, and Cu, it has been observed that the clusters consisting of fewer than 50–60 atoms exhibit dissimilar electronic structures compared to their respective bulk counterparts, and only for larger clusters bulk-like electronic structures appear [19]. In full-shell systems, the interaction between the neighboring groups in bulk is relatively weak, resulting in fast convergence to bulk properties. Among others, systems to be named are noble gases, alkali halides, water, and so on. In contrast to the pure metal clusters and full-shell systems, it is interesting to know how fast the properties of inorganic materials such as oxides converge to the bulk properties as a function of cluster size. Also, it will be interesting to know if there are some properties which are peculiar to the metal-oxide clusters.

Sun et al. [20] examined this problem by considering W_nO_m clusters and demonstrated that in materials characterized by strong covalent or ionic bond, very small clusters have properties similar to their bulk. To shed light on how the properties of oxide clusters evolve as a function of the cluster size, they studied theoretically as well as experimentally $(WO_3)_n$ clusters with $n=1$–4.

In contrast to the generally accepted view for the metallic clusters that more than 50–60 atoms are required for the onset of the bulk structures, geometric and electronic structures of the $(WO_3)_4$ cluster turned out to be nearly identical to those of the WO_3 bulk crystal, indicating that the electronic properties of oxide clusters can converge to the bulk values more rapidly than those in pure metal clusters. They showed that the interatomic bond lengths, atomic arrangements, relative trends in stabilities, electronic structures, and HOMO–LUMO gap (this corresponds to the indirect bandgap in bulk semiconductors) of the $(WO_3)_4$ cluster are nearly same as those in bulk tungsten oxide. In addition, a systematic study of W_2O_m ($m<6$) clusters showed that the W–W bond was broken when the oxygen content reached the bulk composition, namely, at $m=6$. The adiabatic electron affinities (EA) and vertical detachment energies (VDE) of $(WO_3)_n$ ($n=1$–4) clusters are nearly the same, indicating that the bulk bonding characteristics are already developed even in the smallest cluster having the bulk composition.

To further demonstrate how closely these clusters resemble the bulk structure, we compare the geometry of W_4O_{12} cluster with that cut out from the monoclinic structure of bulk tungsten oxide in Figure 6. The internal W–O bond length in the W_4O_{12} cluster is 1.92 Å, which is very close to the bulk value of 1.89 Å. Since in the free cluster the terminal W–O bonds are different from the internal bond due to lack of proper coordination, and in bulk WO_3 the W–O bonds form zigzag chains with varying bond distances, Sun et al. [20] have calculated the average W–O bond length. For a free cluster, it is 1.83 Å while for the monoclinic crystal the same average yields 1.80 Å. The nearest W–W interatomic distance in the W_4O_{12} cluster is 3.80 Å while it is 3.7 Å in bulk WO_3. Similarly, the two O–W–O bond angles in W_4O_{12} cluster agree very well with those in the crystalline phase. These comparisons clearly demonstrate that the geometry of the W_4O_{12} cluster is as if it is a fragment of the bulk.

To investigate the extent to which clusters yield the electronic properties of the bulk, we compare the calculated energy gaps of $(WO_3)_n$ ($n=1$–4) clusters with experimental values. Sun et al. [20] obtained this gap first from

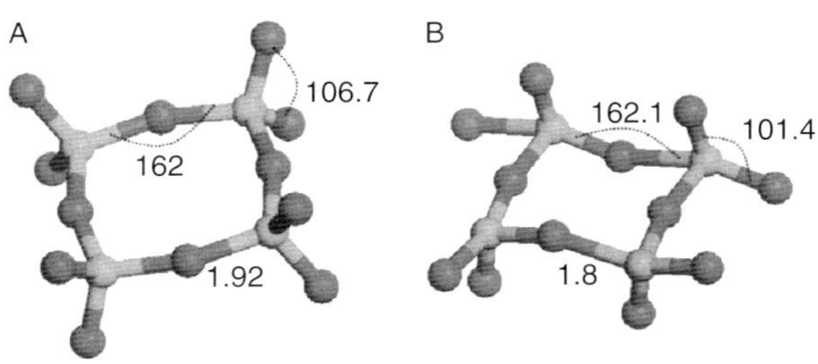

FIGURE 6 Comparison between the equilibrium geometry of (A) W_4O_{12} cluster and (B) the corresponding fragment taken from the monoclinic phase of bulk WO_3 [20].

the difference in the two calculated transition energies as an electron from the doublet anions makes a vertical transition to the singlet and triplet states of the corresponding neutrals. The calculated HOMO–LUMO gaps for $(WO_3)_n$ are 1.10, 2.07, 2.34, and 2.76 eV, respectively, for $n=1$, 2, 3, and 4. The corresponding experimental values are 1.5, 2.5, >1.5, and >1.2 eV, respectively. Note that both calculated gaps increase with cluster size. It is interesting to note that the bulk band gap lies between 2.6 and 3.5 eV. Thus, the HOMO–LUMO gap of the $(WO_3)_4$ cluster already mimics the bulk behavior.

(ii) Alkali Halide Clusters

Alkali halide clusters are classic examples of ionic clusters, written as M_nX_m, where M is an alkali atom (cation) and X is a halogen atom (anion). Because of the dominant ionic nature of the bond between M and X, they are interesting systems to model bulk crystals because the bulk structural motif emerges at rather small cluster sizes [21]. In spite of their bulk structural motif, the cluster size and stoichiometry play important roles in their detailed structural and electronic properties.

Stable alkali halide clusters are formed out of NaCl structure by taking i, j, and k atoms (including cations and anions) along the three cube axes. They are denoted as $(i \times j \times k)$ clusters. While many alkali halide clusters show a cuboid structure with sixfold coordination (corresponding to the NaCl structure), some show a cuboid structure with eightfold-coordinated CsCl structure. Some others also show zinc blende structure [22]. Here, we will confine our discussions to clusters with cuboid structure only, where $i=j=k$. Even for these, there is a structural difference between cuboids with i odd and i even. The total number of atoms (including both alkalis and halogens) $N=i^3$. For odd i, N is odd and can be written as $2n+1$. Examples are $(MX)_nM$ or $(MX)_nX$. These types of clusters (Class I) have been studied extensively because of the simplicity of their structure and excess electron or hole localization problem [23–26]. Some of the interesting electronic properties of these clusters are controlled by the localization characteristics of this excess electron (hole). Clusters for which N is even (Class II) are $(MX)_n$ and can be perceived as $(MX)_nM$ with a cation vacancy or $(MX)_nX$ with an anion vacancy. The smallest such cluster is a $(2\times2\times2)$ cluster, M_4X_4.

From an electronic structure point of view, metal-excess alkali halide clusters (both neutral and charged) have been of great experimental interest. The neutrals M_nX_{n-1} display a wide range of excess-electron-dominated behavior [23–26]. Excess electron and excess hole states have also been investigated in charged clusters [27,28]. As far as we are aware, excess hole states in neutral clusters of the type $X(MX)_n$ have not been investigated extensively. From the experimental side, Rayane et al. were the first to measure the electric dipole susceptibility of one excess electron in Na_nF_{n-1} clusters [29].

a. Atomic Structure

Earlier theoretical investigations aimed at understanding the structures, structural transitions, and melting in alkali halide clusters by employing empirical potentials and molecular dynamics simulations. Not much has been done using *ab initio* methods, particularly for large clusters. One nice thing about these ionic clusters is that one can start with a simple ionic model to understand their structural properties. Once we assign $+1$ charge to the cations and -1 charge to the anions, the parameters that control the structure is the ionic size difference. This can be characterized by the ratio $r = r_+/r_-$ where $r_+(r_-)$ are the ionic radii of the cation (anion). The other factors determining the structure are the values of i, j, and k. Here, we discuss a few examples of sodium halide clusters with cuboid ($i = j = k$) structure. These results were obtained using DFT. These results, however, are generic.

The ground-state structure, particularly for the small clusters, depends sensitively on the ionic size mismatch between the cation and the anion. The ionic radii taken from the bulk sodium halides give values of $r = 0.70$, 0.52, and 0.44 for F, Cl, and I, respectively. Thus one expects the largest deviation from a simple cuboid geometry for the NaI clusters. This is indeed seen in Table 1, where we give different structural parameters. The deviations from the usual 90° for the X–Na–X (defined as α) and Na–X–Na (defined as β) range from approximately 25% to 10% for I to approximately 4% for F, consistent with the values of the r parameter. We also show the structures of Na_4I_4, $Na(NaI)_{13}$, and $I(NaI)_{13}$ in Figure 7, which are the most distorted ones among three different halide clusters.

One can draw several general conclusions regarding the structure of these alkali halide clusters. The bond lengths are shorter than those seen in crystals. When the difference between the ionic radii of the cation and the anion is bigger, the Na–X–Na and X–Na–X angles deviate more from the usual 90°. We find that a cluster with a cation at the center, that is, $Na(NaX)_{13}$, shows more surface distortion compared to $X(NaX)_{13}$, where a anion is at the center.

The BEs per atom are given in Table 2 in units of electronvolts per atom. Clearly the clusters, both stoichiometric and nonstoichiometric ones are weakly bound compared to the bulk. However, the BE increases with increasing size of the cluster. The reduction of the BE on going from F to I in bulk (increasing lattice constant) is also reflected in the clusters.

b. Electronic Structure and Charging-Induced Atom Emission

Theoretical studies on the electronic structure of alkali halide clusters have focused on several aspects. In addition to the usual ionization and affinity studies of clusters of the type M_nX_n and $M_nX_{n-1}^+$, electron localization, insulator–metal transition, and phase separation effects have been studied in alkali halide clusters of the type M_nX_m for different $n-m$ values. For $m < n$, that is, in the halogen deficient case, one looks at the properties of excess

TABLE 1 Bond Lengths (Å) and Bond Angles (in degrees) for Relaxed Clusters

		F	Cl	I
(a) $(NaX)_4$				
Bond length	Na–X	2.16	2.64	3.04
	Na–Na	2.97	3.41	3.71
	X–X	3.14	4.01	4.75
Angle	Na–X–Na	87	80	75
	X–Na–X	93	99	103
(b) $Na_{14}X_{13}$				
Bond length	Na–X	2.17–2.31	2.62–2.83	2.99–3.31
	Na–Na	3.06–3.26	3.68–3.93	4.18–4.51
	X–X	3.22	4.00	4.68
Angle	Na–X–Na	87–92	85–88	83–86
	X–Na–X	92	100	103
(c) $Na_{13}X_{14}$				
Bond length	Na–X	2.19–2.31	2.71–2.87	3.09–3.35
	Na–Na	3.18	3.83	4.36
	X–X	3.10–3.27	3.83–4.06	4.43–4.73
Angle	Na–X–Na	93	90	88
	X–Na–X	88–93	90–97	91–100

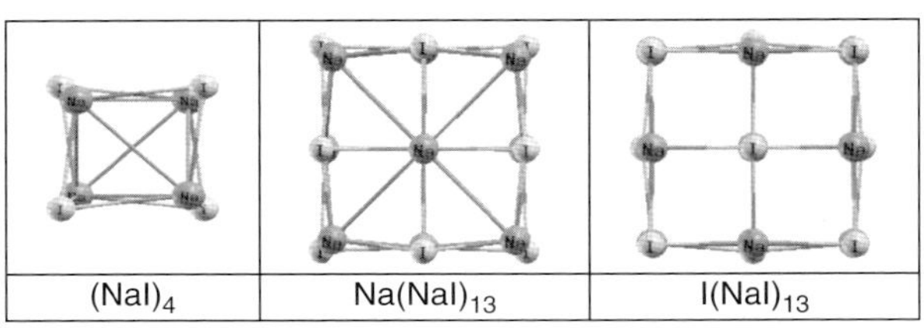

FIGURE 7 The lowest energy structures of NaI clusters for small stoichiometric and non-stoichiometric clusters indicating structural distortions from cubic symmetry.

electrons, whereas for $n<m$ one looks at the properties of excess holes [30–32]. The metal-excess alkali halide clusters (both neutral and charged) have been of great experimental interest, where the neutrals M_nX_{n-1} display

TABLE 2 Binding Energies (eV/Atom) of Solids NaX Are Compared with Those of Clusters

X	Bulk	$(NaX)_4$	$Na(NaX)_{13}$	$X(NaX)_{13}$
F	4.39	3.89	4.00	4.06
Cl	3.42	3.06	3.11	3.14
I	2.70	2.45	2.47	2.49

a wide range of excess-electron-dominated behavior [24,33]. In this review, we focus on the electronic structure of these clusters.

Neutral alkali halide clusters with one extra cation (therefore one extra electron), for example Na_nF_{n-1}, are essentially built around the rock salt structure. There is an interesting correlation between the value of $2n-1$, the number of atoms in the cluster, and the nature of the electronic state associated with the extra electron. For example, when $2n-1=i\times j\times k$, the structure of the cluster is a cuboid portion of the bulk rock salt structure (size $i\times j\times k$), and the excess electron is weakly bound to the surface of the cluster with a small value of the adiabatic ionization potential (AIP). For a $Na_{14}F_{13}$ cuboid, one finds IP approximately 2 eV [30]. On the other hand, if $2n=i\times j\times k$, the structure is a cuboid lattice with an anion vacancy, like a bulk F-center. In this case, the excess electron is localized near the anion vacancy and has a very large (~3.5 eV) AIP. The smallest such cluster is $n=4$, $i=j=k=2$ and the resulting system Na_4F_3 has an AIP value of 3.48 eV. Actually, there is a third class when $2n-2 = i \times j \times k$. In this case, the cluster can be thought of as an extra Na atom attached to a cuboid alkali halide, for example Na_5F_4. The extra electron is strongly localized near the extra Na atom and is much more strongly bound than the other two cases. AIP values for these types of clusters range between 3.1 and 4.0 eV for $n=3$, 5, 7, 10, and 11 [30]. Durand et al. have studied new types of localization sites such as edge states, R centers, and other surface defects for an excess electron in Na_nF_{n-1} clusters [34].

Earlier theoretical attempts to understand the nature of the excess-electron state in alkali halide clusters were made by Landman et al. [35] and by Durand et al. [36]. Using a pseudopotential model to take into account the interaction between the extra electron with closed-shell cations and anions, they came up with the picture of a weakly bound electron hovering over the surface of the cluster [36]. This model was used to derive the ground-state structure, energetics, and electronic properties of one-excess-electron Na_nF_{n-1} clusters in the range $2\leq n\leq 29$. The electron localization and structural characters were closely related. Based on this observation, they classified the clusters into five types: two of them exhibiting rather strong localization namely F-centers and Na-tail structures, the others exhibiting a less bound electron localizing in a surface state, in an edge state, or localizing on an atom-depleted face of the cluster.

In addition to the localization property of the excess electrons and holes, an electronic property of alkali halide clusters that has been extensively investigated is their electrical polarity (EP). EP of clusters, due either to their permanent or to the induced dipole moment, is an important quantity that determines the interaction of a cluster with an external electric field. In metallic clusters, the polarizability α is very close to the classical value, that is, $\alpha = R_c^3$, where R_c is the radius of the cluster. However, in ionic crystals, it is possible to exceed this classical result. An example is $Na_{14}Cl_{13}$, where the excess electron occupies a loosely bound surface state outside a closed shell and the system can undergo a second-order Jahn–Teller instability to a polar state. Rayane et al. [29] and Durand et al. [37] calculated the static electronic polarizability and the permanent electric dipole moment of a $Na_{14}F_{13}$ cluster using a simple one-electron model and found that there was indeed a strong coupling between the deformed electronic charge distribution (in the presence of an external electric field) and both the static and dynamic ionic displacements.

Although earlier calculations using a one-electron picture were successful in giving a qualitative (semiquantitative) picture of the excess-electron states and EP, one has to carry out more realistic calculations using *ab initio* methods for a more accurate description of their physics. DFT-based studies of EP for both closed-shell $(NaF)_n$ and open-shell clusters $(NaF)_nNa$, $n = 13$ have been carried out by Schmidt et al. [38]. They found that in closed-shell clusters the ratio of vibrational to electronic contributions to the total electric polarizability increased dramatically with n. In contrast, due to the loosely bound nature of the extra electron, open-shell clusters show a greatly enhanced electronic polarizability. Also they found that in the open-shell clusters it was important to treat all electrons quantum mechanically, not just the loosely bound excess electron. A distorted polar structure with C_{3v} symmetry, which was found to have the lowest energy when only the loosely bound electron was treated quantum mechanically, becomes cubic with O_h symmetry when all the electrons are treated quantum mechanically. This observation shows clearly the delicate balance between different structures, and one has to therefore carry out careful fully quantum mechanical calculations to understand the geometry of the alkali halide clusters and associated electronic properties, particularly for the open-shell systems. For example, systems with an extra electron behave quite differently from systems with an extra hole. This is illustrated in Figure 8, where we show the charge densities associated with the HOMOs for these two cases. The excess electron in $Na_{14}Cl_{13}$ (Figure 8A) is loosely bound to the surface. In contrast, the excess hole (Figure 8B) in the case of $Na_{13}Cl_{14}$ is localized around the three Cl atoms along the cube diagonal. The same trend is also seen in NaI and NaF clusters (figures not shown).

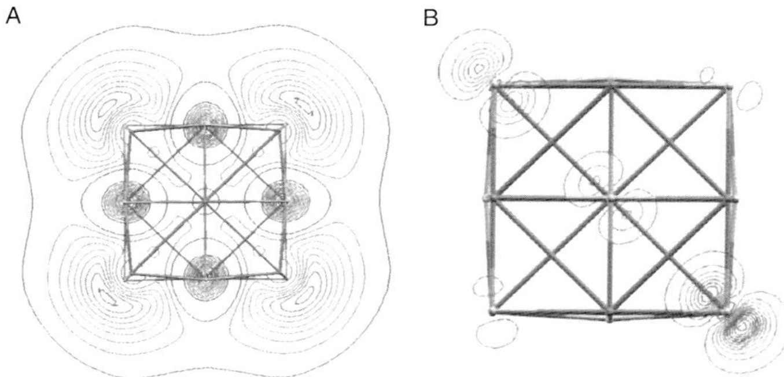

FIGURE 8 Charge densities associated with the HOMO states of (A) $Na_{14}Cl_{13}$ (electron excess) and (B) $Na_{13}Cl_{14}$ (hole excess) clusters.

The other interesting study vis-à-vis the electronic polarity of clusters is the importance of many-body effects [39]. By using a simple model, the authors looked at the polarizability of a few surface electrons in a small cluster. They found a significant increase in the polarizability from its classical value (R_c^3), in agreement with the earlier findings of Durand et al. [37]. The enhancement, however, disappears when the number of surface electrons ("metallic electrons") is increased.

Finally, we would like to point out some interesting recent theoretical studies on the energetics of atom emission from clusters. Ceresoli et al. have carried out *ab initio* DFT-based studies to establish the energetics of extraction of neutralized corner atoms from neutral and charged NaCl nanocubes [40]. In general, detachment of ions, atoms, and molecules from a bulk alkali halide crystal is an energetically expensive process. For example, the energy required to remove a NaCl molecule from bulk NaCl is approximately 1.8 eV because this involves breaking several ionic bonds. In clusters, the physics is somewhat different. Different atoms have different local environments. If by adding an electron or a hole one of the surface or edge or corner atoms can be neutralized locally, then it may be easy to pull this atom away from the cluster. One, of course, has to take into account the energy required to add the extra electron or hole. Ceresoli et al. examined the emission of neutral Cl and Na atoms upon addition of a hole (for Cl) and an electron (for Na) from NaCl clusters. The extraction energy (after taking away or donating an electron) for corner alkali (halogen) is found to be 0.6 eV (0.8 eV). The added electron or hole gets localized near a corner Na^+ or Cl^- ion, thereby neutralizing these ions. From a practical point of view, the electron-induced extraction of the Na atom appears to be more feasible. First, the electron addition process is exothermic. An electron coming from vacuum will lower its energy by attaching to the nanocube corner. This energy is approximately 1 eV, out of which 0.6 eV is required to extract a neutral Na atom. Thus, it is possible to make a good source of neutral alkali atoms by synthesizing NaCl nanocubes and charging them negatively.

(iii) Thermoelectric (Lead Chalcogenide) Clusters

Bulk crystals of lead chalcogenides (PbTe, PbSe, and PbS) have been of great interest over the last several decades because of their applications in thermoelectric, photovoltaic, and infrared devices. Recently, colloidal nanocrystals of these systems have been synthesized in order to see whether embedding these nanoparticles in different types of bulk matrices will give improved performance [41–46].

In contrast to the alkali halides where ionic bonding dominates the physics, in lead chalcogenides, the bonding between the cations and neighboring anions have both ionic and covalent characters; the degree of covalency changes on going from S to Te, the latter being more covalent. In addition, the valence electrons associated with Pb cations are in the p orbitals. Due to the increased strength of covalent bonding, one should expect to see larger structural relaxation effects in lead chalcogenide clusters compared to the alkali halides. In addition, neutral clusters with excess cation such as $Pb_{14}Te_{13}$ have two loosely bound extra electrons which should move on the surface of the cluster, correlating with each other. Their localization properties should also differ from the excess electron in the $Na_{14}Cl_{13}$ cluster.

Nano- and microcrystals of various morphologies, including face-open nanoboxes, microflowers, semimicroflowers, cubic nanoparticles, spherical nanoparticles, nanorods, and so on, have been observed in experiments [41–46]. Theoretical studies have also been carried out for PbS, PbSe, and PbTe (nano) clusters [47–50]. Compared to PbSe and PbS, PbTe has stronger surface relaxation [51,52]. Here we discuss the structural properties and energetics of PbTe clusters of different sizes (up to 343 atoms) and shapes (spherical and cube-shaped). Structural optimization, total energy, and electronic structure calculations were performed within the DFT formalism. We treated the two outermost electron shells of the constituent atoms as valence electrons and the rest as cores. Scalar relativistic effects (mass–velocity and Darwin terms) were included. For computational reasons, spin–orbit interaction was not included in these calculations. Cluster calculations were carried out using a supercell model where the clusters from the neighboring supercells were separated by a vacuum of approximately 10 Å.

In order to study the effects of surface relaxation on the structural properties of PbTe clusters, two classes of cube-shaped clusters, stoichiometric and nonstoichiometric, are considered. Both classes are constructed as cubes $(i \times j \times k)$, $i=j=k$, cut from the bulk. For the stoichiometric clusters, $i=2$, 4, 6,. . .·and for the nonstoichiometric clusters $i=3$, 5,. . .. The compositions of these stoichiometric and nonstoichiometric clusters are Pb_nTe_n and $Pb_{n+1}Te_n$, respectively. We use as starting point the bulk lattice constant $a=6.55$ Å for PbTe. All the surfaces are (001) type.

In Figure 9A–C, structures of several Pb_nTe_n clusters ($n=4$, 32, 108) are shown. We start with Pb_4Te_4, the tetramer of PbTe, which is the smallest

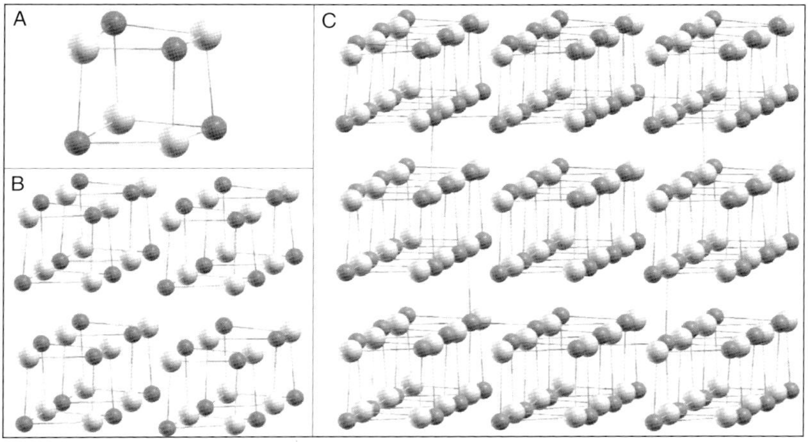

FIGURE 9 Relaxed structures of the stoichiometric cube-shaped Pb_nTe_n clusters: (A) $n=4$, (B) $n=32$, and (C) $n=108$. Large balls are for Pb and small balls are for Te.

cluster under investigation. The relaxed structure of Pb_4Te_4 is a distorted cube (see Figure 9A). The Te–Pb–Te angle (α) and the Pb–Te–Pb angle (β), which are both equal to 90° in the unrelaxed structure, become $\alpha=96.1°$ and $\beta=83.6°$ for the distorted relaxed structure. The deviation of α and β from 90° is quite large, approximately 7%. If we use the ionic radii for Te^{2-} and Pb^{2+} as $r_{2-}=2.21$ Å and $r_{2+}=1.06$ Å, respectively, then the r parameter $(r_{2+}/r_{2-})=0.48$, which is very close to that for Na^+ and I^-. For Na_4I_4, we found $\alpha=103°$ and $\beta=75°$.

The structural distortions seen in PbTe is a result of the asymmetric surface relaxation of cations and anions [51,52]. The Pb–Te bond length is also reduced considerably by approximately 6.8%. Similar results are found for Pb_4Se_4 and Pb_4S_4 (with the calculated bulk lattice constant $a=6.20$ and 5.99 Å, respectively). The reduction in the Pb–Se (Pb–S) bond length is 8.1% (9.1%), and the deviation of α and β from the right angle is approximately 4% (for Pb_4Se_4) or 1.6% (for Pb_4S_4). This indicates that the distortion of the relaxed structure of Pb_4Q_4 decreases in going from Q = Te → Se → S. Similar reduction in distortion was found in NaX, X = I → Cl → F. This reflects the effect of the change in the r parameter on going from Te (0.48) → Se (0.53) → S (0.58). Of course, the physics of the relative importance of ionic and covalent contributions to the bond is reflected in their r parameters.

For larger stoichiometric Pb_nTe_n clusters ($n=32$, 108), there is a local fragmentation which we define as "tertramerization"; the original cubes tend to split into tetramers (Pb_4Te_4), which are the distorted cubes with the eight atoms connected in Figure 9B and C. The intertetramer bond lengths are larger than the intratetramer ones by ~5–12%, where the differences are larger for the outer bonds and decrease toward the center of the cluster. For $Pb_{32}Te_{32}$, the bond lengths for the intra- and intertetramers are, respectively, 3.12 and 3.46 Å, a rather large difference of approximately 11%. The distortion of the tetramers is also larger for those near the surface. In the cluster with $n=108$ (see Figure 9C), the deviation of α and β from the right angle in the tretramer at one of the corners of the cluster is approximately 7.5%, whereas in the central tetramer it is approximately 5%. Without going to larger cluster sizes and based on what one learns about the surface relaxation in the PbTe(001) surface [52], one can now state that the atoms that are near the surfaces of the Pb_nTe_n nanoclusters tend to tetramerize (i.e., split into tetramers of PbTe) whereas those far away from the surfaces preserve their bulk geometry.

The structures of nonstoichiometric $Pb_{n+1}Te_n$ clusters were also studied for $n=13$, 62, and 171. We present the structures for $n=13$ and 62 in Figure 10. The lowest energy structure of $Pb_{14}Te_{13}$ (Figure 10A) is significantly distorted, where one of the corner Pb atoms forms a tetrahedron with the nearest Pb atoms. This tetrahedron is capped by four nearest Te atoms including the one at the center of the cube. These rearrangements result in a large displacement from symmetric cubic positions. For larger nonstoichiometric clusters, we observe "tetramerization" as seen in large stoichiometric clusters. For $Pb_{63}Te_{62}$ the intratetramer bond lengths are approximately 3.12 Å, and the bond length between the tetramers and neighboring ions is

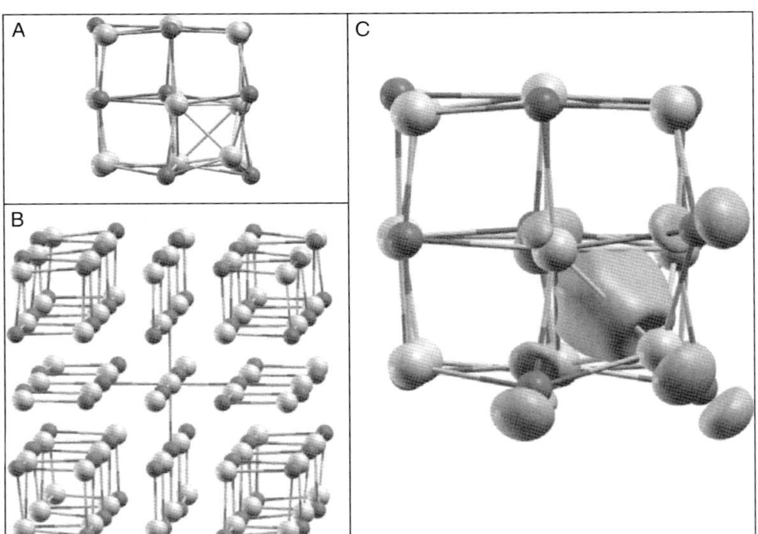

FIGURE 10 Relaxed structures of the nonstoichiometric cube-shaped $Pb_{n+1}Te_n$ clusters: (A) $n=13$, (B) $n=62$, and (C) the partial charge density of HOMO for $n=13$. Larger balls are Pb and smaller balls are Te.

approximately 3.38 Å, a difference of 6% which is a factor of approximately 2 smaller than that seen in stoichiometric $Pb_{32}Te_{32}$ discussed above.

We have also studied the charge distribution associated with the extra two electrons in the $Pb_{14}Te_{13}$ cluster to see whether there is any difference from the $Na_{14}Cl_{13}$ case (see Figures 8A and 10C). The charge density associated with the HOMO is strongly localized near one of the corner Pb atoms, forming a strong bond with the Te at the center. This strong charge localization is accompanied by a large local structural distortion of this Pb and its surrounding three Pb atoms and three Te atoms. Clearly, the PbTe cluster shows a dramatically different behavior compared to the NaX (X = F, Cl, I) clusters.

In addition to the cube-shaped clusters, spherical clusters of PbTe have also been constructed. These are built by considering all the bulk Pb and Te atoms contained in a sphere (centered on the Pb–Te bond). These spherical clusters are, therefore, stoichiometric with the same numbers of Pb and Te atoms; the composition is Pb_nTe_n. In Figure 11, we plot the total energies per formula unit (f.u. = PbTe) of different spherical and cube-shaped stoichiometric Pb_nTe_n clusters (relative to the bulk PbTe energy) with and without relaxation as a function of the cluster size. The energy decreases as the cluster size is increased. As one increases the cluster size further, the energy should reach the bulk value (the zero level in Figure 11). As can be seen clearly from the figure, the cube-shaped clusters are lower in energy than the spherical ones. For $n = 108$, the energy difference is approximately 0.19 eV/f.u., which

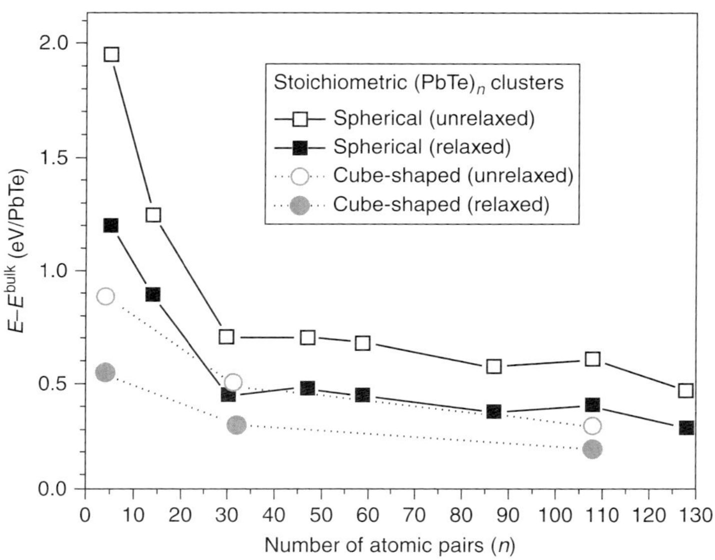

FIGURE 11 Total energy of the stoichiometric (spherical- and cube-shaped) Pb_nTe_n clusters, with respect to the bulk PbTe energy, as a function of cluster size.

is much larger in smaller clusters. This suggests that the cube-shaped Pb_nTe_n clusters with the (001) surface are more stable energetically, which is consistent with the observation of nanoscale PbTe cubes in experiments [42].

The relaxation energy, which is the difference between the "relaxed" and "unrelaxed" curves in Figure 11, also decreases as the cluster size increases. The relaxation energy (per f.u.) goes from 0.75 ($n=5$) to 0.16 eV ($n=128$) for the spherical clusters, whereas for the cubic clusters it goes from 0.34 ($n=4$) to 0.10 eV ($n=108$). The change in the relaxation energy with increasing n for a given type of cluster results from the change in the ratio of the number of atoms on and near the surface to the number of atoms inside the cluster. The relaxation energies in the cube-shaped clusters are smaller than in the spherical ones because the latter usually have atoms on the surface which are loosely bound to the rest of the cluster.

Figure 12A and B gives the total electronic DOS of the cube-shaped Pb_nTe_n ($n=108$; ~1.6 nm in size). The overall DOS of the cluster resembles very well that of the bulk. However, the energy difference between the HOMO and the LUMO is approximately 1.22 eV, which much larger than the bandgap of bulk PbTe (of ~0.8 eV), as expected, because it is well known that quantum confinement effects shift the valence band maximum down to lower energies and the conduction band minimum to higher energies. Note that the true bandgaps will be considerably smaller than their values obtained without spin–orbit interaction. An analysis of the partial charge densities shows that the HOMO is predominantly Te atoms which are at the corners of the cubic cluster. They form, therefore, the surface state(s) of the cluster. These states are associated with the unbonded states of the corner Te atoms. The LUMO, on the other hand, consists of predominantly Pb atoms that are in the middle of the cluster.

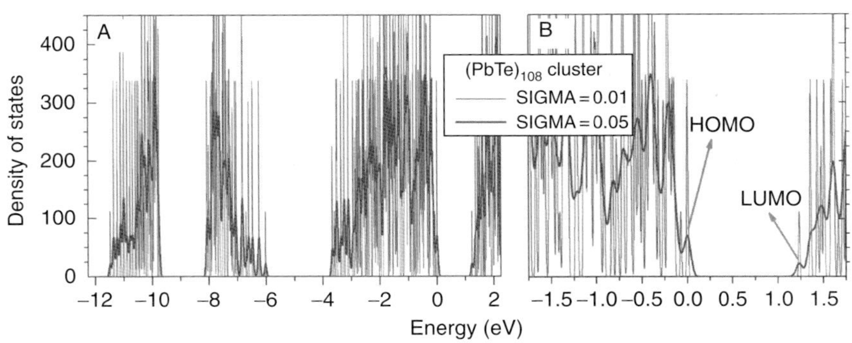

FIGURE 12 (A) DOS of the stoichiometric cube-shaped Pb_nTe_n ($n=108$) cluster and (B) the blowup of the DOS showing the HOMO and LUMO. The states were Gaussian broadened by SIGMA$=0.01$ and 0.05 eV. The zero of the energy is set to the highest occupied level.

III. CLUSTERS AS MODELS FOR UNDERSTANDING COMPLEX MATERIALS

Clusters as models have played an important role for understanding the electronic and magnetic properties of complex materials where local atomic structure dominates the interaction. In this section, we provide two examples that have attracted considerable attention in recent years. One of these is the origin of magnetism in dilute magnetic semiconductors (DMSs) and the other is materials for hydrogen storage. These are illustrated in the following.

A. Spintronics (ZnO)

DMSs offer the possibility of creating new electronic devices where the electron spin as well as its charge can be used. Spintronics, which exploits an electron's spin degree of freedom to store and carry information, has the potential to revolutionize the electronics industry. Among the DMS materials, ZnO and GaN are the semiconductors of choice because of their large bandgaps. For magnetic functional devices, the ultimate goal is to make DMS materials ferromagnetic (FM) at room temperature so that the spin as well as charge of the carriers can be coupled with an external magnetic field to produce new devices.

Until recently, the basic method of introducing magnetism into DMS has been to dope it with transition metals such as V, Co, Cr, Mn, and Fe [53–57]. In these systems, the magnetic moments are localized at the transition-metal sites and carriers are needed to couple these moments. In spite of a great deal of experimental and theoretical work, DMS materials have not lived up to their promise as spintronics materials. Much of the difficulty arises from the strong dependence of experimental results on sample preparation conditions. Another approach that induces ferromagnetism in ZnO has been through the introduction of vacancies or by capping the surface of ZnO nanoparticles with organic ligands. We discuss how clusters have been able to provide an understanding of the origin of ferromagnetism in Mn-doped GaN and ZnO nanoparticles coated with N- and S-containing ligands.

Following the discovery of ferromagnetism in (Ga,Mn)As [58] and the subsequent theoretical prediction [59] that Mn-doped GaN could be FM at or above room temperature, numerous attempts have been made to synthesize this promising DMS material [60–76]. However, the results have been rather confusing. Not only do the reported Curie temperatures [60–71] vary over a wide range (10–945 K), but it is also uncertain whether the ground state of (Ga,Mn)N is FM or antiferromagnetic (AFM) [72–78]. It has been found that at low temperature (< 10 K) the magnetic behavior of the (Ga,Mn)N layers prepared by reactive molecular beam epitaxy shows AFM characteristics with a spin-glass transition [74]. Magnetic measurement at $T = 2$ K using the superconducting quantum interference device magnetometer also shows

AFM coupling between Mn ions in the (Ga,Mn)N sample [75]. An understanding of the controversy between FM and AFM is both important and challenging [77,78]. Clearly, the growth conditions are important as is the proper characterization of the sample. In this connection, it is interesting to note that Overberg et al. [66], by using only the Mn cell and the N plasma under similar conditions, were able to grow samples that showed X-ray diffraction peaks corresponding to the Mn_4N stable phase. This phase is known to be FM [69] with a T_C as high as 745 K. Overberg et al. [66], however, ruled out the possibility of the existence of Mn_4N in the GaMnN film.

An attempt to understand the origin of FM behavior in Mn-doped GaN observed in some experiments was made by Rao and Jena [79] using a cluster model. They noted that the coupling between Mn spins was sensitive to its environment. While bulk Mn is AFM, small clusters of Mn containing less than six atoms are FM and larger clusters exhibit ferrimagnetic behavior. The authors explored the magnetic coupling between Mn atoms interacting through a N atom. To see whether under suitable growth conditions clustering of Mn atoms around N may be favorable and whether such clustering could stabilize the FM phase, Rao and Jena [79] carried out a DFT of Mn_xN ($x = 1$–5) clusters. The geometries of these clusters and the spin density distribution are given in Figure 13. They found that the BE of Mn clusters can be substantially enhanced by N atoms by having hybridized s–d electrons bond with the p electrons of N.

This stabilization is accompanied by FM coupling between the Mn atoms, which in turn are antiferromagnetically coupled to N atoms. This N-mediated FM coupling also gives rise to giant magnetic moments of Mn_xN clusters with total magnetic moments of 4, 9, 12, 17, and 22 μ_B for $x = 1$–5, respectively. They suggested that these giant "cluster magnets" might play a significant role in the observed ferromagnetism in Mn-doped GaN semiconductors. Subsequent experiments have confirmed this hypothesis [80].

Evidence of ferromagnetism in ligand-capped ZnO came from recent experiments where different organic molecules [81] such as dodecylamine ($C_{12}H_{27}N$) and dodecanethiol ($C_{12}H_{25}SH$) were coated on ZnO nanaoparticles of about 10 nm in size. A fundamental understanding of the origin of ligand-induced ferromagnetism requires knowledge of how the ligands bind to the ZnO nanoparticle, how the electronic structure and geometry of the nanoparticles are modified upon ligand capping, which sites carry the magnetic moment, and whether magnetism resides only on the surface of the particle or permeates throughout its interior. Since it is computationally impossible to fully optimize ligated particles of 10 nm in size using first-principles techniques, Wang et al. [82] used a cluster model where Zn_xO_x ($x = 3$, 8, 36) clusters represent ZnO nanoparticles and NH_2 and SCH_3 functional groups represent dodecylamine ($C_{12}H_{27}N$) and dodecanethiol ($C_{12}H_{25}SH$) ligands, respectively. This model may be justified since, as discussed in the above, very small clusters of metal oxides and metal halides

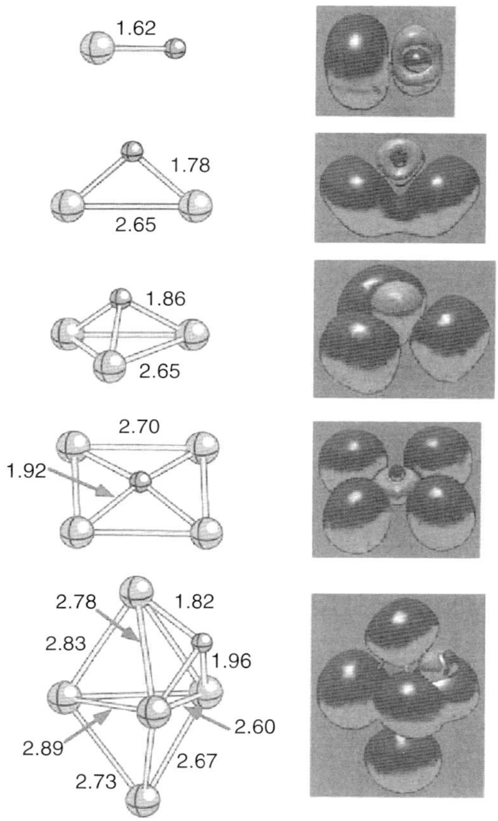

FIGURE 13 Geometries of Mn_xN (left column) clusters in their ground states. The bond lengths are given in angstrom. The spin density surfaces corresponding to 0.005 a.u. for these clusters are plotted in the right column. The green surfaces represent negative spin densities around the *N* site while the blue represents positive spin density around Mn sites (see Ref. [79]). (For interpretation of the references to color in this figure legend, the reader is referred to the Web version of this chapter.)

possess the essential physics and chemistry of their respective bulk due to the strong covalent or ionic bonding.

The optimized structures of Zn_xO_x $x=3$, 8, 36 clusters are shown in Figure 14. The Zn–O bond length in Zn_3O_3, which has a ring-like structure, is 1.868 Å. The Zn–O distance in bulk ZnO is 1.978 Å. Zn_8O_8 has a tube-like structure composed of two eight-membered rings. Here, the Zn–O bond lengths vary between 1.893 and 2.055 Å. In both structures, all the Zn and O atoms are surface atoms and the Zn atoms do not possess the tetrahedral bonding characteristic of bulk ZnO or of a particle of the size of 10 nm. To simulate a more realistic nanoparticle, Wang et al. [82] used a $Zn_{36}O_{36}$ cluster by cutting a small piece from bulk ZnO with wurtzite structure. The cluster was then fully optimized without any symmetry constraints. The optimized geometries are shown in Figure 14.

We see that the structural skeleton of the cluster is still kept, except for bond length contraction of surface atoms due to finite size effect. The average Zn–O bond length on the surface is 1.935 Å, which is contracted by 12% as

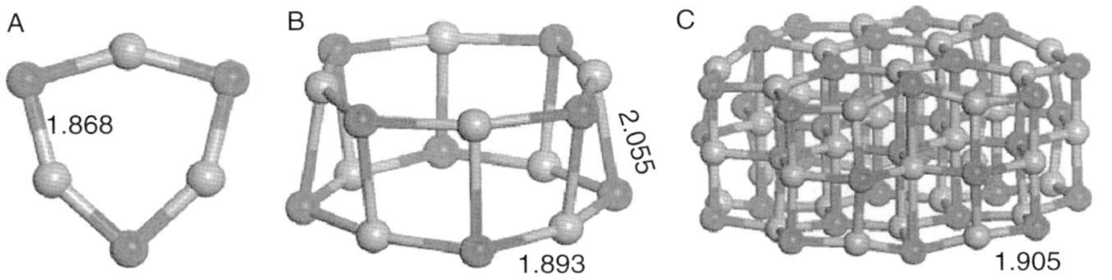

FIGURE 14 Optimized geometry of (A) Zn_3O_3, (B) Zn_8O_8, and (C) $Zn_{36}O_{36}$ (see Ref. [82]).

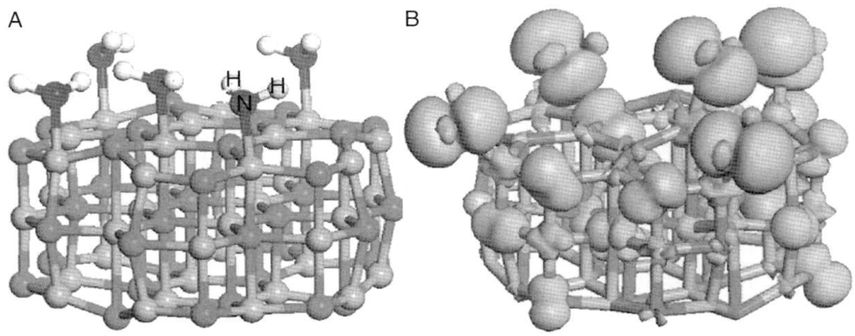

FIGURE 15 (A) Geometry and (B) spin density of $Zn_{36}O_{36}(NH_2)_5$ (see Ref. [82]).

compared to that of inside ones. All the above Zn_xO_x clusters were found to be in a spin singlet state, that is, nonmagnetic as expected.

Wang et al. [82] then studied the binding of NH_2 to Zn_xO_x clusters and found that the lowest energy configuration is one where N binds on top of the Zn atom. The magnetic moment of Zn_3O_3 bound to 1, 2, and 3 NH_2 molecules were found to be 1, 2, and 3 μ_B, respectively. Configurations where the moments on the O sites are coupled antiferromagnetically were higher in energy. Note that the Zn_3O_3 cluster does not fully represent the size of nanoparticles studied experimentally as the Zn atoms do not have the tetrahedral bonding expected in large nanoparticles. Wang et al. [82], therefore, repeated the above calculations for the larger cluster of $Zn_{36}O_{36}$, shown in Figure 14C. They optimized the geometry by capping with five NH_2 functional groups Figure 15A. The magnetic moment of this cluster was found to be 5 μ_B, which is 1 μ_B for each of the NH_2 ligands. The magnetic moment was contributed mainly from the 2p orbitals of N and O surface atoms, as shown by spin density in Figure 15B. These results indicate that magnetism resides on the surface atoms and cluster models provide an understanding of the origin of FM order.

B. Hydrogen Storage (Alanates and Borohydrides)

The limited supply of fossil fuels, its adverse effect on the environment, and the growing worldwide demand for energy have necessitated the search for new and clean sources of energy [83,84]. The possibility of using hydrogen, the third most abundant element on earth, to meet this growing energy need has rekindled [85–87] interest in the study of the interaction of hydrogen in materials. Critical to the hydrogen economy is our ability to find materials for safe, efficient, and economical storage of hydrogen. The conventional method for storing hydrogen as a compressed gas or in the liquid form has its own limitations [88]. While the former requires the use of high-pressure tanks, the latter involves cryogenic temperatures. Safety also remains an important issue with these two forms of storage. An alternate method for hydrogen storage involves metal hydrides [89]. For a viable hydrogen economy, the storage capacity of metal hydrides has to be high, the process should be reversible, and kinetics and thermodynamics should be favorable. Although a database (http://hydpark.ca.sandia.gov) lists over 200 elements, compounds, and alloys that form hydrides, none of these meets the above requirements. To meet the requirement of the transportation industry, materials should be able to store approximately 10 wt% of hydrogen and operate under ambient conditions. For this, the elements have to be lighter than Al. Considerable amount of work has been carried out on light metal hydrides. Among these are alanates and borohydrides.

Alanates [90–92], with the chemical composition $[M^{n+}(AlH_4)^{n-}$, $M=Li$, Na, K, Mg] can store hydrogen up to 11 wt%. However, the desorption temperature of hydrogen in these materials is rather high and not ideal for application in the mobile industry. The recent discovery of Bogdanovic and Schwickardi [90] that the doping of Ti-based catalyst in $NaAlH_4$ can significantly lower the hydrogen desorption temperature has created a great deal of excitement. But why and how Ti accomplishes this task remains a mystery. Jena and Khanna [93] first carried out *ab initio* calculations based on a cluster model, to provide an atomic-level understanding of the role of Ti in lowering the hydrogen desorption temperature in the alanates. Note that, in bulk $NaAlH_4$, the Al atom is encapsulated by four hydrogen atoms forming a tetrahedron, much the same way a carbon atom is encapsulated by four hydrogen atoms in CH_4. The bonding between Al and H is covalent. The AlH_4 unit is then bonded to the Na atom and the two are held together by an ionic bond that results from charge transfer from Na^+ to AlH_4^-. The Na^+ cation is surrounded by eight H atoms. Since small clusters are adequate to model bulk materials characterized by covalent and/or ionic bonding, Jena and Khanna [93] studied the interaction of hydrogen with Al, Na, and Ti atoms systematically using a cluster approach. They first studied the nature of the Al–H bond in clusters of AlH_n ($n \leq 4$), its evolution with successive addition of hydrogen, and the mechanism that stabilizes the AlH_4 unit. The geometry and the electronic structure of $NaAlH_4$ were then calculated to understand the bonding between Na and AlH_4. To study the effect of Ti doping, the geometries and total energies of $TiAlH_n$ and $NaTiH_n$ clusters were calculated, without any symmetry constraint, as it is *a priori* not clear whether Ti replaces the Na or the Al atom in the alanates.

Jena and Khanna [93] noted that the neutral AlH_3 and anion AlH_4^- clusters are among the most stable clusters in the AlH_n series. The stability of $NaAlH_4$ thus derives from the charge transfer from Na to AlH_4 moiety and the ionic bond between Na^+ and AlH_4^-. Once the extra charge from AlH_4^- moiety is removed, it dissociated to AlH_2+H_2. The effect of Ti substitution at the Na or the Al site, therefore, must have been to reduce the charge transfer to the AlH_4^- moiety and hence weakening the Al–H bond leading to hydrogen desorption at lower temperature. This was indeed demonstrated by explicitly calculating hydrogen removal energy $\bar{\Delta}$ from $TiAlH_4$ and $NaTiH_4$ clusters and comparing it with the corresponding energy from $NaAlH_4$. The removal of energy from three different clusters is calculated from the following equations:

$$\begin{aligned}
\bar{\Delta}_1 &= E(\mathrm{NaAlH}_n) - E(\mathrm{NaAlH}_{n-1}) - E(\mathrm{H})\\
\bar{\Delta}_2 &= E(\mathrm{TiAlH}_n) - E(\mathrm{TiAlH}_{n-1}) - E(\mathrm{H})\\
\bar{\Delta}_3 &= E(\mathrm{NaTiH}_n) - E(\mathrm{NaTiH}_{n-1}) - E(\mathrm{H}).
\end{aligned}$$

Here, E is the total energy of respective systems. The results are summarized in Table 3.

We see that the energy to remove a hydrogen atom from $NaAlH_4$ is 3.46 eV, which reduces to 2.68 and 2.82 eV when Ti replaces the Na or the Al site, respectively. For subsequent hydrogen removal, Ti does not have much beneficial effect. It is interesting to note that the energy necessary to remove a H atom from crystalline $NaAlH_4$ is 4.0 eV. This energy reduces to 1.9 and 2.5 eV when Ti replaces the Na or the Al sites, respectively. These results are similar to those given in Table 3 and demonstrate that clusters can serve as a good model to understand the role of catalysts in the dehydrogenation of alanates.

TABLE 3 Energy (in eV) Necessary to Remove H Atoms Successively from $NaAlH_4$, $TiAlH_4$, and $NaTiH_4$ Clusters

n	1	2	3	4
$\bar{\Delta}_1$ (eV)	2.78	3.09	2.61	3.46
$\bar{\Delta}_2$ (eV)	2.35	3.62	3.05	2.68
$\bar{\Delta}_3$ (eV)	2.68	3.04	2.92	2.82

Metal borohydrides [$M(BH_4)_n$] are also potential candidates for hydrogen storage as they can contain large gravimetric density of hydrogen. But they suffer from the same drawback as alanates in that their thermodynamics and kinetics are not ideal. The release of hydrogen from $M(BH_4)_n$ has been thought to follow the following reaction paths depending on whether the final product is a stable binary hydride (MH_x) or an elemental metal (M).

$$M(BH_4)_n \rightarrow MH_x + nB + [2n - x/2]H_2$$
$$M(BH_4)_n \rightarrow M + nB + (2n)H_2.$$

Here, n is the valence of the metal atom M.

In reality, hydrogen desorption from MBH_4 takes place in more complicated ways because of the appearance of intermediate phases. Using a synergistic approach involving first-principles calculations of phase stability and Raman spectroscopy measurements, Orimo and coworkers [94] found that the decomposition of $LiBH_4$ occurs via the formation of one or more polyhedral borane phases (i.e., $Li_2B_{12}H_{12}$ and perhaps similar "B_nH_n"-containing compounds) prior to forming elemental B and LiH as the final products. Recently, Bowman and coworkers [95] used the solid-state NMR method to characterize/identify the amorphous intermediates formed during the decomposition reactions of various metal borohydride systems containing Li, Mg, and Sc metals. They have also identified $[B_{12}H_{12}]^{2-}$ to be the most probable intermediate candidate. The existence of $[B_{12}H_{12}]^{2-}$ species have also been confirmed by others [96–100]. However, it is not clear from experiments whether compounds containing other "B_nH_n" complexes are also present in the intermediate phases. We note that B_4H_6 and $[B_6H_7]^-$ clusters as well as other closo-boranes $[B_nH_n]^{2-}$ with $n \neq 12$ are also known to be stable.

Using DFT, Li and Jena [101] have calculated the relative stability of a number of B_nH_m clusters in neutral and anionic forms with varying composition and studied their interaction with Li, Na, Mg, and Ca. They have shown that compounds containing $[B_{12}H_{12}]^{2-}$ complexes are indeed the most stable species, although other compounds may be present during the dehydrogenation of metal borohydrides. They also showed that studies on clusters can not only provide similar insight into the relative stability of various intermediate phases as calculations based on crystal symmetry can, but also can help predict the possibility of other intermediate products during the dehydrogenation process even without knowing their crystal structure. They showed that, while borane clusters carrying a single negative charge are stable, they react with metal atoms more weakly than the closo-boranes ($B_nH_n^{2-}$) and hence are not likely to exist during the dehydrogenation of metal borohydrides. Their results are summarized in Figure 16. Here, the BEs of various B_nH_m complexes interacting with alkali and alkaline-earth atoms as well as the corresponding HOMO–LUMO gaps are presented. We note that the HOMO–LUMO gaps and the BEs exhibit similar trends. In addition, the BEs of the intermediate

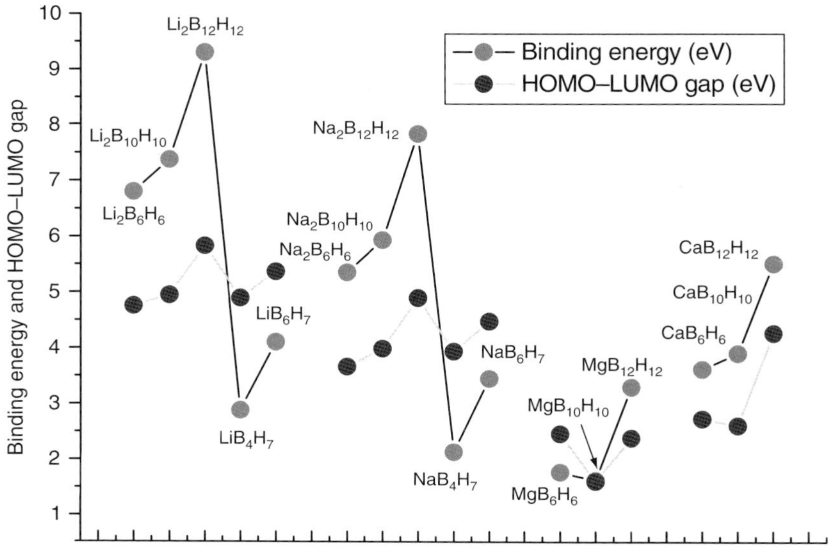

FIGURE 16 The binding energies and HOMO–LUMO gaps for $M_2(B_nH_n)$, and $M(B_nH_m)$ clusters (M = Li, Na, or Mg).

phases of Mg–borane complexes are the lowest among all the systems studied. While $Li_2B_{12}H_{12}$, $Na_2B_{12}H_{12}$, and $MgB_{12}H_{12}$ are preferred intermediates during dehydrogenation of $LiBH_4$, $NaBH_4$, and $Mg(BH_4)_2$, hydriding/dehydriding process in Mg borohydrides may be easier than other hydrides due to the less stable intermediate phases.

IV. CONCLUSIONS

In this chapter, we set out to examine how properties of clusters evolve with size and how many atoms are necessary before a cluster can mimic the properties of their bulk-phase matter. We also examined the role clusters can play as models of bulk matter in accounting for their electronic, magnetic, and energetic properties. Using the jellium model we showed that a metal cluster has to contain more than 100 atoms before bulk properties can emerge. Below this size, the geometry of the clusters bear no resemblance to their crystal structure.

However, the picture is very different when dealing with heteroatomic clusters that are bound by covalent or ionic bonds. We considered clusters of metal oxides, alkali halides, and lead tellurides. We showed that here the geometries, stabilities, and electronic properties of very small clusters containing only a handful of atoms do exhibit bulk properties as long as their compositions are close to their bulk values. Thus, how clusters evolve with size depends upon the nature of their bonding. In addition, when a cluster becomes a crystal also depends upon the properties under investigation. In the case of covalently and ionically bonded clusters such as ZnO, GaN, alanates, and borohydrides, we showed that properties such as magnetic behavior of DMSs and dehydrogentation from complex hydrides can be understood by using small clusters as models.

Although alkali halide clusters where the ionic bonding dominates show bulk-like geometry even for small sizes, there are structural distortions which are cluster specific. The structural distortions decrease when the cluster sizes increase because of the reduced surface-to-volume ratio. The electronic properties of small nonstoichiometric clusters such as $Na_{14}Cl_{13}$ or $Na_{13}Cl_{14}$ show properties unique to clusters. For example, a highly delocalized surface state in the case of the former mimics the lowest conduction band state (or a donor impurity state) of NaCl crystal. Similarly, the relatively localized hole state mimics the acceptor hole state of NaCl. The lead chalcogenide clusters also develop their crystalline structures from very small sizes. However, they show "tetramerization" for small size clusters; this is not seen in the alkali halide clusters. The tetramerization is due to the competing interaction between ionic and covalent bonding between Pb and the chalcogen. Also, due to the increased covalency in lead chalcogenides and the degeneracy of the HOMO state, the excess-electron state in $Pb_{14}Te_{13}$ is highly localized and accompanied by strong structural distortion.

Acknowledgment

P. J. and S. D. M. would like to acknowledge partial support by the Department of Energy.

REFERENCES

[1] (a) Baerends EJ, Ellis DE, Ros P. Chem Phys 1973;2:41. (b) Averill FW, Ellis DE. J Chem Phys 1973;59:6412.
[2] Salahub DR, Messmer RP. Phys Rev B 1977;16:2526.
[3] Jena P, Fradin FY, Ellis DE. Phys Rev B 1979;20:3543.
[4] Raghavachari K, Hallas MD. Mol Phys 2004;102:381.
[5] Hintermann A, Manninen M. Phys Rev B 1983;27:7262.
[6] Knight WD, Clemenger K, de Heer WA, Saunders WA, Chou MY, Cohen ML. Phys Rev Lett 1984;52:2141.
[7] Leuchtner RE, Harms AC, Castleman Jr. AW. J Chem Phys 1991;94:1093.
[8] Khanna SN, Jena P. Phys Rev Lett 1992;69:1664 Phys Rev B 1995;51:13705.
[9] Walter M, Akola J, Lopez-Acevedo O, Jadzinsky PD, Calero G, Ackerson CJ, et al. Proc Natl Acad Sci USA 2008;105:9157.
[10] Claridge SA, Castleman Jr. AW, Khanna SN, Murray CB, Sen A, Weiss PS. ACS Nano 2009;3:244.
[11] Rao BK, Jena P. J Chem Phys 1999;111:1890.

[12] Rao BK, Khanna SN, Jena P. Phys Rev B 2000;62:4666.
[13] Akola J, Häkkinen H, Manninen M. Phys Rev B 1998;58:3601.
[14] Li X, Wu H, Wang XB, Wang LS. Phys Rev Lett 1988;81:1090.
[15] Martin TP, Bergmann T, Gohlich H, Lange T. Chem Phys Lett 1990;172:209.
[16] Raghavachari K, Rohlfing CM. J Chem Phys 1991;94:3670.
[17] Muller J, Liu B, Shvartsburg AA, Ogut S, Chelikowsky JR, Siu KWM, et al. Phys Rev Lett 2000;85:1666.
[18] Kroto HW, Heath JR, O'Brien SC, Curl RF, Smalley RE. Nature 1985;318:162.
[19] Taylor KJ, Pettiette CL, Cheshnovsky O, Smalley RE. J Chem Phys 1992;96:3319.
[20] Sun Q, Rao BK, Jena P, Stolcic D, Kim YD, Gantefor G, et al. J Chem Phys 2004;121:9417.
[21] Whetten RL. Acc Chem Res 1993;26:49.
[22] Frenkel A, Shasha E, Gorodetsky O, Voronel A. Phys Rev B 1993;48:1283.
[23] Ochsenfeld C, Gauss J, Ahlrischs R. J Chem Phys 1995;103:7401.
[24] Honea C, Homer ML, Whetten RL. Phys Rev B 1993;47:7480.
[25] Rajagopal G, Barnett RN, Nitzan A, Landman U, Honea EC, Labastie P, et al. Phys Rev Lett 1990;64:2933.
[26] Chen JL, Wang CS, Jackson KA, Pederson MR. Phys Rev B 1992;45:1927.
[27] Honea C, Homer ML, Whetten RL. Int J Mass Spectrom Ion Process 1990;102:213.
[28] Hudgins RR, Dugourd P, Tenenbaum JM, Jarrold MF. Phys Rev Lett 1997;78:4213.
[29] Rayane D, Compagnon I, Antoine R, Broyer M, Dugourd Ph, Labastie P, et al. J Chem Phys 2002;116:10730.
[30] Rajagopal G, Barnett RN, Landman U. Phys Rev Lett 1991;67:727.
[31] Häkkinen H, Barnett RN, Landman U. Europhys Lett 1994;28:263.
[32] Bader RF, Platts JA. J Chem Phys 1997;107:8545.
[33] Ochsenfeld C, Ahlrischs R. J Chem Phys 1994;101:5977.
[34] Durand G, Spiegelmann F, Labastie P, L'Hermite J-M, Poncharal Ph. Phys Rev Lett 1997;79:633.
[35] Landman U, Scharf D, Jortner J. Phys Rev Lett 1985;54:1860.
[36] Durand G, Giraud-Girard J, Maynau D, Spiegelmann F, Calvo F. J Chem Phys 1999;110:7871.
[37] Durand G, Spiegelman F, Allouche AR. Eur Phys J D 2003;24:19.
[38] Schmidt C, Allen PB, Baruah T, Pederson M. arXiv:cond-mat/0411009v1.
[39] Shytov V, Allen PB. Phys Rev B 2006;74:075419.
[40] Ceresoli D, Zykova-Timan T, Tosatti E. J Phys Condens Matter 2008;20:325236.
[41] Wang C, Zhang G, Fan S, Li Y. J Phys Chem Solids 2001;62:1957.
[42] Poudel B, Wang WZ, Wang DZ, Huang JY, Ren ZF. J Nanosci Nanotechnol 2006;6:1050.
[43] Wu S, Zeng H, Schelly ZA. Langmuir 2005;21:686.
[44] Lu W, Gao P, Jian WB, Wang ZL, Fang J. J Am Chem Soc 2004;126:14816.
[45] Shevchenko EV, Talapin DV, Kotov NA, O'Brien S, Murray CB. Nature 2006;439:55.
[46] Mokari T, Zhang M, Yang P. J Am Chem Soc 2007;129:9864.
[47] Kane RS, Cohen RE, Silbey R. J Phys Chem 1996;100:7928.
[48] Zeng H, Schelly ZA, Ueno-Noto K, Marynick DS. J Phys Chem A 2005;109:1616.
[49] Allan G, Delerue C. Phys Rev B 2004;70:245321.
[50] Hoang K, Lee MS, Mahanti SD, Jena P. to be published.
[51] Satta A, de Gironcoli S. Phys Rev B 2000;63:033302.
[52] Hoang K, Mahanti SD, Jena P. Phys Rev B 2007;76:115432.
[53] Wang Q, Sun Q, Jena P, Hu Z, Note R, Kawazoe Y. Appl Phys Lett 2007;91:063116.

[54] Kittilstved KR, Schwartz DA, Tuan AC, Heald CM, Chambers SA, Gamelin DR. Phys Rev Lett 2006;97:037203.
[55] Venkatesan M, Fitzerald CB, Lunney JG, Coey JM. Phys Rev Lett 2004;93:177206.
[56] Sati P, Deparis C, Morhain C, Schafer S, Stepanov A. Phys Rev Lett 2007;98:137204.
[57] Wang W, Sun Q, Jena P, Kawazoe Y. Appl Phys Lett 2005;87:162509.
[58] Ohno H. Science 1998;281:951.
[59] (a) Dietl T, Ohno H, Matsukura F, Cibert J, Ferrand D. Science 2000;287:1019. (b) Ohno H, Chiba D, Matsukura F, Omiya T, Abe E, Dietl T, et al. Nature (London) 2000;408:944.
[60] Reed ML, Ritums MK, Stadelmaier HH, Reed MJ, Parker CA, Bedair SM, et al. Mater Lett 2001;51:500.
[61] Reed ML, El-Masry NA, Stadelmaier HH, Ritums MK, Reed MJ, Parker CA, et al. Appl Phys Lett 2001;79:3473.
[62] Sonoda S, Shimizu S, Sasaki T, Yamamoto Y, Hori H. J Cryst Growth 2002;237:1358.
[63] Sasaki T, Sonoda S, Yamamoto Y, Suga K-I, Shimizu S, Kindo K, et al. J Appl Phys 2002;91:7911.
[64] Thaler GT, Overberg ME, Gila B, Frazier R, Abernathy CR, Pearton SJ, et al. Appl Phys Lett 2002;80:3964.
[65] Lee JM, Lee KI, Chang JY, Ham MH, Huh KS, Myoung JM, et al. Microelectron Eng 2003;69:283.
[66] Chen PP, Makino H, Kim JJ, Yao T. J Cryst Growth 2003;251:331.
[67] Seo SS, Kim MW, Lee YS, Noh TW, Park YD, Thaler GT, et al. Appl Phys Lett 2003;82:4749.
[68] Shon Y, Kwon YH, Yuldashev ShU, Park YS, Fu DJ, Kim DY, et al. J Appl Phys 2003;93:1546.
[69] Kondo T, Kuwabara S, Owa H, Munekata H. J Cryst Growth 2003;237:1353.
[70] Overberg ME, Abernathy CR, Pearton SJ, Theodoropoulou NA, McCarthy KT, Hebard AF. Appl Phys Lett 2001;79:1312.
[71] Sardar K, Raju AR, Bansal B, Venkataraman V, Rao CNR. Solid State Commun 2003;125:55.
[72] Dhar S, Brandt O, Trampert A, Daweritz L, Friedland KJ, Ploog KH, et al. Appl Phys Lett 2003;82:2077.
[73] Dhar S, Brandt O, Trampert A, Friedland KJ, Sun YJ, Ploog KH. Phys Rev B 2003;67:165205.
[74] Ploog KH, Dhar S, Trampert A. J Vac Sci Technol B 2003;21:1756.
[75] Zajac M, Gosk J, Kaminska M, Twardowski A, Szyszko T, Podsiadlo S. Appl Phys Lett 2001;79:2432.
[76] Ando K. Appl Phys Lett 2003;82:100.
[77] Graf T, Geonnenwein STB, Brandt MS. Phys Status Solidi B 2003;239:277.
[78] Pearton SJ, Abernathy R, Thaler GT, Frazier RM, Norton DP, Ren F, et al. J Phys Condens Matter 2004;16:R209.
[79] Rao BK, Jena P. Phys Rev Lett 2002;18:185504.
[80] (a) Dhar S, Brandt O, Tampert A, Daweritz L, Friedland KJ, Plog KH, et al. Appl Phys Lett 2003;82:2077. (b) Yang R, Yang H, Smith AR. Appl Phys Lett 2006;88:173101.
[81] Garcia MA, Merino JM, Pinel EF, Quesada A, De la Venta J, Ruiz Gonzalez ML, et al. Nano Lett 2007;7:1489.
[82] Wang Q, Sun Q, Jena P. J Chem Phys 2008;129:164714.
[83] Alper J. Science 2003;299:1686.

[84] Cortright RD, Davada RR, Dumesic JA. Nature 2002;418:964.
[85] Chen P, Xiang Z, Luo JZ, Tan KL. Nature 2002;420:302.
[86] Rosi NL, Eckert J, Eddaoudi M, Vodak DT, Kim J, O'Keefe M, et al. Science 2003;300:1127.
[87] Schlappbach L. and Zuttel. Nature 2001;414:353.
[88] Irani RS. MRS Bull 2002;27:680.
[89] Sandrock G. J Alloys Compd 1999;887:293.
[90] Bogdanovic B, Schwickardi M. J Alloys Compd 1997;253:1.
[91] Brinks HW, Hauback BC, Norby P, Fjellvag H. J Alloys Compd 2003;351:222.
[92] Vajeeston P, Ravindran P, Vidya R, Fjellvag H, Kjekshus A. Appl Phys Lett 2003;82:2257–9.
[93] Jena P, Khanna SN. unpublished.
[94] Ohba N, Miwa K, Aoki M, et al. Phys Rev B 2006;74:075110.
[95] Hwang SJ, Bowman RC, Reiter JW, Rijssenbeek J, Soloveichik GL, Zhao J-C, et al. J Phys Chem C 2008;112:3164.
[96] Li HW, Kikuchi K, Nakamori Y, Ohba N, Miwa K, Towata S, et al. Acta Mater 2008;56:1342.
[97] Her JH, Yousufuddin M, Zhou W, Jalisatgi SS, Kulleck JG, Zan JA, et al. Inorg Chem 2008;47:9757.
[98] Purewal J, Hwang SJ, Bowman RC, Rönnebro E, Fultz B, Ahn C. J Phys Chem C 2008;112:8481.
[99] Au M, Jurgensen AR, Spencer WA, Anton DL, Pinkerton FE, Hwang S-J, et al. J Phys Chem C 2008;112:18661.
[100] Soloveichik GL, Gao Y, Rijssenbeek J, Andrus M, Kniajanski S, Bowman RC, et al. Int J Hydrogen Energy 2009;34:916.
[101] Li S, Jena P. to be published.

Chapter 3

Applications of the Cluster Method for Biological Systems

Ralph H. Scheicher*, Minakhi Pujari**, K. Ramani Lata[†], Narayan Sahoo[‡] and Tara Prasad Das[†]

*Department of Physics and Astronomy, Uppsala University, Uppsala, Sweden
[†]Department of Physics, State University of New York at Albany, Albany, New York, USA
**Division of Physical Science, Westwood College, Chicago, Illinois, USA
[‡]Department of Radiation Physics, U.T.M.D. Anderson Cancer Center, University of Texas, Houston, Texas, USA

Chapter Outline Head

Nanoclusters. DOI: 10.1016/S1875-4023(10)01003-X

I. INTRODUCTION

A common problem encountered in first-principles studies of biological systems is that the size of the relevant molecules can be exceedingly large, so that it becomes practically impossible for current state-of-the-art computational resources and algorithms to deal with the enormous number of atoms in their entirety. Fortunately though, such a complete detailed representation is quite often neither required nor desired. Rather, the interest lies on a particular site in the large biomolecule, and that area might encompass only a comparatively small number of atoms. In this context, an important advantageous feature of biological molecules is that they are generally composed of much smaller building blocks or "units." For example, protein chains are polymers of amino acids; DNA and RNA are long chains of nucleotides. These building blocks lend themselves rather perfectly for applications of the cluster method and the main concern is the number of neighboring units to be considered in the calculation for quantities of interest to converge. Also, the termination of the cluster, where dangling bonds may need to be saturated, is a crucial point, and often either hydrogens or methyl groups (CH_3) are found to be suitable choices for this purpose, depending on the location of the dangling bonds and the nature of the biological systems being considered.

In the following sections of this chapter, we will demonstrate how the cluster method has been successfully applied by us to biological molecules with selected examples. Each of the discussed systems plays a vital role in the organism of many living beings. In Section II, we concentrate on hemoglobin (Hb), the molecule which is essential for transporting oxygen in our blood. Both deoxyHb and oxyHb will be considered and the cluster on which we will focus is the heme unit in these molecules as well as heme compounds with halogen atoms as fifth ligands of Fe. In the second section, the biological system of interest is cytochrome *c* (Cyt *c*), an electron transfer protein of vital importance in the breathing cycle. Here, the focus lies not on the heme unit, but on the protein chain and the amino acids from which that polymer chain is built. The heterogeneous polymer cluster taken directly from the actual protein chain will be discussed in Section III. Finally, we come to a biomolecule which acts as a blueprint for all the other molecules in an organism, including the two proteins dealt with in Sections II and III. It is, of course, deoxyribonucleic acid, commonly known by its abbreviation: DNA. We dedicate three sections to this carrier of genetic instructions. In Section IV, the presented studies have been motivated by a rather ingenious experimental exploration of electron transport along the DNA strand, in which two different conformations of DNA, namely, its A-form and B-form (which differ in their base pair separation), were investigated. But, although electron transport *along* the strand is certainly most common, it is not restricted to take place only in that direction: charge transport can also follow a route *perpendicular* to the strand. This approach was proposed as a way to rapidly determine the sequence of a given DNA molecule by scanning the (presumably characteristic) conductance of each of its four bases. We will discuss in Section V how the cluster approach can be once again highly useful in exploring such a scenario from first principles. Finally, more and more hybrid systems have emerged over the past few years in which a combination of DNA molecules with artificially constructed nanomaterials is the focus of interest. In Section VI, we pick two very important related representatives of such nanobiosystems, namely DNA plus graphene and DNA plus carbon nanotubes (CNTs), and demonstrate how cluster methods can be utilized for their exploration.

One aspect that we intentionally left out from our discussion of the cluster method in biological systems is that of the quantum mechanical/molecular mechanical (QM/MM) method, in which a core region of interest is represented by quantum mechanical (*ab initio*) methods, while the effects from a surrounding outer region are simulated using a molecular mechanics (force field) approach. The QM/MM method is undoubtedly very useful in many scenarios, but quite often one can obtain very good results even without it and in the examples discussed here (which often represent a "first look" at novel systems), we have not felt the need to consider this method.

Finally, depending on the quantities one is looking for, different approximations are suitable in the electronic structure calculations. Here, we have used Hartree–Fock (HF) based methods (with correlation effects added via many-body perturbation theory [MBPT]) in those cases where hyperfine parameters were sought. When the focus was more on energetics or forces, or in the case of electronic transport, density functional theory (DFT) was also deemed appropriate.

II. HEME SYSTEMS INCLUDING FIVE-LIGANDED HALOGEN-HEMINS, DEOXY- AND OXY-HEMOGLOBIN

A. Introduction

The electronic structures of deoxyhemoglobin (deoxyHb) and oxyhemoglobin (oxyHb) are of great interest for the understanding of their biological functions, and biochemical and biophysical properties. For providing a thorough background for the first-principles method that is being currently used for the understanding of their electronic structures and associated properties, we shall focus first on the four five-liganded ferric heme compounds with halogens as the fifth ligand, because discussion of these systems allows insights into the accuracy of the method used to study the electronic structures of deoxyHb and oxyHb systems and also because a number of electron-nuclear hyperfine properties are available for these simpler compounds without the proximal imidazole ligand in Hb systems providing linkage to the protein chains. One can make quantitative comparisons of predictions of these hyperfine properties with experimental results that allow an assessment of the accuracy of the calculated electronic wave functions and electronic charge and spin distributions in these systems.

The deoxyHb and oxyHb systems are of course the two most important proteins in the body because of their roles in the respiratory process. The efficiency of this role is characterized by the affinity of oxygen (O_2) for deoxyHb. There is increasing affinity of the heme unit for attachment of O_2 to deoxyHb associated with each of the four protein chains in the systems as O_2 is attached to the other three proteins chains successively, this feature being termed the cooperativity effect [1]. A second feature, termed the Bohr effect [2], is the sensitivity of the O_2 attachment process to the hydrogen ion concentration of the environment of deoxyHb or detachment from oxyHb. This latter feature allows deoxyHb to attach O_2 preferentially in one part of the body and release it efficiently in another part. The understanding of these processes has been in recent years under study using detachments [3] of O_2 and other molecules like CO and NO from six-liganded Hb systems and studying their reattachment to the resulting deoxyHb.

Also of interest is the transport of electrons in deoxyHb and deoxymyoglobin (deoxyMb), which had been studied [4] earlier by attaching an electron accepting atom like ruthenium to the protein chain. One then produces ionization of an electron from the Fe^{2+} ion in the heme in deoxyMb or deoxyHb by the action of light and studies its passage to the ruthenium atom by spectroscopic techniques. This technique allows the study of the speed of passage of the electron but not its pathway. This is however possible as discussed in a later section in this chapter by the study of the motion of electrons leaving trapped muonium atoms in the protein chains through [5] the spin-lattice relaxation of the muons left behind by the fluctuating magnetic field produced at the muon site by the moving electron [6]. A study of the dependence of the muon relaxation rate on the strength of the applied magnetic field [5] then provides insights into the pathway for the electron transport.

The theoretical understanding at a microscopic electronic level of these processes in deoxyHb and deoxyMb requires a very good knowledge of the electronic structures of these systems [7] as also the understanding of hyperfine interactions in these systems that are available from experimental measurements by various techniques including magnetic resonance [8] and Mössbauer spectroscopy [9].

Also, the magnetic properties of oxyHb and oxyMb which were studied simulated by magnetic susceptibility measurements [10] in the 1970s, which were later shown to be incorrect [11–13], have recently become of interest because of muon spin-lattice relaxation measurements on implanted muons in these systems. The understanding of this recent observation makes it important to have a reliable knowledge of the energy difference between singlet and triplet states of oxyMb, and oxyHb both by themselves and in the presence of muon and muonium to be briefly discussed in a later section.

B. Procedure for the Five-Liganded Halo-Heme Compounds and Terminologies for the Properties Involved

The available theoretical hyperfine interaction parameters for fluoro, chloro, bromo, and iodo-heme have been obtained [14–17] using the Gaussian set of programs involving the unrestricted Hartree–Fock (UHF) procedure for the one-electron contributions to magnetic hyperfine and nuclear quadrupole interactions (NQIs) of the nuclei in the halo-heme compounds and isomer shift of ^{57m}Fe. The influence of many-body contributions to these properties can be studied using MBPT for which subroutines are currently available in the Gaussian set of programs. But at the times that the investigations to be discussed in this section were carried out, computing facilities adequate for MBPT investigations were not available, so only the UHF procedure was used, without incorporation of many-body effects.

The variational procedure for implementing the UHF method for these large multicenter systems using the Gaussian set of programs requires the choice of suitable sets of basis functions involving one center Gaussian functions located at the atomic sites in the molecules involved. These basis sets involved both single and contracted Gaussian functions. Thus, the Fe-atom basis functions involve 8s (4, 2, 1, 1, 1, 1, 1, 1), 4p (3, 1, 1, 1) and 3d (3, 1, 1) functions [18], the s basis involving eight variational basis functions, two of them being contracted functions involving linear combinations of respectively four Gaussians functions and two Gaussian functions, and six single Gaussian functions, the same Fe-atom basis sets being used for all four halo-hemes. The basis functions are all normalized and the coefficients of the Gaussians in contracted functions are kept constant while using the Roothaan variational method for the HF procedure. The notations for the p and d basis sets shown are similar to the s-basis functions as just explained. The N, C, and H atom basis functions for chloro-heme (termed hemin) involve [19] for N-atom, the s and p basis sets, 2s (4, 2) and 1p (4), respectively, and for C-atom the basis sets involve 2s (6, 2) and 1p (4), and for H-atom 2s (3, 1). For chlorine atom, the basis sets used involve [20] 4s (6, 2, 2, 2) and 2p (6, 2). For fluoro-heme, the N-atom and C-atom involve [14] basis sets 3s (6, 1, 1) and 1p (4) for N-atom, 2s (4, 2) and 1p (3) for C-atom, and 1s (4) for H-atom [21], and for fluorine atom [14] 3s (4, 1, 1) and 1p (3).

For bromo-heme and iodo-heme, the N-atom involves the same basis set [19] as for chloro-heme; the C-atom involves the same basis set [14] as for fluro-heme and H-atom the same basis set [21] as for fluoro-heme. For the bromine atom, the basis set involves [22] 4s (4, 3, 3, 3), 3p (4, 3, 3), and 1d (4). For iodine atom, the basis set involves [22] 5s (4, 3, 3, 3, 3), 4p (4, 3, 3, 3), and 2d (4, 3). The basis sets involved in deoxyHb and oxyHb will be described in Sections II.D and II.E.

The next feature of the procedure involves the choice of the clusters for the three sets of biological systems that we shall be studying. We shall describe here the clusters of atoms that we shall be using for the halo-hemes. The clusters for deoxyHb and oxyHb will be described in Section II.D. As described in the literature, the cluster used for the HF [23] combined with many-body perturbation procedure (HFMBPT) [24] requires one to choose a cluster of atoms or molecules to represent an infinite solid state system or a large biological molecule with symmetry properties similar to the infinite or large system it is chosen to represent. In the case of energy-dependent properties, one chooses an origin for the cluster which represents a point of symmetry [24], or if the symmetry is low or there is no particular symmetry, a point in the system which has important significance. For the halo-hemin systems as well as deoxy and oxyHb, we shall use Fe as the origin for energy properties and hyperfine properties of ^{57m}Fe nucleus. For properties involving a wave function-dependent property, which is the case for the hyperfine properties of the hemoglobin related systems we are discussing, one chooses as origin an atom [23,24] containing the nucleus whose hyperfine properties are being considered.

Two additional important considerations in deciding on the proper cluster to choose are the size of the cluster and the termination of the cluster. For the first point, one studies the convergence in the calculated value of the property being considered with respect to the size of the cluster [23], or using physical considerations especially for biological systems [25]. As regards the termination of the cluster [23], in the case of solid state systems, the termination is made in a manner dependent on the physical nature of the system. Thus, for an ionic crystal system, one chooses a cluster size that is practicable to handle from computer time considerations for which all the electrons in the cluster are included in the HFMBPT investigation. Additionally, a sizable number of ions outside of the cluster are included handling them as point charges to simulate the influence of ions outside the inner all electron cluster. For semiconductors, the dangling bonds of the atoms at the cluster periphery are terminated by hydrogen atoms. For molecular solids [24], one uses a set of molecules, sufficiently large in number to appropriately represent the infinite solid, leading to a cluster size that is practicable from computation time considerations. Lastly, for metallic or semimetallic systems, not many cluster investigations have been carried out. In these systems, which are neutral in charge, because the number of electrons and nuclear charges are equal, one does not require any special termination, just as in the case of molecular solids. One has however to use a sizable cluster because of the diffuse nature of the electron distribution in most metals.

In biological systems, it has been the practice to terminate carbon, nitrogen, or oxygen atoms by hydrogen atoms as in semiconductors. In the case of the hemoglobin derivatives as in the case of the five liganded heme systems and in deoxyHb and oxyHb discussed in Sections II.D and II.E, the pyrrole rings in the protoporphyrin plane as shown in Figure 1 have CH_2–CH_2–COOH, CH=CH_2, and CH_3 groups attached to the peripheral carbon atoms. In these compounds, for hyperfine and magnetic properties of nuclei in the inner region of the heme unit, the peripheral groups attached to pyrroles are not expected to have important influence on the properties in question and are terminated by H atoms attached to the outer two carbons of each pyrrole ring as in organic molecules or in the case of the dangling bonds in cluster investigations in solid state semiconductor systems [26]. So the protoporphyrin plane in Figure 1 is replaced by the model porphyrin system in Figure 2 for the investigations reported in this chapter. The iron atom position in Figure 2 is taken from crystal structure data in the appropriate systems [27–29].

We shall proceed next to the procedure for evaluation of the hyperfine properties of the various nuclei in the halo-heme systems in terms of the calculated electronic wave functions for these systems in the framework of UHF theory by the HF-Roothaan procedure just described. These hyperfine properties [14–16,23,24] are the isomer shifts (for ^{57m}Fe nucleus) in Mössbauer

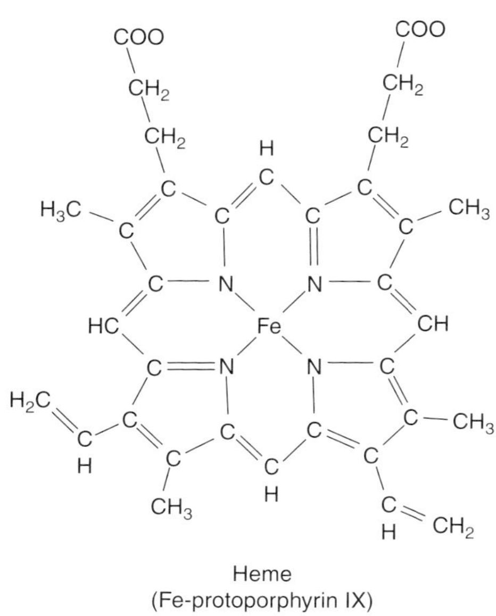

FIGURE 1 Heme unit with pyrrole rings in the protoporphyrin plane having CH_2–CH_2–COOH, CH=CH_2, and CH_3 groups attached to the peripheral carbon atoms.

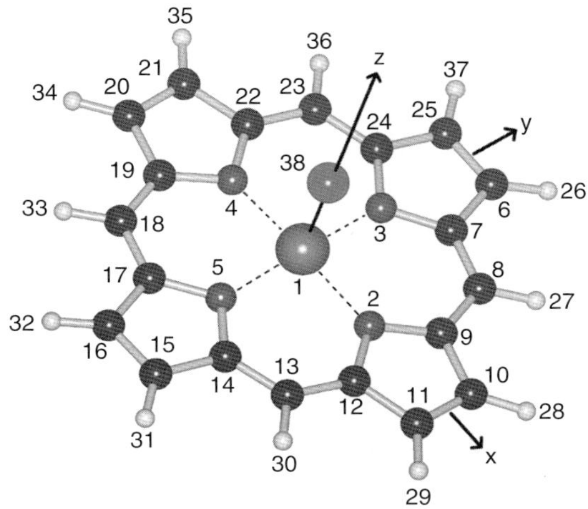

FIGURE 2 Iron porphyrin chloride cluster used for investigations of the five-liganded halogen heme systems.

spectroscopy [30], and the magnetic hyperfine and NQI parameters. The expressions for these hyperfine parameters in UHF theory will now be presented. Using these expressions, the theoretical results that have been obtained in the literature will be discussed later in this section and compared with available experimental results. But before discussing the hyperfine properties, we shall discuss first the evaluation [14–16] of the effective charges and unpaired spin populations on the various atoms in the halo-heme systems which provide valuable insights about the origin of the electronic charge-dependent and spin-dependent hyperfine properties of the halo-heme and related hemoglobin systems.

Using the molecular orbitals for spin up (α) and spin down (β) in UHF theory [14,23,31] given by

$$\psi_{\mu\alpha} = \sum\nolimits_i C_{\mu i\alpha}\chi_{i\alpha} \tag{1a}$$

$$\psi_{\mu\beta} = \sum iC_{\mu i\beta}\chi_{i\beta} \tag{1b}$$

where $\chi_{i\alpha}$ and $\chi_{i\beta}$ are basis functions centered on the various atoms (A) in the cluster with α and β representing up and down spin states, one can obtain the net electronic populations n_A on atom A, effective charges $\xi_{\text{eff},A}$ and unpaired spin populations $n_{\kappa A}$ on them using the Mulliken approximation [32], leading to the expressions

$$\xi_{\text{eff},A} = \xi_A - n_A \tag{2}$$

and

$$n_A = n_{\alpha A} + n_{\beta A} \tag{3a}$$

$$n_{\kappa A} = n_{\alpha A} - n_{\beta A} \tag{3b}$$

where ξ_A is the nuclear charge on atom A and

$$n_{\alpha A} = \sum_{\mu}^{\text{occupied}} \sum_{i \in A} \left[\sum_{j \notin A} \left(C_{\mu i \alpha}^2 + S_{ij\alpha} C_{\mu i \alpha} C_{\mu i \alpha} \right) \right] \quad (4)$$

$$n_{\beta A} = \sum_{\mu}^{\text{occupied}} \sum_{i \in A} \left[\sum_{j \notin A} \left(C_{\mu i \beta^2} + S_{ij\beta} C_{\mu i \beta} C_{\mu i \beta} \right) \right] \quad (5)$$

$$S_{ij\gamma} = \langle \chi_{i\gamma} | \chi_{j\gamma} \rangle \quad \gamma = \alpha, \beta \quad (6)$$

The isomer shift [33–35] represents the difference $\Delta\varepsilon$ in the γ-ray transition frequencies from the excited Mössbauer nuclear level for ^{57m}Fe to the ground nuclear state of ^{57m}Fe, between the biological system B containing iron, like a halo-heme, and the Fe^{3+} ion

$$\Delta\varepsilon = (\varepsilon_{\mathrm{B}} - \varepsilon_{\mathrm{Fe}^{3+}}) = \alpha(\rho_{\mathrm{B}} - \rho_{\mathrm{Fe}^{3+}}) \quad (7)$$

where α is the isomer shift constant [34]

$$(-0.3 \pm 0.03) a_o^3 \, \mathrm{mm/s} \quad (8)$$

The isomer shift is measured in millimeters per second in the standard Mössbauer technique notation on the basis of Doppler shift measurement involving Mössbauer line emitters and absorbers. The quantity ρ_{B} is given by

$$\rho_{\mathrm{B}} = \sum_{\mu\alpha}^{\text{occupied}} |\psi_{\mu\alpha}(Fe)|^2 + \sum_{\mu\beta}^{\text{occupied}} |\psi_{\mu\beta}(Fe)|^2 \quad (9)$$

representing the total electron density at the ^{57m}Fe nucleus of the biological system B. A corresponding expression applies for $\rho_{\mathrm{Fe}^{3+}}$ in Fe^{3+} ion. For numerical accuracy, the same basis set for the Fe atom orbitals should be used for B and Fe^{3+}.

The influence of the magnetic hyperfine interactions, involving the Fermi contact and dipolar hyperfine interactions, is represented in the spin-Hamiltonian used for analyzing the frequencies involved in the transitions in electron spin resonance [36] and in other techniques like the radiative methods such as Mössbauer spectroscopy [33–35] and perturbed angular correlation [37] by the equation [23,38,39]

$$H'_{\mathrm{spin}} = A \vec{I} \cdot \vec{S} + \vec{I} \cdot \overrightarrow{\overrightarrow{B}} \cdot \vec{S} \quad (10)$$

In Equation (10), A is an isotropic scalar term and $\overrightarrow{\overrightarrow{B}}$ is a tensor with components B_{pq}, where $(p, q) = x, y, z$ corresponding to any chosen Cartesian coordinate axis system. The values of A and B_{pq} are obtained [23,38,39] by the usual procedure of equating the expectation value and non-diagonal matrix elements for A and components of B_{pq} of the electron nuclear Fermi-contact hyperfine interaction Hamiltonian $H'_{\text{hfs,F}}$in Equation (11) and the dipolar part of the hyperfine interaction Hamiltonian $H'_{\text{hfs,dip}}$ in Equation (12) over the many-electron determinantal wave functions for the electronic system for (S, M_s) states with S representing the total spin of the system and M_s its Z-component, to the corresponding expectation values and matrix elements of the two terms in Equation (10) involving the spin parts of the (S, M_s) wave functions. This process leads to the expressions for A and B given by Equations (13) and (14).

$$H'_{\text{hfs,F}} = \frac{8\pi}{3}\gamma_e\gamma_N\hbar^2\vec{I}_N \cdot \left[\sum_i \vec{s}_i\delta\left(\vec{r}_i - \vec{R}_N\right)\right] \tag{11}$$

$$H'_{\text{hfs,dip}} = \gamma_e\gamma_N\hbar^2\vec{I}_N \cdot \sum_i \left[\frac{\vec{s}_i}{r_{iN}^3} - 3\left(\frac{\vec{s}_i\vec{r}_{iN}}{r_{iN}^5}\right)\vec{r}_{iN}\right] \tag{12}$$

In Equations (11) and (12), $\vec{I}_N$ and $\vec{s}_i$ represent respectively the nuclear and electron spin angular momentum vectors, in units of $\hbar$, of the nucleus N whose hyperfine interaction is being considered and the electron, $\vec{r}_i$ represents the radius vector of the electron i with respect to a chosen origin, and $\vec{R}_N$ that of the nucleus N. The quantities γ_e and γ_N are the magnetogyric ratios of the electron and the nucleus N given respectively by $\gamma_e = 2(\mu_B/\hbar)$ $\gamma_N = \mu_N/I_N\hbar$, where μ_B and μ_N are respectively the Bohr magneton and the magnetic moment of the nucleus N; also, I_N is the spin of the nucleus and $\hbar$ is the Planck's constant.

$$A = \frac{8\pi}{3}\gamma_e\gamma_N\hbar^2\sum_\mu\left[\left|\psi_{\mu\alpha}\left(\vec{R}_N\right)\right|^2 - \left|\psi_{\mu\beta}(R_N)\right|^2\right] \tag{13}$$

$$B_{pq} = -\gamma_e\gamma_N\hbar^2\sum_\mu\left[\langle\psi_{\mu\alpha}|O_{pq}|\psi_{\mu\alpha}\rangle - \langle\psi_{\mu\beta}|O_{pq}|\psi_{\mu\beta}\rangle\right] \tag{14}$$

$$O_{pq} = \left(\frac{3r_{Np}r_{Nq} - r_N^2\delta_{pq}}{r_N^5}\right) \tag{15}$$

In Equation (14), B_{pq} represents the components of the tensor $\overrightarrow{B}$ in Equation (10).

One can diagonalize the $\overrightarrow{B}$ tensor and obtain the principal components of the tensor $B_{x'x'}, B_{y'y'}$, and $B_{z'z'}$ where the principal axis system is defined by the co ordinate axes X', Y' and Z' whose directions relative to the chosen X, Y, and Z axes are determined in the process of diagonalization. The off-diagonal components $B_{x'y'}$, $B_{y'z'}$, and $B_{z'x'}$ of course vanish in the principal axes system.

The wave functions $\psi_{\mu\alpha}$ and $\psi_{\mu\beta}$ refer to the molecular orbitals for the states with spins parallel and antiparallel respectively to the total spin in the

UHF procedure which is used for the unpaired spin systems. Their values are different because of the difference in exchange potentials experienced by the electrons in the parallel and antiparallel spin states because for the biological systems with finite total spin, there are more electrons occupying the parallel spin valence states than the antiparallel spin valence states and exchange interaction is experienced only between electrons in states with spins parallel to each other. The differences in $\psi_{\mu\alpha}$ and $\psi_{\mu\beta}$, and between properties dependent on parallel and antiparallel spin states are stated in the literature [40,41] to result from this exchange polarization effect. This effect is very important for magnetic hyperfine interactions for both the Fermi contact and dipolar effects as may be seen from Equations (13) and (14). It should be mentioned here that $\vec{r}_N$ represents the difference $\left(\vec{r} - \vec{R}_N\right)$ in the distance vectors for the electron and the nucleus whose hyperfine interaction is under study as in Equations (11) and (12) where summations over electrons i are involved for all the electrons in the system. As is well known in HF theory [42], in taking the matrix elements of Hamiltonian operators like those in Equations (11) and (12) over the many-electron determinantal functions, the summations over electrons in the operators get replaced by summations over one-electron molecular orbital states as in Equations (13) and (14) and the suffixes for electrons are no longer necessary.

For NQI, one has to consider the interactions between the nuclear electric quadrupole moment tensor and the electric field gradient (efg) tensor [43–45] due to the nuclear charges except the charge on the nucleus whose NQI is being considered, and the electrons in the system. On consideration of these interactions and the fact that the nuclear dimensions are of the order of femtometers (10^{-15} m) while the distances between the center of the nucleus and the electron and the nuclear charges are of the order of Angstroms (10^{-10} m), the Hamiltonian for the NQI can be written [44] as

$$H_Q = \frac{1}{6}\sum_{j,k} Q_{jk} V_{jk} \tag{16}$$

where Q_{jk} is the operator corresponding to the jk component of the nuclear quadrupole moment and V_{jk} the jk component of the efg tensor at the nucleus, where $(j, k) = (x, y, z)$. Using the Wigner–Eckart theorem [46] for matrix elements of vectors and tensors in terms of basis sets involving components of the invariant angular momentum vector for a system, which in the case of a nucleus is the spin $\vec{I}$ of the nucleus. Thus, for taking matrix elements of H_Q over basis functions Ψ_{IM}which are eigenfunctions of $\vec{I}^2$ and I_z, where $I\,(I+1)$ is the eigenvalue of $\vec{I}^2$ and M the eigenvalue of I_z, the operator H_Q can be written [44] in the form as shown in Equation 17:

$$H_Q = \frac{eQ}{4I(2I-1)}\sum_{j,k} V_{jk}\left[\frac{3}{2}\left(I_j I_k + I_k I_j\right) - \delta_{jk}\vec{I}^2\right] \tag{17}$$

The components of the efg tensor are given by the relation

$$V_{jk} = \frac{\partial^2 V}{\partial r_j \partial r_k} \tag{18}$$

where V is the potential from the electrons at the nucleus under study and due to the other nuclear charges in the molecule or chosen cluster for the biological molecules like heme compounds and deoxyHb and oxyHb to be discussed in this chapter. The coordinates r_j and r_k are of course the components x, y, or z of the distance vector $\vec{r}$ for a nuclear charge or electron with respect to the nucleus whose efg is being considered. For a point charge $\zeta_N e$, such as a nuclear charge,

$$V = \frac{\zeta_N e}{r_N} \tag{19}$$

where $\vec{r}_N$ is the radius vector of the nuclear charge $\zeta_N e$ with respect to the nucleus at which the efg tensor due to $\zeta_N e$ is being considered. Thus,

$$V_{jk}^{\text{nuclear}} = \frac{\partial^2 V}{\partial r_{jN} \partial r_{kN}} = \zeta_E e \left(\frac{3 r_{jN} r_{kN} - r_N^2 \delta_{jk}}{r_N^5} \right) \tag{20}$$

For an electron of charge $(-e)$ with position rector $\vec{r}$ with respect to the nucleus being considered, similarly $V = -e/r$ and

$$V_{jk}^{\text{electronic}} = \frac{\partial^2 V}{\partial r_{rj} \partial_{rk}} = (-e) \left(\frac{3 r_j r_k - r^2 \delta_{jk}}{r^5} \right) \tag{21}$$

Thus, for a cluster [24,47] with M atoms besides the one at which the efg is being calculated, the total number of electrons is equal to $\sum_N{}^M \zeta_N$ if the cluster is neutral and will be smaller or larger than this number if the cluster used is positively or negatively charged. Thus, for a cluster, the expression for V_{jk} is given by

$$V_{jk} = V_{jk}^{\text{nuclear}} + V_{jk}^{\text{electronic}} \tag{22}$$

where

$$V_{jk}^{\text{nuclear}} = \sum_{N=1}^{m} \zeta_N e \left(\frac{3 r_{jN} r_{kN} - r_N^2 \delta_{jk}}{r_N^5} \right) \tag{22a}$$

$$V_{jk}^{\text{electronic}} = -\sum_{\mu} n_\mu \int e \left| \Psi_\mu \right|^2 \left(\frac{3 r_j r_k - r^2 \delta_{jk}}{r^5} \right) \mathrm{d}\tau \tag{22b}$$

In Equation (22b) the summation in μ is over all the occupied molecular orbital states, and n_μ is the number of electrons in each of the occupied states. For the Restricted Hartree-Fock (RHF) procedure, the space parts of the orbital μ with up spin are the same as for the down spin orbitals and n_μ has the value 2 for each of the occupied paired spin states and 1 for the unpaired spin states. For the UHF procedure, the space parts of the paired orbitals are different for up and down states and thus, the index μ in this case includes both space and spin parts of the occupied molecular orbitals and $n_\mu = 1$ for all occupied states.

It should be stated here that in some cases as in deoxyHb and oxyHb or a hemoglobin derivative [17,48,49] like NOHb and met-hemoglobin [50] where the sixth ligand of Fe in heme is a H_2O molecule, it may be necessary to include in the clusters, for accurate prediction of properties, both the heme system and parts of the adjacent amino acid groups in the protein chains. This may however be impracticable from a computational point of view or too time consuming. In such a case, one could use the influence of some parts of the distant amino acid groups by handling the atoms involved as effective point charges as is often done in ionic crystals [47,51]. With such an approximation, one would have to take V_{jk} as

$$V_{jk} = V_{jk}^{\text{nuclear}} + V_{jk}^{\text{electronic}} + V_{jk}^{\text{distant}} \tag{23}$$

with the expressions for V_{jk}^{nuclear} and $V_{jk}^{\text{electronic}}$ in (22a) and (22b) and

$$V_{jk}^{\text{distant}} = \sum_{N=1}^{M'} \zeta_N^{\text{eff}} e \left(\frac{3r_{jN} r_{kN} - \delta_{jk} r_N^2}{r_N^5} \right) \tag{23a}$$

with M' depending on the number of distant atoms in the proteins that are included. The effective chargesζ_N^{eff} could be obtained from an evaluation of the charges on the atoms in the amino acid molecules from electronic structure calculations on either single amino acid molecules or a cluster made up of a number of amino acid neighbors adjacent to each other. The effective charges on the atoms N in Equation (23a) are obtained using the Mulliken approximation [32], as in Equations (2)–(6).

Once the efg tensor components V_{jk} for the chosen cluster representing the heme systems are calculated using the chosen coordinate system, one can diagonalize the efg tensor and obtain the principal components in the principal axes system (X', Y', Z'). For this system, there are no nondiagonal component such as $V_{x'y'}, V_{y'z'}$ and $V_{z'x'}$and one only has to deal with three principal components $V_{x'x'}, V_{y'y'}$ and $V_{x'x'}$ with the designation of the $X', Y'$and Z'such that the magnitudes of the principal axes components have the order given by

$$|V_{z'z'}| > |V_{y'y'}| > |V_{x'x'}| \tag{24}$$

As these principal components satisfy the Laplace equation

$$V_{x'x'} + V_{y'y'} + V_{z'z'} = 0 \tag{25}$$

only two of them are independent. We can therefore express three principal components by two parameters [44] q and η, where

$$eq = V_{z'z'} \tag{26a}$$

and

$$\eta = \frac{V_{x'x'} - V_{y'y'}}{V_{z'z'}} \tag{26b}$$

Because of Equation (24), one can see that η, the asymmetry parameter, can range from 0 to 1, the former limit occurring when one has axial symmetry described by

$$V_{x'x'} = V_{y'y'} \tag{27}$$

and the value of $\eta = 1$ when

$$V_{x'x'} = 0 \tag{28}$$

Using Equation (17), one can express H_Q in the principal axis system as

$$H_Q = \frac{eQ}{4I(2I-1)}\left[V_{z'z'}\left(3I_z^2 - \vec{I}^2\right) + \left(V_{x'x'} - V_{y'y'}\right)\left(I_{x'}^2 - I_{y'}^2\right)\right] \tag{29a}$$

$$= \frac{e^2qQ}{4I(2I-1)}\left[\left(3I_{z'}^2 - \vec{I}^2\right) + \eta\left(I_{x'}^2 - I_{y'}^2\right)\right] \tag{29b}$$

As the eigenvalue of $\vec{I}^2$ is $I(I+1)$, one can rewrite Equation (29b) as

$$H_Q = \frac{e^2qQ}{4I(2I-1)}\left[\left\{3I_{z'}^2 - I(I+1)\right\} + \eta\left(I_{x'}^2 - I_{y'}^2\right)\right] \tag{29c}$$

One can obtain the eigenvalues and eigenstates of the Hamiltonian H_Q Equation (29c) by diagonalizing the matrix for H_Q in terms of the functions, the basis $\Psi_{I,M}$, where M is the eigenvalue of I_z and has the values

$$M = -I, -(I-1), (I-1), I \tag{30}$$

The choice of these basis sets is suggested by the features of H_Q that $\Psi_{I,M}$ are the eigenstates for $\eta = 0$. The reason for this is that H_Q commutes with $\vec{I}^2$ and I_z and so H_Q, $\vec{I}^2$, and I_z all three have the same eigenstates, namely, $\Psi_{I,M}$. In the case where $\eta \neq 0$, there is mixing [44] of the states $\Psi_{I,M}$ and $\Psi_{I,M\pm2}$. So one can prepare the matrix of H_Q in Equation (29c) in the $\Psi_{I,M}$ basis and then diagonalize it to get the eigenvalues and eigenfunctions of H_Q. As an example [44,52], for the eigenvalues for the case of $I = 3/2$ and M having the values given by Equation (30) for $I = 3/2$, namely, $M = \pm 3/2$ and $\pm 1/2$, one gets

$$E_{3/2,\pm3/2} = 3A\left(1 + \eta^2/3\right)^{1/2} \tag{30a}$$

$$E_{3/2,\pm1/2} = -3A\left(1 + \eta^2/3\right)^{1/2} \tag{30b}$$

where $A = \frac{e^2qQ}{4I(2I-1)} = \frac{e^2qQ}{12}$ for $I = 3/2$ and the eigenfunctions are mixtures of $\Psi_{3/2,\pm3/2}$ with $\Psi_{3/2,\pm1/2}$ but the energy levels are denoted by $I = 3/2$ and

$M = \pm 3/2$, $\pm 1/2$ because these are the quantum numbers for the eigenstates for the same value of $A = e^2qQ/12$ in the limit $\eta = 0$.

From the energy levels in Equations (30a) and (30b), one can obtain expressions for the frequencies in terms of the nuclear quadrupole coupling constant e^2qQ and the asymmetry parameter η which, using the values of the experimentally observed nuclear quadrupole resonance (NQR) frequencies for a nucleus with $I = 3/2$, allows a determination of the values of these two NQI parameters. NQR [44,53] corresponds to resonance experiments carried out in the absence of any static magnetic field with Equation (29c) representing the actual Hamiltonian for the nucleus in a solid state or biological system, or with a weak magnetic field involving Zeeman energy term in the Hamiltonian, small in size compared to the NQI term.

When the value of e^2qQ is weak, NQR measurements are difficult, because the NQR frequencies are small and so the signal/noise ratios are small and the NQR signal is too weak to detect. One has in such cases to carry out nuclear magnetic resonance (NMR) measurements [43,54] in the presence of a strong magnetic field and compare the observed frequencies with the expressions for the expected NMR frequencies combining the magnetic Zeeman interaction term for the nuclei with the NQI Hamiltonian in Equation (29c).

In addition to NMR and NQR methods, the NQI parameters e^2qQ and η can also be obtained by radiative methods like Mössbauer spectroscopy [30,33,35] and the perturbed angular correlation technique [37,55] just as in the case of magnetic hyperfine contact and dipolar parameters. We would also like to remark in conclusion that the hyperfine interaction parameters, isomer shift, contact and dipolar magnetic hyperfine parameters, and the NQI parameters e^2qQ and η, as the expressions for them given earlier in this Section indicate, provide rather extensive and comprehensive information about the nature [23,56] of the electron distributions in biological and condensed matter systems near the nuclear sites. The isomer shift provides information about the isotropic electronic charge density at the nucleus, and the NQI parameters e^2qQ and η about the anisotropic electronic charge distributions near the nuclear site, the former being associated with the departure from spherical or cubic symmetry and the latter with the departure from axial symmetry. The contact magnetic hyperfine interaction parameter describes the isotropic electronic spin density and the dipolar magnetic hyperfine parameters describe the anisotropic electronic spin density near the nucleus, both departures from spherical and axial symmetry.

C. Results for Five-Liganded Heme Systems

The electronic structures and associated properties of five-liganded halohemin systems have been studied extensively since the 1960s. The method that was employed for these systems initially [57,58] was the self-consistent charge extended Hückel (SCCEH) method [59] which utilized the linear combination of atomic orbitals-molecular orbital (LCAOMO) procedure [60]. This method was applied both to spectroscopic properties [57], and magnetic and nuclear quadrupole hyperfine interaction properties [58], isomer shifts [61], and X-ray photoemission properties (XPS) [62], also referred to in the literature as electron spectroscopy for chemical analysis (ESCA) [63]. The SCCEH procedure was a semi-empirical procedure, in which the matrix elements of the electronic Hamiltonian over the atomic orbital basis sets in the variational procedure adopted to solve for the molecular orbital energies and electronic wave functions were obtained from the experimental ionization energies for the atomic orbital basis sets for the atoms involved and their ions, and the effective charges on them obtained using the calculated electronic wave functions in the molecule under study. The effective charges are obtained [57,58] by the Mulliken approximation [32]. It uses a proposed dependence of the matrix elements of the Hamiltonian on the effective charges on the atoms. Instead of looking explicitly for the convergence in the electronic energy eigenvalues and linear combination parameters for the basis functions in the molecular orbital wave functions in the literature [14–17] as the procedure used to study the convergence, the SCCEH procedure [57–59] uses the effective charges on the atoms to indirectly study the convergence in the density matrix involving the atomic orbital coefficients in the molecular orbitals. This method is described extensively in the literature for the properties of heme compounds [57,58,61,62] and hemoglobin derivatives [64–68]. There is also an improvement in the SCCEH procedure in the literature to include better the dependence of the atomic orbital matrix elements on the charges on the atoms. This procedure was called the "charge-coupled self consistent charge extended Hückel" procedure abbreviated as the CCSCCEH procedure [68]. The improvement over the SCCEH procedure involved including the interactions between the electrons on an atom and the effective charges on the neighboring atoms. We shall not be discussing results from the semiempirical procedures but rather from the HF procedure itself or its combination with MBPT [69–73] to include many body effects, which is referred to as the HFMB procedure. We shall however be occasionally referring to how the results from the HFMB procedure compare with those from the SCCEH and CCSCCEH procedures because such a comparison provides useful insights into the nature and accuracy of results by the HF and HFMB procedures, which have been used for heme compounds [14–17] and hemoglobin derivatives [74,75], and other biological systems [76–80] since the 1990s especially for hyperfine and related properties that we shall be focusing on for the five-liganded heme compounds and deoxy- and oxy-Hb systems.

We shall next discuss results on hyperfine interactions and related properties from first-principles investigations in the recent past on the five-liganded halogen heme systems. The results are available for the HF procedure without inclusion of many-body effects. From the results for a number of properties, possible effects of inclusion of many-body effects will be inferred. Investigations are planned for the near future on many-body effects on the properties of these systems.

Before presenting the results for hyperfine and related properties for the five-liganded halo-heme systems, a few remarks about the geometry utilized for the electronic structure investigations on these systems would be helpful. The heme planes of these compounds referred to as the protoporphyrin base shown in Figure 1 involve the four pyrrole groups linked to one another by the mesomethine carbon and hydrogen atoms. The protoporphyrin base involves the chain groups attached to the distant carbon atoms in the four pyrrole ring. The iron atoms, from X-ray structure data [22], in the fluro-heme and chloro-heme compounds were taken to be located at 0.455 Å above the center of the protoporphyrin ring in all four systems. The Fe–halogen distances in the chloro-heme and fluoro-heme compounds, referred to in the literature as hemin and fluoro-hemin compounds, are available from structural data. For the bromo and iodoheme systems, X-ray structural data were not available at the time of the investigations reported in this section, so the iron–halogen distances were obtained by adding the differences between bromine and chorine and iodine and chlorine to the Fe–Cl distance in hemin. Additionally, the heme plane was found from X-ray data in the hemin and fluoro-hemin to be nearly tetragonal and was adjusted [14,15] to be exactly tetragonal for economy of computing effort for the electronic structure investigations. Also, for economy of computing effort, the chain groups at the distant pyrrole carbons were replaced by hydrogen atoms shown in Figure 2 with the CH bond distances being taken to be the same as in organic compounds involving single C–C bonds. This truncation of the chain groups is similar to those in the cluster HF procedure used for study of electronic structures and associated properties for impurity systems in semiconductors. This truncation is not expected to affect hyperfine properties of nuclei in atoms in the interior of the compounds, like the Fe, N, C, and halogen atoms.

The variational Gaussian basis sets used for the HF investigations for all four compounds, fluoro-hemin, chloro-hemin, bromo-hemin, and iodo-hemin are already described in Section II.B.

(i) Results for Electron Distributions and Associated Hyperfine Properties in the Five-Liganded Halogen-Hemin Compounds

The results of the HF investigations of the electronic structures and associated hyperfine properties of the five-liganded halogen-heme compounds will now be discussed. To start out, the results for the electron distributions and in particular the net charges and spin populations on the various atoms in the four compounds are presented in Tables 1 and 2. Table 1 presents these results for fluoro-hemin and chloro-hemin and Table 2 presents those in bromo-hemin and iodo-hemin.

TABLE 1 Effective Charges and Unpaired Spin Populations on Atoms in F-Hemin and Cl-Hemin

	F-Hemin		Cl-Hemin	
Atom[a]	Charge	Unpaired spin population	Charge	Unpaired spin population
Fe_1	2.322	4.625	2.045	4.527
N_3	−0.522	0.066	−0.461	0.091
C_6	−0.191	0.002	−0.206	0.002
C_7	0.048	-0.022	0.055	−0.034
C_8	−0.189	0.024	−0.210	0.038
H_{26}	0.178	0.000	0.193	−0.003
H_{27}	0.190	−0.002	0.206	0.000
R_{38}	−0.522	0.178	−0.518	0.227

R_{38} represents F_{38} and Cl_{38} atoms respectively for F-hemin and Cl-hemin.
[a]Because of the symmetry, the charges and the populations on the rest of the atoms are equal to the appropriate atoms listed here, the terminal H_{26}-atoms (and equivalent H atoms) replacing the side chains at the corresponding sites.

TABLE 2 Effective Charges and Unpaired Spin Populations on Atom in Br-Hemin and I-Hemin

	Br-Hemin		I-Hemin	
Atom[a]	Charge	Unpaired spin population	Charge	Unpaired spin population
Fe_1	2.167	4.371	2.176	4.342
N_3	−0.504	0.090	−0.503	0.114
C_6	−0.191	0.001	−0.189	0.001
C_7	0.046	−0.027	0.048	−0.044
C_8	−0.188	0.028	−0.185	0.049
H_{26}	0.178	0.000	0.181	0.000
H_{27}	0.190	0.002	0.194	−0.003
R_{38}	−0.418	0.375	−0.514	0.356

R_{38} represents Br_{38} and I_{38} atoms, respectively, for Br-hemin and I-hemin.
[a]Because of the symmetry, the charges and the populations on the rest of the atoms are equal to the appropriate atoms listed here, the terminal H_{26}-atoms (and equivalent H atoms) replacing the side chains at the corresponding sites.

(ii) Charge and Spin Distribution in Halogen-Hemin Systems

Before proceeding to discuss the results in Tables 1 and 2, it is important to point out one fact about the electronic structure results, which is of general importance and another that is pertinent for the results presented in these two tables. The first fact is relevant for the UHF procedure used for systems with unpaired spins, as in the case of the five-liganded hemin systems being discussed. This feature has to do with the fact that in the UHF procedure, use of a single determinant many-electron wave function Ψ_{UHF} with the spatial wave functions for paired spin states different from each other leads to the general result that Ψ is not an eigenfunction of $\vec{S}^2$, where $\vec{S}$ is the total spin of the system, which is a requirement for the many-electron wave function as the exact Hamiltonian H for a many-electron system - commutes with $\vec{S}^2$. The root causes [81] of this departure from eigenfunction behavior are well known and ways to remedy it are carefully discussed in the literature. One way [69–72], widely used in MBPT applications in atomic systems, is to start with the restricted HF wave function Ψ_{HF}, which is a single determinant function with the spatial parts of the one-electron wave functions for paired spin states identical, and which is an eigenfunction of $\vec{S}^2$. The exchange core polarization (ECP) feature present in Ψ_{UHF} is then produced by using the difference $H' = H - H_{RHF}$ as the perturbation Hamiltonian. The first order correction to Ψ_{UHF} using H' then introduces the ECP effect through the first-order correction to Ψ_{RHF} which in effect brings in the differences between spatial parts of the paired spin states. Since H' commutes with $\vec{S}^2$, because the exact H and H_{RHF} both commute with $\vec{S}^2$, the first-order perturbed many electron wave function will be an eigenfunction $\vec{S}^2$. Considering higher order perturbation equations, higher order perturbation correction to Ψ_{HF} due to H' which incorporates many-electron correlation effects, can also be shown to be eigenfunctions of $\vec{S}^2$. This is an important feature of the linked cluster many-body perturbation theory [82] which has been successfully applied extensively for atomic systems [69–72,83] and can be applied to biomolecular systems, but is complicated computationally because of the multicenter nature of these systems.

A second method that could be used in many-body procedures involves the use of a multi-determinant many-electron HF wave function Ψ obtained by projecting out [84] from Ψ_{UHF} the unsatisfactory part which makes the latter depart from eigenfunction character with respect to $\vec{S}^2$. This corrected many-electron wave function is termed Ψ_{PUHF}, to which variation principle can be applied as is usual in HF theory to determine the one-electron wave functions of the system. A procedure that is available in computational methods [85] for electronic structure investigations to test how much Ψ_{UHF} departs from eigenfunction character with respect to $\vec{S}^2$ is to use the UHF many-electron wave functions to obtain the expectation value of $\vec{S}^2$ after the one electron wave functions have been obtained variationally. This is a good test, but if one finds that Ψ_{UHF} does not give $S(S+1)$ as the expectation value and then applies [84] the projection operator for projecting out the part of Ψ_{UHF} that is not an eigenfunction of $\vec{S}^2$ after the UHF equations are solved, the resulting many electron wave function ψ_{PUHF} will not be an eigenfunction of the UHF Hamiltonian for the problem, although it is an eigenfunction of $\vec{S}^2$. For this reason, it is important, in general, in trying to obtain many-electron wave functions for non-zero total spin systems in HF theory which are both eigenfunctions $\vec{S}^2$ and the UHF Hamiltonian, to first obtain the multi-determinant form of Ψ_{UHF}, by applying the projection operator to the single determinant analytic form of Ψ_{PUHF} involving different spatial wave functions for different spin states for paired spin electrons. One should then solve the HF equations for the one-electron states that result from minimization of the expectation value of the many-electron Hamiltonian H for the many-electron function Ψ_{PUHF}. These HF equations will be more complicated computationally to solve than the HF equations for the single determinant Ψ_{RHF} with same spatial wave functions for the one-electron states with different spins for paired spin states, but by no means impossible.

It was important to discuss this feature of UHF theory here, as it is expected that UHF or related more satisfactory methods discussed here will continue to be applied to biological systems and molecular and condensed matter systems in general. Fortunately, for molecular systems containing transition metal atoms or ions, the expectation value of using Ψ_{UHF} does not depart significantly from $S(S+1)$ so that the spin contamination effect, as the departure from eigenfunction character with respect to it is termed, is not a serious problem. This feature most likely is connected with the relatively strong binding of the 3d-like electrons in the regions inside, and near, the transition metal atoms in the systems of interest here. This is observed to be the case for the five-liganded heme systems considered here. Thus, for the fluoro, chloro, bromo, and iodo hemin systems the expectation values are respectively 8.76, 8.78, 8.75, and 8.75 very close and in the last two cases exactly equal to the value of 8.75 for $S(S+1)$, corresponding to the spin of 5/2 for these systems. Consequently, the application of MBPT to the heme systems including deoxyHb and oxyHb using UHF procedure as the starting point (or zero order) is justifiable for including many-body effects.

The other feature of Tables 1 and 2 that should be mentioned before discussing the results listed in them is that because of the near tetragonal symmetry of the structures of the five-liganded halo-hemins and the exact tetragonal symmetry assumed here, we have given the results for the charges and spin populations for only one atom for each group of atoms that are equivalent because of the tetragonal symmetry of the cluster chosen for these systems. Thus, using the notations in Figure 2, Fe and the halogen atoms are single atoms, the halogen atoms being denoted by R_{38} in the tables for each of the halo-hemin systems. The entries for N_3 atom in the tables are representative of the family of four equivalent atoms (N_2, N_3, N_4, N_5). Correspondingly, belongs C_6 to the family of eight (C_6, C_{10}, C_{11}, C_{15}, C_{16}, C_{20}, C_{21}, C_{25}), C_7 to the family of eight (C_6, C_9, C_{12}, C_{14}, C_{17}, C_{19}, C_{22}, C_{24}), C_8 to the family of four (C_8, C_{13}, C_{18}, C_{23}), H_{26} to the family of eight (H_{26}, H_{28}, H_{29}, H_{31}, H_{32}, H_{34}, H_{35}, H_{37}), and H_{27} to the family of four (H_{27}, H_{30}, H_{33}, H_{36}) atoms.

From Tables 1 and 2, it can be seen that the charges on the Fe and halogen atoms in all four halo-hemin compounds are sizable, namely, more than 2.0 and less than 3.0 for the Fe atom, that is, between the Fe^{2+} and Fe^{3+} ions, while the halogen atoms carry negative charges lying between -0.4 and -0.5 with magnitudes less than 1.0. This indicates that the Fe–halogen bonds are not fully ionic, in which case the charges would have been respectively $+3$ and -1 for the Fe and halogen atoms. This suggests that there is significant covalent σ-bonding between the two atoms in all four compounds.

The nitrogen and carbon atoms all carry significant charges which are negative in sign for all of them except for those of the C_7 family which are positive. The nitrogen charges are comparable in magnitude to the charges on the halogen atoms. The hydrogen atoms all carry positive charges, both for the H_{27} family bonded to the methine carbons of the C_8 family, and the H_{28} family that were chosen to replace the side chains in each of the two terminal carbons (Figure 1) of the four pyrrole rings in the heme planes in the four compounds. The sizes and signs of the charges on the atoms on the porphyrin systems suggest that there is significant bonding, not only through direct covalent-bonding between the Fe and N atoms, but also partly through π-bonding between Fe and the methine carbon atoms of the porphyrin ring and most likely, almost entirely through the latter atoms for the rest of the atoms in the porphyrin ring. The charges on the hydrogen atoms which are bonded to the methine carbons or the carbons of the C_6 family in the pyrrole rings have their positive charges determined by the electronegativity difference between the hydrogen atoms and the carbon atoms to which they are bonded. They are most likely also indirectly influenced by the charges on their carbon bonding partners which in turn are influenced by π-bonding between Fe and the porphyrin ring.

As regards the unpaired spin populations, as can be seen from both Tables 1 and 2, they are significant for the nitrogen and carbon atoms, except for the C_6 family atoms in the pyrrole rings where they are very small. The populations on the hydrogen atoms are relatively small compared to those of the nitrogen and carbon atoms. The unpaired spin populations seem to have a more short range spread than the charges within the porphyrin ring. Also, it is important that there are negative spin populations for some of the carbon and hydrogen atoms. Both these features are expected, because the unpaired spin populations are determined by the unpaired spin populations on the Fe atom both by direct spreading of the latter population through covalent bonding and by exchange spin polarization effects important in UHF theory, making the paired spin orbitals have different spatial characters for the electronic wave functions for the different spin states. The exchange spin polarization effect can explain the existence of negative spin populations, where it dominates over the direct spread of the unpaired spin population from Fe and its nearest neighboring nitrogen atoms of the N_3 family. The short-range nature of the unpaired spin population is also attributed to the exchange spin polarization effect which is short range in nature. In contrast, the charges on the atoms in the porphyrin ring are determined by the net electronic populations on them, which are not as sensitively dependent on exchange interactions between the electrons as the unpaired spin populations on the atoms, which represent the difference in populations of electrons with spins parallel and antiparallel to the unpaired spins on Fe atom. These differences, as discussed earlier in this paragraph, are strongly influenced by exchange spin polarization effects sensitive to the exchange interactions between electrons with parallel spins.

In examining the trends in charges and unpaired spin populations on the atoms in the four halo-hemins, one observes that there is some irregularity in terms of ionic and covalent characters of the Fe–halogen bonds as compared to what one would have expected from the electronegativity differences between Fe and halogen atoms. There is a drop in the charge on Fe from F-hemin to Cl-hemin as expected from weaker ionic character for the Fe–Cl bond as compared to the Fe–F bond because of the stronger Fe–F electronegativity difference as compared to Fe–Cl, but there are increases from Cl-hemin through Br-hemin and I-hemin in contrast to what would be expected from the decreases in the Fe–Cl, Fe–Br, and Fe–I electronegativity differences over these three pairs. This result is in keeping with our discussion earlier in this section about the electron distributions being not only influenced just by electronegativity differences but also by the π-bonding between Fe and the atoms in the porphyrin ring. This combined influence of both electronegativity differences and π-bonding between Fe and the atoms in the porphyrin ring is probably also responsible for the irregular trend observed in the charges on the halogen and nitrogen atoms among the four halo-hemin compounds. It should be pointed out here that the trends seen from Tables 1 and 2 for the spin populations on Fe atom, the halogens, and other atoms are also influenced by the factors just discussed for the charges on these atoms and by the exchange spin polarization effect as well, as discussed earlier in this section.

The net electron distributions and the unpaired spin distributions over the atoms in the five-liganded halo-hemin compounds are expected to determine the hyperfine properties, namely NQIs and magnetic hyperfine interactions as well as the isomer shifts in ^{57m}Fe Mössbauer spectroscopy which will be presented next. These hyperfine properties can be experimentally measured by Mössbauer spectroscopy and magnetic resonance and NQR techniques. Comparisons between theoretical results and experimental data where available have been made in the literature and will now be discussed.

(iii) Magnetic Hyperfine Interactions in Halogen-Hemin Systems

We shall next discuss the results for hyperfine interactions in halo-hemin systems. The results for the ^{57m}Fe, ^{14}N, ^{1}H and the halogen nuclei, ^{19}F, ^{35}Cl, ^{79}Br, and ^{127}I in all four halogen-hemin systems using HF one electron theory as listed in Table 5 will be analyzed including trends among the four systems and compared with available experimental data. Before we proceed to the results in all the four systems, we shall first discuss in detail the results for the chloro-hemin system, the first system studied thoroughly in the literature [88], to obtain insights into the various contributions to the hyperfine properties including electron–nuclear contact and dipolar interactions in this system, which will be helpful for the understanding of the results in all the systems. Table 3 has the isotropic contact parameters A and dipolar parameters B_{ij} in all the nuclei in the model halo-hemin system with tetragonal symmetry shown in Figure 2, which shows the choice of the Cartesian axis system X, Y, and Z and the notations for all the nuclei (atoms) whose hyperfine interactions will be discussed. For the ^{14}N, ^{13}C, and ^{1}H nuclei as discussed earlier for Tables 1 and 2, the nuclei listed in Table 5 are representative of the corresponding sets of equivalent nuclei. The experimental results listed in the last column of Table 3 for the ^{57m}Fe Mössbauer data [86] refer to $(A+B_{xx})$, as also the corresponding net theoretical result in the adjacent column. For the ^{14}N, ^{13}C, and ^{1}H nuclei, the net theoretical results listed refer to $(A+B_{zz})$ as do the experimental results listed in the last column. The listed experimental results are obtained from electron-nuclear double resonance (ENDOR) data [86,87].

From Table 3, the contact hyperfine parameter A for the ^{57m}Fe nucleus is negative as is the case for most iron compounds [89] and ferromagnetic iron metal [90]. It represents the sum of the contributions from the ECP contributions from the 2s core electrons which is negative and the 3s components of the paired spin orbitals which are positive, the 1s core electron contribution being very small because of the weak exchange interaction between the tightly bound 1s core electrons and the unpaired 3d-type molecular orbitals in chloro-hemin. The direct contributions to A for the ^{57m}Fe nucleus from the unpaired spin d-type molecular orbitals are positive, but much smaller in magnitude than the negative ECP contribution as can be seen from Table 4.

The dipolar components for ^{57m}Fe are seen from Table 3 to be axially symmetric with $B_{xx}=B_{yy}$. They are two orders of magnitude smaller than the contact contributions and can be seen from Table 4 to have comparable magnitudes for the direct and ECP contributions but opposite signs. The calculated net hyperfine constant $(A+B_{xx})$ parallel to the heme plane is -35.25 MHz, corresponding to a hyperfine magnetic field at the ^{57m}Fe nucleus of -639.29 kilogauss (kG). The experimental values from Mössbauer spectroscopic measurements [86] are -26.4 MHz and -480 kG agreeing in their signs with theory but somewhat smaller in magnitude, by about 25%, than the theoretical value.

TABLE 3 Magnetic Hyperfine Parameters (MHz) for the Nuclei in Chloro-Hemin

Nucleus	Contact parameter A	Dipolar parameters						Net hyperfine constant	
		B_{xx}	B_{yy}	B_{zz}	B_{xy}	B_{yz}	B_{zx}	This work	Experimental
$^{57}Fe_1$	−35.514	0.264	0.264	−0.528	0.000	0.000	0.000	−35.251	−26.47[a]
$^{14}N_3$	12.128	−1.221	1.457	−0.236	0.000	−0.449	0.000	11.892	$\|7.55 \pm 0.05\|$[a]
$^{13}C_6$	1.148	0.567	−0.214	−0.353	−0.088	−0.015	0.075	0.795	
$^{13}C_7$	−3.682	1.344	0.051	−1.395	−0.745	−0.101	0.264	−5.077	
$^{13}C_8$	3.804	−0.190	−0.190	0.380	−0.671	−0.140	0.140	4.184	
$^{1}H_{26}$	0.057	−0.413	−0.943	−0.530	0.371	−0.139	−0.037	0.473	
$^{1}H_{27}$	−0.668	0.419	0.419	−0.838	1.486	−0.183	−0.183	−1.505	$\|1.014\|$[b]
$^{35}Cl_{38}$	4.061	−3.140	−3.140	6.280	0.000	0.000	0.000	10.341	

[a] *See Ref. [86].*
[b] *See Ref. [87].*

TABLE 4 Contributions from Unpaired Spin States, Paired Spin States, and Total for Magnetic Hyperfine Parameters (MHz) in Chloro-Hemin

Nucleus	Nature of contribution	Contact parameter	Dipolar parameters					
			B_{xx}	B_{yy}	B_{zz}	B_{xy}	B_{yz}	B_{zx}
$^{57}Fe_1$	Unpaired	1.225	1.013	1.013	−2.026	0.0	0.0	0.0
	Paired	−36.738	−0.749	−0.749	1.498	0.0	0.0	0.0
	Total	−35.514	0.264	0.264	−0.528	0.0	0.0	0.0
$^{14}N_3$	Unpaired	0.608	−0.597	1.246	−0.649	0.0	−0.417	0.0
	Paired	11.520	−0.625	0.212	0.413	0.0	−0.032	0.0
	Total	12.128	−1.221	1.457	−0.236	0.0	−0.449	0.0
$^{13}C_6$	Unpaired	13.039	0.438	−0.098	−0.341	−0.018	−0.012	0.075
	Paired	−11.891	0.129	−0.117	−0.012	−0.069	−0.003	−0.001
	Total	1.148	0.567	−0.214	−0.353	−0.088	−0.015	0.075
$^{13}C_7$	Unpaired	9.571	0.932	−0.190	−0.742	−0.653	−0.076	0.257
	Paired	−13.252	0.412	0.241	−0.653	−0.092	−0.025	0.007
	Total	−3.682	1.344	0.051	−1.395	−0.745	−0.101	0.264
$^{13}C_8$	Unpaired	9.995	0.257	0.257	−0.514	−0.508	−0.123	0.123
	Paired	−6.192	−0.447	−0.447	0.894	−0.163	−0.017	0.017
	Total	3.804	−0.190	−0.190	0.380	−0.671	−0.140	0.140

Continued

TABLE 4 Contributions from Unpaired Spin States, Paired Spin States, and Total for Magnetic Hyperfine Parameters (MHz) in Chloro-Hemin—Cont'd

Nucleus	Nature of contribution	Contact parameter	Dipolar parameters					
			B_{xx}	B_{yy}	B_{zz}	B_{xy}	B_{yz}	B_{zx}
$^{1}H_{26}$	Unpaired	0.363	−0.270	0.949	−0.679	0.548	−0.135	−0.035
	Paired	−0.306	−0.143	−0.007	0.150	−0.178	−0.004	−0.002
	Total	0.057	−0.413	0.943	−0.530	0.371	−0.139	−0.037
$^{1}H_{27}$	Unpaired	0.341	0.465	0.465	−0.930	1.247	−0.175	−0.175
	Paired	−1.009	−0.046	−0.047	0.093	0.239	−0.008	−0.008
	Total	−0.668	0.419	0.419	−0.838	1.486	−0.183	−0.183
$^{35}Cl_{38}$	Unpaired	225.474	−0.571	−0.571	1.142	0.0	0.0	0.0
	Paired	−221.413	−2.569	−2.569	5.138	0.0	0.0	0.0
	Total	4.061	−3.140	−3.140	6.280	0.0	0.0	0.0

For the ^{14}N nucleus, the contact hyperfine term A is seen from Table 3 to be positive. It is seen from Table 4 to be dominated by the ECP term, the direct unpaired spin contribution being small and also positive. Both the direct and ECP terms result from the unpaired spin population transferred from Fe to the pyrrole nitrogen atoms (Table 3), the direct term results from the finite nitrogen atom 2s-character in the unpaired spin orbitals, and the ECP from the exchange interaction between the unpaired spin orbitals and the orbitals in the paired spin orbitals, with spin parallel to the unpaired spin orbitals. The dipolar tensor is no longer axially symmetric with B_{xx}, B_{yy}, and B_{zz} being all unequal and B_{yz} being finite while B_{zx} and B_{xy} have zero value. This is to be expected from Figure 2 because the ^{14}N nucleus, unlike the ^{57m}Fe nucleus, is not at a site of tetragonal symmetry. The appropriate quantity to compare in this case with the experimental value (sign is not obtainable by ENDOR measurements) for the net hyperfine parameter from ENDOR measurements is $(A+B_{zz})$ which has a magnitude [86] of 7.55 MHz. The theoretical value from the results for A and B_{zz} in Table 3 is -11.89 MHz, the experimental value being about 35% lower than theory. It should be noted that the contact contribution to the net hyperfine parameter is only one order of magnitude larger than the dipolar, in contrast to about two orders for the case of ^{57m}Fe. The possible reason for this trend will be remarked on a little later, after considering the $^{1}H_{27}$ results.

The methine protons $^{1}H_{27}$ and its three other equivalent partners in the porphyrin ring are the only other family of nuclei in which magnetic hyperfine properties have been studied in Cl-hemin by the ENDOR technique [87]. From Table 3, the theoretical value of the contact parameter A is seen to be negative, composed of a direct positive contribution from the unpaired spin orbitals which is smaller in magnitude than the negative ECP contribution, the difference in behavior from ^{57m}Fe and ^{14}N being that, as may be seen from Table 4, it has a magnitude significantly closer to the larger negative ECP contribution than were the cases in these other two nuclei. This feature is the result of the fact that the 1s orbital in H is the only occupied orbital with the principal quantum number of unity and picks up the main contribution of the spin transfer from the central Fe atom through the intermediate N and C atoms, to H. The other feature that is different from ^{57m}Fe and ^{14}N hyperfine interactions is that the contact and dipolar parameters are of the same order of magnitude for $^{1}H_{27}$, while in ^{57m}Fe and ^{14}N, the dipolar contributions were respectively two and one order of magnitude smaller than contact. This trend of the increasing order of importance of dipolar interactions over contact as one goes to the more distant nuclei from the central Fe atom, is primarily the result of the classical dipolar contribution from the unpaired spin populations on both Fe and the intervening atoms. This effect is a more long range effect proportional to $1/d^3$, where d is the distance between the internal atoms and the recipient distant ones and is expected to be larger than the local dipolar contribution from electronic orbitals on the atom in question. The net theoretical value of the contact and dipolar hyperfine interaction parameters $(A+B_{zz})$ is -1.50 MHz. In comparison, the corresponding experimental value [87] of 1.014 ± 0.003 MHz is about 35% lower in magnitude as was the case for ^{14}N.

For the rest of the nuclei in Cl-hemin listed in Tables 3 and 4, namely $^{35}Cl_{38}$, $^{13}C_{6,7,8}$, and $^{1}H_{26}$ no experimental data are available from ENDOR measurements to compare with theoretical predictions. The latter have similar trends of decreasing contact contributions as one goes to the more distant atoms because of the reduction in unpaired spin population transfers as listed in Table 1. Also, the increasing relative importance of dipolar interactions as compared to contact contributions as one goes to distant nuclei is seen to hold for the hyperfine interactions from these nuclei from Table 3. The case of $^{35}Cl_{38}$ deserves special comment. This is a nucleus that is a nearest neighbor of Fe, but in this case, the contact and dipolar contributions are already comparable, unlike the other nearest neighbor $^{14}N_3$ nucleus and more like the distant methine protons represented by $^{1}H_{27}$ where the dipolar interaction had major contributions from the Fe and N_3 atoms. This feature of comparable contact and dipolar contributions in $^{35}Cl_{38}$ can be understood as follows. For $^{35}Cl_{38}$ the contact contribution arises mainly from the diffuse Fe 4s orbital mixed with the 3d orbital pointing in the direction of the Cl atom, while the local dipolar interaction arises from the local dipolar contribution from the unpaired spin population transfers to the chlorine 3p orbitals from Fe. Both of these contributions are expected to be comparable because they arise from similar local sources.

Having discussed the case of Cl-hemin in detail where the mechanisms for the origin of the contributions to the hyperfine interactions in the nuclei in Fe and the other atoms were discussed, it is appropriate now to describe briefly the results and trend in the hyperfine interactions of the nuclei in the four halo-hemins. Table 5 lists the contact, dipolar, and total contributions for the four systems, fluro, chloro, bromo, and iodo-hemins, compiled in Ref. [17] on the basis of the theoretical results, and makes comparisons with available experimental results in Refs. [86] and [87] for fluoro-hemin and chloro-hemin, and in Ref. [87] for bromo-hemin. All of the theoretical results listed are on the basis of nonrelativistic UHF investigations. For the dipolar tensor contribution, only the B_{xx} contribution is listed for the ^{57}Fe nucleus for evaluating the net hyperfine constant to compare with available Mössbauer data [86], and B_{zz} contributions are listed for the rest of the nuclei to compare with available ENDOR data [86,87].

The general features of the contact and dipolar contributions are similar for the Fe, halogen, carbon, and nitrogen nuclei and protons, as in the case of hemin described earlier in this section. The special feature of the comparable magnitudes of the contact and dipolar contributions for $^{1}H_{26}$, $^{1}H_{27}$, and ^{35}Cl nuclei in chloro-hemin, in contrast to the rest of the nuclei where the contact contributions are substantially larger in magnitude, is seen to apply to the other three halo-hemins from Table 5. The same arguments as in the case of chloro-hemin can be made for this marked feature for the protons and halogen nuclei in fluoro-, bromo-, and iodo-hemins as well.

Some of the other notable trends over the four halo-hemins that can be seen from Table 5 are the following. For the ^{57m}Fe nuclei, the net hyperfine constants from theory show a continuous decrease in magnitude from

TABLE 5 Magnetic Hyperfine Parameters (MHZ) for the Nuclei in the Halo-Hemins

	F-Hemin[a]				Cl-Hemin[a]				Br-Hemin[b]				I-Hemin[b]			
Nuclei	*A*	*B*[c]	Total	Experimental values	*A*	*B*[c]	Total	Experimental values	*A*	*B*[c]	Total	Experimental values	*A*	*B*[c]	Total	Experimental values[d]
Fe_1	−36286	0.039	−36.247	−28.38[e]	−35.514	0.264	−35.250	−26.470[e]	−31.505	−0.259	−31.764		−32.86	0.667	−32.193	
N_3	9.789	−0.418	9.371	7.530[e]	12.128	−0.236	11.892	7.550[e]	12.398	−0.297	12.101	7.28[e]	13.95	0.085	14.035	
C_6	0.884	−0.308	0.576		1.148	−0.353	0.795		0.899	−0.337	0.562		1.12	−0.376	0.744	
C_7	−2.010	−1.091	−3.101		−3.682	−1.395	−5.077		2.247	−1.193	1.054		−4.269	1.583	−2.686	
C_8	2.203	0.058	2.261		3.804	0.38	4.184		2.247	0.175	2.422		3.819	0.671	4.49	
H_{26}	0.050	−0.5	−0.450		0.057	−0.53	−0.473		0.077	−0.529	−0.452		0	−0.519	−0.519	
H_{27}	−0.428	−0.824	−1.252	0.953[f]	−0.668	−0.837	−1.505	1.014[f]	−0.419	−0.444	−0.863	1.01[f]	−0.225	−0.811	−1.036	
R[g]	140.2	56.68	196.91		4.061	6.281	10.342		11.518	1.064	12.582		11.012	8.088	19.1	
	441.5	178.5	620.0		122.7	189.8	312.5		169.5	15.7	185.2		129.8	95.3	225.1	

[a]*References [14] and [15].*
[b]*References [16] and [17].*
[c]*For* Fe_1, B *refers to the* B_{xx} *component of* B *tensor (laboratory system) and for the rest of the nuclei,* B *refers to the* B_{zz} *component in all halo-hemins.*
[d]*No experimental values are available.*
[e]*Reference [86].*
[f]*Reference [87].*
[g]*The first row refers to the hyperfine interaction parameter in megahertz as for the rest of the results above this arrow. The second row refers to the hyperfine fields in kilo Gauss at the various nuclei which are more convenient to compare for trends.*

fluoro-hemin to bromo-hemin, the result for bromo-hemin being slightly lower than that for iodo-hemin, mirroring the variation of the unpaired spin populations on Fe atom over the four compounds as seen from Tables 1 and 2. This trend is supported by experiment from fluoro- to chloro-hemin. Unfortunately, Mössbauer data for the ^{57m}Fe hyperfine constants are not available for bromo- and iodo-hemin. They would be very helpful to have for comparing with the theoretical trend in these compounds in the future.

For the ^{14}N nuclei, the net theoretical hyperfine constants are seen from Table 5 to have the trend of continuous increase over the fluoro-to iodo-hemins in agreement with the increase in unpaired spin populations over the series, the chloro- and bromo-hemin values for the hyperfine constants being nearly equal, just as the unpaired spin populations in Tables 1 and 2. The experimental values [86] available for fluoro-, chloro-, and bromo-hemins listed in Table 5 support the closeness of the net hyperfine constants for the latter two systems, but for fluoro- and chloro-hemins they show nearly the same value while theory suggests somewhat larger value for chloro-hemin. For the protons, the predicted values of the net hyperfine constants are all small and negative and show little variations over the four compounds. For the methine protons $^{1}H_{27}$, the available experimental ENDOR values [87] of the net hyperfine constants (Table 5) support the trend from theory and even the predicted magnitudes very well, the signs of the experimental results being not determined as is the usual situation with ENDOR measurements.

Overall then, there is good correlation between the net hyperfine constants of the nuclei predicted by UHF theory and the unpaired spin populations on the atoms in the halo-hemins. Experimental results, wherever available, are in reasonable agreement with theoretical predictions. There is need for additional experimental data to test theoretical predictions. Additionally, there is need for incorporation of many-body effects in the theoretical investigations which was not practicable at the time of the UHF investigations in the 1990s. They are more practicable now with the improvements in software [91] and hardware (supercomputer facilities) and should be attempted. It should however be noted that MBPT or configuration interaction procedures will require availability of sizable number of excited molecular orbital states. These will require use of large variational basis sets which for these relatively large biological systems will involve substantial investments in supercomputer time and memory.

(iv) Electronic Charge Density-Dependent Hyperfine Properties in Halogen-Hemin Systems

So far the properties discussed, namely, the magnetic hyperfine properties are dependent on the spin distribution in the five-liganded heme compounds. We shall complete our discussion of the hyperfine properties by considering the other class of such properties that are dependent on the charge distributions around the nuclei. Both the spin distributions and charge distributions associated with the electrons depend on the electron densities from the occupied state eigenfunctions of the molecule. The spin distributions depend on the differences in up spin and down spin electron densities at points within the molecule associated with the eigenfunctions while the charge distributions are associated with the sums of the up and down spin densities. So comparisons of the theoretically predicted magnetic hyperfine and charge-dependent hyperfine properties with experimental values of both these sets of properties provide a reasonably complete test of the accuracy of the electron distributions in the five-liganded heme compounds and hence of the eigenfunctions and indirectly, the energy eigenvalues associated with the molecular orbitals in the HF approximation.

We consider next the charge distribution-dependent hyperfine properties of the excited metastable nuclear state ^{57m}Fe of the iron nucleus which has nuclear spin 3/2, greater than 1/2, and therefore a nuclear quadruple moment, while the ground state ^{57}Fe has spin 1/2 and no nuclear quadrupole moment. There are two electronic charge-dependent hyperfine properties associated with the ^{57m}Fe nucleus [30,33,35] that can be studied using Mössbauer spectroscopy, namely isomer shift and NQI parameters e^2qQ and η. The process of evaluation of these properties has already been described in Section II.B. The isomer shift associated with the four halo-hemins will be discussed first. Before presenting the results for these charge-dependent properties, we shall briefly describe the physical aspects of their origin.

The isomer shift [30,33,35] is associated with the change in the coulomb interaction energy between the electrons and the charge of the nucleus because of the influence of the finite size of the nucleus leading to a distribution in the nuclear charge over the volume of the nucleus. The coulomb interaction energy for iron-containing compounds is different for the nuclear states ^{57m}Fe and ^{57}Fe. So when there is a transition from the excited nuclear state ^{57m}Fe to the ground state, the γ-ray energy is influenced by this difference as discussed in Section II.B. The difference also depends on the electron density at the nuclear site which in turn depends on the nature of the chemical compound containing the iron, for instance in a Fe^{3+} ion and a five-liganded halogen heme compound like the ones being discussed in this section. As mentioned in Section II.B, the difference in the γ-ray energies, or equivalently in the frequencies, in a halogen-heme compound and the Fe^{3+} ion, is called the isomer shift between these two systems. It has been shown by theoretical analysis [30] of the interaction between the electrons and the distributed charge in the nucleus for the ground ^{57m}Fe and excited ^{57m}Fe states that the isomer shift can be described by the formula in Equation (7) in Section II.B, namely

$$\Delta\varepsilon = (\varepsilon_1 - \varepsilon_2) = \alpha(\rho_1(0) - \rho_2(0))$$

where α is termed the isomer shift constant that depends on the difference in the nuclear radii for the excited and ground nuclear states. The value of α is given in Equation (8), namely

$$\alpha = (-0.3 \pm 0.03) a_0{}^3 \mathrm{mm/s}$$

where a_0 is the Bohr radius, 0.5292 Å or 5.292 nm and millimeter per second is the frequency unit in Mössbauer spectroscopy associated with the Doppler shift in γ-ray frequency arising from the source and absorber, with 1 MHz equal to 0.086 mm/s. The other factor in Equation (7) is the difference in electron densities at the nucleus in compounds 1 and 2 in atomic units ($a_0{}^{-3}$)

For Cl-hemin, the calculated value of the electron density difference at the Fe nucleus in this compound and that in Fe^{3+} ion is given by[15]

$$\rho_{\mathrm{ClHm}}(0) - \rho_{\mathrm{Fe}^{3+}}(0) = 1.931 a_o{}^{-3} \tag{31}$$

leading to $\Delta\varepsilon_{\mathrm{ClHm}} = -0.4173$ mm/s, the symbol $\Delta\varepsilon_{\mathrm{ClHm}}$ representing the isomer shift for Cl-hemin with respect to Fe^{3+} ion, which we shall be studying for all four halo-hemins.

The experimental value is not directly available for the isomer shift between Cl-hemin and Fe^{3+}. It is however available for Cl-hemin with respect to ^{57m}Fe in iron–copper alloy [92] of 0.2 mm/s, the latter system with respect to nitroprusside [93] of 0.483 mm/s, ^{57m}Fe in nitroprusside with respect to stainless steel [93] of -0.175 mm/s, the latter system with respect to FeF_2 [94] of -1.4 mm/s, and finally for the latter system with respect to Fe^{3+} ion [95] of -0.0437 mm/s. The addition of all these isomer shifts leads to the experimental value of $\Delta\varepsilon_{\mathrm{ClHm}}$ of -0.399 mm/s, in very good agreement with the theoretical value from Equation (31).

For fluoro-hemin, the calculated value from the electronic densities at the ^{57m}Fe nucleus for both fluoro-hemin and Fe^{2+} leads, for the difference ($\rho_{\mathrm{FHm}} - \rho_{\mathrm{Fe}^{2+}}$), to a value of $3.321 a_0{}^{-3}$. Combining this with the value of α in Equation (8), one obtains for $\Delta\varepsilon_{\mathrm{FHm}}$, the theoretical value of the isomer shift for ^{57m}Fe with respect to Fe^{2+} ion of -0.996 mm/s. This result is also very close as in the case of Cl-hemin to the experimental value of -0.96 mm/s for $\Delta\varepsilon_{\mathrm{FHm}}$, obtained in the same manner as described for Cl-hemin, using the isomer shift of ^{57m}Fe in F-hemin with respect to iron–copper alloy [92].

For Br-hemin, the theoretical value obtained for $\Delta\varepsilon_{\mathrm{BrHm}}$ is -0.858 mm/s, from the UHF value of $2.86 a_0{}^{-3}$ for ($\rho_{\mathrm{IHm}} - \rho_{\mathrm{Fe}^{3+}}$)and the isomer shift constant in Equation (8). No experimental value is available for the isomer shift for Br-hemin with respect to iron-copper alloy, as were available for both F-hemin and Cl-hemin, from which one could obtain the experimental value of $\Delta\varepsilon_{\mathrm{BrHm}}$ to compare with the available theoretical result. But from the very good agreement between theory and experiment for both F-hemin and Cl-hemin, it is not unreasonable to expect good agreement for Br-hemin, when the experimental value for $\Delta\varepsilon_{\mathrm{BrHm}}$ is available.

Lastly, for I-hemin, the UHF value for $\rho_{\mathrm{IHm}} - \rho_{\mathrm{Fe}^{3+}}$ is found to have value of $2.80 a_0{}^{-3}$ leading for $\Delta\varepsilon_{\mathrm{IHm}}$ to a value of -0.84 mm/s. Again, like Br-hemin, no experimental value for $\Delta\varepsilon_{\mathrm{IHm}}$ is available, so it is not possible to compare the theoretical value of the isomer shift for I-hemin with experiment.

From the situations with F-hemin and Cl-hemin, where very good agreement was found between theory and experiment, it is again reasonable to expect good agreement for I-hemin just as in the case of Br-hemin

The other charge-dependent properties of the five-liganded hemin systems that provide information about the electron density at the ^{57m}Fe nucleus are the NQI parameters, namely the ^{57m}Fe quadrupole coupling constant e^2qQ, and the asymmetry parameter η. The procedure for evaluation of these parameters has been briefly discussed earlier in Section II.B. The asymmetry parameter η is given by Equation (26b). As the geometry of the atomic arrangement around Fe in all four halo-hemins is tetragonal, the principal components $V_{x'x'}$ and $V_{y'y'}$ of the efg tensor at the nucleus such as ^{57m}Fe of interest here, are equal to each other and hence η has zero value.

The quadrupole coupling constant e^2qQ involves the principal component of the efg tensor with the largest magnitude, namely $q = V_{z'z'}$. From the tetragonal geometry around the Fe atom in all four halo-hemin systems one expects the principal axis of the efg tensor for the ^{57m}Fe nucleus to be directed perpendicular to the heme plane. There are no theoretical values for the fluoro-hemin and chloro-hemin systems. They could have been obtained from the UHF wave functions that were utilized for the calculation of the isomer shifts and magnetic hyperfine properties for the two compounds, but were not calculated at that time. There is an experimental value available [96] for e^2qQ in chloro-hemin from Mössbauer spectroscopy, the value being 1.52 mm/s.

For bromo-hemin and iodo-hemin, theoretical values are available [17] for e^2qQ for ^{57m}Fe nucleus using [97] $Q = 0.082$ barns or 8.2×10^{-26} (nm)^2. These values are for Br-hemin, e^2qQ (^{57m}Fe) = 0.887 mm/s and for iodo-hemin e^2qQ (^{57m}Fe) = 0.911 mm/s. Unfortunately, no experimental values are available to compare with these values. There is a later, more accurate value [98] for Q (^{57m}Fe) = 0.11 barns. The theoretical values of the ^{57m}Fe coupling constants would be increased by a factor 1.24 associated with the ratio of the latter and the earlier value of Q, and would be the appropriate ones to compare with experimental e^2qQ values when available.

As we have already mentioned in discussing the nature of agreement between theory and experiment for magnetic hyperfine properties, especially for the contact term in magnetic hyperfine interaction, HF theory provides a satisfactory agreement with experimental data. But, the quantitative agreement between theory and experiment is not perfect and there are often significant quantitative differences of 25–35% for the four halo-hemins as discussed earlier in this section. The reason for the differences is that, not only for biological systems, but also for condensed matter and atomic systems [99], there is need for inclusion of many-body correlation effects, especially for contact magnetic hyperfine interaction, because it involves difference in electron density at the nuclear site, between electronic states with opposite spin. Additionally, as contact hyperfine interaction involves a knowledge of the spin density at a single point, namely at the nucleus, there is no possibility for averaging of correlation effect corrections as can occur for properties that involve contributions from a finite region of space.

For the isomer shift [15,17,30,33,35] involved in Mössbauer spectroscopy, the total electron density at the Mössbauer active nuclei is needed in contrast to the differences in electron densities needed at the nuclear site between up and down spin states for the contact magnetic hyperfine interaction. This is the reason for the much better agreement between theory and experiment found for the isomer shift as we have seen in this section for fluoro-hemin and chloro-hemin as compared to magnetic hyperfine properties. For NQIs involving the ^{57m}Fe nucleus, we do not have the opportunity to make comparisons between results from HF theory and experiment. This is because there were no experimental data available for the bromo- and iodo-hemins to compare with theory [17] and in the case of chloro-hemin and fluoro-hemin no theoretical results for the nuclear quadrupole coupling constants are available. Usually for light atoms or molecules, many-body effects are found to be not very important for NQI parameters because the efg operator involves a $1/r^3$ dependence as can be seen from Equation (22b), where r is the distance of the electron from the nucleus. Thus, the components of the efg tensor involve a distribution over spatial coordinates, and not a single point like the contact magnetic hyperfine interaction, thus allowing averaging of many-body effects over space. For transition group elements, the 3d orbitals are somewhat diffuse and electrons in them can have many-body correlation effects [100] among themselves, and with both the 3s and 3p electrons as well as with partially occupied valence 4s and 4p orbitals. Therefore, many-body correlation effects can be important for both magnetic hyperfine and nuclear quadrupole hyperfine interactions. We plan to study with our collaborators the influence of many-body correlation effects on the magnetic and electric quadrupole hyperfine interaction properties in the halo-hemins in the near future.

We shall next discuss the hyperfine properties of deoxyHb within the framework of HF theory. Following this, we shall discuss about the total energies of singlet and triplet states of oxyHb including the influence of many-body effects. The ordering of these states is important for the understanding of the magnetic susceptibility [101] of oxyHb as well as the results of recent experimental studies [102], using the muon spin rotation technique of spin-lattice relaxation effects associated with positive muon interacting with oxyHb.

D. Electronic Structure and Associated Properties of DeoxyHb

This section deals with the features of the electronic structure and associated properties of deoxyHb. The physical, chemical, and biological importance of having a detailed knowledge of the electronic structure of deoxyHb has been already briefly described in Section II.A, which dealt with the introduction and motivation for the investigation of the electronic structures of the three sets of systems described in this section, namely the five-liganded halogen-hemin systems, deoxyHb, and oxyHb.

Section II.B described the general procedure for HF electronic structure investigations of these three sets of systems and that for obtaining the hyperfine parameters in the spin Hamiltonian for the three systems. These descriptions apply to deoxyHb and we shall not repeat the details of these procedures here. We shall only be describing some specific aspects of the procedure, such as the structure of deoxyHb at the atomic level [103] used here, the basis sets for the HF electronic structure investigation, and some special features of the hyperfine properties that have been investigated in the work.

The structure of deoxyHb at the atomic level is shown in Figure 3. The coordinates of the atoms were chosen on the basis of X-ray diffraction studies [28] and of the nearest neighbor nitrogens of Fe and other atoms were on the basis of both X-ray diffraction and Extended X-ray Absorption Fine Structure (EXAFS) data [104]. As in the case of the halo-hemins, for the investigations of the electronic structure and associated properties of deoxyHb, the distant carbon atoms on the pyrrole rings had their attached groups in protoporphyrin replaced by H atoms as is usually done in the cluster approach to adjust the time and memory requirements from practicability considerations for electronic structure investigations of large systems and especially solid state systems [23] which are infinite in size.

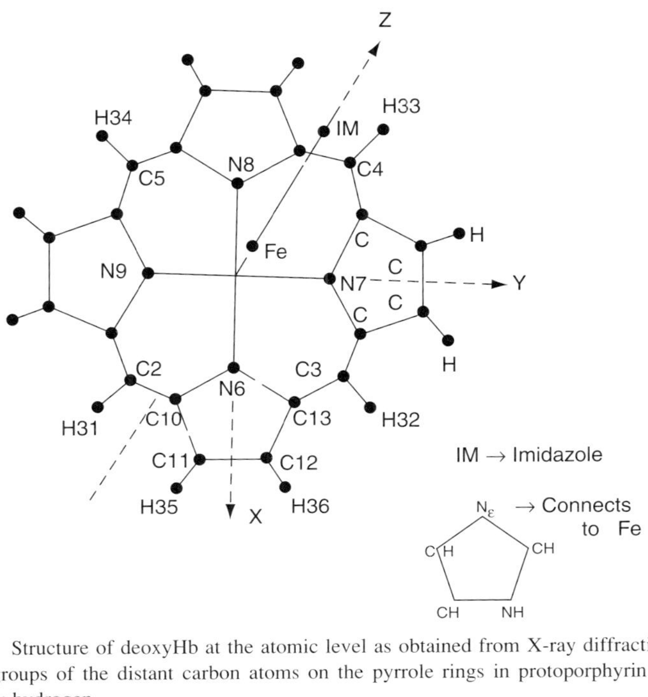

FIGURE 3 Structure of deoxyHb at the atomic level as obtained from X-ray diffraction studies. Attached groups of the distant carbon atoms on the pyrrole rings in protoporphyrin have been replaced by hydrogen.

The bond lengths for the Fe–N bonds in deoxyHb for Fe with the pyrroles were unequal and found to be 2.06 Å on the average [104]. The Fe–N_e bond distance from Fe to the nearest nitrogen, N_ε of the proximal imidazole, was found from the structural data to be 2.12 Å and the Fe atom was located about 0.42 Å from the mean plane for porphyrin, which departed somewhat from planar character. For the investigations that will be described here, the porphyrin plane was chosen as in myoglobin (Mb) [103]. The justification for this was the fact that the Fe–N bond distances in deoxyHb did not differ significantly from those in Mb and the five-liganded halogen-hemin compounds. This approximation of the planar character for porphyrin in deoxyHb is not expected to significantly affect the ^{57m}Fe-hyperfine properties we are interested in. The presence of the imidazole whose plane was found to lie [103] in-between the *XZ* and *YZ* planes, with the *X* and *Y* axes being taken along the Fe–N_6 and Fe–N_7 bonds, removed the tetragonal character [14–17] of the porphyrin electronic charge distribution in the five-liganded halogen-hemin compounds and is expected to produce departures from tetragonal symmetry for the magnetic dipolar and electric quadrupolar hyperfine tensors for the ^{57m}Fe nucleus. This source and another source for departure from tetragonal symmetry for these tensors will be discussed when considering the ^{57m}Fe hyperfine properties. The distal imidazole of the histidine [105] on the other side of the heme planes as compared to the proximal imidazole shown in Figure 3 was ignored because of its sizable distance from Fe and the porphyrin which precludes any significant effect to be expected on the electronic density distribution over the cluster shown in Figure 3.

The molecular orbital investigation for the electronic structure of deoxyHb was carried out as in the case of the halogen-hemin compounds using a Gaussian set of programs. The same Gaussian set of orbitals was used for the Fe and N atoms as for chloro-hemin [14,15] and for the C and H atoms as in fluoro-hemin [14,15], described in Section II.B.

For the deoxyHb system in Figure 3, there were altogether 196 basis functions for all the atoms and 222 electrons with 113 up (parallel) spin electrons and 109 down (antiparallel) spin electrons. The unequal number of electrons with spins parallel and antiparallel to each other required the use of the UHF procedure [23] as in the case of the halo-hemin systems. Also, the latter systems had five extra parallel spins as compared to antiparallels, that is, five unpaired spin electrons, leading to a net electron spin $S=5/2$. In contrast, in deoxyHb systems, there are four extra parallel spin electrons, that is, four unpaired spin electrons. This leads to a net spin of $S=2$. The outermost occupied state electrons involve five parallel spin electrons in the five d-like states, namely d_{z^2}, $d_{x^2-y^2}$, d_{xy}, d_{xz} d_{yz}, and one with spin antiparallel, in one of the down spin d-like states, which comes out from the UHF molecular orbital calculation [74] to be a d_{xy}-like state.

The expectation value of $|\vec{S}|^2$ comes out [15,74] as 6.288 which is quite close to the value of 6.0 for $S=2$, indicating that there is a slight spin contamination of a different total spin state because of the use of the UHF procedure, through a very small admixture of $S=3$ state and the $S=2$ state which is quite insignificant to influence the magnetic hyperfine interaction properties. What is more significant is that the admixture [74] of d_{xy} orbitals with d_{z^2} means that the principal Z' and Z'' axes of the magnetic hyperfine tensor and efg tensor will no longer be along the Z-axis perpendicular to the heme plane as found for the halo-hemin systems [14–17]. Also, the corresponding tensors may no longer be axially symmetric about the principal Z' and Z'' axis for the two cases. We shall be discussing these features when we describe the results [15,74] for the magnetic hyperfine and nuclear quadruple hyperfine interaction tensors.

Turning next to the charge and spin population on the atoms [15,74], these are presented in Tables 6 and 7. The charge on the Fe atom is seen to be very close to 2, as for a ferrous (Fe^{2+}) ion with the electron donation from Fe almost evenly distributed over the four pyrrole nitrogen neighbors and the imidazole nitrogen neighbor. This situation is somewhat different from that in the five-liganded systems like Cl-hemin [14,15] for instance in Table 1 where the Fe atom carried a charge close to $+2$ also, instead of $+3$, as would be expected for a Fe^{3+} ion, which indicated that there was substantial bonding with neighbors, apparently more than in deoxyHb. The spin population is on the other hand somewhat more similar to that for the five-liganded systems like Cl-hemin in Table 1, where the spin population was about 4.5, smaller than 5.0, which would be the case for a Fe^{3+} ion. In deoxyHb, the spin population is less localized, being only about 3.0, about 1.0 short of 4.0 for the fully localized spin population on Fe^{2+} ion. The spin populations on other atoms in deoxyHb also show the more distributed nature as compared to Cl-hemin as would be expected from the lower spin population on Fe in deoxyHb. Unfortunately, it is not possible to measure hyperfine interactions on the other nuclei in deoxyHb because one cannot observe electron spin resonance [106] in deoxyHb because of its integral spin of $S=2$. We shall next analyze the magnetic hyperfine interactions and isomer shifts and NQIs to compare with ^{57m}Fe Mössbauer data to examine the electron spin and charge densities on the iron atom in deoxyHb.

We consider the magnetic hyperfine properties first. We have used the procedure described in Section II.B in discussing the evaluation of the magnetic hyperfine properties for the five-liganded halo-hemins, to evaluate the contact and magnetic dipolar parameters in the hyperfine interaction terms in the spin-Hamiltonian for deoxyHb, namely,

$$H_{\text{hype}} = A\vec{I}\,\vec{S} + B_{x'x'}I_{x'}S_{x'} + B_{y'y'}I_{y'}S_{y'} + B_{z'z'}I_{z'}S_{z'} \tag{32}$$

In Equation (32), $\vec{I}$ refers to the nuclear spin vector for the ^{57m}Fe nucleus and $\vec{S}$ the total spin vector of the electronic system in deoxyHb; $I_{x'}$, $I_{y'}$, $I_{z'}$ refer to the components of $\vec{I}$ in the principal axis system of coordinates,

TABLE 6 The Net Charges on the Various Atoms in DeoxyHb

Nucleus	Charges	Nucleus	Charges
$^{1}Fe^{57}$	1.992	$^{24}C^{13}$ pyrrole	−0.174
$^{2}C^{13}$ methine	−0.213	$^{25}C^{13}$ pyrrole	0.025
$^{3}C^{13}$ methine	−0.252	$^{26}C^{13}$ imidazole	−0.106
$^{4}C^{13}$ methine	−0.206	$^{27}C^{13}$ imidazole	−0.137
$^{5}C^{13}$ methine	−0.192	$^{28}N^{14}$ imidazole	−0.402
$^{6}N^{14}$ pyrrole	−0.448	$^{29}C^{13}$ imidazole	0.056
$^{7}N^{14}$ pyrrole	−0.434	$^{30}N^{14}$ imidazole	−0.401
$^{8}N^{14}$ pyrrole	−0.450	$^{31}H^{1}$ methine	0.144
$^{9}N^{14}$ pyrrole	−0.465	$^{32}H^{1}$ methine	0.163
$^{10}C^{13}$ pyrrole	−0.051	$^{33}H^{1}$ methine	0.170
$^{11}C^{13}$ pyrrole	−0.227	$^{34}H^{1}$ methine	0.163
$^{12}C^{13}$ pyrrole	−0.197	$^{35}H^{1}$ pseudo	0.149
$^{13}C^{13}$ pyrrole	−0.004	$^{36}H^{1}$ pseudo	0.144
$^{14}C^{13}$ pyrrole	0.035	$^{37}H^{1}$ pseudo	0.173
$^{15}C^{13}$ pyrrole	−0.184	$^{38}H^{1}$ pseudo	0.180
$^{16}C^{13}$ pyrrole	−0.191	$^{39}H^{1}$ pseudo	0.147
$^{17}C^{13}$ pyrrole	0.052	$^{40}H^{1}$ pseudo	0.149
$^{18}C^{13}$ pyrrole	0.030	$^{41}H^{1}$ pseudo	0.173
$^{19}C^{13}$ pyrrole	−0.209	$^{42}H^{1}$ pseudo	0.178
$^{20}C^{13}$ pyrrole	−0.202	$^{43}H^{1}$ imidazole	0.244
$^{21}C^{13}$ pyrrole	0.010	$^{44}H^{1}$ imidazole	0.372
$^{22}C^{13}$ pyrrole	0.074	$^{45}H^{1}$ imidazole	0.210
$^{23}C^{13}$ pyrrole	−0.202	$^{46}H^{1}$ imidazole	0.211

X', Y', and Z' of the dipolar magnetic hyperfine tensor and $S_{x'}$, $S_{y'}$, $S_{z'}$ the components of the total electron spin of DeoxyHb. The parameter A refers to the isotropic contact magnetic hyperfine interaction constant and $B_{x'x'}$, $B_{y'y'}$, and $B_{z'z'}$ refer to the principal components of the magnetic dipolar hyperfine interaction tensor.

The calculated value of A using the UHF electronic wave functions obtained by theory [15,74] is −11.763 MHz, composed of a direct contribution 0.148 MHz from the four unpaired electrons in the Fe d_{z^2} , $d_{x^2-y^2}$, d_{xz}

TABLE 7 The Net Unpaired Spin Populations on the Various Atoms in DeoxyHb

Nucleus	Spin populations	Nucleus	Spin populations
$^{1}Fe^{57}$	3.032	$^{24}C^{13}$ pyrrole	0.015
$^{2}C^{13}$ methine	0.409	$^{25}C^{13}$ pyrrole	−0.079
$^{3}C^{13}$ methine	0.255	$^{26}C^{13}$ imidazole	0.011
$^{4}C^{13}$ methine	0.265	$^{27}C^{13}$ imidazole	−0.015
$^{5}C^{13}$ methine	0.110	$^{28}N^{14}$ imidazole	0.071
$^{6}N^{14}$ pyrrole	−0.092	$^{29}C^{13}$ imidazole	−0.014
$^{7}N^{14}$ pyrrole	0.276	$^{30}N^{14}$ imidazole	−0.001
$^{8}N^{14}$ pyrrole	−0.099	$^{31}H^{1}$ methine	−0.033
$^{9}N^{14}$ pyrrole	0.105	$^{32}H^{1}$ methine	−0.021
$^{10}C^{13}$ pyrrole	0.007	$^{33}H^{1}$ methine	−0.021
$^{11}C^{13}$ pyrrole	0.134	$^{34}H^{1}$ methine	−0.007
$^{12}C^{13}$ pyrrole	0.209	$^{35}H^{1}$ pseudo	−0.012
$^{13}C^{13}$ pyrrole	−0.024	$^{36}H^{1}$ pseudo	−0.020
$^{14}C^{13}$ pyrrole	−0.205	$^{37}H^{1}$ pseudo	−0.007
$^{15}C^{13}$ pyrrole	−0.126	$^{38}H^{1}$ pseudo	0.011
$^{16}C^{13}$ pyrrole	0.089	$^{39}H^{1}$ pseudo	−0.009
$^{17}C^{13}$ pyrrole	−0.222	$^{40}H^{1}$ pseudo	−0.012
$^{18}C^{13}$ pyrrole	0.038	$^{41}H^{1}$ pseudo	−0.003
$^{19}C^{13}$ pyrrole	0.119	$^{41}H^{1}$ pseudo	0.006
$^{120}C^{13}$ pyrrole	0.112	$^{43}H^{1}$ imidazole	0.001
$^{21}C^{13}$ pyrrole	−0.010	$^{44}H^{1}$ imidazole	0.000
$^{22}C^{13}$ pyrrole	−0.203	$^{45}H^{1}$ imidazole	−0.001
$^{23}C^{13}$ pyrrole	−0.041	$^{46}H^{1}$ imidazole	0.002

d_{yz}-like molecular orbital states and −11.911 MHz from the d_{xy}-like paired spin states leading to a net value for A of −11.763 MHz. The dipolar magnetic parameters in Equation (32) are obtained using the UHF electronic wave functions for the occupied electronic states in deoxyHb to first obtain the components of the nondiagonal dipolar magnetic tensor $\overrightarrow{B}$ in the chosen coordinate axis XYZ for deoxyHb in Figure 3 using the procedure discussed earlier in Section II.B. It is then diagonalized to obtain the principal components in Equation (32), namely $B_{x'x'} = 6.220$ MHz, $B_{y'y'} = 7.587$ MHz, and $B_{z'z'} = '13.807$ MHz. These values are a little different in the second decimal places for $B_{y'y'}$ and $B_{z'z'}$ in the published paper [74] three years back, because of a slight numerical error in the diagonalization procedure for obtaining the principal components. The principal axes (X' Y' Z') have direction cosines with respect to the XYZ axes in Figure 3 given by

$$\begin{array}{c} \\ X' \\ Y' \\ Z' \end{array} \begin{array}{c} \begin{array}{ccc} \quad X & \qquad Y & \qquad Z \end{array} \\ \begin{pmatrix} -0.010 & 0.162 & 0.989 \\ 0.794 & 0.601 & -0.091 \\ -0.608 & 0.782 & -0.135 \end{pmatrix} \end{array} \tag{33}$$

The Z' axis is seen to make 97.8° with the Z-axis chosen for the heme unit in Figure 3 or about 82.2° with the Z-direction, while the angle with the X is 127.4° or about 52.6° with the X-direction, and 38.5° with the Y-axis. Thus the Z' principal axis for the magnetic hyperfine tensor is closer to the XY plane rather than the Z-axis perpendicular to the heme plane. The net hyperfine tensor components along the X' and Y' axes combining the contact term A and the magnetic dipolar terms $B_{x'x'}$ and $B_{y'y'}$ lead to hyperfine fields along X' and Y'axes corresponding to frequencies $(A + B_{x'x'})$ of -5.543 MHz and $(A + B_{y'y'})$ of -4.176 MHz. The corresponding hyperfine field along the Z' axis corresponds to the frequency for $(A + B_{z'z'})$ of -25.570 MHz.

In comparing with the results of the measurement and analysis by Kent et al. [107], of the Mössbauer hyperfine frequency data, it should be pointed out that the authors assumed for their analysis of the Mössbauer data in the presence of high magnetic fields that the magnetic hyperfine tensor and NQI tensor for ^{57m}Fe had the same principal axes and that the magnetic hyperfine tensor was axially symmetric. From the theoretical results just discussed [74], it is seen that the UHF based result in the preceding paragraph does not show axial symmetry. So it is difficult to make quantitative comparison between the theoretical results and the conclusions from experimental analysis [107]. The only remark one can make is that theoretical results show that the net magnetic hyperfine field components for the X' and Y' principal axes directions are not very different and that the average of the magnitudes of these components is -4.86 MHz, not very different from the single component magnitude of -3.70 MHz (the converted value in MHz from their quoted value in kOe using 1 MHz as the equivalent of 25.412 kOe) that the authors of Ref. [107] have quoted for their predicted single hyperfine field in the heme XY plane.

After discussing next the theoretical and experimental values for the electronic charge distribution-dependent hyperfine properties of isomer shift and NQI parameters for deoxyHb, we shall remark on the additional theoretical and experimental efforts needed for a more detailed comparison of the theoretical and experimental Mössbauer data for the deoxyHb system.

We consider first the isomer shift results that are obtained from the UHF electronic structure investigations. Thus, using the Equations (7) and (8) in Section II.B, one gets

$$\Delta\varepsilon_{\mathrm{deoxyHb}} = \left(\varepsilon_{\mathrm{deoxyHb}} - \varepsilon_{\mathrm{Fe}^{3+}}\right) = -0.3\left(\rho_{\mathrm{deoxyHb}} - \rho_{\mathrm{Fe}^{3+}}\right) \tag{34}$$

where the electron densities at the $^{57\mathrm{m}}$Fe nucleus as in Section II.C for the five-liganded halo-hemin systems are taken in units of a_0^{-3} and $\Delta\varepsilon_{\mathrm{deoxyHb}}$ is obtained in units of millimeter per second. Using the same basis sets for the Fe orbitals for deoxyHb, as described earlier in this Section, for Fe^{3+}, the value of the difference in electron densities at the $^{57\mathrm{m}}$Fe nucleus comes out as $-1.04a_0^{-3}$, leading to $\Delta\varepsilon_{\mathrm{deoxyHb}}$ of 0.312 mm/s.

The experimental value of $\Delta\varepsilon_{\mathrm{deoxyHb}}$ is obtained by the combination of a number of different measurements of the isomer shift between various pairs of systems as for the halo-hemin systems in Section II.C. These isomer shifts between pairs of systems consist of 0.92 mm/s for deoxyHb with respect to iron in rhodium metal [108], 0.125 mm/s for Fe in rhodium metal with respect to iron metal [109], 0.15 mm/s for iron metal with respect to stainless steel [109], −1.4 mm/s for stainless steel with respect to FeF_2 [93], 0.93 mm/s between FeF_2 and K_3FeF_6 [94], and finally −0.437 mm/s between K_3FeF_6 and Fe^{3+} ion [95]. Adding up these six sets of isomer shifts leads to the experimental value of the isomer shift between deoxyHb and Fe^{3+} of 0.288 mm/s in good agreement with the theoretical value of 0.312 mm/s. So again, there is good agreement between HF theory and experiment for the isomer shift as in the case of the halo-hemin systems, the reasons for which have been suggested in Section II.C.

We consider next the electronic charge-dependent hyperfine properties namely the NQI parameters e^2qQ and η for $^{57\mathrm{m}}$Fe nucleus in deoxyHb. The procedure for calculation of the components of the efg tensor has been described in Section II.B, with the efg tensor like the magnetic dipolar hyperfine tensor being first calculated for the chosen X, Y, and Z axes and then diagonalized to obtain the principal components $V_{x''x''}$, $V_{y''y''}$, and $V_{z''z''}$.

From the principal components, one obtains the NQI parameters as in the case of the halo-hemins, the nuclear quadrupole coupling constant e^2Qq and η where $V_{z''z''}=q$, the principal component with the largest magnitude and $\eta=(V_{x'x'}-V_{y'y'})/V_{z'z'}$. Using the calculated value of q and the quadrupole moment $Q=0.11$ barns $=11.0\times10^{-26}$ (nm)^2, we obtain $e^2qQ=23.93$ MHz. The asymmetry parameter η comes out from the calculations as 0.65. Also the orientations of the principal axes (X'', Y'', Z'') with respect to the chosen XYZ axes are given by

$$M_{\mathrm{efg}} = \begin{matrix} \\ X'' \\ Y'' \\ Z'' \end{matrix} \begin{matrix} \begin{matrix} X & \quad Y & \quad Z \end{matrix} \\ \begin{pmatrix} -0.438 & 0.653 & -0.627 \\ -0.287 & 0.754 & -0.591 \\ -0.852 & -0.089 & 0.507 \end{pmatrix} \end{matrix} \tag{35}$$

The asymmetry parameter $\eta = 0.65$ obtained from HF theory is in good agreement with the value of $\eta = 0.7$ from the experimental Mössbauer measurements of Kent et al. [107]. The sizable value of η is a result of the important contribution to the efg from the doubly occupied d_{xy}-like orbital of the Fe in deoxyHb in contrast to the other four d-like orbitals which are singly occupied. The crystal structure of deoxyHb which has the heme unit depart from the tetragonal symmetry for the five-liganded halo-hemin compounds already discussed contributes to the relatively large difference of η from zero because of the inequivalence of the chosen X and Y axes. This contributes to the difference between V_{xx} and V_{yy} and also V_{xz} and V_{yz} which end up in making the principal components $V_{x'x'}$ and $V_{y'y'}$ significantly different from each other. The principal axis Z'' for the efg tensor can be seen, from the transformation matrix in Equation (35), to lead to angles of 60°, 31.6°, and 84.9° with the heme normal along the Z-axis and X and Y respectively. This orientation in particular is also like η, in agreement qualitatively, with the orientation obtained by experiment [107], particularly the fact that Z'' is different from the heme normal direction Z, unlike the result found from earlier Mössbauer measurements [110].

The value of the coupling constant e^2qQ obtained from theory leads to the quadrupole resonance frequency $\nu_Q = (e^2qQ/2)(1+\eta^2)^{1/2}$ for the nuclear spin $I = 3/2$ for ^{57m}Fe of 12.78 MHz which is in reasonable agreement with the experimental value of 25.50 MHz, but almost a factor of two smaller. It is possible that incorporation of many-body correlation effects which are important in transition metal atoms and ions could narrow this difference between HF theory which has no correlation effects at all and experiment.

In completing the descriptions of the theoretical situation regarding the hyperfine properties of deoxyHb using the HF cluster approach, we would like to briefly make some further remarks about the possible sources for improvement of the agreement between theory and experiment for the magnetic hyperfine properties. First of all, it should be mentioned that the excellent agreement found for the asymmetry parameter η for the efg tensor between theory and experiment also has some bearing on the anisotropy of the magnetic hyperfine tensor. The smaller anisotropic effect found for the latter tensor as compared to the much larger value of η for the efg tensor can be understood from the following considerations. While η is substantially influenced by the doubly occupied d_{xy}-like orbital, the magnetic hyperfine dipolar tensor involving the ECP effects involving differences in the up and down spin contributions, in contrast to the sum of these contributions for the efg tensor, is expected to be much less affected by the departure from tetragonal symmetry. This suggests that the assumption of axial symmetry by Kent et al. for the magnetic hyperfine interaction tensor in analyzing the Mössbauer spectroscopy data is not completely valid but this assumption is not as serious as it would have been to assume such a condition for η for the efg tensor. In principle, to bridge the gap between theory and experiment, additional efforts are needed in both areas.

Theoretical side, the need arises for explicit inclusion of many-body effects starting with HF theory and applying MBPT or configuration interaction procedures to explicitly include in the calculation of the magnetic hyperfine interaction and NQI parameters the influence of the difference in the actual electronic Hamiltonian including that in the instantaneous Coulomb interaction between electrons i and j involved in the $1/r_{ij}$ term and the HF one electron potential used for including only the average effects of the pair interaction between the electrons. Many-body effects are needed not only for accurate treatment of magnetic hyperfine interactions but also for NQIs including the determination of quadrupole moments from molecular or solid state data [97,98] as done for Q (^{57m}Fe) using the calculated efg parameters q to extract Q from the measured quadrupole coupling constants. For most nuclei, especially in the nuclear ground state, quadrupole moments have been obtained from atomic nuclear quadrupole coupling data with high accuracy using MBPT [111] in calculating the efg tensor in atoms.

On the experimental side, the analysis of Mössbauer data is needed without any assumptions regarding the symmetry properties of the efg and magnetic hyperfine tensors, particularly, the orientations of the principal axes. Such analyses are now much more practicable with the availability of high speed supercomputers with large storage facilities. Additionally, it would be very helpful for accurate comparison of theoretical and experimental results including many body effects, for instance in deoxyHb, when first-principles calculated quantitative results for the principal components and axes for the NQI and magnetic hyperfine interaction tensors are available. This will allow one to make theoretical predictions for the Mössbauer frequencies and their variations with the orientations of applied magnetic fields and directly compare them with the results of experimental Mössbauer spectroscopic measurements. This would obviate the need for any assumptions in extracting the magnetic hyperfine parameters in the spin Hamiltonian and the orientations of the principal axes of the magnetic hyperfine tensors, as well as the NQI parameters e^2Qq and η and the principal axes for the efg tensors, from experimental Mössbauer spectroscopic data to compare with predictions from theory.

E. Study of the Possibilities of Magnetism at Macroscopic and Microscopic Levels in Oxyhemoglobin

(i) Introduction

a. Motivation for Studying Possible Magnetic Effects Associated with Oxyhemoglobin

We turn next to an example of a hemoglobin related system that has been studied both by purely one-electron HF theory and with both one electron effects and many-body correlation effects included. The molecule we will now be discussing about is oxyHb. The importance of oxyHb in the respiratory process and in the transport of oxygen through the body is well known and vital for the life process of human beings, animals, and birds. It involves, for the efficient operation of the attachment of oxygen to deoxyHb and the detachment of the latter from oxyHb, the cooperativity effect [112] which involves the increased affinity for oxygen by deoxyHb in a non-oxygenated chain when one (or more) of the four chains is (are) oxygenated. Likewise, when one (or more) of the oxygenated chains of oxyHb is (are) deoxygenated, the other oxygenated chains become more susceptible to the process of losing their oxygen molecules. Also, the capability to hold or lose oxygen is influenced by the hydrogen negative ion concentration or pH nature of the environment, an effect named as the Bohr effect [113]. This effect allows the body to have the oxyHb deliver oxygen to the region where it is needed. The origins of both the cooperativity effect and the Bohr effect have still to be understood at an electronic level, which is a real challenge, because of the multicenter and multielectron nature of the massive deoxyHb and oxyHb systems together with their heme units and the protein chains to which they are attached.

We discuss here another aspect of oxyHb, which is the origin of its possible magnetism or absence of it, and the factors that contribute to it. The understanding of this problem of magnetism connected with oxyHb has recently become very important for studying the oxygen content of blood in different parts of the body that is important for understanding functional aspects of processes in different parts of the body in a healthy state or in a diseased state. One example of such a function in a healthy state is the understanding of brain activity induced as a result of an applied stimulation [114]. Another is the problem of understanding the change in oxygen concentration in cancerous cells as compared to healthy cells in the body. These problems are currently being investigated using the functional magnetic resonance imaging technique [115] which can study the amount of deoxyHb in any part of the body because of its magnetic property. The measurement of oxyHb which is believed to be diamagnetic is not directly possible and is estimated indirectly from the total hemoglobin and the deoxyHb content in the system of interest. Any technique that can make a direct measurement of the concentration of oxyHb by inducing magnetic character in it would therefore be very valuable. There appears to be very recent evidence [116] for such a possibility that has made the theoretical analysis of the origin of magnetic character, either direct or induced in oxyHb very important and it will now be discussed.

b. Review of Past Studies of Susceptibility and Singlet–Triplet Separation in OxyHb

Before discussing the results of very current theoretical investigations about the magnetic character of oxyHb, it is helpful to briefly review the past and current status of understanding of magnetism associated with oxyHb. Up to

1978, oxyHb was believed to be diamagnetic, especially since the classic work of [117] C.D. Coryell and Linus Pauling in 1936. In 1977, M. Cerdonio and collaborators [10], on the basis of their bulk susceptibility measurements, reported evidence for paramagnetic susceptibility for oxyHb near room temperature. They proposed that a triplet state existed about 140 cm^{-1} above the ground singlet state indicating that the triplet state was significantly populated near room temperature ($\sim$300 K), leading to the observation of bulk paramagnetic susceptibility. This suggestion was supported by theoretical investigation of the singlet and triplet energy states of oxyHb by Herman and Loew [118] using the Intermediate Neglect of Differential Overlap (INDO) approximation [119] to HF theory combined with configuration interaction for inclusion of many-body correlation effects. This investigation indicated a singlet–triplet separation of 150 cm^{-1}. Pauling has suggested [11] that it was possible that Cerdonio and collaborators' observation of finite paramagnetic susceptibility for oxyHb could be due to the formation of a metastable state of oxyHb in the preparation process for susceptibility measurement, as a result of the water in the aqueous specimen of oxyHb having been frozen and separated from oxyHb. This metastable phase could be partially dissociated at higher temperature leading to a magnetic system like deoxyHb and oxygen both of which could lead to non-vanishing magnetic susceptibility. After this suggestion by Pauling, Cerdonio retracted [12] from his earlier announcement of paramagnetic bulk susceptibility for oxyHb, following careful preparation of oxyHb samples and finding their susceptibilities to be no different than diamagnetic proteins without any metal atoms. A similar conclusion of diamagnetic susceptibility for oxyHb has been reported by another experimental group [120].

c. Recent Results of Muon Rotation Studies in OxyHb Suggesting Microscopic Magnetic Effects Associated with OxyHb

More recently, in 2007, muon spin-lattice relaxation measurements in oxyHb indicated [121] that there was a magnetic contribution to the muon spin lattice relaxation. This relaxation process is different from the muon relaxation process, discussed [5,6] in a later section of this chapter, originating from muonium trapped in the protein chain of Cyt *c* and in strands of DNA. In this latter process, an electron in a trapped muonium in a diamagnetic surrounding leaves and as it moves away, it provides a changing magnetic field interacting with the magnetic moment of the muon left behind which acts as a time-dependent perturbation that leads to a spin-lattice relaxation of the muon magnetic moment. The muon relaxation observed in recent measurements of muon spin rotation in muon trapped in oxyHb [121] appears to be associated with the fluctuating field at a trapped muon as a result of the spin-lattice relaxation of the electron system associated with the heme unit of oxyHb indicating that there is some magnetism associated with oxyHb.

d. Recent First Principles Studies of Singlet–Triplet Energy Separations in OxyHb

The results of this recent muon spin rotation study [121] have reopened the question of magnetism associated with oxyHb either by itself or through the influence of changes in the electron distribution in oxyHb caused by a trapped muon. In trying to understand this situation better, new investigations have been made [122] on the question of the singlet–triplet energy separation in oxyHb by itself using first the HF procedure which has absolutely no many-body correlation effects, as we have emphasized in describing the results of investigations on five-liganded halogen-hemin compounds and deoxyHb system. Following the HF procedure, first-principles MBPT was applied to include many-body correlation effects. The result of these very recent investigations [122] which will now be briefly described provide a clear description of the role of one-electron and many-electron contributions to the factors governing the singlet–triplet separation. They also provide information regarding the charges on the atoms in oxyHb that could be valuable in planning future time-consuming investigations on the trapping of muon by oxyHb and the influence it has on the singlet-triplet energy separation and hence on possible induced magnetization that could occur in the oxyHb system with a trapped muon.

To examine the convergence of recent results of first-principles investigations of singlet and triplet energies in oxyHb, we have listed in Table 8 those energies for three choices of basis sets provided in the Gaussian 09 set of programs. The results are given for recent crystal structure data [123] for oxyHb in the T or tense state with low oxygen affinity. The cluster chosen in Figure 4 for the investigation was similar to deoxyHb in Figure 3. It consisted of the heme unit with the side chains attached to the distant carbons

TABLE 8 Energiesof OxyHb Cluster for Singlet and Triplet States for Three Choices of Gaussian Basis Sets

Basis set	Level of investigation	Singlet energy[a]	Triplet energy[a]
(3-21G)	Hartree–Fock (HF)	−2605.20	−2605.31
(6-31G)	Hartree–Fock (HF)	−2618.10	−2618.20
(6-311G)	Hartree–Fock (HF)	−2618.46	−2618.54
(3-21G)	(HF)+many body	−2608.37	−2608.37[b]
(6-31G)	(HF)+many body	−2621.29	−2621.28
(6-311G)	(HF)+many body	−2621.86	−2621.68

[a]*All energies are in atomic units* (e^2/a_0).
[b]*This total energy is really −2608.367 which is rounded to −2608.37.*

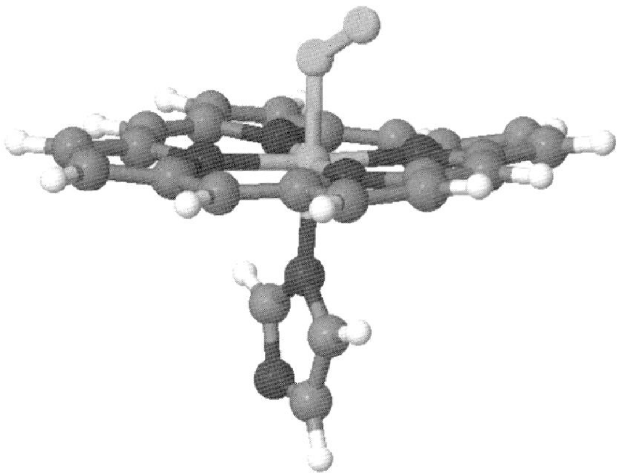

FIGURE 4 Cluster chosen for the investigation of oxyHb. Side chains attached to the distant carbons in the four pyrrole rings have been replaced with hydrogen atoms, as in deoxyHb.

in the four pyrrole rings replaced with hydrogen atoms as in deoxyHb. The proximal imidazole is similar to that in deoxyHb with some differences in the coordinates of the atoms in it in keeping with the structural data for the T-state of oxyHb. The sixth ligand of Fe is the O_2 molecule, with appropriate coordinates of the O atoms taken from the X-ray data [123].

From Table 8, one notices the feature that for all of the basis sets, the HF value for the singlet state energy is always higher than that for the triplet state. This is because, for the singlet state the total spin S is zero and the quantum number corresponding to S_z, the Z-component, $M_s = 0$ and the highest one electron occupied state is doubly occupied with two electrons with opposite spin. For the triplet state the total spin is unity and the maximum value of $M_s = 1$. The highest one electron state cannot be occupied by two electrons with parallel spins to have $M_s = 1$, because of Pauli principle, so they have to go to two different states one in the occupied state in the singlet and the other to the next higher one electron state. This would suggest that the $S = 1$ triplet state would have higher total energy than the singlet. However, for the triplet state, the exchange energy between the two electrons in the highest occupied states which is negative is so strong that it overcomes the positive energy difference between the two states, which leads to a total energy for the $S = 1$ state that is lower than for the $S = 0$ state. When many-body effects are included, the instantaneous correlation from the $1/r_{12}$ coulomb interaction term between the two electrons in the highest occupied state for the $S = 0$ many-electron state is very strong and pushes them apart reducing the total energy. This correlation type of reduction in the total energy is smaller in magnitude for the $S = 1$ many-electron state than for the $S = 0$ state because the two electrons with the highest one-electron energies in the former case in the HF approximation are in two different states and do not get in the way of each other as much as is the case for the $S = 0$ state. The greater magnitude of the negative correlation energy for the $S=0$ state including many-body effects would therefore be lower than the total energy for the $S=1$ state as is observed for all basis sets in Table 8.

The other feature of the results in Table 8 is that the changes in the total energies for both the singlet and triplet states both with and without many-body effects in going from the smallest basis set (3-21G) to (6-31G) are rather substantial. By contrast, the corresponding change in going from (6-31G) basis set to the largest basis set (6-311G) is much smaller indicating good convergence in going to the latter basis set. We shall thus use the results for the largest basis set (6-311G) for our further discussions.

To test if the conclusion about the strong influence of correlation effects in reversing the order of the energies of the triplet and singlet states from purely one-electron HF theory for the T-state or low affinity state of oxyHb is dependent on the structure of oxyHb at the atomic level, the HF and many-body correlation effects have also been studied for the R-state or higher affinity of oxyHb using the structural data [124] for the R state. The results for the two structures are listed in Table 9 using the same choice of (6-311G) basis set in both cases. The results in Table 9 indicate that the triplet-singlet separation is 22% smaller for the R-state than for the T-state, but the other feature of the results for the T state, namely, the stronger correlation effect in the singlet state changing the order of the triplet and singlet states is also clearly demonstrated for the R-state. Additionally, on expressing the energy difference between the triplet and singlet states in terms of kT, where k is the Boltzmann constant and T is expressed in the absolute scale, the values of T for the T and R states, combining HF theory with MBPT, are both two orders of magnitude larger than room temperature 300 K.

There is a value of the triplet-singlet separation in oxyHb available [125] from the use of the BL3YP procedure [126] using a mixture of HF exchange and density functional approximation for exchange and many-body effects approximated by a density functional-dependent correlation potential. The positions of the atoms in the oxyHb cluster were optimized separately for the singlet and triplet states. With this approach, the triplet–singlet energy separation was found [125] to be 1508 K in temperature scale with the same order, with triplet higher than singlet as found from the results in Table 9 and from

TABLE 9 Singlet–Triplet Energy Separations in R and T Structure in OxyHb

	T-state (E_T-E_S)			R-state (E_T-E_S)		
Procedure	Atomic units	eV	T_{Diff}	Atomic units	eV	T_{Diff}
HF	−0.08	−2.18	−25228.99	−0.14	−3.81	−44173
HF+MP2	0.18	4.87	56448.56	0.14	3.81	44173

the first-principles HF+MP2 investigation (MP2 indicating MBPT result at second-order perturbation level). The B3LYP result for T is an order of magnitude lower than the HF+MP2 results for T and R states. Also, the B3LYP result [125] is five times larger than the room temperature of 300 K so that very little triplet population could be found at room temperature as is the case with the HF+MP2 result. The possibility that the substantial quantitative difference between the B3LYP and HF+MP2 results for the triplet–singlet energy gap could be due to the difference in the structural data used in the two cases can be ruled out from the relatively small difference for the R and T states in Table 9. The likely reason seems to be that pointed out by the authors [125] of the chapter on the basis of the B3LYP procedure, namely, that the density functional exchange correlation potential needs to be improved to make the corresponding Hamiltonian approach the actual many-electron Hamiltonian to improve their results in Fe atoms and ions as compared to experiment. A similar conclusion [127] has been reached from recent investigations on energies and hyperfine properties of other atoms by first-principles HF+MP2 investigations based on Gaussian 09 set of programs, the B3LYP procedure, accurate linked cluster many body perturbation theory [128] procedure for atomic systems, and experiment.

It thus appears from the investigations just discussed that for oxyHb system, the singlet state is the ground state. This makes it necessary for first-principles investigations to be carried out for muon trapping in the oxyHb system for singlet and triplet states to try to explain the observed magnetic effects in the muon relaxation rate in oxyHb in muon spin relaxation (μSR) experiments. The analysis of the effective charges obtained from the HF electronic structure investigations on the singlet and triplet states in Table 10 indicates substantial negative charges on the oxygen atoms in the triplet state and much smaller in the singlet state with the oxygen atom distant from the iron in the singlet state even carrying a small positive charge. This means that there is more probability of the muon with the positive charge to be more strongly trapped by the oxygens in the triplet state rather than the singlet. The singlet state had more correlation in energy (in magnitude) than the triplet in the absence of the muon because the paired electrons in the singlet state were closer to each other. With the muon expected to be strongly trapped at the oxygen sites, its positive charge would localize the electrons on the oxygen and make them stay closer to each other and correlate more strongly than in the absence of the muon. This could make the correlation energies for the triplet and singlet states in the presence of the muon closer to each other and reduce the relatively greater lowering of the singlet state in the absence of the muon. This gives the possibility of significant reduction of the triplet–singlet energy separation in the presence of the muon, providing the opportunity for the singlet and triplet states to get close enough to have significant population for the triplet state at room temperature to give the (oxyHb+trapped μ^+)

TABLE 10 Charges on Iron, Pyrrole Nitrogens, and Oxygens in OxyHb

Atom number	Atom	Charge in singlet	Charge in triplet
10	Fe	1.601	1.909
11	O	−0.045	−0.474
16	O	0.042	−0.178
11	N	−1.017	−1.108
12	N	−0.924	−0.892
13	N	−0.990	−1.103
14	N	−0.892	−0.855

system a chance to have magnetic character or even have the triplet state lower in energy to make the entire trapped system magnetic.

The pyrrole and imidazole nitrogens in Table 10 carry substantial negative charges but with no marked differences in their charges for the singlet and triplet states. It would also be worthwhile to explore the influence of trapping of muon at these nitrogens because with the delocalized nature of the electron distribution on the porphyrin ring and the differences in charges on the Fe and O atoms in the triplet and singlet states, it is not unlikely that the trapping of muon could lead to significant changes in the triplet–singlet energy differences as compared to oxyHb in the absence of any muon. In summary, the finding, using the first-principles procedure involving HF combined with MBPT, that the competing effects of exchange and many-body correlation effects determine the relative ordering of singlet and triplet states has provided insights into the possible mechanisms by which muon trapping could conceivably lead to magnetic properties for oxyHb at a microscopic level even though oxyHb by itself is diamagnetic with no bulk susceptibility. If it turns out that muon trapping does not produce any microscopic magnetic property for oxyHb, there is still the possibility that the magnetic contribution to μSR observed experimentally can be explained by a muonium, arising out of a muon in the muon beam capturing an electron, getting trapped on the heme unit in oxyHb in the diamagnetic singlet state, and making it paramagnetic with spin $S=1/2$ as has been found in recent investigations [129,130] in DNA.

III. MUON AND MUONIUM TRAPPING IN THE PROTEIN CHAIN OF CYT *C*

Hemoglobin, discussed in the previous section, is a vitally important protein to transport oxygen through our blood stream. Another heme protein is Cyt *c*, an electron transfer protein. Just as hemoglobin, it is composed of two major parts: its heme unit and its protein chain. In order to reach a better understanding of the dimensionality of the electron transport path, the μSR method [131,132] has been applied to Cyt *c* [133]. The experiment utilizes a 100% spin-polarized pulsed muon beam which is injected into the Cyt *c* sample. The positive muon (μ^+) captures an electron and becomes muonium (Mu) (which can in many ways be considered as a light hydrogen isotope). Once this Mu gets trapped along the protein chain of Cyt *c* and loses the brought-in electron, the latter can move in its characteristic way and thereby cause spin-lattice relaxation to the spin of the μ^+ left behind. A measurement of the relaxation rate as a function of the magnitude of an external magnetic field yields information about the shape of the pathway of the electron transport [134,135]. The important point is that this measurement can only be accurate if the μ^+ stays fixed while acting as a probe for the electron movement and should therefore remain at the original trapping site of the Mu.

Every protein chain, the one of Cyt *c* included, consists of a sequence of amino acid molecules. Following the cluster approach, we began our investigation by carrying out a systematic study of μ^+ and Mu trapped in *individual* amino acid molecules occurring in the protein chain of Cyt *c*. The equilibrium trapping positions for μ^+ and Mu were found by placing the μ^+ or Mu near the site under study and carrying out a geometry optimization of the adduct (μ^+ or Mu), the atom at the trapping site, and its nearest neighbor atoms.

The protein chain of Cyt *c* contains a total of 104 amino acids [136] and is shown schematically in Figure 5. We investigated the trapping properties of μ^+

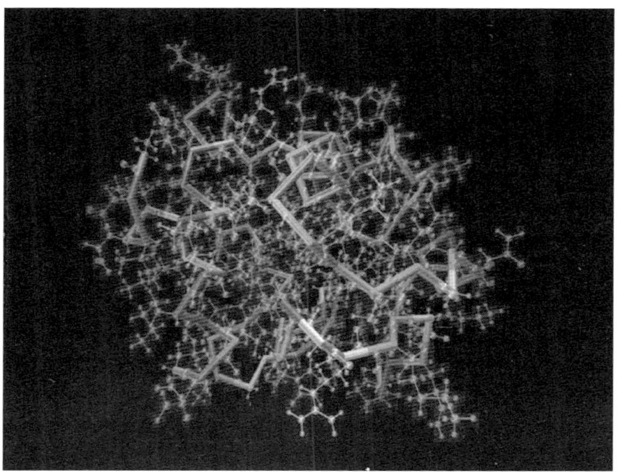

FIGURE 5 A schematic view of the structure of the electron transfer protein cytochrome *c*. The trace of the protein chain is displayed as a sequence of bent rods, while the atoms and chemical bonds that constitute the amino acid building blocks are shown in a partially transparent ball-and-stick model. The colors indicate the particular type of amino acid (e.g., lysine in cyan, glycine in white, and cysteine in yellow).

and Mu in isolated amino acid molecules taken from a sequence that constitutes about 25% of the total length of the chain and includes the amino acids #4 till #27. The results obtained from this sample of amino acids are expected to be somewhat representative for the entire protein. In particular, we were interested in the variation of the binding energies of μ^+ and Mu for different types of amino acids. Furthermore, even in the same type of amino acid, but for different locations in the chain, these parameters will not be necessarily identical. The reason for this lies in the fact that two amino acids of the same type but at different places in the sequence of the protein chain will most likely have different neighboring amino acids. Hence, the local environment depends on the position in the chain, and is not precisely equal for the same type of amino acid. Therefore, before any comparison of μ^+ and Mu trapping properties between different amino acid molecules can be made, we should first establish, how large the range of variations in the same amino acid type is in different parts of the chain, and also what the statistical averages of the properties of the same amino acid type are over the entire chain.

One expects two main effects from the environment of the protein chain on the amino acids under investigation in connection with μ^+ and Mu attachment. The first one is the direct effect that the electron distribution on the neighboring amino acid molecules in the protein chain has on the μ^+/Mu trapping properties. The second one is the indirect effect that this electron distribution has on the structural geometry of the amino acid in the protein chain which is trapping μ^+/Mu, thus leading to a deviation from the geometry of the isolated amino acid. The incorporation of the first effect would require large clusters to be used in the calculations, which is somewhat time-expensive from a computational point of view. In the present investigation, only the second effect was taken into account, by using the structural geometry of Cyt *c* as it has been determined from NMR measurements [136]. By keeping the vast majority of the atoms in our cluster fixed to their experimental positions, we recover the indirect effect of the electron distribution in the much larger chain.

The atomic positions of μ^+/Mu and of carbon and oxygen from the double bond in the carboxyl group of the various amino acids were optimized with respect to minimal energy, while the remaining atoms in the cluster were kept frozen at their respective positions obtained by the analysis of splitting and widths of NMR spectra [136]. In connection with the displacements in the positions of atoms adjacent to the trapped μ^+/Mu, we have found from our study of μ^+ and Mu trapping in the single amino acid molecules cysteine, lysine, and alanine [137] that it is usually sufficient to optimize the positions of the nearest and next nearest atoms to find good convergence. The trapping of μ^+ and Mu was first investigated in the amino acid cysteine in order to obtain general rules for possible trapping sites in amino acid groups. These rules were then confirmed by equivalent investigations on two other amino acids, namely lysine and alanine. We will now describe the results for μ^+ trapping in cysteine.

Our calculations led to positive binding energies for μ^+ at all negatively charged atoms in the molecule while μ^+ was unstable at atoms which carry a positive charge. The atoms in cysteine that carry a negative charge were the nitrogen atom of the amino group, the central carbon atom to which the residual of cysteine, CH_2SH, is connected, the sulfur atom of the residual, and the two oxygen atoms in the carboxyl group. We found that Mu was only trapped by two atoms in cysteine, namely the carbon and the oxygen atom that form a double bond in the carboxyl group of the amino acid. As for the mechanism of the Mu trapping, we observed that the double bond is transformed into a single bond, as shown for example by the elongated bond length between C and O following the trapping of Mu. The Mu then uses one of the two electrons from the dangling bonds that result on C and O, leaving most of an unpaired electron on the other atom with which Mu does not bind. Such a mechanism was also observed in other studies on Mu attachment in covalently bonded systems [138].

As has already been stated above, it is important that μ^+ and Mu are stable at the same site for the μSR measurement to work in the intended way. It appears from our results that of the two sites where Mu can be trapped, namely the C and O forming the double bond in the carboxyl group, μ^+ cannot be bound to the former, as C carries a positive charge. We therefore conclude that the double-bonded oxygen is the only site where both Mu and μ^+ are stable simultaneously, and we therefore predict that the μSR relaxation process is associated with μ^+ trapped at that site after the electron on Mu leaves. The possibility that a moving electron can produce spin-lattice relaxation of the μ^+ if the latter is forced to leave the site where the Mu was trapped (like the carbon in the C=O bond) is rather low. Thus, the main sensing of the electron motion through muon spin-lattice relaxation is expected to occur through the muon at the oxygen site of the C=O in the carboxyl group, which exists in every amino acid.

We studied a selection of single amino acid molecules taken from a sequence of 24 amino acids (#4–27) of the protein chain of Cyt *c*. The binding energies are influenced in various degrees by the environment. The average values and standard deviations were obtained for types of amino acids occurring, such as lysine, glycine, glutamic acid, and cysteine. Clear trends were observed in the binding energies for μ^+ and Mu: as the binding energy of μ^+ increases, that for Mu decreases, and vice versa. A possible reason for this correlation could be the connection between the strength of a double bond and the charges on the atoms in this double bond. More importantly however, even within the statistical fluctuations, the different amino acid molecules show very distinct binding energy values for μ^+ and Mu trapping, and could therefore be identified if a measurement of the binding energy would experimentally be feasible.

We presented here only a very brief account of how the cluster method can be employed to study muon and muonium trapping in the protein chain of Cyt *c*.

For a more detailed presentation and discussion of the quantitative results, we refer the interested reader to References [137–141] listed at the end of this chapter.

IV. ELECTRON TRANSPORT ALONG THE STRAND OF A-FORM AND B-FORM DNA

In the previous section, we have dealt with the electron transfer protein Cyt *c*. Electron transport is normally not considered a biological function of deoxyribonucleic acid (DNA), but may still nevertheless play an important role, for example when it comes to oxidative damage. In any case, the electric conductivity of DNA is a longstanding issue of interest for scientists and the possibility of using DNA as a component in molecular electronic devices has stirred a lot of interest over the past [142,143], and continues to do so [144]. In microdevices based on manufactured organic molecules, DNA could for example be used to transport a signal from one point on the device to another. Scientists in the field of molecular electronics are trying to establish both experimentally and theoretically whether DNA would make for a good conductor of electrical signals. It is however notoriously difficult in such an assessment to distinguish between the genuine resistance of the molecular wire and the resistance due to the contact between the DNA molecule and the (bulk-)electrode used to hook up the molecule to a power supply in a regular circuit, so as to measure the current flowing as a function of the applied voltage.

As the probing of electrical conductance in a molecular wire such as DNA is microscopically based on the transport of single electrons, a promising approach to probe conductance properties of DNA is once again the μSR method [131,132], which has already been used successfully to study electron transfer in Cyt *c* (as discussed in the previous section of the present chapter). In the μSR approach, the positive muon (μ^+) plays two roles: it first brings along an electron into the DNA molecule, and then serves as a probe for the movement of that electron. Such μSR experiments have been carried out [145] on two types of DNA, namely A-form and B-form DNA with the goal in mind to test for possible differences in electrical conductance between these two forms. In the following, some general properties of DNA and its two particular forms (A-form and B-form) will be provided. After that, a quick summary of the experimental findings will be presented which leads to the motivation for our theoretical study. Details of the procedure of our study are followed by a presentation and discussion of the results. We conclude this section with a number of findings that can be drawn and a list of possible future investigations on this topic.

DNA is a double-helical molecule with roughly the form of a spiral staircase [146] and consists of two unbranched polynucleotide chains. A good way to visualize the molecule [147] is to imagine it straightened into a ladder: the side rails of this ladder are made up of alternating sugar and phosphate groups. These groups are linked by so-called 3′- or 5′-phosphodiester bonds. The two sides of the molecular backbone are often described as antiparallel because the sequence of atoms in each rail runs in opposite directions. The rungs of the ladder consist of specific hydrogen-bonded pairs of nitrogenous bases that are attached to the desoxyribose sugars in the backbone's side rails. The hydrophilic sugar phosphates form the external backbone of the DNA molecule while the hydrophobic base pairs are protected inside that backbone from the aqueous environment of the biological cell. The base pairs contain four types of bases, namely adenine (A), guanine (G), cytosine (C), and thymine (T). The pairs of nitrogenous bases that make up the rungs are in the keto (as opposed to enol) tautomeric configuration [146]. The two configurations are distinguished by the location of the hydrogen atoms that provide hydrogen bonding between the two bases.

The smaller single-ringed pyrimidines, cytosine and thymine are always paired with the larger double-ringed pyrimidines, guanine, and adenine. According to the Watson-Crick model [146], the adenine–thymine (AT) base pair is stabilized by two hydrogen bonds, while the guanine–cytosine (GC) base pair is stabilized by three hydrogen bonds. This mechanism leads to identically sized ladder rungs consisting of AT and GC pairs. This consistent pairing is a key element in the function of DNA as a genetic information storage device. The typical configuration of DNA in the cell (i.e., *in vivo*) is called B-form DNA, which suggests that there exists another configuration, called A-form DNA which is only found *in vitro* B-form DNA, can be obtained from A-form DNA when the DNA fibers absorb water until more than 40% of their weight is due to H_2O molecules. In the laboratory, B-form DNA is optimally produced for a relative humidity level of about 90% [147]. The extra hydration helps the DNA molecule to assume the lowest energy configuration. The two helical backbone chains are farther apart than in the A-form and are also elongated by about 30%, so that the B-form has its bases oriented perpendicular to the fiber axis. Due to the elongated form, one rotation of the "staircase" requires only 10 base pairs with a corresponding height of 34 Å, or 3.4 Å between adjacent pairs.

The packing density of base pairs in A-form DNA is about 30% higher than in B-form DNA. It would therefore be very interesting to study and compare electron transport in these two systems because this will yield information on the dependence of electron mobility upon the base pair separation. This in turn, could then lead to a better understanding of the electron transport mechanism in DNA. With that aim in mind, a series of comparative μSR experiments [145] in A-form and B-form DNA have been carried out. The results of this study suggest that electron mobility is enhanced in A-form DNA where base pairs are 2.6 Å apart, and reduced in B-form DNA where the separation of base pairs amounts to 3.4 Å. As this effect is really a result of the moving electron, one could describe it as a *dynamical effect*. However, there is also a *static effect* due to the magnetic hyperfine interaction of the Mu with the electronic environment. Due to the difference in the molecular geometry and level of hydration, this environment is certainly distinct for A-form and B-form DNA, but it is not obvious to what extent this difference would affect the hyperfine interaction of muonium (Mu) in the two systems. It would clearly be beneficial for the interpretation of the μSR data to have an estimate of the importance of this static effect. We therefore set out to study the magnetic hyperfine interaction of Mu in a first-principles manner, to assess how significant the difference in the hyperfine interaction of Mu between A-form and B-form DNA would be [148,149]. Besides the potential usefulness of this investigation for the interpretation of the μSR data, it is also an interesting subject to study in its own right; for example we explored the influence that μ^+ and Mu have on the geometry of a nucleic acid when either one of them is attached to a specific site. Furthermore, when vibrational effects are taken into account, the results on the binding energy of μ^+ can be related to protonation studies on nucleic acids [150,151], as the μ^+ can be regarded as a light isotope of H^+.

We studied the magnetic hyperfine interaction of Mu and general trapping properties of μ^+ and Mu in the nucleic acid adenine in three configurations, namely adenine from A-form DNA, adenine from B-form DNA, and isolated adenine molecule. The structural data of adenine as it occurs in A-form DNA has been taken from experimental X-ray diffraction data [152,153] and analogously for adenine in B-form [154]. Only those atoms that belong to the base and sugar ring of adenine were included in the cluster. The link to the phosphate has been terminated with a methyl group. To evaluate how important the sugar ring is for our results, and also to make the study more general, we furthermore included the isolated adenine molecule (i.e., just the base) in our investigation, with the only modification from its original stoichiometry being the replacement of the H atom on N9 by a methyl group, so that better comparison could be made [155] with adenine in DNA which possesses a link to the sugar-phosphate backbone at N9.

For the isolated adenine molecule, a full optimization of the geometry with respect to minimal energy was carried out at the MP2 level using the 6-31G (d) basis set both with and without μ^+/Mu attached. For adenine from A-form or B-form DNA, only partial optimizations of the geometry with respect to minimum energy of the μ^+/Mu position and of the nearest atoms where μ^+/Mu has been attached to were carried out, while the remaining atoms in the cluster were kept fixed at the geometry determined from experiment [153,154]. The reason why we did not carry out a full optimization of all atomic positions was because we wanted to preserve the characteristic geometry of A-form and B-form DNA, and a full optimization of the cluster would cause any distinction between A-form and B-form to disappear. The thus optimized systems with μ^+/Mu added to the sites under investigation were then studied for their energetic properties and magnetic hyperfine interaction of Mu with its local environment. In particular, following the motivation of the study, we were interested to see how large the difference in the hyperfine interaction would be between adenine from A-form DNA and B-form DNA. In addition, we checked for alterations in the geometry of the isolated adenine molecule when μ^+/Mu was added, especially whether the presence of μ^+/Mu causes any deviation from the planar shape of isolated adenine. For our theoretical investigations, we employed the HF cluster method, as it is implemented in the Gaussian 98 programming package [156]. The UHF procedure was used to allow for spin polarization [149,155]. To incorporate correlation effects [75,141], MBPT [69] has been used in the form of second-order Møller–Plesset perturbation theory (MP2), and the convergence with respect to basis set size was studied using the standard split-valence basis sets 6-31G(d,p), 6-31++G(d,p), 6-311++G(d,p), and cc-pVDZ.

We first report results on the effect that μ^+ and Mu have on the geometry of isolated adenine when trapped at different sites. When Mu or μ^+ is attached to C4 or C5, the geometry of adenine is severely altered from its planar shape. The two rings of adenine appear tilted, with an angle of about 25°. For Mu attached to N3 or N7, the same qualitative behavior is observed; however, quantitatively the change is less drastic; the tilt angle between the rings is about 10° for N3 and about 7° for N7. This kind of change in geometry does not occur when μ^+ is attached to N3 or N7. For those sites, the planar geometry of adenine is preserved. For the trapping sites C2, C6, or C8, adenine keeps its planar geometry for both, μ^+ and Mu attachment. This is due to different reasons: at C2 and C8, the hydrogen atom at these two sites (Figure 6) will move out of the plane of the molecule, and thus compensate the effect that Mu or μ^+ has on the geometry of adenine. Similarly, at C6, the NH_2 group is bending out of the plane, thereby compensating for the effect of Mu or μ^+.

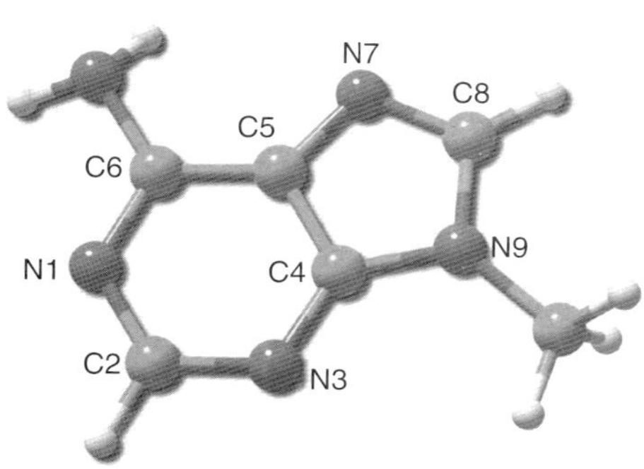

FIGURE 6 Chemical structure of the nucleic acid adenine with atomic labels for reference. The H atom on N9, which is present in the natural form of adenine, was replaced here by a methyl group for reasons discussed in the text.

Our results show that the hyperfine coupling constant is in general significantly lower for Mu trapped in isolated adenine as compared to the case when it is trapped in either A-form or B-form adenine. As there is no sugar ring present in isolated adenine, this implies that the sugar ring has a significant effect on the hyperfine interaction of Mu trapped in the base of adenine. Secondly, comparing the results for A-form DNA and B-form DNA, we see that the numbers are still quite different for most trapping sites, while not as different as in the previous comparison to the isolated adenine system. This means that there is indeed a significant difference in Mu hyperfine interaction between A-form and B-form adenine. Lastly, we note that the results for either A-form adenine or B-form adenine show substantial variation, so that in principle, a distinction of the Mu trapping site on the basis of the measurement of the hyperfine coupling constant could be made.

In conclusion, we have shown from first-principles that the magnetic hyperfine interaction of Mu trapped in the nucleic acid adenine exhibits significant differences depending on whether the geometry of the isolated molecule, the A-form DNA, or the B-form DNA is used for the cluster in our calculations. This result could have important implications for the interpretation of μSR experiments on the electron mobility in A-form and B-form DNA. We also predict that it should in principle be possible to distinguish between different trapping sites in adenine from either A-form DNA or B-form DNA, by measuring the hyperfine coupling constant of Mu, as we found substantial variation in the values for the isotropic hyperfine coupling constant A at different sites. A comparison of our results for the binding energy of μ^+ in adenine with earlier protonation studies in the same nucleic acid yielded very good agreement. We furthermore found that μ^+ and Mu can have profound effects on the geometry of adenine when trapped at various sites in the base, leading to a tilt between the two rings. For some of the sites however, the effect of μ^+ and Mu on the geometry is compensated by the reorientation of an H atom or an NH_2 group bond to that site.

Of course, it is always desirable to use larger clusters, but the large number of electrons that is added when the phosphate group of the backbone is included in the cluster makes this task computationally quite demanding. Nevertheless, a study of μ^+ and Mu trapping properties in a cluster containing several connected base pairs was carried out by us to show how important the direct effect of neighboring nucleic acids is [149]. Finally, the influence of the differing amounts of water that A-form and B-form DNA contain was taken only indirectly into account by using the experimentally determined geometries. It would however be interesting to also study the direct effect that the different hydration environments have on the hyperfine interaction of Mu and on the general trapping properties of μ^+ and Mu by incorporating an appropriate distribution of water molecules in the surrounding of the cluster. The equilibrium positions of these water molecules could be determined using partial geometry optimization for smaller clusters, or, for larger clusters, using molecular dynamics methods.

V. TRANSVERSE ELECTRON TRANSPORT THROUGH DNA FOR RAPID GENOME SEQUENCING

Electron transport in the direction *along* the DNA strand, as discussed for motivation of the cluster studies described in the previous section, has been thoroughly researched. A much less common approach considers the application of a bias voltage *across* the DNA strand in a perpendicular direction. This scenario forms the core of a hypothetical setup with the aim to realize a revolutionary new generation of whole-genome sequencing devices. Once again, we shall see how a cluster approach can be highly valuable in studying such a system from first principles.

Because of their complexity, the existing methods for DNA sequencing are too expensive and time-consuming to be useful for large-scale whole-genome determination. However, the latter is absolutely required in order to fully understand the correlation between particular genetic patterns and certain diseases. Thus, there is a strong drive to drastically reduce the involved cost and time, with a typical aim mentioned in the relevant literature as around US$1000 for one genome in the course of around 24 h. If that goal could be achieved, it would mean that large-scale DNA sequencing would become a feasible option; a significant step toward the vision of "personalized medicine" and a better understanding of hereditary diseases [157–160].

One of the most promising new techniques for next-generation DNA sequencing involves passing a single-stranded DNA (ssDNA) molecule through a so-called "nanopore" with a diameter of only a few nanometers [161–165]. This is accomplished in solution where the phosphate groups on the backbone of ssDNA are negatively charged, compensated by positive counter ions in the solvent. An electric field applied perpendicular to the surface of the separating membrane in which the nanopore is located can eventually capture and drive ssDNA molecules through the nanopore. While originally, monitoring of the time-dependent ionic current was thought to possibly yield the DNA sequence [164,166], it now appears more realistic that embedded electrodes in the walls of the solid-state nanopore are required to reach single-base resolution [167,168]. When applying a bias voltage across these electrodes, a small tunneling current perpendicular to the DNA strand can be recorded and thus, as the ssDNA translocates through the pore, time-dependent current–voltage signals from the electrodes would potentially allow the identification of the nucleotide sequence.

Unfortunately, as it turns out, this "simple approach" leads to a rather broad statistical distribution of the current signals which are on average also of the same order of magnitude [169,170] and so a further improvement is required to actually achieve a robust electrical distinction between the four bases. The idea originally proposed by us [171] to solve this issue was to functionalize the nanopore-embedded electrodes with suitable probing molecules that could weakly couple to the target DNA bases during the translocation process. To test whether such a functionalization can lead to the required accuracy, we calculated the corresponding transport properties, employing DFT as implemented in the Spanish Initiative for Electronic Simulations with Thousands of Atoms (SIESTA) code [172] together with the nonequilibrium Green's Function (NEGF) method provided through the Spin and Molecular Electronics in Atomically-Generated Orbital Landscapes (SMEAGOL) code [173–175].

Figure 7 shows the principal setup of our calculation. We simulated the two gold electrodes (which are supposed to be embedded oppositely from each other in the nanopore). The functionalizing molecule is anchored to the inner surface of one of the gold electrodes via a thiol group. Finally, the target DNA to be scanned is represented by a single base (including the sugar and phosphate group). Here we see how once again, the cluster method can be used to simplify the computational setup and keep the involved elements to a minimum. The functionalizing probing molecule fulfills two functions: (i) dynamic stabilization of the target base in the DNA sequence by formation

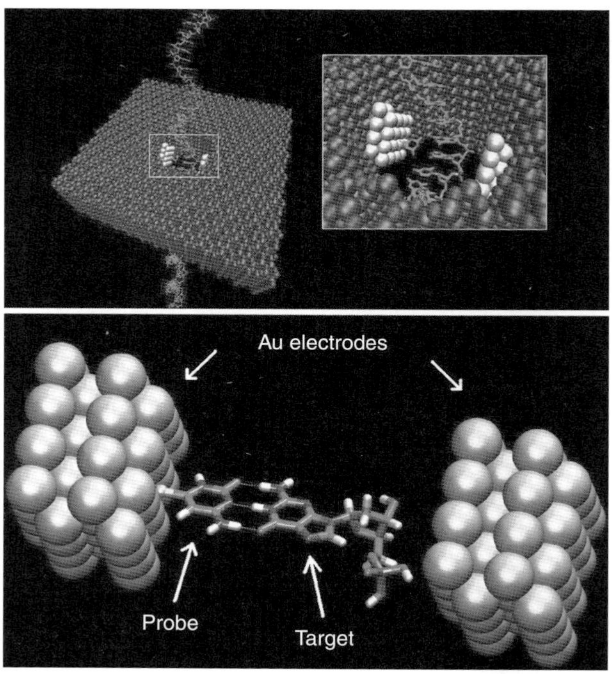

FIGURE 7 The top panel shows a schematic drawing of the idealized investigated system with ssDNA passing through a nanopore formed in a Si_3N_4 membrane with functionalized gold electrodes embedded. A probe molecule (cytosine) is immobilized on the inner surface of the left electrode by a sulfur atom. The bottom panel illustrates the actual cluster used in the calculations. Nanopore walls and remainders of the ssDNA molecule are omitted, and only the gold electrodes, the functionalizing probe, and the target guanosine monophosphate unit are included.

of weak hydrogen bonds and (ii) improved detection of the target base by better coupling electronically to it. The natural property of hybridization in the double-stranded DNA helix has led us to try the four base molecules, namely adenine (A), cytosine (C), guanine (G), and thymine (T), as functionalizing probe molecules in our calculations.

After the most stable pairing geometries between the probe and the target molecule have been determined, the corresponding geometries were used for the electronic transport calculation. Figure 8 shows the *I–V* curves when considering C as the probe, as this choice was found by us to yield the most reliable identification of all four base molecules in DNA. This can be achieved by carrying out three sequencing runs at three different bias voltages, namely 100, 250, and 750 mV. The first set of measurements at 100 mV would result in a series of current signals that fall into two easily distinguishable categories: "high" current values if A or G is the target base, and "low" current values if C or T is the target base. The difference between the two categories is nearly two orders of magnitude, which should make the distinction extraordinarily robust. We now require additional information to resolve the remaining ambiguity between the purines (A and G) on the one hand and the pyrimidines (C and T) on the other hand. In a second measurement at 250 mV, it will be possible to distinguish first between C and T, as their respective current values differ by one order of magnitude at that bias voltage. Thus, any "high" current value would lead to the identification of a T in the sequence, while any "low" current value means that a C is at this position

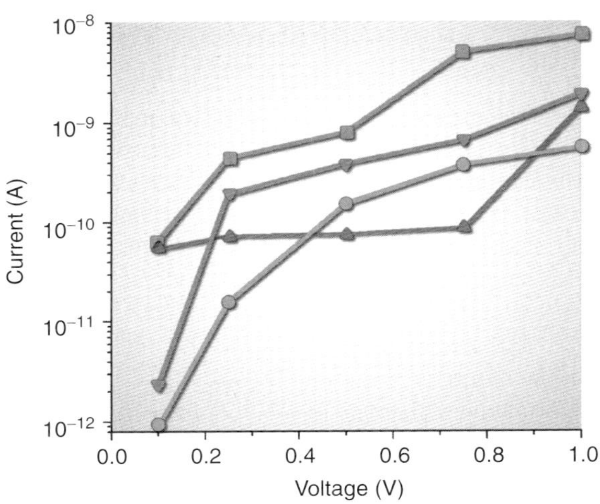

FIGURE 8 The calculated *I–V* curves for the nanopore-embedded gold electrodes functionalized with a cytosine probe, as obtained for all four possible target DNA nucleobases (A: red square; C: green circle; G: blue triangle; T: violet upside-down triangle). Note that the current signals are plotted on a logarithmic scale.

in the sequence. Finally, a third measurement at 750 mV causes the current values for the bases A and G to differ by two orders of magnitude, leading to an easy distinction between the two, where "high" current values correspond to A, while "low" current values correspond to G. In this manner, a series of three current measurements at the three specified voltages leads to the unambiguous determination of the target base identity.

In this section, we have shown that the cluster approach can be very useful in evaluating the feasibility of nanopore-embedded gold electrodes for rapid whole-genome analysis. The main result of our studies is the finding that the addition of functionalizing molecules can lead to a dramatic improvement in the sensitivity toward target base molecules in a pore-translocating DNA sequence. We emphasize that until now, the described setup has only been studied in the realm of theory and is very challenging to implement experimentally, but efforts to fabricate electrically connected nanopores are currently underway. A recently proposed new approach suggests the use of graphene in a double function, namely, as both separating membrane and electrodes (in the form of nanoslits) to sequence DNA [176]. Experimentally, it was demonstrated that DNA can be pulled through a graphene nanopore, as seen from the blockage in the ionic current [177–179]. The atomically thin graphene sheets would be very well suited to achieve single-base resolution. One issue of this setup that requires consideration is the possibility that DNA might stick to unprotected graphene sheets. In the next section, we will further explore such interactions, albeit because of an entirely different motivation.

VI. INTERACTION OF DNA FRAGMENTS WITH GRAPHENE AND CARBON NANOTUBES

There has been a steady increase in interest over the past four years in the non-covalent interaction of DNA with CNTs. This hybrid system at the junction of the biological regime and the nanomaterials world possesses features which make it very attractive for a wide range of applications. Initially, the focus rested on a new way to disperse CNT bundles in aqueous solution [180]. It was also recognized quickly that the combination of DNA and CNTs provides an efficient method to separate the latter according to their electronic properties [181–183]. After this initial discovery, interest was also directed toward applications aimed at electronic sensing of various odors [184]. The probing of conformational changes in DNA *in vivo* triggered by a change in the surrounding ionic concentration [185] shows great potential for new detection mechanisms. This and other envisioned applications of CNTs certainly demand a critical understanding of how such nanomaterials can impact biological systems.

Several theoretical investigations of the interaction between DNA and CNTs have been carried out [186,187], and it is generally understood that the stacking of the π-electron networks of the nucleobases of DNA and of the graphene-like surface of CNTs determines the equilibrium geometry of the hybrid system, while van der Waals forces cause the weak binding between DNA and CNT. It is desirable to understand all aspects of the binding mechanism as precisely as possible, in particular the nature of the relative strength of base-CNT binding, as experiments have shown that many properties of DNA–CNT hybrid systems depend on the base sequence of the utilized DNA strand [183].

As DNA molecules generally represent a very large system for *ab initio* calculations, we have utilized the cluster approach to simplify the calculations, and therefore considered only the interaction of individual nucleobases with the graphitic surface of CNTs. With the above mentioned sequence dependence in mind, our specific interest was focused on an assessment of the subtle differences in the adsorption strength of these nucleobases on large-diameter CNTs (modeled in the form of a graphene sheet, which can be seen as a model system for CNTs with a diameter much larger than the dimensions of the bases, and hence negligible curvature) [188], and small-diameter CNTs (as represented by a (5,0) CNT, which possesses very high curvature) [189].

In both cases (graphene and (5,0) CNT), the supercell approach was utilized. For the graphene system, a 5×5 array of the graphene unit cell in the *X–Y* plane with a separation of 15 Å between adjacent graphene sheets in the *Z*-direction was found to be a suitable choice. The cluster representations of DNA/RNA in the form of base molecules were terminated at the cut bond to the sugar ring with a methyl group in order to generate an electronic environment in the nucleobase more closely resembling the situation in DNA and RNA rather than that of just individual isolated bases by themselves. This has the additional benefit that a small magnitude of steric hindrance can be expected from the methyl group, quite similar to the case in which a nucleobase with attached sugar and phosphate group would interact with graphene.

For the CNT system, we employed a supercell approach which saw the unit cell of a (5,0) single-walled CNT (consisting of a ring of 20 carbon atoms with a diameter of 3.92 Å) repeated three times along the tube axis. In the directions perpendicular to the tube axis, a distance of at least 15 Å was kept between repeated units to avoid unphysical interactions between periodic repetitions of the CNT. Here too, the DNA/RNA clusters consisted of base molecules which were terminated with a methyl group for the reasons mentioned above.

We first carried out an initial force relaxation calculation step for each of the five nucleobases to determine the preferred orientation and optimum height of the planar base molecule relative to the graphene/CNT system surface. The potential energy surface was then explored by translating the relaxed base molecules parallel to the surface (which is flat in the case of graphene; curved in the case of the CNT), covering a mesh of 10×10 scan points. The determination of the minimum total energy configuration was then followed by a 360° rotation of the base molecules in steps of 5° to probe the dependence of the energy on the orientation of the base molecules with respect to the underlying graphene/CNT system. The configuration yielding the minimum total energy was used in the final optimization step in which all atoms in the system were free to relax. This last step possesses, in principle, no restrictions for the arrangement of the atomic positions, but practically, the only way to guarantee that the correct equilibrium configuration corresponding to the global energy minimum is identified, would be to start the geometry optimization process from a large set of plausible trial configurations. In particular, one could try an explicitly bent structure of the base molecules to better accommodate the interaction with the curved surface of the CNT. However, in such a scenario, one should also consider that the resulting distortion of the geometry would require the deformation of relatively stiff covalent bonds which normally keep the base molecule in its native planar geometry. The expected gain in energy from the increase in the comparably weak van der Waals (vdW) interaction may or may not be sufficient to compensate for the energy required to bend the molecule.

An additional set of calculations was performed in the case of the graphene system using the *ab initio* HF approach coupled with MP2 as implemented in the Gaussian03 suite of programs [190]. Due to the use of localized basis sets (rather than plane-waves), the system here consisted of the five nucleobases on top of a patch of nanographene [191,192], that is, a finite sheet containing 28 carbon atoms. The previously optimized configuration and the 6-311++G(d,p) basis sets for C, H, N, and O atoms were used for the MP2 calculations. This additional set of calculations was deemed necessary because Local-density approximation (LDA) is in principle not the optimal choice for investigations of van der Waals bound systems, as it is known that LDA cannot provide an accurate description of dispersion forces. The use of more reliable methods (such as MBPT) throughout would certainly be desirable, but in many cases the high computational cost makes it impossible to apply these methods to systems of larger size. For the particular type of system investigated in our study, it has been reported [193,194] that the LDA approximation appears to give a good (though perhaps fortuitous) description of the dispersive interactions, unlike the generalized gradient approximation (GGA) for which binding is basically nonexistent for van der Waals bound systems. In a study of the adsorption of the base molecule adenine on graphite [195] using LDA and a modified version of the London dispersion formula [196] for vdW interactions in combination with GGA, it was found that LDA, while underbinding the system, does in fact yield a potential energy surface which is almost indistinguishable in its structure from the one obtained via the GGA + vdW approach (cf. Figure 1A and B of Ref. [195]). Furthermore, LDA yields almost the same equilibrium distance of adenine to graphene as GGA + vdW.

From the optimization steps involving the translational scan over the graphene surface of the energy surface, it is apparent that the energy barriers to lateral movement of a given base can range from 0.04 to 0.10 eV (Figure 9), thereby considerably affecting the mobility of the adsorbed nucleobases on the graphene sheet at room temperature, and constricting their movement to certain directions. Also, the rotational scans carried out by us found energy barriers of up to 0.10 eV, resulting in severe hindrance to changes in the orientation of the adsorbed nucleobase. In their equilibrium configuration, three of the five bases tend to position themselves on graphene in a configuration reminiscent of the Bernal's AB stacking of two adjacent graphene layers in graphite (Figure 10). Virtually no changes in the interatomic structure of the nucleobases were found in their equilibrium configurations with respect to the corresponding gas-phase geometries, as it could be expected for a weakly interacting system such as the one studied here. This finding is also in agreement with earlier results reported in the literature for the nucleobase adenine [195]. The stacking arrangement shown in Figure 10 can be understood from the tendency of the π-orbitals of the nucleobases and graphene to minimize their overlap, in order to lower the repulsive interaction. The geometry deviates from the perfect AB base-stacking as, unlike graphene, the six- and five-membered rings of the bases possess a heterogeneous electronic structure because of the presence of both nitrogen and carbon in the ring systems. In addition, there exist different side groups containing CH_3, NH_2, or O, all of which contribute to the deviation from the perfect AB base-stacking as well. Adenine, thymine, and uracil display the least deviation from AB stacking

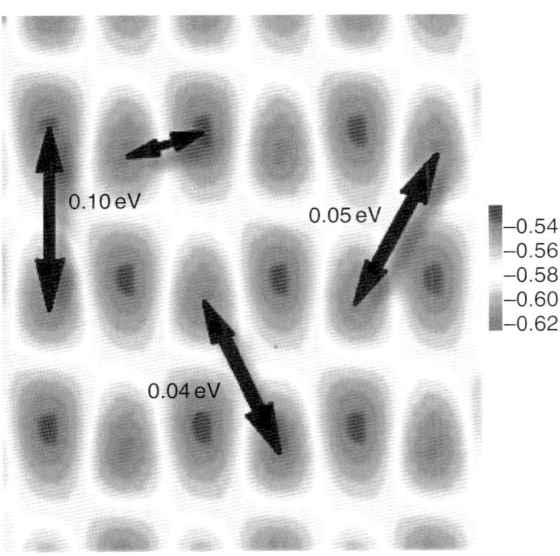

FIGURE 9 Potential energy surface (PES) plot (in eV) for guanine with graphene. The energy range between peak and valley is approximately 0.10 eV, while the energy barrier between adjacent global minima is only around half that high or 0.05 eV.

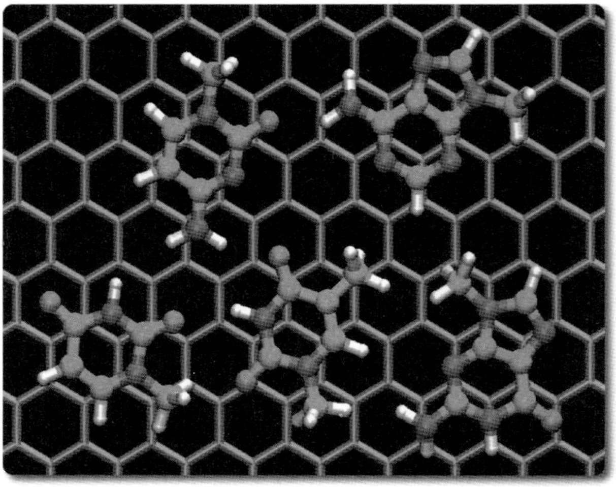

FIGURE 10 Equilibrium geometry of nucleic acid bases on graphene. From left to right, in the top row: cytosine and adenine; in the bottom row: uracil, thymine, and guanine. (We would like to emphasize that the calculation of the equilibrium geometries was carried out individually for each nucleobase isolated on a graphene supercell; the figure merely combines the results for all five nucleobases on a single graphene sheet for better representation.)

out of the five nucleobases. For guanine and cytosine on the other hand, there is almost no resemblance to the AB stacking configuration recognizable.

For the nucleobases adsorbed on CNT, the optimal separation was about 3.2 Å, which is a little less than the characteristic distance for π–π stacked systems, as found in the case of graphene. This is because the high-curvature surface of a tube such as (5,0) allows for the π-orbitals of the nucleic acid base to come closer before the repulsive interaction sets in. From the translational scan of the energy surface, we found an energy barrier of about 0.07 eV for all five molecules. At room temperature, this barrier is sufficiently large to affect the mobility of the base molecules physisorbed on the CNT surface and to constrict their movement to certain directions. The base rotation led to energy barriers of up to 0.12 eV, resulting in severe hindrance to changes in the orientation of the physisorbed nucleic acid base. In their equilibrium configuration, the base molecules A, T, and U tend to position themselves on the CNT in a configuration reminiscent of the Bernal's AB stacking of two adjacent graphene layers in graphite. The base molecules G and C, on the other hand, show a lesser degree of resemblance to the AB stacking. Again, the interatomic structure of the nucleic acid bases in their equilibrium configurations underwent virtually no changes when compared to the corresponding gas-phase geometries.

Next, we calculated the binding energy for all five nucleobases. The binding energy of the system consisting of the nucleobase and graphene/CNT is taken as the energy of the equilibrium configuration with reference to the asymptotic limit obtained by varying the distance between the base and the graphene/CNT sheet in direction perpendicular to their respective surface. For graphene, we found with LDA that adenine, cytosine, and thymine all possess nearly identical binding energies of about 0.49 eV, while guanine with 0.61 eV is bound more strongly, and uracil with 0.44 eV somewhat more weakly. For the systems consisting of nucleic acid bases and the (5,0) CNT, we find guanine bound with 0.49 eV, adenine with 0.39 eV, thymine with 0.34 eV, cytosine with 0.29 eV, and uracil with 0.28 eV.

Comparing the respective results for graphene and CNT, we clearly see that the binding energy of the base molecules is substantially reduced for physisorption on small-diameter CNTs with high curvature. While the curvature allows the nucleic acid base to approach the surface more closely, the majority of the carbon atoms in CNT are actually further removed from the atoms of the bases than in the corresponding case on a graphene sheet. As the attractive interaction falls off as the distance between the carbon nanomaterial (either graphene or a CNT) and the base molecule is increased, the overall binding energy is reduced in the case of CNT.

How can we understand the observed hierarchy of binding energies for the different nucleobases? The answer lies in the nature of the interaction, which is of the van der Waals type. Thus, the binding energy is proportional to the polarizabilities of the interacting entities. To test this idea quantitatively in a first-principles framework, we calculated the polarizabilities of the five nucleobases at the MP2 level of theory. The polarizability of a given nucleobase [197], which represents the deformability of the electronic charge distribution, is known to arise from the regions associated with the aromatic rings, lone pairs of nitrogen, and oxygen atoms. Accordingly, the purine bases guanine and adenine with their five- and six-membered rings possess the largest polarizabilites, whereas the pyrimidine bases with only one six-membered ring exhibit smaller polarizabilities among the five nucleobases. Furthermore, the purine base guanine with its double-bonded oxygen atom and amine group will possess a larger polarizability than the purine base adenine. Our MP2 calculations confirm this trend.

When the molecular polarizabilities of the base molecules are compared with the LDA binding energies, we can see a clear trend (even more so when the binding energies are also determined at the MP2 level of theory, as was done in the case of graphene). The polarizability of a nucleobase is the key factor which governs the strength of interaction with both graphene and CNT. The MP2 binding energies for the system containing graphene are systematically higher than those calculated within the LDA approximation. This is due to the well established fact that MP2 provides a more accurate treatment of the vdW interaction than LDA. We note that the adsystem consisting of the base and the sheet is not bound at the HF level of theory, which underscores the importance of electron correlation in describing the weak vdW interactions in this system.

In the equilibrium configuration, a redistribution of the total charge density within a given base adsorbed on graphene seems to appear. From an analysis of the Mulliken charges for the MP2 calculations, we also find a negligible charge transfer ($<0.02e$) between any of the five nucleobases and patch of nanographene in the equilibrium configuration. This small amount implies that electrostatic interactions in the adsystem are very unlikely to contribute to the interaction energy. Regarding the charge transfer between the bases and the CNT, we find slightly larger amounts of charge transferred. For guanine, for example, the Bader analysis shows that the (5,0) CNT possesses an excess charge of -0.08 e and correspondingly there exists a slight depletion of electrons on guanine by $+0.08e$ (Figure 11). For adenine with CNT, -0.05 e was found to have been transferred from the nucleic acid base to the CNT. Thus, on the one hand, the higher curvature of the (5,0) CNT appears to lead to an increased electronegativitiy which manifests itself in the larger amount of charge transferred to it. The different behavior of guanine and adenine on the other hand becomes understandable when one considers that guanine has a smaller ionization potential than adenine, making it easier to remove an electron from guanine than from adenine. While there are no "whole elementary charges" transferred in this case, but only fractions, it still shows that the CNT is able to withdraw more charge from guanine than from adenine. However, the total amount of transferred charge remains relatively small and the resulting contribution to the binding energy from the attractive Coulomb interaction can be estimated to be at or below the 0.01 eV margin of error in our calculations.

In summary, we have illustrated on this last example of the present chapter how the cluster approach can be utilized to investigate the interaction of the five DNA/RNA base molecules with CNTs, represented in the form of both

FIGURE 11 Isosurface plot of the charge density for guanine physisorbed on a (5,0) single-walled carbon nanotube.

flat graphene and a (5,0) zigzag CNT of very high curvature. From the first-principles calculations, the five nucleic acid bases are found to exhibit significantly different interaction strengths with graphene and CNT. Molecular polarizability of the base molecules is found to play the dominant role in the interaction strength of the base molecules with CNT. This observation could be of importance in understanding the sequence-dependent interaction of DNA with CNTs observed in experiments [182–184]. When comparing the results obtained for nucleobases interacting with flat graphene with those for interaction with the high-curvature (5,0) CNT, we see that the interaction strength of nucleic acid bases is smaller for the tube. Thus, it appears that introducing surface curvature reduces the binding energy between the base molecule and the substrate. The binding energies for the two extreme cases of negligible curvature (flat graphene sheet) and of very high curvature (the (5,0) CNT studied here) represent the upper and lower boundaries, and it is expected that the binding energy of bases for CNTs of intermediate curvature is likely to lie in between these two extremes. On the basis of the results obtained up to this point, the hierarchy of the binding energies of the nucleic acid bases to the graphene-like surfaces of CNTs appears to be universally valid, as long as the interaction is dominated by vdW forces. Further studies on this subject could for example also consider the effect that different chiralities may have on the interaction of nucleic acid bases with high-curvature CNTs.

Acknowledgments

The authors are grateful to Professor Lee Chow, Professor Alfons Schulte, Professor Archana Dubey, Dr. Shyam Raj Badu, Dr. Roger H. Pink for very useful discussions. Also we are grateful to Dr. Pink for valuable help and suggestions in the preparation of the manuscript of this chapter.

REFERENCES

[1] Perutz MF, Fermi G, Luisi B, Shaanan B, Liddington RC. Stereochemistry of cooperative mechanisms in hemoglobin. Acc Chem Res 1987;20:309.

[2] (a) Benesch R, Benesch RE. The chemistry of the bohr effect: i. the reaction of n-ethyl maleimide with the oxygen of hemoglobin. J Biol Chem 1961;236:405. (b) Benesch RE, Ranney GM, Smith GM, Benesch R. The chemistry of the bohr effect: ii. some properties of hemoglobin h. J Biol Chem 1961;236:2926.

[3] Champion PM, Rosca F, Ionascu D, Cao WX, Ye X. Rapid timescale processes and the role of electronic surface coupling in the photolysis of diatomic ligands from heme proteins. Faraday Discuss 2004;127:123.

[4] Mines Gary A, Bjerrum Morten J, Hill Michael G, Casimiro Danilo R, Chang I-Jy, Winkler Jay R, et al. Rates of heme oxidation and reduction in ru (his33) cytochrome *c* at very high driving forces. J Am Chem Soc 1996;118:1961.

[5] Nagamine K, Pratt FL, Ohira S, Watanabe I, Ishida K, Nakamura SN, et al. Intra- and intermolecular electron transfer in cytochrome *c* and myoglobin observed by the muon spin relaxation method. Phys B 2000;289:631.

[6] (a) Scheicher RH, Cammarere D, Sahoo N, Briere TM, Pratt FL, Nagamine K, et al. First-principles study of muon and muonium trapping in the protein chain of cytochrome *c*. Phys B 2003;326:30. (b) Scheicher RH, Das TP, Torikai E, Pratt FL, Nagamine K. First-principles study of muonium in a- and b-form dna. Phys B 2006;374:448.

[7] Ramani Lata K, Sahoo N, Dubey Archana, Scheicher RH, Badu SR, Pink RH, et al. Investigation of the hyperfine properties of deoxy hemoglobin based on its electronic structure obtained by hartree-fock-roothaan procedure. Hyperfine Interactions. Springer; 2008, vol. 181, p. 75, corresponding page number in book of proceedings is 595.

[8] Scholes CP, Isaacson RA, Feher G. Electron nuclear double resonance studies on heme proteins: determination of the interaction of fe^{3+} with its ligand nitrogens in metmyoglobin. Biochim Biophys Acta 1972;263:448. This measurement of the magnetic hyperfine constant for 14N in metmyoglobin was performed using the Electron-Nuclear Double Resonance (ENDOR) technique developed by G. Feher, Phys Rev 1956;103:500.

[9] (a) Gonser U, Grant RW. Mössbauer effect in hemoglobin and some iron-containing biological compounds. Biophys J 1965;5:823. (b) Kent TA, Spartalian K, Lang G, Yonetani CA, Reed CA, Collman JP. High magnetic field mössbauer studies of deoxymyoglobin, deoxyhemoglobin, and synthetic analogues. Biochim Biophys Acta 1979;580:245.

[10] Cerdonio M, Congiu-Castellano A, Mogno F, Pispisa B, Romani GL, Vitale S. Magnetic properties of oxyhemoglobin. Proc Natl Acad Sci USA 1977;74:298.

[11] Pauling Linus. Covalence of atoms in the heavier transition metals. Proc Natl Acad Sci USA 1977;74:2612.

[12] Cerdonio M, Morante S, Torresani D, Vitale S, DeYoung A, Noble RW. Reexamination of the evidence for paramagnetism in oxy- and carbonmonoxyhemoglobins. Proc Natl Acad Sci USA 1985;82:102.

[13] Phils John S, Dreyor Ulrich, Schuster Todd M. Diamagnetism of human apo-, oxy-, and carbon monoxyhemoglobin. Biochemistry 1984;23:865.

[14] Sahoo N, Ramani Lata K, Das TP. Investigation of electron distribution and hyperfine properties of hemin by first-principles hartree-fock self-consistent field procedure. Theor Chim Acta 1992;82:285.

[15] Ramani Lata K. First principles unrestricted hartree-fock calculations for the electronic structure and hyperfine properties of different biological and aquoion systems. PhD Thesis, Department of Physics, State University of New York at Albany, Albany, New York; 1993.

[16] Pujari M, Ramani Lata K, Sahoo N, Das TP. Hartree-fock investigations of magnetic hyperfine interactions in fluoro-, chloro-, bromo-and iodo-hemin systems, american physical society. Annual March Meeting, March 17–22. American Physical Society; 1996, abstract #N39.10.

[17] Pujari Minakhi. First-Principles Investigation of Electronic Structures and Hyperfine Properties of Heme and Nitrosyl-Hemoglobin Systems. Albany, New York: University at Albany, State University of New York at Albany; 1999.

[18] Hay PJ. Gaussian basis sets for molecular calculations. the representation of 3d orbitals in transition-metal atoms. J Chem Phys 1997;66:4377.

[19] Mezey PG, Csizmadi FG. Uniform quality constrained gaussian basis sets. Can J Chem 1977;55:1181.

[20] McLean AD, Chandler GS. Contracted gaussian basis sets for molecular calculations. i. second row atoms, z=11–18. J Chem Phys 1980;72:5639.
[21] Dunning Jr. TH. Gaussian basis functions for use in molecular calculations. contraction of (12s9p) atomic basis sets for the second row atoms. Chem Phys Lett 1970;7:423.
[22] Huzinaga S, Andzelm J, Kobukowski M, Radzio-Andzelm E, Sakai Y, Tatewaki H. Gaussian basis sets for molecular calculations. Amsterdam: Elsevier; 1984.
[23] Das TP. In: Kaplan TA, Mahanti SD, editors. Electronic Properties of Solids Using Cluster Methods. New York: Plenum Press; 1995. p. 1–28.
[24] Aryal MM, Mishra DR, Paudyal DD, Byahut S, Maharjan NB, Adhikari NP, et al. Understanding of nuclear quadrupole interactions of ^{35}cl, ^{79}br and ^{129}i and binding energies of solid halogens at first-principles level. Hyperfine Interact 2007;176:51.
[25] Scheicher RH, Cammarere D, Briere TM, Sahoo N, Das TP, Pratt FL, et al. First-principles theory of muon and muonium trapping in the protein chain of cytochrome *c* and associated hyperfine interactions. Hyperfine Interact 2001;136:755.
[26] Pink RH, Badu SR, Dubey A, Scheicher RH, Jeong J, Byahut SR, et al. First principles hartree-fock cluster study of very dilute transition metal and rare-earth ion systems in silicon, Conference ProceedingsAmerican Institute of Physics; 2008;1003:235.
[27] (a) Koenig DF. The structure of α–chlorohemin. Acta Cryst 1965;18:663. (b) Hoffman R. An extended hückel theory. i. hydrocarbons. J Chem Phys 1963;39:1397. Hemin System.
[28] Fermi G, Perutz MF, Shaanan B. The crystal structure of human deoxyhaemoglobin at 1.74 Å resolution. J Mol Biol 1984;175:159. Deoxy Hemoglobin System.
[29] (a) Paoli M, Liddington R, Tame J, Wilkinson A, Dodson G. Crystal structure of t state haemoglobin with oxygen bound at all four haems. J Mol Biol 1996;256:775. Oxy Hemoglobin R State. (b) Park Sam-Yong, Yokoyama T, Shibayama N, Shiro Y, Tame JRH. 1.25 Å resolution crystal structures of human haemoglobin in the oxy, deoxy and carbonmonoxy forms. J Mol Biol 2006;360:690. Oxy Hemoglobin T State.
[30] Goldanskii VI, Herber RH. Chemical applications of mössbauer spectroscopy. New York, London: Academic Press; 1968.
[31] Szabo A, Ostlund N. Modern quantum chemistry. New York: McGraw Hill; 1989.
[32] Mulliken RS. Electronic population analysis on LCAOMO molecular wave functions. i. J Chem Phys 1955;23:1833.
[33] Wertheim GK. Mössbauer effect: principles and applications. New York: Academic Press; 1964.
[34] Nieuwpoort WC, Post D, van Duijnen PTh. Calibration constant for ^{57}fe mössbauer isomer shifts derived from ab initio self-consistent-field calculations on octahedral fef_6 and $fe(cn)_6$ clusters. Phys Rev B 1978;17:91.
[35] Shenoy GK. Mössbauer isomer shifts. In: Shenoy GK, Wagner FE, editors. Mössbauer Isomer Shifts. Amsterdam: North-Holland; 1978.
[36] (a) Abragam A, Bleaney B. Electron Paramagnetic Resonance of Transition Ions. Clarendon Press; 1970. International Series of Monograms on Physics. (b) Slichter CP. Principles of Magnetic Resonance. New York: Springer Verlag; 1978.
[37] Kaufmann EN, Vianden RJ. The electric field gradient in noncubic metals. Rev Mod Phys 1979;51:161.
[38] Das TP. Relativistic Quantum Mechanics of Electrons. New York: Harper and Row; 1973. Chapter 5.
[39] Sahoo N, Das TP. Hyperfine interactions and relaxivities in divalent and trivalent aquoion systems. J Chem Phys 1989;91:7740.

[40] Cohen MH, Goodings DA, Heine V. Contribution of core polarization to the atomic hyperfine structure and knight shift of li and na. Proc Phys Soc Lond 1959;73:811.

[41] Gaspari GD, Shyu Wei-Mei, Das TP. Theory of core-electron contributions to hyperfine interactions. Phys Rev 1964;134:A852.

[42] Hartree DR. The calculation of atomic structures. New York: John Wiley & Sons, Inc. New York; 1957.

[43] Cohen MH, Reif F. In: Seitz F, Turnbull D, editors. Advances in Solids State Physics. New York: Academic Press; 1957, vol. 5.

[44] Das TP, Hahn EL. In: Seitz F, Turnbull D, editors. Nuclear Quadrupole Resonance Spectroscopy in Advances in Solid State Physics. New York: Academic Press; 1958.

[45] Lucken Edwin Anthony Charles. Nuclear Quadrupole Coupling Constants. Academic Press; 1969.

[46] Eckart C. The application of group theory to the quantum dynamics of monatomic systems. Rev Mod Phys 1929;1:305.

[47] Mitchell Dale W, Sulaiman SB, Sahoo N, Das TP, Potzel W, Kalvius GM. Hartree-fock cluster investigation of ^{67}zn nuclear quadrupole interaction in zinc oxide. Phys Rev B 1991;44:6728.

[48] Mun SK, Chang Jane C, Das TP. Origin of observed changes in ^{14}n hyperfine interaction accompanying r $\rightarrow$ t transition in nitrosylhemoglobin. Proc Natl Acad Sci USA 1979;76:4842.

[49] Chevion M, Stern A, Peisach J, Blumenberg WE, Simon S. Analogous effect of protons and inositol hexaphosphate on the alteration of structure of nitrosyl fetal human hemoglobin. Biochemistry 1978;17:1745.

[50] See Scholes CP et al. Ref. [8].

[51] Kelires PC, Mishra KC, Das TP. Self-consistent field cluster study of the origin of hyperfine interaction and isomer shift in ^{57m}fe$_2$o$_3$. Hyperfine Interact 1987;34:285.

[52] (a) Bersohn R. Nuclear electric quadrupole spectra in solids. J Chem Phys 1952;20:1505. (b) Dean C. Zeeman splitting of nuclear quadrupole resonances. Phys Rev 1954;96:1053. (c) Cohen MH. Nuclear quadrupole spectra in solids. Phys Rev 1954;96:1278.

[53] (a) Dehmelt HG, Kruger H. Kernquadrupolfrequenzen in festem dichloräthylen. Naturwiss 1950;37:111, 1951;38:921. (b) Dehmelt HG. Nuclear quadrupole resonance in rhombic sulfur and the quadrupole moments of s^{33} and s^{35}. Phys Rev 1953;91:313. (c) Barnes PJ, Bray PJ. Nuclear quadrupole resonances of as^{75}. J Chem Phys 1955;23:407.

[54] (a) Lamarche G, Volkoff GM. Further calculations on the nuclear resonance spectrum of al^{27} in spodumene. Can J Phys 1953;31:1010. (b) Volkoff GM, Lamarche G. Can J Phys 1954;32:493.

[55] Karlsson Erik. Solid State Phenomena: As Seen by Muons, Protons, and Excited Nuclei. Oxford University Press; 1995.

[56] Das TP. Theory of origin of hyperfine interactions in atomic systems. Hyperfine Interact 1987;34:149.

[57] Zerner M, Gouterman M, Kobayashi H. Porphyrins viii. extended huckel calculations on iron complexes. Theor Chim Acta Berlin 1966;6:636.

[58] SeeRoy JN, Mishra SK, Mishra KC, Das TP. Theory of orbital contributions to magnetic hyperfine fields in hemoglobin derivatives—application to azidomyoglobin. J Chem Phys 1989;90:7273. Mishra KC, Mishra SK, Scholes CP, Das TP. Origin of asymmetry in nitrogen hyperfine interactions in metmyoglobin. J Am Chem Soc 1983;105:7553 and references to our earlier work therein, in particular the earliest investigations by the

Self-Consistent Charge Extended Hückel procedure, especially Rettig MF, Han PS, Das TP, Theor Chim Acta (Berlin) 1968;12:178 and Erratum 1969;13:432.
[59] (a) Hoffmann R, Lipscomb WN. Theory of polyhedral molecules. i. physical factorizations of the secular equation. J Chem Phys 1962;36:2179. (b) Rettig MF, Drago RS. Electron delocalization in paramagnetic metallocenes. ii. extended hückel molecular orbital calculations. J Am Chem Soc 1962;91:3432, For a detailed description of the Self-Consistent Charge Extended Hückel procedure, see Han PS, Das TP, Rettig MF, Theor Chim Acta 1970;16:1.
[60] Coulson CA. Valence. Oxford: Oxford University Press; 1952; For implementation of LCAOMO procedure in the framework of the Hartree-Fock method see Roothaan CCJ, Rev Mod Phys 1960;32:179 and for applications see for example Wahl AC, Phys Rev A 1971;4:852; Cade PE, Bader RFW, Pelletier J, J Chem Phys 1972;56:573; For inclusion of many-body effects using many-body perturbation theory, see for example, Rodgers JE, Lee T, Das TP, Phys Rev A 1972;7:51.
[61] Chang Jane C, Kim YM, Duff KJ, Das TP. Investigation of electron distribution and hyperfine properties of hemin by first-principles hartree-fock self-consistent field procedure. Theor Chim Acta 1976;41:37.
[62] Chang Jane C, Kim YM, Das TP. Theory of exchange splitting of core levels in x-ray photoemission spectroscopy of hemin. J Chem Phys 1975;62:4350.
[63] Siegbahn K, Hammond D, Fellner-Feldegg H, Barnett EF. Electron spectroscopy with monochromatized x-rays. Science 1972;176:245.
[64] Mun SK, Chang Jane C, Das TP. Critical appraisal of electronic structure of metmyoglobin. ^{14}N and ^{57}fe hyperfine interactions. Biochim Biophys Acta 1977;490:249. Theory of hyperfine fields at iron-57 and nitrogen-14 sites in metmyoglobin and related compounds, J Am Chem Soc 1979;101:5562.
[65] Mishra KC, Mishra SK, Das TP. Electronic structure of ferricytochrome c and associated hyperfine interactions. J Am Chem Soc 1983;105:7729.
[66] Mun SK, Chang Jane C, Das TP. Origin of observed changes in ^{14}N hyperfine interaction accompanying r → t transition in nitrosylhemoglobin. Proc Natl Acad Sci USA 1979;76:4842 and references therein to the study of magnetic and hyperfine properties of six-liganded Hemoglobin derivatives by the SCCEH procedure.
[67] Roy JN, Mishra SK, Mishra KC, Das TP. Variational theory for g tensor in electron paramagnetic resonance experiments: application to ferricytochrome c and azidomyoglobin. J Phys Chem 1989;93:194 and references therein to earlier investigations on g-shift in Hemoglobin derivatives.
[68] Roy JN, Mishra SK, Mishra KC, Das TP. Influence of coulomb corrections in the self-consistent charge extended hückel (scceh) procedure on hyperfine properties of biological molecules. Hyperfine Interact 1987;35:961.
[69] Chang ES, Pu RT, Das TP. Many-body approach to the atomic hyperfine problem. i. lithium-atom ground state. Phys Rev 1968;174:1. Many-body calculation of atomic polarizability-relation of hartree-fock theory. Phys Rev Lett 1968;174:16.
[70] Dutta NC, Matsubara C, Pu RT, Das TP. Brueckner-goldstone many-body theory for the hyperfine structure of phosphorus. Phys Rev Lett 1968;21:1139.
[71] (a) Andriessen J, Raghunathan K, Ray SN, Das TP. Relativistic many-body approach to hyperfine interaction in rare earths: explanation of experimental result in europium. Phys Rev 1977;B15:2533. (b) Vajed-Samii M, Ray SN, Lee T, Das TP. Accurate values of nuclear magnetic moments of francium isotopes. Phys Rev Lett 1982;48: 1330. Errata 1982;49:1466, 1800.

[72] Owusu A, Dougherty RW, Gowri G, Andriessen J, Das TP. Relativistic many-body investigation of hyperfine interactions in excited s states of alkali metals: francium and potassium. Phys Rev 1997;A56:305 and references therein to our earlier investigations in Relativistic Many-Body Theory Studies in other atoms and ions.

[73] Head-Gordon M, Pople JA, Frisch MJ. Mp2 energy evaluation by direct methods. Chem Phys Lett 1988;153:503.

[74] Ramani Lata K et al. Ref. [7].

[75] Badu SR, Dubey Archana, Scheicher RH, Pink RH, Chow Lee, Nagamine K, et al. First principles electronic structure investigations of order of singlet and triplet states of oxyhemoglobin and muon trapping. [Presented at the 3rd Joint International Conference on Hyperfine Interactions and International Symposium on Nuclear Quadrupole Interactions, PS3-11-147, September 13–17, 2010 at CERN/Geneva, Switzerland] [to be published in the Proceedings of the Conference in Hyperfine Interactions (Springer)].

[76] Cammarere D, Scheicher RH, Sahoo N, Das TP, Nagamine K. First-principle determination of muon and muonium trapping sites in horse heart cytochrome *c* and investigation of magnetic hyperfine properties. Phys B 2000;289–290:636.

[77] Scheicher RH, Cammarere D, Sahoo N, Briere TM, Pratt FL, Nagamine K, et al. First-principles study of muon and muonium trapping in the protein chain of cytochrome *c*. Phys B 2003;326:30.

[78] Scheicher RH, Das TP, Torikai E, Pratt FL, Nagamine K. First-principles study of muonium in a- and b-form dna. Phys B 2006;374:448.

[79] Badu SR, Pink RH, Dubey Archana, Scheicher RH, Saha HP, Chow Lee, et al. First principles study of nuclear quadrupole interactions in single and double chain DNA and solid nucleobases. [Presented at the 3rd Joint International Conference on Hyperfine Interactions and International Symposium on Nuclear Quadrupole Interactions, PS3-55-180, September 13–17, 2010 at CERN/Geneva, Switzerland] [Paper under preparation].

[80] Badu SR, Pink RH, Dubey Archana, Scheicher RH, Chow Lee, Huang MB, et al. First Principles Study of Muonium Trapping and Associated Magnetic Hyperfine Interactions in Nucleobases in Single and Double Chain DNA and Solid Nucleobases. [Presented at the 3rd Joint International Conference on Hyperfine Interactions and International Symposium on Nuclear Quadrupole Interactions, PS3-56-182, September 13–17, 2010 at CERN/Geneva, Switzerland] [Paper under preparation].

[81] (a) Nesbet RK. Configuration interaction in orbital theories. Proc R Soc Lond 1955; A230:312 and Phys Rev 1960;118:61. (b) Marshall W. The unrestricted hartree-fock method. Proc Phys Soc Lond 1961;78:113. (c) Bessis N, Lefebvre-Brion H, Moser CM. Calculation of the magnetic hyperfine constant of the nitrogen atom. Phys Rev 1961;124:1124. See also discussions about the question concerning the expectation value of S*S in the UHF and related procedures in pages 860 and 861 in the Conclusion section of Ref. [41] and p. 5 of Chang ES et al. Ref. [69].

[82] Goldstone J. Derivation of the brueckner many-body theory. Proc R Soc Lond 1957; A239:267.

[83] Kelly HP. Correlation effects in atoms. Phys Rev 1963;131:684. See also Chang ES, Pu RT, Das TP, Ref. 69.

[84] Bessis N, Lefebvre-Brion H, & Moser CM. Ref. [81].

[85] Frisch MJ, Trucks GW, Schlegel HB, Scuseria GE, Robb MA, Cheeseman JR, et al. Gaussian 03, Revision C.02. Wallingford, CT: Gaussian, Inc.; 2004.

[86] (a) Johnson CE. Mössbauer effect in some haemoglobin compounds. Phys Lett 1966;21:491. (b) Lang G, Marshall W. Mössbauer spectroscopy of haem proteins. Proc

Phys Soc Lond 1966;87:3. (c) Lang G. Q Rev Biophys 1970;3:1 57mFe Mössbauer data in FluoroHemin and ChloroHemin. (d) Scholes CP, Isaacson RA, Feher G. Electron nuclear double resonance studies on heme proteins: determination of the interaction of fe^{3+} with its ligand nitrogens in metmyoglobin. Biochim Biophys Acta 1972;263:448. (e) Feher G, Isaacson CP, Scholes CP, Nagel R. Electron nuclear double resonance (endor) investigation on myoglobin and hemoglobin. Ann NY Acad Sci 1973;206:86. (f) Van Camp HL, Scholes CF, Mulks CF. Electron nuclear double resonance on heme compounds. endor from the iron ligands in protohemin chloride and protohemin bromide. J Am Chem Soc 1976;98:4094. (g) Mulks CF, Scholes CP, Dickinson LC, Lapidot A. Electron nuclear double resonance from high- and low-spin ferric hemoglobins and myoglobins. J Am Chem Soc 1979;101:1645. (h) Persach J. An. interim report on electronic control of oxygenation of heme proteins. Ann NY Acad Sci 1975;244:187. (i) Peisach J, Blumberg WE, Adler A. Electron paramagnetic resonance studies of iron porphin and chlorin systems. Ann NY Acad Sci 1973;206:310. (j) Chien JCW. Electron paramagnetic resonance study of the stereochemistry of nitrosylhemoglobin. J Chem Phys 1969;51:4220. (k) Doetschmann DC, Utterback SG. Electron paramagnetic resonance study of nitrosylhemoglobin and its chemistry in single crystals. J Am Chem Soc 1981;103:2847. (l) La Mar GN, Overkamp M, Sick K, Gersonde K. Research article proton nuclear magnetic resonance hyperfine shifts as indicators of tertiary structural changes accompanying the bohr effect in monomeric insect hemoglobins. Biochemistry 1978;17:352. (m) Johnson ME, Fung LWM, Ho LWM. Magnetic field and temperature induced line broadening in the hyperfine-shifted proton resonances of myoglobin and hemoglobin. J Am Chem Soc 1977;99:1245. (n) Shulman RG, Glarum M, Karplus M. Electronic structure of cyanide complexes of hemes and heme proteins. J Mol Biol 1971;57:93. ^{14}N hyperfine data.

[87] Reference for ^{14}N hyperfine interaction data also contains ^{1}H hyperfine interaction data.

[88] Ref. [14] and references therein to earlier investigations on the electronic structure and hyperfine interactions in hemin.

[89] Kelires PC, Das TP. Self-consistent field cluster study of the origin of hyperfine interaction and isomer shift in $^{57m}fe_2o_3$. Hyperfine Interact 1987;34:285.

[90] (a) Duff KJ, Das TP. Electron states in ferromagnetic iron. ii. wave-function properties. Phys Rev B 1971;3:2294. (b) Hanna SS, Heberla J, Perlow GJ, Preston RS, Vincent DM. Direction of the effective magnetic field at the nucleus in ferromagnetic iron. Phys Rev Lett 1960;4:513.

[91] See Ref. [85].

[92] Bearden AJ, Moss TH, Caughey WC, Beaudreau CA. Mössbauer spectroscopy of heme and hemin compounds. Proc Natl Acad Sci USA 1965;53:1246.

[93] Danon J. In: Golldanskii VI, Herber RH, editors. Chemical Applications of Mössbauer Spectroscopy. New York: Academic Press; 1968, Chapter 3.

[94] Champion VI, Vaughan RW, Drickamer HG. Effect of pressure on the mössbauer resonance in ionic compounds of iron. J Chem Phys 1967;47:2583.

[95] Duff KJ. Calibration of the isomer shift for ^{57}fe. Phys Rev B 1974;9:66.

[96] See Ref. [17], p. 72.

[97] Duff KJ, Mishra KC, Das TP. Ab initio determination of ^{57m}fe quadrupole moment from mössbauer data. Phys Rev Lett 1981;46:1611.

[98] Hagelberg F, Mishra KC, Das TP. Hartree-fock calculations for $fecl_2$ and $febr_2$: the question of the ^{57m}fe quadrupole moment. Phys Rev B 2002;65:014425.

[99] (a) See for instance the following papers for accurate linked cluster many-body perturbation theory investigations on atoms and ions: Chang ES, Pu RT, Das TP. Many-body approach

to the atomic hyperfine problem. i. lithium-atom ground state. Phys Rev 1968;174:1, Lithium atom. (b) Dutta NC, Matsubara C, Pu RT, Das TP. Lithium atom; "brueckner-goldstone many-body theory for the hyperfine structure of phosphorus. Phys Rev Lett 1968;21:1139, Phosphorous atom. (c) Andriessen J, Van Ormondt D, Ray SN, Das TP. Relativistic configuration interaction using many-body techniques. hyperfine interaction in gd^{3+}. J Phys B 1978;11:2601, Gd3+ Ion.

[100] Ray SN, Das TP. Nuclear quadrupole interaction in the fe^{2+} ion including many-body effects. Phys Rev B 1974;16:4794 57. mFe nuclear quadrupole interaction in Fe^{2+} ion.

[101] See Refs. [10] and [12] for Cerdonio M et al.'s earlier work showing detection of paramagnetic susceptibility in Oxyhemoglobin near room temperature and later work where they disagreed with their earlier results.

[102] Nagamine Kanetada, Shimomura Koichiro, Miyader Haruo, Kim Yong-Jae, Scheicher Ralph Hendrik, Das Tara Prasad, et al. Hemoglobin magnetism in aqueous solution probed by muon spin relaxation and future applications to brain research. Proc Jpn Acad B 2007;83:120.

[103] Fermi G et al. Ref. [28] and references to earlier structural data for deoxyhemoglobin quoted therein. See also Watson HC in "Progress in Stereochemistry" Eds. Aylett BJ, Harris NM, Butterworth, London, 1969;4:299, for structure of deoxymyoglobin.

[104] (a) Eisenberg P, Shulman RG, Kincaid BM, Brown GS, Ogawa S. Extended x-ray absorption fine structure determination of iron nitrogen distances in haemoglobin. Nat Lond 1978;274:30. (b) Perutz MF, Hasnain SS, Duke PJ, Sessler DL, Hahn JB. Stereochemistry of iron in deoxyhaemoglobin. Nat Lond 1982;295:135.

[105] See Ref. [103].

[106] Please see Abragam A, Bleaney B 1970 in Ref. [36].

[107] Kent TA et al. Ref. [9].

[108] Kent TA et al. Ref. [9].

[109] Shenoy GK. Ref. [35].

[110] Gonser U, Maeda Y, Trautwein A, Parak F, Formanek H. Sign and orientation determination of principal axis of electric-field gradient in fe-57 enriched deoxygenated myoglobin single-crystals. Zeits Naturforsch 1974;296:241.

[111] (a) Rodgers JE, Das TP. Many-body theory of the nuclear quadrupole coupling in the boron atom. Phys Rev 1975;A12:353. (b) Rodgers JE, Ray R, Das TP. Many-body calculation of the electric field gradient in the aluminum atom. Phys Rev 1976;A14:543. (c) Ray SN, Lee TP, Das TP. Study of the nuclear quadrupole interaction in the excited (2^3p) state of the beryllium atom by many-body perturbation theory. Phys Rev 1973;A8:1748.

[112] Perutz MF et al. Ref. [1].

[113] Benesch R, Benesch RE. Ref. [2].

[114] (a) Ogawa S, Lee TM, Kay AR, et al. Brain magnetic resonance imaging with contrast dependent on blood oxygenation. Proc Natl Acad Sci USA 1990;87:9868. (b) Kwang K, et al. Proc Natl Acad Sci USA 1992;89:5675.

[115] See Ref. [114] for additional references to the Functional Magnetic Resonance Imaging (FMRI) procedure the Authors of Ref. [114] have used for their investigations on the brain.

[116] Nagamine K et al. Ref. [102].

[117] Pauling L, Coryell CD. The magnetic properties and structure of hemoglobin, oxyhemoglobin and carbonmonoxyhemoglobin. Proc Natl Acad Sci USA 1936;22:210.

[118] Herman ZS, Loew GH. A theoretical investigation of the magnetic and ground-state properties of model oxyhemoglobin complexes. J Am Chem Soc 1980;102:1815.

[119] Pople JA, Beveridge D. Approximate Molecular Orbital Theory. McGraw-Hill: New York; 1970.
[120] Phils JS et al. Ref. [13].
[121] Nagamine K et al. Ref. [102].
[122] Badu SR et al. Ref. [75].
[123] Paoli M et al. Ref. [29 (a)].
[124] Park S-Y et al. Ref. [29 (b)].
[125] Scherlis Damian A, Estrin Dario A. Structure and spin-state energetic of an iron porphyrin model: an assessment of theoretical methods. Int J Quant Chem 2002;87:158.
[126] Becke AD. A new mixing of hartree-fock and local density-functional theories. J Chem Phys 1993;98:1372.
[127] Pink RH, Badu SR, Dubey Archana, Scheicher RH, Raghunathan k, Chow L, Das TP. Test of variational methods for studying hyperfine interactions of molecular and solid state systems by application to atomic systems, PS3-37-145.n third joint HFI-NQI international conference, CERN. Geneva; 2010.
[128] Dutta NC et al. Ref. [99], Lithium Atom Correlation Energy and Hyperfine Interaction. 1968; Dutta NC et al. Ref. [99], Phosphorous Atom Hyperfine Interaction. 1968.
[129] Scheicher RH, Das TP, Torikai E, Pratt FL, Nagamine K. First-principles study of muonium in a- and b- form dna. Phys B 2006;374:448.
[130] Badu SR et al. Ref [80].
[131] Nagamine K. Introductory muon science. Cambridge: Cambridge University Press; 2003.
[132] Schenck A. Muon spin rotation spectroscopy: principles and applications in solid state physics. Boston: Adam Hilger; 1985.
[133] Nagamine K, Pratt FL, Ohira S, Watanabe I, Ishida K, Nakamura SN, et al. Intra- and inter-molecular electron transfer in cytochrome *c* and myoglobin observed by the muon spin relaxation method. Phys B 2000;289–290:631.
[134] Butler MA, Walker LR, Soos ZG. Dimensionality of spin fluctuations in highly anisotropic tcnq salts. J Chem Phys 1976;64:3592.
[135] Risch R, Kehr KW. Direct stochastic theory of muon spin relaxation in a model for *trans*-polyacetylene. Phys Rev B 1992;46:5246.
[136] Banci L, Bertini I, Gray HB, Luchinat C, Reddig T, Rosato A, et al. Solution structure of oxidized horse heart cytochrome *c*. Biochemistry 1997;36:9867.
[137] Cammarere D, Scheicher RH, Sahoo N, Das TP, Nagamine K. First-principle determination of muon and muonium trapping sites in horse heart cytochrome *c* and investigation of magnetic hyperfine properties. Phys B 2000;289–290:636.
[138] Briere TM, Jeong J, Das TP, Ohira S, Nagamine K. Ab initio molecular orbital studies of the positive muon and muonium in 4-arylmethyleneamino-tempo derivatives. Phys B 2000;289–290:128.
[139] Scheicher RH, Cammarere D, Briere TM, Sahoo N, Das TP, Pratt FL, et al. First-principles theory of muon and muonium trapping in the protein chain of cytochrome *c* and associated hyperfine interactions. Hyperfine Interact 2001;136/137:755.
[140] Scheicher RH, Cammarere D, Sahoo N, Nagamine K, Das TP. Theory of ^{14}N and 17O nuclear quadrupole interactions in the single amino acids occurring in the protein chain of cytochrome *c*. Z Naturforsch 2002;57a:532.
[141] Scheicher RH, Cammarere D, Sahoo N, Briere TM, Pratt FL, Nagamine K, et al. First-principles study of muon and muonium trapping in the protein chain of cytochrome *c*, Phys B 2003;326:30.

[142] Fink HW, Schonenberger C. Electrical conduction through dna molecules. Nature 1999;398:407.
[143] Sakamoto K, Gouzu H, Komiya K, Kiga D, Yokoyama S, Yokomuri T, et al. Molecular computation by dna hairpin formation. Science 2000;288:1223.
[144] Cuniberti G, Fagas G, Richter K, editors. Introducing Molecular Electronics. Berlin: Heidelberg, Springer; 2005 [Chapter 15].
[145] Torikai E, Nagamine K, Pratt FL, Watanabe I, Ikedo Y, Urabe H, et al. Probing electron transfer in dna – new life science with muons. Hyperfine Interact 2001;138:509.
[146] Watson JD, Crick FHC. Molecular structure of nucleic acids. Nature 1953;171:737.
[147] Elkin LO. Rosalind franklin and the double helix. Phys Today 2003;56:42.
[148] Scheicher RH, Torikai E, Pratt FL, Nagamine K, Das TP. In: Maier K, Vianden R, editors. Book Chapter in HFI/NQI 2004. Springer: Dordrecht, The Netherlands; 2004. p. 53.
[149] Scheicher RH, Das TP, Torikai E, Pratt FL, Nagamine K. First-principles study of muonium in a-and b-form dna. Phys B 2006;374–375:448.
[150] Russo N, Toscano M, Grand A, Jolibois F. Protonation of thymine, cytosine, adenine, and guanine dna nucleic acid bases: theoretical investigation into the framework of density functional theory. J Comput Chem 1998;19:989.
[151] Smets J, Houben L, Schoone K, Maes G, Adamowicz L. Multiple site proton affinities of methylated nucleic acid bases. Chem Phys Lett 1996;262:789.
[152] Berman HM, Olson WK, Beveridge DL, Westbrook J, Gelbin A, Demeny T, et al. The nucleic acid database: a comprehensive relational database of three-dimensional structures of nucleic acids. Biophys J 1992;63:751. The Nucleic Acid Database at http://ndbserver.rutgers.edu.
[153] Raaijmakers H, Suck D, Mayer C. Crystal Structure Analysis of the A-DNA Dodecamer GACCACGTGGTC. Unpublished, NDB ID 153.
[154] Shakked Z, Guzikevich-Guerstein G, Frolow F, Rabbinovich D, Joachimiak A, Sigler PB. Determinants of repressor/operator recognition from the structure of the trp operator binding site. Nature 1994;368:469, NDB ID 154.
[155] Oganesyan VS, Hubbard PL, Butt JN, Jayasooriya UA. Laying the foundation for understanding muon implantation in dna: ab initio dft calculations of the nucleic acid base muonium adducts. Phys B 2003;326:25.
[156] Frisch MJ, et al. Gaussian 98, Revision A.11.3. Pittsburgh, PA: Gaussian, Inc.; 2002.
[157] Chakravarti A. Single nucleotide polymorphisms:... to a future of genetic medicine. Nature 2001;409:822.
[158] Church GM. Genomes for all. Sci Am 2006;294:47.
[159] Stoneking M. Single nucleotide polymorphisms: from the evolutionary past.... Nature 2001;409:821.
[160] Hood L, Heath JR, Phelps ME, Lin B. Systems biology and new technologies enable predictive and preventative medicine. Science 2004;306:640.
[161] Kasianowicz JJ, Brandin E, Branton D, Deamer DW. Characterization of individual polynucleotide molecules using a membrane channel. Proc Natl Acad Sci USA 1996;93:13770.
[162] Akeson M, Branton D, Kasianowicz JJ, Brandin E, Deamer DW. Microsecond time-scale discrimination among polycytidylic acid, polyadenylic acid, and polyuridylic acid as homopolymers or as segments within single rna molecules. Biophys J 1999;77:3227.
[163] Meller A, Nivon L, Brandin E, Golovchenko J, Branton D. Rapid nanopore discrimination between single polynucleotide molecules. Proc Natl Acad Sci USA 2000;97:1079.
[164] Deamer W, Akeson M. Nanopores and nucleic acids: prospects for ultrarapid sequencing. Trends Biotechnol 2000;18:147 and references therein.

[165] Khanna VK. Existing and emerging detection technologies for dna (deoxyribonucleic acid) finger printing, sequencing, bio- and analytical chips: a multidisciplinary development unifying molecular biology, chemical and electronics engineering. Biotechnol Adv 2007;25:85.
[166] Vercoutere WA, Winters-Hilt S, DeGuzman VS, Deamer D, Ridino SE, Rodgers JT, et al. Discrimination among individual watson-crick base pairs at the termini of single dna hairpin molecules. Nucleic Acids Res 2003;31:1311.
[167] Li J, Gershow M, Stein D, Brandin E, Golovchenko JA. Dna molecules and configurations in a solid-state nanopore microscope. Nat Mater 2003;2:611.
[168] Rhee M, Burns MA. Nanopore sequencing technology: nanopore preparations. Trends Biotechnol 2007;25:174.
[169] Zwolak M, Di Ventra M. Electronic signature of dna nucleotides via transverse transport. Nano Lett 2005;5:421.
[170] Lagerqvist J, Zwolak M, Di Ventra M. Fast dna sequencing via transverse electronic transport. Nano Lett 2006;6:779.
[171] He H, Scheicher RH, Pandey R, Rocha AR, Sanvito S, Grigoriev A, et al. Functionalized nanopore-embedded electrodes for rapid dna sequencing. J Phys Chem 2008;C112:3456 see also the pre-print arXiv:0708.4011.
[172] Soler JM, Artacho E, Gale JD, García A, Junquera J, Ordejón P, et al. The siesta method for *ab initio* order-*n* materials simulation. J Phys Condens Matter 2002;14:2745.
[173] A.R. Rocha, V. M. García-Suárez, S. W. Bailey, C. J. Lambert, J. Ferrer, S. Sanvito, et al. Computer code SMEAGOL (Spin and Molecular Electronics in an Atomically Generated Orbital Landscape). www.smeagol.tcd.ie. 2006.
[174] Rocha AR, García-Suárez VM, Bailey SW, Lambert CJ, Ferrer J, Sanvito S. Towards molecular spintronics. Nat Mater 2005;4:335.
[175] Rocha AR, García-Suárez VM, Bailey S, Lambert C, Ferrer J, Sanvito S. Spin and molecular electronics in atomically generated orbital landscapes. Phys Rev B 2006;73:085414.
[176] Postma HWCh. Rapid sequencing of individual dna molecules in graphene nanogaps. Nano Lett 2010;10:420.
[177] Schneider GF, Kowalczyk SW, Calado VE, Pandraud G, Zandbergen HW, Vandersypen LMK, et al. Dna translocation through graphene nanopores. Nano Lett 2010;10:3163.
[178] Merchant CA, Healy K, Wanunu M, Ray V, Peterman N, Bartel J, et al. Dna translocation through graphene nanopores. Nano Lett 2010;10:2915.
[179] Garaj S, Hubbard W, Reina A, Kong J, Branton D, Golovchenko JA. Graphene as a subnanometre trans-electrode membrane. Nature 2010;467:190.
[180] Nakashima N, Okuzono S, Murakami H, Nakai T, Yoshikawa K. Dna dissolves single-walled carbon nanotubes in water. Chem Lett 2003;32:456.
[181] Zheng M, Jagota A, Semke ED, Diner BA, Mclean RS, Lustig SR, et al. Dna-assisted dispersion and separation of carbon nanotubes. Nat Mater 2003;2:338.
[182] Zheng M, Jagota A, Strano MS, Santos AP, Barone P, Chou SG, et al. Structure-based carbon nanotube sorting by sequence-dependent dna assembly. Science 2003;302:1545.
[183] Tu X, Manohar S, Jagota A, Zheng M. Dna sequence motifs for structure-specific recognition and separation of carbon nanotubes. Nature 2009;460:250.
[184] Staii C, Johnson Jr. AT, Chen M, Gelperin A. Dna-decorated carbon nanotubes for chemical sensing. Nano Lett 2005;5:1774.

[185] Heller DA, Jeng ES, Yeung T-K, Martinez BM, Moll AE, Gastala JB, et al. Optical detection of dna conformational polymorphism on single-walled carbon nanotubes. Science 2006;311:508.

[186] Enyashin AN, Gemming S, Seifert G. Dna-wrapped carbon nanotubes. Nanotechnology 2007;18:245702.

[187] Johnson RR, Charlie Johnson AT, Klein ML. Probing the structure of dna– carbon nanotube hybrids with molecular dynamics. Nano Lett 2008;8:69.

[188] Gowtham S, Scheicher RH, Ahuja R, Pandey R, Karna SP. Physisorption of nucleobases on graphene: density-functional calculations. Phys Rev B 2007;76:033401.

[189] Gowtham S, Scheicher RH, Pandey R, Karna SP, Ahuja R. First-principles study of physisorption of nucleic acid bases on small-diameter carbon nanotubes. Nanotechnology 2008;19:125701.

[190] Frisch MJ, et al. In: GAUSSIAN 03, Revision C.02. Wallingford, CT: Gaussian, Inc.; 2004.

[191] Shibayama Y, Sato H, Enoki T, Endo M. Disordered magnetism at the metal-insulator threshold in nano-graphite-based carbon materials. Phys Rev Lett 2000;84:1744.

[192] Harigaya K, Enoki T. Mechanism of magnetism in stacked nanographite with open shell electrons. Chem Phys Lett 2002;351:128.

[193] Simeoni M, De Luca C, Picozzi S, Santucci S, Delley B. Interaction between zigzag single-wall carbon nanotubes and polymers: a density-functional study. J Chem Phys 2005;122:214710.

[194] Tournus F, Latil S, Heggie MI, Charlier JC. π-stacking interaction between carbon nanotubes and organic molecules. Phys Rev B 2005;72:075431.

[195] Ortmann F, Schmidt WG, Bechstedt F. Attracted by long-range electron correlation: adenine on graphite. Phys Rev Lett 2005;95:186101.

[196] London F. Zur theorie und systematik der molekularkräfte. Z Phys 1930;63:245 Some characteristics and uses of molecular force, Z Phys Chem Abstr B 1930;11:222.

[197] Seela F, Jawalekar AM, Münster I. Replacement of Canonical DNA Nucleobases by Benzotriazole and 1,2,3-Triazolo[4,5-d]pyrimidine: Synthesis, Fluorescence, and Ambiguous Base Pairing. Helv Chim Acta 2005;88:751.

Chapter 4

Cluster Structures: Bridging Experiment and Theory

F. Janetzko*, A. Goursot†, T. Mineva†, P. Calaminici‡, R. Flores-Moreno‡,§, A. M. Köster‡ and D. R. Salahub*,||,¶

*Department of Chemistry, University of Calgary, Alberta, Canada

†ICGM, UMR 5253 CNRS - Ecole de Chimie - UM1, Montpellier, Cédex 5, France

‡Departamento de Química CINVESTAV, Avenida Instituto Politécnico Nacional 2508 A.P. 14–740 México, Mexico

§Chemistry and Biochemistry Department, COSAM, Auburn University, Auburn, Alabama, USA

||IBI, Institute for Biocomplexity and Informatics, University of Calgary, Alberta, Canada

¶ISEEE, Institute for Sustainable Energy, Environment and Economy, University of Calgary, Alberta, Canada

Chapter Outline Head

Nanoclusters. DOI: 10.1016/S1875-4023(10)01004-1

I. INTRODUCTION

Density functional theory (DFT) has progressed remarkably over the years, developing from fringe applications to its current state as a mainstay of computational chemistry, physics, and biology. Many frontiers have been traversed: structure optimization, excited states, spectroscopies of various types, reactivity, and dynamics. Much current effort is focused on hybrid methods that combine DFT with force fields of various types in order to treat a quantum mechanical core with an adequate description of micro- and macro-environmental effects. The field of clusters has a similar history and has shown similar remarkable development over about the same period. Cluster science emerged from the confluence of molecular spectroscopy, molecular beam techniques, catalysis, and low-temperature matrix spectroscopy on the "chemistry" side and solid state electronics, magnetic, and surface science on the "physics" side. Clusters have fascinating properties in their own right and they also serve as models for bulk systems in a wide variety of contexts. Indeed, one can view the whole field of nanoscience and nanotechnology as dealing precisely with the size dimension where clusters, properly embedded in micro and macro environments, are relevant. In this chapter, we look at two current frontiers of both DFT and cluster science, areas where progress has been made but much remains to be done. We look first at transition-metal clusters where the compact d shell with its strong spin-dependent correlation provides severe challenges, which are at present only partially met. The other "last frontier of DFT" lies in the area of weak intermolecular interactions. Roughly speaking, good results can be obtained for hydrogen-bonded systems with contemporary functionals of the generalized gradient approximation (GGA) or so-called meta-GGA functionals. Weaker non-bonded interactions, such as those found in the stacking of DNA bases, or mid-range interactions in peptides, are presently not under control; GGA and meta-GGA functionals fail. An intense search for improved functionals is under way.

The chapter reviews both theoretical/computational approaches and the experimental techniques for a number of examples in these two classes of cluster systems. We wish to build a bridge between theory and experiment, as well as the bridges that clusters provide, between "chemistry" and "physics" and among the "picoscopic," "nanoscopic," microscopic, and macroscopic worlds. We have chosen to include considerable detail on both the theoretical methodology and the experiments in the hope that it will help experimentalists to better judge the quality of the calculations and that it will allow theorists to better appreciate and relate to the experiments they are trying to explain. Some of the material is at a textbook level and numerous references are provided. We will gladly accept criticism from more senior readers on these points, if the material helps students to find their way into this exciting field of research.

A. Theoretical Methods

The theoretical study of clusters is challenging due to the paucity of empirical rules to guide the choice of structures, to the complicated electronic structures often found, and to the importance of weak interactions between the building blocks in many cases. Often, theoreticians face the formidable task of locating the global minimum on the potential energy surface of clusters without any further information except the types and numbers of building blocks. Moreover, low-lying excited states, theoretically difficult-to-describe interactions, symmetry breakings, and other obstacles complicate the situation in many cluster studies. Thus, it is not surprising that cluster science is a strong driving force for the development of new theoretical methods, nowadays most often electronic structure methods, as well as local and global optimization algorithms. Despite all this effort, global minimization [1–4] employing electronic structure methods are still in a very early stage of development. The computational demands of reliable first-principle electronic structure methods represent the main drawback.

The development of molecular density functional theory (DFT) methods over the last four decades [5] raises the hope that this situation may change in the not so distant future. In particular, auxiliary density functional theory (ADFT), which employs auxiliary densities for the calculation of the Coulomb and exchange-correlation potential, is very promising in this respect. It combines the reliability of first-principles DFT methods with a highly efficient implementation. The auxiliary density perturbation theory (ADPT) that was recently developed [6] represents a natural extension of the ADFT method. With the ADPT, polarizabilities, electronic excitations, vibrational frequencies, and many other molecular properties can be calculated in the framework of ADFT. We now describe the ADFT implementation in the program deMon2k [7].

(i) Kohn–Sham Density Functional Theory

Based on the Hohenberg–Kohn theorem [8], the ground-state energy of a many-electron system with the external potential $v(\mathbf{r})$ can be expressed by the following energy functional [9,10]:

$$E[\rho] = T[\rho] + \int \rho(\mathbf{r}) v(\mathbf{r})\, d\mathbf{r} + V_{ee}[\rho]. \tag{1}$$

Here $T[\rho]$ is the kinetic energy and $V_{ee}[\rho]$ collects all electron–electron interaction energies. A major technical problem is the accurate description of the kinetic energy functional $T[\rho]$. In the Kohn–Sham method, this problem is avoided by introducing orbitals of a noninteracting reference system. Levy and Perdew [11,12] have shown that these orbitals are delivered by the following procedure:

$$T_{KS}[\rho] = \min_{\Psi \mapsto \rho_0} \langle \Psi | \hat{T} | \Psi \rangle. \tag{2}$$

Here Ψ is the Slater determinant composed from the Kohn–Sham orbitals $\psi_i(\mathbf{r})$ and $T_{KS}[\rho]$ is the corresponding Kohn–Sham kinetic energy. This energy can be calculated from the Kohn–Sham orbitals:

$$T_{KS}[\rho] = \sum_i^{occ} \left\langle \psi_i \middle| -\tfrac{1}{2}\nabla^2 \middle| \psi_i \right\rangle. \tag{3}$$

The Kohn–Sham orbitals are functionals of the density and can be derived from it as shown by Parr and coworkers [13–15]. Imposing the constraint

$$\rho_0(\mathbf{r}) = \sum_i^{occ} |\psi_i(\mathbf{r})|^2, \tag{4}$$

to the variation in Equation (2) using a local Lagrange multiplier $v_{KS}(\mathbf{r})$, we obtain

$$\left(-\frac{1}{2}\nabla^2 + v_{KS}(\mathbf{r})\right)\psi_i(\mathbf{r}) = \varepsilon_i \psi_i(\mathbf{r}) \tag{5}$$

These are the Kohn–Sham equations and $v_{KS}(\mathbf{r})$ determines (within a trivial constant) the external potential of the Kohn–Sham reference system.

In order to find a more explicit representation of $v_{KS}(\mathbf{r})$, we now rewrite the energy functional, Equation (1), using the Kohn–Sham kinetic energy term (Equation (3)) and the explicit expression for the electronic Coulomb energy

$$J[\rho] = \frac{1}{2}\int\int \frac{\rho(\mathbf{r})\rho(\mathbf{r}')}{|\mathbf{r}-\mathbf{r}'|}\,\mathrm{d}\mathbf{r}\,\mathrm{d}\mathbf{r}' = \frac{1}{2}\langle \rho \parallel \rho \rangle, \tag{6}$$

where the symbol $\parallel$ represents the $1/|\mathbf{r} - \mathbf{r}'|$ operator. It follows that

$$E[\rho] = T_{KS}[\rho] + \int \rho(\mathbf{r})v(\mathbf{r})\,\mathrm{d}\mathbf{r} + J[\rho] + E_{xc}[\rho], \tag{7}$$

where the newly introduced exchange-correlation energy functional is defined as

$$E_{xc}[\rho] \equiv T[\rho] - T_{KS}[\rho] + V_{ee}[\rho] - J[\rho]. \tag{8}$$

This quantity collects all nonclassical interactions between the electrons and the difference of the kinetic energies of the interacting and noninteracting system.

The accuracy of the Kohn–Sham method is mainly determined by the quality of the approximation used for the calculation of $E_{xc}[\rho]$. The kinetic energy difference in $E_{xc}[\rho]$ is in the order of magnitude of the correlation energy [16].

Based on the Hohenberg–Kohn theorem, the ground-state density minimizes the energy functional Equation (7) and hence satisfies the Euler equation

$$\mu = \frac{\delta E[\rho]}{\delta \rho(\mathbf{r})}, \tag{9}$$

where μ is the Lagrange multiplier associated with the normalization of the electronic density to the number of electrons n in the system:

$$\int \rho(\mathbf{r})\,\mathrm{d}\mathbf{r} = n. \tag{10}$$

The functional derivative of the energy functional, Equation (7), yields

$$\frac{\delta E[\rho]}{\delta \rho(\mathbf{r})} = \frac{\delta T_{\mathrm{KS}}[\rho]}{\delta \rho(\mathbf{r})} + v(\mathbf{r}) + \int \frac{\rho(\mathbf{r}')}{|\mathbf{r}-\mathbf{r}'|}\,\mathrm{d}\mathbf{r}' + \frac{\delta E_{\mathrm{xc}}[\rho]}{\delta \rho(\mathbf{r})} = \frac{\delta T_{\mathrm{KS}}[\rho]}{\delta \rho(\mathbf{r})} + v_{\mathrm{KS}}(\mathbf{r}). \tag{11}$$

Thus, the Kohn–Sham potential has the following explicit form:

$$v_{\mathrm{KS}}(\mathbf{r}) = v(\mathbf{r}) + \int \frac{\rho(\mathbf{r}')}{|\mathbf{r}-\mathbf{r}'|}\,\mathrm{d}\mathbf{r}' + v_{\mathrm{xc}}[\rho], \tag{12}$$

where the exchange-correlation potential is defined as

$$v_{\mathrm{xc}}[\rho] \equiv \frac{\delta E_{\mathrm{xc}}[\rho]}{\delta \rho(\mathbf{r})}. \tag{13}$$

Inserting Equation (12) in Equation (5) yields the canonical Kohn–Sham orbital equations:

$$\left(-\frac{1}{2}\nabla^2 + v(\mathbf{r}) + \int \frac{\rho(\mathbf{r}')}{|\mathbf{r}-\mathbf{r}'|}\,\mathrm{d}\mathbf{r}' + v_{\mathrm{xc}}[\rho]\right)\psi_i(\mathbf{r}) = \varepsilon_i \psi_i(\mathbf{r}). \tag{14}$$

These equations have to be solved iteratively. They can be cast in the matrix form, yielding Roothaan–Hall-like equations [17,18].

(ii) The LCGTO Kohn–Sham Method

In the linear combination of Gaussian-type orbital (LCGTO) ansatz, the Kohn–Sham orbitals are expanded into the following atomic orbitals:

$$\psi_i(\mathbf{r}) = \sum_{\mu} c_{\mu i}\mu(\mathbf{r}). \tag{15}$$

Here, $\mu(\mathbf{r})$ represents an atomic orbital and $c_{\mu i}$ the corresponding molecular orbital coefficient. To avoid unnecessary complications in the presentation, we restrict ourselves to the closed-shell case. The extension to the open-shell formalism [19–21] is straightforward. With the LCGTO expansion, we find for the electronic density (Equation (4))

$$\rho(\mathbf{r}) = \sum_{\mu,\nu} P_{\mu\nu}\mu(\mathbf{r})\nu(\mathbf{r}). \tag{16}$$

$P_{\mu\nu}$ represents an element of the closed-shell density matrix, defined as

$$P_{\mu\nu} = 2\sum_{i}^{occ} c_{\mu i}c_{\nu i}. \tag{17}$$

Using the above expansions for the Kohn–Sham orbitals (Equation (15)) and the electronic density (Equation (16)), the Kohn–Sham energy expression (Equation (7)) can be rewritten in the following form:

$$E = E_{\mathrm{SCF}} + \sum_{A>B}^{\mathrm{atoms}} \frac{Z_A Z_B}{|\mathbf{A} - \mathbf{B}|}, \tag{18}$$

$$E_{\mathrm{SCF}} = \sum_{\mu,\nu} P_{\mu\nu}H_{\mu\nu} + \frac{1}{2}\sum_{\mu,\nu}\sum_{\sigma,\tau} P_{\mu\nu}P_{\sigma\tau}\langle\mu\nu \parallel \sigma\tau\rangle + E_{\mathrm{xc}}[\rho], \tag{19}$$

$$H_{\mu\nu} = \left\langle \mu \left| -\frac{1}{2}\nabla^2 \right| \nu \right\rangle - \sum_{C}^{\mathrm{atoms}} \left\langle \mu \left| \frac{Z_C}{|\mathbf{r} - \mathbf{C}|} \right| \nu \right\rangle, \tag{20}$$

$$E_{\mathrm{xc}}[\rho] = \int \rho(\mathbf{r})\varepsilon_{\mathrm{xc}}[\rho]\,d\mathbf{r}. \tag{21}$$

Here, E_{SCF} collects all terms depending on the electronic density, which changes during the self-consistent field (SCF) iterations. The SCF convergence is based on this energy expression. The total energy (Equation (18)) is the sum of E_{SCF} and the nuclear repulsion energy which is calculated from the nuclear charges Z_A and Z_B and atomic position vectors **A** and **B**. The $H_{\mu\nu}$ (Equation (20)) are elements of the core Hamiltonian matrix. They are built from the kinetic and nuclear attraction energy of the electrons and describe the movement of an electron in the nuclear framework. The computation of this matrix has a formal quadratic scaling with the number of basis functions N of the system. The second term in Equation (19) represents the-Coulomb repulsion energy of the electrons. In contrast to the Hartree–Fock theory, the calculation of the Coulomb and exchange energies are separated in DFT. The Coulomb term introduces a formal N^4 scaling. For the calculation of the exchange-correlation energy (Equation (21)), a numerical integration has to be performed. This integration scales formally as $N^2 \times G$, where G is the number of grid points necessary for the numerical integration.

From the above discussion, it follows that the calculation of the Coulomb repulsion energy represents the most demanding computational task in

Equation (19). The introduction of the variational approximation of the Coulomb potential [22–24] reduces the formal scaling of this term to $N^2 \times M$, where M is the number of auxiliary functions which is usually three to five times N. This technique is nowadays used in most LCGTO-DFT programs. It is identical to the so-called resolution of the identity (RI) [25] that cropped up in wavefunction methods, too. At this point, it should be mentioned that only the RI Coulomb fitting is size-consistent. The variational approximation of the Coulomb potential, as implemented in deMon2k, is based on the minimization of the following self-interaction error:

$$\mathcal{E}_2 = \frac{1}{2}\int\int \frac{[\rho(\mathbf{r}) - \widetilde{\rho}(\mathbf{r})][\rho(\mathbf{r}') - \widetilde{\rho}(\mathbf{r}')]}{|\mathbf{r} - \mathbf{r}'|}\, \mathrm{d}\mathbf{r}\, \mathrm{d}\mathbf{r}'. \tag{22}$$

The approximated density $\widetilde{\rho}(\mathbf{r})$ is expanded in primitive Hermite Gaussians which are centered at the atoms:

$$\widetilde{\rho}(\mathbf{r}) = \sum_{\bar{k}} x_{\bar{k}} \bar{k}(\mathbf{r}). \tag{23}$$

The primitive Hermite Gaussian auxiliary functions are indicated by a bar. An (unnormalized) auxiliary function $\bar{k}(\mathbf{r})$ centered at atom K with exponent ζ_k has the form

$$\bar{k}(\mathbf{r}) = \left(\frac{\partial}{\partial K_x}\right)^{\bar{k}_x} \left(\frac{\partial}{\partial K_y}\right)^{\bar{k}_y} \left(\frac{\partial}{\partial K_z}\right)^{\bar{k}_z} \mathrm{e}^{-\zeta_k(\mathbf{r}-\mathbf{K})^2}. \tag{24}$$

In deMon2k, these auxiliary functions are normalized with respect to the Coulomb norm and grouped into s, spd, and spdfg sets. The exponents are shared within each of these sets [26,27]. Based on the exponent range of the atomic basis set, an automatic generation of auxiliary function sets, which are indicated by the abbreviation GEN, is available in deMon2k. A detailed description is given in Ref. [28].

With the LCGTO expansion for $\rho(\mathbf{r})$ and $\widetilde{\rho}(\mathbf{r})$, we obtain the following representation for $\mathcal{E}_2$:

$$\begin{aligned}
\varepsilon_2 &= \frac{1}{2}\langle \rho - \widetilde{\rho} \parallel \rho - \widetilde{\rho}\rangle \\
&= \frac{1}{2}\langle \rho \parallel \rho\rangle - \langle \rho \parallel \widetilde{\rho}\rangle + \frac{1}{2}\langle \widetilde{\rho} \parallel \widetilde{\rho}\rangle \\
&= \frac{1}{2}\sum_{\mu,\nu}\sum_{\sigma,\tau} P_{\mu\nu}P_{\sigma\tau}\langle \mu\nu \parallel \sigma\tau\rangle \\
&\quad -\sum_{\bar{k}}\sum_{\mu,\nu} P_{\mu\nu}\langle \mu\nu \parallel \bar{k}\rangle x_{\bar{k}} + \frac{1}{2}\sum_{\bar{k},\bar{l}} x_{\bar{k}} x_{\bar{l}}\langle \bar{k} \parallel \bar{l}\rangle.
\end{aligned} \tag{25}$$

The expansion coefficients $x_{\bar{k}}$ of the approximated density are determined by the minimization of the self-interaction error $\mathcal{E}$:

$$\frac{\partial \mathcal{E}_2}{\partial x_{\bar{m}}} = -\sum_{\mu,\nu} P_{\mu\nu}\langle \mu\nu \parallel \bar{m}\rangle + \sum_{\bar{k}} x\bar{k}\langle \bar{k} \parallel \bar{m}\rangle = 0 \; \forall \; \bar{m}. \tag{26}$$

At this point, it is useful to introduce the Coulomb matrix

$$\mathbf{G} = \begin{pmatrix} \langle \bar{1} \parallel \bar{1}\rangle & \langle \bar{1} \parallel \bar{2}\rangle & \ldots & \langle \bar{1} \parallel \bar{m}\rangle \\ \langle \bar{2} \parallel \bar{1} & \langle \bar{2} \parallel \bar{2}\rangle & \ldots & \langle \bar{2} \parallel \bar{m}\rangle \\ \vdots & \vdots & \ddots & \vdots \\ \langle \bar{m} \parallel \bar{1}\rangle & \langle \bar{m} \parallel \bar{2}\rangle & \ldots & \langle \bar{m} \parallel \bar{m}\rangle \end{pmatrix}, \tag{27}$$

and the Coulomb vector

$$\mathbf{J} = \begin{pmatrix} \sum_{\mu,\nu} P_{\mu\nu}\langle \mu\nu \parallel \bar{1}\rangle \\ \sum_{\mu,\nu} P_{\mu\nu}\langle \mu\nu \parallel \bar{2}\rangle \\ \vdots \\ \sum_{\mu,\nu} P_{\mu\nu}\langle \mu\nu \parallel \bar{m}\rangle \end{pmatrix}. \tag{28}$$

With **G** and **J,** the following inhomogeneous equation system for the determination of the fitting coefficients, collected in x, can be formulated:

$$\mathbf{G}x = \mathbf{J}. \tag{29}$$

A straightforward solution is obtained by the inversion of the Coulomb matrix **G**:

$$\mathbf{x} = \mathbf{G}^{-1} \; \mathbf{J}. \tag{30}$$

Because **G** is symmetric and positive definite, its inversion can be very efficiently performed via Cholesky decomposition [29]. However, if very large auxiliary function sets are used, the Coulomb matrix tends to become ill-conditioned. As a consequence, tight SCF convergence cannot be reached. Therefore, a singular value decomposition (SVD) is used in deMon2k (keyword MATINV) in such cases.

After the description of the calculation of the fitting coefficients, we now turn to the energy and SCF calculation. Because $\mathcal{E}_2$ is positive definite, the following inequality holds:

$$\frac{1}{2}\sum_{\mu,\nu}\sum_{\sigma,\tau} P_{\mu\nu}P_{\sigma\tau}\langle \mu\nu \parallel \sigma\tau\rangle \geq \sum_{\bar{k}}\sum_{\mu,\nu} P_{\mu\nu}\langle \mu\nu \parallel \bar{k}\,\rangle x_{\bar{k}} - \frac{1}{2}\sum_{\bar{k},\bar{l}} x_{\bar{k}}x_{\bar{l}}\langle \bar{k} \parallel \bar{l}\,\rangle.$$

With this inequality, an approximate SCF energy, which is based on Equation (19), can be derived:

$$E_{\rm SCF} = \sum_{\mu,\nu} P_{\mu\nu} H_{\mu\nu} + \sum_{\bar{k}} \sum_{\mu,\nu} P_{\mu\nu} \langle \mu\nu \parallel \bar{k} \rangle x_{\bar{k}} - \frac{1}{2} \sum_{\bar{k},\bar{l}} x_{\bar{k}} x_{\bar{l}} \langle \bar{k} \parallel \bar{l} \rangle + E_{\rm xc}[\rho]. \tag{31}$$

The variation of this energy expression with respect to the molecular orbital coefficients, constraining the Kohn–Sham orbitals to be orthonormal,

$$\sum_{\mu,\nu} c_{\mu,i} S_{\mu\nu} c_{\nu j} = \delta_{ij} \ \forall \ i,j, \tag{32}$$

yields

$$\frac{\partial E_{\rm SCF}}{\partial c_{\mu i}} = 2 \sum_{\nu} \left(H_{\mu\nu} + \sum_{\bar{k}} \langle \mu\nu \parallel \bar{k} \rangle x_{\bar{k}} + \langle \mu | v_{\rm xc}[\rho] | \nu \rangle \right) c_{\nu i} - 2 \sum_{\nu} \sum_{j} S_{\mu\nu} c_{\nu j} \varepsilon_{ji} \ \forall \ \mu, i. \tag{33}$$

Here, the derivative of the exchange-correlation energy, restricting ourselves to local functionals for the clarity of the presentation, was developed as [30]

$$\frac{\partial E_{\rm xc}[\rho]}{\partial c_{\mu i}} = \int \frac{\delta E_{\rm xc}[\rho]}{\delta \rho(\mathbf{r})} \frac{\partial \rho(\mathbf{r})}{\partial c_{\mu i}} \, d\mathbf{r} = \sum_{\nu} \langle \mu | v_{\rm xc}[\rho] | \nu \rangle c_{\nu i}. \tag{34}$$

The ε_{ji} in Equation (33) are the undetermined Lagrange multipliers. Their transformation to diagonal form yields the canonical Kohn–Sham equations

$$\mathbf{K} \ \mathbf{c} = \mathbf{S} \ \mathbf{c} \boldsymbol{\varepsilon}. \tag{35}$$

Here, the elements of the Kohn–Sham matrix **K** are defined as

$$K_{\mu\nu} = H_{\mu\nu} + \sum_{\bar{k}} \langle \mu\nu \parallel \bar{k} \rangle x_{\bar{k}} + \langle \mu | v_{\rm xc}[\rho] | \nu \rangle. \tag{36}$$

In Equation (35), **S** represents the overlap matrix, **c** the molecular orbital coefficient matrix, and $\boldsymbol{\varepsilon}$ the diagonal matrix of the Lagrange multipliers, that is the Kohn–Sham orbital energies.

As can be seen from Equation (36), the Kohn–Sham matrix elements depend on the fitting coefficients $x_{\bar{k}}$ and the molecular orbital coefficient $c_{\mu i}$ via the dependence of the exchange-correlation potential from the orbital density $\rho(\mathbf{r})$. In deMon2k, this theoretical model is selected by the keyword specification VXCTYP BASIS and is often referred to as the BASIS approach. Due to efficient three-center electron repulsion integral algorithms [31], the numerical integration of the exchange-correlation energy and potential is the computationally most demanding task in this approach. Therefore, a more efficient approach for the calculation of the exchange-correlation energy and potential is desirable. In fact, the use of auxiliary functions for the calculation of the exchange-correlation energy and potential has a long history in DFT methods [32,33]. In programs like deMon-KS [34], DGAUSS [35], or GTOFF [36], the exchange-correlation potential is expanded in auxiliary functions as proposed by Sambe and Felton [33]. The expansion coefficients are obtained by a least-squares fit on a grid. Because this fit and the corresponding energy expression are not variational, only approximate gradients and higher energy derivatives are available.

(iii) Auxiliary Density Functional Theory

As an alternative to the fitting of the exchange-correlation potential by auxiliary functions, the direct use of the auxiliary function density from the variational fitting of the Coulomb potential for the calculation of the exchange-correlation potential has been investigated over the last few years [37–39]. The resulting energy expression, from now on named the ADFT energy, is variational and has the form [39]

$$E_{\mathrm{SCF}} = \sum_{\mu,\nu} P_{\mu\nu} H_{\mu\nu} + \sum_{\bar{k}} \sum_{\mu,\nu} P_{\mu\nu} \langle \mu\nu \parallel \bar{k} \rangle x_{\bar{k}} - \frac{1}{2} \sum_{\bar{k},\bar{l}} x_{\bar{k}} x_{\bar{l}} \langle \bar{k} \parallel \bar{l} \rangle + E_{\mathrm{xc}}[\widetilde{\rho}]. \tag{37}$$

In deMon2k, this theoretical model is selected by the keyword specification VXCTYP AUXIS, which is the default setting, and often referred to as the AUXIS approach. Because the approximated density is a linear combination of auxiliary functions, the density calculation at each grid point scales linear. Instead, with the orbital density the products of basis functions have to be evaluated. Obviously, this represents a considerable simplification of the grid work. In particular, using auxiliary function sets that share the same exponents reduces considerably the number of expensive exponential function evaluations at each grid point. Moreover, by using Hermite Gaussian auxiliary functions, the Hermite polynomial recurrence relations [40] can be used for the function calculations on the grid.

The variation of the ADFT energy expression (Equation (37)) with respect to the molecular orbital coefficients, again constraining the Kohn–Sham orbitals to orthonormality (Equation (32)), yields

$$\frac{\partial E_{\mathrm{SCF}}}{\partial c_{\mu i}} = 2 \sum_{\nu} \left(H_{\mu\nu} + \sum_{\bar{k}} \langle \mu\nu \parallel \bar{k} \rangle (x_{\bar{k}} + z_{\bar{k}}) \right) c_{\nu i} - 2 \sum_{\nu} \sum_{j} S_{\mu\nu} c_{\nu j} \varepsilon_{ji} \;\forall\; \mu, i. \tag{38}$$

The derivative of the exchange-correlation energy, again restricting ourselves to local functionals, now contains the fitted density

$$\frac{\partial E_{\mathrm{xc}}[\widetilde{\rho}]}{\partial c_{\mu i}} = \int \frac{\delta E_{\mathrm{xc}}[\widetilde{\rho}]}{\delta \widetilde{\rho}(\mathbf{r})} \frac{\partial \widetilde{\rho}(\mathbf{r})}{\partial c_{\mu i}} \, d\mathbf{r}. \tag{39}$$

The resulting functional derivative defines the exchange-correlation potential calculated with the fitted density:

$$v_{\mathrm{xc}}[\widetilde{\rho}] \equiv \frac{\delta E_{\mathrm{xc}}[\widetilde{\rho}]}{\delta \widetilde{\rho}(\mathbf{r})}. \tag{40}$$

The derivative of the fitted density with respect to molecular orbital coefficients is given by

$$\frac{\partial \widetilde{\rho}(\mathbf{r})}{\partial c_{\mu i}} = \sum_{\bar{l}} \frac{\partial x_{\bar{l}}}{\partial c_{\mu i}} \bar{l}(\mathbf{r}). \tag{41}$$

For the fitting coefficients, it follows from Equation (30)

$$x_{\bar{l}} = \sum_{\bar{k}} G_{\bar{l}\bar{k}}^{-1} J_{\bar{k}} = \sum_{\bar{k}} \sum_{\mu,\nu} G_{\bar{l}\bar{k}}^{-1} \langle \mu\nu \parallel \bar{k} \rangle P_{\mu\nu}. \tag{42}$$

With

$$\frac{\partial P_{\mu\nu}}{\partial c_{\mu i}} = 2c_{\nu i}, \tag{43}$$

it follows for the derivative of the fitted density

$$\frac{\partial \widetilde{\rho}(\mathbf{r})}{\partial c_{\mu i}} = 2 \sum_{\bar{k},\bar{l}} \sum_{\nu} c_{\nu i} \langle \mu\nu \parallel \bar{k} \rangle G_{\bar{k}\bar{l}}^{-1} \bar{l}(\mathbf{r}). \tag{44}$$

Inserting this expression and the definition of the approximate exchange-correlation potential Equation (40) into Equation (39) yields

$$\frac{\partial E_{\mathrm{xc}}[\widetilde{\rho}]}{\partial c_{\mu i}} = 2 \sum_{\bar{k}} \sum_{\nu} c_{\nu i} \langle \mu\nu \parallel \bar{k} \rangle z_{\bar{k}} \tag{45}$$

with

$$z_{\bar{k}} = \sum_{\bar{l}} G_{\bar{k}\bar{l}}^{-1} \langle \bar{l} | v_{\mathrm{xc}}[\widetilde{\rho}] \rangle. \tag{46}$$

In order to distinguish the $x_{\bar{k}}$ and $z_{\bar{k}}$ coefficients, we name them Coulomb and exchange-correlation coefficients, respectively. The corresponding Kohn–Sham matrix elements are defined as

$$K_{\mu\nu} = H_{\mu\nu} + \sum_{\bar{k}} \langle \mu\nu \parallel \bar{k} \rangle (x_{\bar{k}} + z_{\bar{k}}). \tag{47}$$

They depend only on the fitting coefficients. In contrast to the traditional fitting of the exchange-correlation potential, ADFT fits the density and hence, the numerical integration of v_{xc} is not avoided. However, the work on the grid is considerably reduced because only one-center terms are involved in the integrals in Equation (46). These terms are evaluated numerically. Because the ADFT energy is variational, analytic gradients [39] and higher energy derivatives can be formulated.

(iv) Auxiliary Density Perturbation Theory

For second energy derivatives that appear in the calculation of polarizabilities, vibrational frequencies, and other second-order properties, the perturbed density matrix is required. McWeeny's self-consistent perturbation (SCP) theory [41–46] represents a direct approach for the calculation of this matrix. For the clarity of the presentation, we assume perturbation-independent basis and auxiliary functions and restrict ourselves to closed-shell systems. Under these conditions, the elements of the perturbed density matrix are given by the SCP formalism as [45]

$$P_{\mu\nu}^{(\lambda)} \equiv \frac{\partial P_{\mu\nu}}{\partial \lambda} = 2 \sum_{i}^{occ} \sum_{a}^{uno} \frac{\mathcal{K}_{ia}^{(\lambda)}}{\varepsilon_i - \varepsilon_a} \left(c_{\mu i} c_{\nu a} + c_{\mu a} c_{\nu i} \right). \tag{48}$$

Here λ denotes the perturbation parameter, for example, an electric field component in the calculation of polarizabilities, ε_i and ε_a orbital energies of the ith occupied and ath unoccupied orbital and $\mathcal{K}_{ia}^{(\lambda)}$ the perturbed Kohn–Sham matrix in the molecular orbital representation.

$$\mathcal{K}_{ia}^{(\lambda)} = \sum_{\mu,\nu} c_{\mu i} c_{\nu a} K_{\mu\nu}^{(\lambda)} \tag{49}$$

with

$$K_{\mu\nu}^{(\lambda)} = H_{\mu\nu}^{(\lambda)} + \sum_{\bar{k}} \langle \mu\nu \parallel \bar{k} \rangle \left(x_{\bar{k}}^{(\lambda)} + z_{\bar{k}}^{(\lambda)} \right). \tag{50}$$

The perturbation of the exchange-correlation coefficients is given by

$$z_{\bar{k}}^{(\lambda)} = \sum_{\bar{l}} G_{\bar{k}\bar{l}}^{-1} \left\langle \bar{l} \,|\, v_{\text{xc}}^{(\lambda)}[\tilde{\rho}] \right\rangle. \tag{51}$$

Since $v_{\text{xc}}[\tilde{\rho}]$ itself is a (local) functional of the density, it follows

$$\left\langle \bar{l} | v_{\text{xc}}^{(\lambda)}[\tilde{\rho}] \right\rangle = \iint \bar{l}(\mathbf{r}) \frac{\delta v_{\text{xc}}[\tilde{\rho}]}{\delta \tilde{\rho}(\mathbf{r}')} \frac{\partial \tilde{\rho}(\mathbf{r}')}{\partial \lambda} \, d\mathbf{r} \, d\mathbf{r}' = \sum_{\bar{m}} \langle \bar{l} | f_{\text{xc}}[\tilde{\rho}] | \bar{m} \rangle x_{\bar{m}}^{(\lambda)} \tag{52}$$

with the exchange-correlation kernel defined as

$$f_{\text{xc}}(\mathbf{r}, \mathbf{r}') = \frac{\delta^2 E_{\text{xc}}[\tilde{\rho}]}{\delta \tilde{\rho}(\mathbf{r}') \delta \tilde{\rho}(\mathbf{r})} = \frac{\delta v_{\text{xc}}[\tilde{\rho}]}{\delta \tilde{\rho}(\mathbf{r}')}. \tag{53}$$

Compared to the standard LCGTO kernel $\langle \mu\nu | f_{\text{xc}} | \sigma\tau \rangle$, the scaling of the kernel calculation is reduced by almost two orders of magnitude in the ADPT approach. With this result, we now rewrite Equation (49) as

$$\mathcal{K}_{ia}^{(\lambda)} = \mathcal{H}_{ia}^{(\lambda)} + \sum_{\bar{k}} \langle ia \parallel \bar{k} \rangle x_{\bar{k}}^{(\lambda)} + \sum_{\bar{k},\bar{l}} \langle ia \parallel \bar{k} \rangle F_{\bar{k}\bar{l}} x_{\bar{l}}^{(\lambda)} \tag{54}$$

with

$$F_{\bar{k}\bar{l}} = \sum_{\bar{m}} G_{\bar{k}\bar{m}}^{-1} \langle \bar{m} | f_{xc}[\tilde{\rho}] | \bar{l} \rangle. \tag{55}$$

On the other hand, we find as the derivative of the fitting equation system (Equation (29))

$$\sum_{\bar{k}} G_{\bar{m}\bar{k}} x_{\bar{k}}^{(\lambda)} = \sum_{\mu,\nu} P_{\mu\nu}^{(\lambda)} \langle \mu\nu \parallel \bar{m} \rangle. \tag{56}$$

Substituting Equations (54) and (56) into Equation (48) yields

$$\begin{aligned}\sum_{\mu,\nu} P_{\mu\nu}^{(\lambda)} \langle \mu\nu \parallel \bar{m} \rangle &= 4 \sum_{i}^{occ} \sum_{a}^{uno} \frac{\langle \bar{m} \parallel ia \rangle \mathcal{H}_{ia}}{\varepsilon_i - \varepsilon_a} + 4 \sum_{i}^{occ} \sum_{a}^{uno} \sum_{\bar{k}} \frac{\langle \bar{m} \parallel ia \rangle \langle ia \parallel \bar{k} \rangle}{\varepsilon_i - \varepsilon_a} x_{\bar{k}}^{(\lambda)} \\ &+ 4 \sum_{i}^{occ} \sum_{a}^{uno} \sum_{\bar{k},\bar{l}} \frac{\langle \bar{m} \parallel ia \rangle \langle ia \parallel \bar{k} \rangle}{\varepsilon_i - \varepsilon_a} F_{\bar{k}\bar{l}} x_{\bar{l}}^{(\lambda)} \\ &= \sum_{\bar{k}} G_{\bar{m}\bar{k}} x_{\bar{k}}^{(\lambda)}.\end{aligned} \tag{57}$$

We now define the elements of the Coulomb and exchange-correlation coupling matrix as

$$\begin{aligned} A_{\bar{k}\bar{l}} &= \sum_{i}^{occ} \sum_{a}^{uno} \frac{\langle \bar{k} \parallel ia \rangle \langle ia \parallel \bar{l} \rangle}{\varepsilon_i - \varepsilon_a}, \\ B_{\bar{k}\bar{l}} &= \sum_{i}^{occ} \sum_{a}^{uno} \sum_{\bar{m},\bar{n}} \frac{\langle \bar{k} \parallel ia \rangle \langle ia \parallel \bar{m} \rangle}{\varepsilon_i - \varepsilon_a} G_{\bar{m}\bar{n}}^{-1} \langle \bar{n} | f_{xc}[\tilde{\rho}] | \bar{l} \rangle = \sum_{\bar{m}} A_{\bar{k}\bar{m}} F_{\bar{m}\bar{l}}. \end{aligned} \tag{58}$$

Similarly, the elements of the perturbation vector are given by

$$b_{\bar{k}}^{(\lambda)} = \sum_{i}^{occ} \sum_{a}^{uno} \frac{\langle \bar{k} \parallel ia \rangle \mathcal{H}_{ia}}{\varepsilon_i - \varepsilon_a}. \tag{59}$$

With these matrices and vectors, Equation (57) can be now recast to

$$(\mathbf{G} - 4\mathbf{A} - 4\mathbf{B})\mathbf{x}^{(\lambda)} = 4\mathbf{b}^{(\lambda)} \Leftrightarrow \mathbf{x}^{(\lambda)} = \left(\frac{1}{4}\mathbf{G} - \mathbf{A} - \mathbf{B} \right)^{-1} \mathbf{b}^{(\lambda)}. \tag{60}$$

Thus, the calculation of the perturbed fitting coefficients is reduced to the solution of the above inhomogeneous equation system. In contrast to the traditional coupled-perturbed Kohn–Sham equation system [47,48], the dimension of Equation (60) is M^2 and, therefore, the memory requirement for the solution of this system of equations is similar to the corresponding SCF calculation. With the perturbed fitting coefficients, the perturbed Kohn–Sham matrix can be constructed via Equation (54) and the perturbed density matrix can then be calculated by Equation (48).

B. Experimental Techniques

In the following subsections, we review briefly the principles of selected experimental techniques that can be used to study clusters. An estimate of the accuracy of the experimentally obtainable data is given for each technique, and we describe which conclusions can be drawn from these data concerning the structural and electronic properties of clusters. We also discuss how results of quantum chemical calculations can assist experiments, for example by providing necessary information for the analysis of raw experimental data, verify experimental results, or complement and link different experimental techniques and findings.

(i) Polarizabilities

A good overview of experimental polarizability measurement techniques is given in Ref. [49]. Here, we will give a brief description of the most common experimental methods used for the polarizability measurements of clusters and molecules in the gas phase. The average polarizability can be accurately determined by measuring the dielectric constant ε. For this measurement, a capacitor is constructed in such a way that the volume between the plates can be filled with the material to be measured or completely evacuated. The dielectric constant is determined by taking the ratio between the vacuum capacitance C_0 and εC_0, the capacitance from the studied gas. The dielectric constant can be measured by making the capacitor part of a resonant circuit and observing the change in resonance frequency when material is introduced between the capacitor plates [50]. A more common way of measuring the dielectric constant is by using an alternating current bridge circuit to compare the unknown capacitance to a capacitance standard [51–54]. In these methods, an alternating current frequency in the range of 1–20 kHz is used. This frequency is small enough to obtain static polarizabilities. Measurements of the dielectric constant provide the most accurate experimental values for static or near-static polarizabilities of a gas. The refractive index is the ratio between the speed of light in vacuum and in a material. A measurement of the refractive index η of a substance provides information about the polarizability of the constituent molecules because of the relation

$$\eta = \sqrt{\varepsilon_r}. \tag{61}$$

Only one single refractive index can be measured for a gas, liquid, or amorphous solid if the molecules are not optically active, and if the sample is not in an applied electric or magnetic field. Accurate measurements of the refractive index of a gas can be made interferometrically. These measurements are useful for substances for which the capacitance measurement of the dielectric constant is impractical.

Rayleigh scattering is the nonresonant scattering of light by a polarizable particle much smaller than the light wavelength. The oscillating electric field of the incident light induces an oscillating dipole moment which reradiates light in a dipole pattern. For a spherically symmetric molecule and linearly polarized light, the induced dipole moment is parallel to the incident polarization and is proportional to the polarizability of the particle. If the molecule is not spherically symmetric, then the induced dipole moment depends on the orientation of the molecule, and the Rayleigh scattering causes depolarization of the incident light. The amount of Rayleigh scattering depends on both the mean polarizability $\bar{\alpha}$ and the anisotropy $|\Delta\alpha|$ of the polarizability tensor:

$$\bar{\alpha} = \frac{1}{3}\left(\alpha_{xx} + \alpha_{yy} + \alpha_{zz}\right), \tag{62}$$

$$|\Delta\alpha|^2 = \frac{1}{2}\left[\left(\alpha_{xx} - \alpha_{yy}\right)^2 + \left(\alpha_{xx} - \alpha_{zz}\right)^2 + \left(\alpha_{yy} - \alpha_{zz}\right)^2\right]. \tag{63}$$

Here α_{xx}, α_{yy}, and α_{zz}, are the principal polarizabilities of the molecule.

In the most common setup for observing Rayleigh scattering, a laser linearly polarized along **e** is used to illuminate a gaseous sample. Scattered radiation of polarization **q** is detected after it passes through a polarizer. The mean radiation rate per unit solid angle (the steradiance) for a single particle, averaged over molecular orientation, is

$$\langle S_0\rangle = I(2\pi v/c)^4 \frac{1}{5}\bar{\alpha}^2\left[3\kappa^2 + \cos^2\theta\left(5 + \kappa^2\right)\right] \text{with}\, \kappa = \frac{|\Delta\alpha|^2}{3\bar{\alpha}^2}, \tag{64}$$

where I is the incident intensity and θ is the angle between **e** and **q**. For a dilute gas, the scattered light from individual particles is incoherent and the steradiance from each particle can be added together. If the number of illuminated particles is known, it is possible to measure both $\bar{\alpha}^2$ and κ^2. The limitations of this classical expression are discussed in detail by Bridge and Buckingham [55]. The depolarization ρ_0 is defined as the ratio of the minimum to maximum steradiances in one direction. If the light is observed in a direction which is perpendicular to both the propagation direction and the polarization of the incident light, then the depolarization is

$$\rho_0 = \frac{3\kappa^2}{5 + 4\kappa^2}. \tag{65}$$

The depolarization is independent of the number density and determines $|\Delta\alpha|$ via κ^2 independent of $\bar{\alpha}$. Rayleigh scattering is most often used to determine the depolarization of a gas in a cell [55–57]. However, Rayleigh scattering from a beam of particles can be used to measure the polarizability, too [58].

An excellent way to determine the polarizability of a molecule is to observe the deflection of a collimated molecular beam which is passed through a static, inhomogeneous electric field. Generally, the electric field is produced using a two-wire deflecting field. The result is a relatively large region over which the product of the electric field and the field gradient is constant. Defining the field to be along **z**, the force on an atom in this field is

$$\mathbf{F} = \alpha_{zz}\mathcal{E}\frac{d\mathcal{E}}{dz}\mathbf{z}, \tag{66}$$

where **F** is the electric field force on a molecule, α_{zz} is the zz component of the polarizability tensor, and ε is the magnitude of the electric field. The spatial deflection of a molecule is approximately proportional to $1/v^2$, where v is the velocity of the molecule [49]. A typical setup for this experimental technique needs a source, that is an oven with a narrow rectangular orifice. The particles pass through a rotating-disk velocity selector in order to minimize errors associated with an imperfect knowledge of the velocity distribution. A slit placed before the deflecting field, in conjunction with the opening of the oven, limits the transverse velocity and position distribution of the beam. Deflection is measured by a hot wire detector. The oven and hot-wire detector are useful for measurements involving alkali metals. Other materials might require a different source and/or detector.

Dispersive Fourier transform spectroscopy (DFTS) is an interferometric technique used to measure the complex refractive index of gases and liquids, which can be related to the mean polarizability via the Lorentz–Lorentz equation [49].

$$\bar{\alpha} = \frac{3}{4\pi n}\left(\frac{\eta^2 - 1}{\eta^2 + 2}\right). \tag{67}$$

Here, n denotes the number density of the gas [59]. The ability of DFTS to determine simultaneously both the real and the imaginary parts of the refractive index (or polarizability) is its key feature. The accuracies of DFTS polarizabilities depend on the molecular species. A disadvantage of this technique is that it can only be applied to measure the refractive index and absorption coefficient of macroscopic quantities. Determinations with single molecules or beams are not possible.

Position-sensitive time-of-flight mass spectrometry is a very good technique for measuring static polarizabilities of clusters. It was successfully applied to Al_n clusters [60–62]. The method is based on the traditional approach of deflecting a neutral beam using a nonuniform electric field. The novel aspect of this technique is its detection scheme using a position-sensitive time-of-flight spectrometer and laser photoionization. The experimental setup is illustrated in Figure 1. A cluster source (1) based on laser vaporization is used to generate a cluster beam with clusters of up to several thousand atoms. The clusters traverse a long region with an inhomogeneous direct current field (2). Each cluster is deflected by an amount that is directly proportional to its polarizability and inversely proportional to its mass. After deflection, the clusters enter a long drift tube that allows the deflected clusters to further spread out in the transverse direction (3). At the end of the tube, the clusters enter a region where they are ionized by absorbing photons produced by an excimer laser (4).

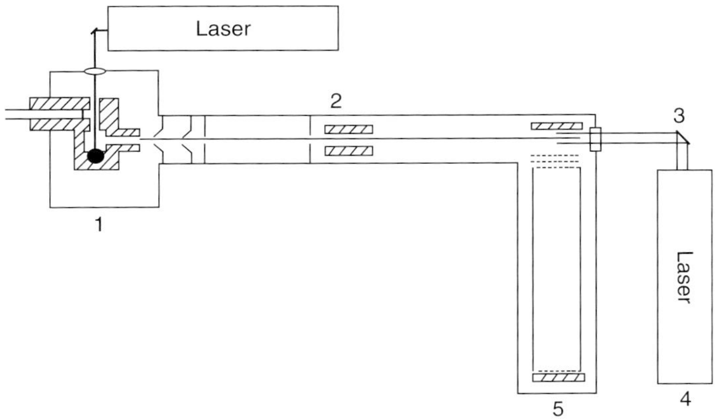

FIGURE 1 Position sensitive time-of-flight apparatus.

The ionization volume is adjusted in order to generate the maximal single-shot signal. After ionization occurs, an electric field is pulsed on to accelerate the ionized clusters orthogonal to the original drift direction. This electric field and two others are used to form a time-of-flight mass spectrometer. After another drift path, the cluster ions are detected by a Daly ion detector (5). It is important to understand how the position-of-flight spectrometer affects the cluster beam. The transverse spatial profile of the deflected beam in the ionization region is converted into a time profile by the time-of-flight spectrometer. In addition, the time-of-flight spectrometer plays its role of separating out different clusters in time according to their charge to mass ratio. The measurement consists determining the average shift in the signal due to a given cluster mass when the deflecting field is on compared to the signal when the deflecting field is off. For each beam pulse, this time shift is determined for all the different cluster masses. The shift in time Δt for a particular cluster mass can be related to the transverse deflection d of that cluster when the deflection field is on versus off. The deflection of a given cluster mass is given by

$$d = \left(\bar{\alpha} \mathcal{E} \frac{d\mathcal{E}}{dz} \right) \frac{K}{mv^2}, \tag{68}$$

where K is a geometric factor and v is the longitudinal velocity of the cluster. The quantity d is the displacement of an object of mass m on which acts the force

$$F = \bar{\alpha} \mathcal{E} \frac{d\mathcal{E}}{dz} \tag{69}$$

for a time $t = L/v$ where L is the length over which the force acts. The other factor of v in the denominator is due to the time the beam is allowed to propagate before the effects of the force are observed ($t_{obs} = D/v$) where D is the distance travelled from the interaction region to the point of observation. Due to the fact that it is very difficult to accurately determine K and $\mathcal{E}/\mathrm{d}\mathcal{E}/\mathrm{d}z$, these quantities are calibrated by deflecting lithium. In this way, the polarizability of a cluster A, α_A can be deduced by comparing its deflection to that of lithium, that is,

$$\frac{\alpha_A}{\alpha_{Li}} = \frac{d_A}{d_{Li}} \frac{m_A}{m_{Li}} \left(\frac{v_A}{v_{Li}}\right)^2. \tag{70}$$

The deflection of each cluster is determined by measuring the time shift in the centroid of the pulse corresponding to the arrival of clusters of a particular mass at the Daly ion detector. This technique offers several advantages that should be noted. It uses a universal detector that can be applied to any particle species that remains stable upon ionization. It is a position-sensitive measurement that uses differences in timing to deduce differences in position. In addition, the deflections of all the different cluster types in the beam are measured for each pulse. This has the advantage that systematic errors in the experiment are minimized. Moreover, the entire deflection profile for the beam is recorded for each shot, in contrast with deflection experiments where a slit is slowly scanned to map out the beam profile.

The experimentally measured average polarizabilities can be directly compared with their calculated counterparts. The experimental errors are usually too large to establish an unequivocal mapping to the calculated structures. However, the trend of the polarizabilities with cluster size can be well reproduced by theory in some cases. In deMon2k, the elements of the polarizability tensor are either calculated by a finite-field method [63,64] or analytically via the ADPT described in Section I.A.4. Using the ADPT, the perturbed density matrix the polarizability tensor elements are given by

$$\alpha_{uv} = \sum_{\mu,v} P_{\mu v}^{(u)} \langle \mu \mid u \mid v \rangle; u, v = x, y, z. \tag{71}$$

With the polarizability tensor elements, the mean polarizability and polarizability anisotropy can be calculated according to Equations (62) and (63).

(ii) Collision-Induced Dissociation

In a collision-induced dissociation (CID) experiment, a cluster ion A_n^+ is dissociated through collision with a chemically inert reactant R.

$$A_n^+ + R \rightarrow A_{n-1}^+ + A + R \tag{72}$$

The energy threshold E_0 of the endothermic reaction (72) corresponds to the bond dissociation energy (BDE) [65]:

$$D\left(A_{n-1}^+ - A\right) = E_0. \tag{73}$$

The cluster ions can be generated by laser vaporization followed by mass selection using a mass spectrometer. In most cases, cations are used, although anions can be generated and measured in CID experiments, too [65]. Before they enter the reaction zone, the ions are brought to a certain kinetic energy, for example through a supersonic expansion process or using a thermal flow tube [66]. The reactant R is usually a rare gas, preferably Xe [67]. The reactants and all products are collected after the collision and detected by a second mass spectrometer. The total reaction cross section σ_{tot} and the individual reaction cross sections of the products σ_{P} can be calculated from the measured intensity I_{R} and I_{P} of the reactants and products according to Equation (74).

$$I_{\mathrm{R}} = I_0 \mathrm{e}^{-\sigma_{\mathrm{tot}}\rho l}, \qquad \sigma_{\mathrm{P}} = \sigma_{\mathrm{tot}} \frac{I_{\mathrm{P}}}{\sum_{\mathrm{P}'} I_{\mathrm{P}'}}. \tag{74}$$

Here, $I_0 = I_{\mathrm{R}} + \sum_{\mathrm{P}} I_{\mathrm{P}}$ is the incident ion intensity and ρ and l are the gas density and the effective path length, respectively [68]. Values of σ_{P} for the products of interest are measured for different kinetic energies of the reacting species. No product will be observed until the kinetic energy of the reactants is increased to the energy threshold of the reaction. The data can be fitted to a mathematical expression (75) to yield the cross section σ as a function of the relative kinetic energy E of the particles.

$$\sigma(E) = \frac{\sigma_0}{E} \sum_i g_i (E + E_i - E_0)^N \tag{75}$$

The sum runs over all internal rovibronic states i with energies E_i and populations $g_i \left(\sum_i g_i = 1\right)$, which are obtained from a Maxwell–Boltzmann distribution. N and σ_0 are an adjustable parameter and an energy-independent scaling parameter, respectively, and E_0 is the energy threshold for the reaction at 0 K [65].

Equation (73) is valid only under the assumption that there occurs no activation barrier additional to the energy threshold E_0 of the dissociation reaction (72). For noncovalent bonds, this is in general the case [66,69]. However, a careful examination of the CID results and comparison with other experimental or theoretical data should be performed. Furthermore, it is assumed that the reaction probability (i.e., σ_0 and N) is equal for all vibrational states of the reactants. The internal energy E_i entering Equation (75) can be calculated from the vibrational frequencies of the clusters, and also the Rice–Ramsperger–Kassel–Marcus (RRKM) ansatz can be incorporated into Equation (75) to account for lifetime effects of slower reactions [65,66,70–72].

CID experiments provide bond dissociation energies of ions with an absolute accuracy of about 1–20 kcal/mol (in general about 5–15% relative error) [65,66] depending on the kind of molecules or clusters. However, further information can be obtained from the results of CID experiments on cluster ions, especially in combination with other experimental results [65,66].

If the ionization energies (IE) of A_n and A_{n-1} are known, the BDE of the neutral cluster can be calculated according to [65]

$$D(A_{n-1} - A) = D\left(A_{n-1}^{+} - A\right) + IE(A_n) - IE(A_{n-1}). \quad (76)$$

By measuring the BDE for clusters A_n^+ of different sizes n, relative stabilities of the clusters can be revealed. The reason for a higher stability of a certain size can be the electronic structure (closed electronic shells) or the geometric structure (compact or highly symmetrical structures). This information can be used as guidance, for example, for a theoretical structure investigation of these clusters. For more insight into the electronic structure, for example, of transition-metal clusters, their reaction with hydrogen (H_2) or deuterium (D_2) can be studied.

$$A_n^+ + D_2 \rightarrow A_nD^+ + D \quad (77)$$

The BDE $D(A_n^+ - D) = D(D_2) - E_0$, where E_0 is the energy threshold of reaction (77), can be compared with the BDE $D(A_n^+ - A)$. If both energies are similar, the bonding is mostly due to s-like orbitals: otherwise, d orbitals contribute considerably [65]. Further details and examples will be given in Section II.

Quantum chemical calculations can be valuable in combination with CID experiments in two ways. On the one hand, the obtained BDE and predictions about relative stabilities of different clusters can be tested against calculated BDE and relative stabilities from high level calculations. In cases where certain electronic and geometrical properties of clusters can be concluded from the CID experiments, these data can be used as starting points for more detailed theoretical studies in order to obtain refined structure information. Further properties, for example, polarizabilities, can then be calculated from the theoretical structures, which can be compared with results from other experimental techniques. In this way, quantum chemical simulations can act as a link between different experimental methods.

On the other hand, for the analysis of the raw data in CID experiments certain molecular information, such as frequencies or rotational constants, and assumptions (e.g., no reaction barrier) of the clusters are needed to obtain the final results. These data can be provided by quantum chemical calculations, too. In general, an estimate of these properties, for example of the frequencies, is sufficient. Therefore, less accurate theoretical methods can be employed and the exact geometric structures of the clusters need not be known.

(iii) Vibrational and Rotational Spectroscopy

Vibrational spectroscopy [73] can be used to obtain structural information for molecules and clusters. In analogy to classical mechanics, the force F between two atoms in a molecule can be described in the harmonic approximation by Hooke's law.

$$F = -k(r - r_0) \tag{78}$$

The factor k describes the strength of the bond between the atoms and is the quadratic force constant, and r_0 denotes the equilibrium distance of the atoms. The potential associated with Equation (78) is the harmonic potential (Equation (79)).

$$V(r) = \frac{1}{2}kr^2 \tag{79}$$

According to quantum mechanics, the energy of a molecule is quantized, that is, a molecule can only possess discrete vibrational energy states. By solving a one-dimensional Schrödinger equation with the harmonic potential (Equation (79)), one obtains a simple expression for the energy E_v and vibrational terms G_v associated with each state v of the normal mode ν.

$$E_v = h\nu\left(v + \frac{1}{2}\right); \quad v = 0, 1, 2, \ldots \tag{80}$$

$$G_v = \nu\left(v + \frac{1}{2}\right). \tag{81}$$

Here, v is the vibrational quantum number. With $v = 0$, the zero-point energy $E_0 = \frac{1}{2}h\nu$ of the vibrational ground state is obtained. The vibrational energy states of the harmonic potential are equidistant ($\Delta E = h\nu$). In general, the harmonic potential describes the lower vibrational states quite well, while it shows the wrong behavior (e.g., no dissociation limit) for $v \to \infty$. Therefore, anharmonic potentials such as the Morse potential function (Equation (82)) [74] can be used to obtain a more realistic picture. The Morse function yields energy levels with decreasing distance (Equation (83)) and a dissociation energy $D_0 = D - E_0$.

$$V(r) = D\{1 - \exp[-a(r - r_0)]\}^2, \tag{82}$$

$$E_v = h\nu\left(v + \frac{1}{2}\right) - h\nu x\left(v + \frac{1}{2}\right)^2, \tag{83}$$

$$G_v = \nu\left(v + \frac{1}{2}\right) - \nu x\left(v + \frac{1}{2}\right)^2. \tag{84}$$

Here, x is the anharmonicity constant. For polyatomic molecules, the expression for the vibrational term (Equation (84)) can be approximated in the following form [75]:

$$G = \sum_i v_i \left(v_i + \frac{d_i}{2} \right) + \sum_{i,j} x_{ij} \left(v_i + \frac{d_i}{2} \right) \left(v_j + \frac{d_j}{2} \right). \tag{85}$$

The indices i and j label different normal modes, and d_i and d_j are the degeneracies of the modes. The x_{ij} $(i \neq j)$ are the so-called off-diagonal anharmonicity constants. They are usually small for low vibrational levels and their contribution can be neglected. However, they can lead to a mixing of normal modes, indicating that the harmonic approximation is no longer valid in those cases.

The standard approach for the quantum chemical calculation of force constants, vibrational frequencies, and infrared (IR) or Raman intensities is based on the double-harmonic approximation [76]. In most cases, the DFT frequencies are in good agreement with the experimental data. For the intensities, only a rough qualitative match with experiment (strong, medium, weak) can be expected. The calculation of the anharmonicity constants x_{ij} involves the evaluation of the third derivatives of the potential, which is very time consuming and therefore usually avoided.

a. Infrared Vibrational Spectroscopy

A vibrational excitation $v' \leftarrow v''$ for a normal mode v occurs in IR spectroscopy by absorption of a photon with an energy $E = E_{v'} - E_{v''} = \Delta E = h\nu$. However, this absorption can take place only if the dipole moment of the molecule or cluster changes due to vibration in this mode [77]. The absorption probability, and therefore the intensity of the absorption peak in the IR - spectrum, is proportional to the transition moment $\mathbf{R}^{v'v''}$.

$$\mathbf{R}^{v'v''} = \int \psi^*_{v'} \boldsymbol{\mu} \psi_{v'} \, d\tau \tag{86}$$

$\psi_{v'}$ and $\psi_{v''}$ are the vibrational eigenfunctions of the two vibrational states and $\boldsymbol{\mu}$ is the molecular dipole moment. The transition moment (Equation (86)) vanishes except if $\Delta v = v' - v'' = \pm 1$ under the assumption of the harmonic potential (Equation (79)) [77]. Because the potential in real systems is not strictly harmonic, the selection rule $\Delta v = \pm 1$ is not rigorously valid and (usually weak) absorption with $|\Delta v| > 1$ can be observed.

Typical IR absorptions occur between 4000 and 100 cm^{-1} (IR spectrum) and in the range of 100–10 cm^{-1} (far-infrared (FIR) spectrum). The region 4000–14,000 cm^{-1} is called the near-infrared (NIR) region. Spectrometers have in general a resolution between 0.002 and 2 cm^{-1} [73] and allow in many cases to record the rotational fine structure of vibrational bands.

b. Raman Vibrational Spectroscopy

In Raman spectroscopy, a vibrational mode v is excited by an inelastic scattering process. The molecule is irradiated by monochromatic light $hv_0 \gg hv$ of the ultraviolet (UV), visible (VIS), or NIR region. The exciting radiation undergoes an inelastic scattering, leading to radiation with an energy E_R.

$$E_R = hv_0 \pm hv \tag{87}$$

This scattering is called Raman scattering. Additionally, the irradiating light is scattered elastically (Rayleigh scattering, see Section 1.B.1), leaving its energy unchanged. The lines occurring for $hv_0 - hv$ and $hv_0 + hv$ are called the Stokes and anti-Stokes lines, respectively. The Rayleigh scattering has the highest probability and therefore the highest intensity in the Raman spectrum. The intensity of the Stokes line is larger than that of the anti-Stokes line, because the vibrational ground state is more populated than the excited states. Raman scattering can be observed only for normal modes that change the polarizability tensor $\boldsymbol{\alpha}$ of a molecule. In analogy to the transition moment (Equation (86)), a scattering moment $\mathbf{P}^{v'v''}$ can be used in order to describe the transition probability for Raman scattering [77].

$$\mathbf{P}^{v'v''} \sim \int \psi_{v'}^{*} \boldsymbol{\alpha} \psi_{v'} \, \mathrm{d}\tau \tag{88}$$

The scattering moment (Equation (88)) is different from zero only if $\boldsymbol{\alpha}$ changes due to the vibration and if $\Delta v = \pm 1$, leading to the same vibrational selection rule as for IR spectroscopy. Like in the IR spectrum, a rotational fine structure can be observed in the Raman spectrum, too. However, the rotational selection rule is $\Delta J = 0, \pm 2$ here, in contrast to IR spectroscopy [77]. IR and Raman spectroscopy are complementary techniques, since IR-active normal modes are usually Raman-forbidden and vice versa.

c. Microwave Rotational Spectroscopy

Rotational spectroscopy deals with the absorption or emission of an electromagnetic radiation typically in the microwave region of the electromagnetic spectrum. The FIR region lying adjacent to the microwave region may be used for rotational spectra as well. The microwave spectrum of a molecule is associated with a corresponding change in the rotational quantum number in the molecule [78]. The analysis of a rotational spectrum yields the elements of the inertia tensor ($\mathbf{I}$), which are directly related to the molecular rotational levels or the molecular angular momentum.

A transition between two rotational levels of a molecule is determined by the symmetry of the molecule. For convenience, molecules are classified into four different types according to their symmetry and rotational behavior: linear molecules, symmetric tops (C_6H_6, NH_3), spherical tops (P_4, CCl_4, NH_4^+), and asymmetric tops (H_2O, NO_2). The structure of a rotational spectrum is then analyzed adopting the rigid rotor (rotating top) model. The simplest example is a diatomic molecule that, in the rigid-rotor model, is characterized by two equal eigenvalues of the momentum of inertia tensor and the third one equal to zero. The rotational energy levels are obtained from the Schrödinger equation, with the Hamiltonian expressed in terms of the angular momentum operator $\hat{L}$:

$$H\Psi = \frac{\hat{\mathrm{L}}^2}{2I}\Psi. \tag{89}$$

Solving the above equation gives the rotational energy levels E_J:

$$E_J = J(J+1)\frac{h^2}{2I} = J(J+1)B, \quad J = 0, 1, 2, \ldots, \tag{90}$$

where B is the rotational constant. Determination of the spectroscopic constant B allows one to analyze the rotational spectra in terms of B, hence giving the I values. It is then straightforward to compute the bond lengths in diatomic molecules, for example. For the general molecules, the analysis of the rotational spectrum becomes more complicated as the rotational energy is a function of the three nonequivalent principal axes of the momentum of inertia.

To facilitate the analysis of complex microwave spectra, isotopic substitutions are often used. The substitution of an atom in a molecule with its isotope does not change the corresponding bond lengths but changes the momentum of inertia, providing sufficient information to resolve the molecular structures. Microwave spectroscopy is considered to be the most accurate method giving information about the structure of a molecule in the gas phase.

The transitions between two rotational levels are active in the rotational spectrum if $\Delta J = \pm 1$ [79]. Rotational spectroscopy is applicable only to the gas phase. In the case of solids or liquids, the rotational motions are quenched as a result of collisions. In addition to the rotational structures, further details are usually present due to the rotational motion of the molecules, giving rise to magnetic and electrostatic interactions in the molecule.

The electronic structure of a cluster, in a given electronic state, depends on its vibrational and rotational degrees of freedom. This dependence explains the interest in measuring electric dipole moments for various vibrational and rotational quantum numbers. Accurate dipole moment measurements for specific rovibrational levels are performed via the Stark effect. This effect designates any change induced in a molecular spectrum by the presence of an external electric field. Coupling this effect with microwave and FIR spectroscopy aids the assignment of the observed transitions to rovibronic symmetries and enables the evaluation of the dipole moment of molecules and clusters in their electronic ground state and in their ground and excited vibrational states [78,80]. Various refinements of microwave spectroscopy settings have appeared in the last ten years, enabling high resolution and allowing the study of weakly bound complexes. In particular, molecular beam electric resonance [81] and Fourier transform microwave [82,83] spectroscopy are of interest to us. The former method provides the components of the dipole moment, whereas spectra with detailed nuclear–quadrupole and nuclear spin–spin fine structures can be obtained by both techniques. The development of microwave-optical double resonance experiments allows one to study microwave transitions in excited vibrational states of the electronic ground state and in excited electronic states. Detection of the rovibrationally excited states can be done with laser-induced fluorescence [84–86], multiphoton ionization [87,88], mass spectrometry [89–91], and zero-kinetic-energy (ZEKE) spectroscopy [92,93].

The techniques designed to observe rotational and vibrational spectra provide valuable information about structures and are well suited for the determination of intermolecular separations in weakly bonded systems. Assignment of the IR, far IR, or microwave spectra is strongly aided by quantum mechanical (QM) methods. These spectroscopies have provided very detailed and accurate data for small systems such as dimers, trimers, and small complexes. These data are used as benchmarks for the most sophisticated QM methods, too.

(iv) Photoelectron Spectroscopy

The photon energy of the radiation source used in photoelectron spectroscopy (PES) exceeds the ionization energy of a particle A. As a result, the following reaction is observed:

$$\mathrm{A} + h\nu \rightarrow \mathrm{A}^{+} + e^{-}. \tag{91}$$

Here, A as well as A^{+} can be a neutral or a charged species. The superscript "+" just indicates that A^{+} has one electron less than A.

The kinetic energy T_{e} of the released photoelectron corresponds to the excess energy (Equation (92)), because the change in the kinetic energy of the ion A^{+} can be neglected because of the large difference in mass between the ion and the electron.

$$T_{\mathrm{e}} = h\nu - \mathrm{IP}_i \tag{92}$$

IP_i denotes the ionization energy of the ith electron. Weakly bound electrons will possess a higher kinetic energy, and strongly bound electrons smaller ones. The photoelectrons pass through an analyzer and are collected according to their kinetic energy. The ionization energies IP_i and therefore the energy of the bands in the photoelectron spectrum can be calculated according to Equation (92).

If the particle A is neutral, then the principal bands in the photoelectron spectrum correspond to the different ionization potentials of A. If, on the other hand, A is an anion, the bands provide the (negative) electron affinities of the corresponding neutral species [94]. Furthermore, information about the electronic state of a particle might be obtained. For example, if A^+ is a linear cation with a spin quantum number $S = 1/2$ (doublet), and assuming that the coupling between the angular momentum and the spin is rather weak (Russell–Saunders type of coupling) [77], the spin–orbit coupling can lead to a doublet splitting of the corresponding band in the photoelectron spectrum, indicating a nonzero angular momentum number $L \neq 0$ of A^+, that is, a $^2\Lambda$ state ($\Lambda = \Pi, \Delta, \ldots$). From the absence of such a splitting (and given a sufficient resolution of the spectrum), a $^2\Sigma$ state can be assumed for A^+. In PES, the total angular momentum number J can change by more than $\Delta J = J^+ - J = \pm 1$ in contrast to electronic transitions between two bound states. Furthermore, ΔJ can be a half-integer [95]. Information about the molecular structure of the particle can be obtained from the vibrational fine structure of the principal bands in the photoelectron spectrum. Let us consider first the case of a diatomic particle having one (totally symmetric) normal mode. When an electron is ejected during the photoionization process (Equation (91)), a part of the excess energy can lead to vibronic excitations. Under the experimental conditions in PES, the particle A is in general in its vibronic ground state. Therefore, the resulting vibronic structure is caused mainly by $v_{A^+} \leftarrow 0_A$ excitations, where v_{A^+} and 0_A are quantum numbers of vibrational states of A^+ and of the vibrational ground state of A, respectively. For the analysis of the vibronic structure, the Franck–Condon (FC) principle [96–98] can be used. The basic assumption of this principle is that the electronic excitation is much faster than the movement of the nuclei, in accordance with the Born–Oppenheimer approximation [99]. Therefore, the position of the nuclei before and right after the excitation will be practically identical. The intensity of the band for a vibronic excitation $v_{A^+} \leftarrow 0_A$ is proportional to the overlap integral (Equation (93)), which is the so-called FC factor [77].

$$\left(S_{v,0}^{\mathrm{FC}}\right)^2 = \left[\int \psi_v^{A^+} \psi_0^{A}\, \mathrm{d}\tau\right]^2 \tag{93}$$

${\psi_v}^{A^+}$ denotes the vibrational eigenfunction of A^+ with the vibrational quantum number v, and ${\psi_0}^{A}$ is the corresponding function for A. Note, that the molecular dipole moment does not occur in Equation (93) in contrast to the transition moment in Equation (86). Since ${\psi_v}^{A^+}$ and ${\psi_0}^{A}$ belong to different electronic states and the underlying potentials are not exactly harmonic, the overlap integral does not vanish in general for $v_{A^+} \neq 0$. Two principal cases can be distinguished for the prediction of vibronic intensity patterns using the FC principle [94]. In the first case, the geometry of A^+ is very similar to that of A. The maximum overlap S^{FC} and therefore the highest intensity will be expected to occur for the $0_{A^+} \leftarrow 0_A$ excitation ($\Delta v = 0$). The intensities of excitations $\Delta v = 1, 2, \ldots$ will drop rapidly with increasing Δv, and the adiabatic excitation ($\Delta v = 0$) and the vertical excitation (excitation with the highest intensity) are the same. (Note that, in case A is not in its vibronic ground state, in general excitations $\Delta v = 0, \pm 1, \pm 2, \ldots$ can occur.) In the second case, the geometries of A^+ and A are different. Now the adiabatic excitation $\Delta v = 0$ is not equal to the vertical excitation and in general a vibrational progression will be expected to occur in the spectrum. Such a progression can be used, for example, in combination with the anharmonic Morse potential [74] to obtain normal modes, anharmonicity constants, dissociation energies, and equilibrium structures of molecules [94]. This approach is in general straightforward for diatomic molecules.

For polyatomic molecules, not all excitations $\Delta v = 1,2, \ldots$ might be allowed, depending on the symmetry of the corresponding normal modes and vibrational eigenfunctions [75]. The general selection rule is that the integrand in the FC factor (Equation (93)) must be totally symmetric (and therefore, only normal modes of the same kind can combine with each other); otherwise the excitation is forbidden. If a normal mode is totally symmetric, then the intergrand will also be totally symmetric for all $\Delta v = 0,1,2, \ldots$. On the other hand, if a normal mode is not totally symmetric, only excitations with even quantum number changes are allowed ($\Delta v = 0,2,4, \ldots$). Depending on how different the geometries of A and A^+ are, either the $\Delta v = 0$ transitions will dominate (similar geometries) or a vibrational progression occurs in the spectrum (different geometries) as outlined above. However, even in the latter case, experience shows that the $\Delta v = 0$ transition is in general the strongest. Vibrations can mix over the off-diagonal anharmonic constants, as described in Equation (85), if normal modes of the same symmetry are energetically close to each other (Fermi resonance) [94,100]. This indicates a breakdown of the harmonic oscillator approximation. Therefore, the resulting photoelectron spectrum can become quite complicated in this case and the interpretation of the spectrum can be cumbersome or even impossible without further support from other experiments and/or more sophisticated theoretical models.

Nowadays, photoelectron spectra have a resolution of typically 10–30 meV (80–240 cm^{-1}) [94], although in some cases even a better resolution can be obtained [101,102]. While this is sometimes already the limit for resolving the vibrational structure, it is in general insufficient to obtain reliable rotational information, which requires a resolution in the range of 0.5 meV or less. The reason for the limited resolution in PES is the difficulty in measuring the kinetic energy of the photoelectrons [101,103]. Other effects such as the Doppler broadening of bands (due to the speed distribution of the particles) and pressure broadening (life time of excited states is shortend due to collision) play minor roles. Their influence can be diminished, for example, by the use of supersonic jets [94,100]. The so-called natural broadening of the bands, which is due to the limited life time of excited states, leads in general to FWHM (full-width at half-maximum) of the order of 10^{-6}–10^{-4} cm^{-1} and can be neglected here.

a. Zero-Kinetic-Energy Pulsed-Field-Ionization (ZEKE-PFI) Spectroscopy

Zero-kinetic-energy (ZEKE) photoelectron spectroscopy [101,103–106] is based on the principles of photoelectron spectroscopy. It overcomes the limitations in resolution of PES, which are mainly due to the difficulties of measuring the kinetic energy of the photoelectrons with high accuracy as stated in the previous section. Instead of using a radiation with a fixed wavelength for the simultaneous ionization of electrons with different ionization thresholds and measuring subsequently their kinetic energy, a tunable radiation source is used to produce threshold photoelectrons (TPE), that is, electrons with zero kinetic energy. In this way, the different ionization potentials of a particle can be scanned step by step according to the wavelength of the ionizing radiation. By providing the possibility to collect the TPE efficiently and to discriminate them from so-called "hot electrons," that is, electrons with some kinetic energy due to ionization from lower ionization thresholds, the resolution achievable with this technique depends in the first place on the bandwidth of the radiation source.

In contrast to the threshold photoelectron spectroscopy (TPES) [107] where electrons are generated with energies slightly above the corresponding ionization energy, ZEKE-PFI spectroscopy uses electrons with slightly less energy [101,103]. An electron is excited by a short laser pulse (in the range of picoseconds to nanoseconds) with an energy just below the corresponding ionization potential. In the resulting state, the electron occupies an orbital with a high principal quantum number η. Such a state is called a Rydberg state and lies energetically only a few cm^{-1} below the ionization threshold. In fact, for each rovibronic energy level of the corresponding ionized species A^+, there is a series of Rydberg states with decreasing energy distance, eventually converging to the ionization threshold of that level. After a short time delay, usually between a few hundred nanoseconds and a few microseconds, a pulsed electric field is applied to finally ionize the particle (pulsed field ionization) and to create the ZEKE electrons.

The time delay between the laser pulse and the field pulse serves two purposes. First, it allows the separation of the ZEKE electrons from the hot electrons, since the latter ones are leaving the ionization zone before the ZEKE electrons are produced. Furthermore, it enhances the resolution of the corresponding ZEKE spectrum because it allows autoionization of short-lived, energetically low-lying Rydberg states ($n < 150$) leaving only higher Rydberg states ($n > 150$) for the PFI. These high-n Rydberg states, also called core-non-penetrating Rydberg states [104] or "magic" Rydberg states [103], have unusually long life times τ mainly due to the mixing of the orbital angular momentum quantum number l (Stark mixing) and its projection m_l [108]. The mixing is caused by electric fields present in the ionization zone, which originate, for example, from electrons produced in autoionization processes.

Another effect of these fields is that they accelerate the decay of the low-lying Rydberg states. To obtain resolutions of about 0.2 cm^{-1}, the ZEKE-PFI technique can be combined with an additional electric field pulse [103]. If the field strength of the pulse is slowly increased, for example in a stepwise form [104], only a part of the high-n Rydberg states is ionized in each PFI. The ejected electrons can be discriminated by their time of flight (TOF). Further refinements can be achieved by applying a pre-pulse with a field of opposite direction [109].

The same selection rules for the total angular momentum J and the vibrational quantum number v apply as discussed in the previous section for the PES. Additionally, the rotational fine structure can be resolved. Detailed rovibronic symmetry selection rules for photoionization processes of polyatomic molecules can be found in the literature [95]. They will be discussed where applicable in the corresponding examples in Section II.

For the theoretical simulation of ZEKE spectra, a simple FC model, assuming that the neutral and ionic potential are both harmonic, can be used to calculate the multidimensional FC factors. Although for some molecules there is evidence that coupling between highly excited Rydberg states has an important effect on ZEKE intensities, for many molecules a simple FC model appears to yield reasonably accurate intensities. For the spectral simulation, we can take into account the fact that the normal coordinates, frequencies, and molecular structures are different for the neutral and cationic cluster. The normal coordinates $\mathbf{q}$ of the neutral and $\mathbf{q}'$ of the ion are related by

$$\mathbf{q}' = \mathbf{S}\mathbf{q} + \mathbf{Q}, \tag{94}$$

where $\mathbf{S}$ is an orthonormal matrix representing the rotation of the normal coordinate $\mathbf{q}'$ with respect to $\mathbf{q}$, and $\mathbf{Q}$ is a displacement vector. The recursion relations of Doktorov et al. [110,111] are used to calculate the multidimensional FC factors [112]. With some approximations (see, for example, Refs. [113–116] for details), it is possible to obtain an excellent cold spectrum of the FC structure using displacement parameters expressed in terms of the final state (cation for ZEKE and neutral for the PES spectroscopy) [117,118]. In this case, the intensity of transitions between two different vibrational states is due to the displacement of the two electronic surfaces. This treatment reduces the FC integral to a product of one-dimensional integrals.

II. STRUCTURE DETERMINATION BY COMBINING EXPERIMENT AND THEORY

In the second part of this chapter, we describe how experimental and theoretical results can be combined to obtain information about the geometrical and electronic structure of clusters. For each experimental technique described in the previous section, results for example systems are discussed in combination with data obtained from calculations. Some of the systems occur more than once to show how information from different experiments can be combined and bridged by quantum chemical simulations to gain finally reliable insight into the structural and electronic properties of clusters.

A. Polarizabilities

The experimental study of sodium cluster polarizabilities [119–121] reveals a strong size dependence, which was also observed in almost all accompanying theoretical studies [122–130]. In Figure 2, the experimental and theoretical polarizabilities of sodium clusters with less than ten atoms are compared. The data are normalized to the number of atoms in the cluster. In particular, we present Vosko, Wilk, and Nusair (VWN) [131] (—×—) and Perdew, Burke, Ernzerhof (PBE) [132] (—+—) polarizabilities by employing a DFT-optimized triple-zeta valence polarization (TZVP) basis set [133] augmented with field-induced polarization (FIP) functions [126]. The cluster structures were optimized with the PBE functional and are topologically identical to those published in Ref. [126].

The qualitative trend of the experimental polarizabilities is well reproduced at both levels of theory. The comparison of the VWN and PBE results show that gradient corrections reduce the deviation between theoretical and experimental sodium cluster polarizabilities. Nevertheless, the theoretical polarizabilities are in general too small. Similar results are found with other DFT and correlated wave-function methods [130]. Due to this situation, it is

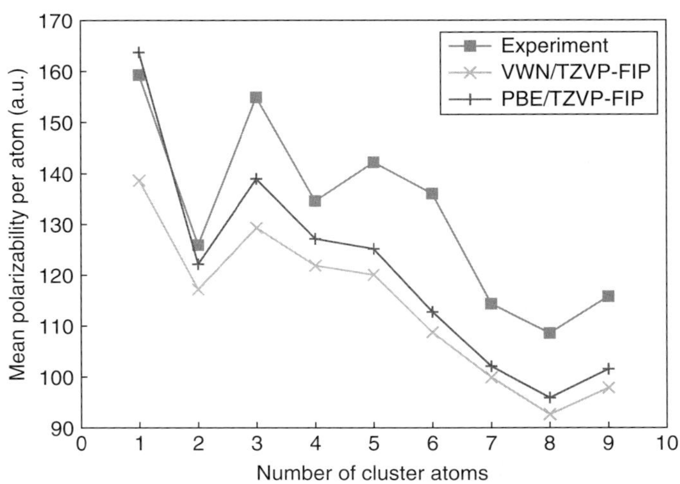

FIGURE 2 Correlation between calculated and experimental polarizability per atom of sodium clusters.

speculated that temperature effects are not negligible for the theoretical description of sodium cluster polarizabilities.

The electronic configuration of the noble metals Cu, Ag, and Au is characterized by a closed d shell and a single valence electron and, therefore, is closely related to that of the alkali metals. In view of this prominent characteristic, clusters of noble metals are expected to exhibit certain similarities to alkali-metal clusters. Only very recently static electric polarizabilities of Cu_n clusters with $n \geq 9$ have been measured [134]. Therefore, the polarizabilities of smaller noble-metal clusters obtained with theoretical methods are very valuable. Furthermore, the comparison of reliable theoretical polarizabilities of noble metal clusters with those obtained experimentally for alkali-metal clusters can give insight into the electronic structure of these systems. For the calculation of the copper cluster polarizabilities, a newly developed copper FIP basis set for density functional calculations [135] in combination with the generalized gradient approximation for the exchange by Becke [136] and the correlation by Perdew (B88-P86) [137] was employed. The copper cluster structures were optimized at the local level of theory. They are reported in Ref. [138].

In Figure 3A, the mean polarizability per atom of copper clusters is plotted. For comparison the experimental polarizability per atom of sodium and lithium, clusters up to the nonamer are shown in Figure 3B and C, respectively. The values of the experimental polarizability per atom of sodium clusters are those reported by Knight et al. [119], while the experimental polarizability data of lithium clusters are those reported by Benichou et al. [139].

The experimental work of Knight et al. [119] by electric deflection techniques has shown that the optical properties of alkali-metal aggregates, such as sodium and potassium clusters, follow a general trend toward the bulk value. Knight et al. have also shown the existence of a pronounced size dependence in the polarizability of small clusters. In fact, for the sodium aggregates, from the atom to the pentamer, the polarizability per atom has an oscillating behavior. After the pentamer, it decreases to a minimum for the octamer, and then increases again for the nonamer (Figure 3B). Potassium cluster polarizabilities follow closely the same pattern [119]. More recently, Benichou et al. [139] have measured static electric polarizabilities of lithium clusters up to 22 atoms by deflecting a well-collimated beam through a static, inhomogeneous transverse electric field. In order to avoid any systematic error in their experiment, Benichou et al. carried out also measurements of sodium cluster polarizabilities. The values they have obtained were in close agreement with those previously measured by Knight. The work of Benichou et al. shows that the trend of the polarizability per atom of small lithium clusters differs from those of small sodium and potassium clusters. In fact, a sharp decrease by about a factor of two from the monomer to the trimer is observed. For larger sizes, $n \geq 4$, the polarizability per atom decreases slowly (Figure 3C).

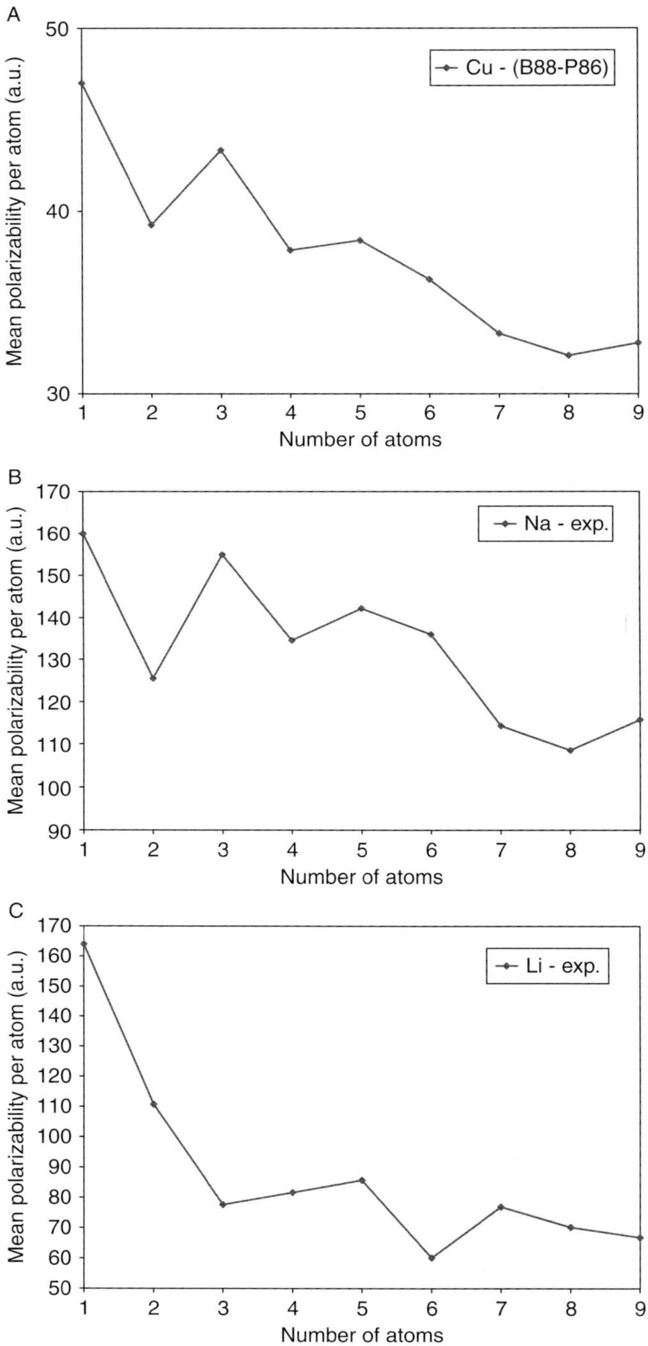

FIGURE 3 Comparison of the calculated polarizability per atom of copper clusters (A) with the corresponding experimental data for sodium (B) and lithium (C) clusters.

As Figure 3 shows, going from the atom up to the pentamer, the mean polarizability per atom of small copper clusters has an oscillating behavior. From the pentamer, the polarizability per atom decreases, with a minimum value for the octamer. It increases then again, going from the octamer to the nonamer. The comparison of Figure 3A–C shows that the calculated polarizability per atom of copper clusters, going from the atom to the nonamer, presents the same trend as experimentally observed by Knight for sodium. However, the mean polarizability per atom of copper clusters is about three times smaller. This result indicates that in copper clusters the electrons are more strongly attracted by the nuclei than in the sodium clusters. Therefore, their electronic structure is more compact.

We are confident that the reliability of the absolute values for the calculated polarizabilities of the copper clusters is in the same range as for the previous study we have carried out on small sodium clusters. Thus, deviations of less than 10% for the absolute values can be expected. The theoretical prediction of the trend of the mean polarizabilities per copper atom is believed to be reliable, too. However, the comparison of the calculated static polarizability of Cu_9 with the reported experimental value [134] is not very satisfying.

Other first-principles investigations of the polarizability of small and intermediate-sized copper clusters were also carried out later on [140,141]. They reveal similar discrepancies to the reported experimental data.

Polarizabilities give insight into the electronic structure of clusters. In the case of the alkali-metal clusters, the closing of the shell structure is found. This trend is reflected by the calculated polarizabilities.

B. Collision-Induced Dissociation

In order to compare the results of CID experiments with results from quantum chemical studies, calculations on a relatively high theoretical level have to be performed in general. Su et al. [142] have investigated cationic vanadium clusters V_n^+ with $n = 2$–20 by measurements of the CID thresholds. The BDEs of the cations were obtained. By combining the CID results with values for the ionization energies (IE) from the literature [143,144], they also provided BDEs of the corresponding neutral clusters. As a result, they found alternating BDEs for the neutral and the cationic clusters for $n = 2$–10, where clusters with an odd number of atoms are less strongly bound than the neighboring clusters with an even number of atoms. The BDEs for the neutral and cationic dimers and trimers are $D(\mathrm{V{-}V}) = 2.74 \pm 0.29$ eV and $D(\mathrm{V^+{-}V}) = 3.13 \pm 0.14$ eV and $D(\mathrm{V_2{-}V}) = 1.42 \pm 0.10$ eV, and $D(\mathrm{V_2^+{-}V}) = 2.27 \pm 0.09$ eV, respectively. The result for $D(\mathrm{V{-}V})$ agrees very well with the value of 2.752 ± 0.002 eV obtained from a resonant two-photon ionization spectrum of V_2 [145,146], while the value for the vanadium dimer cation could be confirmed by measuring the threshold for the photodissociation of V_2^+ ($D(\mathrm{V^+{-}V}) = 3.143 \pm 0.003$ eV) [142].

Several theoretical studies have dealt with small and medium-sized vanadium clusters [114,147–150]. Grönbeck and Rosén [149] performed DFT calculations for vanadium clusters V_n and V_n^+ with $n = 2$–9 by employing a triple-zeta basis set with polarization functions using the local spin density (LSD) approximation proposed by Vosko, Wilk, and Nusair (VWN) [131], with and without gradient corrections (GC) according to Becke [136] and Perdew [137]. The Coulomb potential was evaluated via a least-squares fit procedure. While the BDEs obtained with the gradient correction showed differences in the corresponding experimental CID values of less than ± 0.4 eV, the VWN results were found to be up to more than 1 eV too high. The alternating trend of the BDEs was only reproduced for $n \leq 3$. For the neutral dimer, Grönbeck and Rosén obtained a bond dissociation energy of $D(V-V) = 3.11$ eV, which is 0.36 eV higher than the CID value. However, they found a triplet $3d^4 4s^1$ state as the ground state of the vanadium atom in contrast to the experimental $3d^3 4s^2$ triplet ground state, leading to a wrong atomic reference. This was found earlier by Salahub and Baykara [148] using the local JMW [151] exchange-correlation potential. A reason for this behavior is the inadequately described correlation of the s^2 shell [148,152]. As shown by Wu and Ray [150] and by Calaminici et al. [114], the relative energies of the electronic states of the vanadium atom are very sensitive to the basis set. Wu and Ray [150] obtained a BDE of $D(V-V) = 2.91$ eV using the Becke (B88) [136] exchange and the Perdew–Wang (PW91) [153] correlation approximation together with the 6-311G basis set [154]. They found the correct atomic ground state, and their calculations reproduced the oscillating behavior of the CID BDEs for $n \leq 9$. However, all their BDEs were still overestimated by up to several tenths of an electronvolt.

Calaminici et al. have optimized a double zeta plus valence polarization (DZVP) basis set for use with the generalized gradient approximation (DZVP-GGA) for all 3d and 4d transition metals [114,155] with the AllChem [156] and deMon2k [7] LCGTO-KS DFT programs. They used this basis set together with the GGA of Perdew and Wang (PW86) [157] and Perdew (P86) [137,158] and the A2 auxiliary function set [133] for the variational fitting of the Coulomb potential [22,23] and obtain the correct electronic ground state for the vanadium atom and a BDE for the neutral and cationic dimer of $D(V-V) = 2.77$ eV and $D(V^+-V) = 3.24$ eV, respectively, in very good agreement with the experimental CID values. The calculations were performed first with the BASIS approach described in Section I.A.2 [114] and confirmed later using the ADFT method (AUXIS approach, Section I.A.3) [155]. The optimized V–V bond length using the PW86-P86/DZVP-GGA method was found to be 1.802 Å, while the VWN/DZVP-GGA-optimized bond length had a value of 1.760 Å. Assuming the typical bracketing of the experimental length by the too small VWN and too large GGA value, a bond length of 1.78 ± 0.02 Å can be assigned to V_2 from the results of the calculations. For the neutral dimer, a $^3\Sigma_g$ ground state was found, and for the cationic one a $^4\Sigma_g$ ground state.

The calculated BDEs for the neutral trimer (1.59 eV) and cationic trimer (2.10 eV) are in good agreement with the CID values. Both structures were found to be equilateral triangles with V–V bond lengths of 2.169 and 2.180 Å, respectively. As electronic ground states, $^2A_1'$ and $^3A_2$ states were found. The very good agreement of the calculated BDEs with the experimental CID values indicates that the calculated electronic and geometric structures are reasonably accurate. Table 1 summarizes the different theoretical and experimental results for the neutral vanadium dimer. As can be seen, the BDEs alone are not a sufficient indicator of the accuracy of the electronic and geometric structure obtained with quantum chemical methods since, although the BDEs are quite different for the different methods, the bond lengths, for example, are quite similar. Therefore, in order to verify the structures, further experiments are necessary as will be shown in Section II.D.

Small neutral and charged copper clusters Cu_n with $n \leq 5$ were investigated by Calaminici et al. [159] using the the local VWN [131] and the gradient-corrected PW86-P86 [137,157,158] exchange-correlation contributions in combination with the DZVP basis set and the A2 auxiliary function set [133] with the LCGTO-KS DFT program deMon-KS [34,160–162]. The structures have been fully optimized from different starting geometries by means of the Broyden–Fletcher–Goldfarb–Shanno (BFGS) algorithm [163]. For the most stable structures, vibrational analyses have been

TABLE 1 Results for the Neutral Vanadium Dimer V2: Bond Length $r_{V–V}$/Å and Bond Dissociation Energy D(V–V)/eV

Method	$r_{V–V}$	D(V–V)
CAS SCF[a]	1.77	0.33
LSD[b]	1.75	1.71–3.85
LSD[c]	1.77	4.40
LSD-GC[c]	1.78	3.11
BPW91/6-311G[d]	1.73	2.91
PW86-P86/DZVP-GGA[e]	1.77	2.77
CID[f]	–	2.74 ± 0.29
Two-photon ionization spectrum[g]	–	2.752 ± 0.002

[a]*Ref. [147].*
[b]*Ref. [148].*
[c]*Ref. [149].*
[d]*Ref. [150].*
[e]*Ref. [114].*
[f]*Ref. [142].*
[g]*Refs. [145,146].*

performed in order to identify these structures as true minima on the potential energy surface (PES). Later, this study was extended to neutral and charged copper clusters with up to 10 atoms [138,164] using the same level of theory with the LCGTO-KS DFT program ALLCHEM [156].

A planar structure was found for all neutral Cu_n clusters with $n \leq 6$; in the case of the cationic and anionic clusters, only clusters were planar with up to 4 and 5 atoms, respectively. [138]. The binding energies were calculated and compared with the results from CID experiments of cationic [165] and anionic [166] clusters using electron affinities or threshold detachment energies [167] and ionization potentials [168–171] from the literature. The calculated binding energies per atom for the neutral clusters are shown, together with the corresponding values from CID experiments of anions (CIDA) and cations (CIDC), in Figure 4. The calculated values have been corrected for basis-set superposition errors [172] and include the calculated zero-point energies for the clusters. They are in good agreement with the CID experiments of anionic clusters; all values are within the experimental error bounds [138]. Due to large uncertainties in the ionization potentials, the theoretical values are also within the experimental errors for the cationic clusters. However, the latter are considerably smaller than both the theoretical and anionic CID binding energies. Furthermore, separation energies and fragmentation channels for cationic clusters have been investigated [138], and the results of calculations can be compared with CID experiments [173]. The separation energies $D(n^+, m^0)$ for a cationic cluster with n atoms under cleavage of a monomer ($m^0 = 1$) or dimer ($m^0 = 2$) have been calculated from the corresponding binding energies E_B as follows:

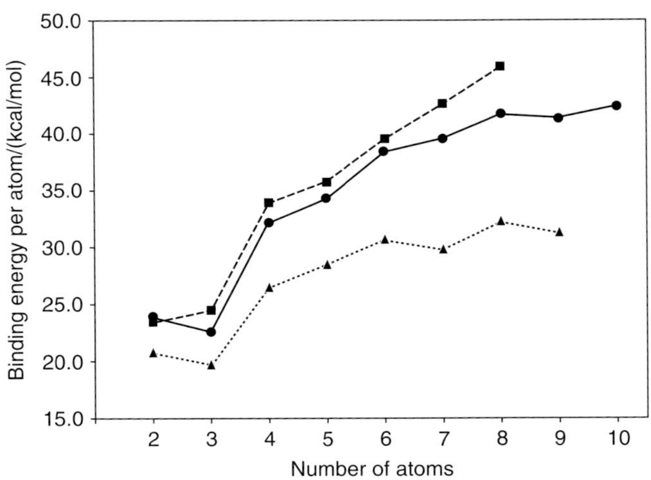

FIGURE 4 Comparison of the calculated binding energies per atom for the neutral clusters (●) with values from CID experiments of anions (■) and cations (▲). All values are in kcal/mol.

$$D(n^+, 1) = E_B(n^+) - E_B([n-1]^+), \quad (95)$$

$$D(n^+, 2) = E_B(n^+) - E_B([n-2]^+) - E_B(2) \quad (96)$$

The calculated separation energies for monomer and dimer cleavage are shown in Figure 5, together with the error bounds of the experimental values [173]. The even-numbered copper cation clusters show a lower separation energy for monomer cleavage than for dimer fragmentation. This preference for monomer cleavage is in line with the experimental results [173]. The case for the odd-numbered clusters is less satisfactory, however. While the CID experiments find monomer cleavage for the Cu_9^+ cluster, monomer and dimer cleavage for the Cu_7^+ cluster, and a preference for dimer cleavage for the Cu_3^+ and Cu_5^+ clusters, the theoretical results agree only in the case of the Cu_5^+ cluster [138]. From calculations, the energy difference for monomer and dimer cleavage for Cu_3^+ is only 0.05 eV. Therefore, both types of fragmentation could take place. For the Cu_9^+ cluster, the theoretical results clearly suggest a dimer fragmentation (Figure 5), while monomer cleavage is predicted for the Cu_7^+ cluster. The discrepancies between theoretical and experimental results can be due to the fact that the experiments deal with highly excited clusters, where consecutive reactions cannot be excluded, whereas the calculations refer to ground states.

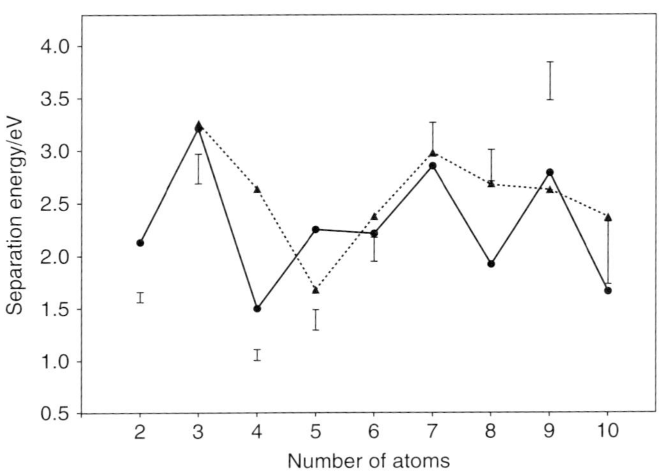

FIGURE 5 Comparison of experimental and theoretical separation energies per eV of copper clusters for monomer (●) and dimer (▲) cleavage. The error bounds of the experimental values are shown.

C. Vibrational and Rotational Spectroscopies

(i) Transition-Metal Clusters

Vibrational frequencies are linked to the structure of clusters via the force constants and therefore give indirect insight into bond strengths and bond lengths. The higher the frequency, the stronger the bond and, in general, the shorter the bond length. However, care must be taken when deriving structural information of clusters based only on the comparison of frequencies as will be shown in this section.

The vibrational modes of the first-row transition-metal dimers from Sc_2 to Cu_2 have been the subject of a number of theoretical and experimental studies. Recently, Calaminici et al. [155] have investigated the influence of different basis sets on the calculation of 3d transition-metal dimer properties using the KS-DFT program deMon2k [7] with the DZVP-LDA [133], DZVP-GGA, and TZVP-GGA basis sets [155] and the VWN [131], PW86-P86 [137,157,158], and B88-LYP [136,174] exchange-correlation functionals. They found only a small dependence of the vibrational frequencies on the basis set. For the scandium dimer, a frequency of 240 cm^{-1} is found by Raman spectroscopy [175], whereas the frequency of the copper dimer is experimentally determined by Fourier transform emission spectroscopy [176] to be 266 cm^{-1}. For both dimers, the calculated frequencies [155] are in very good agreement (less than 40 cm^{-1} difference) regardless of the functional and basis set used. A similar behavior is found for the titanium dimer (experimental frequency 408 cm^{-1} [177], resonance Raman progression in a solid argon matrix). It is worth noting that frequencies calculated with the local VWN functional are in general larger than the corresponding GGA values. The VWN functional overestimated the bonding, leading to slightly too short bond lengths and too large force constants. On the other hand, GGA functionals tend to yield too large bond lengths (so-called "bracketing" of the bond lengths by local and GGA functionals).

No experimental values for the bond lengths are known for the scandium dimer up to now. Hence, no direct comparison of the theoretically predicted range of 2.53–2.68 Å for the dimer bond length obtained by Calaminici et al. [155] can be made, although these results confirm earlier theoretical studies (see Ref. [155] and references therein). In this case, experimentalists are challenged to take the next step.

The bond lengths for Ti_2 (1.943 Å [178]) and Cu_2 (2.219 Å [176]) are experimentally well determined. The calculated bond length for the titanium dimer using the GGA-optimized basis sets is in fair agreement with experiment. The VWN/DZVP-GGA value is 1.900 Å, which is about 4 pm too short, while the PW86-P86/DZVP-GGA overestimates the bond length by approximately the same amount. In the case of Cu_2, the GGA/TZVP-GGA results show a fair agreement with experiment, too (2.169 Å for VWN and 2.260 Å for B88-LYP). Here, the good agreement of the frequencies is in line with the behavior of the geometric structure of the dimers.

The nickel dimer was studied experimentally by Moskovits and Hulse [179]. They obtained a frequency of 330 cm^{-1} from UV–vis measurements. Ahmed and Nixon [180] measured 381 cm^{-1} for Ni_2 in an Ar matrix. However, a more recent negative photoelectron study [181] suggests a much smaller value of 280 ± 20 cm^{-1}, which is supported by the 259.2 ± 3.0 cm^{-1} of Raman measurements by Wang et al. [182]. Castro et al. [183] have investigated the neutral iron, cobalt, and nickel dimers with the LCGTO DFT code deMon-KS [160] using the DZVP2 basis set [26]. The calculated frequencies for Ni_2 are 354 and 337 cm^{-1} using the VWN and PW86-P86 functionals, respectively. While the agreement with the UV–vis results is very good, the frequencies are slightly too high compared with the more recent measurements. The calculated frequencies of Calaminici et al. [155] of 286–300 cm^{-1} using GGA functionals in combination with the DZVP-LDA and DZVP-GGA basis sets are in good agreement with more recent experimental results. The corresponding TZVP-GGA values and the VWN results give higher frequencies. The VWN/TZVP-GGA values of 358 cm^{-1} and the VWN/DZVP-LDA (DZVP-GGA) value of 335 (336) cm^{-1} are similar to the ones obtained by Castro et al. (see above). The experimental Ni_2 bond length is 2.1545 Å. Like in the case of the dimers discussed above, the typical bracketing of the experimental values by the VWN and GGA bond lengths is obtained by Calaminici et al. with the DZVP-GGA basis sets with a deviation of less than 4 pm for the GGA functionals. The GGA value of Castro et al. (2.10 Å) is a little too short.

The transition-metal dimers V_2, Cr_2, Mn_2, Fe_2, and Co_2 proved to be very challenging systems for both experimental and theoretical methods. In the following, we will discuss the iron and the cobalt dimers as two examples in more detail. Starting with Co_2, the frequencies calculated by Castro et al. [183] with the VWN and PW86-P86 functionals are 444 and 421 cm^{-1}, respectively, which are much too high with respect to the experimentally determined values [184,185] of 300 and 297 cm^{-1}, both obtained by Raman spectroscopy. The VWN values of Calaminici et al. [155] show a similar behavior, although the VWN/TZVP-GGA frequency of 355 cm^{-1} is much closer to the experiment than the VWN/DZVP-LDA value (425 cm^{-1}) or the VWN/DZVP-GGA value (411 cm^{-1}). However, their GGA values of 249 cm^{-1} (B88-LYP/DZVP-GGA) and 290 cm^{-1} (PW86-P86/DZVP-GGA) are in good to excellent agreement with the experimental values. In this case, the GGA calculations lead to broken symmetry states for the cobalt dimer. As for the case of the scandium dimer, no experimental value is available for Co_2.

Fe_2 is one of the most investigated 3d transition-metal dimers using both experimental and theoretical methods. However, the assignment of the electronic ground state is still subject to controversial discussions in the literature, and a comprehensive presentation of all studies and findings is far beyond the scope of this chapter. Nevertheless, we will give a short overview of some experimental and theoretical results and discuss some of the conclusions with emphasis on the consequences for the bond length and frequency of Fe_2. The iron dimer bond length $r_{Fe–Fe}$ was measured with the extended X-ray absorption fine structure (EXAFS) approach in argon [186] and neon [187] matrices giving $r_{Fe–Fe} = 1.87$ Å and $r_{Fe–Fe} = 2.02 \pm 0.02$ Å, respectively, showing quite a large uncertainty of 0.15 Å. The frequency of 299.6 cm^{-1} found by Raman spectroscopy in an argon matrix [184], however, is in excellent agreement with the value of 300 ± 15 cm^{-1} from the one-electron detachment PES studies of Fe_2^- [188] of Leopold and Lineberger. Based on their photoelectron spectrum, they suggest a $^9\Sigma_g^-$ ground state for Fe_2 [189]. However, this conclusion contradicts the finding of Baumann et al. [190], who were unable to observe Fe_2 in their electron spin resonance (ESR) experiments. They support a $^7\Delta_u$ ground state for the iron dimer.

As the experimental findings contradict each other in part, so do the results of theoretical studies. DFT calculations [155,191,192] yield, in general, for a variety of functionals and basis sets, a $^7\Delta_u$ ground state for Fe_2, and the calculated bond length is in reasonable to good agreement with the experimental results (keeping in mind the large uncertainty of the experiments, as pointed out above), mostly favoring the larger $r_{Fe–Fe} = 2.02 \pm 0.02$ Å result. For example, Chrétien and Salahub [191] found a bond length of 2.008 Å using the PW86-P86 GGA functional; Gutsev and Bauschlicher [192] obtained a B88-PW91 [136,153] value of 2.011 Å, and recently Calaminici et al. [155] calculated $r_{Fe–Fe} = 2.033$ Å with the PW86-P86 exchange-correlation functional in combination with the DZVP-GGA basis set. However, the calculated frequencies in these studies (415, 397, and 399 cm^{-1}, respectively) are clearly too large compared with the experimental values. Castro and Salahub [193] pointed out earlier that there is a delicate balance between exchange and correlation, between bonding and magnetism, and between occupation of d and s orbitals in highly spin-polarized systems, which might be insufficiently represented by the corresponding functionals. Recently, Rollmann et al. [194] ascribe this difference between calculated and experimental frequencies to an improper description of electronic correlation effects of partly filled d or f shells within DFT LDA and GGA methods. They employed a simple model, GGA-U, which corrects the GGA total energy expression with a two-parameter correction term. The so-called Hubbard parameter $U = 2.20$ eV was fitted according to experimental results for the Fe_2 dissociation energy, ionization potential, and electron affinity. This modification leads to a stabilization of the $^9\Sigma_g^-$ state compared to the $^7\Delta_u$ state for the iron dimer, with a frequency of 346 cm^{-1}. Although this frequency is lower by about 55–70 cm^{-1} compared with the DFT results without the correction term, it is still larger than the experimental values by at least 30–75 cm^{-1}. Furthermore, the bond length calculated by Rollmann et al. is 2.143 Å and therefore much too large compared to the experimental range.

While pure DFT calculations find the $^7\Delta_u$ ground state, multireference configuration interaction (MRCI) methods generally obtain the $^9\Sigma_g^-$ state as the electronic ground state. For example, both recent MRCI studies of Hübner and Sauer [195] and of Bauschlicher and Ricca [196] agree that the calculated $^9\Sigma_g^-$ state for Fe_2 is energetically favored compared to the $^7\Delta_u$ state. The calculated frequencies (298.6 and 301 cm^{-1}, respectively) are in excellent agreement with the experiment. However, both studies find bond lengths much longer than experiment (2.179 and 2.190 Å, respectively). Finally, a $^9\Sigma_g^-$ ground state contradicts the electron spin resonance (ESR) results, as discussed above.

One reason from these contradictory results might be the assumption that some of the theoretical and/or experimental findings are incorrect. For example, the electronic ground state of Fe_2 could be a symmetry-broken state. This would lower the frequency in the DFT calculations as shown for the Co_2 dimer discussed above [155]. Also, Bauschlicher and Ricca demonstrated that the relative stabilities of the two states in the MRCI calculations strongly depend on the extent of the active space in the configuration interaction (CI) calculations [196]. They conclude that the inclusion of more extensive correlation might lower the $^7\Delta_u$ state in their calculation. On the other hand, the experimental range for the bond length (showing a large uncertainty anyway) might not be correct and the Fe–Fe distance is actually larger. However, the ESR results remain contradictory in this case. Another conclusion was given by Bauschlicher and Ricca [196], assuming all experimental results are correct. In this case, maybe the $^9\Sigma_g^-$ state was indeed observed in the PES experiment [188]. Still, the ground state of Fe_2 might be the $^7\Delta_u$ state. It might not have been observed in the PES experiment because it is not connected to the state of Fe_2^- in the one-electron detachment process. In combination with the remarks concerning the extent of the active space in the MRCI calculations, this would lead to a consistent picture.

The example of Fe_2 shows clearly that transition-metal systems are challenging for both theoretical and experimental methods. The results of both have to be checked very carefully and, in general, no comprehensive structure information can be obtained based on the frequencies alone. Additional data, both experimental and theoretical, have to be used.

(ii) Van der Waals Clusters

Van der Waals (vdW) clusters can be defined as supermolecules formed from the interaction of permanent, induced, and instantaneous multipoles in their constituent molecules. These clusters, studied in the gas phase, can be studied as isolated molecules. They have unique characteristics such as very low binding energies, large intermolecular distances, and low-frequency intermolecular vibrational vdW modes, with, in general, a very small perturbation of the structures and properties of their constituent molecules.

Patterns of the rotational transitions, obtained from microwave spectra and analyzed with the consideration of the adequate symmetry-based rotation model, provide the value of the separation between the centers of mass of the two consitutents of the dimer or their average separation, if several isomers are present. The low intermolecular vibrational frequency modes of the complexes are also studied using Raman vibronic double-resonance spectroscopy, which provides information about the shifts in the individual monomer vibrations due to the complexation, as well as their possible coupling and the intermolecular vibrations. In addition, the binding energy of the complex can be estimated from the excess of the vibrational energy in its excited state needed to induce its vibrational predissociation state. It is worth noting that most of the techniques adopted to study these complexes use direct excitation of the vibrational vdW states in their electronically excited S1 potential energy surface (laser-induced fluorescence, vibronic Raman spectroscopy, two-color mass spectroscopy, etc.), in order to obtain high resolution of the intermolecular vibrational spectra. A recently developed technique, coherent ion dip spectroscopy (CIS), reaches this goal by investigating high overtones in the electronic ground states and has been applied to vdW systems.

Information gathered from the experimental spectra of various vdW complexes have been reviewed earlier by Dyke [197], Legon and Millen [198], and Leopold et al. [199], and more recently by Xu et al. [200]. Despite the recent developments in the experimental microwave, or other rovibrational spectroscopic techniques, the characterization and description of the vdW systems remain impractical, especially for clusters containing more than 10 atoms. Therefore, providing quantum chemical calculations to the experimentalists is necessary. Indeed, the crucial role this type of cluster plays in many biological and chemical processes has prompted the development of various theoretical methodologies aiming mainly at accurate description of the intermolecular potentials.

The theoretical progress made for describing weakly bonded systems has been summarized and discussed in relation to the experimental data in the comprehensive review articles of Buckingham et al. [201], Hobza [202], and Müller-Dethlefs and Hobza [93]. However, comparison between experiment and theory is not trivial, even for small complexes, due to the complexity of the rovibrational spectra and the very flat PES. Usually, the experimental analyses are carried out assuming rigid monomer structures, thus providing results only about the intermolecular distances and angles averaged over the possible isomers of the complex. Moreover, only structures with dipole moments (or induced dipole moments) are observed through microwave experiments.

One must stress that vdW interactions are a challenge for theoretical methods. Indeed, the intermolecular distances between the constituents imply that the wave function and the electron density are distributed through a large portion of space with areas of very small densities, requiring very large basis sets with polarization functions, very large grids, and tight numerical accuracy criteria. The interaction energy ΔE of a vdW cluster is calculated from two approaches: (1) the supermolecular variational method, which gives the ΔE as the difference between the energy of the cluster and the sum of the energies of the isolated subsystems forming a cluster; and (2) the perturbation method, which directly gives the interaction energy as a sum of different energy terms with well-defined physical meaning, such as Coulombic, exchange-repulsion, dispersion, induction, etc. For critical and detailed theoretical analyses of the interaction forces and energies in the vdW clusters, the review of Hobza [202] can be recommended among the numerous review papers on this subject. In most cases, ΔE is obtained from post-Hartree–Fock methods (MP2, MP4, and CCSD/CCSD(T)) due to the importance of accurate correlation energy computations.

Another important point for theoretical studies of vdW is the choice of basis set. Large basis sets including diffuse polarization functions are needed to reproduce correctly the properties of both monomers and the supermolecule. Indeed, correct description especially of the monomer polarizability allows a correct estimation of the intermolecular dispersion energy.

At the same time, the vdW, the short-range repulsive, and the long-range attractive forces between the constituents in the cluster should be properly computed. Moreover, the interaction energy is a subject of large basis set superposition error (BSSE), if less extensive basis sets are employed. Dunning's correlation consistent polarized valence XZ (cc-pvXZ, X = D, T, Q) or Dunning's augmented (aug-cc-pvXZ) bases [203,204], containing diffuse functions of all types are recommended [202]. The (at first glance contradictory) observation that DFT calculations of hydrogen-bonded systems employing double-zeta basis sets without BSSE corrections lead to reasonable results for energies and structures is due to the fact that the BSSE is of the same size and direction as the dispersion interactions, as pointed out by Jurečka et al. recently [205]. Therefore, such calculations benefit from the cancellation of errors: the missing dispersion energy is compensated by the BSSE.

Combination of highly correlated post-Hartree–Fock methods with large basis set sizes is often limited because of the required unacceptable computational time. Therefore, methodologies based on DFT for weak interactions are now starting to be accessible.

One way of extending DFT to weakly bonded systems is the reparameterization of existing functionals or the development of new exchange-correlation approximations capable of describing hydrogen-bonded and dispersion-bonded interactions. For example, Truhlar and coworkers have introduced recently the M05-2X [206] hybrid and the M06-L [207] local meta-GGA functional. These exchange-correlation functionals contain adjustable parameters which were optimized using a training set of molecules including hydrogen-bonded and dispersion-bonded complexes. Zhang and

Salahub [208,209] have reparameterized the $\tau 1$ correlation functional [210] in combination with the TPSS exchange functional [211]. The resulting TPSS$\tau 3$ method described hydrogen- and dispersion-bonded complexes better compared to the original Bm$\tau 1$ [210] and TPSS$\tau 1$ methods.

The applicability of the DFT method augmented by a damped empirical dispersion term E_{disp} (Equation (97)), added simply to the usual DFT energy [212] in order to improve the description of weak interactions, will be demonstrated below.

$$E_{\text{disp}} = -\sum_{i=1}^{N}\sum_{j=i+1}^{N-1}\frac{C_6^{ij}}{r_{ij}^6} f_{\text{damp}}(r_{ij}) \tag{97}$$

E_{disp} is a sum of all possible contributions of the atom pair (i, j) with the interatomic distance r_{ij}, in an N-atomic system, damped with a function

$$f_{\text{damp}}(r_{ij}) = \frac{1}{1 + \mathrm{e}^{-\alpha\left(\frac{r_{ij}}{s_R r_0} - 1\right)}} \tag{98}$$

with α and s_R as adjustable parameters, and r_0 being the sum of the atomic van der Waals radii [213]. The dispersion coefficients C_6^{ij} are computed from the atomic C_6^{i} as proposed by Wu and Yang [212]:

$$C_6^{ij} = \frac{2C_6^i C_6^j}{C_6^i + C_6^j}. \tag{99}$$

By adding ∇E_{disp} to the DF energy gradient, structure optimizations can also be carried out within this DFT scheme. Jurečka et al. [205] have recently optimized the parameters α and s_R in combination with different density functionals for a set of systems with weak interactions. Their fit set included in total 22 complexes, out of which 7 were hydrogen-bonded, 8 dispersion-bonded, and 7 of mixed character. They found a dependence of the two fitting parameters on the density functionals used. The TPSS exchange-correlation functional [214] gave the best overall results among the functionals tested. Furthermore, it was shown that the best results could be obtained if different parameters s_R for the hydrogen-bonded complexes and for the dispersion-bonded complexes were used. In other words, parameters that minimize the errors for the hydrogen-bonded complexes will not give minimal errors for the dispersion-bonded systems and vice versa. There could be two reasons for this behavior [205]. On one hand, the electron densities of the moieties in the dispersion-bonded complexes are essentially unperturbed compared to the isolated moieties in contrast to the densities in the hydrogen-bonded systems. Therefore, the correlation effects can reach further, mediated by the shared electron density in the case of the hydrogen-bonded systems. On the other hand, the wrong asymptotic behavior of the exchange-correlation functionals might be an explanation.

We will denote the DFT dispersion scheme based on Equations (3)–(5) with $\alpha = 23.0$ and $s_R = 1.0$ in the subsequent sections as DFT-D. Here, the discussion will be restricted to a few examples of combined experimental and theoretical studies, chosen among the most difficult and debated cases.

a. Carbon Monoxide Dimer

The carbon monoxide dimer is one example of a small vdW complex that has been widely studied by means of microwave, millimeter-wave [215–220], and infrared spectroscopy [216,219–223] mainly because of its substantial astrophysical relevance. The flat intermolecular potential surface of CO···CO consisting of many nearly equally deep minima, separated by rather small barriers (10–20 cm^{-1}) [217], makes this cluster a real challenge for both experiment and theory [216,223]. The carbon monoxide dimer shows a rich rotational structure (with about 46 rotational energy levels known presently [223]) in its ground-state C···O stretching vibration ($v_0 = 0$). For an unambiguous assignment of these transitions, very accurate computations are highly desired but the quantum chemical methods encounter also the difficulties of describing properly such anisotropic intermolecular potential surfaces. Recent computations [224] that attempted to predict the lowest rovibrational states agreed only semiquantitatively with the observed microwave and millimeter-wave spectra, thus providing some guidance for their assignments. Therefore, the technique of combination differences, based on the previously known energy levels, is the most frequently adopted approach [216]. The rotational energy levels (grouped in stacks with known symmetry) are derived (if possible) by unambiguous assignments of the observed transitions. They are used in the simple semirigid rotor energy level equation [218] to obtain the rotational constants and hence the effective intermolecular separation values. Following this path and considering that the rotational stacks tend to fall into two groups, the hypothesis of the existence of two "isomers" of the carbon monoxide dimer, proposed in the early study [215], is supported also presently [216,223]. The experimentally estimated intermolecular separation distances of 4.4 and 4.0 Å have been attributed to a C-bonded complex as ground state and to an O-bonded complex as the first excited state, respectively [216]. These data agreed nicely with the structural CCSD(T) calculations [225,226] reporting intermolecular distances of 4.34 Å (ground state) and 3.86 Å (first excited state). The reported binding energies from the CCSD(T) computations are 0.45 and 0.36 kcal/mol for the ground and first excited state, respectively. This small energy difference explains why the commonly used MP4 and CCSD(T) supermolecule methods are of insufficient quality for the intermolecular potential surface of the CO dimer [226]. A combination of DFT and CCSD(T) methods [227] to calculate a four-dimensional PES assuming rigid molecules led to results comparable with those reported in Ref. [224]. To our knowledge, no other study of the CO dimer based on DFT or DFT-D is available.

b. Methane–Water

The intriguing methane–water complex represents a prototype for hydrophobic systems [228–232]. In contrast to the carbon monoxide dimer, much less is known about the potential energy surface of $CH_4 \cdots H_2O$ from the available rotational and FIR spectra [232,233]. The FIR [232] and microwave [233] spectra were assigned with the aid of additional Stark effect and nuclear spin hyperfine measurements. Hence, an estimated zero-point center-of-mass separation between carbon and oxygen of 3.7024 Å was obtained from the rotational constants, assuming the pseudoatom model. The estimated harmonic vibrational C···O stretching frequency of 55 cm^{-1} allowed unambiguous assignment of the observed lower modes (in the range of 18–35.5 cm^{-1}) to vibration-rotation-tunneling modes. *Ab initio* calculations (MP2, CCSD, and DFT-D) [234] carried out with at least the cc-PVQZ basis set [203,204] and considering full structural optimization agree that two isomers of the complex exist (Figure 6A and B). In the DFT-D calculations, the GGA PBE exchange functional (revised version [235] from 1998) combined with the LYP correlation [174] (revPBE-LYP-D) and the meta-GGA TPSS (TPSS-D) exchange and correlation functional [214] were used. The isomer **a** is found to be slightly more stable than isomer **b** by a maximum of 0.4 eV, according to the self-consistent field method. More details on optimized structural parameters, such as the C–O distance ($r_{C\text{–}O}$) and the distances between the water hydrogen atoms and the carbon atom ($r_{C\text{–}H1}$ and $r_{C\text{–}H2}$), and the BSSE-corrected binding energies (ΔE) are presented in Table 2. When medium basis sets, such as cc-PVTZ and cc-PVQZ, are used, a large BSSE is obtained (in the range of 40–60% of ΔE for both isomers). The BSSE vanishes for the aug-cc-PVQZ basis set. In isomer **a,** one O–H bond of the water points towards one of the faces of the CH_4 tetrahedron and forms an

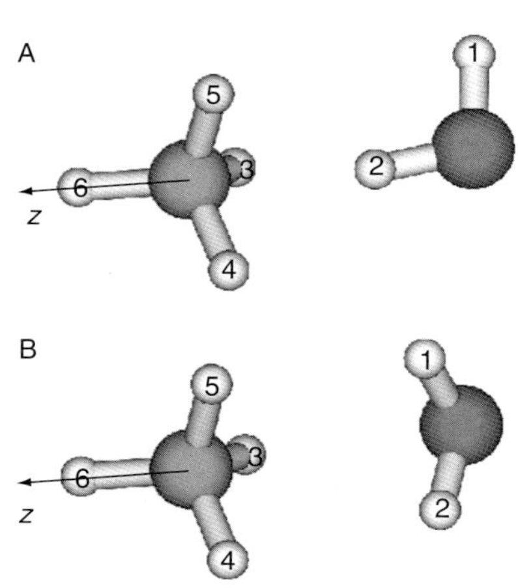

FIGURE 6 Schematic presentation of the most stable minimum (A) and the second minimum (B) structures of $CH_4 \cdots H_2O$.

TABLE 2 BSSE-Corrected Interaction Energies ΔE/(kcal/mol) and Distances $r_{C-...O}$, $r_{C-...H1}$, and $r_{C-...H2}$ in Å for Isomers a and b of $CH_4 \cdots H_2O$ Computed at Different Theoretical Levels

Method	$-\Delta E^a$	$-\Delta E^b$	$r_{C-...O}{}^a$	$r_{C-...H1}{}^a$	$r_{C-...H2}{}^a$	$r_{C-...O}{}^b$	$r_{C-...H1}{}^b$	$r_{C-...H2}{}^b$
revPBE-LYP-D[c]	–	0.80	–	–	–	3.667	3.506	3.505
revPBE-LYP-D[d]	1.17	0.76	3.579	2.697	3.641	3.671	3.512	3.505
revPBE-LYP-D[e]	0.82	0.54	3.605	2.648	3.859	3.777	4.065	4.327
TPSS-D[d]	0.81	0.65	3.569	2.814	3.498	3.741	3.464	3.472
TPSS-D[e]	0.96	0.83	3.671	2.709	3.948	3.572	3.093	3.052
MP2[c]	0.58	0.30	3.456	2.650	3.493	3.672	3.569	3.573
MP2[d]	0.75	0.40	3.497	2.553	3.289	3.724	3.317	3.312
CCSD[c]	–	0.92	–	–	–	3.743	3.678	3.678

[a]*Interaction energies and structural parameters for isomer a.*
[b]*Interaction energies and structural parameters for isomer b.*
[c]*cc-PVTZ basis.*
[d]*cc-PVQZ basis.*
[e]*aug-cc-PVQZ basis.*

O–H2–C angle (see Figure 6 for the atom numbering) in the range of 163–168°, depending on the computational method. The H_2O molecule in the isomer **b** is displaced towards the CH_4 moiety in such a way that both H atoms are at almost equal distances to the carbon atom. In the case of this second isomer, the water molecule is differently rotated, depending on the computational method, around an axis perpendicular to the C_2 water axis. These rotations do not influence significantly the C–H1 and C–H2 distances and the interaction energies. Similar structural configurations as those depicted in Figure 6, with different mutual rotations between CH_4 and H_2O in the complex, were reported as the two lowest lying minima in an earlier *ab initio* study [236] and in a more recent one [237] where the symmetry adapted perturbation theory (SAPT) was employed to describe the $CH_4 \cdots H_2O$ PES. The C–O distances and the interaction energies agree well with the reported CCSD(T)-PES minima [237].

It is worth noting that the DFT-D methodologies applied to the structural intermolecular study of the methane–water complex give results quite comparable with the MP2/cc-PVQZ and the CCSD/cc-PVTZ values, with a trend to somewhat larger C···O distances (Table 2). The individual structures of the H_2O and CH_4 moieties in the supermolecule remain almost unchanged with respect to their optimized molecular structures. The existence of other minima on the PES of this ideal model of the water–methane interaction cannot be excluded, as also suggested from the analysis of the microwave spectra, leading to the conclusion of a nearly free rotation of the water molecule in the complex [232,233].

Theoretically obtained accurate low-frequency modes are indispensable information for an unequivocal assignment of the observed transitions in the microwave and FIR region, but their precision depends dramatically on the choice of the basis set and method. This explains the lack of vibrational spectra reported in most of the theoretical work devoted to the study of weakly bound clusters. Our computed low-frequency modes for the$CH_4{\cdots}H_2O$ complex (isomer **a**) are comparable at the rev-PBE-D/cc-PVQZ and MP2/cc-PVTZ levels. Increasing the size of the basis set did not change the frequency values: we found differences of only 1–2 cm^{-1} for the low modes and of 2–5 cm^{-1} for the higher bands.

From the analysis of the six normal vectors in the FIR, we find that the five lowest ones couple internal rotations between the CH_4 and H_2O moieties in the supermolecular complex. The highest one is the only normal vector with nonzero components on O and C atoms, and is easily assigned to the intermolecular C···O stretching mode. Its computed value is 90/91 cm^{-1} and 107/118 cm^{-1} for the isomers **a**/**b**, using the MP2 and revPBE-D methods, respectively. These values are higher than the estimated experimental C···O stretching of 55 cm^{-1} [232,233]. To our knowledge, other experimental information about FIR frequencies of the methane–water complexes are not available. However, a direct comparison between the computed and experimental FIR vibrations is hampered by the harmonic approximation and the 0 K temperature assumption in the computations.

c. Benzene Dimer

A very large amount of literature on weakly bound complexes deals with studies related to vdW complexes involving benzene. Benzene complexes with rare gas or small molecules like H_2O or NH_3 have also been studied experimentally and theoretically [202,238,239]. Among the benzene complexes, the benzene dimer has raised the most interest, both for theoreticians and experimentalists, as the prototype of weakly interacting π systems. Here, the challenge for traditional *ab initio* theory is to deal with a molecular size already too large for a highly correlated treatment of the PES including all degrees of freedom. The challenge is no less for DFT methods, due to their problems in handling damped dispersion interactions. On the other hand, the numerous experimental studies performed up to now have only given clues about the structure of this complex.

Experiments on the benzene dimer have been performed for a long time using various techniques. Molecular beam electric resonance, performed as early as the 1970s, yielded the important information that the dimer has a permanent dipole moment [240,241]. If the dimer structure is assumed to be rigid, this result means that the two benzene moieties are not equivalent, and a T-shape structure (T) was proposed. This assignment has been confirmed later from hole-burning experiments on mixtures of benzene dimer isotopomers, although weaker bands could suggest the presence of another structure [242]. However, the analyses of different features in the mass-resolved excitation spectra of benzene dimer isotopomers concluded that the complex contains two equivalent sites with a symmetry lower than threefold [243,244].

More recently, the study of the low rotational transitions in the microwave spectra [245] has provided the value for the average distance of 4.96 Å between the two centers of mass of the benzene moieties, assuming that the dimer has a T-shape structure and using the symmetric-top Hamiltonian to fit the data. Another conclusion is that the dimer is far from rigid, with strong tunneling interchanging the benzene moieties. Similar conclusions were drawn from Raman vibronic double-resonance spectroscopy measurements, with the complementary result of a free rotation of the top benzene moiety around the C_6 axis [246]. From all these diverse experiments, one can thus conclude that there may be several possible structures for the benzene dimer, which may transform dynamically, although the most probable conformation, sampled through experiments, corresponds to a T-shape structure. As has been pointed out previously in a review article on the structure and rigidity of aromatic vdW clusters [239], the interpretation of the experimental spectra, to be fully correct, should take into account the molecular symmetry of the whole dimer, without assuming *a priori* the restriction of a quite large number of its degrees of freedom.

The first theoretical studies devoted to the benzene dimer used empirical potentials and suggested that the complex had a PD (parallel-displaced) [242] or a V-shaped (herringbone) [247,248] structure. The dependence of the final structure (T-shaped, PD, or V-shaped) on the adopted atomic net charges was shown in different studies [249–251]. Quantum chemical calculations have been performed on various possible structures of the benzene dimer [252,253]. The PD and T-shapes are the most stable, and their relative stability depends on the level of theory utilized in these calculations. As for the other examples given above, the results are very much dependent on the number of orbitals and polarization functions, as well as on the treatment of correlation.

As a general rule, the MP2 method (and MP4 to a lesser extent) provides strongly overestimated stabilization energies with respect to CCSD(T). However, full CCSD(T) calculations cannot be performed with the largest bases including polarization functions and thus a theoretical energy limit is not within computational range. In the present state of the art, it can be inferred that the T-shaped and PD structures have equivalent stabilization energies. With benzene monomer geometries optimized at the MP2/DZP2 level and energies calculated at the CCSD(T)/ aug-cc-pVDZ(C)/cc-pvDZ(H), Hobza et al. [254] obtained a stabilization of 2.3 kcal for the T-shaped and 2.0 kcal/mol for the PD structures. More recently, Tsuzuki et al., using their AIMI model developed for calculating the interaction energies between aromatic molecules [255], found T-shaped and PD stabilization energies at 2.46 and 2.48 kcal/mol, respectively. The benzene monomer geometries were optimized at the MP2/6-31G* level and the energies estimated at the CCSD (T) level with extrapolation to the basis set limit. This study concludes that the benzene dimer binding energy is essentially due to dispersion. All the *ab initio* results reported include BSSE corrections and were obtained by fixing all degrees of freedom of the dimer except the distance between the two benzene moieties. Finally, the points of the PES of the benzene dimer calculated from CCSD(T) interaction energies have been used to parameterize an intermolecular potential model [256]. This potential has been used, in conjunction with microwave and Raman spectra, to study the rotational and vibrational properties of the dimer. Among the various structures, only the T-shaped one is found as a minimum. A nearly free rotation around the axis connecting the two benzene centers is found, and a barrier of only 0.1 kcal/mol for the rotation of the proton donating benzene around its C_6 axis. Despite a qualitative agreement with experiment, the latter barrier is two times smaller than the experimental evaluation, and the theoretical distance between the two centers of mass is 0.25 Å too short.

DFT results have appeared quite recently in the literature compared to *ab initio* studies, due to the well-known problem of this method to account for long-range dispersion. Most DFT functionals tested for the parallel benzene dimer showed repulsive PES and strongly underestimated values for the other structures [257]. Recent attempts to augment the DFT energies with an empirical damped dispersion correction have shown that this method can be a solution to solve the DFT "dispersion problem" [258]. More work is certainly necessary in this domain, because of the large sensitivity of the computed values to the functionals chosen, the basis extension, and the grid size.

D. Photoelectron Spectroscopy

Recently, vibrationally resolved negative ion PES was applied to the group V metal trimer monoxides V_3O, Nb_3O, and Ta_3O yielding insight into the bonding of these early transition-metal oxides [259]. The experimental data reported in this work provide a benchmark for computational studies of early transition metal clusters. Because 4d and 5d elements are involved, it is interesting to analyze the role that relativistic effects play in these systems.

In 2003 Calaminici et al. assigned the ground state structure of V_3O and V_3O^- by simulating the vibrationally resolved negative ion photoelectron spectrum of the vanadium trimer monoxide [115,260]. In that study, all-electron DFT calculations were presented. The new program deMon2k includes a novel pseudo-potential integrator [261] and, therefore, it is possible to study heavier elements like Nb and Ta including scalar relativistic corrections via effective core potentials (ECPs). Thus, the Nb and Ta oxides were recently investigated using ECPs [116,262].

The molecular structures were optimized using the local density approximation with the VWN [131] correlation functional. Scalar quasirelativistic ECPs [263,264] were used for all atoms. A valence space of 6 and 13 electrons was used for the oxygen and group V elements, respectively. For the fitting of the auxiliary density, the GEN-A2 auxiliary set was automatically generated. A description of the automatic generator of auxiliary functions in pseudopotential calculations has been published in Ref. [262]. For the optimization of geometries, different initial geometries and multiplicities have been considered. To avoid spin contamination, the ristricted open Kohn-Sham (ROKS) method was used. An harmonic vibrational analysis was performed in order to discriminate between minima and transition states. In all cases, a low spin planar C_{2v} structure with an edge-bound oxygen atom was found as the ground state. The three metal atoms form an isosceles triangle ($r_{12} = r_{13}$) with the oxygen attached to atoms 2 and 3 (see Figure 7).

In Table 3, the values of the corresponding structural parameters are reported. The results for vanadium systems were obtained from all-electron calculations. The assigned ground state symmetry symbols are also given in Table 3. For this assignment, the molecular plane was taken as the *yz* plane with the oxygen atom lying in the *z*-axis. The molecular topologies and the assigned ground states for the neutral systems are in agreement with experimental predictions. However, the calculated anion ground states are assigned as $^1A_1$ instead of the $^3B_1$ assignment suggested from the experiments [259].

The deviations between the AUXIS and BASIS results are of a few tenths of a picometer. The accuracy of the ADFT theory in geometry optimizations, previously validated only for all-electron calculations [265], is also very good for ECPs. There is no comparison to experiment because no experimental structure data exist. Therefore, the theoretical results provide the structural

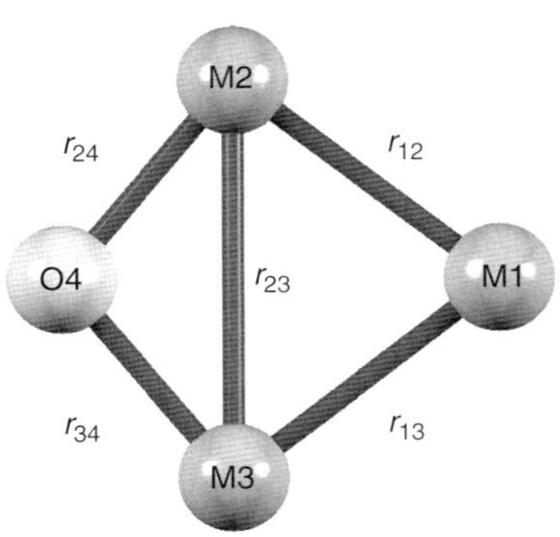

FIGURE 7 Schematic C_{2v} structure of M_3O with bond lengths r_{ab}, metal atoms M1, M2, and M3 and the bridging oxygen atom O4.

TABLE 3 M_3O Ground State Structures

		Structure parameters (Å)					
		AUXIS			BASIS		
Molecule	State	r_{24}	r_{23}	r_{12}	r_{24}	r_{23}	r_{12}
Neutral							
V_3O^a	$^2B_2$				1.794	2.527	2.051
Nb_3O	$^2B_2$	1.915	2.752	2.293	1.912	2.750	2.296
Ta_3O	$^2B_2$	1.917	2.769	2.363	1.918	2.772	2.370
Anion							
V_3O^{-a}	$^1A_1$				1.805	2.544	2.040
Nb_3O^-	$^1A_1$	1.926	2.773	2.293	1.926	2.771	2.294
Ta_3O^-	$^1A_1$	1.935	2.787	2.358	1.934	2.790	2.367

[a]*All-electron results [115].*

information required to confirm the experimental predictions from spectroscopy. For these systems, planar structures with double bridging oxygen atoms were predicted [259]. The present calculations confirm these predictions. Furthermore, given the success of DFT in the determination of molecular geometries in transition-metal systems, the computed structures combined with the corresponding electronic states support the assignment of the reported spectrum.

By comparison of the anionic structures with the corresponding neutral ones, it can be seen that the addition of the extra electron to the neutral system does not affect r_{12} in an appreciable way independently of the transition-metal involved. The structural change is mainly observed in r_{24} and r_{23}. These interatomic distances are increased upon electron addition. This fits nicely with the experimental prediction, which suggests that the extra electron occupies an antibonding orbital in the case of V and Nb. However, it also has been experimentally predicted that the anionic electron occupies a bonding orbital in the case of Ta_3O.

Schematic orbital correlation diagrams such as those shown in Figure 8 give insight into the electronic structure change due to the electron addition and can be used to validate the experimental predictions. From these diagrams, it can be seen that the anionic electron occupies an antibonding orbital in the V and Nb system. For the Ta system, the half-filled orbital of the neutral molecule is antibonding, too. However, in the anion the highest occupied molecular orbital (HOMO) and the lowest unoccupied molecular orbital (LUMO), which is a bonding orbital, interchange their position. This indicates

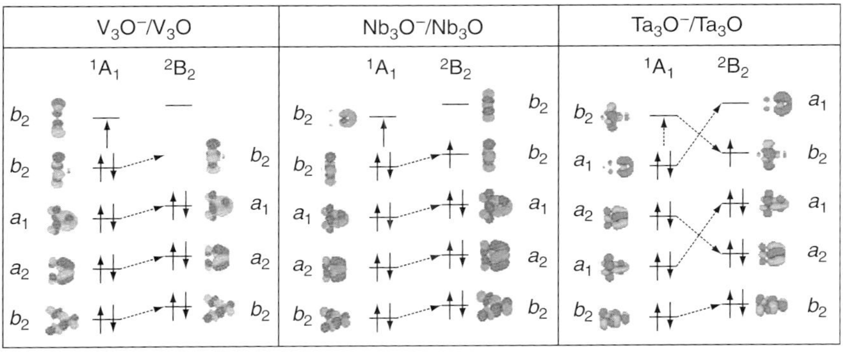

FIGURE 8 Schematic Kohn–Sham molecular orbital correlation diagrams from the anionic ($^1A_1$) ground state to the neutral ($^2B_2$) ground state. Both neutral and anionic ground state structures are lying in the plane of the paper with the oxygen atom on the left side of each structure. The dashed arrow indicates the ionization.

a breakdown of the simple one-electron picture. Structural changes in the Ta_3O system cannot be explained solely in terms of orbital diagrams. It is interesting to note that the HOMO of Ta_3O^- is topologically equivalent to the LUMO of Nb_3O^-. A similar orbital is not found in the V_3O^- orbital diagram. Based on this observation, one can speculate that scalar relativistic effects, which increase down the column of the periodic table, stabilize this type of orbital.

Table 4 shows the calculated vibrational frequencies for normal modes of the reported neutral and anionic ground states together with the experimental frequencies determined by negative ion PES. The calculated vibrational frequencies for the total symmetric modes agree well with the experimental values [259]. Note that the differences between AUXIS and BASIS results (see Sections I.A.2 and I.A.3) are less than 6 cm^{-1}. This is within the accuracy of the finite difference method used for the calculation of the harmonic frequencies. This suggests that ADPT (see Section I.A.4) should be well suited for the calculation of reliable analytic vibrational frequencies. In contrast to V and Nb, the b_1 mode is the third normal mode in the Ta system. This normal mode corresponds to the out-of-plane deformation of the oxygen atom. The interchange of the first a_1 mode with the out-of-plane b_1 mode reflects the stronger oxygen bond in Ta_3O.

From the calculated frequencies and the corresponding normal mode vectors, harmonic multidimensional FC factors required for the simulation of photoelectron spectra can be calculated. As described in Section I.B.4, the fact was taken into account that the normal coordinates, frequencies, and molecular structures are different for the ionic and neutral clusters. In this way, the vibrationally resolved negative ion photoelectron spectra were simulated. Results for the niobium system are presented here.

TABLE 4 Calculated and Experimental Vibrational Frequencies of M_3O and M_3O^-

			Frequency (cm^{-1})		
System	Type[a]	Symmetry	AUXIS	BASIS	Experimental[b]
V_3O (V_3O^-)	ν	b_2		161 (149)	
	π	b_1		309 (324)	
	ν	a_1		361 (368)	340 ± 15 (355 ± 20)
	ν	a_1		410 (416)	415 ± 15
	ν	b_2		549 (554)	
	ν	a_1		730 (723)	750 ± 20 (770 ± 20)
Nb_3O (Nb_3O^-)	ν	b_2	236 (219)	235 (218)	
	π	b_1	289 (291)	288 (296)	
	ν	a_1	317 (313)	315 (310)	320 ± 15 (300 ± 20)
	ν	a_1	372 (372)	371 (370)	
	ν	b_2	575 (558)	576 (559)	
	ν	a_1	729 (712)	732 (715)	710 ± 15
Ta_3O (Ta_3O^-)	ν	b_2	186 (175)	190 (170)	
	ν	a_1	220 (220)	216 (214)	225 ± 15 (215 ± 10)
	π	b_1	254 (265)	246 (260)	
	ν	a_1	270 (269)	269 (266)	
	ν	b_2	588 (554)	584 (555)	
	ν	a_1	715 (691)	714 (691)	710 ± 15

[a] ν, stretch; π, out-of-plane deformation.
[b] Negative ion photoelectron spectra results from Ref. [259].

In Table 5, the calculated spacing from the origin and the relative intensities for the most relevant peaks are compared with experiment. The experimental values are given in parentheses. As Table 5 shows, the calculated spacing of all peaks is in good agreement with the observed data. Also, the intensities of the hot band and the fundamentals show fair agreement with experiment. The intensity of the non-FC transition $1_0^1 3_0^1$, however, is much too large in the simulation. Thus, further investigations are required [116]. The adiabatic electron affinity can also be determined from negative ion photoelectron spectra. The reported value is 1.393 ± 0.006 eV [259].

TABLE 5 Peak Assignments, Relative Intensities, Given as $v_{ion}^{neutral}$, and Positions of the Simulated $Nb_3O^- \rightarrow Nb_3O + e^-$ Photoelectron Spectrum at 300 K

Assignment	Relative intensity	Spacing from origin (cm^{-1})
3_1^0	0.017 (0.036)	−310 (−295)
0_0^0	1	0
3_0^1	0.135 (0.164)	316 (320)
3_0^2	0.001 (0.001)	631 (635)
1_0^1	0.037 (0.033)	738 (710)
$1_0^1 3_0^1$	0.077 (0.012)	1054 (1025)

Experimental data from Ref. [259] are given in parentheses.

The adiabatic electron affinity calculated is 1.39 eV in excellent agreement with the experimental one.

(i) ZEKE-PFI

The possibility of combining theoretical calculations with experimental data from ZEKE-PFI PES to resolve the structure of small TM clusters has attracted much attention in the past few years. A combination of DFT and harmonic FC factor calculations, as described in Section I.B.4, has been used to determine the structures of several free and substituted TM clusters [113–116,266–268]. The ZEKE-PFI technique was also used to determine accurate adiabatic ionization potentials of the bare V_3 and V_4 clusters [269]. The observation of a ZEKE-PFI spectrum of V_3 with well-resolved vibrational bands has opened the possibility for a structure determination via the above described technique.

Different theoretical works exist on vanadium clusters. For a brief review of some of them, we direct the reader to Refs. [114,149,150,270–276]. This literature shows the main difficulties in the determination of the ground state structures of small TM clusters based only on calculations. In this section, we review a combined theoretical and experimental approach for the determination of the V_3 ground state structure presented in Ref. [114]. The neutral and cationic vanadium trimers were optimized with the DFT program AllChem [156]. The experimental ZEKE-PFI spectrum was simulated using multidimensional FC factors calculated from the geometries and harmonic frequencies of the obtained structures of V_3 and V_3^+. The comparison between the experimental and theoretical ZEKE-PFI spectra establishes unequivocally the ground-state structure of the V_3 cluster as an equilateral doublet. This represented the first work in which the structure of a pure TM metal cluster has been determinated from the combination of DFT FC simulations and ZEKE photoelectron spectroscopy [114]. A newly developed DZVP basis set optimized for gradient-corrected functionals [277] was used for the vanadium. This kind of basis set was later on optimized for all 3d TM elements [155,278,279]. Compared to the DZVP basis set optimized for local functionals [133], the new basis set improves the accuracy of the low-lying atomic energy levels of the 3d TM atoms. The exponents and contraction coefficients of the newly developed GGA-optimized DZVP basis set for the vanadium atom are given in Ref. [114]. For neutral V_3, an equilateral triangle $^2A_1'$ ground state was found. The optimized bond distance is 2.17 Å. The D_{3h} vanadium trimer ground state differs from the C_{2v} structures reported in previous DFT studies [149,150]. Only 0.03 eV above the $^2A_1'$ ground state lies an acute triangle $^4A_2$ state and, similar to the neutral trimer, an equilateral triangle ground state for V_3^+ was found [114]. This structure can be assigned to a $^3A_2'$ state.

The optimized bond distances of 2.18 Å in the cationic ground state are very close to those of the neutral trimer. The calculated adiabatic ionization potential of 5.61 eV is in good agreement with the experimental value of 5.49 eV [143,269]. Due to the very small energy difference between the calculated neutral ground state and the first excited state, the vanadium trimer ground state structure cannot be assigned on the basis of these calculations. Fortunately, Hackett and coworkers recorded a well-resolved ZEKE-PFI spectrum of V_3 [269]. The simulation of this spectrum from DFT results [114] allowed the assignment of the V_3 ground state structure. For this purpose, the harmonic frequencies of the optimized neutral and cationic vanadium trimers were calculated. From these frequencies and the corresponding normal mode vectors, the harmonic FC factors were calculated and the ZEKE-PFI spectrum of V_3 was simulated [114].

Figure 9 shows the experimental spectrum recorded at room temperature (top) and the simulated one for the transition from the neutral $^2A_1'$ to the cationic $^3A_2'$ state. The simulation was performed at room temperature (middle) and at 700 K (bottom). The theoretical frequencies are reported relative to the experimental adiabatic ionization potential, which corresponds to the origin of the ZEKE-PFI spectrum at 44,342 cm^{-1}. Therefore, the experimental ZEKE-PFI spectrum was assigned to the transition from the neutral $^2A_1'$ to the cationic $^3A_2'$ state, and it can be concluded that the vanadium trimer possesses a $^2A_1'$ ground state [114]. Both neutral and cationic ground states are equilateral triangles with very similar structural parameters. Thus, the ZEKE-PFI spectrum is dominated by the 0–0 band. The good agreement between the simulated and measured spectrum allowed also an assignment of the observed satellite bands. A detailed discussion of the band assignments is presented in Ref. [114].

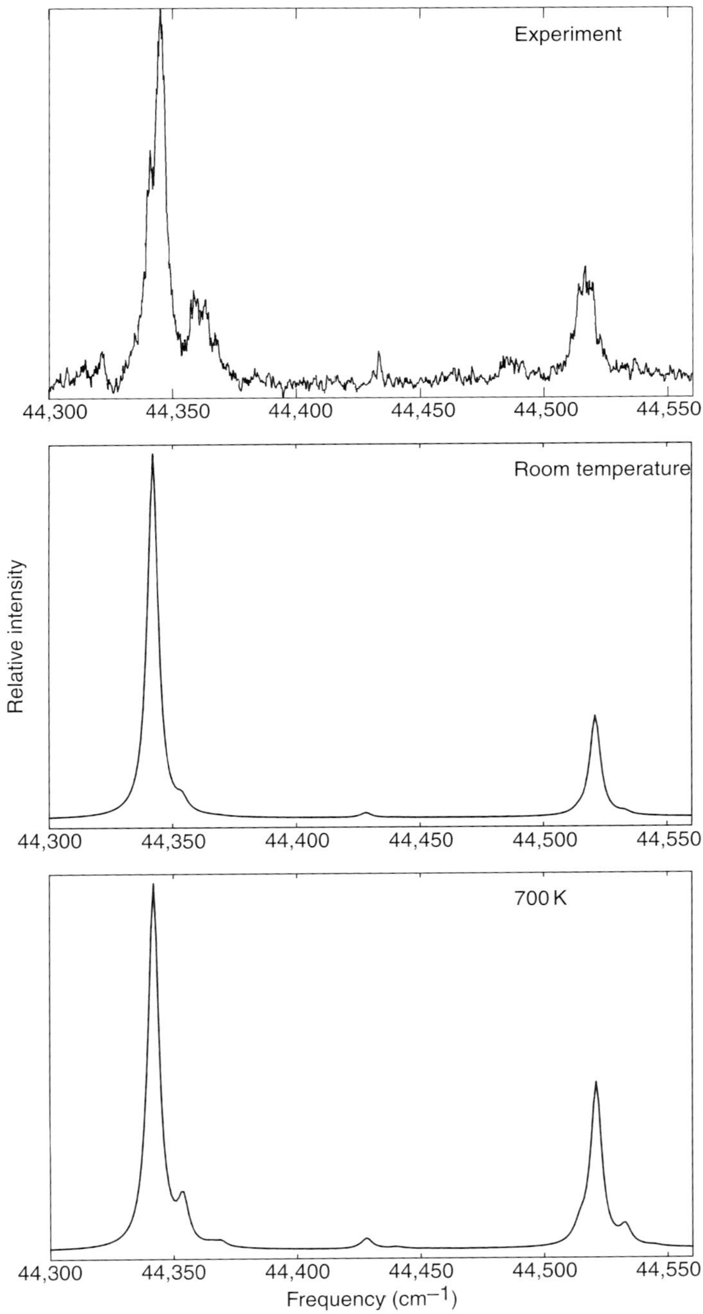

FIGURE 9 Experimental (top) and simulated V3 ZEKE-PFI spectrum for the $^2A_1' \rightarrow {}^3A_2'$ transition at room temperature (middle) and 700 K (bottom). Both simulated spectra were convoluted with a 5 cm^{-1} FWHM Lorentzian line shape to simulate the (rotational) width of the experimental bands.

Another example for the combination of experimental and theoretical work in the field of ZEKE-PFI spectroscopy is represented by the study of transition-metal oxides [113,115,116,280–291]. These studies are of broader interest because the properties related to the metal–oxygen bond are crucial for the understanding of chemisorption and the catalytic activities of metal oxides. In particular, niobium oxides are very complex transition-metal clusters and, therefore, the investigation of the niobium–oxygen chemical bonding is of great interest. On the other hand, the theoretical description of molecules containing transition metals is rather challenging due to the open d shells. Simple transition-metal oxide molecules provide ideal systems for the investigation of the reliability of theoretical methods. Recently, negative ion PES was applied to the group V metal trimer monoxides V_3O, Nb_3O, and Ta_3O, yielding insight into the bonding of these early transition-metal cluster oxides [259]. The structures of Nb_3O and Nb_3O^+ have been determined by vibrational resolved ZEKE-PFI spectroscopy and DFT calculations [113]. The good agreement between experimental and theoretical data allowed the assignment of the most stable structure for Nb_3O and Nb_3O^+ as planar structures with C2v symmetry and low, i.e. doublet and singlet, multiplicity [113]. The topology and geometrical parameters of the ground state structures obtained for Nb_3O and Nb_3O^+ are reported in Fig. 1 of Ref. [116]. From the harmonic vibrational frequencies of the neutral and cationic ground-state structures and the corresponding normal mode vectors, harmonic FC factors were calculated.

The vibrationally resolved ZEKE-PFI spectrum for Nb_3O was simulated and was reported in Ref. [113]. Details of this work are given in Ref. [116]. Figure 10 shows the comparison between (a) the experimental ZEKE-PFI spectrum recorded at 300 K (top) and at 100 K (bottom) and (b) the corresponding spectra obtained by simulation. As Figure 10 shows, the agreement between the experimental and theoretical ZEKE-PFI is very satisfying. We notice that the simulated spectra in Ref. [116] are even in better agreement with the experimental ones than the simulated spectra reported in Ref. [113]. This is probably due to the difference in bond lengths calculated for the neutral Nb_3O system (see Fig. 1 of Ref. [116] and Fig. 4 of Ref. [113]).

III. MATCHING EXPERIMENT AND THEORY—CONDITIONS AND IMPROVEMENTS

Over the last 40 years, DFT has changed the face of computational chemistry in an impressive way. This is certainly true for theoretical cluster studies. They have now reached an accuracy that allows detailed comparison with experiments. This chapter shows some examples. Despite the huge success of DFT in cluster science, many problems still remain. In particular, the reliable calculation of atomization energies and reaction barriers for clusters are nontrivial tasks. For van der Waals systems, this even holds for simple structure optimization and characterization. Therefore, the development of DFT

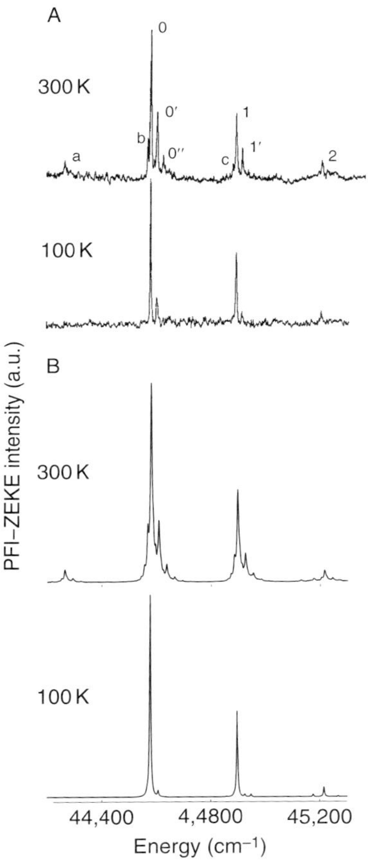

FIGURE 10 Experimental (A) and simulated (B) ZEKE-PFI spectra of Nb_3O at 300 and 100 K. The position of the 0–0 band has been shifted to the experimental value of 44,578 cm^{-1}.

functionals that improve these drawbacks of the current methodology represents one of the most challenging topics in theoretical chemistry and physics. A promising avenue lies in the merger between the conventional Kohn–Sham method and limited correlation interaction or propagator expansions. However, the double counting of interactions is a fundamental problem along this road. To the best of our knowledge, a satisfying solution remains to be reached.

In view of this situation, more pragmatic approaches, such as DFT-optimized basis sets or the DFT-D method described here, are suggested to extend the application range and accuracy of the current Kohn–Sham methodology. Even though these approaches are empirical in their formulation, they are extremely useful to gather more physical insight into our current approximations. Crucial ingredients for the further development are reliable experimental reference data that can be used to test and validate new theoretical methods.

Another important issue in the calculation of clusters of heavier elements is the inclusion of relativistic effects. The use of pseudo-potential methods has two main advantages in the treatment of these systems: easy inclusion of relativistic corrections and efficient evaluation of valence properties. Since bonding is mainly a valence property, pseudo-potentials are very useful for the study of clusters containing heavier elements as demonstrated in the applications to group V trimer monoxides.

For the more direct comparison with experiment, the inclusion of temperature and pressure seems to be one of the most logical next steps in DFT cluster studies. The simulation of vibrationally resolved photoelectron spectra based on DFT calculations represents a first attempt in this direction. Here, temperature is included via a Boltzmann distribution over the calculated harmonic vibrational levels. Despite the simple model, qualitative temperature effects can be observed. These simulations are at the moment also our most powerful technique to determine cluster structures.

The importance of cluster isomers is closely linked to the above discussion of temperature. In fact, in many experimental studies the question arises whether the recorded data are associated to one cluster structure or to an isomer mixture. Polarizability measurements as discussed in Section I.B.1 are typical examples. Not only is the polarizability itself temperature dependent but also the isomer mixture. To match experiment and theory in those cases, molecular dynamics studies become mandatory. The ADFT described here seems a promising avenue to realize Born–Oppenheimer molecular dynamics simulation on finite systems. A welcome byproduct of such a development could be a first-principles global optimization method. It is hoped that these developments will shift tedious human work to CPU load. In this way a reliable and representative dataset could be generated that will help to gain further insight into the basic rules of cluster chemistry and physics.

Acknowledgments

F. J. thanks the Deutsche Forschungsgemeinschaft (DFG) for a fellowship (Forschungsstipendium). P. C. acknowledges financial support from the CONACYT project U48775. R. F.-M. gratefully acknowledges a CONACYT Ph.D. fellowship 163442. D. R. S. is grateful to NSERC-Canada for support of this work over the years.

REFERENCES

[1] Johnston RL. Dalton Trans 2003;4193.
[2] Wales DJ, Scheraga HA. Science 1999;285:1368.
[3] Doye JPK, Wales DJ, Miller MA. J Chem Phys 1998;109:8143.

[4] Garzón IL, Michaelian K, Beltrán MR, Posada-Amarillas A, Ordejón P, Artacho E, et al. Phys Rev Lett 1998;81:1600.
[5] Salahub DR, Goursot A, Weber J, Köster AM, Vela A. In: Dykstra CE, Frenking G, Kim GE, Scuseria GE, editors. Theory and application of theoretical chemistry: the first forty years. Amsterdam: Elsevier; 2005.
[6] Flores-Moreno R. Ph.D. thesis. CINVESTAV.
[7] Köster AM, Calaminici P, Casida ME, Flores-Moreno R, Geudtner G, Goursot A, et al. The deMon developers, CINVESTAV, See also: http://www.demon-software.com; 2006.
[8] Hohenberg P, Kohn W. Phys Rev 1964;136:B864.
[9] Parr RG, Yang W. Density–functional theory of atoms and molecules. New York: Oxford University Press; 1989.
[10] Dreizler RM, Gross EKU. Density functional theory. Berlin: Springer-Verlag; 1990.
[11] Levy M. Proc Natl Acad Sci USA 1979;76:6062.
[12] Levy M, Perdew JP. In: Dreizler RM, da Providencia J, editors. Density functional methods in physics. New York: Plenum; 1985. 11pp.
[13] Zhao Q, Parr RG. Phys Rev A 1992;46:2337.
[14] Zhao Q, Parr RG. J Chem Phys 1993;98:543.
[15] Zhao Q, Morrison RC, Parr RG. Phys Rev A 1994;50:2138.
[16] Almbladh CO, Pedroza AC. Phys Rev A 1984;29:2322.
[17] Roothaan CCJ. Rev Mod Phys 1951;23:69.
[18] Hall GG. Proc R Soc Ser A 1951;205:541.
[19] Roothaan CCJ. Rev Mod Phys 1960;32:179.
[20] Binkley JS, Pople JA, Dobosh PA. Mol Phys 1974;28:1423.
[21] Pople JA, Nesbet RK. J Chem Phys 1954;22:571.
[22] Dunlap BI, Connolly JWD, Sabin JR. J Chem Phys 1979;71:4993.
[23] Mintmire JW, Dunlap BI. Phys Rev A 1982;25:88.
[24] Mintmire JW, Sabin JR, Trickey SB. Phys Rev B 1982;26:1743.
[25] Vahtras O, Almlöf J, Feyereisen MW. Chem Phys Lett 1993;213:514.
[26] Andzelm J, Radzio E, Salahub DR. J Comput Chem 1985;6:520.
[27] Andzelm J, Russo N, Salahub DR. J Chem Phys 1987;87:6562.
[28] Köster AM, Calaminici P, Escalante S, Flores-Moreno R, Goursot A, Patchkovskii S, et al. The deMon User´s Guide, Version 1.1 The deMon Developers, CINVESTAV, See also: http://www.demon-software.com; 2004.
[29] Press WH, Teukolsky SA, Vetterling WT, Flannery BP. Numerical recipes in FORTRAN. 2nd ed. Cambridge University Press; 1992.
[30] Gel´fand IM, Fomin SV. Calculus of variations. Englewood Cliffs, NJ: Prentice Hall; 1963.
[31] Köster AM. J Chem Phys 2003;118:9943.
[32] Baerends EJ, Ellis DE, Ros P. Chem Phys 1973;2:41.
[33] Sambe H, Felton RH. J Chem Phys 1975;62:1122.
[34] Casida ME, Daul C, Goursot A, Köster AM, Petterson LGM, Proynov E, et al. deMon-KS Version 3.4. deMon Software, Montréal.
[35] Andzelm J, Wimmer E. J Chem Phys 1992;96:1280.
[36] Trickey SB, Alford JA, Boettger JC. In: Leszcynski J, editor. Computational materials science, theoretical and computational chemistry, vol. 15. Amsterdam: Elsevier; 2004. 171pp.
[37] Laikov DN. Chem Phys Lett 1997;281:151.
[38] Köster AM. Habilitation thesis. Universität Hannover; 1998.
[39] Köster AM, Reveles JU, del Campo JM. J Chem Phys 2004;121:3417.

[40] Saunders VR. In: Diercksen GHF, Wilson S, editors. Methods in computational physics. Dordrecht: Reidel; 1983. 1pp.
[41] McWeeny R. Phys Rev 1962;126:1028.
[42] Diercksen G, McWeeny R. J Chem Phys 1966;44:3554.
[43] McWeeny R, Diercksen G. J Chem Phys 1968;49:4852.
[44] Raynes WT, Dodds JL, McWeeny R, Riley JP. Mol Phys 1977;33:611.
[45] McWeeny R, Dodds JL, Sadlej AJ. Mol Phys 1977;34:1779.
[46] McWeeny R. Methods of molecular quantum mechanics. 2nd reprinting London: Academic Press; 2001.
[47] Fournier R. J Chem Phys 1990;92:5422.
[48] Komornicki A, Fitzgerald G. J Chem Phys 1993;98:1398.
[49] Bonin KD, Kresin VV. In: Electric-dipole polarizabilities of atoms, molecules and clusters. Singapore: World Scientific; 1997.
[50] Narnes ANM, Turner DJ, Sutton LE. Trans Faraday Soc 1971;67:2902.
[51] Sutter H, Cole RH. J Chem Phys 1970;52:132.
[52] Bose TK, Sochanski JS, Cole RH. J Chem Phys 1972;57:3592.
[53] Meyer C, Morrison G. J Phys Chem 1991;95:3860.
[54] Goodwin ARH, Morrison G. J Phys Chem 1992;96:5521.
[55] Bridge NJ, Buckingham AD. Proc R Soc 1966;295:334.
[56] Alms GR, Burnham AK, Flygare WH. J Chem Phys 1975;63:3321.
[57] Bogaard MP, Buckingham AD, Pierens RK, White AH. J Chem Soc Faraday Trans I 1978;74:3008.
[58] Maltsev VA, Nerushev OA, Novopashin SA, Selivanov BA. Chem Phys Lett 1993;212:480.
[59] Boetter C, van Belle O, Bordewijk P, Rip A. Theory of electric polarization, vol. 1. In: 2nd ed. Amsterdam: Elsevier; 1973.
[60] Heer W, Milani P, Chatelain A. Phys Rev Lett 1989;63:2834.
[61] Milani P, Moullet I, Heer W. Phys Rev A 1990;42:5150.
[62] Heer W, Milani P. Rev Sci Instrum 1991;62:670.
[63] Stewart JJP, Kurtz HA, Dieter KM. J Comp Chem 1990;11:82.
[64] Calaminici P, Jug K, Köster AM. J Chem Phys 1998;109:7756.
[65] Armentrout PB. Annu Rev Phys Chem 2001;52:423.
[66] Rodgers MT, Armentrout PB. Mass Spec Rev 2000;19:215.
[67] Hales DA, Armentrout PB. J Cluster Sci 1990;1:127.
[68] Ervin KM, Armentrout PB. J Chem Phys 1985;83:166.
[69] Armentrout PB, Simons J. J Am Chem Soc 1992;114:8627.
[70] Khan FA, Clemmer DE, Schultz RH, Armentrout PB. J Phys Chem 1993;97:7978.
[71] Rodgers MT, Ervin KM, Armentrout PB. J Chem Phys 1997;106:4499.
[72] Martínez A, Calaminici P, Köster AM, Mitchell SA. Chem Phys Lett 1999;299:630.
[73] Schrader B, editor. Infrared and Raman spectroscopy. Weinheim: VCH Verlagsgesellschaft; 1995.
[74] Morse PM. Phys Rev 1929;34:57.
[75] Herzberg G. Molecular spectra and molecular structure III. Electronic spectra and electronic structure of polyatomic molecules. 4th ed. New York: Van Nostrand Reinhold Company; 1966.
[76] Komornicki A, McIver JW. J Chem Phys 1979;70:2014.
[77] Herzberg G. Molecular spectra and molecular structure I. Spectra of diatomic molecules. 2nd ed, New York: Van Nostrand Reinhold Company; 1950.

[78] Townes CH, Schawlow AL. Microwave spectroscopy. New York: Courier Dover Publications; 1975.
[79] Kroto HW. Molecular rotation spectra. New York: Courier Dover Publications; 2003.
[80] Nelson RD, Lide DR, Maryott AA. Technical report. Natl. Stand. Ref. Data. Ser., U.S. National Bureau of Standards, Washington; 1967.
[81] Dyke TR, Muenter JS. International review of science, physical chemistry II. Molecular structure and properties, volume II. London and Boston: Butterwoths; 1971. p. 27.
[82] Gillies JZ, Cillies CW, Lovas FL, Matsumura K, Suenram RD, Kraka E, et al. J Am Chem Soc 1991;113:6408.
[83] Andrews AM, Nemes L, Marucas SL, Hillig W, Kuczkowski RL, Muenter JS. J Mol Spectrosc 1993;160:422.
[84] Childs WJ. Physics Reports—Rev Sect Phys Lett 1992;211:113.
[85] Namiki KC, Robinson JS, Steimle TC. J Chem Phys 1998;109:5283.
[86] Chakraborty T, Lim EC. J Phys Chem 1993;97:11151.
[87] Bak A, Hutchinson JM, Holiday RJ. J Chem Phys 2002;116:4955.
[88] Akagi H, Yokoyama K, Yokoyama A. J Chem Phys 2003;118:3600.
[89] Ebenstein WL, Muenter JS. J Chem Phys 1984;80:3989.
[90] Shostak SL, Ebenstein WL, Muenter JS. J Chem Phys 1991;94:5875.
[91] Pino T, Parneix P, Douin S, Bréchignac P. J Phys Chem 2004;108:7364.
[92] Watkins WJ, Cockett MCR. J Chem Phys 2003;119:11636.
[93] Müller-Dethlefs K, Hobza P. Chem Rev 2000;100:143.
[94] Ellis A, Feher M, Wright T. Electronic and photoelectron spectroscopy fundamentals and case studies. Cambridge: Cambridge University Press; 2005.
[95] Signorelli R, Merkt F. Mol Phys 1997;92:793.
[96] Condon E. Phys Rev 1926;28:1182.
[97] Franck J. Trans Faraday Soc 1926;21:536
[98] Birge RT. Phys Rev 1926;28:1157.
[99] Born M, Oppenheimer JR. Ann. Physik 1927;84:457.
[100] Struve WS. Fundamentals of molecular spectroscopy. New York: Wiley; 1989.
[101] Held A, Schlag EW. Acc Chem Res 1998;31:467.
[102] Weinkauf R, Lehrer F. Faraday Discuss 2000;115:363.
[103] Müller-Dethlefs K, Schlag EW. Angew Chem Int Ed 1998;37:1347.
[104] Cockett MCR. Chem Soc Rev 2005;34:935.
[105] Müller-Dethlefs K, Sander M, Schlag EW. Chem Phys Lett 1984;112:291.
[106] Müller-Dethlefs K, Schlag EW. Annu Rev Phys Chem 1991;42:109.
[107] Peatman WB, Borne TB, Schlag EW. Chem Phys Lett 1969;3:492.
[108] Chupka WA. J Chem Phys 1993;98:4520.
[109] Dietrich HJ, Müller-Dethlefs K, Baranov LY. Phys Rev Lett 1996;76:3530.
[110] Doktorov EV, Malkin IA, Man'ko VI. J Mol Spectrosc 1975;56:2.
[111] Doktorov EV, Malkin IA, Man'ko VI. J Mol Spectrosc 1977;64:302.
[112] Note that there is an error in the recursion relation of Refs. [110] and [111]. SeeCromp B, Triest T, Carrington Jr. T, Reber C. Spectrochimica Acta A 1999;55:575.
[113] Yang D-S, Zgierski MZ, Rayner DM, Hackett P, Martínez A, Salahub DR, et al. J Chem Phys 1995;103:5335.
[114] Calaminici P, Köster AM, Carrington Jr. T, Roy P-N, Russo N, Salahub DR. J Chem Phys 2001;114:4036.
[115] Calaminci P, Köster AM, Salahub DR. J Chem Phys 2003;118:4913.
[116] Calaminici P, Flores-Moreno R, Köster AM. J Chem Phys 2004;121:3558.

[117] Zgierski MZ. Chem Phys 1986;108:61.
[118] Zgierski MZ, Zerbetto F. J Chem Phys 1993;99:3721.
[119] Knight WD, Clemenger K, de Heer WA, Saunders WA. Phys Rev B 1985;31:2539.
[120] Rayane D, Allouche AR, Benichou E, Antoine R, Aubert-Frecon M, Dugourd P, et al. Eur Phys J D 1999;9:243.
[121] Tikhonov G, Kasperovich V, Wong K, Kresin VV. Phys Rev A 2001;64:063202.
[122] Manninen M, Nieminen RM, Puska MJ. Phys Rev B 1986;33:4289.
[123] Moullet I, Martins JL, Reuse F, Buttet J. Phys Rev Lett 1990;65:476.
[124] Guan J, Casida ME, Köster AM, Salahub DR. Phys Rev B 1995;52:2184.
[125] Harbola MK. Solid State Commun 1996;98:629.
[126] Calaminici P, Jug K, Köster AM. J Chem Phys 1999;111:4613.
[127] Kronik L, Vasiliev I, Chelikowsky JR. Phys Rev B 2000;62:9992.
[128] Kümmel S, Berkus T, Reinhard P-G, Brack M. Eur Phys J D 2000;11:239.
[129] Solov'yov IA, Solov'yov AV, Greiner W. Phys. Rev. A 2002;65:053203.
[130] Chandrakumar KRS, Ghanty TK, Ghosh SK. J Chem Phys 2004;120:6487.
[131] Vosko SH, Wilk L, Nusair M. Can J Phys 1980;58:1200.
[132] Perdew JP, Burke K, Ernzerhof M. Phys Rev Lett 1996;77:3865.
[133] Godbout N, Salahub DR, Andzelm J, Wimmer E. Can J Chem 1992;70:560.
[134] Knickelbein MB. J Chem Phys 2004;120:10450.
[135] Calaminici P, Köster AM, Vela A, Jug K. J Chem Phys 2000;113:2199.
[136] Becke AD. Phys. Rev. A 1988;38:3098.
[137] Perdew JP. Phys Rev B 1986;33:8822.
[138] Jug K, Zimmermann B, Calaminici P, Köster AM. J Chem Phys 2002;116:4497.
[139] Benichou E, Antoine R, Rayane D, Vezin B, Dalby FW, Dugourd P, et al. Phys Rev A 1999;59:R1.
[140] Jaque P, Toro-Labbe A. J Chem Phys 2002;117:3208.
[141] Yang ML, Jackson KA. J Chem Phys 2005;122:184317.
[142] Su C-X, Hales DA, Armentrout PB. J Chem Phys 1993;99:6613.
[143] Cox DM, Whetten RL, Zakin MR, Trevor DJ, Reichmann KC, Kaldor A. In: Stwalley WC, Lapp M, editors. Opticla Science and Engineering Ser. 6, Advances in Laser Science, AIP Conf. Proc. 146, volume I. New York: AIP; 1986. p. 527.
[144] Sugar JS, Corliss C. J Phys Chem Ref Data 1985;14:202.
[145] Spain EM, Morse MD. Int J Mass Spectrom Ion Process 1990;102:183.
[146] Spain EM, Morse MD. J Phys Chem 1992;96:2479.
[147] Walch SP, Bauschlicher Jr. CW, Roos BO. Chem Phys Lett 1983;103:175.
[148] Salahub DR, Baykara NA. Surf Sci 1985;156:605.
[149] Grönbeck H, Rosén A. J Chem Phys 1997;107:10620.
[150] Wu X, Ray AK. J Chem Phys 1999;110:2437.
[151] Janak JF, Moruzzi VL, Wiliams AR. Phys Rev B 1975;12:1257.
[152] Boch BH, Dunning Jr TH, Harrison JF. J Chem Phys 1981;75:3466.
[153] Perdew JP, Wang Y. Phys Rev B 1992;45:13244.
[154] Wachters AJH. J Chem Phys 1970;52:1033.
[155] Calaminici P, Janetzko F, Köster AM, Mejia-Olvera R, Zuniga-Gutierrez B. J Chem Phys 2007;126:044108.
[156] Köster AM, Krack M, Leboeuf M, Zimmermann B. ALLCHEM. Univeristät Hannover; 1998.
[157] Perdew JP, Wang Y. Phys Rev B 1986;33:8800.
[158] Perdew JP. Phys Rev B 1986;34:7406.
[159] Calaminici P, Köster AM, Russo N, Salahub DR. J Chem Phys 1996;105:9546.

[160] St-Amant A, Salahub DR. Chem Phys Lett 1990;169:387.
[161] Salahub DR, Fournier R, Mlynarski P, Papai I, St-Amant A, Ushio J. In: Labanowski L, Andzelm J, editors. Density functional methods in chemistry. New York: Springer; 1991.
[162] St-Amant A. Ph.D. thesis. Université de Montréal; 1991.
[163] Schlegel HB. In: Lawley KP, editor. Ab initio methods in quantum chemistry-I. New York: Wiley; 1987.
[164] Jug K, Zimmermann B, Köster AM. Int J Quant Chem 2002;90:594.
[165] Ingólfsson O, Busolt U, Sugawara KI. J Chem Phys 2000;112:4613.
[166] Spasov VA, Lee TH, Ervin KM. J Chem Phys 2000;112:1713.
[167] Ho J, Ervin KM, Lineberger WC. J Chem Phys 1990;93:6987.
[168] Moor CE. Atomic energy levels, vol. II. Nat. Bur. Standards; 1971.
[169] James AM, Lemire GW, Langridge-Smith PRR. Chem Phys Lett 1994;277:503.
[170] Knickelbein MB. Chem Phys Lett 1992;192:129.
[171] Powers DE, Hansen GS, Geusic ME, Michalopulos DL, Smalley RE. J Chem Phys 1983;78:2866.
[172] Jansen HB, Ross P. Chem Phys Lett 1969;3:140.
[173] Krückeberg S, Schweikhard L, Ziegler J, Dietrich G, Lützenkirchen K, Walther C. J Chem Phys 2001;114:2955.
[174] Lee C, Yang W, Parr RG. Phys Rev B 1988;37:785.
[175] Moskovits M, DiLella DP, Limm W. J Chem Phys 1984;80:626.
[176] Ram RS, Jarman PF. C N and Bernath. J Mol Spectrosc 1992;156:468.
[177] Cossé C, Fouassier M, Mejean T, Tranquille M, DiLella DP, Moskovits M. J Chem Phys 1980;73:6076.
[178] Doverstål M, Karlsson L, Lindgren B, Sassenberg U. Chem Phys Lett 1997;270:273.
[179] Moskovits M, Hulse JE. J Chem Phys 1977;66:3988.
[180] Ahmed F, Nixon ER. J Chem Phys 1979;71:3547.
[181] Ho J, Polak ML, Ervin KM, Lineberger WC. J Chem Phys 1993;99:8542.
[182] Wang H, Haouari H, Craig R, Lombardi JR, Lindsay DM. J Chem Phys 1996;104:3420.
[183] Castro M, Jamorski C, Salahub DR. Chem Phys Lett 1997;271:133.
[184] Moskovits M, DiLella DP. J Chem Phys 1980;73:4917.
[185] Dong J-G, Hu Z, Craig R, Lombardi JR, Lindsay DM. J Chem Phys 1994;101:9280.
[186] Montano PA, Shenoy GK. Solid State Commun 1980;35:53.
[187] Purdum H, Montano PA, Shenoy GK, Morrison T. Phys Rev B 1982;25:4412.
[188] Leopold DG, Lineberger WC. J Chem Phys 1986;85:51.
[189] Leopold DG, Almlöf J, Lineberger WC, Taylor PR. J Chem Phys 1988;88:3780.
[190] Baumann CA, Van Zee RJ, Weltner Jr W. J Phys Chem 1984;88:1815.
[191] Chrétien S, Salahub DR. Phys Rev B 2002;66:155425.
[192] Gutsev GL, Bauschlicher Jr. CW. J Phys Chem A 2003;107:7013.
[193] Castro M, Salahub DR. Phys Rev B 1994;49:11842.
[194] Rollmann G, Herper HC, Entel P. J Phys Chem A 2006;110:10799.
[195] Hübner O, Sauer J. Chem Phys Lett 2002;358:442.
[196] Bauschlicher Jr. CW, Ricca A. Mol Phys 2003;101:93.
[197] Dyke TR. Top Curr Chem 1984;120:85.
[198] Legon AC, Millen DJ. Chem Rev 1986;86:635.
[199] Leopold KR, Fraser GT, Novick SE, Klemperer W. Chem Rev 1994;94:1807.
[200] Xu Y, Van Wijngaarden J, Jäger W. Int Rev Phys Chem 2005;24:301.
[201] Buckingham AD, Fowler PW, Hutson JM. Chem Rev 1988;88:963.
[202] Hobza P. Annu Rep Prog Chem Sect C: Phys Chem 1997;93:257.

[203] Dunning Jr TH. J Chem Phys 1989;90:1007.
[204] Woon DE, Dunning Jr TH. J Chem Phys 1995;103:4572.
[205] Jurečka P, Černý J, Hobza P, Salahub DR. J Comp Chem 2006;28:556.
[206] Zhao Y, Schultz NE, Truhlar DG. J Chem Theory Comput 2006;2:364.
[207] Zhao Y, Truhlar DG. J Chem Phys 2006;125:194101.
[208] Zhang Y, Salahub DR. Chem Phys Lett 2007;436:394.
[209] Zhang Y, Vela A, Salahub DR. Theor Chem Acc 2007;118:693.
[210] Proynov EI, Chermette H, Salahub DR. J Chem Phys 2000;113:10013.
[211] Perdew JP, Tao JM, Staroverov VN, Scuseria GE. J Chem Phys 2004;120:6898.
[212] Wu Q, Yang W. J Chem Phys 2002;116:515.
[213] Bondi A. J Phys Chem 1964;68:441.
[214] Tao J, Perdew JP, Staroverov VN, Scuseria GE. Phys Rev Lett 2003;91:146401.
[215] Vanden Bout PA, Steed JM, Bernstein LS, Klemperer W. Astrophys J 1979;234:503.
[216] Surin LA, Fourzikov DN, Lewen F, Dumesh BS, Winnewisser G, McKeller ARW. J Mol Spectrosc 2003;222:93 and references therein.
[217] Roth DA, Hepp M, Pak I, Winnewisser G. Chem Phys Lett 1998;298:381.
[218] Roth DA, Pak I, Surin LA, Dumesh BS, Winnewisser G. J Chem Phys 2000;113:3034.
[219] Tang J, McKeller ARW, Surin LA, Fourzikov DN, Dumesh BS, Winnewisser G. J Mol Spectrosc 2002;214:87.
[220] Walker KA, Xia C, McKeller ARW. J Chem Phys 2000;113:6618.
[221] Brooks MD, McKeller ARW. Chem Phys Lett 1998;287:365.
[222] Brooks MD, McKeller ARW. J Chem Phys 1999;111:7321.
[223] McKeller ARW. J Mol Spectrosc 2004;226:190 and references therein.
[224] Vissers GWM, Wormer PES, van der Avoird A. Phys Chem Chem Phys 2003;5:4767.
[225] Rode M, Sadlej J, Moszynski R, Wormer PES, van der Avoird A. Chem Phys Lett 2001;334:424.
[226] Rode M, Sadlej J, Moszynski R, Wormer PES, van der Avoird A. Chem Phys Lett 1999;314:326.
[227] Vissers GWM, Heßelmann A, Jansen G, Wormer PES, van der Avoirda A. J Chem Phys 2005;122:054306.
[228] Sloan Jr. ED. Ind Eng Chem Res 2000;39:3123.
[229] Stern A, Kirby SH. J Phys Chem A 2001;105:4664.
[230] Moon C, Taylor PC, Rodger PM. Can J Phys 2003;81:451.
[231] Radhakrishnan R, Trout BL. J Chem Phys 2002;117:1786.
[232] Dore L, Cohen RC, Schmuttenmaer CA, Busarow KL, Elrod MJ. J Chem Phys 1994;100:863.
[233] Suenram RD, Fraser GT, Lovas FJ, Kawashima Y. J Chem Phys 1994;101:7230.
[234] Mineva T, Goursot A, Talbi D, Köster AM, del Campo JM, Zhang Y, Salahub DR. (unpublished results).
[235] Zhang Y, Yang W. Phys Rev Lett 1998;80:890.
[236] Szczęśniak MM, Chalasiński G, Cybulski SM, Cieplak P. J Chem Phys 1993;98:3078.
[237] Akin-Ojoa O, Szalewicz K. J Chem Phys 2005;123:134311.
[238] Menapace JA, Bernstein ER. J Phys Chem 1987;91:2533.
[239] Sun S, Bernstein ER. J Phys Chem 1996;100:13348.
[240] Janda KC, Hemminger JC, Win JS, Novick SE, Harris SJ, Klemperer W. J Chem Phys 1975;63:1419.
[241] Steed JM, Dixon TA, Klemperer W. J Chem Phys 1979;70:4940.
[242] Schauer M, Bernstein ER. J Chem Phys 1985;82:3722.

[243] Law KS, Schauer M, Bernstein ER. J Chem Phys 1984;81:4871.
[244] Börnstein KO, Selzle HL, Schlag EW. J Chem Phys 1986;85:1726.
[245] Arunan E, Gutowsky HS. J Chem Phys 1993;98:4294.
[246] Henson BF, Hartland GV, Venturo VA, Felker PM. J Chem Phys 1992;97:2189.
[247] Schlag EW, Selzle HL. J Chem Soc Faraday Trans 1990;86:1.
[248] Williams DE. Acta Crystallogr 1980;A36:715.
[249] Van der Waal BW. Chem Phys Lett 1986;123:69.
[250] Fraga S. J Comp Chem 1982;3:329.
[251] Rubio M, Torrens F, Sanchez-Maryn J. J Comp Chem 1993;14:647.
[252] Hobza P, Selzle HL, Schlag EW. J Phys Chem 1993;97:3937.
[253] Hobza P, Selzle HL, Schlag EW. J Am Chem Soc 1994;116:3500.
[254] Hobza P, Selzle HL, Schlag EW. J Phys Chem 1996;100:18790.
[255] Tsuzuki S, Honda K, Uchimaru T, Mikami M, Tanabe K. J Am Chem Soc 2002;124:104.
[256] Spirko V, Engkvist O, Soldan P, Selzle HL, Schlag EW, Hobza P. J Chem Phys 1999;111:572.
[257] Johnson ER, Wolkow RA, Dilabio GA. Chem Phys Lett 2004;394:334.
[258] Grimme S. J Comp Chem 2004;25:1463.
[259] Green SME, Alex S, Fleischer NL, Millam EL, Marcy TP, Leopold DG. J Chem Phys 2001;114:2653.
[260] Calaminici P, Köster AM. Int J Quant Chem 2003;91:317.
[261] Flores-Moreno R, Alvarez-Mendez RJ, Vela A, Köster AM. J Comp Chem 2006;27:1009.
[262] Calaminici P, Flores-Moreno R, Köster AM. Comput Lett 2005;1:164.
[263] Bergner A, Dolg M, Küchle W, Stoll H, Preuß H. Theor Chim Acta 1990;77:123.
[264] Andrae D, Häußermann U, Dolg M, Stoll H, Preuß H. Theor Chim Acta 1990;77:123.
[265] Reveles JU. Ph.D. thesis. CINVESTAV; 2004.
[266] Yang D-S, Zgierski MZ, Berces A, Hackett PA, Roy P-N, Martínez A, et al. J Chem Phys 1996;105:10663.
[267] Yang D-S, Zgierski MZ, Berces A, Hackett PA, Martínez A, Salahub DR. Chem Phys Lett 1997;227:71.
[268] Yang D-S, Zgierski MZ, Hackett PA. J Chem Phys 1998;108:3591.
[269] Yang D-S, James AM, Rayner DM, Hackett PA. Chem Phys Lett 1994;231:177.
[270] Salahub DR, Messmer RP. Surf Sci 1981;106:415.
[271] Liu F, Khanna S, Jena P. Phys Rev B 1991;43:8179.
[272] Walch SP, Bauschlicher Jr. CW. J Chem Phys 1985;83:5735.
[273] Dreyssé H, Dorantes-Davila J, Vega A, Balbas L-C, Bouarab S, Nait-Laziz H, et al. J Appl Phys 1993;73:6207.
[274] Lee K, Callaway J. Phys Rev B 1993;48:15358.
[275] Lee K, Callaway J. Phys Rev B 1994;49:13906.
[276] Zhao J, Chen X, Sun Q, Liu F, Wang G, Lain KD. Physica B 1995;215:177.
[277] Krack M, Köster AM. NATO ASI on Metal-Ligand Interactions in Chemistry, Physics and Biology. 1998, Cetraro, Italy, September 01–12.
[278] Köster AM, Calaminici P, Gómez Z, Reveles U. In: Sen K, editor. Reviews of modern quantum chemistry, a celebration of the contribution of Robert G Parr. Singapore: World Scientific Publishing Co; 2002.
[279] Mejia-Olvera R. Licenciatura thesis. CINVESTAV; 2005.
[280] Fontijin A, editor. Gas-phase metal reactions. Amsterdam: Elsevier Science; 1992.
[281] Armentrout PB, Kickel BL. In: Freiser BS, editor. Organometallic ion chemistry. Dordrecht: Kluwer Academic; 1996.

[282] Fournier R, Papai I. In: Chong DP, editor. Recent advances in density functional methods. Singapore: World Scientific; 1995.
[283] Balducci G, Gigli G, Guido M. J Chem Phys 1983;79:5616.
[284] Loock HP, Simard B, Wallin S, Linton C. J Chem Phys 1998;109:8980.
[285] Jones LH. Inorganic vibrational spectroscopy. New York: Dekker; 1971.
[286] Bauschlicher Jr CW, Maitre P. Theor Chim Acta 1995;90:189.
[287] Bauschlicher Jr. CW, Langhoff SR. J Chem Phys 1986;85:5936.
[288] Somorjai GA. Introduction to surface chemistry and catalysis. New York: Wiley-Interscience; 1994.
[289] Cox PA. Transition metal oxides. Oxford: Claredon; 1992.
[290] Marcy TP, Leopold DG. Int J Mass Spectrom 2000;195/196:653.
[291] Simard B, Mitchell SA, Rayner DM, Yang DS. In: Russo N, Salahub DR, editors. Metal-ligand interactions in chemistry, physics and biology. Dordrecht: Kluwer Academic; 2000.

Multiple Aromaticity, Multiple Antiaromaticity, and Conflicting Aromaticity in Planar Clusters

Dmitry Yu. Zubarev and Alexander I. Boldyrev
Department of Chemistry and Biochemistry, Utah State University, Logan, Utah, USA

Chapter Outline Head

Nanoclusters. DOI: 10.1016/S1875-4023(10)01005-3

I. INTRODUCTION

Clusters of atoms represent a very special class of chemical species, which are different from solid state substances and conventional individual molecules. As an intermediate state of matter, they play a significant role in chemistry [1]. In spite of enormous amount of work in the past three decades, there is no simple chemical bonding model, beyond the jelly model, that can describe chemical bonding in metal and nonmetal clusters as efficiently and robustly as the Lewis model in organic chemistry.

However, certain progress has been made in recent years in developing chemical bonding models for clusters (see Refs. [2–14] and references therein). The major success in advancing the theory of chemical bonding in clusters is based on the application of the aromaticity and antiaromaticity concepts. Since the introduction of the aromaticity/antiaromaticity in aluminum clusters [2b,c], these ideas received significant development and it was understood that the concepts of aromaticity and antiaromaticity in clusters have very specific flavor if compared with organic chemistry where they came from. The many-fold nature of aromaticity, antiaromaticity, and conflicting aromaticity is a common feature of bonding in clusters [2a,f,g]. In addition to the delocalized bonding one also must account for the localized two-center two-electron (2c–2e) bonds and lone pairs in chemical bonding models.

In this chapter, we present a comprehensive analysis of chemical bonding in planar clusters on the basis of aromaticity/antiaromaticity. First, we would like to introduce all the possible modes of aromaticity on the basis of the simplest X_3 clusters.

II. POSSIBLE TYPES OF AROMATICITY AND ANTIAROMATICITY IN X_3 CLUSTERS

When only s-atomic orbitals (AOs) are involved in chemical bonding, one may expect only σ-aromaticity/σ-antiaromaticity in X_3 systems. If p-AOs are involved in chemical bonding, σ-tangential (σ_t-), σ-radial (σ_r-), and π-aromaticity/antiaromaticity could occur. In this case, there can be multiple (σ- and π-) aromaticity, multiple (σ- and π-) antiaromaticity, and conflicting aromaticity (simultaneous σ-aromaticity and π-antiaromaticity or σ-antiaromaticity and π-aromaticity). If d-AOs are involved in chemical bonding, σ-tangential (σ_t-), σ-radial (σ_r-), π-tangential (π_t-), π-radial (π_r-), and δ-aromaticity/antiaromaticity could occur. In this case, there can be multiple (σ-, π-, and δ-) aromaticity, multiple (σ-, π-, and δ-) antiaromaticity, and conflicting aromaticity (simultaneous σ-aromaticity, π-aromaticity, and δ-antiaromaticity; σ-aromaticity, π-antiaromaticity, and δ-aromaticity; σ-antiaromaticity, π-aromaticity, and δ-aromaticity; σ-aromaticity, π-antiaromaticity, and δ-antiaromaticity; σ-antiaromaticity, π-aromaticity, and δ-antiaromaticity; σ-antiaromaticity, π-antiaromaticity, and δ-aromaticity).

A. s-AO-Based σ-Aromaticity and σ-Antiaromaticity in X_3 Clusters

The σ-aromaticity in the simplest metal cluster Li_3^+ was initially discussed by Alexandrova and Boldyrev [15], and then by Havenith et al. [16] and Wang et al. [14d]. The global minimum structure of Li_3^+ is a perfect triangle [15–17]. The $1a_1'$-valence MO is a sum of the 2s-AOs of three lithium atoms (Figure 1A). It is completely bonding and in this sense similar to the completely bonding π-MO in the prototypical π-aromatic $C_3H_3^+$ cation (Figure 1B) [17b,c].

The only difference is that the π-MO is a sum of $2p_z$-AOs of carbons. The delocalized π-MO in $C_3H_3^+$ renders its π-aromaticity according to the famous $4n+2$ Huckel rule. On the basis of the analogy between the π-delocalized

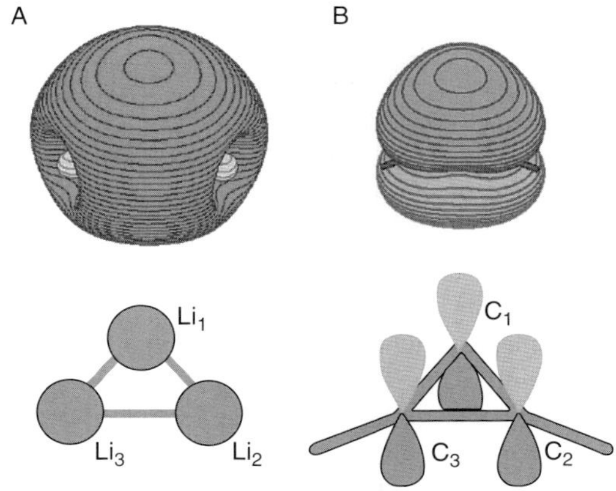

FIGURE 1 (A) $1a_1'$-HOMO of Li_3^+ and its schematic representation as a linear combination of 2s-AOs of Li atoms, (B) $1a_2''$-HOMO of $C_3H_3^+$ and its schematic representation as a linear combination of $2p_z$-AOs of C atoms.

MO in $C_3H_3^+$ and the σ-delocalized MO in Li_3^+, we would like to call the latter cation σ-aromatic. Wang et al. [17a] confirmed aromaticity in Li_3^+, as well as in Na_3^+, K_3^+, and Cu_3^+ cations on the basis of negative values of nucleus-independent chemical shift (NICS) [18]. However, Havenith et al. [16] on the basis of absence of ring current in Li_3^+ concluded that it is not a σ-aromatic system. We believe that the final judgment in establishing aromaticity in our Alk_3^+ cations should be on the basis of the evaluation of the resonance energy, high symmetry, and the $4n+2$ electron counting rule for cyclic structures. It is however inconvenient to evaluate the resonance energy for a cation. We therefore optimized geometry for the Li_3Cl neutral molecule, containing the Li_3^+ cation and the Cl^- anion [15]. The bidentate structure of Li_3Cl (C_{2v}, I, $^1A_1$) with Cl^- coordinated to the edge of the Li_3^+ triangular ion is the global minimum according to calculations by Alexandrova and Boldyrev [15], Dugourd et al. [19a], and Durand et al. [19b] The Cl^- anion only slightly perturbs the σ-aromatic HOMO in Li_3Cl when compared to the isolated Li_3^+ cation [15]. The σ-resonance energy in the Li_3^+ cation can be calculated as the energy of the reaction (1)

$$Li_3Cl\ (C_{2v},{}^1A_1) \rightarrow Li_2 + LiCl, \qquad (1)$$

where Li_2 and LiCl are reference classical molecules. According to our calculations the energy of the reaction (1) was found to be 35.7 kcal/mol (CCSD(T)/6-311+G(2df)//CCSD(T)/6-311+G*+ZPE/CCSD(T)/6-311+G*). The calculated resonance energy is certainly very high compared to the Li_2 dissociation energy (23.1 kcal/mol at the same level of theory). Thus, we believe that the use of the σ-aromaticity for the description of the Li_3^+ cation as well as for all other alkali metal cations is justified [15].

The aromatic counting rule for σ-electrons is $4n+2$ (singlet coupling) if only the s-AOs are responsible for bonding. Then, for σ-antiaromatic species, the counting rule is $4n$ (singlet coupling). The Li_3^- anion is a good example of σ-antiaromatic system with 4σ-electrons. The electronic configuration for the singlet state of Li_3^- at the D_{3h} symmetry is $1a_1'^2 1e'^2$, and the triangular structure with the singlet electronic state must undergo the Jahn–Teller distortion toward linear $D_{\infty h}$ structure with the $1\sigma_g^2 1\sigma_u^2$ valence electronic configuration. Indeed, the linear structure for Li_3^- is well documented [20]. Two σ-delocalized MOs can be approximately localized into two 2c–2e bonds and the linear structure of Li_3^- can be formally considered as a classical structure [15]. This situation is similar to the antiaromatic cyclobutadiene structure, which can be considered as having two double and two single carbon–carbon bonds, and, thus, being formally a classical structure. The antiaromaticity should manifest itself in the reduction of stability of the molecule. Two reactions below show that the atomization energy of Li_3^- (reaction (2), CCSD(T)/6-311+G(2df)//CCSD(T)6-311+G*+ZPE/CCSD(T)/6-311+G*) is indeed substantially lower than the atomization energy of Li_3^+ (reaction (3), CCSD(T)/6-311+G(2df)//CCSD(T)6-311+G*+ZPE/CCSD(T)/6-311+G*).

$$Li_3^-(D_{\infty h},{}^1\Sigma_g^+) \rightarrow 2Li\ ({}^2S) + Li^-({}^1S)\Delta E = +42.2\ \text{kcal/mol}, \quad (2)$$

$$Li_3^+(D_{3h},{}^1A_1') \rightarrow 2Li\ ({}^2S) + Li^+({}^1S)\Delta E = +65.0\ \text{kcal/mol}. \quad (3)$$

When two more electrons are added to Li_3^- leading to Li_3^{3-} cluster, the number of σ-electrons again satisfies the $4n+2$ rule and the corresponding cluster is expected to be aromatic again. In order to avoid dealing with the highly unstable triply charged Li_3^{3-} anion, let us switch to the isoelectronic Mg_3 cluster. It was found that Mg_3 (D_{3h}, ${}^1A_1'$, $1a_1'^2 1e'^4$) is a weakly bound van der Waals complex [21a] with the atomization energy of 5.2 kcal/mol (CCSD(T)/6-311+G(2df)//CCSD(T)6-311+G*+ZPE/CCSD(T)/6-311+G*). In order to have aromaticity, we need to have delocalized bonding, but when all the orbitals formed out of the same AOs are completely occupied this leads to formation of lone pairs. The bonding, nonbonding, and antibonding MOs for Mg_3 are of the same type as MOs of Li_3 (Figure 2).

Apparently, when all bonding, nonbonding, and antibonding MOs composed of the 3s AOs in Mg_3 are occupied the result is formation of three $3s^2$ lone pairs at each Mg atom with almost zero net bonding effect. The same

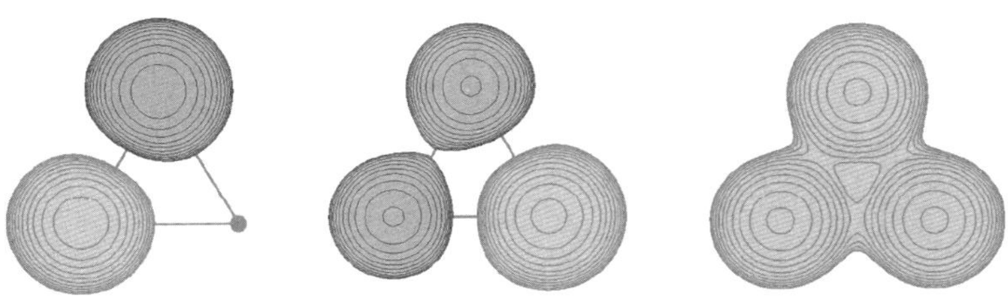

FIGURE 2 All valence molecular orbitals for the Mg_3 (D_{3h}, ${}^1A_1'$) cluster.

holds true for the reference hydrocarbon aromatic molecules. If we were able to make a $C_5H_5^{5-}$ pentaanion with all bonding, nonbonding, and antibonding π-MOs completely occupied (with 10 π-electrons satisfying the $4n+2$ rule for $n=2$) we would have five $2p_z$ lone pairs with close to zero contribution to bonding.

B. p-AO-Based π-Aromaticity in X_3 Clusters

It was shown [21a] that Mg_3^{2-} within either $NaMg_3^-$ or Na_2Mg_3 has the $1a_1'^2 1e'^4 1a_2''^2$ electronic configuration with the highest occupied molecular orbital (HOMO) being completely bonding π-MO, which is similar to the π-bonding MO of $C_3H_3^+$ (Figure 1B). The first three MOs ($1a_1'$ and $1e'$) are primarily formed by 3s-AOs of Mg atoms. Because bonding ($1a_1'$) and antibonding ($1e'$) orbitals are completely occupied, these MOs can be localized into three $3s^2$ lone pairs located at each Mg atom. Thus, bonding in these species comes primarily from this π-bonding MO. In these clusters, π-aromaticity occurs without formation of the σ-bonding framework, which is highly unusual for chemical species.

C. p-AO-Based Double (π- and σ-) Aromaticity in X_3 Clusters

Analysis of the chemical bonding in the B_3^-, Al_3^-, and Ga_3^- species revealed that these species are doubly aromatic [17b,c]. All these clusters have the $1a_1'^2 1e'^4 1a_2''^2 2a_1'^2$ electronic configuration (valence MOs for Ga_3^- are shown in Figure 3).

Although the first three MOs ($1a_1'$ and $1e'$) can be localized into three 2c–2e B–B bonds in B_3^- (the occupation number [ON] for the localized by natural bond analysis (NBO) [22] B–B bond is 1.88 |e| compared to the ideal number 2.00 |e|), their localization would lead to three lone pairs in Al_3^- (ON = 1.78 |e|) and Ga_3^- (ON = 1.84 |e|). The $2a_1'$-HOMO is completely bonding σ-radial (σ_r-) orbital formed by *n*p-AOs and this MO is responsible for σ-aromaticity in these three clusters. The $1a_2''$-HOMO-1 is a completely bonding π-orbital (similar to π-HOMO in $C_3H_3^+$) and this MO is responsible for the π-aromaticity in the B_3^-, Al_3^-, and Ga_3^- species. Thus, these clusters are doubly (σ- and π-) aromatic. The double aromaticity leads to rather high atomization energies (all at the CCSD(T)/6-311+G(2df)//CCSD(T)/6-311+G*+ZPE/CCSD(T)/6-311+G*) for all three species:

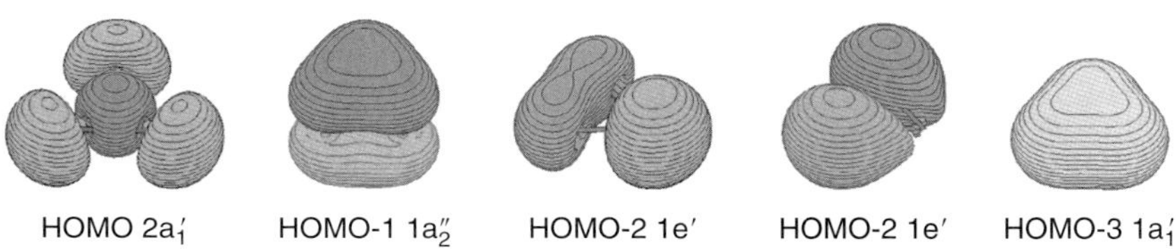

FIGURE 3 Valence molecular orbitals of Ga_3^- (similar to those of B_3^- and Al_3^-).

$$D_0(B_3^-(D_{3h},{}^1A_1') \rightarrow 2B\ (^2P) + B^-(^3P)) = +247.6\ \text{kcal/mol}, \quad (4)$$

$$D_0(Al_3^-(D_{3h},{}^1A_1') \rightarrow 2Al(^2P) + Al^-(^3P)) = +118.6\ \text{kcal/mol}, \quad (5)$$

$$D_0(Ga_3^-(D_{3h},{}^1A_1') \rightarrow 2Ga\ (^2P) + Ga^-(^3P)) = +107.7\ \text{kcal/mol}, \quad (6)$$

From these dissociation energies, it is very clear that the significantly higher D_0 for the B_3^- anion is due to the presence of the three 2c–2e B–B bonds, which are absent in the Al_3^- and Ga_3^- anions.

In addition to the high D_0, the double aromaticity in these clusters leads to high symmetry (D_{3h}), high resonance energy (see Ref. [10] for the resonance energy in Al_3^-), high first vertical electron detachment energy [17b] (VDE (B_3^-) 2.72 eV, VDE(Al_3^-) = 1.73 eV, and VDE = 1.69 eV all at OVGF/6-311+G(2df)//CCSD(T)/6-311+G*), and significantly negative NICS values (B_3^-: NICS(0.0) = −73.6 ppm, NICS(0.5) = −57.9 ppm, NICS(1.0) = −28.2 ppm; Al_3^-: NICS(0.0) = −35.8 ppm, NICS(0.5) = −33.9 ppm, NICS (1.0) = −26.7 ppm; Ga_3^-: NICS(0.0) = −27.3 ppm, NICS(0.5) = −26.8 ppm, NICS(1.0) = −22.6 ppm; all at B3LYP/6-311+G*).

D. p-AO-Based Conflicting Aromaticity in X_3 Clusters

The joint experimental and theoretical study of $AlSi_2^-$, $AlSiGe^-$, and $AlGe_2^-$ anions [21b] revealed that these species have unexpected global minimum C_s ($^1A'$) structures shown in Figure 4A–C.

The structures of these anions as well as the well documented C_{2v} ($^1A_1$) structure of Si_3 [23] (Figure 4D) were rationalized using the concept of conflicting aromaticity. In Figure 5A and B, we presented MOs of the Si_3 and $AlGe_2^-$ species, respectively.

Bonding in Si_3 can be understood starting with Si_3^{2+} (which is isoelectronic to the previously considered Al_3^-, see Section II.C) doubly aromatic species. The 2e′-LUMO in Si_3^{2+} is a doubly degenerate antibonding σ-MO. Its partial occupation in the Si_3 cluster leads to the Jahn–Teller distortion of the D_{3h} structure towards the C_{2v} structure. The HOMO-1 $2b_2$ of Si_3 (Figure 5A) adds

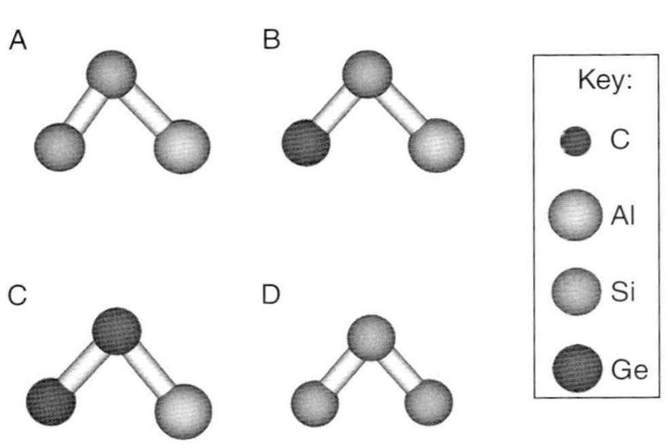

FIGURE 4 Calculated geometric structures of (A) $AlSi_2^-$, (B) $AlSiGe^-$, (C) $AlGe_2^-$, and (D) Si_3.

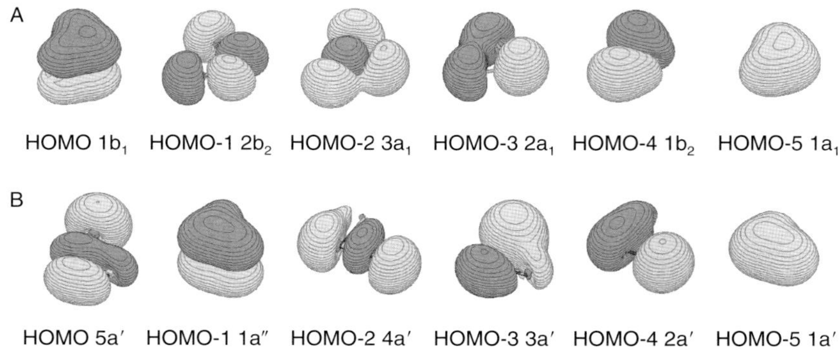

FIGURE 5 Molecular orbitals of (A) Si_3 and (B) $AlGe_2^-$.

antibonding character in the σ-framework, resulting in substantial elongation of one of the Si–Si bonds (Figure 4D). With four σ-electrons, Si_3 is σ-antiaromatic and this results in the structural distortion from D_{3h} Si_3^{2+} to C_{2v} Si_3, similar to the transition from the aromatic $C_4H_4^{2+}$ to the antiaromatic C_4H_4. Antiaromaticity in C_4H_4 manifests itself in the localization of π-electrons. In Si_3 σ-antiaromaticity leads to the localization of σ-electrons. As it was shown above, in the pure σ-antiaromatic triatomic Li_3^- anion, four σ-electrons lead to the linear structure. Although Si_3 is a σ-antiaromatic system with four σ-electrons, it is not linear because of the important influence from the π-electrons. The HOMO $1b_1$ of Si_3 is a completely delocalized π orbital, making the cluster π-aromatic. Because Si_3 has σ-antiaromaticity and π-aromaticity, it is a system with conflicting aromaticity. Substitution of one of the silicon atoms in Si_3 by Al^- (isoelectronic to a Si atom) results in a very similar $AlSi_2^-$ C_s structure (Figure 4A). The isoelectronic $AlSiGe^-$ (Figure 4B) and $AiGe_2^-$ (Figure 4C) also have very similar structures as Si_3. Their valence molecular orbitals are also rather similar to those of Si_3 (Figure 5A). Thus, all four species should be considered as having conflicting aromaticity, similar to the prototypic Al_4^{4-} cluster [2a,c], which is σ-aromatic and π-antiaromatic. In systems with the conflicting aromaticity, it is difficult to make a judgment about the net aromaticity or antiaromaticity [24]. We believe that the geometric criterion (lowering of the high symmetry upon the transition from aromatic to antiaromatic system) should be considered to be paramount relative to other criteria of aromaticity or antiaromaticity. Thus, the structural distortion in $AlSi_2^-$, $AlSiGe^-$, and $AlGe_2^-$ makes them net antiaromatic, again similarly to Al_4^{4-} [13].

E. d-AO-Based Aromaticity in the $Ta_3O_3^-$ Cluster

It was shown [25] that the $Ta_3O_3^-$ global minimum has a perfect D_{3h} ($^1A_1'$) planar triangular structure (Figure 6A). The structure and bonding in $Ta_3O_3^-$

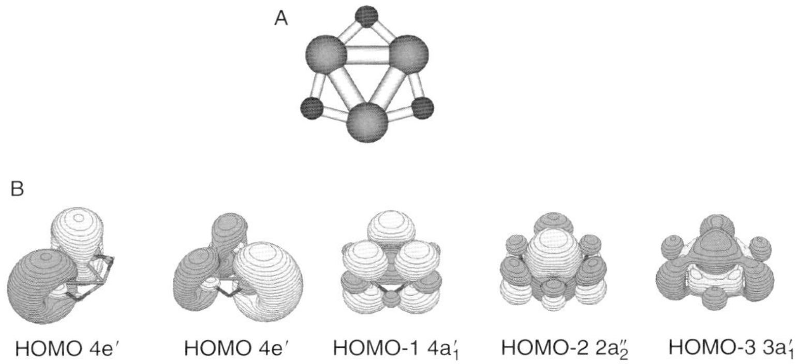

FIGURE 6 Calculated (A) geometric structure and (B) five upper molecular orbitals of $Ta_3O_3^-$.

can be understood by analyzing molecular orbitals (Figure 6B). Of 34 valence electrons in $Ta_3O_3^-$, 24 belong to either pure oxygen lone pairs or those polarized toward Ta (responsible for the covalent contributions to Ta–O bonding). Other 10 valence electrons are responsible for the direct metal–metal bonding, as shown in Figure 6B. Among the five upper MOs, three MOs are of σ-type: the partially bonding/antibonding doubly degenerate 4e′ HOMO and the completely bonding $3a_1'$ HOMO-3. The antibonding nature of the completely occupied doubly degenerate HOMO significantly reduces the bonding contribution of completely bonding HOMO-3 to the σ-bonding in the Ta_3 framework. If the HOMO (4e′) and the HOMO-3 ($3a_1'$) were composed of the same s–d hybrid functions, they would completely cancel each other. However, the hybridization in the 4e′ and $3a_1'$ orbitals is somewhat different. Therefore, there should remain some σ-aromatic character in $Ta_3O_3^-$. For example, the three-center delocalization in the aromatic $W_3O_9^{2-}$ molecule due to a d–d σ-bond was estimated previously to provide about 1 eV additional resonance energy, similar to that estimated for benzene [26].

In the $Ta_3O_3^-$ anion, the HOMO-2 ($2a_2''$) is a completely bonding π orbital composed primarily of the 5d orbitals of Ta, giving rise to π-aromatic character according to the $(4n+2)$ Hückel rule for π-aromaticity. Here, we apply the $(4n+2)$ counting rule separately for each type of aromaticity encountered in a particular planar system, that is, separately for σ-, π-, δ-, and ϕ-type molecular orbitals.

The HOMO-1 ($4a_1'$), which is a completely bonding orbital mainly coming from the overlap of the d_{z^2} AOs on each Ta atom is in fact a δ-aromatic orbital. This orbital has the "appearance" of a π-orbital with major overlaps above and below the molecular plane, but it is not a π-type MO because it is symmetric with respect to the molecular plane. This MO possesses two nodal surfaces perpendicular to the molecular C_3 axis, and thus it is a δ orbital (see detailed discussion in Ref. [25]). Therefore, the $Ta_3O_3^-$ cluster possesses an unprecedented multiple (δ and π) aromaticity, which is responsible for the metal–metal bonding and the perfect triangular Ta_3 framework. The energy ordering of σ (HOMO-3) < π (HOMO-2) < δ (HOMO-1) [25] indicates that the strength of the metal–metal bonding increases from δ to π to σ, in agreement with the intuitive expectation that σ-type overlap is greater than π-type overlap, and δ-type overlap is expected to be the weakest.

Atomic f-AOs offer additional possibility to form ϕ-bonds and thus could lead to systems with a ϕ-aromaticity. However, at this point, such systems have not yet been reported.

III. POSSIBLE TYPES OF AROMATICITY AND ANTIAROMATICITY IN X_4 CLUSTERS

A. s-AO-Based σ-Aromaticity and σ-Antiaromaticity in X_4 Clusters

The Li_4^{2+} dication is the simplest tetra-atomic metal cluster with two valence electrons and it is known to have a tetrahedral structure [15]. Thus, this σ-aromatic dication is not planar.

When two more electrons are added, the neutral Li_4 planar molecule has four σ-electrons and according to the $4n$ rule for antiaromaticity it is antiaromatic. Four atomic σ-antiaromatic molecules are rhombus rather than rectangular because of the first-order Jahn–Teller effect in the singlet $1a_g^2 1e_u^2$ electronic configuration in the D_{4h} structure [2a,15]. Figure 7 shows these orbitals for the D_{4h} (Figure 7B) and their counterparts for the rhombus D_{2h} (Figure 7C) structures.

Antiaromaticity in Li_4 can be also interpreted in terms of the island aromaticity. The bonding HOMO-1 $1a_g$ and antibonding HOMO $1b_{1u}$ (Figure 7C) could be localized into two three-center two-electron (3c–2e) bonds located over two triangular fragments of Li_4 with the ONs 1.82 |e| each (Figure 7D). This localization was performed using the NBO method (NBO 5.0 program) [21]. The obtained 3c–2e bonds are similar to the 3c–2e bond in the Li_3^+ σ-aromatic cluster, so they form two islands of aromaticity in the globally antiaromatic Li_4 cluster. The calculated negative out-of-plane component of the NICS tensor ($NICS_{zz}$) [27] at the center of the triangular fragment of Li_4 is -13.5 ppm and it clearly supports the presence of the island aromaticity. The formation of the island aromaticity in the globally antiaromatic Li_4 cluster is the reason of the significant dimerization energy (CCSD(T)/6-311+G(2df)//CCSD(T)/6-311+G*+ZPE/CCSD(T)/6-311+G*) in the reaction (7):

$$Li_2 + Li_2 \rightarrow Li_4\,,\quad \Delta E = -19.5\ \text{kcal/mol}. \tag{7}$$

Li_4^{2-} and Mg_4^{2+} are examples of systems with six valence σ-electrons. Both of these systems are metastable (because of the charge -2 or $+2$), although they have square-planar local minimum structures. In order to avoid working with the metastable highly charged species, Alexandrova and Boldyrev

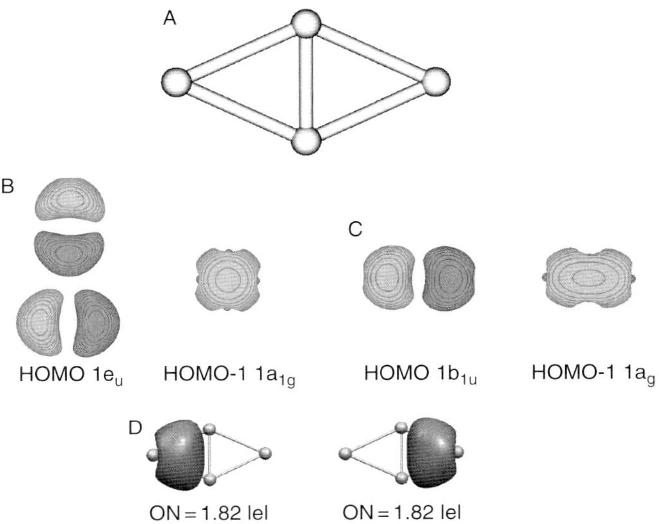

FIGURE 7 Calculated (A) geometric structure of Li_4, D_{2h}, (B) canonic valence molecular orbitals of Li_4, D_{4h}, (C) canonic valence molecular orbitals of Li_4, D_{2h}, and (D) localized 3c–2e natural bonding orbitals of Li_4, D_{2h}.

considered valence isoelectronic Li_2Mg_2 cluster [15]. The Li_2Mg_2 global minimum structure is in fact the cyclic σ-aromatic (with the six σ-electrons) structure (Figure 8). It is 11.3 kcal/mol (CCSD(T)/6-311+G(2df)//CCSD(T)/6-311+G*) lower in energy than the "classical" linear Li–Mg–Mg–Li structure. The higher stability of the σ-aromatic cyclic structure over the classical linear one provides us with an additional justification for the introduction of the σ-aromaticity into the description of metal clusters.

Another class of interesting aromatic clusters, namely square-planar Cu_4^{2-}, Ag_4^{2-}, and Au_4^{2-} within the Cu_4Li_2, Ag_4Li_2, and Au_4Li_2, respectively, has been initially studied by Schleyer and coworkers [28]. According to their calculations, there is a significant charge transfer from Li to the coinage-metal clusters. For example, in Cu_4Li_2, the natural population analysis (NPA) charge on Li is +0.8 |e|. These authors also reported the NICS values in the centers of Cu_4Li_2 (−14.5 ppm), Ag_4Li_2 (−14.1 ppm), and Au_4Li_2 (−18.6 ppm) clusters. They stated that the participation of p-orbitals in the bonding (and cyclic electron delocalization) of these clusters is negligible. Instead, these clusters benefit strongly from the delocalization of d and to some extent s-orbitals. They also pointed out that d-orbital aromaticity of Cu_4Li_2 is indicated by its high (243.2 kcal/mol) atomization energy.

In the follow-up paper, Sundholm, Wang, and coworkers [29] reported a joint experimental and theoretical paper on Cu_4Li^-, Ag_4Li^-, Au_4Li^-, as well as theoretical results on Cu_4Li_2, Ag_4Li_2, Au_4Li_2, and Cu_4^{2-}. They found that the Cu_4Li^- and Ag_4Li^- anions have a pyramidal structure consistent with the

FIGURE 8 Optimized global minimum structure of Li_2Mg_2.

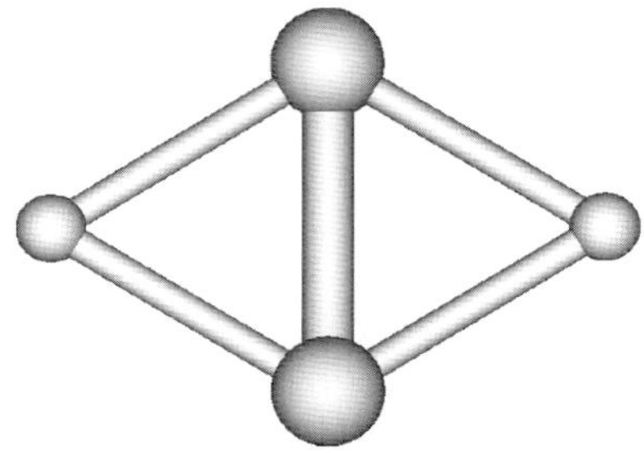

bipyramidal structure reported by Schleyer and coworkers, while the Au_4Li^- anion was found to be planar. Using the Gauge-Including Magnetically Induced Current (GIMIC) method, Sundholm, Wang, and coworkers [29] concluded that strong ring currents are sustained mainly by the highest occupied molecular orbital derived from the Cu 4s-AOs. Thus, the GIMIC calculations show that the Cu_4^{2-} ring is σ-aromatic and that the d-orbitals do not play any significant role for the electron delocalization effects. This study did not support the notion that the square-planar Cu_4^{2-} is the first example of d-orbital aromatic molecules.

If bonding in Cu_4^{2-} and Ag_4^{2-} rings is primarily because of σ-orbitals, then these dianions are similar to Li_4^{2-} and Mg_4^{2+} aromatic systems with six σ-electrons as discussed above.

B. p-AO-Based Aromaticity and Antiaromaticity in X_4 Clusters

The multiple aromaticity due to p-AOs has been initially discovered for the Al_4^{2-} dianion [2a,b]. The extensive search for the global minimum structure of Al_4^{2-} using three theoretical methods: B3LYP/6-311+G*, MP2/6-311+G*, and CCSD(T)/6-311+G* revealed that the planar-square structure D_{4h} ($^1A_{1g}$) is the most stable one (Figure 9A).

The perfect square structure of the Al_4^{2-} global minimum is unexpected, because all the alternative structures would present better charge separation [2a], which is expected to be important in determining the relative stability of the doubly charged anion. There must be some unique features of chemical bonding in Al_4^{2-} that give rise to the stability of the favored square-planar structure. MO pictures for Al_4^{2-} are presented in Figure 9B. The HOMO ($1a_{2u}$), HOMO-1 ($2a_{1g}$), and HOMO-2 ($1b_{2g}$) are completely bonding orbitals formed out of 3p-AOs of Al and represent p_π (π-orbitals formed from the p_z-AOs directed perpendicularly to the molecular plane), $p_{\sigma-r}$ (σ-orbitals formed from the $p_{x,y}$-AOs directed radially toward the center), and $p_{\sigma-t}$ (σ-orbitals formed from the $p_{x,y}$-AOs directed tangentially with respect to the cycle), respectively. The remaining four MOs are bonding, nonbonding, and antibonding orbitals formed primarily from the filled valence 3s orbitals of Al

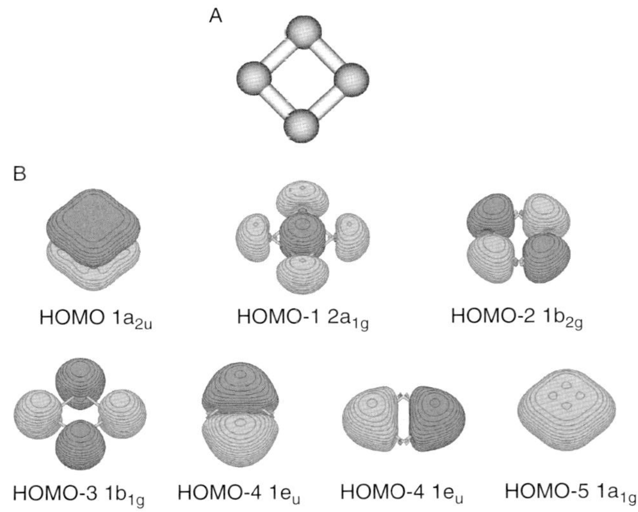

FIGURE 9 Calculated (A) geometric structure and (B) valence molecular orbitals of Al_4^{2-}.

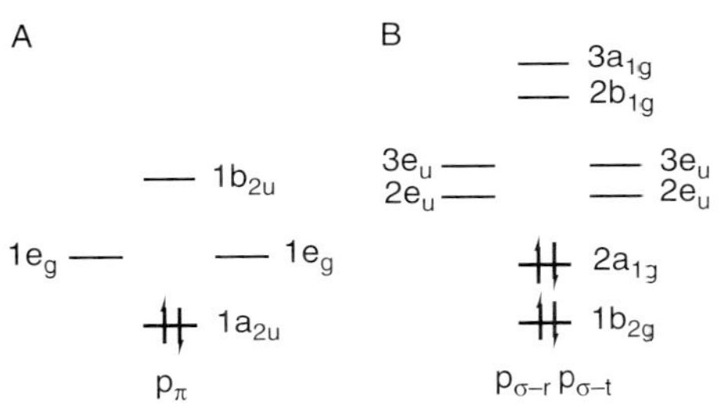

FIGURE 10 Molecular orbital diagram (A) π-MOs and (B) σ-MOs.

and can be viewed as atomic lone pairs. Thus, the upper three MOs are primarily responsible for the chemical bonding in Al_4^{2-}. If we split the σ- and π-orbitals into two separate sets, we can represent the MOs formed by 3p-AOs of Al with the MO diagram shown in Figure 10.

The lowest-lying π-MO and the two lowest-lying σ-MOs (Figure 10) are completely bonding, whereas the highest-lying ones are completely antibonding. The two MOs in the π-set and the four MOs in the σ-set that are located in between the completely bonding and completely antibonding MOs are doubly degenerate with bonding/antibonding (or nonbonding) characters. The $2e_u$- and $3e_u$-MOs are composed out of $p_{\sigma-r}$- and $p_{\sigma-t}$-AOs. This is the reason why $p_{\sigma-r}$- and $p_{\sigma-t}$-MOs are presented as one set in Figure 10. On the basis of this mixing of $p_{\sigma-r}$- and $p_{\sigma-t}$-AOs, the counting rule for σ-electrons for cyclic systems with the even number of vertices should be $4n+4/4n+6$ for aromaticity/antiaromaticity, but they are $4n+2/4n$ for the cyclic systems with the odd number of vertices. On the basis of two distinct types of MOs, we can introduce two types of aromaticity: π-aromaticity from the p_{π} MOs, and σ-aromaticity including the $p_{\sigma-r}$ and $p_{\sigma-t}$ MOs. The occupation of all three bonding MOs in Al_4^{2-} makes its shape a perfect square and renders it doubly aromatic in nature.

The double aromaticity (σ- and π-) found in Al_4^{2-} should result in a high "resonance" or stabilization energy (RE) [2d,10]. Boldyrev and Kuznetsov evaluated RE to be 48 kcal/mol according to the procedure described in Ref. [2]. Dixon and coworkers [10] arrived to the RE(Al_4^{2-}) value of 73 kcal/mol. While the RE values found in both studies are dispersed, they are consistently very high as compared to the RE of C_6H_6 (20 kcal/mol). The high RE value for Al_4^{2-} clearly supports the double aromaticity proposed on the basis of the MO analysis.

Fowler et al. [8] constructed current density maps for Al_4^{2-} separately for the σ- and π-orbitals. They found that the contribution to the current from the π-electrons is rather weak compared to the contribution from the σ-electrons. The strong diatropic σ-ring current confirms the presence of σ-aromaticity in Al_4^{2-}, whereas the combination of σ- and π-diatropic ring current (although the latter is the weak one) confirms the multiple aromatic character of Al_4^{2-}. Results of Sundholm and coworkers [9] suggested that the Al_4^{2-} dianion sustains a large diatropic ring current in an external magnetic field and thus on the basis of this criterion it should be considered aromatic. Schleyer and coworkers [25] tested aromaticity in the Al_4^{2-} dianion using NICS and confirmed that not only the diatropic π-MO (−17.8 ppm) but also the σ-MOs (sum −11.1 ppm) contribute to the considerable aromaticity. The total NICS $(0)=-30.9$ ppm value for Al_4^{2-} is appreciably higher than the NICS$(0)=-9.7$ ppm for the prototypical aromatic benzene molecule. Thus, all magnetic probes (NICS and ARCS) as well as the maps of the total and σ- and π-induced ring currents clearly indicate the simultaneous presence of π-aromaticity and σ-aromaticity in the Al_4^{2-} dianion.

We would like to mention, that there were also attempts to explain the enhanced stability of $NaAl_4^-$ and Na_2Al_4 using the jellium model [30], which was very successful in explaining extra stability of so-called "magic" clusters of alkali metal atoms. However, Dhavale et al. [30] concluded that the enhanced stability of the $NaAl_4^-$ and Na_2Al_4 species cannot be explained on the basis of a simple jellium model.

Several computational studies have also been reported focusing on the aromaticity in other group III valence isoelectronic dianions, X_4^{2-} [2d,9a,31,32]. The most accurate *ab initio* data were obtained by Sundholm et al. [9a] for the B_4^{2-}, Ga_4^{2-}, In_4^{2-}, and Tl_4^{2-} doubly charged anions. In addition, a series of $X'X_3^-$ (X = B, Al, Ga, In, Tl, X′ = C, Si, Ge, Sn, Pb) [33–35], X″ X_3 (X = B, Al, Ga, In, Tl, X″ = N, P, As, Sb, Bi) [36–41], B_2Si_2 [9], Al_2Si_2 [9], Ga_2Si_2 [9], and Si_4^{2+} [42] have also been probed for aromaticity. They all have been found to possess multiple aromaticity similar to Al_4^{2-}, except for the CAl_3^- and B_2Si_2 species (see Ref. [2] for details).

The concept of the multiple aromaticity introduced initially for Al_4^{2-} is capable of explaining the advantages of the cyclic structure for a variety of tetra-atomic molecules with 14 valence electrons. If all atoms in a molecule have similar electronegativity, which favors delocalization, the aromatic cyclic structure can be predicted to be the global minimum.

C. p-AO-Based Multiple Aromaticity in the Hg_4^{6-} Cluster

Mercury has a closed shell electron configuration ($6s^2$) and therefore a neutral Hg_4 cluster is expected to be a van der Waals complex. However, it was recently shown, that one particular sodium–mercury amalgam Na_3Hg_2 contains Hg_4^{6-} square units as its building blocks [43]. The reason why Hg_4^{6-} is a particularly stable building block can be understood if we recognize that it is valence isoelectronic to the doubly aromatic cluster Al_4^{2-} if we ignore completely occupied d-AOs on Hg [44].

Figure 11 displays the seven valence MOs of the square-planar Hg_4^{6-}. Comparing Figure 11 to Figure 9B, one sees clearly that the same set of occupied MOs is present in both cases. Hence, the structural and electronic stability of the square-planar Hg_4^{6-} and Al_4^{2-} should be attributed not only to π-aromaticity due to the occupation of $1a_{2u}$ π-orbital by two electrons but also to σ-aromaticity due to the occupation of the two σ-bonding orbitals, $1b_{2g}$ and $2a_{1g}$. The finding of the multiple aromaticity in Hg_4^{6-} establishes a solid bridge between our gas-phase studies of multiply aromatic clusters and bulk materials containing such species. It is surprisingly unexpected that such ancient materials as amalgams can be rationalized on the basis of multiple aromaticity initially discovered in the gas-phase studies of the Al_4^{2-} cluster.

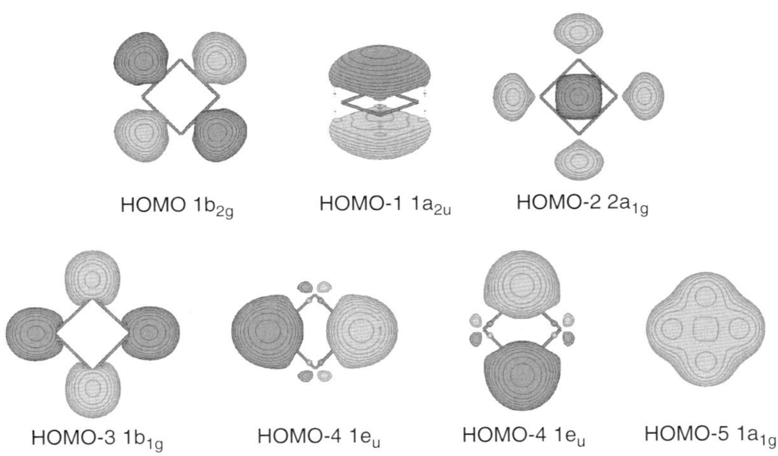

FIGURE 11 Valence molecular orbitals for Hg_4^{6-}.

D. p-AO-Based Conflicting Aromaticity in Al_4^{4-} and Si_4 Clusters

As discussed above, the Al_4^{2-} dianion is an all-metal doubly aromatic system. The double aromaticity comes from three types of MOs that can be separated in a square structure into two sets: σ-MOs and π-MOs (Figure 10). It was shown that Al_4^{4-} in the isolated form (though being electronically unstable) or as a part of the either $Li_3Al_4^-$ or Li_4Al_4 clusters is the first net antiaromatic all-metal cluster [2], because the additional electron pair occupies one of the degenerate π-($1e_g$) MO (Figure 10).

The partial occupation of the degenerate π-($1e_g$) MO makes this cluster π-antiaromatic but it is still a σ-aromatic system because of the occupied $1b_{2g}$-MO and $2a_{1g}$-MO (Figure 10). These MOs are $p_{\sigma-r}$ (σ orbitals formed from the $p_{x,y}$-AOs directed radially toward the center), and $p_{\sigma-t}$ (σ orbitals formed from the $p_{x,y}$-AOs directed tangentially with respect to the cycle). Thus, the Al_4^{4-} tetra-anion, as either isolated Al_4^{4-} or decorated by Li^+ cations in the anionic $Li_3Al_4^-$ or in the neutral Li_4Al_4, is a system with conflicting aromaticity. The π-antiaromaticity in the Al_4^{4-} tetra-anion inside of $Li_3Al_4^-$ (Figure 12A) or Li_4Al_4 (Figure 12B) species makes it rectangular, because the π HOMO of Al_4^{4-} is bonding within the two shorter Al–Al bonds, but antibonding within the longer Al–Al bonds. Both the π bonding and the rectangular shape of Al_4^{4-} are analogous to the prototypical antiaromatic organic molecule cyclobutadiene. Thus Al_4^{4-} is the first all-metal π-antiaromatic species ever made, even though it still possesses σ-aromaticity derived from the two delocalized σ bonding MOs. The conflicting aromaticity in this system results in the net antiaromaticity as can be judged on the basis of the geometric criterion.

The conflicting aromaticity of Al_4^{4-} in $Li_3Al_4^-$ generated wide discussion in the literature [2a,c,3a,8d,9b,12a,b,c,13a,b,45–51]. Schleyer and coworkers [45] confirmed that both the $Li_3Al_4^-$ and Li_4Al_4 ground states contain the rectangular Al_4^{4-} tetra-anion with four π-electrons, that gives positive (paramagnetic) contributions to the NICS indices, that is, it is strongly π-antiaromatic. They also obtained negative NICS indices for the σ electrons,

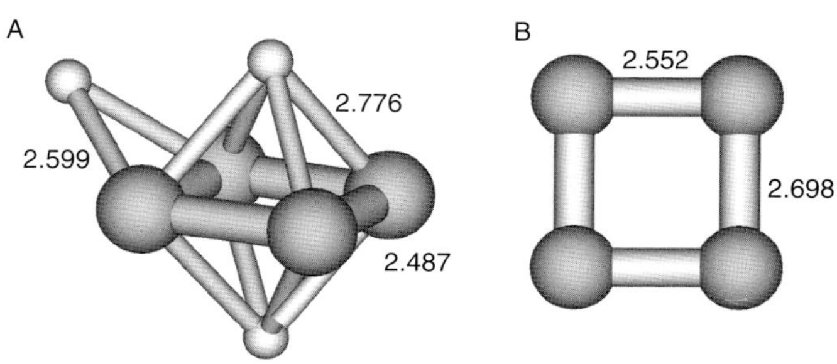

FIGURE 12 Optimized molecular structure of (A) $Li_3Al_4^-$ and (B) Li_4Al_4; Li atoms are removed for clarity.

confirming the σ-aromaticity in these all-metal clusters. However, when they added all the NICS indices from both π and σ electrons, they obtained an overall negative value because the magnitude of the NICS index from the σ electrons is higher than that from the π electrons. The total NICS values that obtained are $NICS(Li_3Al_4^-) = -4.8$ ppm and $NICS(Li_4Al_4) = -11.4$ ppm, from which they concluded that both systems should be considered net aromatic rather than net antiaromatic [45]. This controversy was extended into a discussion on the overall aromaticity or antiaromaticity in all-metal systems when both σ(π)-aromaticity and π(σ)-antiaromaticity are present simultaneously and was reported in a feature article by Ritter in the C & E News [51].

We maintained that the singlet $Li_3Al_4^-$ isomer with the rectangular Al_4^{4-} tetra-anion should be considered net antiaromatic on the basis of the rectangular distortion, which is exactly analogous to the structural distortion in the prototypical antiaromatic C_4H_4 molecule. Most of the participants in the discussion agree with our assessment.

The Si_4^{2+} dication is isoelectronic to the Al_4^{2-} dianion, which was shown previously to be a doubly aromatic system (π-aromatic and σ-aromatic) [2,3]. When a pair of electrons is added to Si_4^{2+}, the resulting neutral Si_4 is expected to be either π-antiaromatic, or σ_r-antiaromatic, depending on which MO is occupied by the extra electrons. Unlike isoelectronic Al_4^{4-}, where additional pair of electrons occupies the partially bonding/antibonding doubly degenerate π-orbital, in Si_4 the extra pair of electrons partially occupies the doubly degenerate σ-orbital resulting in the distortion of the rectangular structure into the rhombus structure (Figure 13A).

The rhombus structure first predicted by Raghavachari and Logovinski [52] is the global minimum for Si_4, and the σ-antiaromaticity of Si_4 provides the natural explanation for the rhombus distortion. However, the rhombus Si_4 can still be viewed as π-aromatic and therefore it is a system with conflicting aromaticity. As in the case of Al_4^{4-}, we believe that the Si_4 is a net antiaromatic cluster.

E. p-AO-Based Multiple Antiaromaticity in the Si_4^{2-} Cluster

The most remarkable transformation occurs upon addition of an electron pair to the neutral Si_4 to form the Si_4^{2-} dianion. Now we have several choices. We can add the electron pair to the σ_r-system of the σ_r-antiaromatic Si_4, thus

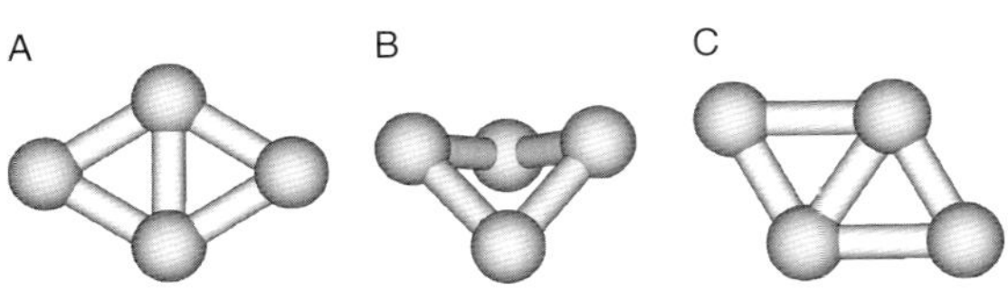

FIGURE 13 Optimized structures of (A) Si_4, D_{2h}, (B) Si_4^{2-}, D_{2d}, and Si_4^{2-}, C_{2h}.

completing the formation of four 2c–2e Si–Si bonds (Figure 10). This Si_4^{2-} dianion should be considered as a 2π aromatic system and should be square-planar, but the second-order Jahn–Teller distortion leads to a butterfly structure (Figure 13B), analogous to the structural distortion in another classical π-aromatic system, $C_4H_4^{2-}$. The butterfly structure turned out to be the global minimum for Si_4^{2-} according to our calculations. Alternatively, we can add the pair of electrons to the π-system in the σ_r-antiaromatic rhombus Si_4 structure, yielding a doubly (σ_r- and π-) antiaromatic Si_4^{2-} (Figure 13C). Double antiaromaticity has an extremely interesting structural consequence. As mentioned above, the σ_r-antiaromaticity induces a rhombus distortion and the π-antiaromaticity has a rectangular distortion. Hence, for a system with both σ_r- and π-antiaromaticity, one would expect a geometry that represents a compromise between the rhombus and rectangular distortions, that is, a parallelogram structure. Calculations indeed yielded such a doubly antiaromatic structure (Figure 13), which is 4.2 kcal/mol higher in energy than the global minimum butterfly structure [53].

These two structures of Si_4^{2-} with the doubly antiaromatic parallelogram geometry or the aromatic butterfly geometry stabilized by one Na^+ cation have been observed experimentally using photoelectron spectroscopy. The presence of these two isomers of $NaSi_4^-$ in molecular beams has been confirmed through comparison of the theoretical and experimental photoelectron spectra.

F. p-AO-Based π-Aromaticity in the X_4^{2-} (X = N, P, As, Sb, Bi) and X_4^{2+} (X = O, S, Se, Te) Clusters

A remarkable compound 2,2,2-crypt-potassium terabismuthide (−2), $(C_{18}H_{36}N_2O_6K^+)_2Bi_4^{2-}$, containing a metal Bi_4^{2-} cluster was reported in 1977 by Cisar and Corbett [54]. Follow-up theoretical studies [55–59] confirmed that it is valence isoelectronic to the classical aromatic $C_4H_4^{2-}$ dianion [60] with a completed σ-framework (four 2c–2e C–C bonds and four 2c–2e C–H bonds) and six delocalized π-electrons. In 1984, Critchlow and Corbet [61] reported another potassium-crypt salt of tetraantimonide (2-) $(2,2,2\text{-crypt-K}+)_2Sb_4^{2-}$ containing the aromatic Sb_4^{2-} dianion.

The isolated tetraatomic group V aromatic clusters (Pn_4^{2-}), P_4^{2-}, As_4^{2-}, and Sb_4^{2-}, have been studied theoretically and experimentally [62]. Experimentally, the Pn_4^{2-} species have been stabilized by one Na^+ cation in the form of $NaPn_4^-$ and their photoelectron spectroscopy (PES) spectra in molecular beams have been recorded. Excellent agreement between *ab initio* and experimental PES spectra for NaP_4^-, $NaAs_4^-$, and $NaSb_4^-$ confirmed the theoretically obtained pyramidal structures of these species, on the basis of the square-planar Pn_4^{2-} unit capped by Na^+. Molecular orbital analysis showed that indeed all the Pn_4^{2-} species possess six π-electrons (Figure 14). Jin et al. [63] reported a theoretical study of MP_4 (M = Be, Mg, and Ca) and M'_2P_4

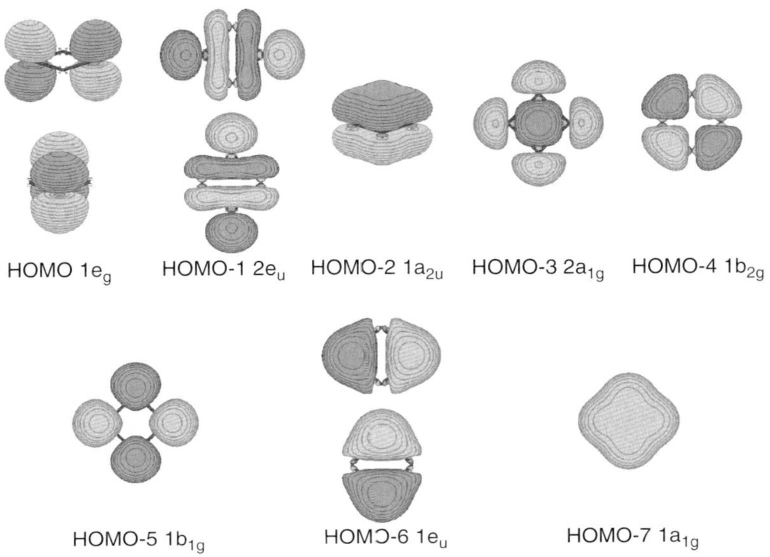

FIGURE 14 Valence molecular orbitals of As_4^{2-}.

(M′ = Li, Na, and K) clusters and confirmed the aromatic nature of P_4^{2-} in all these species. Recently, the first compound $Cs_2P_4 \cdot 2NH_3$ containing the aromatic P_4^{2-} cluster was synthesized by Kraus et al. [64].

The series of group VI isoelectronic chalcogen dications, Ch_4^{2+} (Ch = Te, Se, S, O), has also been synthesized in solid state and studied theoretically [65–86]. The existence of the Se_4^{2+} and Te_4^{2+} dications in solution was reported by Gillespie et al. in 1968 [69–71]. In 1972, Corbett and coworkers [72] reported the synthesis and crystal structure of $Te_4^{2+}(AlCl_4^-)_2$ and showed directly that the Te_4^{2+} unit indeed has a nearly perfect square structure. Compounds containing planar-square aromatic S_4^{2+} have been also reported [73,74]. Theoretical analyses [76–84] of the chemical bonding in S_4^{2+}, Se_4^{2+}, and Te_4^{2+} have confirmed that all these species are aromatic with six π electrons. In 1989, Burford et al. [85] reviewed experimental and theoretical studies of the Ch_4^{2+} (Ch = Te, Se) dications. Very recently the lightest member of this family, the O_4^{2+} dication, has been studied [86] theoretically and has been shown to be the first all-oxygen aromatic species.

The above discussion shows that tetraatomic aromatic clusters of groups V and VI elements with six π electrons and a classical σ-framework (four lone pairs located at every atom and four 2c–2e bonds) are well established in chemistry. They can be metal clusters such as Bi_4^{2-}, metalloid clusters such as Sb_4^{2-}, As_4^{2-}, and Te_4^{2+}, or nonmetal clusters such as P_4^{2-}, Se_4^{2+}, S_4^{2+}, and O_4^{2+}.

IV. POSSIBLE TYPES OF AROMATICITY AND ANTIAROMATICITY IN X_5 CLUSTERS

A. p-AO-Based Multiple Aromaticity in the B_5^+ Cluster

The slightly distorted planar pentagonal structure for B_5^+ was reported as the global minimum by Kato et al. [87] and Rica and Bauschlicher [88]. This distortion from the perfect pentagon to the C_{2v} $^1A_1$ structure is due to the second-order Jahn–Teller effect. However, as it was shown by Zubarev and Boldyrev [89], the global minimum C_{2v} ($^1A_1$) structure is only 0.365 kcal/mol lower in energy than the second-order saddle point D_{5h} ($^1A_1'$) planar pentagon structure and after ZPE correction the vibrationally averaged D_{5h} ($^1A_1'$) structure is actually lower in energy than the vibrationally averaged C_{2v} ($^1A_1$) structure by 0.010 kcal/mol (all data at CCSD(T)/6-311+G*). Thus, for all practical purposes, we can consider the B_5^+ cluster as a planar pentagon.

The molecular orbital analysis (Figure 15) can readily explain the beautiful planar pentagonal structure of B_5^+. The HOMO ($2a_1'$) in B_5^+ is a globally bonding σ-MO and HOMO-1 ($1a_2''$) is a globally bonding π-MO. Thus, they make the cation doubly (σ- and π-) aromatic. On the other hand, the NBO analysis showed that HOMO-2 and HOMO-2′ ($1e_2'$), HOMO-3 and HOMO-3′ ($1e_1'$), and HOMO-4 ($1a_1'$) can be localized into five peripheral 2c–2e B–B bonds. Double aromaticity in conjunction with the presence of five 2c–2e B–B peripheral bonds is responsible for the vibrationally averaged highly symmetric D_{5h} structure of the B_5^+ cluster. The double aromaticity in B_5^+ manifests itself in the high first singlet vertical excitation energy (2.97 eV at TD-B3LYP/6-311+G*), highly negative values of NICS (NICS(0) = −36.2 ppm; NICS (0.5) = −31.0 ppm, and NICS(1.0) = −18.8 ppm), and most importantly it

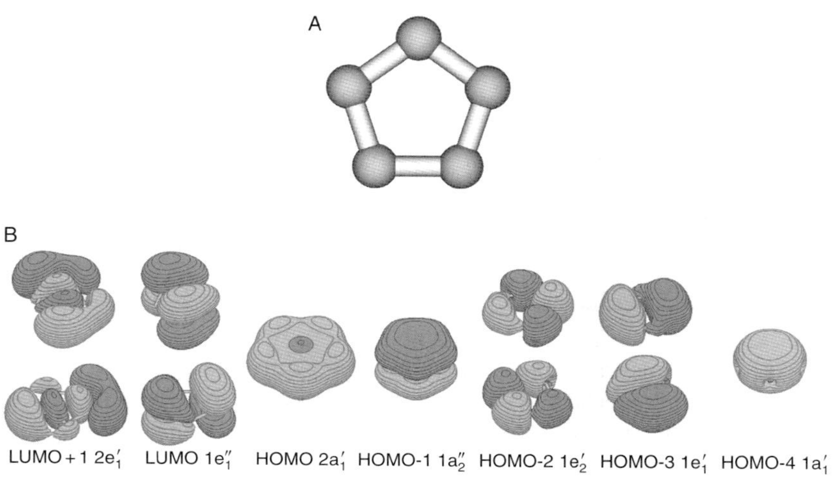

FIGURE 15 (A) Molecular structure of B_5^+, D_{5h} and (B) valence molecular orbitals of B_5^+, D_{5h}.

explains why B_5^+ has high stability and low reactivity in collision-induced dissociation (CID) experiments by Anderson and coworkers [90].

B. p-AO-Based Conflicting Aromaticity in the B_5^- Cluster

Let us start with the MOs of the B_5^+ cluster in order to explain the global minimum structure of the B_5^- cluster. The LUMO+1 ($2e_1'$-MO) in B_5^+ is a partially bonding/antibonding σ-orbital related to the completely bonding σ-HOMO ($2a_1'$) (Figure 15B). These three MOs are a part of the set of five MOs formed by the 2p-radial AOs of B and responsible for the global σ-bonding. In B_5^-, this LUMO+1 ($2e_1'$-MO) is partially occupied instead of the LUMO ($1e_1''$). The singlet $1a_1'^2 1e_1'^4 1e_2'^4 1a_2''^2 2a_1'^2 2e_1'^2$ electronic configuration of the D_{5h} structure of B_5^- undergoes the first-order Jahn–Teller distortion leading to the C_{2v} ($^1A_1, 1a_1^2 1b_2^2 2a_1^2 3a_1^2 1b_1^2 2b_2^2 4a_1^2 2b_2^2$) planar structure which is the B_5^- global minimum structure. Thus, B_5^- has four electrons on the globally delocalized σ-HOMO-1 ($4a_1$) and σ-HOMO ($3b_2$) and two electrons on the globally delocalized π-HOMO-3 ($1b_1$). This makes B_5^- a system with conflicting aromaticity (σ-antiaromatic and π-aromatic). NBO analysis for the B_5^{5+} cation at the geometry of B_5^- and with the $1a_1^2 1b_2^2 2a_1^2 3a_1^2 2b_2^2$ electronic configuration shows that there are five 2c–2e B–B peripheral bonds (ON = 1.76–1.91 |e|). The C_{2v} $^1A_1$ structure of B_5^- has been experimentally established in the joint photoelectron and *ab initio* study by Zhai et al. [91]. We believe that the B_5^- anion is "net antiaromatic" because of its low symmetry: the σ-antiaromaticity apparently overwhelms the π-aromaticity. The low first singlet vertical excitation energy (1.10 eV, at TD-B3LYP/6-311+G*) supports our overall assignment of antiaromaticity. This anion is a remarkable example showing that limiting chemical bonding analysis to π-electrons only does not allow one to explain why π-aromatic (with two π-electrons) B_5^- cluster has low C_{2v} symmetry and low first singlet vertical excitation energy.

C. Pentaatomic π-Aromatic Species of Groups IV and V Elements: M_5^- (M = N, P, As, Sb, Bi) and M_5^{6-} (Ge, Sn, Pb)

In 1994, Gausa et al. [92,93] reported an experimental and theoretical study of the Bi_5^- metal cluster. On the basis of isovalent and isostructural analogy between Bi_5^- and $C_5H_5^-$, they concluded that Bi_5^- may be viewed as aromatic with six π-electrons. They pointed out that the high first VDE is the proof that this system is aromatic. In 2002, Zhai et al. [94] obtained the PES spectra of four Pn_5^- species (Pn = P, As, Sb, Bi) and examined their electronic structure and chemical bonding in more detail. De Proft et al. [95] reported ring current and NICS calculations for the planar pentagonal structure of Pn_5^-. They found that conclusions based on the map of the ring current and NICS agree with the presence of aromaticity in the Pn_5^- anions.

Scherer [96,97] synthesized a sandwich compound $[CpFe(As_5)Fe(C_5Me_5)]PF_6$ containing a planar aromatic As_5^- and the cyclic P_5^- has been isolated in alkali metal salts [98,99] and incorporated in several mixed sandwich complexes, $[(\eta^5\text{-}C_5Me_5)Fe(\eta^5\text{-}P_5)]$, $[(\eta^5\text{-}C_5Me_5)_2Fe(\eta^5\text{-}P_5)]$, $[(\eta^5\text{-}C_5Me_5)Cr_2(\eta^5\text{-}P_5)]$ [100–102]. Recently, the carbon free sandwich complex $[Ti(\eta^5\text{-}P_5)_2]^-$, was prepared and characterized [103].

Christe and coworkers [104] reported mass-spectroscopic identification of the N_5^- cluster. This all-nitrogen aromatic species has been extensively studied computationally [105], but experimental spectroscopic data are not yet available.

Several very interesting compounds such as Na_8BaPb_6, Na_8BaSn_6, and Na_8EuSn_6 have been reported by Todorov and Sevov [106]. On the basis of the stoichiometric formulas, one may think that these compounds contain hexaatomic Pb_6 or Sn_6 building blocks. However, structural analyses revealed that these three compounds are built out of columns of pentagonal rings of Sn_5^{6-} or Pb_5^{6-} stacked exactly on top of each other. The alkaline- or rare-earth cations are found exactly halfway between the ring planes in a ferrocene-like geometry. The isolated Sn^{4-} or Pb^{4-} monoatomic anions are positioned within the plane of the pentagonal rings. The pentagonal rings Sn_5^{6-} or Pb_5^{6-} are valence isoelectronic to the previously discussed Pn_5^- anion and therefore they should be considered as π-aromatic systems with six delocalized π-electrons in addition to a lone pair on each Sn or Pb atom and five 2c–2e Sn–Sn or Pb–Pb σ-bonds. These compounds are examples of the importance of aromaticity in the formation of materials in the solid state.

V. POSSIBLE TYPES OF AROMATICITY AND ANTIAROMATICITY IN PLANAR AND QUASI-PLANAR BORON CLUSTERS

Small and medium size boron clusters are known to be planar or quasi-planar [2,89f,89]. Their planarity or quasi-planarity has been explained on the basis of the following bonding model:

- Boron atoms in the peripheral n-membered ring are bound with 2c–2e B–B bonds. The number of 2c–2e peripheral B–B bonds in all planar or quasi-planar clusters is equal to the number of peripheral edges. This takes $2n$ electrons out of all the valence electrons available for bonding.
- The rest of the valence electrons participate in the globally delocalized bonding between all boron atoms.
- There are globally delocalized π-MOs that make a cluster either globally π-aromatic if it has $4n+2$ π-electrons or globally antiaromatic if it has $4n$ π-electrons for singlet coupled electrons. For triplet coupled π-electrons the number of electrons should satisfy the inverse $4n$ rule for aromaticity.

- There are globally delocalized σ-MOs that make a cluster either globally σ-aromatic if it has $4n+2$ σ-electrons or globally antiaromatic if it has $4n$ σ-electrons for singlet coupled electrons. For triplet coupled σ-electrons the number of electrons should satisfy the inverse $4n$ rule for aromatic systems.
- As the size of the system increases, the species emerge that form an internal ring within the external ring of the 2c–2e B–B bonds. In addition to these 2c–2e bonds and globally delocalized bonding between all boron atoms, one should consider bonding within the internal ring, which can be localized or delocalized.

A. Doubly Aromatic Boron Clusters

We have already discussed B_3^- and B_5^+ doubly aromatic clusters above (see Sections II.C and IV.A, respectively). The other doubly aromatic clusters: B_4, B_4^{2-}, B_5^+, B_6^{2+}, B_6, B_7^+, B_7^-, B_8, B_8^{2-}, B_9^-, B_{10}, B_{11}^+, B_{12}, and B_{13}^+ are shown in Figure 16.

(i) B_4

The D_{2h} ^1A_g rhombus global minimum structure (Figure 16) of the neutral B_4 cluster comes from the second-order (or "pseudo") Jahn–Teller effect, as it

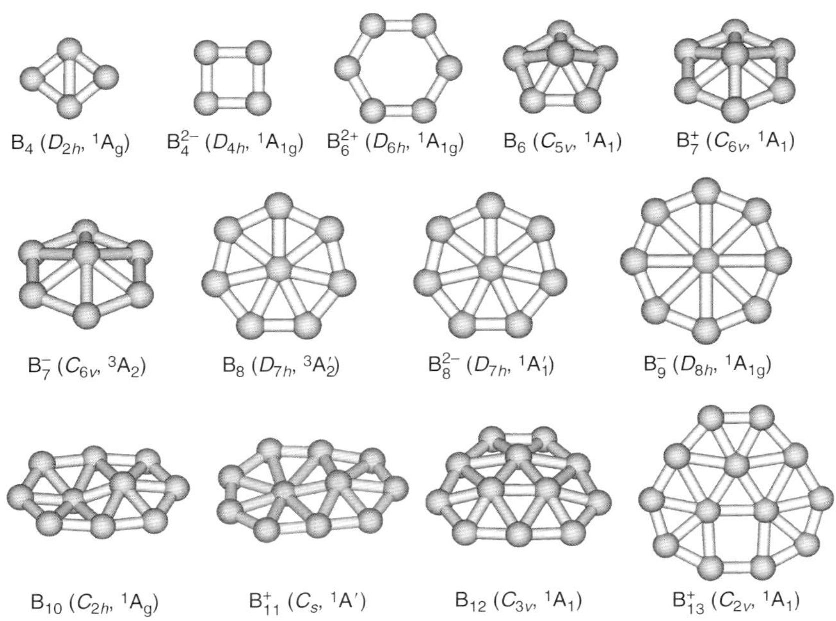

FIGURE 16 Optimized global minimum structures of: B_4, B_4^{2-}, B_6^{2+}, B_6, B_7^+, B_7^-, B_8, B_8^{2-}, B_9^-, B_{10}, B_{11}^+, B_{12}, and B_{13}^+.

was discussed by Martin et al. [107]. The barrier for "squareness" is rather small (0.7–0.8 kcal/mol [108,109]), therefore, at moderate temperature the B_4 cluster is effectively square. The electronic configuration of the D_{2h} structure of B_4 is $1a_g^2 1b_{1u}^2 1b_{2u}^2 1b_{3g}^2 1b_{3u}^2 2a_g^2$. The lowest four MOs (HOMO-2 ($1b_{3g}$), HOMO-3 ($1b_{2u}$) HOMO-4 ($1b_{1u}$), and HOMO-5 ($1a_g$)) can be localized, as it has been shown by NBO analysis, into four classical peripheral 2c–2e B–B bonds. The remaining two MOs are globally delocalized, and participate in the global bonding in the cluster. The HOMO-1 ($1b_{3u}$) is a completely bonding π-MO formed by the out-of-plane overlap of $2p_z$-AOs on the B atoms. The two electrons populating this MO make the cluster π-aromatic. The HOMO ($2a_g$) of B_4 is a σ-radial MO, just like the HOMO ($2a_1'$) of B_3^-, formed by the radial overlap of 2p-AOs. The system thus can be characterized as σ-aromatic. The assignment of double (σ and π)-aromaticity in B_4 first made by Zhai et al. [108] is supported by effective high-symmetric (square) structure, calculated first singlet vertical excitation energy (3.08 eV at TD-B3LYP/6-311+G*), and calculated NICS index, which is highly negative at the center of the cluster (−35.6 ppm).

(ii) B_4^{2-}

According to Sundholm and coworkers [9] the doubly charged B_4^{2-} cluster has a square-planar D_{4h} $^1A_{1g}$ ($1a_{1g}^2 1e_u^4 1b_{1g}^2 1b_{2g}^2 2a_{1g}^2 1a_{2u}^2$) structure. They used the isoelectronic analogy with Al_4^{2-}, where extensive search for the global minimum structure has been performed [2a] and the D_{4h} $^1A_{1g}$ square-planar structure was found to be the most stable isomer. The Al_4^{2-} dianion has been discussed above (Section III.B). It was shown that it is a doubly (σ- and π-) aromatic system and therefore, the isoelectronic, isostructural B_4^{2-} is also doubly (σ- and π-) aromatic system with four 2c–2e peripheral B–B bonds.

(iii) B_6^{2+}

The B_6^{2+} dication has the perfect hexagonal structure (Figure 16) with the electronic configuration: $1a_{1g}^2 1e_{1u}^4 1e_{2g}^4 1b_{2u}^2 1a_{2u}^2 2a_{1g}^2$. The NBO localization leads to six peripheral 2c–2e B–B bonds formed out of HOMO-2 ($1b_{2u}$), HOMO-3 and HOMO-3′ ($1e_{2g}$), HOMO-4 and HOMO-4′($1e_{1u}$), and HOMO-5 ($1a_{1g}$). The two other completely bonding σ-HOMO ($2a_{1g}$) and π-HOMO-1 ($1a_{2u}$) are responsible for σ- and π-aromaticity. Large first vertical excitation energy (1.94 eV) and large negative NICS (−29.6 ppm) confirm our assignment of double aromaticity in B_6^{2+}.

(iv) B_6

The B_6 cluster is the first example of a structure where the cavity inside the cycle is occupied by an atom. However, the cavity inside of the pentagon is too small to favorably accommodate a boron atom; therefore, the planar pentagonal structure with the boron atom located at the center of the five-atomic ring is not a minimum. According to Niu et al. [110] and Alexandrova et al. [111] calculations, the global minimum structure of B_6 is the pentagonal pyramid C_{5v} ($^1A_1$, $1a_1^21e_1^42a_1^21e_2^43a_1^22e_1^4$) with the boron atom located 0.94 Å above the center of the B_5 perfect pentagon. In order to simplify interpretation of molecular orbitals, let us first perform MO analysis for the D_{5h} ($^1A_1'$, $1a_1'^21e_1'^41a_2''1e_2'^22a_1'^22e_1'^4$) structure (Figure 17A), in which the central atom is pushed into the center.

Five MOs (HOMO-2 and HOMO-2′ ($1e_2'$), HOMO-4 and HOMO-4′ ($1e_1'$), and HOMO-5 ($1a_1'$)) can be localized into five 2c–2e B–B bonds, so they are responsible for the peripheral bonding. The HOMO-3 ($1a_2''$) is formed by $2p_z$-AOs and it is responsible for the global π-bonding. The HOMO and HOMO′ ($2e_1'$), and HOMO-1 ($2a_1'$) are formed from 2p-radial AOs and they are responsible for the global σ-bonding in the B_6 cluster. Thus, B_6 in the D_{5h} $^1A_1'$ configuration is a doubly (σ- and π-) aromatic system with two π- and six σ-electrons. While in the pyramidal structure σ- and π-MOs are mixed (Figure 17B), we believe that the bonding picture developed for the D_{5h} $^1A_1'$ structure is still qualitatively valid and can explain why B_6 clusters adopt such a structure. The large first vertical excitation energy (2.34 eV) and large negative NICS (−59.1 ppm) confirm the assignment of double aromaticity in B_6.

(v) B_7^+

The C_{6v} $^1A_1$ ($1a_1^21e_1^41e_2^42a_1^21b_1^23a_1^22e_1^4$) hexagonal pyramid [112–116] with the central boron atom located 0.72 Å above the plane (Figure 16) is the

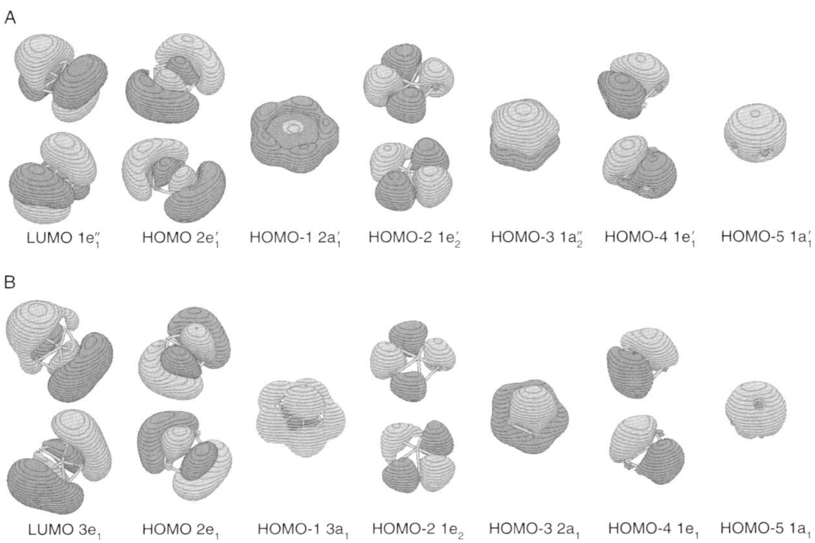

FIGURE 17 Valence molecular orbitals of (A) B_6, D_{5h} and (B) B_6, C_{5v}.

global minimum structure of the cationic B_7^+ cluster. Again, for simplicity let us perform MO analysis for the D_{6h} ($^1A_{1g}$, $1a_{1g}^2 1e_{1u}^4 1e_{2g}^4 1b_{2u}^2 2a_{1g}^2 1a_{2u}^2 2e_{1u}^4$) structure, in which the central atom is pushed into the plane. Six MOs (HOMO-3 ($1b_{2u}$), HOMO-4 and HOMO-4′ ($1e_{2g}$), HOMO-5 and HOMO-5′ ($1e_{1u}$), and HOMO-6 ($1a_{1g}$)) can be localized into six 2c–2e B–B bonds. The HOMO-1 ($1a_{2u}$) is formed by $2p_z$-AOs and is responsible for the global π-bonding. The HOMO and HOMO′ ($2e_{1u}$), and HOMO-2 ($2a_{1g}$) are formed from 2p-radial AOs and they are responsible for the global σ-bonding in the B_7^+ cluster. Thus, B_7^+ in the D_{6h} $^1A_{1g}$ configuration is a doubly (σ- and π-) aromatic system with two π- and six σ-electrons. The central cavity in the B_6 hexagon is still too small to favorably accommodate the central boron atom and therefore, the C_{6v} $^1A_1$ pyramidal structure corresponds to the global minimum. The doubly aromatic nature of B_7^+ is consistent with findings of Anderson and coworkers [90] that this cluster is highly stable and has low reactivity. Also, the high first vertical excitation energy (1.99 eV) and highly negative NICS (−42.3 ppm) confirm this assignment.

(vi) B_7^-

Alexandrova et al. [117] have shown that the B_7^- cluster has a very flat triplet C_{6v} $^3A_2$ ($1a_1^2 1e_1^4 1e_2^4 2a_1^2 3a_1^2 1b_1^2 2e_1^4 3e_1^2$) [118] pyramidal structure similar to the B_7^+ structure as the global minimum. MO analysis was performed for the planar B_7^- D_{6h} $^3A_{2g}$ ($1a_{1g}^2 1e_{1u}^4 1e_{2g}^4 1b_{2u}^2 2a_{1g}^2 1a_{2u}^2 2e_{1u}^4 1e_{1g}^2$) [117] model system. The HOMO-7 ($1a_{1g}$), HOMO-6 and HOMO-6′ ($1e_{1u}$), HOMO-5 and HOMO-5′ ($1e_{2g}$), and HOMO-4 ($1b_{2u}$) can be localized into six peripheral 2c–2e B–B bonds, forming the hexagonal bonding framework. Double occupation of the completely bonding HOMO-2 ($1a_{2u}$) and partial occupation of the partially antibonding HOMO and HOMO′ ($1e_{1g}$) make this cluster π-aromatic with four π-electrons according to the inverse 4n rule for triplet states. HOMO-3 ($2a_{1g}$) and HOMO-1 and HOMO-1′ ($2e_{1u}$) are delocalized σ-MOs which make this cluster σ-aromatic. Thus, the B_7^- cluster is a doubly (σ- and π-) aromatic system with six peripheral B–B bonds.

(vii) B_8

The neutral B_8 cluster has a triplet perfect heptagon (Figure 16) structure D_{7h} $^3A_2'$ ($1a_1'^2 1e_1'^4 1e_2'^4 2a_1'^2 1e_3'^4 1a_2''^2 2e_1'^4 1e_3''^2$) [119–122] (correction to the erroneous ^3A_g state reported in Ref. [119]). In the triplet D_{7h} $^3A_2'$ structure, HOMO-3 and HOMO-3′ ($1e_3'$), HOMO-5 and HOMO-5′ ($1e_2'$), HOMO-6 and HOMO-6′ ($1e_1'$), and HOMO-7 ($1a_1'$) (Figure 18) can be localized into seven 2c–2e B–B bonds. The π-aromaticity in this cluster is due to the partially occupied HOMO and HOMO′ ($1e_1''$) and doubly occupied HOMO-2 ($1a_2''$), which are formed from $2p_z$-AOs. The σ-aromaticity is due to HOMO-1 and HOMO-1′ ($2e_1'$) and HOMO-4 ($2a_1'$) which are formed from

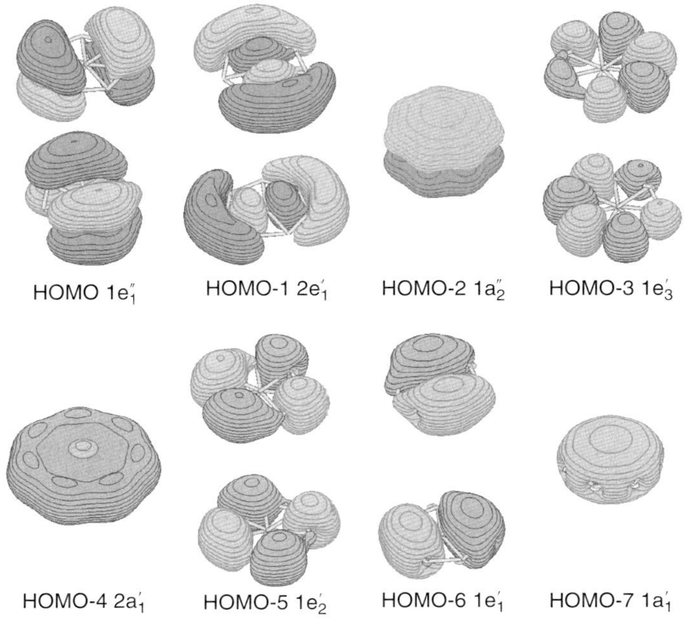

FIGURE 18 Valence molecular orbitals of B_8^{2-}, D_{7h}.

2p-radial AOs. Thus, the D_{7h} $^3A_2'$ structure is a doubly (σ- and π-) aromatic system with four π-electron (satisfying the inverse 4*n* rule for aromaticity for triplet coupled electrons), with six σ-electrons (satisfying the $4n+2$ rule for aromaticity for singlet coupled electrons), and with seven 2c–2e peripheral B–B bonds.

(viii) B_8^{2-}

The doubly charged B_8^{2-} anion has a planar D_{7h} $^1A_1'$ ($1a_1'^2 1e_1'^4 1e_2'^4$-$2a_1'^2 1e_3'^4 1a_2''^2 2e_1'^4 1e_1''^4$) singlet global minimum structure (Figure 16). Molecular orbitals of this dianion are similar to the orbitals of the neutral B_8 (Figure 18). The contribution of the valence orbitals to bonding in B_8^{2-} is the same as in B_8 (see above) with just one difference: in the dianion the doubly degenerate HOMOs are completely occupied. Thus, the singlet D_{7h} $^1A_1'$ structure of B_8^{2-} is a doubly (σ- and π-) aromatic system with six π-electron, six σ-electrons, and seven 2c–2e peripheral B–B bonds. This analysis of chemical bonding for B_8^{2-} was first proposed by Zhai et al. [119]. Its double aromaticity is also confirmed by highly negative values of NICS: NICS(0) = −84.7 ppm, NICS(0.5) = −27.0 ppm, and NICS(1.0) = −24.8 ppm. The isolated dianion was not studied experimentally, but its high-symmetric structure was experimentally confirmed in a joint study of the LiB_8^- cluster including photoelectron spectroscopy and *ab initio* calculations by

Alexandrova et al. [123]. In that study, it was shown that the calculated photoelectron spectrum of the half-sandwich structure of LiB_8^- in which Li^+ cation is located above the slightly distorted B_8^{2-} heptagon agrees well with the experimentally recorded spectra of the anion in the molecular beam.

(ix) B_9^-

The joint study of B_9^- using photoelectron spectroscopy and *ab initio* calculations [119] established the perfect octagonal structure of this cluster (Figure 16). The remarkable shape of B_9^- can be easily rationalized on the basis of the presence of double (σ- and π-) aromaticity. The eight MOs HOMO-3 ($1b_{2g}$), HOMO-5, HOMO-5′ ($1e_{3u}$), HOMO-6, HOMO-6′ ($1e_{2g}$), HOMO-7, HOMO-7′ ($1e_{1u}$), and HOMO-8 ($1a_{1g}$) can be localized into eight 2c–2e B–B peripheral bonds [2]. The three π-MOs HOMO, HOMO′ ($1e_{1g}$), and HOMO-2 ($1a_{2u}$) are responsible for π-aromaticity and the three σ-MOs HOMO-1, HOMO-1′ ($2e_{1u}$), and HOMO-4 ($2a_{1g}$) are responsible for σ-aromaticity in B_9^-. This chemical bonding analysis for B_9^- was first proposed by Zhai et al. [119]. The double (σ- and π-) aromaticity in B_9^- is supported by high symmetry, high first singlet vertical excitation energy (2.79 eV at TD-B3LYP/6-311+G*), and highly negative NICS values: NICS(0) = −28.3 ppm, NICS(0.5) = −23.3 ppm, and NICS(1.0) = −13.7 ppm. The presence of eight 2c–2e B–B peripheral bonds together with the double (σ- and π-) aromaticity is responsible for the unprecedented example of octacoordinated atom in planar environment.

Minkin and coworkers [124,125] reported the octacoordinated planar structures for the CB_8, SiB_8, and PB_8^+ species and their π-aromatic character with six π-electrons. We would like to stress that on the basis of our analysis of chemical bonding in B_9^-, the valence isoelectronic CB_8, SiB_8, and PB_8^+ species are also σ-aromatic with six σ-electrons and they also have eight 2c–2e peripheral B–B bonds. π- and σ-aromaticity together with eight peripheral B–B bonds is responsible for the beautiful octagonal shape of these species.

(x) B_{10}

Zhai et al. [126] and Boustani [120] determined the global minimum structure of B_{10} as C_{2h}, ^1A_g (Figure 16). This structure is nonplanar with eight boron atoms forming a planar cycle around two atoms at the center. One of the central atoms is located above the plane and the other one below the plane. Zhai et al. [126] first performed the analysis of chemical bonding but only for the π-system. They showed that this structure has six π-electrons and thus it is π-aromatic. More detailed chemical bonding analysis in B_{10} was performed by Zubarev and Boldyrev [2]. They considered the C_{2h} ^1A_g structure flattened into the planar D_{2h} ^1A_g structure for simplicity and performed MO analysis (Figure 19) for it.

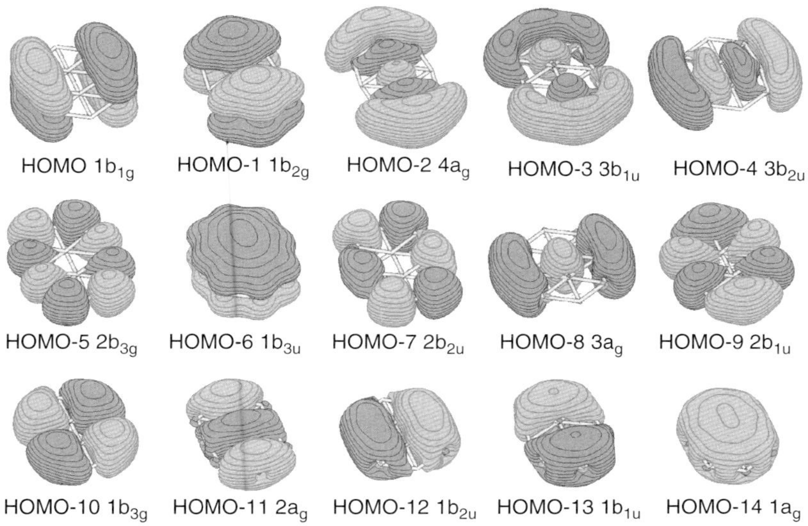

FIGURE 19 Valence molecular orbitals of the flattened D_{2h} structure of B_{10}.

The NBO analysis leads to the localization of the HOMO-5 ($2b_{3g}$), HOMO-7 ($2b_{2u}$), HOMO-9 ($2b_{1u}$), HOMO-10 ($1b_{3g}$), HOMO-11 ($2a_g$), HOMO-12 ($1b_{2u}$), HOMO-13 ($1b_{1u}$), and HOMO-14 ($1a_g$) into eight peripheral 2c–2e B–B bonds (ON = 1.87–1.93 |e|) and HOMO-8 ($3a_g$) into one 2c–2e B–B bond (ON = 1.52 |e|) between the central atoms. Rather low ON for the central B–B bond shows that we should treat the existence of this bond with caution. Yet, assuming the presence of the 2c–2e bond between the central atoms as the result of localization of HOMO-8 ($3a_g$), we are left with six more MOs. Three MOs: HOMO ($1b_{1g}$), HOMO-1 ($1b_{2g}$), and HOMO-6 ($1b_{3u}$) are responsible for the global π-bonding. Finally, last three MOs: HOMO-2 ($4a_g$), HOMO-3 ($3b_{1u}$), and HOMO-4 ($3b_{2u}$) are responsible for the global σ-bonding and σ-aromaticity. Thus, B_{10} can be tentatively considered as a doubly (σ- and π-) aromatic cluster with eight peripheral 2c–2e B–B bonds.

(xi) B_{11}^+

The quasi-planar C_s $^1A'$ structure for the B_{11}^+ cluster (Figure 16) was reported by Ricca and Bauschlicher [88]. In this structure, the nine-atomic ring is too small to accommodate two boron atoms within the plane. The electronic configuration of the flattened B_{11}^+ C_{2v} $^1A_1$ is $1a_1^2 2a_1^2 1b_2^2 3a_1^2 2b_2^2 4a_1^2 3b_2^2 5a_1^2 6a_1^2 4b_2^2 1b_1^2 7a_1^2 5b_2^2 8a_1^2 2b_1^2 1a_2^2$ [2]. Ten lowest canonical MOs can be localized into nine 2c–2e peripheral B–B bonds (originating from HOMO-6, HOMO-7, HOMO-9, HOMO-10, HOMO-11, HOMO-12, HOMO-13, HOMO-14, and HOMO-15; with ON = 1.81–1.94 |e|) and one central B–B bond (originating from HOMO-8; with ON = 1.53 |e|).

The remaining three σ-MOs: HOMO-2 ($8a_1$), HOMO-3 ($5b_2$), and HOMO-4 ($7a_1$), and HOMO-8 ($5a_1$) are responsible for the global σ-bonding and σ-aromaticity [2]. Three π-MOs: HOMO ($1a_2$), HOMO-1 ($2b_1$), and HOMO-5 ($1b_1$) are responsible for the global π-bonding and π-aromaticity. Thus, the B_{11}^+ cation can be tentatively considered as a doubly (σ- and π-) aromatic system.

(xii) B_{12}

The quasi-planar convex C_{3v} $^1A_1$ structure of B_{12} (Figure 16) was reported by Zhai et al. [126] and Boustani [113]. Again, three central boron atoms cannot fit into the plane of the nine-membered ring. The MO analysis was again performed for the planar structure for simplicity [2]. NBO analysis showed nine peripheral 2c–2e B–B bonds (ON = 1.89–1.94 |e|). However, the NBO analysis did not show 2c–2e B–B bonds between three central atoms. Instead, it showed the presence of three "lone pairs" with the average ON about 1.1 |e| and with the total accumulation of 3.2 |e| on each of three central atoms. This result could be a deficiency of the employed NBO method and a hint that there are bonds formed within the three-center internal cycle and between the nine-center external and three-center internal cycles. The nine 2c–2e B–B peripheral bonds are formed out of HOMO-4 ($3e'$), HOMO-7 ($1a_2'$), HOMO-8 ($2a_1'$), HOMO-9 ($2e'$), HOMO-10 ($1e'$), and HOMO-11 ($1a_1'$). These bonds account for 18 valence electrons out of 36. Molecular orbital picture showed the presence of three globally delocalized π-MOs: HOMO-5 ($1a_2''$), HOMO-1, and HOMO-1′ ($1e''$), which reveals π-aromaticity, as it was initially reported by Zhai et al. [126]. Out of the remaining six σ-MOs, one (HOMO-6 ($3a_1'$)) can be associated with delocalized σ-bonding between three central atoms, and five (HOMO ($5e'$), HOMO-2 ($4a_1'$), and HOMO-3 ($4e'$)) with delocalized σ-bonding between three central and nine peripheral atoms. Each set of orbitals satisfies $4n+2$ rule (with $n=0$ for bonding between the central atoms and $n=2$ for bonding between the central and peripheral atoms). Thus, in addition to the π-aromaticity, this cluster is also σ-aromatic. At this point, the B_{12} cluster can be considered as being doubly (σ- and π-) aromatic [2].

(xiii) B_{13}^+

Anderson and coworkers [90] reported that the B_{13}^+ cationic cluster has anomalously high stability and low reactivity in comparison with other cationic boron clusters in collision-induced dissociation experiment. They initially attributed this high stability to a filled icosahedral structure. In the follow-up theoretical study by Kawai and Weare [127], it was shown using Car-Parrinello *ab initio* molecular dynamics simulations that a filled icosahedron of B_{13}^+ is not even a minimum on the potential energy surface. The global minimum planar C_{2v} $^1A_1$ structure of B_{13}^+ was established by

Ricca and Bauschlicher [88] (Figure 16). In this structure, three boron atoms can fit perfectly into the plane of the 10-membered ring. On the basis of plotted MOs, Fowler and Ugalde concluded that three doubly occupied π-MOs give six π-electrons to the cyclic system, a situation reminiscent of benzene and the Huckel aromaticity and thus they were the first who proposed that the exceptional stability and low reactivity of B_{13}^+ is related to its aromatic character [128]. In the follow-up paper Aihara [7] evaluated the topological resonance energy (TRE) for π-electrons using the graph theory of aromaticity. He concluded that B_{13}^+ is much more aromatic than polycyclic aromatic hydrocarbons of the similar size.

Like in case of other large boron clusters, the σ-bonding has not been initially taken into account. Zubarev and Boldyrev were the first who performed comprehensive molecular orbital analysis for the B_{13}^+ cation [2]. Valence MOs of B_{13}^+ are plotted in Figure 20. The NBO analysis showed 10 2c–2e σ-B–B peripheral bonds (ON = 1.89–1.93 |e|). The 10 2c–2e σ-B–B peripheral bonds are formed out of HOMO-8 ($7a_1$), HOMO-9 ($6a_1$), HOMO-10 ($4b_2$), HOMO-11 ($5a_1$), HOMO-12 ($3b_2$), HOMO-14 ($2b_2$), HOMO-15 ($3a_1$), HOMO-16 ($2a_1$), HOMO-17 ($1b_2$), and HOMO-18 ($1a_1$). These 10 bonds take 20 of 38 valence electrons. HOMO-6 ($1b_1$), HOMO-2 ($1a_2$), and HOMO-1 ($2b_1$) are globally delocalized π-MOs responsible for π-aromaticity, as it was previously reported by Fowler and Ugalde [128] and Aihara [7]. The 12 remaining electrons on HOMO ($10a_1$), HOMO-3 ($6b_2$), HOMO-4 ($9a_1$), HOMO-5 ($8a_1$), HOMO-7 ($5b_2$), and HOMO-13 ($4a_1$), are responsible for globally delocalized σ-bonding and σ-aromaticity. The HOMO-13 ($4a_1$) is responsible for delocalized σ-bonding between three central atoms and the five remaining delocalized σ-MOs are responsible for delocalized σ-bonding between three central atoms and the peripheral atoms. Considering these two sets of delocalized σ-MOs, we see that they separately satisfied the $4n+2$

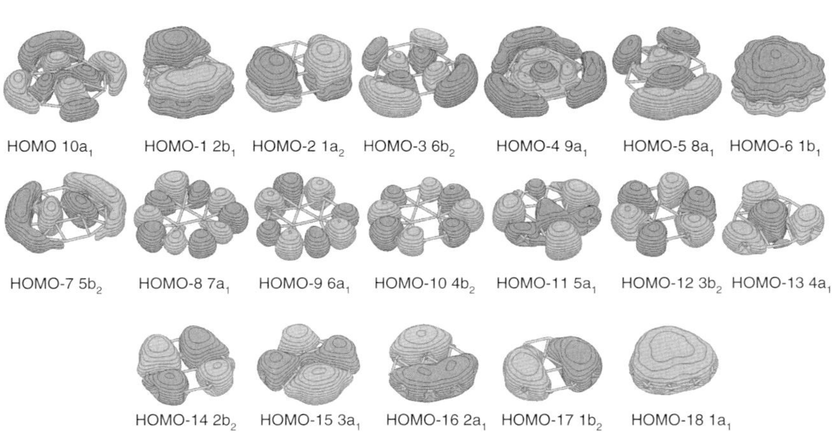

FIGURE 20 Valence molecular orbitals of B_{13}^+.

rule (with $n=0$ for bonding between the central atoms and with $n=2$ for bonding between the central and peripheral atoms). Thus, the B_{13}^{+} is doubly (σ- and π-) aromatic and that can explain high first singlet vertical excitation energy (2.09 eV, at TD-B3LYP/6-311+G*), highly negative NICS values, and most importantly the anomalously high stability and low reactivity of B_{13}^{+} in comparison to other cationic boron clusters observed by Anderson and coworkers [90].

B. Doubly Antiaromatic Boron Clusters

The B_6^{2-} dianion is the best explained doubly antiaromatic boron cluster [2f,g,2,8,111b,2,8,111]. B_6^{2-} has the planar D_{2h} ^1A_g ($1a_g^2 1b_{1u}^2 2a_g^2 1b_{2u}^2 1b_{3g}^2 1b_{3u}^2 3a_g^2 2b_{2u}^2 2b_{1u}^2 1b_{2g}^2$) global minimum structure. This structure can be traced back to the B_6^{2+}, D_{6h} hexagon (see Section V.A), which underwent the first-order Jahn–Teller distortion upon the addition of two extra electron pairs, that occupy one of the doubly degenerate π-MO ($1e_{1g}$-LUMO in B_6^{2+}) and one of the doubly degenerate σ-MO MO ($2e_{1u}$-LUMO+1 in B_6^{2+}). Molecular orbital analysis helps us to interpret chemical bonding in B_6^{2-} (Figure 21A).

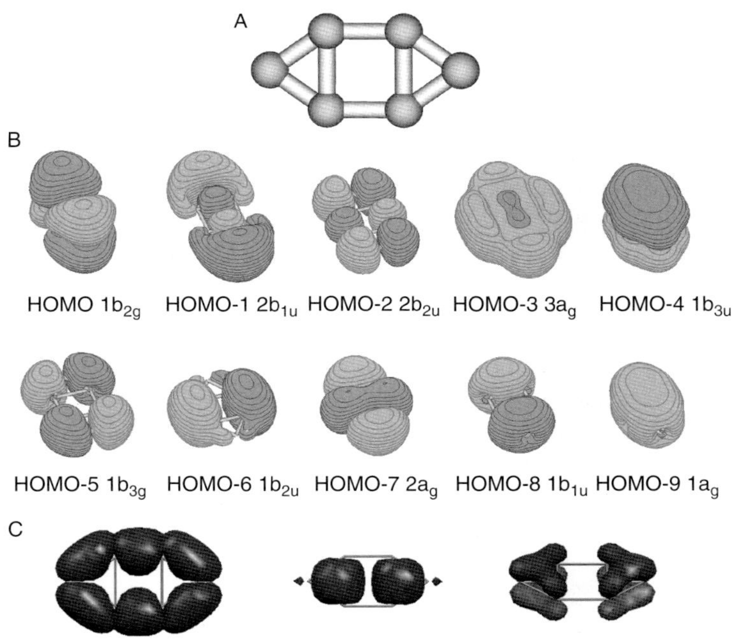

FIGURE 21 (A) Optimized molecular structure of B_6^{2-}, (B) valence canonical molecular orbitals of B_6^{2-}, and (C) six NBO localized 2c–2e B–B bonds, two NBO localized 3c–2e σ-bonds, and two NBO localized 3c–2e π-bonds.

Six MOs (HOMO-2 ($2b_{2u}$), HOMO-5 ($1b_{3g}$), HOMO-6 ($1b_{2u}$), HOMO-7 ($2a_g$), HOMO-8 ($1b_{1u}$), and HOMO-9 ($1a_g$)) can be localized into six 2c–2e B–B bonds, so they are responsible for the peripheral bonding in this cluster. The HOMO-1 ($2b_{1u}$) and HOMO-3 ($3a_g$) are σ-radial MOs, with the HOMO-3 being completely bonding, and the HOMO-1 being partially antibonding. Thus, B_6^{2-} has two globally delocalized σ-MOs, which makes this dianion σ-antiaromatic. The two other delocalized orbitals HOMO-4 ($1b_{3u}$) and HOMO ($1b_{2g}$) are π-MOs. The HOMO-4 is completely bonding, and the HOMO is a partially bonding orbital. The B_6^{2-} dianion has four σ- and four π-electrons on the globally delocalized MOs and six 2c–2e peripheral B–B bonds. Thus, we can assign B_6^{2-} to doubly (σ- and π-) antiaromatic systems.

Aihara et al. [129] claimed that the boron B_6^{2-} cluster is highly aromatic on the basis of TRE. The calculated TRE for B_6^{2-} in terms of the resonance integral between two bonded boron atoms ($|\beta_{BB}|$) is 0.549 $|b_{BB}|$. For the reference, the TRE for benzene is 0.273 $|b_{CC}|$. However, Zubarev and Boldyrev [2g] pointed out that this large resonance energy does not contradict the assignment of double antiaromaticity to B_6^{2-}. We see this cluster as being *antiaromatic globally*. It does not, however, mean that this cluster cannot have positive resonance energy. This globally antiaromic system can be also viewed as composed out of two islands of aromaticity. Indeed, π-MOs of B_6^{2-} cluster can be viewed as composed of two π-aromatic B_3^- clusters (see Refs. [2]f,g, [111] for the detailed discussion). The $1b_{2g}$-HOMO and $1b_{3u}$-HOMO-4 are a pair of bonding and antibonding π-MOs in B_3^- (Figure 21B). Therefore, π-MOs essentially do not contribute to chemical bonding between two B_3^- groups. The π-MOs in B_6^{2-} are effectively localized over B_3 triangular fragments and give rise to the island π-aromaticity in this cluster. Similar analysis for the delocalized σ-MOs reveals that we also have the island σ-aromaticity in B_6^{2-}. Thus, the globally antiaromatic B_6^{2-} dianion consists of two island aromatic subunits. The island π-aromaticity is responsible for the positive TRE in B_6^{2-} in Aihara et al. [129] calculations. Alexandrova et al. [130] demonstrated that for Li_2B_6 molecule in the gas-phase the global minimum structure is C_{2h} ($^1A_1$) with two Li^+ ions located above and below the B_3^- triangular areas in B_6^{2-}. These results confirmed the presence of the island aromaticity in the globally antiaromatic system.

The question may arise, why a doubly antiaromatic structure is the global minimum. We believe that the B_6^{2-} cannot favorably support six delocalized electrons in either σ- or π-subsystems because of the weak electrostatic field from the screened boron nuclei. Another example of a doubly antiaromatic boron cluster is B_{15} (see Ref. [2]g for the detailed discussion).

C. Boron Clusters with Conflicting Aromaticity

There are two identified boron clusters with conflicting aromaticity: B_5^- and B_{14} [2]. We discussed the π-aromatic and σ-antiaromatic B_5^- cluster in detail in Section IV.B. The B_{14} cluster is σ-aromatic and π-antiaromatic according to the chemical bonding analysis performed by Zubarev and Boldyrev [2]. One can see that boron clusters with conflicting aromaticity are also quite rare.

VI. POSSIBLE TYPES OF AROMATICITY AND ANTIAROMATICITY IN PLANAR CARBON CLUSTERS

Medium size pure carbon clusters are known to be the most stable in the cyclic form [131]. Chemical bonding in terms of aromaticity and anti-aromaticity has been discussed by Schleyer et al. [132], Martin-Santamaria and Rzepa [132], Deleuze et al. [132], and most recently by Wodrich et al. [27]. In our review we will briefly discuss these results.

A. Doubly Aromatic Carbon Clusters

(i) C_6^{4+}

Up till now, C_6^{4+} is the smallest carbon cluster that was identified by Wodrich et al. [27] as doubly (σ- and π-) aromatic. This tetracation has the high symmetry D_{6h} singlet structure (Figure 22).

According to the analysis in Ref. [22c], this cluster has six delocalized π-electrons and two delocalized σ-electrons (occupying radial delocalized orbital). The highly negative $NICS_{\pi zz} = -38.5$ ppm and $NICS_{radzz} = -16.0$ pmm confirmed the presence of double aromaticity in this cluster. We repeated C_6^{4+} calculation at the B3LYP/6-311+G* level of theory and found that indeed this structure is a local minimum. The valence electronic configuration for C_6^{4+} is $1a_{1g}^2 1e_{1u}^4 1e_{2g}^4 1b_{2u}^2 1a_{2u}^2 2a_{1g}^2 1e_{1g}^4$. Indeed, $1a_{2u}$ and $1e_{1g}$ are π-delocalized MOs responsible for the π-aromaticity in this system and $2a_{1g}$ is the σ-delocalized MO responsible for the σ-aromaticity as was stated by Wodrich et al. [27]. However, we thought that the lowest six valence MOs ($1a_{1g}$, $1e_{1u}$, $1e_{2g}$, and $1b_{2u}$) should be responsible for the formation of the six 2c–2e C–C bonds, similar to the cluster B_6^{2+}, where these MOs are responsible for the formation of the six B–B bonds (see Section V.A). In order to test this, we performed the NBO analysis for the C_6^{12+} cluster at the optimized geometry of C_6^{4+} with only these six MOs being occupied. At two different levels of theory, B3LYP/6-311+G* and RHF/6-311+G*, we found six localized C–C bonds with the ON 1.97 |e|, thus confirming our bonding interpretation of these MOs. We believe that like in planar boron clusters, these peripheral 2c–2e C–C bonds together with double aromaticity are responsible for the planarity of this and other carbon clusters.

(ii) C_{10}

The D_{5h} singlet structure of C_{10} is more stable than the lowest energy linear isomer (Figure 22) [133]. The D_{5h} structure with 10 delocalized π-electrons and 10 delocalized σ-electrons (occupying radial delocalized orbital) was

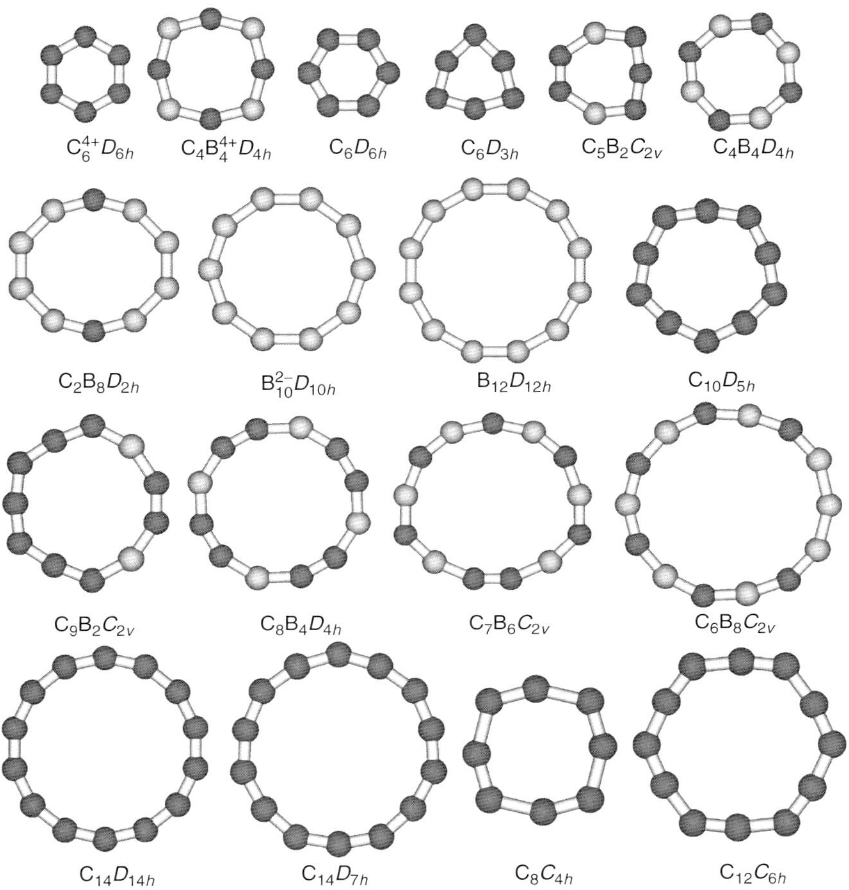

FIGURE 22 Molecular structures of the monocyclic carbon, boron, and borocarbon rings (boron is blue and carbon is brown). (For interpretation of the references to color in this figure legend, the reader is referred to the Web version of this chapter.)

recognized as a doubly aromatic system by Martin-Santamaria and Rzepa [132], and by Wodrich et al. [27]. The deviation from the D_{10h} high symmetry in the C_{10} cluster occurs due to the second-order Jahn–Teller effect. We calculated the D_{10h} and D_{5h} structures of C_{10} at the B3LYP/6-311+G* and the D_{10h} structure has one imaginary frequency. Geometry optimization following the imaginary normal mode led to the D_{5h} minimum structure. The most accurate barrier value for the interconvertion of the equivalent D_{5h} minima through the D_{10h} transition state structure is only 1.0 kcal/mol (CCSD(T)/cc-pvTZ) [134]. Therefore, for all practical purposes, the doubly aromatic C_{10} cluster can be considered as having D_{10h} symmetry after the vibrational averaging. The distortion from the high symmetry D_{10h} structure occurs because of the second-order Jahn–Teller effect. We performed MO analysis for the both D_{10h} and D_{5h} structures. The 10 lowest valence MOs ($1a_{1g}^2 1e_{1u}^4 1e_{2g}^4 1e_{3u}^4 1e_{4g}^4 1b_{2u}^2$ in D_{10h} and $1a_1'^2 1e_1'^4 1e_2'^4 2e_2'^4 2e_1'^4 1a_2'^2$ in D_{5h}) can be localized into 10 2c–2e C–C bonds (ON is 1.98 |e| in the C_{10}^{20+} cation at either D_{10h} or D_{5h} optimized geometry of corresponding neutrals). The five delocalized π-MOs ($1a_{2u}^2 2e_{2g}^4 1e_{4u}^4$ in D_{10h} and $1a_2''^2 1e_1''^4 1e_2''^4$ in D_{5h}) make this cluster π-aromatic and remaining five delocalized σ-MOs ($2a_{1g}^2 1e_{2u}^4 2e_{2g}^4$ in D_{10h} and $2a_1'^2 3e_1'^4 3e_2'^4$ in D_{5h}) make this cluster σ-aromatic. Again, we believe that formation of 10 peripheral 2c–2e C–C bonds is as important for the stability of the cyclic planar structure of C_{10} cluster as its double (σ- and π-) aromaticity. The highly negative $NICS_{\pi zz} = -39.0$ ppm and $NICS_{radzz} = -13.9$ pmm calculated for the D_{5h} singlet structure confirm the assignment of the cluster as doubly aromatic.

(iii) C_{14}

The D_{7h} singlet structure of C_{14} was recognized as doubly aromatic by Martin-Santamaria and Rzepa [132], and by Wodrich et al. [27]. The deviation from the D_{14h} high symmetry in the C_{14} cluster occurs because of the second-order Jahn–Teller effect. The equivalent D_{7h} minima and the D_{14h} transition state structure are essentially isoenergetic at B3LYP/6-311+G* [27]. Again, for all practical purposes, the doubly aromatic C_{14} cluster can be considered as having D_{14h} symmetry after vibrational averaging. As in the previous cases, we believe that 14 2c–2e C–C peripheral bonds are also responsible for the stability and planarity of the cluster. The highly negative $NICS_{\pi zz} = -56.7$ ppm and $NICS_{radzz} = -64.0$ pmm calculated for the D_{14h} singlet structure confirm the assignment of the cluster as doubly aromatic.

B. Doubly Antiaromatic Carbon Clusters

(i) C_8

Currently, the cyclic C_8 is the smallest carbon cluster that was identified by Wodrich et al. [27] as doubly (σ- and π-) antiaromatic. According to their analysis, this cluster has eight delocalized π-electrons and eight delocalized σ-electrons (occupying radial delocalized orbitals). The highly positive $NICS_{\pi zz} = +151.5$ ppm and $NICS_{radzz} = +28.9$ pmm confirm the assignment of the cluster as doubly antiaromatic. Martin and Taylor [134] reported the cyclic C_{4h} singlet structure with the alteration of C–C bonds as being more stable than the linear structures (Figure 22). We calculated the D_{8h}, D_{4h}, and C_{4h} structures at the B3LYP/6-311+G* level of theory and obtained their relative energies 68.4 kcal/mol (D_{8h}, fourth-order saddle point) and 24.1 kcal/mol (D_{4h}, second-order saddle point) with respect to the most stable minimum structure C_{4h}. The distortion from the D_{8h} structure occurs due to the first-order

Jahn–Teller effect and has significantly more pronounced energetic effect compared to the second-order Jahn–Teller effects in the aromatic systems.

The second-order Jahn–Teller effect in aromatic carbon clusters does not affect the bond lengths in the carbon clusters and distorts angles only, while in the doubly antiaromatic C_8 cluster, the first-order Jahn–Teller effect leads to the significant bond alteration with $\Delta R(\text{C–C}) = 0.13$ Å (B3LYP/6-311+G*).

C. Carbon Clusters with Conflicting Aromaticity

Wodrich et al. [27] identified the monocyclic planar C_7 and C_9 clusters as having mixed aromaticity. The cyclic C_7 cluster has six delocalized π-electrons and eight delocalized σ-electrons in the C_{2v} structure according to these authors. The cyclic C_9 cluster has 10 delocalized π-electrons and eight delocalized σ-electrons in the C_{2v} structure according to Wodrich et al. [27]. However, as these authors pointed out, the structures of C_7 and C_9 tested for aromaticity are not minima on the potential energy surface and therefore will not be discussed in any detail here.

Our chemical bonding analysis of carbon clusters reveals that in addition to aromaticity/antiaromaticity the formation of 2c–2e C–C bonds in the peripheral ring is essential for stability and planarity of carbon cyclic clusters. In fact, we believe that the planar cyclic carbon clusters are possible because of the presence of these peripheral C–C bonds. For example, in silicon clusters, where lone pairs are encountered instead of 2c–2e Si–Si bonds, the three-dimensional structures become global minima beyond the tetramers.

VII. POSSIBLE TYPES OF AROMATICITY AND ANTIAROMATICITY IN MONOCYCLIC BOROCARBON CLUSTERS

Aromaticity and antiaromaticity in borocarbon rings (Figure 22) have been recently considered by Wodrich et al. [27]. According to the electron counting, cyclic $C_4B_4^{4+}$ (six π- and two σ delocalized electrons), C_5B_2 (six π- and two σ delocalized electrons), C_4B_4 (six π- and six σ delocalized electrons), C_2B_8 (six π- and six σ delocalized electrons), C_9B_2 (10 π- and 10 σ delocalized electrons), C_8B_4 (10 π- and 10 σ delocalized electrons), C_7B_6 (10 π- and 10 σ delocalized electrons), and C_6B_8 (10 π- and 10 σ delocalized electrons) are doubly aromatic systems. Wodrich et al. [27] confirmed the σ- and π-aromaticity in these clusters by calculating NICS [135].

The CB_6^{2-} D_{6h} and CB_7^- D_{7h} singlet anions have been proposed as examples of hexa- and hepta-coordinated carbon species [3b,c,8b,136]. The CB_7^- species is isoelectronic to B_8^{2-}, which as we have shown previously possesses the global minimum D_{7h} structure with a hepta-coordinated boron (see Section V.A). The D_{7h} CB_7^- (Figure 23A) can be viewed as replacing

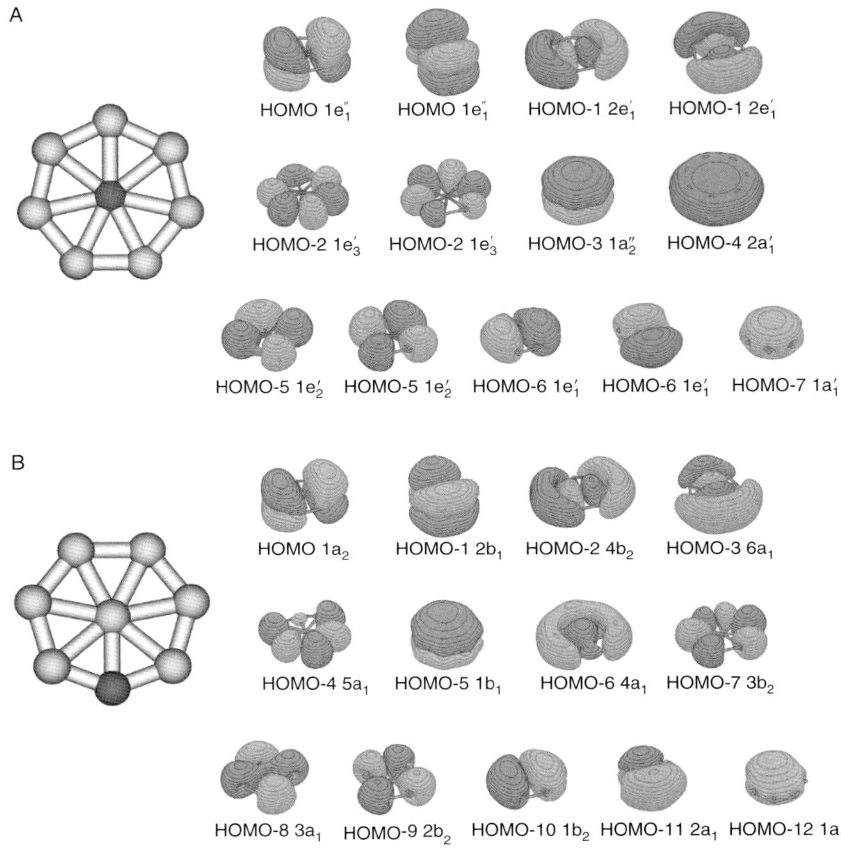

FIGURE 23 (A) The optimized D_{7h} structure of CB_7^- and its valence molecular orbitals and (B) the optimized C_{2v} global minimum structure of CB_7^- and its valence molecular orbitals.

the center B^- ion in B_8^{2-} by a C atom. However, in a joint photoelectron spectroscopy and *ab initio* study it was shown that the observed species is a C_{2v} CB_7^- (Figure 23B), in which the C atom replaces a B^- ion in the rim of the D_{7h} B_8^{2-} molecular wheel [137].

The difference in stability and chemical bonding in the D_{7h} and C_{2v} molecular wheel structures of CB_7^- was explained on the basis of the analysis of valence molecular orbitals, as shown in Figure 23. The MOs of the D_{7h} CB_7^- (Figure 23A) are identical to those of the B_8^{2-} molecular wheel (Figure 18), that is, it is doubly aromatic with six totally delocalized π electrons (HOMO $1e''_2$ and HOMO-3 $1a''_2$) and six totally delocalized σ electrons (HOMO-1 $2e'_1$ and HOMO-4 $2a'_1$), and seven MOs (HOMO-2 $1e'_3$, HOMO-5 $1e'_2$, HOMO-6 $1e'_1$, and HOMO-7 $1a'_1$) which can be localized into seven 2c–2e B–B peripheral bonds. The MOs of the C_{2v} global minimum (Figure 23B) are very similar to those of the D_{7h} isomer. The structure is also π aromatic with six delocalized π electrons (HOMO $1a_2$, HOMO-1 $2b_1$, and HOMO-5 $1b_1$) and it is also σ-aromatic with six delocalized σ-electrons (HOMO-2 $4b_2$, HOMO-3 $6a_1$, and HOMO-6 $4a_1$). There are also seven MOs (HOMO-4 $5a_1$, HOMO-7 $3b_2$, HOMO-8 $3a_1$, HOMO-9 $2b_2$, HOMO-10 $1b_2$, HOMO-11 $2a_1$, and HOMO-12 $1a_1$), that could be localized into five 2c–2e peripheral B–B and two 2c–2e C–B peripheral bonds, similar to those in the D_{7h} isomer.

The central carbon atom in the D_{7h} isomer is involved in completely delocalized (doubly σ- and π- aromaticity) bonding only, while in the C_{2v} global minimum structure, the carbon atom is involved in the two 2c–2e B–C peripheral bonds, in addition to the participation in the delocalized σ- and π-bonding. Carbon is known to form strong 2c–2e σ-bonds because of its high valence charge. This makes the peripheral position of carbon atom significantly more preferable compared to the central position. On the other hand, boron is known to participate in delocalized σ-bonding because of its relatively low valence charge, making the doubly aromatic C_{2v} structure the most stable.

Thus, the experimental and theoretical study performed by Wang et al. [137] showed that the hepta-coordinated carbon in the C–B system is extremely unfavorable. We expect that the central position of carbon in other planar cyclic borocarbon clusters with coordination number of six and higher would be also highly unfavorable.

VIII. OVERVIEW

The discussion presented above clearly demonstrated the importance of aromaticity/antiaromaticity in describing chemical bonding in planar clusters. For the main group atoms, the delocalized bonding can be characterized by double (σ- and π-) aromaticity, double (σ- and π-) antiaromaticity, and conflicting aromaticity (σ-aromaticity and π-antiaromaticity or σ-antiaromaticity and π-aromaticity). In transition metal clusters, participation of d-AOs in chemical bonding could lead to σ-tangential (σ_t-), σ-radial (σ_r-), π-tangential (π_t-), π-radial (π_r-), and δ-aromaticity/antiaromaticity. In this case, there can be multiple (σ-, π-, and δ-) aromaticity, multiple (σ-, π-, and δ-) antiaromaticity, and conflicting aromaticity (simultaneous σ-aromaticity, π-aromaticity, and δ-antiaromaticity; σ-aromaticity, π-antiaromaticity, and δ-aromaticity; σ-antiaromaticity, π-aromaticity, and δ-aromaticity; σ-aromaticity, π-antiaromaticity, and δ-antiaromaticity; σ-antiaromaticity, π-aromaticity, and δ-antiaromaticity; σ-antiaromaticity, π-antiaromaticity, and δ-aromaticity). There is also a possibility to find ϕ-aromaticity/antiaromaticity in clusters composed out of f-elements. In this case, even more complicated cases of conflicting aromaticity can be encountered. The discussion of the viability of ϕ-aromaticity/antiaromaticity is beyond the scope of this review.

The globally antiaromatic systems could also be treated as systems with island aromaticity. As we have shown, antiaromatic σ-system of Li_4 delocalized over the entire cluster can be represented as two islands of σ-aromaticity delocalized over two triangular fragments (see Section III.A). Analogously, antiaromatic σ-system and antiaromatic π-system of the globally doubly antiaromatic B_6^{2-} cluster can be viewed as two islands of σ- and π-aromaticity delocalized over two triangular fragments (see Section V.B).

The electron counting rule for aromaticity/aniaromaticity (singlet coupling case) should be applied separately for each of σ-, π-, and δ-systems. If the aromatic subsystem includes radial and tangential parts, for example, p-AOs-based σ-aromaticity/antiaromaticity and d-AOs based π-aromaticity/antiaromaticity, the counting rules for the cyclic structures with even number of vertices are $4n+4$(aromaticity)/$4n+6$(antiaromaticity). For the cyclic structures with odd number of vertices they are $4n+2$(aromaticity)/$4n$ (antiaromaticity).

Previously, we developed a chemical bonding model for boron clusters, which combines σ- and π-aromaticity/antiaromaticity and peripheral 2c–2e B–B bonds. This model is also applicable to pure carbon and borocarbon cyclic planar clusters. As we mentioned above, we believe that the peripheral 2c–2e C–C bonds are responsible for the stability of medium carbon clusters. The same is true for the borocarbon clusters. Whenever lone pairs are encountered instead of these peripheral bonds, the deviation from planarity occurs in favor of three-dimensional structures. Such examples are found for most of the clusters of the main group and transition metal atoms with rather low atomic electronegativity. Aluminum and silicon clusters illustrate this point, because they become three-dimensional beyond five and four atoms, respectively.

At this point, we believe that the chemical bonding model for the planar clusters is well developed and can be used to explain their structures, stability, reactivity, and other properties. The next challenge is to develop a similar simple "pencil and paper" chemical bonding model for the three-dimensional clusters, which would be as successful as the chemical bonding model for the planar clusters.

Acknowledgments

This work was supported by the National Science Foundation (CHE-0714851). Computer time from the Center for High Performance Computing at Utah State University is gratefully acknowledged. The computational resource, the Uinta cluster supercomputer, was provided through the National Science Foundation under Grant CTS-0321170 with matching funds provided by Utah State University.

REFERENCES

[1] Cluster Chemistry and Dynamics. Special Feature. Proc Natl Acad Sci USA 2006;103:10552–605.

[2] (a) Boldyrev AI, Wang LS. All-metal aromaticity and antiaromaticity. Chem Rev 2005;106:3716. (b) Li X, Kuznetsov AE, Zhang HF, Boldyrev AI, Wang LS. Observation of all-metal aromatic molecules. Science 2001;291:859. (c) Kuznetsov AE, Birch KA, Boldyrev X, Li X, Zhai H-J, Wang LS. All-metal antiaromatic molecule: Rectangular Al-4 (4−) in the Li3Al4− anion. Science 2003;300:622. (d) Boldyrev AI, Kuznetsov AE. On the resonance energy in new all-metal aromatic molecules. Inorg Chem 2002;41:532. (e) Alexandrova AI, Boldyrev AI. Sigma-aromaticity and Sigma-antiaromaticity in alkali metal and alkaline earth metal small clusters. J Phys Chem A 2003;107:554. (f) Alexandrova AN, Boldyrev H-J, Zhai H-J, Wang LS. All-boron aromatic clusters as potential new inorganic ligands and building blocks in chemistry. Coord Chem Rev 2006;250:2811. (g) Zubarev DY, Boldyrev AI. Comprehensive analysis of chemical bonding in boron clusters. J Comput Chem 2007;28:251.

[3] (a) Chen Z, Wannere CS, Corminboeuf C, Puchta R, Schleyer PvR. Nucleus-independent chemical shifts (NICS) as an aromaticity criterion. Chem Rev 2005;106:3842. (b) Exner PvR, Schleyer PvR. Planar hexacoordinate carbon: A viable possibility. Science 2000;290:1937. (c) Wang ZX, Schleyer PvR. Construction principles of "hyparenes": Families of molecules with planar pentacoordinate carbons. Science 2001;292:2465. (d) Tanaka H, Neukermans E, Janssens E, Silverans RE, Lievens P. Sigma aromaticity of the bimetallic Au5Zn+ cluster. J Am Chem Soc 2003;125:2862.

[4] (a) Tsipis CA. DFT study of "all-metal" aromatic compounds. Coord Chem Rev 2005;249:2740. (b) Tsipis AC, Tsipis CA. Hydrometal analogues of aromatic hydrocarbons: A new class of cyclic hydrocoppers(I). J Am Chem Soc 2003;125:1136.

[5] (a) Jemmis ED, Jayasree EG. Analogies between boron and carbon. Acc Chem Res 2003;36:816. (b) Balakrishnarajan MM, Hoffmann R, Pancharatna PD, Jemmis ED. Magic electron counts and bonding in tubular boranes. Inorg Chem 2003;42:4650. (c) Jemmis MM, Balakrishnarajan MM, Pancharatna PD. Electronic requirements for macropolyhedral boranes. Chem Rev 2002;102:93.

[6] (a) Fowler JE, Ugalde JM. The curiously stable B-13(+) cluster and its neutral and anionic counterparts: The advantages of planarity. J Phys Chem A 2000;104:397. (b) Mercero JM, Ugalde JM. Sandwich-like complexes based on "all-metal" (Al-4(2−)) aromatic compounds. J Am Chem Soc 2004;126:3380.

[7] Aihara J. B-13(+) is highly aromatic. J Phys Chem A 2001;105:5486.

[8] (a) Fowler PW, Havenith RWA, Steiner E. Unconventional ring currents in an 'all-metal aromatic', Al-4(2−). Chem Phys Lett 2001;342:85. (b) Havenith RWA, Fowler PW, Steiner E. Ring currents in a proposed system containing planar hexacoordinate carbon, CB62−. Chem Eur J 2002;8:1068. (c) Fowler PW, Havenith RWA, Steiner E. Ring current and electron delocalisation in an all-metal cluster, Al-4(2−). Chem Phys Lett 2002;359:530. (d) Havenith JH, van Lenthe JH. A valence bond study of the sigma and pi aromatic species Al-4(2−). Chem Phys Lett 2004;385:198. (e) Havenith RWA, De Proft F, Fowler PW, Geerling P. Sigma-aromaticity in H-3(+) and Li-3(+): Insights from ring-current maps. Chem Phys Lett 2005;407:391.

[9] (a) Juselius J, Straka M, Sundholm D. Magnetic-shielding calculations on Al-4(2−) and Analogues. A new family of aromatic molecules? J Phys Chem A 2001;105:9939. (b) Lin Y-C, Juseluis J, Sundholm D. Magnetically induced current densities in Al-4(2−) and

Al-4(4−) species studied at the coupled-cluster level. J Chem Phys 2005;122:214308. (c) Lin Y-C, Sundholm D, Juseluis J, Cui L-F, Li X, Zhai H-J, Wang LS. Experimental and computational studies of alkali-metal coinage-metal clusters. J Phys Chem A 2006;110:4244.

[10] Zhan C-G, Zheng F, Dixon DA. Electron affinities of Al-n clusters and multiple-fold aromaticity of the square Al-4(2−) structure. J Am Chem Soc 2002;124:14795.

[11] Santos JC, Tiznado W, Contreras R, Fuentealba P. Sigma-pi separation of the electron localization function and aromaticity. J Chem Phys 2004;120:1670.

[12] (a) Datta A, Pati SK. Charge-transfer induced large nonlinear optical properties of small Al clusters: Al4M4 (M = Li, Na, and K). J Phys Chem. A 2004;108:9527. (b) Datta A, Pati SK. Stable transition metal complexes of an all-metal antiaromatic molecule (Al4Li4): Role of complexations. J Am Chem Soc 2005;127:3496. (c) Datta A, Pati SK. Rationalization of the pi-sigma (anti)aromaticity in all metal molecular clusters. J Chem Theor Comput 2005;1:824. (d) Datta A, John NS, Kulkarni GU, Pati SK. Aromaticity in stable tiara nickel thiolates: Computational and structural analysis. J Phys Chem A 2005;109:11647. (e) Rehaman A, Datta A, Mallajosyulu SS, Pati SK. Quantifying aromaticity at the molecular and supramolecular limits: Comparing homonuclear, heteronuclear, and H-bonded systems. J Comput Theor Comput 2006;2:30.

[13] (a) Havenith RWA, Fowler PW, Steiner E, Shetty S, Kanhere D, Pal S. Aromaticity and antiaromaticity of LixAl4 clusters: Ring current patterns versus electron counting. Phys Chem Chem Phys 2004;6:285. (b) Shetty S, Kar R, Kanhere DG, Pal S. Intercluster reactivity of metalloaromatic and antiaromatic compounds and their applicationsin molecular electronics: A theoretical investigation. J Phys Chem A 2006;110:252.

[14] (a) Hu XB, Li HR, Liang WC, Han SJ. Theoretical study on all-metal aromatic complexes: interaction with small gas molecule. Chem Phys Lett 2004;397:180. (b) Hu X, Li H, Liang WC, Han SJ. Theoretical study on all-metal aromatic complexes: all-metal aromatic hydrogen bond. Chem Phys Lett 2005;402:539. (c) Hu XB, Li HR, Liang WC, Han SJ. Exploring a new kind of aromatic hydrogen bond: hydrogen bonding to all-metal aromatic species. New J Chem 2005;29:1295. (d) Wang F-F, Li Z-R, Wu D, Sun X-Z, Chen W, Li Y, et al. Novel superalkali superhalogen compounds (Li-3)(+)(SH)(−) (SH=LiF2, BeF3, and BF4)with aromaticity: New electrides and alkalides. Chem Phys Chem 2006;7:1136.

[15] (a) Alexandrova AN, Boldyrev AI. Sigma-aromaticity and sigma-antiaromaticity in alkali metal and alkaline earth metal small clusters. J Phys Chem A 2003;107:554. (b) Yong L, Wu XX, Chi XX. Theoretical study of aromaticity in small hydrogen and metal cation clusters X-3(+) (X = H, Li, Na, K, and Cu). Int J Quant Chem 2007;107:722.

[16] Havenith RWA, De Proft F, Fowler PW, Geerlings P. Sigma-Aromaticity in H-3(+) and Li-3(+): Insights from ring-current maps. Chem Phys Lett 2005;407:391.

[17] (a) Kuznetsov AE, Boldyrev AI. Theoretical evidence of aromaticity in X-3(−) (X = B, Al, Ga) species. Struct Chem 2002;13:141. (b) Yang L-M, Ding Y-H, Sun C-C. Design of the sandwich-like compounds based on the all-metal aromatic unit Al-3(−). Chemphyschem 2006;7:2478.

[18] Schleyer PvR, Maerker C, Dransfeld A, Jiao HJ, Hommes NJRvE. Nucleus-independent chemical shifts: A simple and efficient aromaticity probe. J Am Chem Soc 1996;118:6317.

[19] (a) Dugourd Ph, Rayane D, Labastie P, Vezin B, Chevaleyre J, Broyer M. Measurements of lithium cluster ionization-potentials. Chem Phys Lett 1992;197:433. (b) Durand G, Heitz M-C, Spiegelman F, Meier C, Mitric R, Bonacic-Koutecky V, Pittner J. Different approaches for the calculation of electronic excited states of nonstoichiometric alkali halide clusters: The example of Na3F. J Chem Phys 2004;121:9898.

[20] Bishop DM, Chaillet M, Larrieu K, Pouchan C. Some electrical-properties of Li-3+ and Li-3−. Mol Phys 1984;51:179.
[21] (a) Weinhold F, Landis C. Valency and Bonding. Canbridge, UK: Cambridge University Press; 2005. (b) Glendening ED, Badenhoop JK, Reed AE, Carpenter JE, Bohmann JA, Morales CM, et al. NBO 5.0. Madison: Theoretical Chemistry Institute, University of Wisconsin; 2001.
[22] (a) Kuznetsov AE, Boldyrev AI. A single pi-bond captures 3, 4 and 5 atoms. Chem Phys Lett 2004;388:452. (b) Zubarev DY, Boldyrev AI, Li X, Wang LS. Observation of triatomic species with conflicting aromaticity: AlSi2− and AlGe2−. J Phys Chem B 2006;110:9743.
[23] (a) Kafafi ZH, Hauge RH, Fredin L, Margrave JL. Infrared matrix-isolation spectrum of the Si2C molecule. J Phys Chem 1983;87:797. (b) Grev RS, Schaefer III HF. The ground-state of Si-3 – 2 near degenerate isomers. Chem Phys Lett 1985;119:111. (c) Raghavachari K. Theoretical-study of small silicon clusters – cyclic ground-state structure of Si-3. J Chem Phys 1985;83:3520. (d) Ragchavachari K. Theoretical-study of small silicon clusters – equilibrium geometries and electronic-structures of Si-2-7, Si-10. J Chem Phys 1986;84:5672.
[24] (a) Ritter SK. Deciphering metal antiaromaticity. Chem Eng News 2003;81:23. (b) Chen Z, Corminboeuf T, Heine T, Bohmann J, Schleyer PvR. Do all-metal antiaromatic clusters exist? J Am Chem Soc 2003;125:13930. (c) Shetty S, Kanhere DG, Pal S. Metallo-antiaromatic Al4Na4 and Al4Na3− compounds: A theoretical investigation. J Phys Chem A 2004;108:628. (d) Chacko S, Kanhere DG, Paranjape VV. Evolution of the structural and bonding properties of aluminum-lithium clusters. Phys Rev A 2004;70:023204. (e) Santos JC, Andres J, Aizman A, Fuentealba P. An aromaticity scale based on the topological analysis of the electron localization function including sigma and pi contributions. J Chem Theor Comput 2005;1:83. (f) Lin Y-C, Juselius J, Sundholm D. Magnetically induced current densities in Al-4(2−) and Al-4(4−) species studied at the coupled-cluster level. J Chem Phys 2005;122:214308.
[25] Zhai H-J, Averkiev BB, Zubarev DY, Wang LS, Boldyrev AI. Delta aromaticity in [Ta3O3](−). Angew Chem Int Ed 2007;46:4277–80.
[26] Huang X, Zhai HJ, Kiran B, Wang LS. Observation of d-orbital aromaticity. Angew Chem Int Ed 2005;44:7251.
[27] (a) Fowler PW, Steiner E. Ring currents in six-membered heterocycles: the diazaborinines (CH)(2)B2N2. Mol Phys 2000;98:945. (b) Steiner E, Fowler PW, Jennesskens LW. Counter-rotating ring currents in coronene and corannulene. Angew Chem Int Ed 2001;40:362. (c) Wodrich MD, Corminboeuf C, Park PvR, Schleyer PvR. Double aromaticity in monocyclic carbon, boron, and borocarbon rings based on magnetic criteria. Chem Eur J 2007;13:4582–93.
[28] Wannere CS, Corminbouef C, Wang Z-X, Wodrich MD, King RB, Schleyer PvR. Evidence for d orbital aromaticity in square planar coinage metal clusters. J Am Chem Soc 2005;127:5701.
[29] Lin Y-C, Sundholm D, Juselius J, Cui L-F, Li X, Zhai H-J, et al. Experimental and computational studies of alkali-metal coinage-metal clusters. J Phys A 2006;110:4244.
[30] Dhavale A, Kanhere DG, Blundell SA, Zope R. Density-functional investigation of the size dependence of the electronic structure of mixed aluminum-sodium clusters. Phys Rev B 2002;65:085402.
[31] Zhao C, Balasubramanian K. Spectroscopic properties of novel aromatic metal clusters: NaM4 (M=Al,Ga,In) and their cations and anions. J Chem Phys 2004;120:10501.
[32] Kuznetsov AE, Boldyrev AI, Li X, Wang LS. On the aromaticity of square planar Ga-4(2−) and In-4(2−) in gaseous NaGa4− and NaIn4− clusters. J Am Chem Soc 2001;123:8825.

[33] Li X, Zhang HF, Wang LS, Kuznetsov AE, Cannon NA, Boldyrev AI. Experimental and theoretical observations of aromaticity in heterocyclic XAI(3)(−) (X = Si, Ge, Sn, Pb) systems. Angew Chem Int Ed 2001;40:1867.
[34] Chi XX, Li XH, Chen XJ, Yuang ZS. Ab initio studies on the aromaticity of bimetallic anionic clusters XGa3− (X = Si, Ge). J Mol Struct (Theochem) 2004;677:21.
[35] Boldyrev AI, Simons J, Li X, Chen W, Wang LS. Combined photoelectron spectroscopy and ab initio study of the hypermetallic Al3C molecule. J Chem Phys 1999;110:8980.
[36] Feng PY, Balasubramanian K. Electronic states of Al3P and AlP3 and their positive ions. Chem Phys Lett 1999;301:458.
[37] Archibong EF, St-Amant A, Goh SK, Marynick D. Structure and electron detachment energies of Al3P− and Al3P3−. J Phys Chem A 2002;106:5932.
[38] Ping YF, Balasubramanian K. Electronic states of Ga3P and GaP3 clusters. Chem Phys Lett 1998;288:1.
[39] Archibong EF, St-Amant A, Goh SK, Marynick D. On the structure and electron photodetachment spectra of Ga3P− and Ga3As−. Chem Phys Lett 2002;361:411.
[40] Archibong EF, St-Amant A. An ab initio and density functional study of Al3As, Al3As−, AlAs3, and AlAs3−. J Phys Chem A 2002;106:7390.
[41] Balasubramanian K, Zhu X. Spectroscopic properties of mixed gallium arsenide tetramers: GaAs3 +/−, GaAs3, Ga3As +/−, and Ga3As. J Chem Phys 2001;115:8858.
[42] Zhai H-J, Kuznetsov AE, Boldyrev AI, Wang LS. Multiple aromaticity and antiaromaticity in silicon clusters. ChemPhysChem 2004;5:1885.
[43] Nielsen JW, Baenziger NC. The crystal structures of NaHG2, NaHg and Na3Hg2. Acta Cryst 1954;7:277.
[44] Kuznetsov AE, Corbett JD, Wang LS, Boldyrev AI. Aromatic mercury clusters in ancient amalgams. Angew Chem Int Ed 2001;40:3369.
[45] Chen Z, Corminboeuf C, Heine T, Bohmann J, Schleyer PvR. Do all-metal antiaromatic clusters exist? J Am Chem Soc 2003;125:13930.
[46] Santos JC, Andres J, Aizman A, Fuentealba P. An aromaticity scale based on the topological analysis of the electron localization function including sigma and pi contributions. J Chem Theor Comput 2005;1:83.
[47] Shetty S, Kanhere DG, Pal S. Metallo-antiaromatic Al4Na4 and Al4Na3− compounds: A theoretical investigation. J Phys Chem A 2004;108:628.
[48] Chattaraj PK, Roy DR, Elango M, Subramanian V. Stability and reactivity of all-metal aromatic and antiaromatic systems in light of the principles of maximum hardness and minimum polarizability. J Phys Chem A 2005;109:9590.
[49] Mallajosyula SS, Datta A, Pati SK. Aromatic superclusters from all-metal aromatic and antiaromatic monomers, [Al-4](2−) and [Al-4](4−). J Phys Chem B 2006;110:20098.
[50] Datta A, Mallajosyula SS, Pati SK. Nonlocal electronic distribution in metallic clusters: A critical examination of aromatic stabilization. Acc Chem Res 2007;40:213.
[51] Ritter SK. Deciphering metal antiaromaticity. Chem Eng News 2003;81:23.
[52] Raghavachari K, Logovinsky V. Structure and bonding in small silicon clusters. Phys Rev Lett 1985;55:2853.
[53] Zhai H-J, Kuznetsov AE, Boldyrev AI, Wang L-S. Multiple aromaticity and antiaromaticity in silicon clusters. Chemphyschem 2004;5:1885.
[54] Cisar A, Corbett JD. Polybismuth anions – synthesis and crystal-structure of a salt of tetrabismuthide(2−) ion, Bi-4(2−) basis for interpretation of structure of some complex intermetallic phases. Inorg Chem 1977;16:2482.

[55] Clark T, Wilhelm D, Schleyer PvR. Y vs cyclic delocalization in small ring dications and dianions – the dominance of charge repulsion over huckel aromaticity. Tetrahedron Lett 1982;23:3457.
[56] Hess Jr. BA, Ewig CS, Schaad LJ. An abinitio study of the cyclobutadiene dianion and dication. J Org Chem 1985;50:5869.
[57] Skancke A. Destabilization of the planar huckeloid cyclobutadiene dianion – diagonal repulsions. Nouv J Chim 1985;9:577.
[58] Zandwijk Gv, Janssen RAJ, Buck HM. 6-pi aromaticity in 4-membered rings. J Am Chem Soc 1990;112:4155.
[59] Balci M, McKee ML, Schleyer PvR. Theoretical study of tetramethyl- and tetra-tert-butyl-substituted cyclobutadiene and tetrahedrane. J Phys Chem A 2000;104:1246.
[60] Sekiguchi A, Matsuo T, Watanabe H. Synthesis and characterization of a cyclobutadiene dianion dilithium salt: Evidence for aromaticity. J Am Chem Soc 2000;122:5652.
[61] Critchlow SC, Corbett JD. Homopolyatomic anions of the post transition-elements – synthesis and structure ofpotassium-crypt salts of the tetraantimonide(2−) and heptaantimonide (3−) anions, Sb4 2− and Sb7 3−. Inorg Chem 1984;23:770.
[62] Kuznetsov AE, Zhai H-J, Wang LS, Boldyrev AI. Peculiar antiaromatic inorganic molecules of tetrapnictogen in Na(+)Pn(4)(−)(Pn = P, As, Sb) and important consequences for hydrocarbons. Inorg Chem 2002;41:6062.
[63] Jin Q, Jin B, Xu WG. Aromaticity of the square P-4(2−) dianion in the P4M (M = Be, Mg, and Ca) and P4M2 (M = Li, Na, and K) clusters. Chem Phys Lett 2004;396:398.
[64] Kraus F, Aschenbrenner JC, Korber N. P-4(2−): A 6 pi aromatic polyphosphide in dicesium cyclotetraphosphide-ammonia (1/2). Angew Chem Int Ed 2003;42:4030.
[65] Gillespie RJ, Passmore J. Polycations of group-6. Acc Chem Res 1971;4:413.
[66] Brown ID, Crump DB, Gillespie RJ, Santry DF. Structure of Se4 2+ ion. Chem Commun 1968;(15):853.
[67] Brown ID, Crump DB, Gillespie RJ. Crystal structure of tetraselenium bis(hydrogen disulfate). Inorg Chem 1971;10:2319.
[68] Gillespie RJ, Pel GP. Infrared and raman spectra of Se4 2+ ion. Inorg Chem 1969;8:1229.
[69] Gillespie RJ, Barr J, Kapoor R, Malhotra KC. New polyatomic cations of group 6 b elements .1. solutions of selenium in highly acidic solvents. Can J Chem 1968;46:149.
[70] Gillespie RJ, Barr J, Crump DB, Kapoor R, Ummat PK. Polyatomic cations of group vib elements .2. compounds of Se4 2+ cation. Can J Chem 1968;46:3607.
[71] Gillespie RJ, Barr J, Crump DB, Kapoor R, Pez GP. Te4 2+cation. J Am Chem Soc 1968;90:6855.
[72] Couch TW, Lokken DA, Corbett JD. Crystal-structures of tetratellurium(2+) tetrachloraluminate and heptachlorodialuminate, Te4 2+(AlCl4−)2 and Te4 2+(Al2Cl7−)2. Inorg Chem 1972;11:357.
[73] Gillespie RJ, Passmore J, Ummat PK, Vaidya OC. Polyatomic cations of sulfur .1. preparation and properties of S16 2+, S8 2+, and S4 2+. Inorg Chem 1971;10:1327.
[74] Passmore J, Sutherland G, White PS. Preparation and crystal-structures of (S7I)4s4(AsF6)6 and S4(AsF6)2.0.6SO2 – a convenient synthesis of hexafluoroarsenate salts of chalcogen homoatomic cations. J Chem Soc Chem Commun 1980;330.
[75] Cameron TS, Dionne I, Jenkins HDB, Parson S, Passmore J, Roobottom HK. Preparation, X-ray crystal structure determination, lattice potential energy, and energetics of formation of the salt S-4(AsF6)(2)center dot ASF(3) containing the lattice-stabilized tetrasulfur [2+] cation. Implications for the understanding of the stability of M-4(2+) and M-2(+) (M = S, Se, and Te) crystalline salts. Inorg Chem 2000;39:2042 and references therein.

[76] Tanaka K, Yamabe T, Terama EH, Fukui K. MO theoretical-study on the dications of tetrasulfur, tetraselenium, and tetratellurium. Inorg Chem 1979;18:3591.

[77] Kao J. S2 +4 and S2 −4 – an abinitio study. J Mol Struct 1980;63:293.

[78] Tang T-H, Bader RFW, MacDougall PJ. Structure and bonding in sulfur nitrogen-compounds. Inorg Chem 1985;24:2047.

[79] Skrezenek FL, Harcourt RD. Abinitio valence bond calculations and the spin-paired diradical character of S-4(2+). Theor Chim Acta 1985;67:271.

[80] Saethre LJ, Gropen O. Structure and bonding in square-planar chalcogen rings. Can J Chem 1992;70:348.

[81] Sannigrahi M, Grein F. On the structure of S4(2+) and its formation from 2S2(+). ab-initio SCF and CASSCF studies. Can J Chem 1994;72:298.

[82] Krossing I, Passmore J. Preference of np(pi)-np(pi) bonding (n = 3, 4) over purely sigma-bonded species in M-4(2+) (M = S, Se): Geometries, bonding, and energetics of several M-4(2+) isomers. Inorg Chem 1999;38:5203.

[83] Jenkins HDB, Jitariu LC, Krossing I, Passmore J, Suontamo R. Basis set and correlation effects in the calculation of accurate gas phase dimerization energies of two M-2(+) to give M-4(2+) (M = S, Se). J Comput Chem 2000;21:218.

[84] Tuononen HM, Suontamo R, Valkonen J, Laitinen RS. Electronic structures and spectroscopic properties of 6 pi-electron ring molecules and ions E2N2 and E-4(2+) (E = S, Se, Te). J Phys Chem A 2004;108:5670.

[85] Burford N, Passmore J, Sanders JCP. Liebman JF, Greenburg A, editors. From Atoms to Polymers. Isoelectronic Analogies. New York: VCH; 1989. p. 53–108.

[86] (a) Elliott BM, Boldyrev AI. Ab initio probing of the aromatic oxygen cluster O-4(2+). J Phys Chem A 2005;109:236. (b) Havenith RWA, Fowler PW. Cyclic O-4(2+): metastable and aromatic. Mol Phys 2006;104:3161.

[87] Kato H, Yamashita K, Morokuma K. Ab initio MO study of neutral and cationic boron clusters. Chem Phys Lett 1992;190:361.

[88] Ricca A, Bauschlicher Jr. CW. The structure and stability of B-n(+) clusters. Chem Phys 1996;208:233.

[89] Zubarev DYu, Boldyrev AI. Comprehensive analysis of chemical bonding in boron clusters. J Comput Chem 2007;28:251.

[90] Hanley L, Whitten JL, Anderson SL. Collision-induced dissociation and abinitio studies of boron cluster ions – determination of structures and stabilities. J Phys Chem 1988;92:5803.

[91] Zhai H-J, Wang LS, Alexandrova AN, Boldyrev AI. Electronic structure and chemical bonding of B-5(−) and B-5 by photoelectron spectroscopy and ab initio calculations. J Chem Phys 2002;117:7917.

[92] Gausa M, Kascher R, Lutz HO, Seifert G, Meiwes-Broer K-H. Photoelectron and theoretical investigations on bismuth and antimony pentamer anions – evidence for aromatic structure. Chem Phys Lett 1994;230:99.

[93] Gausa M, Kascher R, Seifert G, Faehmann J-H, Lutz HO, Meiwes-Broer K-H. Photoelectron investigations and density functional calculations of anionic Sb-n(−) and Bi-n(−) clusters. J Chem Phys 1996;104:9719.

[94] Zhai H-J, Wang LS, Kuznetsov AE, Boldyrev AI. Probing the electronic structure and aromaticity of pentapnictogen cluster anions Pn(5)(−) (Pn = P, As, Sb, and Bi) using photoelectron spectroscopy and ab initio calculations. J Phys Chem A 2002;106:5600.

[95] De Proft F, Fowler PW, Havenith RWA, Schleyer PvR, Lier GV, Geerling P. Ring currents as probes of the aromaticity of inorganic monocycles: P5-As5-S2N2, S3N3−, S4N3+, S4N42+, S5N5+S42+ and Se-4(2+). Chem Eur J 2004;10:940.

[96] Scherer OJ. Complexes with substituent-free acyclic and cyclic phosphorus, arsenic, antimony, and bismuth ligands. Angew Chem Int Ed 1990;29:1104.
[97] Scherer OJ. P-n and As-n ligands: A novel chapter in the chemistry of phosphorus and arsenic. Acc Chem Res 1999;32:751.
[98] Baudler M, Akpapoglou S, Ouzounis D, Wasgestian F, Meinigke B, Budzikiewicz H, et al. Contributions to the chemistry of phosphorus .185. on the pentaphosphacyclopentadienide ion, P5(−). Angew Chem Int Ed 1988;27:280.
[99] Baudler M, Duster D, Ouzounis DZ. Contributions to the chemistry of phosphorus .172. existence and characterization of the pentaphosphacyclopentadienide anion, P5−, the tetraphosphacyclopentadienide ion, P4CH−, and the triphosphacyclobutenide ion, P3CH2−. Anorg Allg Chem 1987;544:87.
[100] Scherer OJ, Bruck T. [(ETA-5-P5)FE(ETA-5-C5ME5)], a pentaphosphaferrocene derivative. Angew Chem Int Ed 1987;26:59.
[101] Scherer OJ, Schwalb J, Wolmershauser G, Kaim W, Gross R. Cyclo-P5 as complex ligand – the phosphorus analog of the cyclopentadienyl ligand. Angew Chem Int Ed 1986;25:363.
[102] Chamizo JA, Ruizmazon M, Salcedo R, Toscano RA. P10Fe, a phosphorus analog of ferrocene. Inorg Chem 1990;29:879.
[103] Urnezius E, Brennessel WW, Cramer CJ, Schleyer PvR. A carbon-free sandwich complex [(P-5)(2)Ti](2−). Science 2002;295:832.
[104] Vij A, Pavlovich JG, Wilson WW, Vij V, Christe KO. Experimental detection of the pentaazacyclopentadienide (pentazolate) anion, cyclo-N-5(−). Angew Chem Int Ed 2002;41:3051.
[105] Dixon DA, Feller D, Christe KO, Wilson WW, Vij A, Vij V, et al. Enthalpies of formation of gas-phase N-3, N-3(−), N-5(+), and N-5(−) from ab initio molecular orbital theory, stability predictions for N5+N3− and N5+N5−, and experimental evidence for the instability of N5+N3−. J Am Chem Soc 2004;126:834 and references there in.
[106] Todorov I, Sevov SC. Heavy-metal aromatic rings: Cyclopentadienyl anion analogues Sn-5 (6) and Pb-5(6−) in the zintl phases Na8BaPb6, Na8BaSn6, and Na8EuSn6. Inorg Chem 2004;43:6490.
[107] Martin JML, François JP, Gijbels R. Abinitio study of boron, nitrogen, and boron-nitrogen clusters potential-energy surface of B4 and total atomization energies of B2, B3, and B4. Chem Phys Lett 1992;189:529.
[108] Zhai H-J, Wang LS, Alexandrova AN, Boldyrev AI, Zakrzewski VG. Photoelectron spectroscopy and ab initio study of B-3(−) and B-4(−) anions and their neutrals. J Phys Chem A 2003;107:9319.
[109] Gillery C, Linguerri R, Rosmus P, Maier JP. Theoretical study of the electronically excited states of B-4, B-4(+) and B-4(−). Z Phys Chem 2005;219:467.
[110] Niu J, Rao BK, Jena P. Atomic and electronic structures of neutral and charged boron and boron-rich clusters. J Chem Phys 1997;107:132.
[111] Alexandrova AN, Boldyrev AI, Zhai AH-J, Wang LS, Steiner E, Fowler PW. Structure and bonding in B6(−) and B6: Planarity and antiaromaticity. J Phys Chem A 2003;107:1359.
[112] Kato H, Yamashita K, Morokuma K. Ab initio mo study of neutral and cationic boron clusters. Chem Phys Lett 1992;190:361.
[113] Boustani I. Systematic LSD investigation on cationic boron clusters – BN+ (N=2–14). Int J Quant Chem 1994;52:1081.
[114] Boustani I. Structure and stability of small boron clusters – a density-functional theoretical-study. Chem Phys Lett 1995;240:135.

[115] Ricca A, Bauschlicher Jr. CW. The structure and stability of B-n(+) clusters. Chem Phys 1996;208:233.

[116] Li Q-S, Gong L-F, Gao Z-M. Structures and stabilities of B-7, B-7(+) and B-7(−) clusters. Chem Phys Lett 2004;390:220.

[117] Alexandrova AN, Boldyrev AI, Zhai H-J, Wang L-S. Electronic structure, isomerism, and chemical bonding in B-7(−) and B-7. J Phys Chem A 2004;108:3509.

[118] In Ref. [117] the A1 spectroscopic state was reported for the $1a_{12}1e_1{}^41e_2{}^42a_1{}^23a_1{}^21b_1{}^22e_1{}^43e_1{}^2$ electronic configuration of the B_7^- C_{6v} structure as determined by the Gaussian 03 program. However, the correct spectroscopic state for such electronic configuration is 3A2. Similarly, in Ref. [117] the 3A1g was reported for the $1a_{1g}{}^21e_{1u}{}^41e_{2g}{}^41b_{2u}{}^22a_{1g}{}^21a_{2u}{}^22e_{1u}{}^41e_{1g}{}^2$ configuration of the B_7^- D_{6h} structure, while the correct spectroscopic state is $^3A_{2g}$.

[119] Zhai HJ, Alexandrova AN, Birch KA, Boldyrev AI, Wang L-S. Hepta- and octacoordinate boron in molecular wheels of eight- and nine-atom boron clusters: Observation and confirmation. Angew Chem Int Ed Engl 2003;42:6004.

[120] Boustani I. Systematic ab initio investigation of bare boron clusters: Determination of the geometry and electronic structures of B-n (n=2–14). Phys Rev B 1997;35:16426.

[121] Bonacic-Koutecky V, Fantucci P, Koutecky J. Quantum-chemistry of small clusters of elements of group-IA, group-IB, and group-IIA – fundamental-concepts, predictions, and interpretation of experiments. Chem Rev 1991;91:1035.

[122] Li Q, Zhao Y, Xu W, Li N. Structure and stability of B-8 clusters. Int J Quant Chem 2005;101:219.

[123] Alexandrova AN, Zhai H-J, Wang LS, Boldyrev AI. Arachno, nido, and closo aromatic isomers of the Li6B6H6 molecule. Inorg Chem 2004;43:3588.

[124] Minkin VI, Minyaev RM. Hypercoordinate carbon in polyhedral organic structure. Mendelleev Commun 2004;14:43.

[125] Minyaev RM, Gribanova TN, Starikov AG, Minkin VI. Octacoordinated main-group element centres in a planar cyclic B-8 environment: an ab initio study. Medeleev Commun 2001;11:213.

[126] Zhai H-J, Kiran B, Li J, Wang LS. Hydrocarbon analogues of boron clusters – planarity aromaticity and antiaromaticity. Nat Mater 2003;2:827.

[127] Kawai R, Weare JH. Anomalous stability of B-13(+) clusters. Chem Phys Lett 1992;191:311.

[128] Fowler JE, Ugalde JM. The curiously stable B-13(+) cluster and its neutral and anionic counterparts: The advantages of planarity. J Phys Chem A 2000;104:397.

[129] Aihara J, Kanno H, Ishida T. Aromaticity of planar boron clusters confirmed. J Am Chem Soc 2005;127:13324.

[130] Alexandrova AN, Zhai H-J, Wang LS, Boldyrev AI. Arachno, nido, and closo aromatic isomers of the Li6B6H6 molecule. Inorg Chem 2004;43:3588.

[131] Orden AV, Saykally RJ. Small carbon clusters: Spectroscopy, structure, and energetics. Chem Rev 1998;98:2313 and references therein.

[132] (a) Schleyer PvR, Jiao HJ, Glukhovtsev MN, Chandrasekhar J, Kraka E. Double aromaticity in the 3,5-dehydrophenyl cation and in cyclo[6]carbon. J Am Chem Soc 1994;116:10129. (b) Martin-Santamaria S, Rzepa HS. Double aromaticity and anti-aromaticity in small carbon rings. Chem Commun 2000;(16):1503. (c) Deleuze MS, Giuffreda JP, Francois JP, Cederbaum LS. Valence one-electron and shake-up ionization bands of carbon clusters. II. The C-n (n=4,6,8,10) rings. J Chem Phys 2000;112:5325.

[133] (a) Raghavachari K, Binkley JS. Structure, stability, and fragmentation of small carbon clusters. J Chem Phys 1987;87:2191. (b) Liang C, Schaefer III HF. Carbon clusters – the structure of C-10 studied with configuration-interaction methods. J Chem Phys 1990;93:8844.
[134] Martin JML, Taylor PR. Structure and vibrations of small carbon clusters from coupled-cluster calculations. J Phys Chem 1996;100:6047.
[135] Martin JML, El-Yazal J, Francois J-P. Structure and relative energetics of C-2n+1 (n=2–7) carbon clusters using coupled cluster and hybrid density functional methods. Chem Phys Lett 1996;252:9.
[136] (a) Erhardt S, Frenking G, Chen Z, Schleyer PvR. Aromatic boron wheels with more than one carbon atom in the center: C2B8, C3B93+, and C5B11+. Angew Chem Int Ed 2005;44:1078. (b) Ito K, Chen Z, Corminboeuf C, Wannere CS, Zhang XH, Li QS, Schleyer PvR. Myriad planar hexacoordinate carbon molecules inviting synthesis. J Am Chem Soc 2007;129:1510. (c) Minyaev RM, Gribanova TN, Starikov AG, Minkin VI. Heptacoordinated carbon and nitrogen in a planar boron ring. Dokl Chem 2002;382:41. (d) Minkin VI, Minyaev RM. Hypercoordinate carbon in polyhedral organic structure. Mendeleev Commun 2004;14:43.
[137] Wang L-M, Huang W, Averkiev BB, Boldyrev AI, Wang LS. CB7-: Experimental and theoretical evidence against hypercoordinate planar carbon. Angew Chem Int Ed 2007;46:4550.

Chapter 6

Reactivity and Thermochemistry of Transition Metal Cluster Cations

P. B. Armentrout
Department of Chemistry, University of Utah, Salt Lake City, Utah, USA

Chapter Outline Head

I. INTRODUCTION

Studies of gas-phase metal clusters have many motivations. First, clusters may serve as experimentally tractable models for surfaces and heterogeneous catalysts as well as tailored catalysts produced by cluster deposition on a support. Because many industrial catalysts involve highly dispersed metals and surface defect sites are often the active sites for chemical reactions [1,2],

Nanoclusters. DOI: 10.1016/S1875-4023(10)01006-5

the relationship between model clusters and real catalysts is plausible. Second, certain sized clusters may have unusual properties and reactivity. If the details underlying these unusual properties can be understood, the prospects of designing efficient and selective catalysts may be enhanced. Third, gas-phase cluster research serves as an ideal interface between experimental and theoretical work. Because of the myriad spin states accessible for open shell transition metal clusters, such systems are among the most challenging computational targets, but the finite size of cluster systems allows for specific comparison of experimental and theoretical data.

In this review, studies in our group to examine the stability and reactivity of transition metal cluster cations are recounted. By using powerful mass spectrometric techniques, reactions of metal clusters with a variety of small molecules can be examined in detail and their cross sections and rate constants measured. Examination of the kinetic energy dependence of these reactions allows thermochemistry for a variety of ligands bound to transition metal cluster cations to be measured as a function of cluster size. Such thermodynamic data provide useful benchmarks against which theoretical calculations can be tested. Further, it is found that even modest sized clusters reach a limiting adsorbate binding energy. Surprisingly, this limiting energy correlates well with the bond energies of these adsorbates to the bulk metal surfaces in those cases where such information is available. This correlation suggests that cluster–adsorbate binding energies can then be extended to more complicated systems for which surface science has yet to reveal quantitative thermochemical data. The available thermodynamic data from such cluster studies are fully reviewed here.

II. EXPERIMENTAL METHODS

The experiments described in this review are conducted on a guided ion beam tandem mass spectrometer equipped with a laser ablation/supersonic expansion cluster ion source. Only a brief description is provided here, as the experimental apparatus and techniques have been detailed previously [3]. A laser vaporization/supersonic expansion source is used to generate thermalized metal cluster cations [4,5]. The output (511 and 578 nm) of an Oxford ACL 35 copper vapor laser operating at 7 kHz is tightly focused onto a continuously translating and rotating metal rod inside an aluminum source block. The optimum pulse energy for metal cluster ion production generally ranges between 3 and 4 mJ/pulse. The vaporized material is entrained in a continuous flow ($5–6 \times 10^3$ standard cm^3/min) of He passing over the ablation surface. Many collisions and rapid mixing lead to the formation of thermalized clusters as they travel down a 2 mm diameter × 63 mm long condensation tube. Although direct measurements of the internal temperatures of the clusters are not possible, our studies suggest that the clusters are not internally excited and are likely to be near room temperature [3].

This seeded helium flow then undergoes a mild supersonic expansion in a field free region that is skimmed, and passes through two differentially pumped regions. Positively charged ions are accelerated and injected into a 60° magnetic sector momentum analyzer. The mass-selected ions are decelerated and focused into an rf octopole ion guide [6,7] that extends through a reaction cell. The octopole beam guide is biased with dc and rf voltages. The former allows accurate control of the translational energy of the incoming ions, whereas the latter establishes a radial potential that efficiently traps the parent and product ions that travel through the octopole. The pressure of the neutral reactant gas in the reaction cell is kept relatively low to reduce the probability of multiple collisions with the ions. To test this, studies are generally conducted at both ~0.2 and ~0.4 mTorr pressures for the neutral reagent. When necessary, the resultant cross sections are extrapolated to zero pressure, such that all results further analyzed correspond to products formed in single ion-molecule collisions. The product and reactant ions drift to the end of the octopole, where they are extracted, and injected into a quadrupole mass filter for mass analysis. Ion intensities are measured with a Daly detector [8] coupled with standard pulse counting techniques. Reactant ion intensities generally range from 10^5 to 10^6 ions/s. Observed product intensities are converted to absolute reaction cross sections as discussed in detail elsewhere [7]. Absolute errors in the cross sections are on the order of ±30%.

The absolute zero in the kinetic energy of the ions and their energy distributions (0.7–2.0 eV, gradually increasing with cluster size) are measured using the octopole as a retarding energy analyzer [7].The error associated with the zero of the absolute energy scale is 0.05 eV in the lab frame. Kinetic energies in the laboratory frame are converted to center-of-mass (CM) energies using the stationary target approximation, $E(\mathrm{CM}) = E(\mathrm{lab})m/(m+M)$ where m and M are the masses of the neutral and ionic reactants, respectively. The data at the lowest energies are corrected for truncation of the ion beam energy distribution [7].

A. Threshold Analysis and Thermochemistry

The energy dependences of cross sections for endothermic processes in the threshold region can be modeled using Equation (1) [9–11]

$$\sigma(E) = \left(\frac{N\sigma_0}{E}\right)\sum_i g_i \int_{E_0-E_i}^{E} \left[1 - e^{-k(E^*)\tau}\right](E-\varepsilon)^{N-1}\,\mathrm{d}\varepsilon, \qquad (1)$$

where σ_0 is an energy independent scaling parameter, E is the relative kinetic energy, and E_0 is the threshold for reaction at 0 K. The summation is over the rovibrational states of the clusters having energies E_i and populations g_i, where $\sum g_i = 1$. Vibrational frequencies for the bare metal clusters are obtained by using an elastic cluster model suggested by Shvartsburg et al. [12]. The integration is over the range of impact parameters leading to reaction as characterized by the parameter N [13]. This leads to a distribution in ε, the energy transferred from the reactants' translational energy into internal energy of the ion complex during the collision. The term in square brackets represents the probability for reaction on the experimental time scale of the reaction, τ, approximately 10^{-4} s in our apparatus. Because metal clusters have many low frequency vibrational modes, the lifetime of the transient intermediate formed during reaction can exceed τ. Such lifetime effects become increasingly important as the size of the cluster increases. [14] The unimolecular rate constant for dissociation, $k(E^*)$, is incorporated in Equation (1) as detailed elsewhere [15], where E^* is the internal energy of the energized molecule (EM) after the collision, that is, $E^* = \varepsilon + E_i$. This rate constant is estimated using statistical Rice–Ramsperger–Kassel–Marcus (RRKM) theory [16–18] as shown in Equation (2).

$$k(E^*) = \frac{dN_{\mathrm{vr}}^{\dagger}(E^* - E_0)}{h\rho_{\mathrm{vr}}(E^*)}. \tag{2}$$

Terms in Equation (2) include d, the reaction degeneracy; h, Planck's constant; $N_{\mathrm{vr}}^{\dagger}(E^* - E_0)$, the sum of rovibrational states of the transition state (TS) at an energy $E^* - E_0$; and $\rho_{\mathrm{vr}}(E^*)$, the density of rovibrational states of the EM at the available energy, E^*.

We have previously pointed out that the use of RRKM theory is not entirely appropriate for species like transition metal clusters that have an appreciable density of electronic states [3]. Nevertheless, there are no reliable means of accurately estimating the density of electronic states nor are more appropriate models yet available. Errors associated with the neglect of the electronic state density should largely cancel because both reactants and products share this high density of electronic states.

To extract thermodynamic information, kinetic energy dependent cross sections are analyzed using Equation (1). The model cross section is compared to the data after convolution with the kinetic energy distributions of the ion and neutral reactants [7]. The parameters σ_0, N, and E_0 are adjusted using a least-squares criterion until the data are reproduced over extended ranges of energy and magnitude. Uncertainties in the derived E_0 values include errors associated with variations in E_0 over the range of N values that adequately reproduce several data sets, variations in the vibrational frequencies of the reactant cluster ions, EMs, and TSs by factors of 1/2 and 2, and the absolute uncertainty in the energy scale. Thus, the uncertainties include all known random and systematic uncertainties in the data acquisition and analysis.

III. THE STABILITIES OF BARE METAL CLUSTERS

The stabilities of the bare metal cluster ions can be examined by collision-induced dissociation (CID), reaction (3).

$$M_n^+ + Xe \rightarrow M_{n-1}^+ + M + Xe. \tag{3}$$

Threshold energies for these reactions can be obtained using the methods described above. Given the long-range attractive interaction between the cluster ion and the metal fragment, there should be no barriers in excess of the endothermicity of this process. Thus, the experimentally observed threshold energy, E_0, equals the bond dissociation energy (BDE), DM_{n-1}^+–M. Our work in this area now includes the first row transition metal clusters of Ti_n^+ ($n=2$–22) [19], V_n^+ ($n=2$–20) [20], Cr_n^+ ($n=2$–21) [21,22], Mn_2^+ [23,24], Fe_n^+ ($n=2-19$) [3,14,25], Co_n^+ ($n=2$–18) [26,27], Ni_n^+ ($n=2$–18) [28,29], and Cu_2^+ [30], the second row transition metal clusters of Nb_n^+ ($n=2$–11) [31,32], and the third row transition metal clusters of Ta_n^+ ($n=2$–4) [33].

These experimental studies, which have been reviewed thoroughly before [34–36], show interesting variations in the stability of the clusters with their size. To examine how these cluster stabilities approach bulk-phase properties, the cohesive energy, $E_c(n)$, defined as the total energy necessary to atomize the cluster (i.e., the sum of the BDEs), divided by the number of atoms in the cluster can be examined. Miedema [37,38] has proposed a spherical drop model that predicts that the only difference between the energy of a metal cluster (assumed to be spherical) and that of the same group of atoms in the bulk is the surface energy of the cluster, that is,

$$E_c(n) = \Delta_{vap}H_0 - \left(\frac{36\pi}{n}\right)^{1/3} \gamma^0 V_a{}^{2/3}, \tag{4}$$

where $\Delta_{vap}H_0$ is the bulk-phase heat of vaporization, γ^0 is the surface free energy of the bulk metal, V_a is the atomic volume in the bulk, and all thermodynamic values are at 0 K. The predictions of this model using the bulk-phase values for iron ($\Delta_{vap}H=4.28$ eV, $\gamma^0=1.6\times10^{19}$ eV/m^2, and $V_a{}^{2/3}=5.2\times10^{-20}$ m^2) [37,38] are compared with the cluster results ($n=2$–19) in Figure 1. In all metal systems we have examined so far (Ti, V, Cr, Fe, Co, Ni, Nb), the cohesive energies plotted versus $n^{-1/3}$ extrapolate to a value at $n=\infty$ within 10% of the $\Delta_{vap}H_0$ values [34,36]. Furthermore, the slope of this extrapolation provides a value for $\gamma^0V_a{}^{2/3}$ that lies within ~20% of the bulk-phase value suggested by Miedema [37,38]. Overall, these comparisons show that the properties of the transition metals that control the binding energies of small clusters are the same properties that control the bulk-phase thermodynamic stabilities.

It is also worth noting that even though the cluster stabilities extrapolate smoothly to the bulk phase, even the largest clusters examined here ($n\sim20$) have a cohesive energy that is only about 70% of the bulk-phase value. This type of extrapolation behavior is typical of many cluster properties, which makes the results discussed below for adsorbate interaction energies surprising.

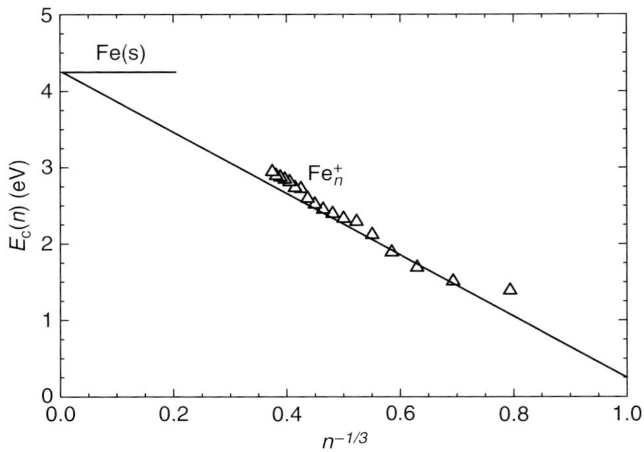

FIGURE 1 Cohesive energies of iron cluster cations (symbols) compared to the spherical drop model, Equation (4) (line). The horizontal line indicates the bulk-phase heat of vaporization. Adapted from Ref. [14].

IV. REACTIVITY STUDIES WITH DIATOMS

Following this initial work to establish the stabilities of bare metal cluster cations, the reactivity of these species has been studied to systematically evaluate the reaction thermochemistry and to begin to correlate reactivity with electronic structure and geometry of the clusters. To date, we have studied vanadium, chromium, iron, cobalt, and nickel cluster cations reacting with D_2 [39–43] and O_2 [44–48], as well as iron cluster cations reacting with N_2 [49]. Thermochemistry is obtained by examining the kinetic energy dependence of bimolecular reactions, such as processes (5)–(7).

$$M_n{}^+ + L_2 \rightarrow M_nL^+ + L, \tag{5}$$

$$M_{n+1}{}^+ + L_2 \rightarrow M_nL^+ + M + L, \tag{6}$$

$$M_{n+x}{}^+ + L_2 \rightarrow M_nL_2{}^+ + xM. \tag{7}$$

Again the data are analyzed using the methods outlined above to yield the threshold for the reaction, E_0, which can be related to bond energies of interest through Equations (8)–(10).

$$D(M_n{}^+ - L) = D(L - L) - E_0(5), \tag{8}$$

$$D(M_n{}^+ - L) = D(L - L) + D(M_n{}^+ - M) - E_0(6), \tag{9}$$

$$D(M_n{}^+ - 2L) = D(L - L) + D(M_n{}^+ - xM) - E_0(7). \tag{10}$$

Clearly, the latter two equations rely on knowledge of the bond energies of the bare clusters, as established using the CID approach discussed above. Early results from these reactivity studies have been reviewed elsewhere [36,50,51].

A. Reactions with D_2

The dominant reaction for metal cluster cations reacting with D_2 at elevated energies leads to the formation of $M_nD^+ + D$, reaction (5). In all cases examined, this reaction is endothermic, reflecting the fact that the cluster deuteride bond energy is weaker than the D_2 bond energy. Figure 2 shows typical data for the reaction of metal cluster cations with D_2. The cross section reaches a maximum because the M_nD^+ product dissociates further at higher energies to $M_n^+ + D$, a process that can begin at $D(D_2) = 4.54$ eV. It can be seen that this corresponds well to the energy of the peak for the smaller cluster, whereas for larger clusters, the peak moves to higher energies because the lifetime for dissociation of the M_nD^+ product can exceed the experimental flight time of the ion. Dissociation could ostensibly occur by losing a metal atom instead to form $M_{n-1}D^+ + M$, but this product ion is never observed. This demonstrates that M_n^+–D BDEs must be lower than $M_{n-1}D^+$–M BDEs, as verified by the quantitative measurements discussed further below.

Figure 2 also illustrates that the $M_nD_2^+$ adduct can also be formed. This only occurs if the cluster is sufficiently large ($n \geq 6$ for V, $n \geq 5$ for Cr, $n \geq 9$ for Fe, $n = 4$, 5, and ≥ 9 for Co, and $n \geq 5$ for Ni). In most cases (e.g., Fe_{15}^+ in Figure 2), this process appears to be an activated one requiring energy to occur efficiently, whereas a few systems exhibit barrierless formation of these product ions: Cr_n^+ ($n = 6$–8), Fe_9^+, Co_n^+ ($n = 4$, 5, ≥ 10), and Ni_n^+ ($n \geq 5$). Adduct formation is not observed for smaller clusters because the lifetime of the $M_nD_2^+$ complex is too short to observe experimentally. However, lifetimes of $M_n(D_2)^+$ adducts in which the D_2 molecule is physisorbed to the cluster are calculated to be much shorter than the experimental time available for dissociation back to reactants ($\sim 10^{-4}$ s), which strongly suggests that the $M_nD_2^+$ complexes involve dissociative chemisorption of D_2 on the cluster surface. Thus, the activation barrier observed for most of these product ions can be assigned to the barrier required to induce dissociative chemisorption on the cluster.

M_n^+–D BDEs are obtained from the thresholds for formation of $M_nD^+ + D$ using Equation (8). The values obtained are compared to $D(M_n^+–M)$ values in Figure 3. For chromium, we find that the Cr_n^+–D and Cr_n^+–Cr BDEs are similar in magnitude and both exhibit even–odd oscillations, reflecting the similarity of the electronic structure of D atoms, $^2S(1s^1)$, and Cr atoms, $^7S(4s^13d^5)$. This also demonstrates that the stable, half-filled 3d shell of chromium is not particularly active in the metal–metal bonding. In contrast, the V_n^+–V BDEs exceed those for V_n^+–D by a considerable amount, 1.5 ± 0.2 eV ($58 \pm 11\%$) for $n \geq 5$ where the two sets of bond

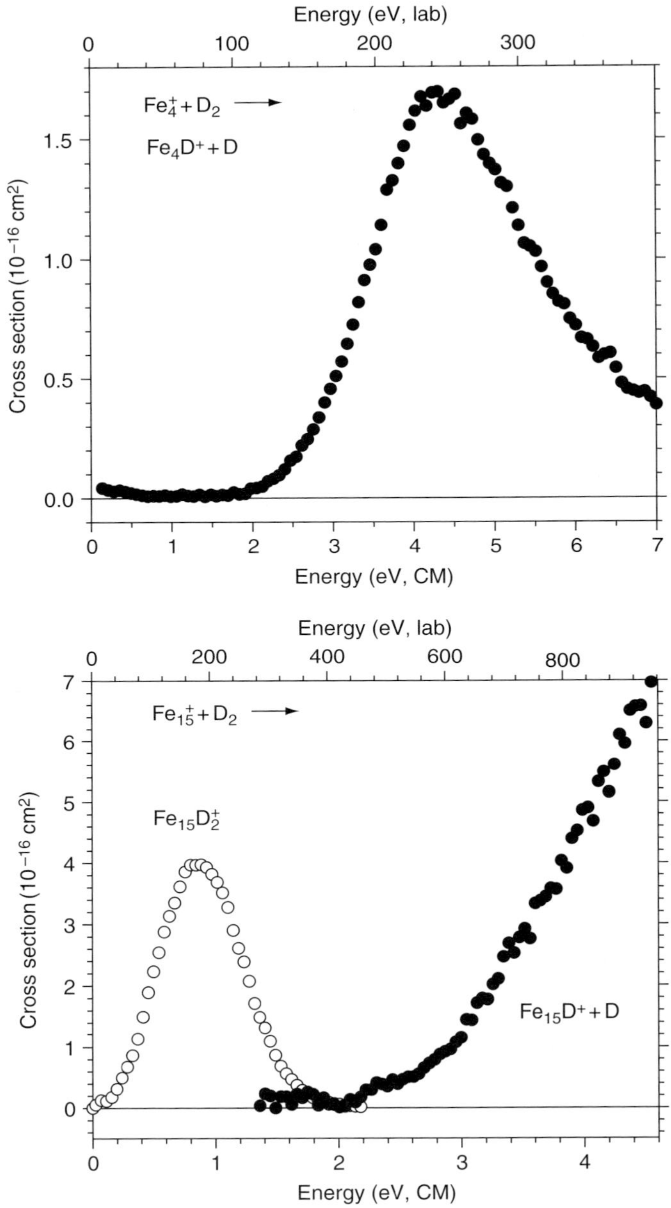

FIGURE 2 Cross sections for the reactions of Fe_4^+ and Fe_{15}^+ with D_2 as a function of collision energy in the center-of-mass (lower x axis) and laboratory (upper x axis) frames. $Fe_nD_2^+$ and Fe_nD^+ products are indicated by open and closed circles, respectively. Reproduced with permission from Ref. [39].

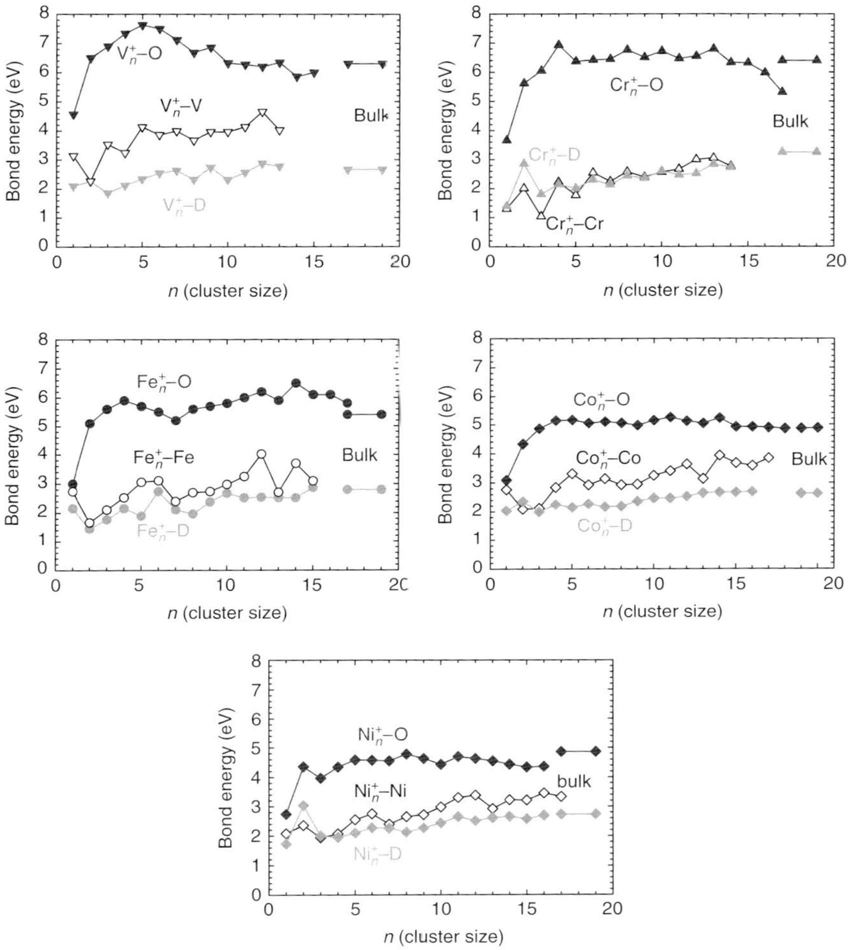

FIGURE 3 Bond energies (eV) for M_n^+–D, M_n^+–M, and M_n^+–O as a function of cluster size. Parts (A)–(E) show results for vanadium [20,41,46], chromium [22,40,45], iron [14,39,44], cobalt [27,42,47], and nickel [29,43,48], respectively. Bulk-phase values for metal surfaces binding D [54,65] and O [60] are also shown. Parts (A)–(C) and (E) are adapted from Ref. [51]. (For interpretation of the references to color in this figure legend, the reader is referred to the Web version of this chapter.)

energies parallel one another rather closely. Clearly, V_n^+–V bonding must have considerable contributions from both 4s–4s and 3d–3d interactions.

Likewise in the iron, cobalt, and nickel systems, most M_n^+–M BDEs are stronger than the M_n^+–D BDEs. For Fe, the average difference is 0.30 ± 0.07 eV for $n = 2$–4, 6, 7, 9, 10, 13 and 15; for Co, the average difference is 0.66 ± 0.10 eV for $n = 1$, 4, 6, 8–10, and 13; and for Ni, $n = 5$, 6, 8–11, and 14–16, the difference averages to 0.56±0.10 eV. These differences can again be attributed to 3d–3d interactions in the metal–metal bonds, but this contribution is not as extensive as those of the early transition metal, vanadium. For other sized clusters, notably $n=12$ for Fe, Co, and Ni, and $n=5$ and 14 for Fe and Co, the metal–metal bonds are stronger than the metal–deuteride bonds by a larger amount. In rationalizing why these M_{n+1}^+ clusters have particularly high stability, it is significant that these clusters can have a highly symmetric geometrical structure compared to neighboring clusters: M_6^+ (octahedral), M_{13}^+ (icosahedral or octahedral with fcc or bcc packing), and M_{15}^+ (bcc rhombic dodecahedral). Substitution of D for M in these clusters breaks the symmetry, changing the molecular orbital ordering, thereby leading to a less strongly bound system.

B. Reactions with O_2

Typical data for M_n^+ reacting with O_2 are shown in Figure 4. In almost all cases, metal clusters oxidize efficiently at low energies with reaction efficiencies approaching 100%. For all metal systems, the dominant products at low energies are $M_mO_2^+$ where the exothermicity of the oxidation process determines the number of M atoms lost $(n-m)$. As the clusters get larger, more energy can be dissipated in the larger cluster and fewer M atoms are lost until the $M_nO_2^+$ adduct is observed for the largest clusters studied. In all cases, the $M_mO_2^+$ product ions lose additional metal atoms as the energy available is increased. Competing with these processes is loss of an oxygen atom to form M_nO^+, with considerably smaller reaction probabilities. This product then undergoes further decomposition by sequential loss of metal atoms as the available energy increases. Small amounts of CID yielding M_m^+ products are also observed. Unlike the D_2 system, the observation that $M_mO_2^+$ and M_mO^+ product ions dissociate preferentially by metal atom loss indicates that M_n^+–O BDEs must be larger than M_n^+–M BDEs.

For large clusters, the ionization energies of the M_mO_2 and M_mO products are smaller than those of M and MO, such that $M_mO_2^+$ and M_mO^+ are always the product ions observed. For small clusters, however, alternate pathways can be seen. Hence, the two features observed in the Fe_3O^+ cross section (Figure 4B) are associated with $Fe_3O^+ + FeO$ and $Fe_3O^+ + Fe + O$ at higher energies. Likewise, the observation of Fe^+ and Fe_2^+ at low energies in Figure 4A must be associated with the formation of stable neutral iron dioxide cluster products.

In the deuterium system, the endothermic reaction (5) leads straightforwardly to thresholds that are related directly to the BDEs of interest via Equation (8). In contrast, in the O_2 reaction system, thresholds for formation of the $M_nO^+ + O$ product channels do not correspond to the thermodynamic limit. This is believed to be primarily a consequence of the severe competition with the dominant products observed, $M_mO_2^+$. However, because of the extensive

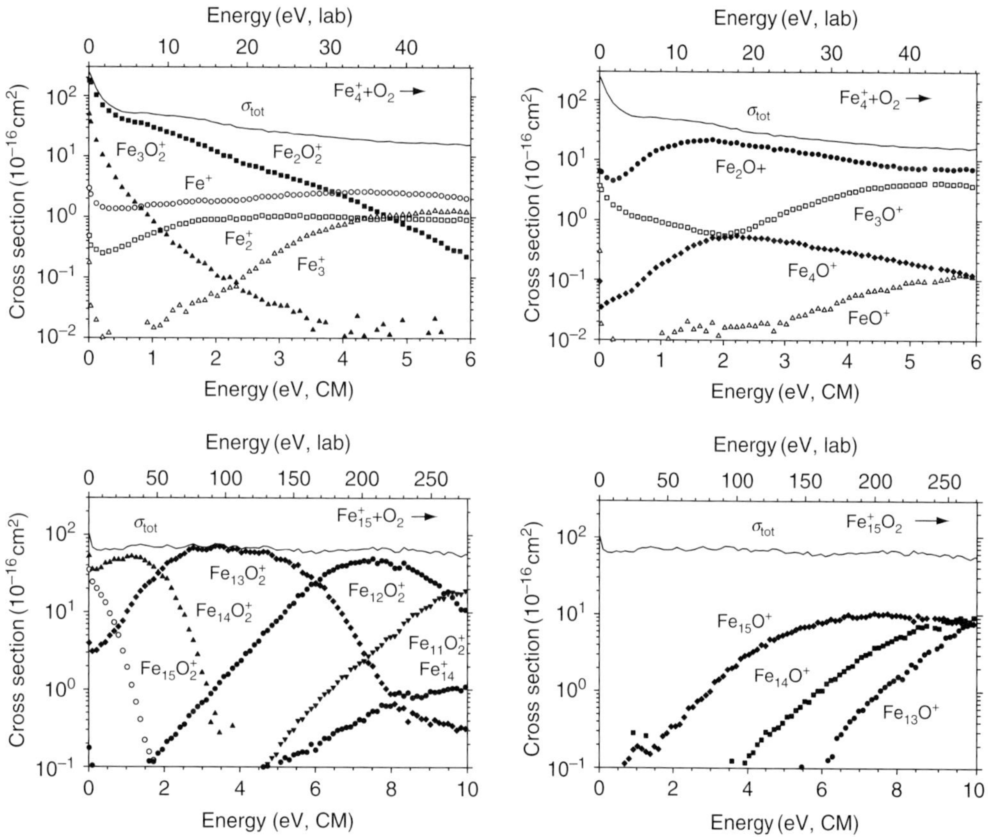

FIGURE 4 Cross sections for the reactions of Fe_4^+ and Fe_{15}^+ with O_2 as a function of collision energy in the center-of-mass (lower x axis) and laboratory (upper x axis) frames. The first panels exhibit the cluster dioxide and cluster fragment products. The second panels show the cluster monoxide products. σ_{tot} is the sum of all products from both panels. Reproduced with permission from Ref. [44].

loss of metal atoms observed, M_nO^+ BDEs can be determined from the thresholds for the secondary process (6) using Equation (9) as well as higher order reactions in which additional metal atoms are lost [44–48]. Verification that the thermochemistry obtained in this way is accurate comes from measurements of these bond energies from reactions of chromium and iron cluster cations with CO_2 [52,53], which is a good donor of a single oxygen atom. Further, for all five metals, these results can be compared with those for binding two oxygen atoms to the clusters, M_n^+–2O BDEs. These values are determined from an analysis of the cross sections for formation of $M_mO_2^+$ products in endothermic reactions in the O_2 systems, where thresholds for reactions (7) are converted to bond energies using Equation (10). For all metals, the M_n^+–O BDEs are comparable to half the M_n^+–2O BDEs [44–48]. As these determinations are completely independent, the agreement substantiates the accuracy of both measurements. Furthermore, the latter observation indicates that the first and second oxygen atom can find similar binding sites on all but the smallest clusters.

The M_n^+–O BDEs are shown in Figure 3 and compared with the M_n^+–D and M_n^+–M BDEs. In agreement with the qualitative behavior of the reaction cross sections, the M_n^+–O BDEs exceed the M_n^+–M BDEs by a considerable amount. Furthermore, the values reach an asymptotic limit at much smaller cluster sizes than the M_n^+–D BDEs. This is believed to be a consequence of the very strong metal–oxygen bonds disrupting the metal–metal bonding, essentially annealing the cluster oxide complexes, and thereby allowing the strongest possible cluster bond to oxygen.

C. Comparison of Cluster Hydride and Oxide Bond Energies to Bulk-Phase Values

Figure 3 also compares the M_n^+–D and M_n^+–O BDEs to values for H and O atom binding to bulk surfaces of the same metals. These bulk-phase adsorbate energies are usually confined to a fairly narrow range. For vanadium, the average binding energy of H to the V(100) surface is 2.66 ± 0.08 eV [54,55]. For chromium, the binding energy to polycrystalline surfaces is about 3.21 eV [56,57]. In the iron case, the bulk-phase BDE is 2.80 ± 0.10 eV for Fe(100), Fe(110), and Fe(111) [58–60], whereas for cobalt, hydrogen binds to Co (0001) by 2.60 eV and to Co(1010) by 2.65 eV [61,62]. For nickel, the binding energies of hydrogen atoms on Ni(111) are 2.70 [63] and 2.74 eV [64,65], 2.74 eV on Ni(100) [64], and 2.70 eV on Ni(110) [64]. Figure 3 shows that the M_n^+–D BDEs for the largest n have generally reached a fairly constant value that is within experimental error of the bulk-phase value for all five metal systems. Some smaller clusters have considerably weaker BDEs and can be almost half the bulk value, but other small clusters (notably Cr_2^+–D, Fe_6^+–D, and Ni_2^+–D) can have very strong bond energies. These variations must be related to the geometric and electronic structures of the metal cluster cations.

For the oxygen systems, the bulk-phase oxidation energies vary considerably with metal. For convenience, we utilize the atomic adsorption BDEs for oxygen listed by Benziger [60]: 6.3, 6.5, 5.4, 5.0, and 4.9 eV for V, Cr, Fe, Co, and Ni surfaces, respectively. The first four of these values are estimated from the enthalpies of formation of bulk compounds and only the latter value has been measured directly. In calorimetry experiments [56,66–72], values of 6.3 ± 0.2 eV for chromium, 4.0–5.5 eV for iron, 4.7–5.1 for cobalt, and 5.0 and 4.8 eV for Ni(111), 5.0 eV for Ni(110), and 5.4 eV for Ni(100) have been measured for oxygen atom binding. In all cases, Figure 3 shows that these bulk-phase values compare reasonably well to the cluster oxide BDEs. In particular, the cluster BDEs properly reflect the strong changes observed with the identity of the metal. It is particularly surprising that this bulk-phase limit is reached even for very small clusters, generally by $n = 4$. In part, this occurs because the M_n^+–O BDEs are much stronger than the M_n^+–M BDEs, thereby allowing adsorption of an oxygen atom on the cluster to disrupt metal–metal bonding in order to form strong metal–oxide bonds. In contrast, M_n^+–D BDEs are slightly weaker than the M_n^+–M BDEs, such that larger clusters are required to reach the bulk-phase limit in the deuteride systems.

Ultimately, the interesting observation is that the adsorbate interaction energies reach bulk-phase values for modestly sized clusters. This contrasts with the approach to bulk-phase behavior illustrated in Figure 1. This is rationalized on the basis that the adsorbates can only bind to one, two, three, perhaps four metal atoms on a surface, such that small clusters can mimic the binding. The strength of the interaction surely depends on the ability of the surface (that of the bulk or the cluster) to move electron density around, so once the cluster is big enough to provide sufficient electronic "flexibility", it can form as strong a bond as it ever will.

D. Reactions with N_2

The reactions of N_2 with iron cluster cations have been studied and representative cross sections are shown in Figure 5. Unlike the D_2 and O_2 systems where reactions (5)–(7) dominate the products observed, the CID reaction (3) accounts for the majority of the products observed. This is not surprising given the very strong N_2 bond energy, 9.76 eV, compared to 4.54 and 5.12 eV for D_2 and O_2, respectively. Nevertheless, activation of the N_2 bond is observed at higher collision energies with the formation of both mononitride and dinitride cluster ions, Fe_mN^+ and $Fe_mN_2^+$. For both the mononitride and dinitride cluster ions, these species then dissociate by sequentially losing iron atoms as the energy increases. As in the oxygen system, this demonstrates that M_n^+–N BDEs must exceed the M_n^+–M BDEs.

The monomer and dimer ions react to form Fe_nN^+, whereas larger clusters all form $Fe_{n-1}N^+$ product ions, until $n \geq 15$ where the Fe_nN^+ product is again observed. Cluster dinitride product ions begin to be observed at $n = 6$, where

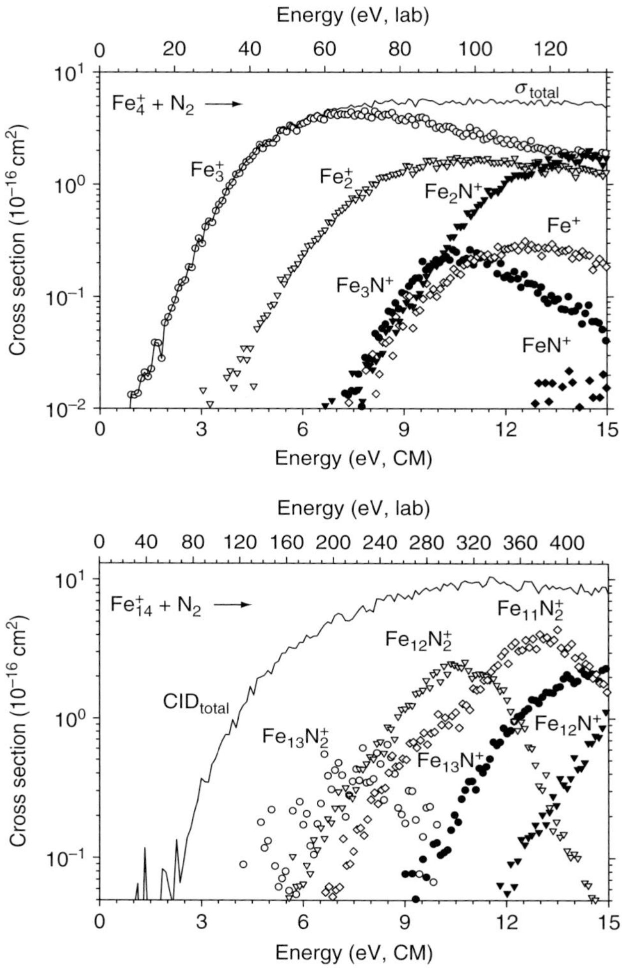

FIGURE 5 Cross sections for the reactions of Fe_4^+ and Fe_{14}^+ with N_2 as a function of collision energy in the center-of-mass (lower *x* axis) and laboratory (upper *x* axis) frames. Fe_nN^+ products are indicated by closed symbols. $Fe_nN_2^+$ products are indicated by open symbols in the second panel. Collision-induced dissociation products are shown by open symbols in the first panel and their sum by the line in the second panel. Reproduced with permission from Ref. [49].

the $Fe_4N_2^+$ product ion is observed, whereas the $Fe_{n-1}N_2^+$ product ion begins to be formed at $n=9$. The $Fe_nN_2^+$ adduct is first observed for reaction of $n=12$, is not observed for $n=13$ and 14, and becomes fairly prominent for $n\geq 16$. These reactivity trends with cluster size can be understood by realizing that the lifetime of the product ions increases with the size of the cluster. Therefore, small clusters need to lose two iron atoms to stabilize the $Fe_{n-2}N_2^+$ species such that they can be observed on our experimental time scale of 10^{-4} s. Larger clusters have sufficiently long lifetimes that $Fe_{n-1}N_2^+$ is observed, and the largest clusters don't need to lose any Fe atoms to extend their lifetime past the experimental limit. As formation of a physisorbed $Fe_n(N_2)^+$ adduct should exhibit no barrier to its formation because of the long-range ion-induced dipole potential between Fe_n^+ and N_2, the observation of appreciable thresholds for formation of the $Fe_nN_2^+$ adducts (5–7 eV) indicates that nitrogen must be dissociatively chemisorbed on the iron cluster in these species.

Analysis of the bond energies for this system is complicated by the competition between the mononitride and dinitride product ions. Nevertheless, there is good agreement among bond energies derived from different sequential processes, that is, from the thresholds for generation of $Fe_{n-x}N^+$ and $Fe_{n-x}N_2^+$ where x varies between 0 and 3. Furthermore, there is generally good agreement between the values obtained for $D(Fe_{n-x}^+{-}N)$ and $D(Fe_{n-x}^+{-}2N)/2$, indicating that the first and second nitrogen atoms bind to the clusters with similar strengths. The exceptions are $n=4$ and 5 where $D(Fe_{n-x}^+{-}N)$ exceeds $D(Fe_{n-x}^+{-}2N)/2$ by about 0.6 eV. Clearly, these small clusters may not be able to accommodate formation of strong bonds to two nitrogen atoms.

Another significant result from this study was the thresholds obtained for formation of the $Fe_nN_2^+$ adduct for $n=12$ and 15–18. Analysis of the cross sections yielded consistent thresholds for these reactions with an average value of $E_0=0.48\pm0.03$ eV. This value can be favorably compared with previous work that estimates that the activation barrier for dinitrogen dissociation on Fe surfaces is greater than 0.35–0.43 eV [60]. Other measurements place this activation energy lower, 0.0–0.3 eV [59,73], but it has been suggested that these measured barriers actually correspond to diffusion away from surface defect sites where the activation actually occurs [60]. Furthermore, an activation energy of 0.48 eV corresponds to nitrogen atom adsorption energies on iron surfaces of 5.8 ± 0.1 eV, which agrees nicely with a value of about 5.7 eV as obtained from a reinterpretation of the work of Boszo et al. [73] by Stoltze and Nørskov [74].

The bond energies measured here for single nitrogen atoms are compared to the bulk-phase value in Figure 6. Clearly, the variations in the cluster bond energies as a function of size extend to larger clusters than for the D and O cases, and the cluster bond energies have not yet reached an asymptotic value. This appears to be a result of the sensitivity of this system to the large activation barrier required to break the very strong N_2 bond, whereas activation of D_2, O_2, and the molecular systems described below either have no activation barriers or small ones. The unactivated processes in these latter systems allow the energy released upon dissociative chemisorption to help anneal the transient intermediate formed, thereby allowing the intermediate to find the most stable structure of the resulting products. The high activation energy needed for N_2 doesn't allow this, leading to the variations observed. Nevertheless,

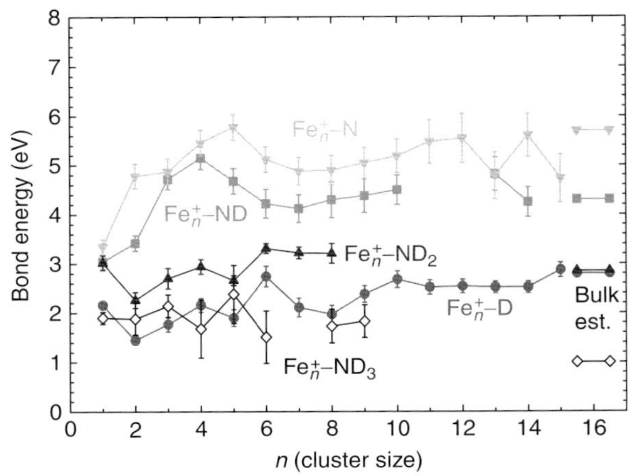

FIGURE 6 Comparison of bond energies for Fe_n^+–D (solid circles) [39], Fe_n^+–ND_3 (open diamonds) [77], Fe_n^+–ND_2 (solid triangles) [77], Fe_n^+–ND (solid squares) [77], and Fe_n^+–N (inverted triangles) [49]. Bulk-phase values for iron surfaces binding D [58–60] and ND_3 [86] and estimates for N [74], ND, and ND_2 [85] are also shown to the right. (For interpretation of the references to color in this figure legend, the reader is referred to the Web version of this chapter.)

the strongest Fe_n^+–N BDEs measured here ($n = 5$, 11, 12, 14) agree nicely with the bulk-phase value from Stoltze and Nørskov [74], but lie somewhat below the value obtained by Boszo et al. [73], helping to validate the reinterpretation of their data.

V. REACTIVITY STUDIES WITH LARGER MOLECULES

A key advantage to the gas-phase methods for thermodynamic measurements introduced above is that they can be applied to molecular fragments bound to metal clusters as well as the simple atomic adsorbates discussed above. Our first studies involving larger molecules focused on reactions of chromium and iron clusters with CO_2 [52,53]. As alluded to above, because CO_2 is a good oxygen atom donor (because of the stability of the CO product formed), the ionic products observed were almost exclusively M_mO^+ species (along with simple CID of the bare cluster ions). These studies provided valuable thermochemistry that verified measurements of the cluster monoxide bond energies made in the O_2 systems.

More complex reaction chemistry and richer thermodynamic data have been acquired for the reactions of iron and nickel clusters with CD_4 [75,76], and iron clusters reacting with ND_3 [77]. Data for reactions of cobalt clusters with CD_4 are also available [78]. In these studies, thermodynamic data are acquired in analogy with the diatom reaction systems in which both the primary reaction (11) and the secondary reaction (12) are generally observed.

As above, the data are analyzed to yield E_0, which is related to BDEs through Equations (13) and (14).

$$M_n^+ + RL \rightarrow M_nL^+ + R, \tag{11}$$

$$M_n^+ + RL \rightarrow M_{n-1}L^+ + M + R, \tag{12}$$

$$D(M_n^+ - L) = D(R - L) - E_0(11), \tag{13}$$

$$D(M_{n-1}^+ - L) = D(R - L) + D(M_{n-1}^+ - M) - E_0(12). \tag{14}$$

A. Reactions with Methane

In the reactions of methane, CD_4 is used instead of CH_4 to enhance our ability to separate products closely spaced in mass. Figure 7 shows an example of the complex data obtained in these systems with reactions (15)–(18) being observed in most systems.

$$M_n^+ + CD_4 \rightarrow M_nD^+ + CD_3, \tag{15}$$

$$\rightarrow M_nCD_2^+ + D_2, \tag{16}$$

$$\rightarrow M_nCD^+ + D + D_2, \tag{17}$$

$$\rightarrow M_nC^+ + 2D_2. \tag{18}$$

The lowest energy process observed is reaction (16) largely because it involves expulsion of the stable D_2 molecule. At somewhat higher energies, a second dehydrogenation in reaction (18) to form the M_nC^+ product is observed. The correspondence between the peak in the $M_nCD_2^+$ cross section and the onset for M_nC^+ production indicates that these are sequential processes. Reaction (15) dominates the product spectrum at elevated kinetic energies. At higher energies, the primary $M_nCD_2^+$ product can also dissociate by D atom loss in reaction (17). In addition, each of these primary products can dissociate further by sequentially losing metal atoms to form $M_{n-1}D^+$, $M_{n-1}CD_2^+$, $M_{n-1}CD^+$, and $M_{n-1}C^+$, as in Figure 7B and D. A minor reaction pathway is CID to form M_{n-1}^+. All of the products exhibit thresholds for all clusters, indicating that activation of methane on iron and nickel clusters is an activated process. This is consistent with observations on these bulk-phase metal surfaces. For the largest iron clusters, $n \geq 10$, the $Fe_nCD_4^+$ adduct is also observed (Figure 7C) but this product is not observed for the nickel clusters. For reasons similar to those discussed above, we assign this species to a dissociative chemisorption process with a threshold determined by the barrier to the bond activation necessary for this reaction (as discussed further below).

The determination of cluster BDEs to D, C, CD, and CD_2 can be carried out in the same manner outlined above for the atomic systems. Because the

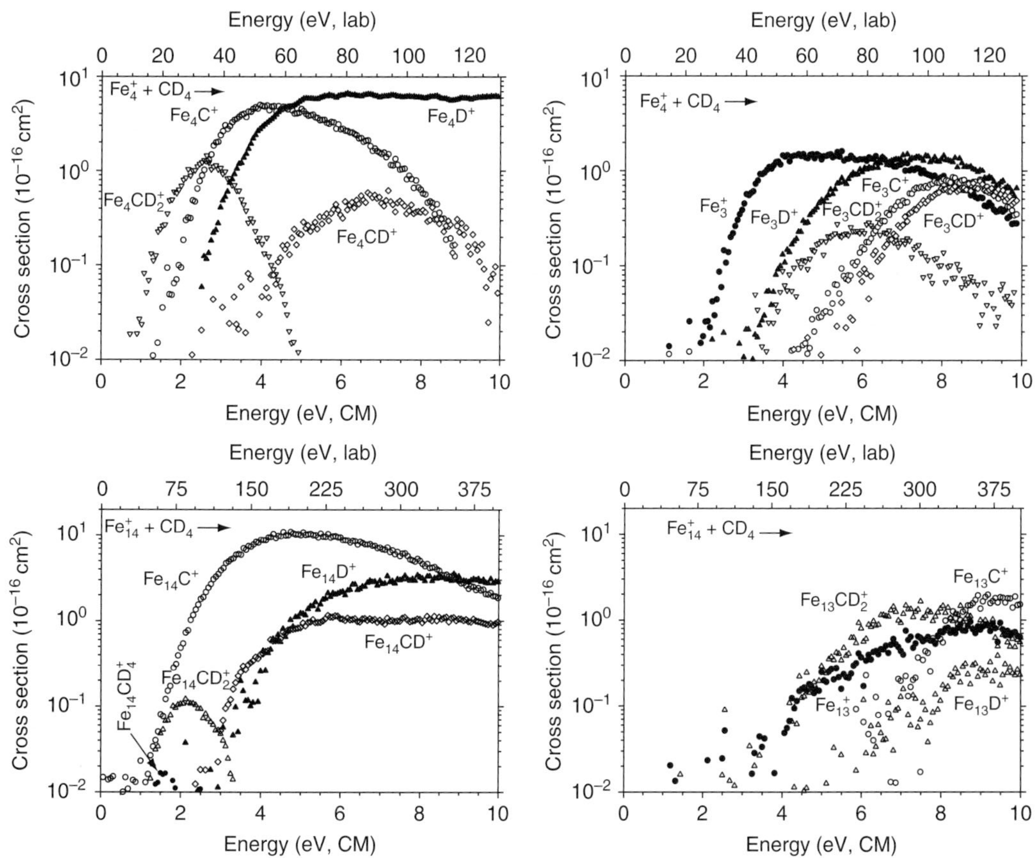

FIGURE 7 Cross sections for the reactions of Fe_4^+ and Fe_{14}^+ with CD_4 as a function of collision energy in the center-of-mass (lower x axis) and laboratory (upper x axis) frames. The first panels exhibit the primary product ions and the second panels show the secondary product ions containing one iron atom fewer than the reactant cluster. Reproduced with permission from Ref. [75].

product ions are observed to lose metal atoms, the threshold energies for the primary and secondary reactions provide an internal check on the derived thermochemistry. Specifically, almost all ions of interest ($Ni_nCD_2^+$ species being a notable exception) are produced in at least two independent pathways, for example, M_nC^+ is formed in the primary reaction (18) of M_n^+ and also in the secondary reaction of M_{n+1}^+ (yielding $M_nC^+ + M + 2D_2$). BDEs for D, C, and CD to both iron and nickel clusters obtained from analyses of these primary and secondary reactions agree with one another within experimental error. Further, the M_nD^+ BDEs are consistent with those obtained from reactions with D_2 [39,43]. In contrast, the $Fe_nCD_2^+$ BDEs obtained from analyses of reactions (16) fall below those from the secondary reaction except for $n = 3$ and 4. For reaction of Fe^+, this discrepancy has been demonstrated to be a result of a barrier along the potential energy surface that exceeds the asymptotic energy of the product ions [79]. It is likely that a similar explanation holds for larger clusters. For Fe^+, the barrier lies in the exit channel [79] and it is anticipated that this is probably true for reaction of the Fe_2^+ dimer as well. For larger clusters, $n \geq 5$, evidence suggests that the barriers must lie in the entrance channel, with the $n = 3$ and 4 clusters providing a region that switches between these two extremes. The key experimental observation suggesting this entrance channel barrier is the observation that the threshold for formation of $Fe_nCD_2^+$ matches that for formation of the $Fe_nCD_4^+$ adduct ($n \geq 10$). This correspondence indicates that these thresholds cannot correspond to the thermodynamic limit for either channel, but must be attributed to a rate limiting TS leading to both products. Because formation of a physisorbed (intact) methane adduct would not have a barrier (as a result of the long-range ion-induced dipole attractive interactions), the $Fe_nCD_4^+$ species must involve activated dissociative chemisorption, clearly a necessary first step in the dehydrogenation reaction. Another indication that this interpretation is reasonable is that for the $n \geq 6$ clusters, the differences in the BDEs derived from the primary and secondary reactions are fairly constant and average to 0.70 ± 0.30 eV. This energy can be assigned to the activation barrier for dissociative chemisorption of methane on the iron clusters. This barrier can be thought of as the crossing point between the potential energy surface for physisorption and chemisorption when viewed using a classic surface model; see Figure 8. This energy also agrees relatively well with calculated estimates of an equivalent methane activation barrier on bulk-phase iron surfaces, approximately 0.3–0.9 eV [80]. Unfortunately, no comparable thermodynamic information could be obtained from the analogous reactions with nickel clusters because the secondary reaction was not observed in this system. Nevertheless, several arguments suggest that the $Ni_n^+–CD_2$ bond energies obtained from analysis of reactions (16) may be low by about 0.3 ± 0.2 eV, as indicated in Figure 9 by the dashed line.

Figure 9 shows the cluster BDEs to the fragments of methane measured in this work. As for the atomic systems, the values vary considerably with cluster

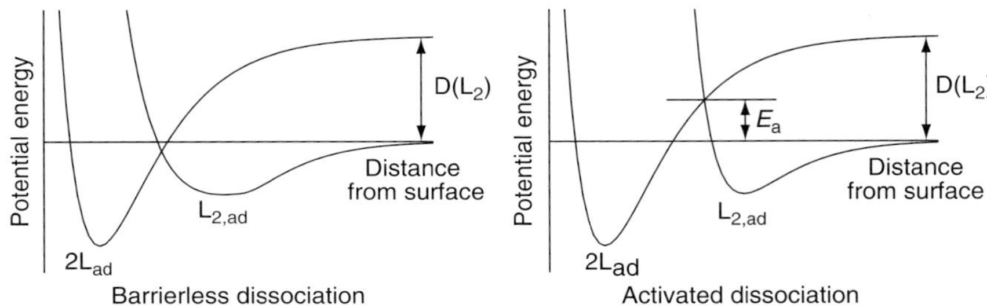

FIGURE 8 Schematic potential energy curves for barrierless and activated dissociative chemisorption on a metal surface.

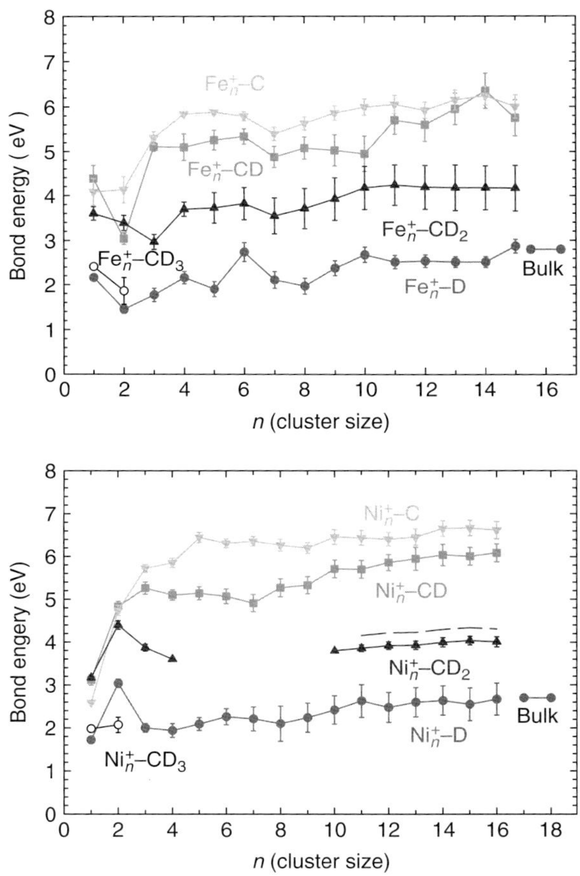

FIGURE 9 Comparison of bond energies for M_n^+–D (solid circles), M_n^+–CD_3 (open circles), M_n^+–CD_2 (solid triangles), M_n^+–CD (solid squares), and M_n^+–C (inverted triangles) for M = Fe [75] and Ni [76]. Bulk-phase values for metal surfaces binding D [58–60,63–65] are shown to the right. The dashed line indicates an estimate of the true Ni_n^+–CD_2 BDEs; see text. Adapted from Refs. [75] and [76]. (For interpretation of the references to color in this figure legend, the reader is referred to the Web version of this chapter.)

size for small n, consistent with substantial changes in the electronic and geometric structures of the metal cluster cations. However, for larger clusters, the BDEs tend to level out to nearly constant values, just as they did for the atomic adsorbates discussed above, as in Figure 3. Just as D and O reach their asymptotic values at different values of n, the asymptotic BDE for the molecular fragments is reached for smaller clusters as the adsorbate BDE increases. The relative values of these asymptotic cluster binding energies to D, CD_2, and CD fall roughly into a reasonable order, specifically, the relative values are consistent with the formation of one, two, and three bonds, respectively. In the few cases where a CD_3 BDE could be measured, the values are close to the BDEs of D, consistent with single bond formation for the methyl group to metal clusters. For the carbon atom, the BDEs are close to those of the CD adsorbate, indicating formation of a triple bond between C and the cluster. This can occur because the carbon atom forms two covalent bonds using its valence p orbitals, and then accepts electron density into the empty p orbital, thereby forming a third bond, similar to that in carbon monoxide.

Comparisons to bulk-phase values cannot be made in these cases because no such experimental values exist for the molecular fragments. The only species for which experimental information does exist (besides the hydrides discussed above) is nickel carbide, which has been measured as about 7.37 eV in a molecular beam experiment [81]. This value has been questioned by Siegbahn and Wahlgren [82] who suggest that the value is "overestimated" because "this value is suspiciously close to the sublimation energy of graphite" at 7.42 eV.

B. Reactions with Ammonia

Ammonia is found to be more reactive with iron cluster cations than methane, a consequence of the lone pair of electrons on nitrogen. The types of reactions observed parallel those found for the methane system, with reactions (19)–(23) being observed in most systems; see Figure 10.

$$M_n^+ + ND_3 \rightarrow M_nD^+ + ND_2, \tag{19}$$

$$M_n^+ + ND_3 \rightarrow M_nND_2^+ + D, \tag{20}$$

$$M_n^+ + ND_3 \rightarrow M_nND^+ + D_2, \tag{21}$$

$$M_n^+ + ND_3 \rightarrow M_nN^+ + D + D_2, \tag{22}$$

$$M_n^+ + ND_3 \rightarrow M_nND_3^+. \tag{23}$$

For most clusters with $n \geq 6$, adduct formation in reaction (23) is the only exothermic process observed, as previously observed by Irion and Schnabel using ion cyclotron resonance mass spectrometry at thermal energies only [83]. For $n = 3$–5, dehydrogenation, reaction (21), is also observed to be exothermic, but is efficient for only the $n = 4$ cluster. Irion and Schnabel observed this

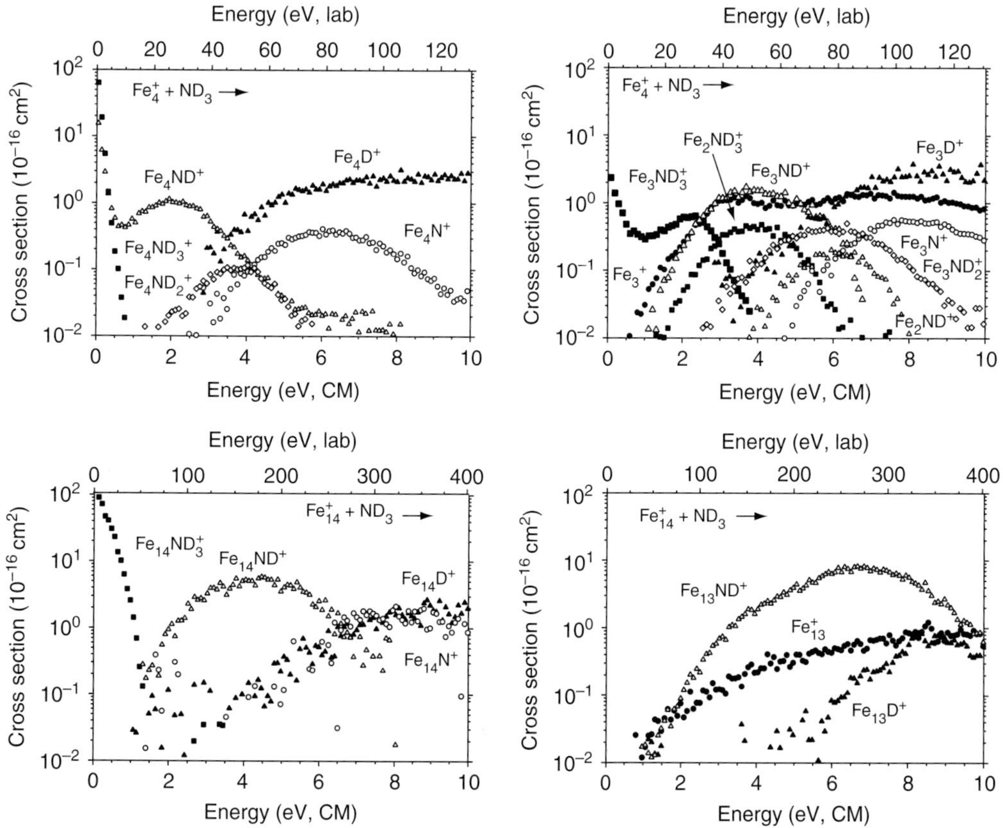

FIGURE 10 Cross sections for the reactions of Fe_4^+ and Fe_{14}^+ with ND_3 as a function of collision energy in the center-of-mass (lower x axis) and laboratory (upper x axis) frames. The first panels exhibit the primary product ions and the second panels show the secondary product ions containing one iron atom fewer than the reactant cluster. Reproduced with permission from Ref. [77].

process only for Fe_4^+, as their sensitivity did not permit observation of the much less efficient processes for Fe_3^+ and Fe_5^+. At somewhat higher energies, reactions (19), (20), and (22) are observed, as well as processes in which all of the product ions lose iron atoms; see Figure 10B and D.

One of the more interesting observations in this system is that the reactions of Fe_n^+ ($n=3$–5) show dual features in the cross sections for dehydrogenation, for example, Figure 10A. As noted above, the exothermic features for $n=3$ and 5 are much smaller than for $n=4$, by about 2 orders of magnitude, whereas the endothermic feature remains about the same magnitude in all three systems and is observed for all other clusters as well. Analysis of these endothermic features yields Fe_n^+–ND BDEs that lie 0.95 ± 0.10 eV below BDEs derived from the secondary reactions, which also agree with BDEs derived from tertiary reactions. As in the methane system, this indicates that activation of the ND bond in ammonia must overcome a barrier of this magnitude, and that this barrier height is relatively independent of cluster size. In the $n=3$–5 systems, however, dehydrogenation occurs not only by this relatively efficient activated process but also by an alternative process that is barrierless and unavailable to smaller and larger clusters. Although there are several speculative explanations for this low energy process (alternate product isomer, alternate surface of differing spin, concerted D_2 elimination vs. sequential deuteride shifts), an appealing one relies in part on the theoretical results of Fossan and Uggerud [84]. They find that reaction with ammonia induces the Fe_4^+ reactant to transform from a planar rhomboid to a tetrahedral structure in which the NH is bound to a threefold site on one face. It is plausible that this structural rearrangement provides the low energy path for dehydrogenation. Because such a rearrangement becomes increasingly difficult for larger clusters because there are more metal–metal bonds, this pathway might reasonably disappear with cluster size.

Figure 6 shows the bond energies for Fe_n^+–ND_x derived in this work and compared to Fe_n^+–D. Examination of the Fe_n^+–ND_2 BDEs shows that they have the same pattern for $n=1$–6 as the Fe_n^+–D BDEs, which is consistent with the fact that both D and ND_2 can form a single covalent bond. This is additional evidence that these patterns reflect changes in the electronic character of the clusters. Although $D(Fe_n^+$–$ND_2)$ and $D(Fe_n^+$–D) parallel each other, the former BDEs average 0.78 ± 0.13 eV stronger than the latter. This can be attributed to a dative interaction of the lone pair of electrons on nitrogen donating back to the metal cluster cation. For $n=6$–8, the Fe_n^+–ND_2 BDEs have reached an asymptotic value, 3.25 ± 0.10 eV. This occurs for smaller clusters than in the case of the Fe_n^+–D BDEs for the same reasons as discussed above for D versus O ligands. This asymptotic value is greater than that previously estimated for the bulk phase, which was assumed to equal that for H atoms [85]. The present work demonstrates that the electron lone pair on nitrogen is capable of contributing to the bonding in these systems.

For the Fe_n^+–ND BDEs, it is found that these exceed the Fe_n^+–CD_2 BDEs by 0.5 ± 0.1 eV for $n\geq6$. As both ND and CD_2 can form two covalent bonds, this enhancement can again be attributed to a dative bond formed by donation of the lone pair of electrons on nitrogen to the cluster cation. The Fe_n^+–ND BDEs for $n=3$–5 are much stronger than that for larger and smaller clusters, consistent with the observation of exothermic dehydrogenation of ammonia by these three clusters exclusively. Apparently, these three clusters have geometric/electronic structures that are able to bind ND more strongly than larger clusters. It can be seen in Figure 6 that the estimated bulk-phase binding energy of ND [85] is close to the values for $n=6$–10 and 14, which serves to validate the previous estimate. It can also be noted that the Fe_n^+–O bonds are stronger still than Fe_n^+–ND (see Fig. 9 in Ref. [77] for a direct comparison), even though O, ND, and CD_2 can all form two covalent bonds. Again this can be attributed to the ability of the lone pairs of electrons on oxygen and nitrogen to contribute to the bonding. Oxygen does this more effectively than ND, effectively forming a triple bond with iron cluster cations.

Bond energies for Fe_n^+–ND_3 were also obtained for the smaller clusters and vary appreciably with cluster size. The patterns differ from those for Fe_n^+–D, which reflects the fact that D is a one electron donor whereas ND_3 forms bonds by donating two electrons to the clusters. The patterns in these two BDEs therefore reflect the ability of the cluster to have a singly occupied orbital available for bonding versus an empty orbital to accept the lone pair of electrons on ammonia. These cluster ammonia BDEs generally exceed the bulk-phase value [86], which can be attributed to enhanced interaction resulting from the positive charge on the clusters.

VI. CONCLUSION

Virtually all technologically important catalytic reactions (the Haber process, water gas shift reactions, Fischer–Tropsch synthesis, hydrocarbon processing, etc.) involve molecular (CH_3, CH_2, CH, NH_2, NH, OH) and atomic (H, C, N, O) fragments bound to metal surfaces. Although thermodynamic data for some atomic ligands are widely available (especially H and O, and only rarely for C and N), information for molecular systems is still quite sparse. The experimental methods that work so well for measuring the adsorption energies of atomic and stable molecular species, generally fail for molecular fragments, which often decompose under such conditions. Unexpectedly, some of the first experimental information regarding such bond energies may be available from the cluster studies described in this review. This relies on the observation that these adsorbate bond energies approach bulk-phase behavior much differently than many other cluster properties, for example, the cohesive energies illustrated in Figure 1. In contrast, the adsorbate bond energies rapidly reach an asymptotic value at modest sized clusters and even more rapidly as the binding interaction gets stronger. In the cases where a comparison between these asymptotic values and bulk-phase adsorbate binding energies can be made (H and O), there is a good correspondence, as

TABLE 1 Average Asymptotic Adsorbate (L) Binding Energies (eV) to Large Metal Cluster Cations. Uncertainties in parentheses

L	V	Cr	Fe	Co	Ni
D	2.8 (0.1)[a]	2.8 (0.1)[b]	2.6 (0.1)[c]	2.7 (0.3)[d]	2.6 (0.1)[e]
O	6.2 (0.2)[f]	6.4 (0.4)[g]	5.8 (0.3)[h]	5.1 (0.3)[i]	4.6 (0.2)[j]
N			5.6 (0.3)[k]		
C			6.1 (0.2)[l]		6.5 (0.1)[m]
CD			5.9 (0.4)[l]		5.9 (0.2)[m]
CD_2			4.2 (0.4)[l]		4.2 (0.3)[m]
CD_3			2.6 (0.2)[l]		2.6 (0.1)[m]
ND			4.5 (0.3)[n]		
ND_2			3.3 (0.2)[n]		

[a]Ref. [41].
[b]Ref. [40].
[c]Ref. [39].
[d]Ref. [42].
[e]Ref. [43].
[f]Ref. [46].
[g]Ref. [45].
[h]Ref. [44].
[i]Ref. [47].
[j]Ref. [48].
[k]Ref. [49].
[l]Ref. [75].
[m]Ref. [76].
[n]Ref. [77].

illustrated in Figure 3. Hence, values for the molecular species shown in Figures 6 and 9 can plausibly be utilized as reasonable experimental estimates of the bulk-phase binding energies as well. In this regard, the studies reviewed here provide not only unique thermodynamic information for clusters, but also plausibly extend these data to surfaces of much wider applicability. For convenience, these asymptotic values have been collected in Table 1, although the reader is referred to the individual papers for more comprehensive discussions of the limitations and assumptions used in deriving these values.

Acknowledgment

This research is supported by the Chemical Sciences, Geosciences, and Biosciences Division, Office of Basic Energy Sciences, U.S. Department of Energy.

REFERENCES

[1] Somorjai G. Chemistry in two dimensions: Surfaces. Ithaca: Cornell University Press; 1981.

[2] Over H, Kim YD, Seitsonen AP, Wendt S, Lundgren E, Schmid M, et al. Atomic-scale structure and catalytic reactivity of the $RuO_2(110)$ surface. Science 2000;287:1474.

[3] Loh SK, Hales DA, Lian L, Armentrout PB. Collision-induced dissociation of Fe_n^+ (n = 2–10) with Xe: Ionic and neutral iron cluster binding energies. J Chem Phys 1989;90:5466.

[4] Deitz TG, Duncan MA, Powers DE, Smalley RE. Laser production of supersonic metal cluster beams. J Chem Phys 1981;74:6511.

[5] Loh SK, Hales DA, Armentrout PB. A continuous source for production of cold, mass-selected transition metal cluster ions. Chem Phys Lett 1986;129:527.

[6] Teloy E, Gerlich D. Integral cross sections for ion-molecule reactions. The guided beam technique. Chem Phys 1974;4:417. (b) Gerlich D. Adv Chem Phys 1992;82:1. (c) Gerlich D. Inhomogeneous rf fields: A versatile tool for the study of processes with slow ions. In: Ng C-Y, Baer M, editors. State-selected and state-to-state ion–molecule reaction dynamics. New York: Wiley; 1992. p. 1.

[7] Ervin KM, Armentrout PB. Translational energy dependence of $Ar^+ + XY \rightarrow ArX^+ + Y$ ($XY = H_2, D_2$, HD) from thermal to 30 eV c.m. J Chem Phys 1985;83:166.

[8] Daly NR. Scintillation type mass spectrometer ion detector. Rev Sci Instrum 1959;31:264.

[9] Chesnavich WJ, Bowers MT. Theory of translationally driven reactions. J Phys Chem 1979;83:900.

[10] Armentrout PB. Thermochemical measurements by guided ion beam mass spectrometry. In: Adams LM, Babcock LM, editors. Advances in gas phase metal ion chemistry, vol. 1. Greenwich: JAI; 1992. p. 83.

[11] Armentrout PB. The kinetic energy dependence of ion-molecule reactions: Guided ion beams and threshold measurements. Int J Mass Spectrom 2000;200:219.

[12] Shvartsburg AA, Ervin KM, Frederick JH. Models for statistical decomposition of metal clusters: Vibrational frequency distributions. J Chem Phys 1996;104:8458.

[13] Muntean F, Armentrout PB. Guided ion beam study of collision-induced dissociation dynamics: Integral and differental cross sections. J Chem Phys 2001;115:1213.

[14] Lian L, Su CX, Armentrout PB. Collision-induced dissociation of Fe_n^+ (n = 2–19) with Xe: Bond energies, geometric structures, and dissociation pathways. J Chem Phys 1992;97:4072.

[15] Rodgers MT, Ervin KM, Armentrout PB. Statistical modeling of collision-induced dissociation thresholds. J Chem Phys 1997;106:4499.

[16] Gilbert RG, Smith SC. Theory of unimolecular and recombination reactions. Oxford: Blackwell Scientific; 1990.

[17] Holbrook KA, Pilling MJ, Robertson SH. Unimolecular reactions. 2nd ed. New York: Wiley; 1996.

[18] Truhlar DG, Garrett BC, Klippenstein SJ. Current status of transition-state theory. J Phys Chem 1996;100:12771.

[19] Lian L, Su CX, Armentrout PB. Collision-induced dissociation of Ti_n^+ (n = 2–22) with Xe: Bond energies, geometric structures, and dissociation pathways. J Chem Phys 1992;97:4084.

[20] Su CX, Hales DA, Armentrout PB. Collision-induced dissociation of V_n^+ (n = 2–20) with Xe: Bond energies, dissociation pathways, and structures. J Chem Phys 1993;99:6613.

[21] Su CX, Hales DA, Armentrout PB. The bond energies of Cr_2^+ and Cr_2. Chem Phys Lett 1993;201:199.

[22] Su CX, Armentrout PB. Collision-induced dissociation of Cr_n^+ ($n = 2$–21) with Xe: Bond energies, dissociation pathways, and structures. J Chem Phys 1993;99:6506.

[23] Ervin K, Loh SK, Aristov N, Armentrout PB. Metal cluster ions: The bond energy of Mn_2^+. J Phys Chem 1983;87:3593.

[24] Armentrout PB. Transition metal cluster ion chemistry. In: El-Sayed M, editor. Laser applications in chemistry and biophysics. Proceedings of SPIE, vol. 620. 1986. p. 38.

[25] Loh SK, Lian L, Hales DA, Armentrout PB. Collision-induced dissociation of Fe_2^+. J Phys Chem 1988;92:4009.

[26] Hales DA, Armentrout PB. Effect of internal excitation on the collision-induced dissociation and reactivity of Co_2^+. J Cluster Science 1990;1:127.

[27] Hales DA, Su CX, Armentrout PB. Collision-induced dissociation of Co_n^+ ($n = 2$–18) with Xe: Bond energies of cationic and neutral cobalt clusters, dissociation pathways, and structures. J Chem Phys 1994;100:1049.

[28] Lian L, Su CX, Armentrout PB. The bond energy of Ni_2^+. Chem Phys Lett 1991;180:168.

[29] Lian L, Su CX, Armentrout PB. Collision-induced dissociation of Ni_n^+ ($n = 2$–18) with Xe: Bond energies, geometrical structures, and dissociation pathways. J Chem Phys 1992;96:7542.

[30] Lian L, Schultz RH, Armentrout PB. (unpublished results).

[31] Hales DA, Lian L, Armentrout PB. Collision-induced dissociation of Nb_n^+ (n = 2–11): Bond energies and dissociation pathways. Int J Mass Spectrom Ion Processes 1990;102:269.

[32] Loh SK, Lian L, Armentrout PB. Collision-induced dissociation of niobium cluster ions: Transition metal cluster binding energies. J Am Chem Soc 1989;111:3167.

[33] Hales DA. CID of transition metal cluster ions. Ph.D. Thesis, University of Utah, 1990; 204pp.

[34] Armentrout PB, Hales DA, Lian L. Collision-induced dissociation of transition metal cluster ions. In: Duncan MA, editor. Advances in metal and semiconductor clusters, vol. II. Greenwich: JAI; 1994. p. 1.

[35] Armentrout PB. Guided-ion beam studies of ionic transition metal clusters and complexes. In: Russo N, Salahub DR, editors. Metal-ligand interactions — structure and reactivity. Dordrecht: Kluwer-Academic Publishers; 1996. p. 23.

[36] Armentrout PB, Griffin JB, Conceição J. Thermochemistry and reactivity of transition metal cluster ions. In: Chuev GN, Lakhno VD, Nefedov AP, editors. Progress in physics of clusters. Singapore: World Scientific; 1999. p. 198.

[37] Miedema AR. Z Metallkunde 1978;69:287.

[38] Miedema AR. Faraday Symp R Soc Chem 1980;14:136.

[39] Conceiçao J, Loh SK, Lian L, Armentrout PB. Guided ion beam studies of the reactions of Fe_n^+ ($n = 2$–15) with D_2. cluster-deuteride bond energies as a chemical probe of cluster structures. J Chem Phys 1996;104:3976.

[40] Conceiçao J, Liyanage R, Armentrout PB. Guided ion beam studies of the reactions of Cr_n^+ ($n = 2$–14) with D_2. Cluster-deuteride bond energies as a chemical probe of cluster electronic structure. Chem Phys 2000;262:115.

[41] Liyanage R, Conceiçao J, Armentrout PB. Guided ion beam studies of the reactions of V_n^+ ($n = 2$–13) with D_2. cluster-deuteride bond energies as a chemical probe of cluster electronic structure. J Chem Phys 2002;116:936.

[42] Liu F, Armentrout PB. Guided ion beam studies of the kinetic energy dependent reactions of Co_n^+ ($n = 2$–16) with D_2: Cobalt cluster-deuteride bond energies. J Chem Phys 2005;122:194320–1-12.

[43] Liu F, Liyanage R, Armentrout PB. Guided ion beam studies of the reaction of Ni_n^+ ($n =$ 2–16) with D_2: Nickel cluster-deuteride bond energies. J Chem Phys 2002;117:132.

[44] Griffin JB, Armentrout PB. Guided ion beam studies of the reactions of Fe_n^+ ($n = 2$–18) with O_2: Iron cluster oxide and dioxide bond energies. J Chem Phys 1997;106:4448.
[45] Griffin JB, Armentrout PB. Guided ion beam studies of the reactions of Cr_n^+ ($n = 2$–18) with O_2: Chromium cluster oxide and dioxide bond energies. J Chem Phys 1998;108:8062.
[46] Xu J, Rodgers MT, Griffin JB, Armentrout PB. Guided ion beam studies of the reactions of V_n^+ ($n = 2$–17) with O_2: Bond energies and dissociation pathways. J Chem Phys 1998;108:9339.
[47] Liu F, Li F-X, Armentrout PB. Guided ion beam studies of the reactions of Co_n^+ ($n = 2$–20) with O_2: Cobalt cluster-oxide and dioxide bond energies. J Chem Phys 2005;123:064304–1-15.
[48] Vardhan D, Liyanage R, Armentrout PB. Guided ion beam studies of the reactions of Ni_n^+ ($n = 2$–18) with O_2: Nickel cluster oxide and dioxide bond energies. J Chem Phys 2003;119:4166.
[49] Tan L, Liu F, Armentrout PB. Thermochemistry of the activation of N_2 on iron cluster cations: Guided ion beam studies of the reactions of Fe_n^+ ($n = 1$–19) with N_2. J Chem Phys 2006;124:084302–1-14.
[50] Armentrout PB. Reactions and thermochemistry of small transition metal cluster ions. Ann Rev Phys Chem 2001;52:423.
[51] Armentrout PB. The thermochemistry of adsorbates on transition metal cluster ions: Relationship to bulk phase properties. Eur J Mass Spectrom 2003;9:531.
[52] Griffin JB, Armentrout PB. Guided ion beam studies of the reactions of Fe_n^+ ($n = 1$–18) with CO_2: Iron cluster oxide bond energies. J Chem Phys 1997;107:5345.
[53] Griffin JB, Armentrout PB. Guided ion beam studies of the reactions of Cr_n^+ ($n = 1$–18) with CO_2: Chromium cluster oxide bond energies. J Chem Phys 1998;108:8075.
[54] Krenn G, Eibl C, Mauritsch W, Hebenstreit ELD, Varga P, Winkler A. Adsorption kinetics and energetics of atomic hydrogen (deuterium) on oxygen and carbon covered V(100). Surf Sci 2000;445:343.
[55] Beutl M, Lesnik J, Laundgren E, Konvicka C, Varga P, Rendulic KD. Interaction of H_2, CO and O_2 with a vanadium (111) surface. Surf Sci 2000;447:245.
[56] Toyoshima I, Somorjai G. Catalysis Rev Sci Eng 1979;19:105.
[57] Frese KW. Calculation of surface binding energy for hydrogen, oxygen, and carbon atoms on metallic surfaces. Surf Sci 1987;182:85.
[58] Boszo F, Ertl G, Grunze M, Weiss M. Appl Surf Sci 1977;1:103.
[59] Boszo F, Ertl G, Weiss M. Interaction of nitrogen with iron surfaces: II. Fe(110). J Catal 1977;50:519.
[60] Benziger JB. Thermochemical methods for reaction energetics on metal surfaces. In: Shustorovich E, editor. Metal-surface reaction energetics. New York: VCH; 1991. p. 53.
[61] Bridge ME, Comrie CM, Lambert RM. Hydrogen chemisorption and the carbon monoxide-hydrogen interaction on cobalt (0001). J Catal 1979;58:28.
[62] Ernst KH, Schwarz E, Christmann K. The interaction of hydrogen with a cobalt(1010) surface. J Chem Phys 1994;101:5388.
[63] Christmann K, Behm RJ, Ertl G, Van Hove MA, Weinberg WH. Chemisorption geometry of hydrogen on Ni(111): Order and disorder. J Chem Phys 1979;70:4168.
[64] Christmann K, Schober O, Ertl G, Neuman M. Adsorption of hydrogen on nickel single crystal surfaces. J Chem Phys 1974;60:4528.
[65] Lapujoulade J, Neil KS. Chemisorption of hydrogen on the (111) plane of nickel. J Chem Phys 1972;57:3535.
[66] Brennan D, Hayward DO, Tradnell BMW. The calorimetric determination of the heats of adsorption of oxygen on evaporated metal films. Proc Roy Soc 1960;A256:81.

[67] Bragg J, Tomkins TC. Calorimetric heats of sorption of gases on evaporated iron films. Trans Faraday Soc 1955;51:1071.
[68] Wedler G. Z Phys Chem 1961;27:388.
[69] Benziger JB, Preston RE. Recombination reactions on Ni(111). Surf Sci 1984;141:567.
[70] Brown WA, Kose R, King DA. Femtomole adsorption calorimetry on single-crystal surfaces. Chem Rev 1998;98:797.
[71] Rao CNR, Kamath PV, Yashonath S. Molecularly adsorbed oxygen on metals: electron spectroscopic studies. Chem Phys Lett 1982;88:13.
[72] Ostrovskii VE. Russ J Phys Chem 1988;62:330.
[73] Boszo F, Ertl G, Grunze M, Weiss M. Interaction of nitrogen with iron surfaces: I. Fe(100) and Fe(111). J Catal 1977;49:18.
[74] Stoltze P, Nørskov J. Bridging the "pressure gap" between ultrahigh-vacuum surface physics and high-pressure catalysis. Phys Rev Lett 1985;55:2502.
[75] Liyanage R, Zhang X-G, Armentrout PB. Activation of methane by size-selected iron cluster cations, Fe_n^+ (n = 2–15): Cluster-CH_x (x = 0–3) bond energies and reaction mechanisms. J Chem Phys 2001;115:9747.
[76] Liu F, Zhang X-G, Liyanage R, Armentrout PB. P. B. methane activation by nickel cluster cations, Ni_n^+ (n = 2–16): Reaction mechanisms and thermochemistry of cluster-CH_x (x = 0–3) complexes. J Chem Phys 2004;121:10976.
[77] Liyanage R, Griffin JB, Armentrout PB. Thermodynamics of ammonia activation by iron cluster cations. Guided ion beam studies of the reactions of Fe_n^+ (n = 2–10, 14) with ND_3. J Chem Phys 2003;119:8979.
[78] Citir M, Liu F, Armentrout PB. Methane activation by cobalt cluster cations, Co_n^+ (n = 2–16): Reaction mechanisms and thermochemistry of cluster-CH_x (x = 0–3) complexes. J Chem Phys 2009;130:054309–NaN.
[79] Haynes CL, Chen Y-M, Armentrout PB. The reaction of $FeCH_2^+ + D_2$: Probing the $[FeCH_4]^+$ potential energy surface. J Phys Chem 1996;100:111.
[80] Hong Y, Whitten JL. Reaction of CH_4 with substitutional Fe/Ni(111). Surf Sci 1993;289:30.
[81] Isett LT, Blakely JM. Binding energies of carbon to Ni(100) from equilibrium segregation studies. Surf Sci 1975;47:645.
[82] Siegbahn PEM, Wahlgren U. Cluster modeling of chemisorption energetics. In: Shustorovich E, editor. Metal-surface reaction energetics. New York: VCH; 1991. p. 1.
[83] Irion MP, Schnabel P. FT-ICR studies of sputtered metal cluster ions. 5. The chemistry of iron cluster cations with ammonia and hydrazine, J. Phys Chem 1991;95:10596.
[84] Fossan KO, Uggerud E. Reactions of cationic iron clusters with ammonia, models of nitrogen hydrogenation and dehydrogenation. Dalton Trans 2004;892.
[85] (a) Ertl G. In: Anderson JR, Boudart M, editors. Catalysis: science and technology, vol. 4. Berlin: Springer; 1983. (b) Ertl G. Catal Rev Sci Eng 1980;21:201. (c) Ertl G. In: Jennings JR, editor. Catalytic ammonia synthesis. New York: Plenum; 1991. p. 109.
[86] Evdokimova ZhA, Valitov NKh. Zh Prikl Khim (S-Peterburg) 1985;58:2121 (In Russian).

Chapter 7

Hydrogen and Hydrogen Clusters Across Disciplines

I. Cabria, M. Isla, M. J. López, J. I. Martínez, L. M. Molina and J. A. Alonso

Departamento de Física Teórica, Atómica y Óptica, Universidad de Valladolid, Valladolid, Spain

Chapter Outline Head

Nanoclusters. DOI: 10.1016/S1875-4023(10)01007-7

I. INTRODUCTION

Hydrogen may play an important role as an alternative to the present fuels and sources of electricity for the massive production of energy. An economy based on hydrogen has two pillars. The first is the production of electricity by means of nuclear fusion of hydrogen isotopes and the second pillar is the production of electricity by means of hydrogen fuel cells. The second mode could be employed for use in buildings or for cars, which will require *in situ* or onboard hydrogen storage, respectively. For these reasons, the study of hydrogen becomes an interesting subject from both the basic and the technological points of view.

It is expected that some day hydrogen isotopes will be the fuel in nuclear fusion reactors (the first pillar), once the outstanding problem of sustaining and controlling the reaction is solved. Experiments performed at the Livermore Laboratory have shown that the irradiation of a dense molecular beam of large deuterium clusters by an ultrafast high intensity laser leads to a violent Coulomb explosion of the clusters [1,2]. Nuclear fusion reactions originating from the collisions between the flying deuterium nuclei have been observed in those experiments. Neutrons are produced in those fusion reactions and table-top neutron sources have been constructed on the basis of cluster beam technique [3]. The first stages of the process, that is, the irradiation of the deuterium clusters, and the Coulomb fragmentation that follows, are extremely interesting, but the details of the energy absorption by the cluster and the energy transfer between electrons and nuclear motion are not well known.

In the near future, hydrogen could replace gasoline in cars, using fuel cells (the second pillar). The main remaining problem in this area is to develop a way of storing the required amount of hydrogen in the tank of the car. Motivated by this need, an intensive effort is nowadays dedicated to investigate materials capable of providing an efficient hydrogen storage for on-board automotive applications. Well-defined targets have been established in order to have a hydrogen-based vehicle whose performance equals that of the present fossil fuel based vehicles: the goals established by the U.S. Department of Energy for the year 2010 are at least 6 wt% of hydrogen and a hydrogen molar volume lower than 44.8 cm^3/mol. Porous materials based on graphitized forms of carbon have been proposed as hydrogen containers.

Hydrogen clusters also have attracted attention because of their peculiar properties [4,5] that arise from the coexistence of strong intramolecular bonding (with a high intramolecular binding energy of 4.8 eV) and weak intermolecular forces. Many investigations have focused on clusters of the family $H_3^+(H_2)_N$ with $N = 1,2, \ldots$. Their easiness to be handled experimentally [6], and the fact that the majority of the hydrogen clusters in the universe belong to this family, justify the study of the ionized species. The caloric curve of size-selected $H_3^+(H_2)_N$ clusters has been determined in high energy collision experiments and has been interpreted as indicating the transition from a bound cluster to the gas phase [7,8]. In contrast, there are few studies of neutral clusters. The hydrogen molecule behaves as a closed-shell spherical unit, and the hydrogen clusters show some similarities with the clusters of the inert gases. The quantum nature of large hydrogen clusters has been another area of interest [9,10]. Atomic hydrogen can also serve as a chemical probe of the reactivity of other clusters and nanostructures [11].

Hydrogen and hydrogen clusters give rise to an area which moves across disciplines, with interesting scientific aspects and firm technological possibilities. In this work we review recent work showing this interdisciplinarity.

II. STRUCTURE AND GROWTH OF NEUTRAL HYDROGEN CLUSTERS

The potential use of hydrogen as a fuel for automobiles calls for a deeper understanding of the weak bonding interactions between hydrogen molecules. Accurate calculations have been performed for $(H_2)_2$, $(H_2)_3$, and $(H_2)_4$ using Möller–Plesset (MP2) and coupled cluster (CC) methods [12,13]. The dimer, $(H_2)_2$, has a T-shaped structure, with a separation of 6.55 a.u. between the mass centers of the two molecules. The relative orientation of the two molecules is associated with the optimum interaction between their permanent electric quadrupoles. The binding energy E_b is between 4 and 5 meV and the potential is very flat. The trimer is planar and cyclic (with C_{3h} symmetry), and the tetramer is slightly nonplanar. For larger sizes, only approximate calculations have been performed, using empirical force fields fitted to the *ab initio* results [12,13]. As the size grows, an increasing number of isomers with similar energies are found. The ground state of $(H_2)_5$ is a pyramid (a nearly flat tetramer plus a molecule overhead), and that of $(H_2)_6$ is a bipyramid. Then, a pentagonal bipyramid is found for $(H_2)_7$. The zero point energy reduces further the already small binding energy of these clusters.

Small *para*-H_2 clusters produced in cryogenic free jets have been studied by Raman spectroscopy [14]. The intensity of the Raman line corresponding to the internal vibration of the H_2 molecules reveals enhanced intensity for some particular cluster sizes, $(H_2)_{13}$, $(H_2)_{32}$, and $(H_2)_{55}$, and these were tentatively interpreted as indicating the formation of successive shells. The existence of those magic numbers motivated a theoretical study of the structures of $(H_2)_N$ clusters in that size range [15] using the density functional theory (DFT) and the local density approximation (LDA) for exchange and correlation effects [16,17]. A molecular bond length of 1.38 a.u., close to the experimental value of 1.40 a.u. for the isolated H_2 molecule [18], is obtained independent of the cluster size. In the optimized structure of $(H_2)_2$, the two molecules are in a perpendicular configuration; a triangle is obtained for $N=3$, and a bent rhombus for $N=4$. The structures obtained belong to the

D_{3h} symmetry group for $N=5$ (triangular bipyramid), to O_h for $N=6$ (square bipyramid), and to D_{5h} for $N=7$ (pentagonal bipyramid). The optimized structures agree in most cases, but not in all, with the structures described above, obtained from the MP2 and CC methods [13]. Larger clusters grow following a pattern of icosahedral growth and a complete icosahedron is obtained for $N=13$. In this cluster, one molecule occupies the central position and the other 12 molecules occupy the 12 vertices of the icosahedron. The growth pattern obtained is similar to that followed by clusters of the inert gases [4] because the H_2 molecule has a filled electronic shell. Calculations for larger clusters were restricted to well-defined growth models. An icosahedron is formed by 20 triangular faces joined by 30 edges and 12 vertices. Molecules can be added on top of the icosahedron in two different ways. In a first type of decoration, molecules are added on top of edge (E) and vertex (V) positions. These provide a total of 42 sites (30+12) to cover $(H_2)_{13}$, and a cluster with 55 molecules is obtained after full covering. This is called multilayer icosahedral (MIC) growth. Alternatively, one can cover V sites and the 20 sites at the center of the triangular faces (F sites). In that way, the covering of $(H_2)_{13}$ by 32 additional molecules (12+20) leads to a cluster with 45 molecules. This second type of decoration is often called face centered (FC) growth. Other growth models, such as the formation of face centered cubic (fcc) cuboctahedral structures, lead to structures of higher energy.

Calculations were performed up to $(H_2)_{35}$. The binding energies per molecule, $e_b(N)$,

$$e_b(N) = \frac{E_b(N)}{N} = \frac{E((H_2)_N) - NE(H_2)}{N} \tag{1}$$

calculated from the ground state energies of the clusters and the free molecule, show that the FC structures are more stable than the MIC structures up to about $N=27$; for N between 27 and 35 the two structures are almost degenerate, and a transition to MIC structures is expected to occur soon for larger clusters. FC growth is also characteristic of the inert gas clusters, where a transition to MIC structures [4] occurs for clusters with 27–28 atoms. The relative stability of the $(H_2)_N$ clusters following the two growth paths can be appreciated in Figure 1, where the quantity

$$\Delta e_b(N) = 2e_b(N) - [e_b(N+1) + e_b(N-1)] \tag{2}$$

is plotted as a function on N for the FC and MIC structures. Peaks in $\Delta e_b(N)$ correspond to the most stable clusters, and these occur for $N=5$, 7, 13, 19, 23, 26, and 32 for FC structures. The peaks for N larger than 13 correspond to the progressive formation and filling of caps. For instance, the peak at $(H_2)_{19}$ corresponds to the decoration of $(H_2)_{13}$ with a cap formed by six molecules. Addition of an adjacent cap leads to the structure of $(H_2)_{23}$ and so on.

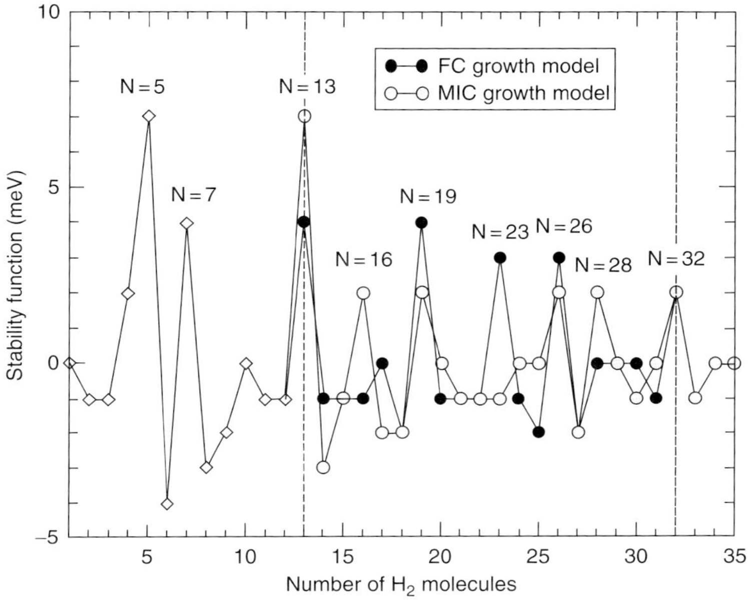

FIGURE 1 Stability function, $\Delta e_b(N)$, of $(H_2)_N$ clusters for MIC and FC growth models. Peaks correspond to clusters with enhanced stability. Vertical lines indicate features observed in a Raman spectroscopy experiment. Reproduced from Ref. [15], with permission of EDP Sciences.

It is known that the LDA overestimates the binding energies. A binding energy E_b between 4 and 5 meV is obtained for $(H_2)_2$ in the MP2 and CC methods [12,13], while the LDA calculation gives 16.5 meV. Considering the tiny binding energy, this is quite successful for DFT, and it is probably the most one can expect from a simple density functional. Because all clusters are subject to similar errors, trends in the binding energy as a function of N are more trustable than the absolute binding energies. However, zero point vibrational corrections reduce the binding energies substantially. The intermolecular potential of $(H_2)_2$ is very anharmonic. Accounting for anharmonicity effects in first order [19], a corrected value of 2.75 meV is obtained for the LDA binding energy E_b, and similar reductions affect other clusters. Zero point effects also reduce the binding energies in the *ab initio* calculations [13]. Because of those small binding energies, hydrogen is a molecular gas except at very low temperatures and high pressures. The small binding energies have other important consequences, as we discuss below.

A part of the findings of Figure 1 are consistent with the results of the Raman spectroscopy experiments [14]. Local maxima in the Raman intensities were observed for $N \approx 13$, 32, and 55, and were ascribed to a large abundance of those clusters. The present theoretical analysis provides an interpretation of the first two magic sizes. In addition, the calculations indicate that an FC to MIC transition is likely to occur soon after $(H_2)_{35}$. This explains the observation of the magic cluster $(H_2)_{55}$, having the structure of an icosahedron with two complete shells. One can notice in Figure 1 that $(H_2)_{32}$ corresponds to a stability peak for both FC and MIC structures. On the other hand, the other predicted stability peaks of Figure 1 between $N=13$ and $N=32$ have not been detected in the Raman experiments. The DFT calculations discussed above correspond to clusters with a well-defined structure. However, zero point energies reduce so much the binding energies that many of these clusters are liquid even at very low temperatures. For instance, in the vibrational ground state of $(H_2)_2$, the calculated amplitude of the intermolecular vibrations is a substantial fraction of the H_2–H_2 intermolecular distance, and the same occurs for other small $(H_2)_N$ clusters. According to the Lindeman melting rule, those clusters are liquid. This is confirmed by the path integral Montecarlo simulations of Mezzacapo [20]. Those simulations, performed with model intermolecular potentials, have predicted that most $(H_2)_N$ clusters with N smaller than 27 are liquid at 1 K. Thus, effects due to particularly stable solid-like structures can only be observed for clusters larger than $N=27$–28. This explains why $(H_2)_{32}$ and $(H_2)_{55}$ are the only magic solid clusters detected in the Raman experiments. Further confirmation of these ideas comes from the fact that clear magic peaks at $N=32$ and 55 are seen in the experiments only at the lowest temperatures studied. The case of $(H_2)_{13}$ is special. A structure of a central molecule surrounded by 12 molecules is so favorable that this structure is expected even if the cluster is liquid; in this case the surface shell is melted.

The photoabsorption spectrum of the $(H_2)_N$ clusters reflects the dominant effect of intramolecular bonding [15]. The spectrum of the isolated H_2 molecule, calculated using the time-dependent density functional theory (TDDFT) [21] shows two absorption peaks in the UV region at 11.2 and 14.3 eV. The high energies arise from the large gap between the highest occupied and the lowest unoccupied electronic levels of the cluster (HOMO–LUMO gap). Then the spectra of $(H_2)_N$ clusters essentially display the same features, broadened by the effect that different molecules in a given cluster experience different environments.

III. IONIZED HYDROGEN CLUSTERS

Ionization of a hydrogen molecule in a $(H_2)_N$ cluster is followed by the exothermic reaction

$$H_2^+ + H_2 \rightarrow H_3^+ + H \tag{3}$$

and the released energy, 1.7 eV, is enough to eject the neutral H atom from the cluster. The H_3^+ cation is stabilized by the other H_2 molecules that form one or more solvating shells around the trimer [22–24]. The cluster cations thus have the composition $H_3^+(H_2)_N$. These clusters can grow easily [25] and those with $N=3$, 6, 8, and 10 exhibit special stability, indicating atomic shell closing effects [26–28]. Ionized clusters are important active species in the nucleation phenomena in the stratosphere and interstellar clouds [29]; being charged makes them easier to handle experimentally [6].

The H_3^+ trimer becomes the core of the $H_3^+(H_2)_N$ structure of those clusters. This trimer ion and its deuterated isotopomers have attracted the attention of experimentalists and theorists because of their fundamental nature, astrophysical significance, and dynamical richness. Its unusual bonding leads to an exceptional roto-vibrational spectrum. The experiments [30–33] and *ab initio* simulations of Tennyson [34–40] have shed light on the electronic structure, the infrared photodissociation spectrum, and the classical and quantal behavior of the molecule at its dissociation limit. Enhanced ionization has been observed for this molecule [41,42]. On the other hand, since the first quantitative model of interstellar chemistry by Herbst and Klemperer [43], it is known that H_3^+ is the main agent responsible for the formation of complex molecules in the reaction network of the interstellar medium. In fact, H_3^+ is present in any environment where molecular hydrogen gas is ionized [44,45]: it has been detected in the atmospheres of the giant planets such as Jupiter [46], where it is distributed widely across the planet's auroral regions, Saturn [47] and Uranus [48]. It has also been identified in the supernova SN1987A [49] and in the interstellar medium, and it took 16 years before H_3^+ could be detected in dense [50] and diffuse [51] interstellar clouds.

The charge has influence on the binding energy of the $H_3^+(H_2)_N$ clusters. Taking $H_3^+(H_2)_5$ as an example, the calculated average LDA binding energy of the H_2 molecules in the cluster, that can be defined as

$$e_b^{av}(\text{charged}) = \frac{E\left(H_3^+(H_2)_5\right) - E\left(H_3^+\right) - 5E(H_2)}{5}, \tag{4}$$

is equal to 60 meV. For comparison, the average binding energy of five H_2 molecules attached to a sixth H_2 molecule to form the neutral $(H_2)_6$ cluster,

$$e_b^{av}(\text{neutral}) = \frac{E\left((H_2)_6\right) - 6E(H_2)}{5}, \tag{5}$$

is 25 meV. So, the effect of the charge in the cluster center is to increase the binding energy substantially. Zero point corrections lower those binding energies.

IV. LIQUID TO GAS PHASE TRANSITION IN HYDROGEN CLUSTERS

The best signature of a first or second order phase transition in a finite system is provided by the specific shape of the caloric curve, that is, the thermodynamic temperature as a function of the total energy. The caloric curve of size-selected hydrogen clusters has been determined in high energy collision experiments [7,8] and has been interpreted as indicating the transition from a bound cluster to the gas phase. In those experiments, hydrogen clusters are formed in a cryogenic cluster expansion source and then ionized and size-selected using an ion accelerator [6].

In the experiments of Gobet and coworkers [7,8], the collisions between size-selected $H_3^+(H_2)_N$ clusters with N up to 14, accelerated to kinetic energies of 60 keV/amu, and a helium gas target were analyzed. A multidetector device records for each collision event simultaneously the number (multiplicity) of each mass-identified fragment ion resulting from the reaction. The fragmentation reactions have the general form

$$H_3^+(H_2)_N + He \rightarrow H_3^+(H_2)_k + bH_3^+ + cH_2^+ + dH^+ + eH_2 + fH \qquad (6)$$

with $a-f=0, 1, \ldots$ (neutral species larger than the dimer are absent). The construction of the caloric curve requires the simultaneous determination of the energy and the temperature. The cluster energy, that is the energy deposited into the cluster by the collision with a He atom, is determined by the nature and multiplicity of the products in the reaction (6). On the other hand, the temperature of the cluster prior to decay is obtained using a relationship [52], tested successfully in nuclear physics, between the characteristic shape of a fragment mass distribution and the temperature of decaying nuclei [53]. The caloric curves for $H_3^+(H_2)_N$ with $N=6$, 8, 9, 11, 12, and 14 are collected in Figure 2. The caloric curves have three parts: after an initial rise, a plateau occurs before the curve rises again. The curves show the characteristic features of a first order phase transition.

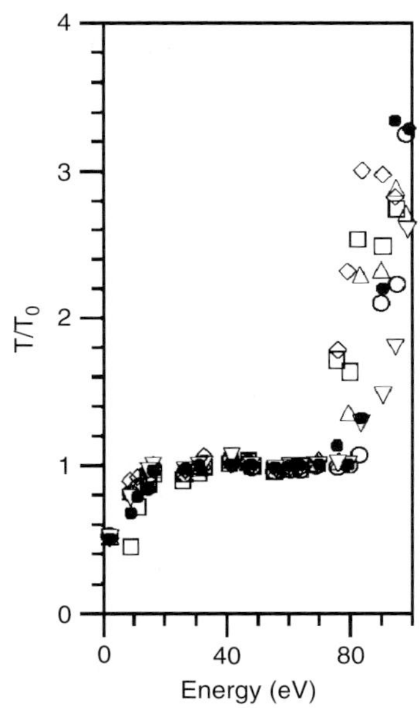

FIGURE 2 Caloric curves for fragmentation of $H_3^+(H_2)_N$ clusters, with $N=6$ (open squares), $N=8$ (open circles), $N=9$ (triangles), $N=11$ (diamonds), $N=12$ (inverted triangles), and $N=14$ (filled circles). Reduced temperature (with T_0 the temperature in the plateau of the curve) is given versus the energy deposited on the clusters. Reproduced from Ref. [8], with permission of the American Physical Society.

V. LASER IRRADIATION OF DEUTERIUM CLUSTERS

Motivated by the advances in laser technology, there is a great interest in the study of the interaction between matter and ultrafast lasers, that is, lasers with intensities higher than 10^{14} W/cm^2 and pulse duration below 100 fs. The intensity of the currently available lasers can exceed the electric field created by an atomic nucleus and the time scale of femtoseconds is also typical of the electron motion. This new research field allows for the exploration of the nonlinear response of atoms to intense laser pulses, leading to the observation of new processes, like above-threshold ionization (ATI) [54], correlated double ionization [55] and high order harmonic generation (HHG) [56]. The theoretical understanding for the atomic case is nowadays well known. The behavior of molecules and clusters under similar laser conditions offers a new challenge because of the existence of additional degrees of freedom, such as the nuclear motion or the presence of intermolecular and intramolecular forces. This gives rise to a broad range of complex phenomena: above-threshold dissociation (ATD), bond softening and hardening, interatomic coulombic decay (ICD) of excited molecules, and enhanced ionization. Some of these phenomena are followed by a Coulomb explosion. When molecules or clusters are multiply ionized by laser pulses of very short duration, the unbalanced positive charges are sufficiently close together to cause a repulsion-induced explosion of the nuclear skeleton [57,58].

In experiments performed by Ditmire and coworkers [1,2], the irradiation of a dense molecular beam of large deuterium clusters, $(D_2)_N$, by an intense femtosecond laser has driven nuclear fusion reactions. Laser irradiation produces the multiple ionization of the clusters, which then explode under the action of the repulsive Coulomb forces between the bare nuclei of the ionized atoms. Some of those flying nuclei collide with nuclei ejected from other clusters in the plasma, and if the kinetic energies are higher than a few keV, nuclear D–D fusion can occur with high probability. This occurs for beams formed by large deuterium clusters, as in that case the nuclei ejected have kinetic energies of many keV. Apart from the obvious interest for future thermonuclear devices, this technique has lead to the development of table-top neutron sources [3]. As the reaction $D+D \rightarrow {}^3He + n$ produces a neutron with an energy of 2.45 MeV, those neutrons could potentially be used in neutron radiography [59] and in materials research [60]. Motivated by those works, Suraud and coworkers [61,62] and Isla and Alonso [63,64] have used TDDFT to study the dynamical response of charged deuterium clusters irradiated by an intense femtosecond laser. This simulates the first stages in the experiments of Ditmire and coworkers [1,2].

A. Density Functional Molecular Dynamics

The method used in those computer simulations is the TDDFT [21], with an implementation in which the response of the system to a time-dependent external perturbation $v_{ext}(\mathbf{r}, t)$ is obtained by directly solving the time-dependent Kohn–Sham equations. This method was originally introduced to study nuclear reactions [65], and has been later applied to clusters [66–71]. The starting point is the ground state of the system in the equilibrium configuration. A laser pulse is then applied and the electronic orbitals are propagated in time:

$$i\frac{\partial}{\partial t}\psi_j(\mathbf{r},t) = \hat{H}_{\mathrm{KS}}\psi_j(\mathbf{r},t) = \left[-\frac{\hbar^2}{2}\nabla^2 + v_{\mathrm{KS}}(\mathbf{r},t)\right]\psi_j(\mathbf{r},t)(j=1,\ldots,N). \tag{7}$$

Here, $v_{\mathrm{KS}}(\mathbf{r}, t)$ is the time-dependent effective Kohn–Sham (KS) potential acting on the electrons,

$$v_{\mathrm{KS}}[n](\mathbf{r},t) = v_{\mathrm{ext}}(\mathbf{r},t) + v_{\mathrm{ionic}}(\mathbf{r},t) + v_{\mathrm{Hartree}}[n](\mathbf{r},t) + v_{\mathrm{xc}}[n](\mathbf{r},t), \tag{8}$$

that is, the sum of the applied external potential (the laser field), the ionic (or nuclear) potential acting on the electrons, the electrostatic Hartree potential, and the quantum-mechanical exchange–correlation (xc) potential. The last two are explicit functionals of the time-dependent electron density

$$n(\mathbf{r},t) = \sum_{j=1}^{N}|\psi_j(\mathbf{r},t)|^2. \tag{9}$$

One advantage of explicitly propagating the time-dependent Kohn–Sham equations is that it permits to couple the electronic system to the ionic background, which can be, for many purposes, treated classically. It is then possible to perform a combined dynamics of electrons and nuclei. The set of equations to be solved is formed by the time dependent Kohn–Sham equations (7), together with the classical Newton's equations for the motion of the nuclei (or the ions):

$$m_\alpha\frac{d^2\mathbf{R}_\alpha}{dt^2} = \mathbf{F}_\alpha(\mathbf{R}_\alpha,t). \tag{10}$$

In these equations, $\mathbf{R}_\alpha$ indicates the position of the ion labeled α, m_α its mass, and $\mathbf{F}_\alpha$ $(\mathbf{R}_\alpha, t)$ is the instantaneous force on that ion

$$\mathbf{F}_\alpha(\mathbf{R}_\alpha,t) = -\langle\Psi(t)|\nabla_{\mathbf{R}_\alpha}\hat{H}(\mathbf{R},t)|\Psi(t)\rangle - \sum_{\alpha'}\frac{(\mathbf{R}_\alpha-\mathbf{R}_{\alpha'})Z_\alpha Z_{\alpha'}}{|\mathbf{R}_\alpha-\mathbf{R}_{\alpha'}|^3} + Z_\alpha\mathbf{E}(t). \tag{11}$$

The first term in the force is the electronic contribution, calculated using the Ehrenfest theorem. This is just the extension of the Hellmann–Feynman theorem to the time-dependent case. The second term is the force exerted by the other ions and the last term is the force due to the external electric field $\mathbf{E}(t)$.

In that expression for the force, $|\Psi(t)\rangle$ is the interacting many body wavefunction, which TDDFT does not provide. However, the previous equation can be applied because the gradient $\nabla_{\mathbf{R}_\alpha}\hat{H}(\mathbf{R},t)$ results in a one-body local operator

$$\nabla_{\mathbf{R}_\alpha}\hat{H}(\mathbf{R},t)=\sum_{j=1}^{N}\sum_{\alpha}\nabla_{\mathbf{R}_\alpha}v_{\text{ion}}^{\alpha}\left(\hat{\mathbf{r}}_j-\mathbf{R}_\alpha\right), \tag{12}$$

and, consequently, the knowledge of the time-dependent density is enough to calculate the Ehrenfest expression

$$\left\langle\Psi(t)|\nabla_{\mathbf{R}_\alpha}\hat{H}(\mathbf{R},t)|\Psi(t)\right\rangle=\int \mathrm{d}\mathbf{r}n(\mathbf{r},t)\sum_{\alpha=1}\nabla_{\mathbf{R}_\alpha}v_{\text{ion}}^{\alpha}(\mathbf{r}-\mathbf{R}_\alpha). \tag{13}$$

This approach is nonadiabatic. The electrons are permitted to populate several energy surfaces (ground state and excited), and to change the value of these populations. This is important for nonlinear excitations as the ones treated next.

B. Laser Irradiation of D_{13}^{+}

The optimized ground state structure of the D_{13}^{+} cluster contains a charged triangular trimer D_3^{+} at the cluster center. This trimer is surrounded by a solvation shell formed by five D_2 molecules, three of them with their centers of mass placed in the plane of the trimer and their axes oriented perpendicular to that plane. The other two molecules are at a larger distance and displaced from that plane. The cluster can be viewed as $D_3^{+}(D_2)_5$ and its structure is shown in the leftmost panel of Figure 3. The system is then perturbed by a classical laser pulse, whose electric field has a cosinoidal envelope,

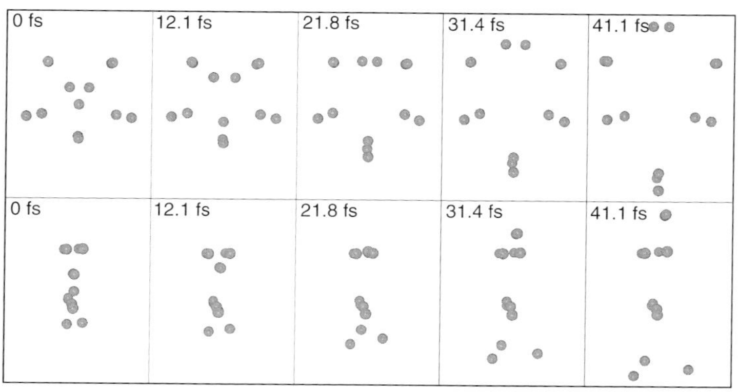

FIGURE 3 Snapshots of the cluster structure at different times after application of a laser pulse in the slow fragmentation case. Two mutually perpendicular views are presented for each snapshot. Reproduced from Ref. [63], with permission of the American Physical Society.

$$E(t) = A_0 \cos\left(\frac{\pi}{2}\frac{t - 2\tau_0 - t_0}{\tau_0}\right)\sin(\omega t)\hat{\mathbf{e}}, |t - t_0| < \tau_0 \tag{14}$$

where ω is the frequency of the field and $\hat{\mathbf{e}}$ is the polarization vector: $2\tau_0$ is the total pulse duration, and A_0 is the amplitude. The frequency ω of the laser is an important parameter in order to achieve an efficient coupling of the laser radiation with the cluster. The first peak in the photoabsorption spectrum of the cluster occurs at a frequency $\omega = 0.352$ a.u. ($\hbar\omega = 9.58$ eV), and the computer simulations were performed at this frequency, for varying intensities of the laser pulse [63].

The results of a simulation in which a laser pulse of 9.6 fs is applied to the cluster are presented first. The amplitude of the field is 0.02 a.u., giving a pulse intensity of 1.4×10^{13} W cm^{-2}. The evolution of the cluster structure with time is shown in Figure 3, where a few snapshots at different times have been selected. The central trimer absorbs energy from the laser pulse and splits in two fragments: a molecule and an atom are emitted in opposite directions, toward the top and the bottom, respectively, in the views presented in Figure 3. The cluster roughly maintains its structure during the time of the applied pulse, the initial 9.6 fs of the simulation, because of the inertia of the atoms and the time the central trimer needs to absorb the necessary energy to break the intramolecular bonds. As the emitted molecule moves upward, it passes near two H_2 molecules, and these two molecules are also set in motion. The bond lengths of these molecules just oscillate about their equilibrium value but the intramolecular bonds remain intact, that is, the molecules do not dissociate. On the other hand, the atom moving downward collides with a third molecule of the solvation shell, giving rise to an exchange collision. As the cluster dissociates, the two D_2 molecules originally most distant from the trimer remain little affected. This dissociation mode of the cluster can be characterized as slow fragmentation.

Before irradiation, all the cluster energy is potential energy stored as binding energy in the intramolecular bonds mainly, and a little as intermolecular interaction energy. When the laser pulse is applied, the binding energy decreases. The energy absorbed by the cluster from the laser pulse is about 28 eV. A substantial fraction, 21 eV, is transformed into potential energy, and a smaller fraction, of 7 eV, is transformed into kinetic energy of the molecules of the fragmenting cluster. The change in potential energy begins almost immediately but the change in the kinetic energy of the molecules is delayed by a few femtoseconds. The laser energy is first absorbed in the form of electronic excitations of the cluster. A part of this excitation energy is employed in dissociating the central trimer and in breaking the weak attractive bonds between the trimer and the surrounding molecules, and another part is transformed into kinetic energy of the molecules. However, this does not exhaust the absorbed energy and a part still remains during the simulation as electronic excitations (although ionization does not occur) and intramolecular vibrations. In spite of the cluster dissociation, the potential energy still remains very large because the system maintains most of its initial binding energy in the intramolecular bonds. The influence of the laser frequency was also explored. For pulses of the same intensity as above, but half of the resonance frequency, that is $\omega = 0.176$ a.u., the atoms oscillate around their equilibrium positions, but the cluster does not dissociate.

The results of the computer simulation change drastically for pulses of frequency $\omega = 0.352$ a.u. and intensity five times larger than the one leading to slow fragmentation. The duration of the pulse is again 9.6 fs. The results of the evolution of the cluster structure are shown in Figure 4. The main feature is that not only the central trimer dissociates now, but all the intramolecular bonds of the other molecules are also broken. The atoms are ejected like in an explosion. This occurs simultaneously in all parts of the cluster. All the molecules dissociate, including the trimer, as the atoms repel each other as a consequence of the massive ionization.

The potential energy quickly increases (decreases in absolute value) during the interaction between the cluster and the laser pulse. This is a manifestation of the dissociation of the trimer and all the D_2 molecules. The potential energy even becomes repulsive and it reaches a maximum value at 7–8 fs. This arises from the electrostatic repulsion between the nuclei of the ionized atoms. A Coulomb explosion of the cluster then occurs and the nuclei fly away. During the explosion, the repulsive potential energy accumulated at the top of the Coulomb barrier is transformed into kinetic energy of the flying nuclei.

Coulomb explosion is a violent dissociation process that occurs in molecules or clusters when these are multiply ionized by femtosecond laser pulses [57]. In small clusters, stripping two electrons may be sufficient to cause Coulomb fragmentation [4,72]. In fact, cluster size is an important parameter to determine whether the cluster will follow this decay channel [73–75]. This is the mechanism corresponding to the fast dissociation process displayed in Figure 4. The irradiation of the cluster creates a plasma, and in a

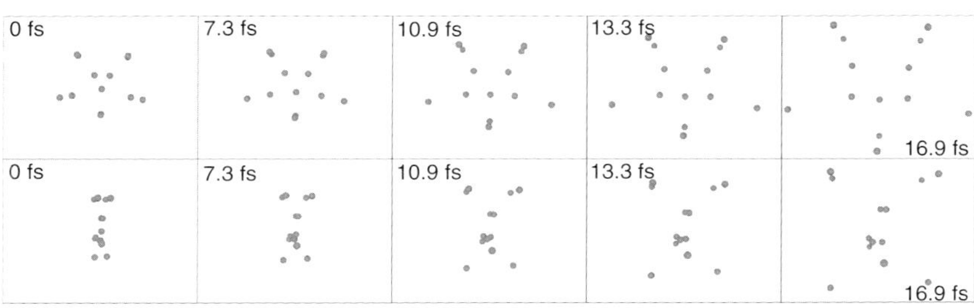

FIGURE 4 Snapshots of the evolution of the cluster structure after application of a laser pulse in the fast fragmentation case. Two mutually perpendicular views are presented for each snapshot. Reproduced from Ref. [63], with permission of the American Physical Society.

short interval, roughly corresponding to the duration of the laser pulse, the plasma loses five electrons. Moreover, the loss of electrons continues afterward, although at a slightly slower rate. Two kinds of Coulomb repulsions occur between the deuterium nuclei: Coulomb repulsion between the two nuclei of each molecule and repulsion between nuclei of different molecules. The first one is stronger because the two atoms in a molecule are at a closer distance. Besides the size of the cluster, there is another requirement for a pure Coulomb explosion to happen: the cluster must be almost stripped of all their electrons in a time scale of only a few femtoseconds. This can be achieved with the very high power femtosecond lasers that have become available in recent years [76].

The kinetic energy of the fastest nuclei resulting from the Coulomb explosion of D_{13}^+ is about 13 eV. In most nuclear fusion processes, from controlled fusion reactors to solar reactions, the reacting particles have kinetic energies in the range of 1–10 keV [77], enough to overcome the Coulomb barrier for fusion. This indicates that the Coulomb explosion of D_{13}^+ delivers energies that are small compared to those required to produce D–D fusion. Of course, this is expected, as the cluster is very small, whereas the clusters in the experiments of Ditmire et al. [1,2] are much larger: the average number of atoms is one thousand times larger. The maximum kinetic energy is proportional to R^2, where R is the radius of the cluster, and a beam of deuterium clusters with radii greater than 50 a.u. is necessary to produce the multi-keV ions required for nuclear fusion [2]. Last and Jortner [78,79] have proposed, and shown by molecular dynamics simulations, that very energetic deuterium or tritium nuclei (D^+ or T^+) can be produced by the multielectron ionization and Coulomb explosion of clusters formed by D_2O and T_2O molecules, similar to water clusters. These clusters will provide substantially higher fusion reaction yields than the homonuclear clusters of the same size.

C. Laser Irradiation of D_3^+

The ground state geometry of D_3^+ is a near equilateral triangle with two sides of length 0.945 Å and one side of length 0.908 Å [64]. The calculated ionization potential of D_3^+ is 35.35 eV. This high value arises from the unscreened attraction of the electrons by the deuterium nuclei. The electronic density exhibits features characteristic of the covalent bonding, with some accumulation of charge between the nuclei. The photoabsorption spectrum shows two dominant features in the UV range. The first is formed by two close excitations at 17.5 and 17.8 eV, forming a double-peak. The average value is in very good agreement with the lowest excitation energy of H_3^+ (17.8 eV) found in the experiments of Wolff et al. [80]. The second feature is a peak at 20.1 eV with a lower absorption strength. Other peaks at higher energies have little strength.

The results of several simulations [64] for varying laser frequency and intensity are now discussed. In all cases, pulses of 9.6 fs are employed to excite the cluster. Two intensities are selected for each frequency, 1.0×10^{12} W cm^{-2} (low intensity) and 1.0×10^{15} W cm^{-2} (high intensity). First, the laser frequency is tuned to the absorption peak at 17.65 eV. In response to a low intensity pulse, the atoms of the trimer oscillate in the plane of the cluster and the motion is close to a breathing mode. On the other hand, when a high intensity pulse is applied, the cluster undergoes a fast dissociation with the characteristics of a Coulomb explosion. This occurs in two steps. First, the system reaches a transient nanoplasma-like state; this is a short lived state, and the two electrons quickly escape. Ionization produces a skeleton of unbalanced positive nuclear charges sufficiently close together to cause the Coulomb explosion of the cluster. When the laser frequency is tuned to resonate with the absorption peak at 20.1 eV, the dynamical response of the cluster to a low intensity pulse is again an oscillatory motion of the atoms in the plane of the trimer, but the amplitude of the oscillations is small compared to the case when the laser was tuned to the peak at 17.65 eV discussed above. For a high intensity pulse, a Coulomb explosion occurs. The same qualitative explanation applies here, but some quantitative differences can be noticed. The velocities of the flying nuclei are 25% smaller compared to the first case, resulting in a slower explosion. This occurs because the initial plasma-like state of the system has a longer life and ionization is slower. A last example corresponds to a nonresonant laser frequency, 5 eV, far from the absorption peaks. Atomic vibrations are noticed in this case for low and high intensities. However, the absorption of energy is not large enough to break the bonds.

The electronic response in the linear domain, which is the case for the low intensity laser field, can be analyzed through the time-dependent dipole moment, shown in Figure 5. For a frequency corresponding to the absorption peak at 17.65 eV, the dipole moment is greatly amplified by resonance with the laser field, and strong dipole oscillations continue long after the laser has been switched off. The high amplification benefits from the fact that both the electrons and the nuclei oscillate in the plane of the nuclei. The behavior for a laser frequency corresponding to the excitation peak at 20.1 eV is different. The dipole moment follows closely the oscillations of the laser field as well as its sinusoidal envelope, but only during the time that the field is acting. A field with this frequency induces oscillations of the electronic cloud perpendicular to the plane of the nuclei. So, the electrons follow closely the excitation field during the 10 fs that the field is acting, although the electronic response is small. The amplitude of the dipole oscillations is much less than in the previous case because the absorption strength of this peak is only 60% of the other, and there is no enhancement due to the nuclear vibrations. The extreme situation is observed in the nonresonant case at 5 eV (right panel of Figure 5), where the dipole moment first follows closely the laser field, returning practically to its initial value when the field is turned off.

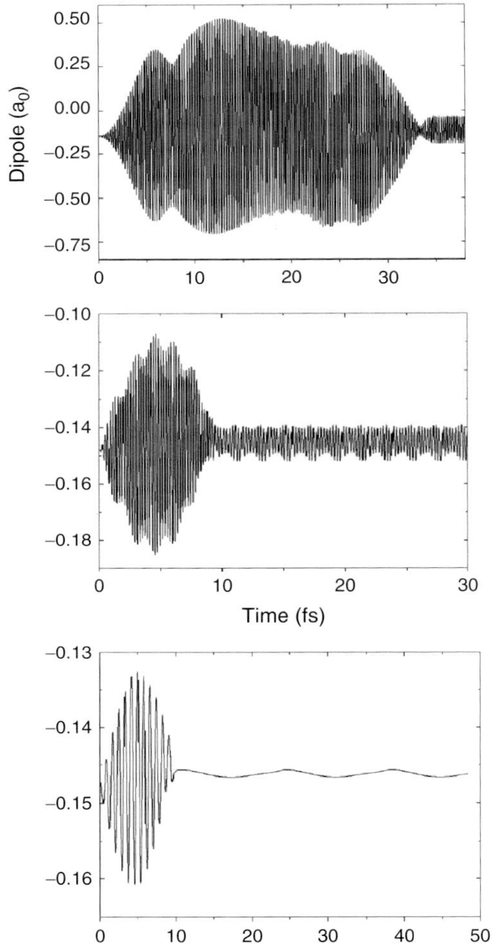

FIGURE 5 Time evolution of the electric dipole moment of D_3^+ in the linear domain (laser intensity = 10^{12} W cm^{-2}). Upper panel: Laser frequency corresponding the excitation peak at 17.65 eV in the absorption spectrum. Central panel: Frequency corresponding the excitation peak at 20.0 eV. Lower panel: Nonresonant frequency, at 5.0 eV. Notice the different scales in the dipole axis. Reproduced from Ref. [64], with permission of the American Chemical Society.

VI. HYDROGEN STORAGE

Hydrogen is a firm candidate to replace gasoline as a fuel in cars. Prototypes of electric cars in which the electric power is generated by the reaction of the stored hydrogen with atmospheric oxygen in a fuel cell have already been built by several car manufacturers. The process is noncontaminant: it only produces water. Hydrogen has a high energy density by mass, 120 MJ/g, nearly three times that of gasoline (44 MJ/g). However, hydrogen is a gas, and consequently its energy density by volume is much smaller than the value of 35 MJ/l of gasoline. Even liquid hydrogen has a volumetric energy density that is only about one fourth of that value. In the existing prototype cars the hydrogen fuel is stored in the tank of the car in the form of compressed gas.

This storage method is not very efficient: it requires a strong and heavy container, and the autonomy of the car is about 200 km, well below the 500 km of the usual gasoline cars. The U.S. Department of Energy (DOE) has established for the year 2010 a hydrogen storage target of 6% of the storage system weight for onboard automotive applications. This has motivated the search for light materials that could store enough hydrogen to fulfill that target. One type of materials that is currently investigated, and which is the focus here, is the class of graphitized carbons. This class comprises different graphitic materials with the common characteristic of having a large specific surface area (of several 100 m^2/g): graphene, carbon nanotubes, porous carbons, activated carbons, etc.

A. The Interaction of Molecular Hydrogen With Graphene

The potential of graphitic materials to store hydrogen is based on the physisorption of molecular hydrogen on the graphitic surface. Yip and coworkers [81] have estimated the optimum adsorption enthalpy for the occurrence of reversible adsorption/desorption under the operating conditions required for automotive applications. The difference in chemical potential $\Delta\mu = \Delta H - T\Delta S$, where ΔH is the change in specific enthalpy and ΔS is the change in specific entropy between the gas and adsorbed phases of hydrogen, controls the direction of adsorption/desorption. One H_2 molecule in the gas at room temperature and 1 atm of pressure has an entropy close to 15.6 k_B [82]. On the other hand, the entropy in the adsorbed phase is much less. Therefore, in order to achieve reversible adsorption/desorption near room temperature, ΔH should be about 10–15 k_BT/molecule, which amounts to a binding enthalpy per H_2 molecule of about 0.30 eV. Another requirement is that ΔH should depend weakly on the hydrogen coverage X. This arises because hydrogen should stay adsorbed to near saturation up to temperatures near 330 K, but should be totally desorbed at $T = 400$ K, which is the upper temperature limit provided by the fuel-cell heat recycling system. This will occur if $\Delta H(X)$ varies from 0.30 to 0.40 eV/molecule when X varies from saturation to zero.

The adsorption of hydrogen in different forms of carbon has been studied by many experimental groups. Dillon and coworkers [83] were the first to study the storage of hydrogen by assemblies of single-wall carbon nanotubes (SWCNTs) and porous activated carbon. They pointed out that the attractive potential of the walls of the pores makes possible a high density storage. Other groups have later performed similar studies in SWCNTs, graphite nanofibers (slitpores), and activated carbons [84–103]. The results are controversial, with reports of high absorption capacity alternating with less optimistic ones.

Some authors have also performed computer simulations of the adsorption of molecular hydrogen inside, outside, and in the interstices of arrays of SWCNTs and in idealized carbon slitpores using model pair potentials to describe the interactions. Wang and Johnson [90,91] used a hybrid path integral-Montecarlo method. They adopted the semiempirical pair potential of Silvera and Goldman [104] for the H_2–H_2 interaction and the H_2–C interaction was modeled by a potential derived by Crowell and Brown [105]. Stan and Cole [106] performed calculations using Lennard-Jones interactions between the hydrogen molecule and the carbon atoms of the nanotubes. Those simulations give useful insight to interpret the results of the experiments. A simple description of the interactions is necessary for massive simulations involving hundreds of H_2 molecules and an assembly of SWCNTs of realistic size, but one can expect more realistic results if the interaction potential is derived from first principles calculations.

The adsorption of H_2 on a planar graphene layer has been studied [107,108] using DFT and the LDA. Figure 6 shows the calculated interaction potential for several positions and orientations of the hydrogen molecule. The equilibrium distances between the center of mass of the molecule and the graphere layer vary between 5.07 and 5.50 a.u., and the binding energies vary between 0.070 and 0.086 eV. The most favorable position for the molecule is above the center of a carbon hexagon, and the configuration with the molecular axis parallel to the layer is slightly more favorable than the perpendicular

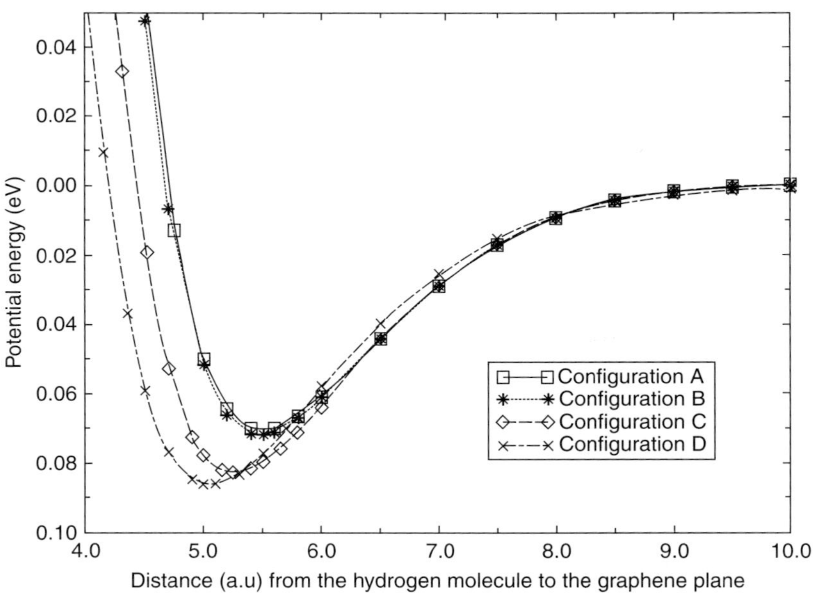

FIGURE 6 Potential energy curves for the interaction of H_2 with a graphene layer. The axis of the molecule is perpendicular (A, B, C) or parallel (D) to the layer. In the perpendicular orientation the molecule is above a carbon atom (A), above the center of a C–C bond (B), and above the center of a hexagon (C). In the parallel orientation (D) the molecule is above the center of a hexagon. Reproduced from Ref. [107], with permission of the American Institute of Physics.

configuration. The zero-point vibrational energy of the molecule in that potential well is about 15 meV. The interaction energy curves of Figure 6 can be interpreted as arising from two main contributions, one attractive and one repulsive. The sharp repulsive wall is due to the short-range repulsion that develops when the close electronic shell of the hydrogen molecule overlaps substantially with the electron gas of the substrate. This contribution is very sensitive to the local electron density sampled by the hydrogen molecule in its approach to the graphene layer and correlates the positions of the different minima with the local values of the electron density of the graphene layer. The attractive contribution is rather similar for all the configurations (notice the similarity of the potential energy curves beyond 5.8 a.u.), and is mainly due to electronic exchange and correlation effects. The exchange energy in the LDA is given by

$$E_x^{\mathrm{LDA}}[\rho] = -C_x \int \rho(\mathbf{r})^{4/3}\, \mathrm{d}^3 r \tag{15}$$

where C_x is a known constant (correlation is neglected in this discussion only for the purposes of simplicity, but it is included in the calculations). This functional is nonlinear in the density. Consequently, in the regime of weakly overlapping densities, there is an attractive contribution between the hydrogen molecule and the graphene layer even in the absence of any rearrangements of the electron density

$$\Delta E_x[\rho] = -C_x \left[\int \left[\rho_{\mathrm{H}_2}(\mathbf{r}) + \rho_g(\mathbf{r}) \right]^{4/3} \mathrm{d}^3 r - \int \rho_{\mathrm{H}_2}(\mathbf{r})^{4/3}\, \mathrm{d}^3 r - \int \rho_g(\mathbf{r})^{4/3}\, \mathrm{d}^3 r \right]. \tag{16}$$

Here, ρ_{H_2} and ρ_g represent the tail densities of the H_2 molecule and the graphene layer, respectively. In addition, there is a density rearrangement (or electronic relaxation) that, although small, also contributes to the binding. The LDA usually overestimates the binding, an effect that in this case serves to compensate for the insufficient account of the van der Waals interaction. The LDA interaction energy curve for the system H_2–graphene is in good agreement with the results of second-order Möller–Plesset (MP2) calculations: Okamoto and Miyamoto [108] obtained a binding energy of 0.086 eV, and Ferre-Vilaplana [109] obtained a value a little lower, 0.064 eV. In contrast, the generalized gradient approximation, GGA, gives a purely repulsive interaction between H_2 and graphene or carbon nanotubes [110]. Girifalco and Hodak [111] have noticed that LDA calculations reproduce well the empirical interaction potentials in graphitic systems for distances near the equilibrium distance, although, of course, the LDA is not able to reproduce the long range dispersion interaction, which arises from nonlocal correlation effects [112].

Figure 6 describes well the essence of the adsorption of molecular hydrogen on the surface of graphitic materials, giving the magnitude of the physisorption energies. It will become clear in what follows that the differences for adsorption on various graphitic materials are just due to the different curvature of the surface, and consequently are due to the amount of available graphitic surface at distances from the hydrogen molecule consistent with the minimum of the potentials of Figure 6.

B. Adsorption of Hydrogen on the Surface of Carbon Nanotubes

The interaction potential between a hydrogen molecule and a (5,5) SWCNT, whose radius is 6.44 a.u., is plotted in Figure 7. The different curves correspond to three configurations of the molecule relative to the nanotube: in its approach to the nanotube, the center of mass of the molecule is upon a C–C bond with the molecular axis parallel to that bond (circles), or upon the center of a hexagon with the molecular axis parallel (crosses) or perpendicular (diamonds) to the hexagon surface [113,114]. Weak physisorption wells are obtained for the molecule inside and outside the nanotube. For adsorption outside, the binding is largest for configurations with the molecule upon the center of a hexagon, $E_b = 0.068$ eV. The energy minimum occurs at equilibrium distances of 5.1 and 5.0 a.u. from the nanotube wall, for the two axis orientations, respectively. These results are consistent with the measured

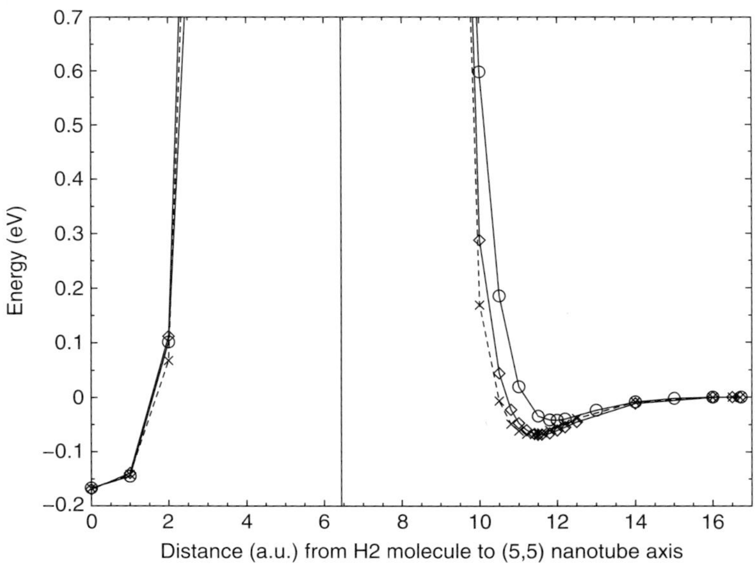

FIGURE 7 Interaction energy of the H_2 molecule and a (5,5) SWCNT. In the approach to the nanotube, the center of mass of the molecule is upon the midpoint of a C–C bond with the molecular axis parallel to that bond (circles) or upon the center of a hexagon with the molecular axis parallel (crosses) or perpendicular (diamonds) to the hexagon surface. Reproduced from Ref. [113], with permission of IEEE.

binding energy of 0.062 eV obtained by Brown et al. [92] by thermally activated H_2 desorption from nanotubes with radii of about 13 a.u. and with the MP2 calculations of Ferre-Vilaplana [115], which predict a binding energy of 0.052 eV for a (5,5) nanotube. The results are also consistent with those for adsorption on planar graphene (equilibrium distance of 5.07 a.u. and binding energy of 0.086 eV). The slightly lower binding energy on the outside of the nanotube is due to the convex surface seen by the molecule. The largest binding energy, 0.17 eV, occurs for the molecule inside the SWCNT, and is due to the curvature effect, a concave surface in this case. The dependence of the binding energy with the radius of the nanotube is small when the H_2 molecule is outside. For a (6,6) SWCNT, with radius $R = 7.78$ a.u., the binding energy for the optimum configuration is $E_b = 0.07$ eV. The binding for the molecule inside is more sensitive to the radius: for the same (6,6) SWCNT, $E_b = 0.12$ eV, and Li and coworkers [81] calculated $E_b = 0.11$ eV for a (7,7) SWCNT. The differences with respect to the narrower (5,5) nanotube give a clear hint of the idea that narrow pores will provide the optimum container for hydrogen in graphitic materials.

Table 1 shows the results of recent calculations of the adsorption energy on the external surface of different carbon nanotubes [116,117]. Results for full external coverage are also indicated. Full coverage means that the outer surface of the nanotubes is covered with one molecule adsorbed on each hexagon. The binding energies per molecule for full coverage of the (5,5) and (6,4) nanotubes, 66 and 68 meV/molecule, respectively, are lower than the binding energies of isolated molecules because for full coverage the distances between some neighbor molecules lie on the repulsive region of the H_2–H_2 potential and the net effect of the interaction between neighbor molecules contributes to decrease a little the binding to the nanotube. The (5,5) nanotube is metallic and the (6,4) and (16,2) nanotubes are semiconducting. The binding energies for single molecule adsorption are, nevertheless, very similar, in spite of early expectations based on the higher polarizability of the metallic nanotubes; then, the electrical character of the nanotube does not affect the single-molecule adsorption energies. A more complex situation occurs for full coverage. The binding energies are again similar for the metallic (5,5) and semiconducting (6,4) nanotubes; however, those for the semiconducting (8,1) and quasimetallic (8,2) nanotubes are larger compared to (5,5) and (6,4). This is a consequence of the different distances between neighbor adsorbed molecules: the distances between a H_2 molecule and its six neighbor molecules in the (8,1) and (8,2) nanotubes lie in the attractive part of the intermolecular potential. Configurations with more than one adsorbed molecule per graphitic hexagon imply reduced intermolecular distances, which give rise to strongly repulsive interactions, and those configurations are not stable.

It has been suggested that physisorption would be more intense on defects in nanotubes, in particular on Stone-Wales defects, and there is a theoretical study of the physisorption and chemisorption of hydrogen on Stone-Wales

TABLE 1 Calculated Binding Energies E_b (meV Per Adsorbed Molecule), and Equilibrium Molecule-Nanotube Wall Distances D_{eq} (in a.u.) for the Physisorption of a H_2 Molecule on the Surface of Hexagonal Carbon Nanotubes and Pentaheptite Carbon Nanotubes (PHNT) Whose Radii (in a.u.) Are Indicated

Nanotube	Radius	Site	E_b	D_{eq}
Single molecule				
(5,5)	6.48	Hexagon	89	4.9
(6,4)	6.52	Hexagon	101	5.3
(8,2)	6.86	Hexagon	88	5.1
(16,2)	12.68	Hexagon	90	5.3
PHNT(8,2)	6.25	Pentagon	78	5.1
PHNT(8,2)	6.25	Heptagon	69	4.7
Full coverage				
(5,5)	6.48	–	66	4.9
(6,4)	6.52	–	68	4.9
(8,1)	6.40	–	98	4.9
(8,2)	6.86	–	96	4.9
PHNT(8,2)	6.25	–	62	5.1

Full coverage means one hydrogen molecule adsorbed on each hexagon (pentagons and heptagons in the PHNT) [116,117].

defects in graphene [118]. Pentaheptite carbon nanotubes (PHNTs) can be formed by rolling up a two-dimensional threefold coordinated carbon network composed of pentagons and heptagons [119], which in turn can be obtained by the creation of Stone-Wales defects on a graphene sheet. These PHNTs are metallic. The optimized (8,2) PHNT ((8,2) in the usual notation of hexagonal carbon nanotubes, and (4,1) if we follow the notation of Crespi et al. [119]) has a slightly buckled surface (the relative difference between the shortest and largest radius is 6%) and is the equivalent of the (8,2) hexagonal nanotube, which has a flat surface. The most favorable adsorption sites are above the centers of the carbon pentagons and heptagons with the molecular axis perpendicular to the nanotube surface. The physisorption energies of a single molecule on the (8,2) PHNT [116], shown in Table 1, are slightly smaller than those on the hexagonal (8,2) nanotube. For full coverage, some intermolecular distances lie on the repulsive part of the intermolecular potential and contribute to a lowering of the binding energy (per molecule) with respect to the case of a single molecule.

The possibility of increasing the storage capacity by the adsorption of a second hydrogen layer around hexagonal nanotubes has also been explored [117]. The second layer is at a different distance from the nanotube wall, in such a way that only the hydrogen molecules of the first layer are in direct contact with the nanotube surface. The binding energies per hydrogen molecule for full covering of the nanotube surface with two hydrogen layers are 20–40% smaller than the binding energies for a single covering layer. This indicates that the outer layer is more weakly bound to the nanotube than the inner layer. Consequently, covering the nanotube with a second hydrogen layer would only be possible at very low temperatures or at high pressures.

C. Molecular Physisorption Versus Atomic Chemisorption

Above we have concentrated on the physisorption of molecular hydrogen. But one could think in adsorbing atomic hydrogen as a way to increase the binding energy. Hydrogen is a molecular gas and dissociating the molecule into its two constituent hydrogen atoms costs an energy of 4.8 eV. Dissociating the molecule adsorbed on the surface of a nanotube is less difficult but there is still a sizable dissociation barrier. The most stable chemisorbed state on a (6,6) PHNT is achieved for the two atoms of the dissociated molecule attached, respectively, to the carbon atoms of a pair between a pentagon and a heptagon. The binding energy of each H atom is 2.6 eV, and this dissociated chemisorption state is, in fact, 150 meV more stable than the lowest molecular physisorption state. Although some dissociated chemisorption states are slightly more stable than all the physisorbed states on a pentaheptite nanotube, an initially physisorbed hydrogen molecule must overcome a substantial energy barrier to reach the dissociated state with two chemisorbed atoms. The calculated barrier is about 2.4 eV. Duplock et al. [118] have studied physisorption and chemisorption on a Stone-Wales defect in a graphene sheet modulated to produce armchair-like curvature, and they also have found that the difference between the total energies of physisorbed and chemisorbed states is very small, about 150 meV, although the physisorbed state was more stable in that case.

For comparison, the molecular physisorption energy on the external surface of a hexagonal (6,6) nanotube is 0.081 eV/molecule [116], very similar to the physisorption energy on a pentagon of the (6,6) PHNT. The binding energy per H atom in the dissociated chemisorbed state on the hexagonal nanotube is 2.25 eV. However, the most stable state of the system is molecular physisorption, which is 0.48 eV more stable than the dissociative chemisorbed state. The energy barrier separating molecular physisorption from dissociative chemisorption in this hexagonal carbon nanotube is 1.8 eV. Other estimations [113] for (5,5) and (6,6) carbon nanotubes give dissociation barriers of 2.5–2.6 eV. The result that the chemisorption energy per H atom is larger on pentaheptites (2.6 eV) than on hexagonal nanotubes

(2.25 eV) is in agreement with the findings of Duplock et al. [118], who also obtained larger binding energies on a Stone-Wales defect on a graphene sheet with curvature than on a regular flat graphene sheet.

D. Adsorption of Hydrogen on Boron Layers and Nanotubes

Many layered materials have demonstrated their ability to form nanotubular structures. These nanostructures could be potential candidates to store hydrogen because of their large surface area. Among the light-weight elements, besides C, boron would be a possible candidate. Although the natural phases of bulk boron do not show laminar structures, the possibility of forming boron nanotubes, BNTs, was predicted by Boustani [120,121]. The prediction was confirmed by Ciuparu et al. [122], who reported the synthesis of single-walled BNTs with a radius of approximately 34 a.u. Boron is lighter than carbon, and hence the adsorbed hydrogen weight percent could be larger than on graphene and carbon nanotubes. Moreover, the atomic polarizability of boron is larger than that of carbon, which could increase the physisorption energies. On the other hand, the van der Waals radius of boron is about 0.43 a.u. larger than that of carbon, and this could decrease the strength of the van der Waals interactions.

A boron sheet consists of a periodic triangular network of B atoms, as shown in Figure 8. Optimization of the geometry leads to a buckled structure formed by parallel alternating up and down (or hill and valley) rows of B atoms. Boron nanotubes can be constructed [123] by rolling up a boron sheet. Flat and buckled structures of boron sheets and nanotubes have been studied [124]. The sheets and nanotubes with a buckled surface are about 0.20 eV more stable, per boron atom, than the corresponding flat structures. The helicity of some B nanotubes does not allow for buckling and those nanotubes exhibit flat surfaces. Flat and buckled boron sheets and nanotubes have metallic character.

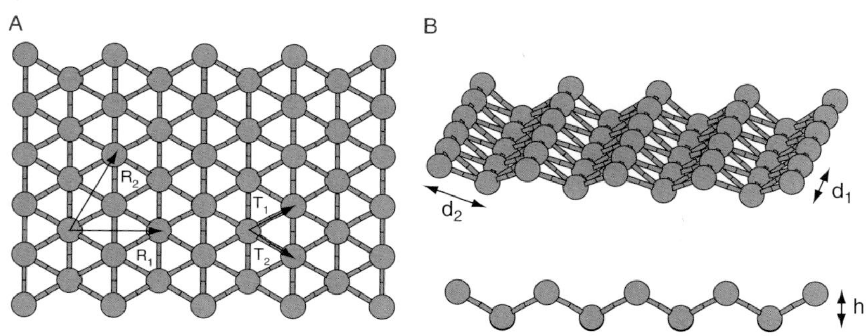

FIGURE 8 (A) Triangular network of B atoms. The primitive vectors of the triangular lattice, T_1 and T_2, and of the graphene unit cell, R_1 and R_2, are shown. (B) Optimized buckled boron sheet.

For full coverage of the external surface of carbon nanotubes, achieved by placing a H_2 molecule per hexagon of the C network, the hydrogen content would be 7.6 wt%. For a similar coverage of one molecule per hexagon in B sheets and BNTs, the hydrogen content is 5.8 wt%. Nevertheless, because of the buckling and of the larger distance between neighboring B atoms in B sheets and BNTs, the distance between physisorbed molecules would be larger than in SWCNTs, with the consequent reduction in the repulsive interactions between molecules. The calculated physisorption energies on BNTs are in the range of 30–60 meV/molecule, depending on the site and orientation of the molecule with respect to the surface. The most stable adsorption sites are on top of a B atom in a valley and on top of a B–B bond in a valley. There are no energy barriers between the hill and the valley sites and, therefore, hydrogen will tend to adsorb on the valley sites. Despite the larger atomic polarizability of the B atom compared to the C atom, hydrogen physisorption energies in B sheets and BNTs are smaller than the corresponding energies (of 70–100 meV/molecule) in graphene and carbon nanotubes [81,107–109,114–116,125–127]. Consistent with these smaller physisorption energies, the equilibrium molecule-surface distances in the boron nanostructures, 5.7–7.0 a.u., are larger than the distances, 4.7–5.3 a.u., on graphene and SWCNTs.

In the chemisorbed state, the molecule is broken and each H atom is strongly attached to a B atom of a B–B bond. Calculations for a (4,4) boron nanotube show that chemisorption on a hill is more stable than chemisorption on a valley. In fact, chemisorption on a hill is even more stable, by 0.76 eV/molecule, than the most stable physisorption configuration (on a valley). Both, molecular physisorption and dissociative chemisorption (in the form of B–H and bridge B–H–B chemical bonds) have been observed in amorphous boron treated by mechanical milling under hydrogen atmosphere [128]. The small fraction of physisorbed hydrogen desorbs at room temperature. However, the two types of chemically bonded hydrogen require higher temperatures to desorb: temperatures of 500 and 720 K are needed to break the B–H and B–H–B bonds, respectively. Turning to the calculations, the energy barriers between the molecular physisorption and the dissociative chemisorption states are quite close to 1 eV. These barriers are two to three times smaller than those found in carbon nanotubes; however, they are still large.

E. Enhancement of the Hydrogen Physisorption Energy in Nanopores

From the previous sections, it becomes clear that the energy of physisorption of molecular hydrogen to the surface of graphitic nanostructures, a graphene layer, or the external surface of a carbon nanotube is close to 100 meV/molecule. This binding energy falls short of the value of 300–400 meV per H_2

molecule required for reversible adsorption/desorption near room temperature and normal pressures. The same conclusion applies to boron or BN nanostructures [129]. This explains why the measured hydrogen storage capacities of carbon nanotubes at room temperature are less than 1% [98]. A strategy for increasing the physisorption energy consists in allowing the hydrogen molecule to interact with two surfaces at the same time, like for a molecule between two closely spaced graphene surfaces, or with a concave surface. This can be achieved inside nanocavities and nanopores [85,101–103]. A graphene slitpore, consisting of two parallel graphene layers separated by a certain distance *d*, gives a simple model for the pores existing in nanoporous carbon materials. The minimum of the interaction potential energy in Figure 6 occurs at 5 a.u., so in a slitpore of width $d = 10$ a.u. the hydrogen molecule would interact optimally with the two parallel graphene surfaces, increasing the binding energy by a factor of two. That is, the binding energy in the slitpore of that size will be about 200 meV. The measured adsorption properties of carbon nanoporous materials, like activated carbons, are usually analyzed by assuming the pores in the material to be slitpores. Another model of a nanopore is the inner channel of a nanotube. For (5,5) and (6,6) carbon nanotubes, the binding energies of an endohedral hydrogen molecule are 0.17 and 0.12 eV, respectively (see Figure 7). The value for the (5,5) nanotube more than doubles the binding energy in the exohedral position obtained in a similar calculation, and the reason is that the diameter of the nanotube, 12.9 a.u., optimizes the interaction between the hydrogen molecule and the graphitic walls (the most stable position occurs for the molecule at the center of the inner channel).

Patchkovskii and coworkers [130] have applied a thermodynamical model to study the hydrogen storage in graphene slitpores. The model takes into account the quantum behavior of the hydrogen molecules confined in the volume of the slitpore. In this way, they have made predictions for the storage capacities of carbon nanoporous materials. Cabria et al. [129] have recently improved that model, arriving at the conclusion that the new model reproduces well the hydrogen storage properties measured for different samples of porous carbons, at low (77 K) and room temperature assuming slitpore widths of about 9.6 a.u. The comparison between the results of the model for storage at 298 K as a function of the external pressure and the experimental measurements of Jorda–Beneyto for activated carbons [103] are given in Figure 9. The experimental results for two selected samples, called KUA1 and KUA5 in the original paper [103], are fitted very well by applying the model with slitpores of widths 9.45 and 9.64 a.u. According to the calculations, the DOE goal for the gravimetric storage capacity appears to be accessible with nanoporous materials at temperatures of 77 K. Of course, those temperatures prevent applications in the car industry, although other applications which require fuel cells operating at low temperatures may be more accessible. This might be the case of satellites in space.

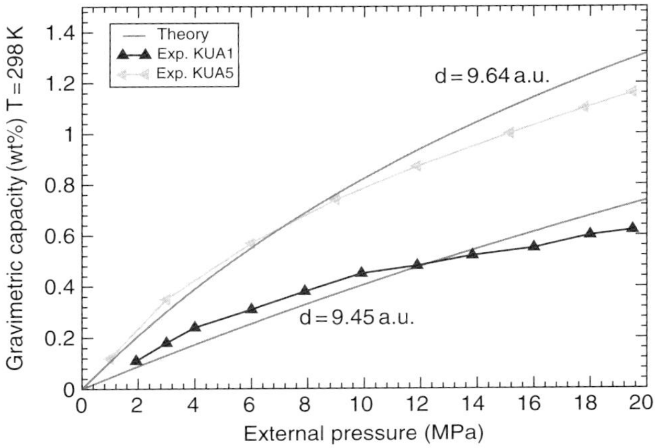

FIGURE 9 Experimental and calculated gravimetric capacities at 298 *K* as a function of the external pressure. The measurements correspond to two samples, KUA1 and KUA5, in the experiments of Jorda–Beneyto [103]. The calculations [129] are for slitpores of widths 9.45 and 9.63 a.u.

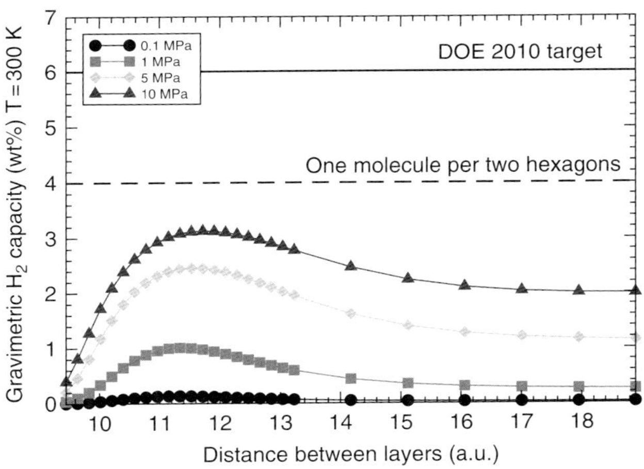

FIGURE 10 Gravimetric capacities of graphene slitpores as a function of the pore width at 300 K, and different pressures: 0.1, 1, 5, and 10 MPa. The DOE target is indicated by the continuous horizontal line. The dashed horizontal line represents coverage of one H_2 molecule per two hexagons.

At room temperature, and for optimized slitpore sizes near $d = 11.3$ a.u., a gravimetric storage capacity of 3.1% is predicted at pressures of 10 MPa (see Figure 10). The trend in the Figure indicates that the storage will increase at higher pressures. Although high pressures may appear as a technological limitation, some car companies have presented prototype cars using hydrogen gas compressed in tanks at 70 MPa.

F. Enhancement of Hydrogen Physisorption Energy by Doping

Doping with appropriate impurities represents another strategy to enhance the binding energies of molecular hydrogen to graphitized surfaces. Chen et al. [131] reported hydrogen storage capacities of 20 wt% on Li-doped nanotubes at 1 atm and temperatures of 473–673 K, 14 wt% on K-doped nanotubes at near room temperature, and capacities of 14 and 5 wt% on Li- and K-doped graphite, respectively. Under similar conditions they obtained a hydrogen storage capacity of 0.4 wt% on clean nanotubes. However, these result are controversial. Other authors made the observation that the increase in the weight of the stored material was likely due to water impurities present in the experiments [132,133]. Nevertheless, Yang [132] pointed out that in dry hydrogen, alkali-doped nanotubes can adsorb about 2 wt% hydrogen, that is, more than pure nanotubes.

Some theoretical studies of hydrogen adsorption on alkali-doped nanotubes and graphene have been published. Dubot and Cenedese [134] performed semiempirical Hartree–Fock calculations of the physisorption of molecular hydrogen on Li-doped nanotubes. Froudakis [125,135] did first-principles calculations of H_2 physisorption on K-doped (5,5) SWCNTs, and obtained a binding energy of 148 meV/molecule. Lee et al. [136] studied the effect of Li doping on the barrier for dissociative chemisorption on SWCNTs. Zhu et al. [137] did first principles calculations of hydrogen physisorption on Li- and K-doped graphene. Maresca et al. [138] have studied the physisorption of the hydrogen molecule intercalated in the region between two parallel coronene molecules with and without Li atoms; those Li atoms were arranged forming a distribution similar to that in the LiC_6 solid intercalated compound. Molecular-dynamics simulations of the interaction of H_2 with pillared graphene sheets and nanotubes, both intercalated with Li, have also been performed [139]. Cabria et al. [140] and Cho and Park [141] have performed extensive calculations explicitly comparing the physisorption of H_2 on pure and Li-doped nanotubes and graphene using DFT with the LDA and GGA approximations, respectively.

The Li atom can be adsorbed on three different sites on a graphene layer: above the center of a carbon hexagon, on top of a carbon atom, and above a C–C bond. The most stable position is above the center of the hexagon. The Li atom transfers some electronic charge (0.5–0.6 e) to the graphene layer [140,142,143]. That charge concentrates in the region between the Li atom and the graphene layer, especially above the perimeter of the carbon hexagon. The adsorption energies of the Li atom in DFT calculations using the LDA and GGA are 0.88 and 0.52 eV, respectively, and the corresponding equilibrium distances between the Li atom and the graphene layer are 3.46 and 3.52 a.u. The binding energies are a few tenths of an eV larger in other calculations [134,143].

The results for the physisorption of molecular hydrogen on a graphene layer with a Li impurity adsorbed above the center of a carbon hexagon [140] are compared in Table 2. The physisorption energy is given by

$$E_b = E(\text{graphene} + \text{Li}) + E(\text{H}_2) - E(\text{graphene} + \text{Li} + \text{H}_2), \qquad (17)$$

The results for the physisorption on pure graphene are given for comparison. The relevant configurations have the Li atom and the hydrogen molecule on the same side of the graphene layer (the effect on the binding energy is very small when Li and H_2 are on opposite sides of the layer). Those configurations with the molecular axis parallel to the graphene layer are shown in Figure 11: the H_2 molecule on top of the Li atom (configuration C), above a C–C bond of the hexagon hosting the Li atom (configuration B), above a carbon atom (configuration A), and above the center of a neighbor hexagon (configuration D). Table 2 also includes results for similar configurations with the molecular axis perpendicular to the layer. The physisorption energies on doped graphene differ substantially from the physisorption energies for analogous configurations on undoped graphene. This can also be appreciated in the plot of the interaction energies in Figure 12 for configuration D with the

TABLE 2 Binding Energies E_b (meV/molecule), Equilibrium Molecule–Graphene Plane Distances D_{eq} (a.u.), Equilibrium Molecule–Li Atom Distances D_{Li-H_2} (a.u.) and Mulliken Charges on the Li Impurity for the Physisorption of a H_2 Molecule on a Graphene Layer Doped with a Li Impurity [140]

	Undoped			Doped				
Orientation	Site	E_b	D_{eq}	Site	E_b	D_{eq}	D_{Li-H_2}	$Q_{Li}(e)$
Perpendicular	C	89	5.3	C	NB	NB	NB	+0.36
Perpendicular	B	82	5.5	B	51	6.8	4.1	+0.40
Perpendicular	A	83	5.5	A	95	6.2	3.9	+0.43
Perpendicular	C	89	5.3	D	184	4.9	4.9	+0.51
Parallel	C	92	5.1	C	160	7.2	3.8	+0.33
Parallel	B	72	5.5	B	178	6.2	3.7	+0.37
Parallel	A	69	5.5	A	145	6.2	3.9	+0.38
Parallel	C	92	5.1	D	174	4.7	4.8	+0.50

The perpendicular or parallel orientation of the molecular axis with respect to the graphene layer is indicated. The molecule is on top of the Li atom (configuration C), above a carbon–carbon bond (configuration B), on top of a carbon atom (configuration A), and above the center of a neighbor hexagon (configuration D). Similar notation is used for adsorption on a clean graphene layer. NB means not bound.

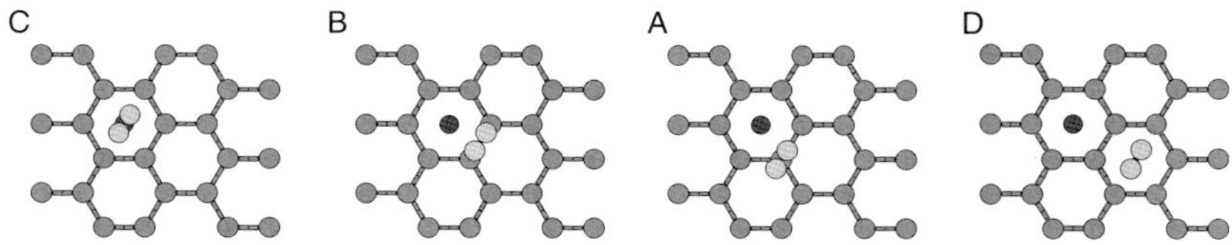

FIGURE 11 Selected physisorption configurations for a hydrogen molecule on a Li-doped graphene layer. From left to right: on top of the Li impurity (configuration C), above a near carbon–carbon bond (configuration B), on top of a carbon atom (configuration A) and above the center of a neighbor hexagon (configuration D). The parallel orientation of the hydrogen molecule is shown in all these configurations.

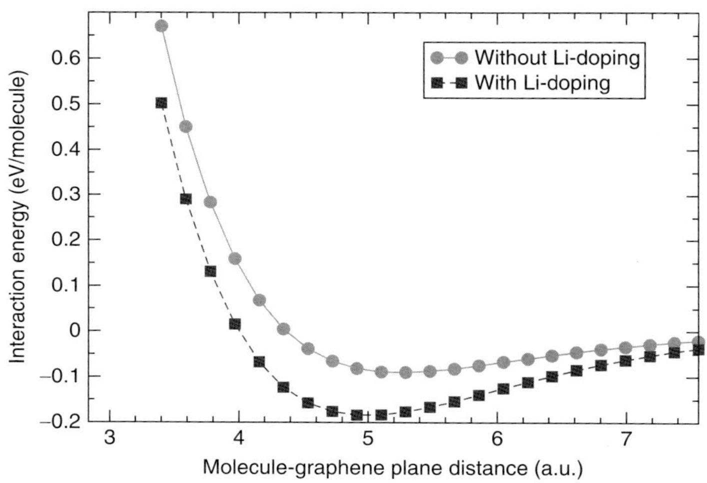

FIGURE 12 Interaction energy of a hydrogen molecule with a clean (circles) and with a Li-doped graphene layer (squares). The Li impurity is adsorbed above the center of a carbon hexagon and the hydrogen molecule is adsorbed above a neighbor carbon hexagon, with the molecular axis perpendicular to the graphene plane.

molecular axis perpendicular to the layer. This indicates that the presence of the nearby Li atom, which allows for the direct interaction between Li and H_2, is crucial to enhance the physisorption energy of the hydrogen molecule. The trend when the molecular axis is perpendicular to the surface is an increase of the binding energies—unbound, 51, 95, and 184 meV/molecule—as the distance *d* between the Li atom and the H_2 molecule, projected on a plane parallel to the layer, increases: $d=0$ for the molecule on top the Li impurity (configuration C), $d=2.3$ a.u. for the molecule above a C–C bond (configuration B), $d=3.0$ a.u. for the molecule above a carbon atom (configuration A), and $d=4.7$ a.u. for the molecule above a neighbor hexagon (configuration D). Consistent with this trend, the separation D_{eq} between the molecule and the graphene layer decreases when *d* increases: D_{eq} = unbound,

6.8, 6.2, and 4.9 a.u. for configurations C, B, A, and D, respectively. The true Li-molecule equilibrium distances D_{Li-H_2} are rather similar in configurations A and B, 3.91 and 4.12 a.u., respectively. These facts are explained by an excluding volume built around the Li atom when the H_2 molecule is nearby. This does not apply to configuration D, as the molecule in a neighbor hexagon is already outside the Li excluding volume.

When the molecular axis is parallel to the surface, the physisorption energies increase by a factor of 2 with respect to the undoped case for all configurations. For this orientation of the axis, the molecule–graphene distance D_{eq} decreases when the distance between the physisorption site and the impurity, projected in the layer plane, increases. The values of D_{eq} are 7.2, 6.2, 6.2, and 4.7 a.u. for configurations C, B, A, and D, respectively. This behavior is the same as for the molecular axis perpendicular to the layer. The equilibrium Li–H_2 distances have similar values when the molecule is near the impurity: 3.78, 3.66, and 3.91 a.u., for configurations C, B, and A. These facts suggest again an exclusion–volume effect. However, this time the binding energies are not affected. The enhancement of the H_2 physisorption energies is due to the charge transfer from the Li atom to the graphene layer. The charge transferred from a Li atom to the graphene layer is approximately 0.4–0.5 e, and this charge remains localized near the Li atom [143]. When the H_2 molecule is not too close to the Li atom, that is, when the molecule is on a neighbor hexagon, the charge transferred to the graphene layer increases the binding energy by a factor of 2 with respect to adsorption on pure graphene for both the perpendicular and the parallel orientations. On the other hand, if the molecule is close to the Li atom, the volume excluded by the impurity leads to a Li–H_2 distance of 3.8—4.2 a.u. and this effect tends to reduce the physisorption energy. Nevertheless, if the molecular axis is parallel to the surface, the molecule is still able to profit from the local enhancement of the electron density of the graphene layer and this causes a bonding contribution that overcompensates the effect due to the exclusion volume of the impurity.

The GGA calculations of Cho and Park [141] give support to the LDA calculations of Cabria et al. [140]. The calculated GGA binding energy of H_2 on a clean graphene layer is 78 meV, and the binding energy on a clean (10,0) SWNT is 95 meV, both on the external and the internal surfaces (in all cases the axis of the molecule is perpendicular to the surface). These binding energies increase when the nanotube is doped with lithium: 122 and 197 meV for H_2 adsorption far from the Li impurity and near the impurity, respectively.

Lee et al. [136] have argued that the reported higher hydrogen adsorption on alkali-doped graphite and nanotubes might be due to the occurrence of atomic chemisorption. However, the theoretical simulations show that the hydrogen molecule is stable in the presence of the Li impurity and does not dissociate [137,138,140].

The most stable position of a lithium impurity adsorbed on the external surface of a (4.4) carbon nanotube is above the center of a hexagon [140].

The distance to the nanotube surface is 3.31 a.u., similar to the distance in graphene, and the adsorption energy of a Li atom is 1.09 eV. Dubot and Cenedese [134] also found this position as the most stable site on an (8,0) nanotube, and obtained a distance of 3.97 a.u. Lee et al. [136] obtained an equilibrium distance of 3.38 a.u. When the Li atom is adsorbed inside the nanotube, the Li adsorption energy is about 0.5 eV. The physisorption energies and equilibrium distances of a hydrogen molecule on the external surface of clean and Li-doped (4,4) nanotubes are similar to those obtained for adsorption on clean and doped graphene (actually, the binding energies are just a little lower because of the curvature of the nanotube) [140]. Again, an enhancement of the physisorption energies with Li doping is found. On the basis of semiempirical calculations, Dubot and Cenedese [134] also predicted that the physisorption of molecular hydrogen on carbon nanotubes becomes stabilized by lithium doping. Froudakis [125] used a mixed quantum-mechanics/molecular-mechanics model (in this approach a small part of the system is treated quantum mechanically while the rest is treated by simple molecular mechanics) to study the adsorption of molecular hydrogen on the external surface of a K-doped (5,5) carbon nanotube, and obtained a binding energy of 148 meV/molecule. From a Mulliken population analysis, Froudakis found a charge transfer of 0.6 *e* from the potassium atom to the nanotube and suggested that the increased hydrogen storage observed in the experiments was due to a dipole interaction between the polarized molecule and the alkali atom. The enhanced hydrogen storage capacity of Li-doped carbon nanotubes reported by Chen et al. [131] was initially assigned to the dissociation of the hydrogen molecules in the presence of Li. However, Yang [132] found that the infrared (IR) spectrum of LiOH–H_2O was very similar to the IR spectrum obtained by Chen et al. [131]. From this and other facts, Yang concluded that there was moisture in the experiments of Chen et al., because the reaction of Li with moisture could produce alkali hydroxides and additional hydrates. Hence, the increased hydrogen storage was only apparent and the increase in weight actually came from the hydroxides and hydrates. Besides, Lee et al. [136] found in their first-principles calculations that the energy barrier for the dissociation of the hydrogen molecule was lowered by about 0.3—0.5 eV on Li-doped nanotubes, although the barrier was still too high, about 2.5—2.7 eV. Calculations by Cabria et al. [140], allowing for the dissociative chemisorption of molecular hydrogen adsorbed on a Li-doped nanotube, did not reveal any tendency for molecular dissociation, in agreement with the results of Lee et al. [136].

VII. HYDROGEN INTERACTION WITH GOLD CLUSTERS

Gold has been usually regarded as an inert element, with scarce application in the field of catalysis because of its noble character when it is in bulk form. However, the discovery of high catalytic activity for oxide-supported small

Au nanoparticles has opened a whole new research area. A number of chemical reactions have been identifed for which Au nanoparticles are active catalysts (low temperature CO oxidation, propene epoxidation, water-gas-shift reaction, and so on), and consequently a large effort is nowadays dedicated to the study of the influence of size, shape, support, and chemical state of gold nanoparticles on their reactivity. One of the possible applications for these new catalysts is as active elements in fuel cells, catalyzing the reaction between hydrogen and oxygen to produce electric power and water as residual element. The special features of these catalysts, namely their high selectivity and the ability of running reactions at relatively low temperatures, make them possible candidates for this task. Therefore, it is interesting to evaluate the intrinsic features of the interaction of gold nanoparticles with hydrogen.

Parallel to the research on supported nanoparticles, the study of the catalytic properties of small Au_n clusters attracts an increasing interest; as the nanoparticles, the clusters are chemically active, and it is easier to control fundamental parameters such as their size or the charge state. Besides, clusters show remarkable effects, the most important being the occurrence of planar structures up to relatively large sizes, $n=12$ for anions [144] and $n=7$ for cations [145]. Such preference for planar structures has been interpreted as being due to relativistic effects that change the chemistry of gold clusters as compared to copper or silver clusters: the 6s and 5d states are unusually close in energy in gold, and their hybridization enhances the tendency for planarity in the small clusters [146].

In this section, we review the results from a DFT study of the interaction between atomic hydrogen and small neutral Au clusters [11]. A GGA functional was used for exchange–correlation effects. The intention is twofold: first, using hydrogen as a chemical probe of the intrinsic reactivity of each cluster, insight can be gained on the local site variation of the reactivity; second, a detailed analysis of the electronic structure upon hydrogen adsorption gives information on the character of the H–Au chemical interaction. The results show that the Au–H bonds are highly directional, strong around the perimeter of the planar cluster and much weaker for H on top of the cluster plane. Such directionality is an intrinsic property of the small gold clusters, and can be generalized to other types of adsorbates and to the interaction of the cluster with an oxide surface (upright conformations of the cluster relative to the surface compete with parallel ones, and are often more stable) [11,147].

The binding energies of atomic hydrogen on different sites of the cluster are shown in Figure 13 for planar Au_n clusters with $n=4$–10. The highest binding energies, highlighted with a circle, occur at sites on the perimeter (rim) of the cluster plane. In contrast, adsorption in the perpendicular direction, above the cluster plane, is weaker by several tenths of an eV (Au_8 is an exceptional case, as the large hole at its center allows for the insertion of the hydrogen atom; such hole can be considered as part of the rim of the cluster). Such differences indicate a strong directionality of the H-cluster bonding.

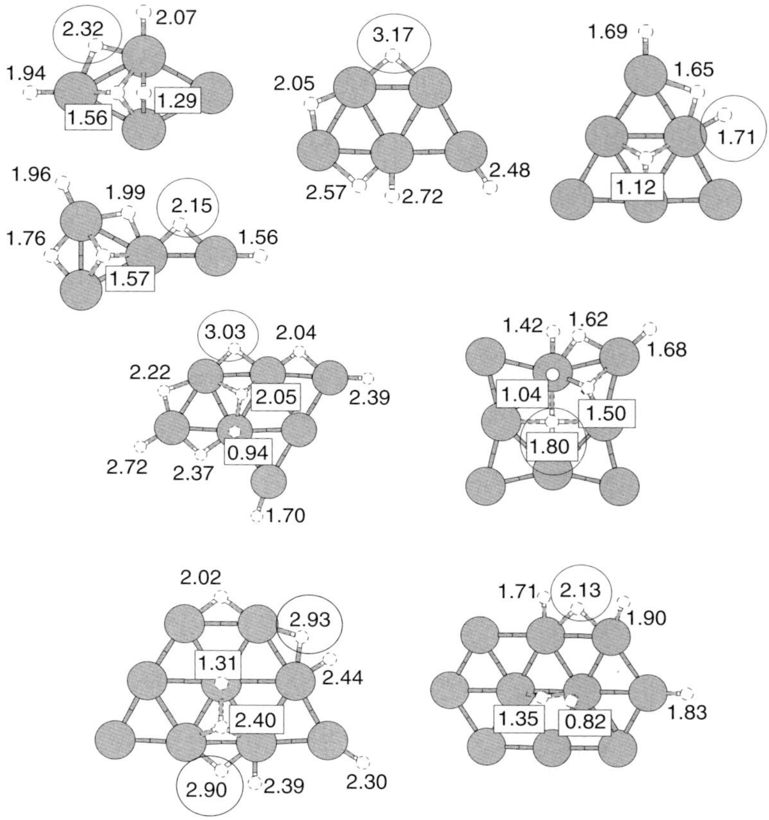

FIGURE 13 Lowest-energy isomers of Au_n clusters ($n = 4$–10), and binding energies (in eV) of a single hydrogen atom at different sites of the cluster. A circle highlights the most stable adsorption sites. A second, low-lying isomer, is included for Au_4. Reproduced from Ref. [11], with permission of the American Chemical Society.

Besides, there is a preference for Au–H–Au bridging configurations; this suggests that H tries to maximize coordination, in order to obtain a better overlap between the 1s orbitals of H and the 5d orbitals of Au. The H atom then probes the local site reactivity of the gold clusters.

To analyze the features of the H–Au bonding, it is useful to make a comparison with results for H adsorption on Al_{13} clusters [148]. The electronic structure of the H atom adsorbed on Al_{13} is analogous to that of a H^- anion embedded into the electron gas of the host, and preferentially binds at regions with the highest values of the Fukui functions (these reactivity indexes serve to characterize the uptake/donation of electronic charge from/to the metallic cluster; see discussion below). The upper part of Figure 14 shows a comparison between the densities of states (DOS) of Au_{10} and HAu_{10} (with H in the

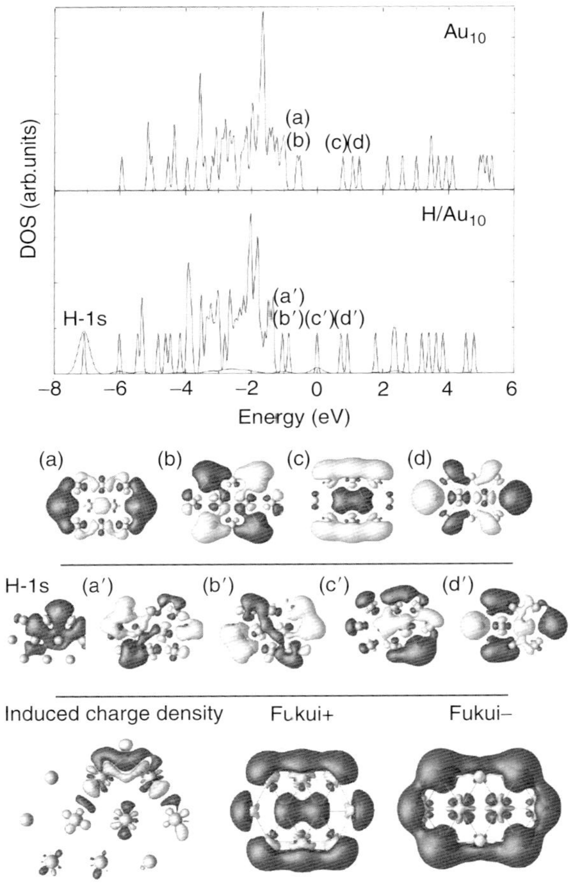

FIGURE 14 Electronic structure analysis of binding of H to Au_{10}. Upper part: densities of electronic states (DOS) of Au_{10} and HAu_{10}. Middle part: plots of the real part of selected Kohn–Sham orbitals (those states are labeled in the DOS plots). States b and c are the HOMO and LUMO of Au_{10}, respectively. State c′ is the HOMO of HAu_{10}. Lower part: induced charge density and Fukui functions of Au_{10}. Values of ± 0.01 eV/A^3 are chosen for the isosurfaces of the induced density, and values of ± 0.003 eV/A^3 are used in the case of the Fukui functions. In the middle and lower panels, the two different colors indicate positive and negative values of the corresponding functions. Reproduced from Ref. [11], with permission of the American Chemical Society.

most stable adsorption configuration). Au_{10} has a closed-shell electronic structure with an energy gap of 1.3 eV between the highest occupied and the lowest unoccupied molecular orbitals (HOMO–LUMO gap). The adsorption of hydrogen results in the formation of bonding and antibonding combinations of H-1s and Au-5d orbitals, leading to a strongly bound, bonding combination at -7 eV, relative to the Fermi energy. This doubly occupied state has a strong H-1s character (see the insets in Figure 14, where plots of the real part of some electronic orbitals are shown). Therefore, the chemistry of the binding between H and the gold clusters closely resembles the picture found in H–Al_{13}. The analysis of the electronic structure indicates that H interacts with one of the states in the middle of the Au-5d band. That state is transformed into the strongly bound H 1s-like state. Two electrons fill that state and the third electron is transferred to the frontier orbitals of the cluster. A more simplified interpretation, focussing on the transfer of the electron of the H atom to the "valence pool" of the cluster, has been proposed by Buckart et al. [149]; as electrons are indistinguishable, the two interpretations are consistent with each other.

In the case of the HAu_{10} cluster, the extra electron is donated to the LUMO of Au_{10} (labeled c in Figure 14), which becomes distorted by the influence of the proton potential in HAu_{10} (orbital c′). Both the HOMO and LUMO orbitals of Au_{10} are localized around the cluster's rim, and the preference for binding sites around the rim can be attributed to the localization of the frontier orbitals in this region. Additional insight is obtained by plotting the Fukui functions, which give information of the local reactivity around the cluster. For a system with N electrons, two Fukui functions can be defined as

$$f^{+}(\mathbf{r}) = n_{N+1}(\mathbf{r}) - n_{N}(\mathbf{r}) \tag{18}$$

$$f^{-}(\mathbf{r}) = n_{N}(\mathbf{r}) - n_{N-1}(\mathbf{r}) \tag{19}$$

where n_N, n_{N+1}, and n_{N-1} represent the ground state densities of the cluster with N, $N+1$, and $N-1$ electrons, respectively. These functions provide a measure of the change in the chemical potential of the system with a variation of the number of electrons, and this gives an accurate estimation of the nucleophilic ($f^{+}(\mathbf{r})$) or electrophilic ($f^{-}(\mathbf{r})$) character: regions where $f^{+}(\mathbf{r})$ is large will stabilize an uptake of electronic charge, whereas regions with high values of $f^{-}(\mathbf{r})$ will stabilize the donation of electronic charge. Assuming a small change in the shape of the Kohn–Sham orbitals upon addition or removal of charge, the Fukui funcions can be approximated by the electron densities of the LUMO and HOMO orbitals, respectively:

$$f^{+}(\mathbf{r}) \approx n_{\mathrm{LUMO}}(\mathbf{r}); \quad f^{-}(\mathbf{r}) \approx n_{\mathrm{HOMO}}(\mathbf{r})$$

In fact, Figure 14 shows that $f^{+}(\mathbf{r})$ of Au_{10} is very similar to the density of the LUMO orbital (the square of the wave function), and also that $f^{-}(\mathbf{r})$ looks like a superposition of the densities of the nearly degenerated a and b (HOMO) orbitals. As both the positive and negative Fukui functions are concentrated on the cluster´s rim, it can be concluded that this is the most reactive region, stabilizing both the uptake and donation of charge. This is a general property of the family of planar gold clusters [11].

The strength of the binding of H to the gold cluster is highly sensitive to the number of atoms in the cluster. Observation of Figure 13 gives evidence of a marked odd–even oscillation in the optimal binding energy, which is higher for odd-numbered clusters compared to neighboring even-numbered ones. This feature can be understood by comparing the electronic structure of two prototypical cases, namely HAu_{10} and HAu_9, the DOS of the first one being shown in Figure 14. As discussed above, when when H interacts with Au_{10}, the extra electron is transferred to the LUMO of the cluster. In contrast, Au_9 is a cluster with a half-filled HOMO, and the filling of the HOMO by the extra electron results in a closed-shell electronic configuration for HAu_9, which results in extra stabilization as compared to HAu_{10}. This explanation is valid for all the clusters in the series, as sizable HOMO–LUMO energy gaps are opened for HAu_n with $n=5$, 7, and 9, whereas open-shell electronic configurations result for $n=4$, 6, 8, and 10. The calculated binding energy per atom of the isolated H_2 molecule is 2.3 eV, so it is clear that the dissociative chemisorption of H_2 on neutral gold clusters will only be thermodinamically feasible for odd-numbered Au_n clusters (of course, one has to take barriers into account).

In conclusion, adsorption of atomic hydrogen is capable of probing the local reactivity of small planar gold clusters: directional bonding preferentially occurs along the cluster rim, and a pronounced inertness is observed perpendicular to the cluster plane. The directionality of the bonds formed by the planar gold clusters is general, and can be used to interpret other types of unusual chemical behavior of the same clusters, such as the preferential upright conformations on oxide surfaces [11,147], or the recent discovery of high stabilities for cage-like gold clusters [150,151].

VIII. SUMMARY

The interdisciplinarity of the studies of hydrogen has been illustrated by reviewing recent work on hydrogen and hydrogen clusters. Hydrogen clusters present interesting characteristics derived from their molecular character. The Coulomb explosion of large deuterium clusters has been induced by laser irradiation of molecular beams, and nuclear fusion of the colliding deuterium nuclei, with production of neutrons, has been observed in those experiments. The first stage of those experiments, that is, the dynamical response of the deuterium clusters to the laser irradiation, has been studied by computer simulation. The dissociation paths, fragmentation or ionization, followed by Coulomb explosion are sensitive to the cluster size and to the frequency and intensity of the laser. An efficient storage of hydrogen in light materials is a crucial requirement to make hydrogen the preferred fuel in the cars of the future. Nanoporous carbon materials doped with impurities like lithium appear as promising candidates in this task. Atomic hydrogen is a reactive species that can be used to probe the local reactivity of different materials, and the case of planar gold clusters has been explored here.

Acknowledgments

Work supported by MEC of Spain (Grant MAT2005-06544-C03-01), and by Junta de Castilla y León (Grant VA039A05).

REFERENCES

[1] Ditmire T, Zweiback J, Yanovsky VP, Cowan TE, Hays G, Wharton KB. Nuclear fusion from explosions of femtosecond laser-heated deuterium clusters. Nature 1999;398:489.

[2] Zweiback J, Smith RA, Cowan TE, Hays G, Wharton KB, Yanovsky VP, et al. Nuclear fusion driven by Coulomb explosions of large deuterium clusters. Phys Rev Lett 2000;84:2634.

[3] Hartke R, Symesa DR, Buersgens F, Ruggles LE, Porter JL, Ditmire T. Fusion neutron detector calibration using a table-top laser generated plasma neutron source. Nucl Instrum Methods A 2005;540:464.

[4] Alonso JA. Structure and properties of atomic clusters. London: Imperial College Press; 2005. See Chapter 3.

[5] Castleman AW, Keese RG. Gas-phase clusters: Spanning the states of matter. Science 1998;241:36.

[6] Farizon B, Farizon M, Gaillard MJ, Gobet F, Carré M, Buchet JP, et al. Experimental evidence of critical behavior in cluster fragmentation using an event-by-event data analysis. Phys Rev Lett 1998;81:4108.

[7] Gobet F, Farizon B, Farizon M, Gaillard MJ, Buchet JP, Carré M, et al. Probing the liquid-to-gas phase transition in a cluster via a caloric curve. Phys Rev Lett 2001;87:203401.

[8] Gobet F, Farizon B, Farizon M, Gaillard MJ, Buchet JP, Carré M, et al. Direct experimental evidence for a negative heat capacity in the liquid-to-gas phase transition in Hydrogen cluster ions: Backbending of the caloric curve. Phys Rev Lett 2002;89:183403.

[9] Sindzingre P, Ceperley DM, Klein ML. Superfluidity in clusters of p-H_2 molecules. Phys Rev Lett 1991;67:1871.

[10] Whaley KB. Structure and dynamics of quantum clusters. Int Rev Phys Chem 1994;13:41.

[11] Molina LM, Alonso JA. Chemical properties of small Au clusters: An analysis of the local site reactivity. J Phys Chem C 2007;111:6668.

[12] Diep P, Johnson JK. An accurate H_2-H_2 interaction potential from first principles. J Chem Phys 2000;112:4465.

[13] Carmichael M, Chenoweth K, Dykstra CE. Hydrogen molecule clusters. J Phys Chem A 2004;108:3143.

[14] Tejeda G, Fernández JM, Montero S, Blume D, Toennies JP. Raman spectroscopy of small para-H_2 clusters formed in cryogenic free jets. Phys Rev Lett 2004;92:223401.

[15] Martínez JI, Isla M, Alonso JA. Theoretical study of molecular hydrogen clusters. Eur Phys J D 2007;43:61.

[16] Kohn W, Sham LJ. Self-consistent equations including exchange and correlation effects. Phys Rev 1965;140:A1133.

[17] Parr RG, Yang W. Density functional theory of atoms and molecules. Oxford: Oxford University Press; 1989.

[18] Lange NA, Forker GM. CRC handbook of physical chemistry. New York: McGraw-Hill; 1969.

[19] Bransden BH, Joachain CJ. Physics of atoms and molecules. Essex: Longman; 1999 p. 391.

[20] Mezzacapo F, Boninsegni M. Superfluidity and quantum melting of p-H_2 clusters. Phys Rev Lett 2006;97:045301.

[21] Gross EKU, Dobson JF, Petersilka M. Density-functional theory of time-dependent phenomena. In: Nalewajski RF, editor. Topics in current chemistry, vol. 81. Heidelberg: Springer-Verlag; 1996. p. 81.

[22] Chermette H, Ymmud IV. Structure and stability of hydrogen clusters up to H_{21}^+. Phys Rev B 2001;63:165427.

[23] Howard IA, Leys FE, March NH, Van Alsenoy C, Alonso JA, Rubio A. Molecular clusters of Hydrogen, deuterium and tritium: especially cationic species $H_3^+(H_2)_m$: m=2,5 and 14. In: Vajtai R, Aimerich X, Kish LB, Rubio A, editors. Proceedings of SPIE Nanotechnology, vol. 5118. Bellingham: SPIE; 2003. p. 331.

[24] Prosmiti R, Villareal P, Delgado-Barrio G. Structures and energetics of H_n^+ clusters (n = 5−11). J Phys Chem A 2003;107:4768.

[25] Clampitt R, Towland L. Clustering of cold Hydrogen gas on protons. Nature 1969;223:815.

[26] Hiraoka K, Kebarle P. A determination of the stabilities of H_5^+, H_7^+, H_9^+, and H_{11}^+ from measurement of the gas phase ion equilibria $H_n^+ + H_2 = H_{n+2}^+$ (n = 3,5,7,9). J Chem Phys 1978;62:2267.

[27] Hiraoka K. A determination of the stabilities of $H_3^+ (H_2)_n$ with n=1–9 from measurements of the gas phase ion equilibria $H_3^+ (H_2)_{n-1} + H_2 = H_3^+ (H_2)_n$. J Chem Phys 1987;87:4048.

[28] Hiraoka K, Mori T. Thermochemical stabilities of $D_3^+(D_3)_n$ with n = 1–10. Chem Phys Lett 1989;157:467.

[29] Ferguson EE, Fehsenfeld FC, Albritton DL. Ion chemistry of the earths's atmosphere. In: Bowers MT, editor. Gas phase ion chemistry, vol. 1. New York: Academic Press; 1979. p. 45.

[30] Henderson JR, Tennyson J. Calculated spectrum for near-dissociation H_3^+: a first attempt. Mol Phys 1996;89:953.

[31] Neale L, Miller S, Tennyson J. Spectroscopic properties of the H_3^+ molecule: A new calculated line list. Astrophys J 1996;464:516.

[32] Dinelli BM, Polyanskya OL, Tennyson J. Spectroscopically determined Born-Oppenheimer and adiabatic surfaces for H_3^+, H_2D^+, D_2H^-, and D_3^+. J Chem Phys 1995;103:10433.

[33] Kozin IN, Roberts RM, Tennyson J. Symmetry and structure of rotating H_3^+. J Chem Phys 1999;111:140.

[34] Dinelli BM, Lesueur CR, Tennyson J, Amos RD. Ab-initio ro-vibrational levels of H_3^+ beyond the Born-Oppenheimer approximation. Chem Phys Lett 1995;232:295.

[35] Tennyson J, Miller S. H_3^+ – from first principles to Jupiter. Contemp Phys 1994;35:105.

[36] Tennyson J, Polyansky OL. Non-Born-Oppenheimer correction to the H_3^+ potential from esperimental-data. Phys Rev A 1994;50:314.

[37] Dinelli BM, Miller S, Tennyson J. A spectroscopically determined potential-energy surface for H_3^+. J Mol Spectr 1994;163:71.

[38] Kreckel H, Tennyson J, Schwalm D, Zajfman D, Wolf A. Rovibrational relaxation model for H_3^+. New J Phys 2004;6:151.

[39] Kostin MA, Polyansky OL, Tennyson J, Mussa HY. Rotation-vibration states of H_3^+ at dissociation. J Chem Phys 2003;118:3538.

[40] Prosmiti R, Polyansky OL, Tennyson J. A global potential energy surface for the H_3^+ molecule. Chem Phys Lett 1997;273:107.

[41] Kawata I, Kono H, Bandrauk AD. Mechanism of enhanced ionization of linear H_3^+ in intense laser fields. Phys Rev A 2001;64:043411.

[42] Yu Hengtai, Bandrauk AD. Molecules in intense laser fields: Enhanced ionization in one- and two-electron linear triatomic molecules. Phys. Rev. A 1997;56:685.

[43] Herbst E, Klemperer W. Formation and depletion of molecules in dense interstellar clouds. Astrophys J 1973;185:505.
[44] Tennyson J. H_3^+ from chaos to the cosmos. Phys World 1995;8:33.
[45] Tennyson J. Spectroscopy of H_3^+ - planets, chaos, and the universe. Rep Prog Phys 1995;58:421.
[46] Drossart P, Maillard JP, Caldwell J, Kim SJ, Watson JKG, Majewski WA, et al. Detection of H_3^+ on Jupiter. Nature 1989;340:539.
[47] Geballe TR, Jagod MF, Oka T. Detection of H_3^+ infrared-emission lines in Saturn. Astrophys J 1993;408:L109.
[48] Trafton LM, Geballe TR, Miller S, Tennyson J, Ballester GE. Detection of H_3^+ from Uranus. Astrophys J 1993;405:761.
[49] Miller S, Tennyson J, Lepp S, Dalgarno A. Identification of features due to H_3^+ in the infrared-spectrum of supernova 1987A. Nature 1992;355:420.
[50] Geballe TR, Oka T. Detection of H_3^+ in interstellar space. Nature 1996;384:334.
[51] McCall BJ, Geballe TR, Hinkle KH, Oka T. Detection of H_3^+ in the diffuse interstellar medium toward Cygnus OB2 No. 12. Science 1998;279:1910.
[52] Fisher ME. Theory of equilibrium critical phenomena. Rep Progr Phys 1967;30:615.
[53] Belkacem M, Latora V, Bonasera A. Critical evolution of a finite system. Phys Rev C 1995;52:271.
[54] Agostini P, Fabre F, Mainfray G, Petite G, Rahman N. Free-free transitions following six-photon ionization of Xenon atoms. Phys Rev Lett 1979;42:1127.
[55] van der Hart HW. The recollision model applied to strong-field double ionization of C^+. J Phys B 2003;36:3477.
[56] Ferray M, L'Huillier A, Li XF, Lempre LA, Mainfray G, Manus C. Multiple-harmonic conversion of 1064 nm radiation in rare gases. J Phys B 1988;21:L31.
[57] Poth L, Wisniewsky ES, Castleman AW. Cluster dynamics: Fast reactions and Coulomb explosion. Am Sci 2002;90:342.
[58] Heidenreich A, Jortner J, Last I. Cluster dynamics transcending chemical dynamics toward nuclear fusion. Proc Natl Acad Sci USA 2007;103:10589.
[59] Loveman R, Bendahan J, Gozani T, Stevenson J. Time of flight fast neutron radiography. Nucl Instrum Methods B 1995;99:765.
[60] Izyumov YuA, Chernoplekov NA. Neutron spectroscopy. Dordrecht: Kluwer; 1994.
[61] Ma LM, Suraud E, Reinhard PG. Laser excitation and ionic motion in small clusters. Eur Phys J D 2001;14:217.
[62] Ma LM, Reinhard PG, Suraud E. Dynamics of H_9^+ in intense laser pulses. Eur Phys J D 2005;33:49.
[63] Isla M, Alonso JA. Fragmentation and Coulomb explosion of deuterium clusters by the interaction with intense laser pulses. Phys Rev A 2005;72:023201.
[64] Isla M, Alonso JA. Interaction of the charged deuterium cluster D_3^+ with femtosecond laser pulses. J Phys Chem C 2007;111:17765.
[65] Flocard H, Koonin SE, Weiss MS. Three-dimensional time-dependent Hartree-Fock calculations: Application to $^{16}O + {}^{16}O$ collisions. Phys. Rev. C 1978;17:1682.
[66] Yabana K, Bertsch GF. Time-dependent local-density approximation in real time. Phys Rev B 1996;54:4484.
[67] Yabana K, Bertsch GF. Optical response of small carbon clusters. Z Phys D 1997;42:219.
[68] Yabana K, Bertsch GF. Optical response of small silver clusters. Phys Rev A 1999;60:3809.
[69] Castro A, Marques MAL, Alonso JA, Bertsch GF, Rubio A. Excited states dynamics in time-dependent density functional theory. Eur Phys J D 2004;28:211.

[70] Castro A, Marques MAL, Alonso JA, Rubio A. Optical properties of nanostructures from time-dependent density functional theory. J Comput Theor Nanosci 2004;1:231.
[71] Alonso JA, Rubio A. Models for optical properties clusters and nanostructures. Handbook of theoretical and computational nanotechnology. In: Rieth M, Schommers W, editors. Magnetic nanostructures and nanooptics, vol.7. Stevenson Ranch: American Scientific Publishers; 2006. p. 209.
[72] Howard IA, Alonso JA, March NH, Rubio A, Van Alsenoi C. Coulomb explosion of deuterium cationic clusters. Phys Rev A 2003;68:065201.
[73] Näher U, Bjornholm S, Frauendorf S, Garcías F, Guet C. Fission of metal clusters. Phys Rep 1997;285:245.
[74] Lezius M, Dobosz S, Normand D, Schmidt M. Explosion dynamics of rare gas clusters in strong laser fields. Phys Rev Lett 1998;80:261.
[75] Garcías F, Lombard RJ, Barranco M, Alonso JA, López JM. Semiempirical model for the fission of multiply charged metal clusters. Z Phys D 1995;33:301.
[76] Seres J, Muller A, Seres E, Keffe KO, Lenner M, Herzog RF, et al. Sub-10-fs, terawatt-scale Ti:sapphire laser system. Opt Lett 2003;28:1832.
[77] Krane KS. Introductory nuclear physics. New York: Wiley; 1988 See Chapter 14.
[78] Last I, Jortner J. Nuclear fusion induced by Coulomb explosion of heteronuclear Clusters. Phys Rev Lett 2001;87:033401.
[79] Last I, Jortner J. Nuclear fusion driven by Coulomb explosion of homonuclear and heteronuclear deuterium and tritium containing clusters. Phys Rev A 2001;64:063201.
[80] Wolff W, Coelho LFS, Wolf HE. Collisional destruction of fast H_3^+ ions in noble gases. Phys Rev A 1992;45:2978.
[81] Li J, Furuta T, Goto H, Ohashi T, Fujuwara Y, Yip S. Theoretical evaluation of hydrogen storage capacity in pure carbon nanostructures. J Chem Phys 2003;119:2376.
[82] Schroeder DV. An introduction to thermal physics. Reading, MA: Addison-Wesley; 2000.
[83] Dillon AC, Jones KM, Bekkedhal TA, Kiang CH, Bethune DS, Heben MJ. Storage of hydrogen in single-walled carbon nanotubes. Nature 1997;386:377.
[84] Hynek S, Fuller W, Bentley J. Hydrogen Storage by Carbon Sorption. Int J Hydrogen Energy 1997;22:601.
[85] Rzepka M, Lamp P, de la Casa-Lillo MA. Physisorption of Hydrogen on Microporous Carbon Nanotubes. J Phys Chem B 1998;102:10894.
[86] Darkrim F, Levesque D. Monte Carlo simulations of hydrogen adsorption in single-walled carbon nanotubes. J Chem Phys 1998;109:4981.
[87] Darkrim F, Vermesse J, Malbrunot P, Levesque D. Monte Carlo simulations of nitrogen and hydrogen physisorption at high pressures and room temperature. Comparison with experiments. J Chem Phys 1999;110:4020.
[88] Ye Y, Ahn CC, Whitham C, Fultz B, Liu J, Rinzler AG, et al. Hydrogen adsorption and cohesive energy of single-walled carbon nanotubes. Appl Phys Lett 1999;74:2397.
[89] Liu C, Fan YY, Liu M, Kong HT, Cheng HM, Dresselhaus MS. Hydrogen storage in single-walled carbon nanotubes at room temperature. Science 1999;286:1127.
[90] Wang Q, Johnson JK. Molecular simulation of hydrogen adsorption in single-walled carbon nanotubes and idealized carbon slit pores. J Chem Phys 1999;110:577.
[91] Wang Q, Johnson JK. Optimization of carbon nanotube arrays for hydrogen adsorption. J Phys Chem B 1999;103:4809.
[92] Brown CM, Yildirim T, Newman DA, Heben MJ, Gennet T, Dillon AC, et al. Quantum rotation of hydrogen in single-wall carbon nanotubes. Chem Phys Lett 2000;329:311.

[93] Bunger U, Zittel W. Hydrogen storage in carbon nanostructures - still a long road from science to commerce? Appl Phys A 2001;72:147.
[94] Cao A, Zhu H, Zhang X, Li X, Ruan D, Xu C, et al. Hydrogen storage of dense-aligned carbon nanotubes. Chem Phys Lett 2001;342:510.
[95] Hirscher M, Becher M, Haluska M, Dettlaff-Weglikowska U, Quintel A, Duesberg GS, et al. Hydrogen storage in sonicated carbon materials. Appl Phys A 2001;72:129.
[96] Dillon AC, Heben MJ. Hydrogen storage using carbon adsorbents: past, present and future. Appl Phys A 2001;72:133.
[97] Khare BN, Meyyappan M, Cassell AM, Nguyen CV, Han J. Functionalization of carbon nanotubes using atomic hydrogen from a glow discharge. Nano Lett 2002;2:73.
[98] Zuttel A. Materials for hydrogen storage. Mater Today 2003;6:24.
[99] Orimo S, Zuttel A, Schlapbach L, Majer G, Fukunaga T, Fujii H. Hydrogen interaction with carbon nanostructures: current situtation and future prospects. J Appl Crystallogr 2003;356:716.
[100] Panella B, Hirscher M, Roth S. Hydrogen adsorption in different carbon nanostructures. Carbon 2005;43:2209.
[101] Gogotsi Y, Dash RK, Yushin G, Yildirim T, Laudisio G, Fischer JE. Tailoring of Nanoscale Porosity in Carbide-Derived Carbons for Hydrogen Storage. J Am Chem Soc 2005;127:16006.
[102] Yushin G, Dash R, Jagiello J, Fischer JE, Gogotsi Y. Carbide-Derived Carbons: Effect of Pore Size on Hydrogen Uptake and Heat of Adsorption. Adv Funct Mater 2006;16:2288.
[103] Jorda–Beneyto M, Suarez–García F, Lozano–Castelló D, Cazorla–Amorós D, Linares-Solano A. Hydrogen Storage on chemically Activated Carbons and Carbon Nanomaterials at high pressures. Carbon 2007;45:293.
[104] Silvera IF, Goldman VV. The isotropic intermolecular potential for H_2 and D_2 in the solid and gas phases. J Chem Phys 1978;69:4209.
[105] Crowell AD, Brown JS. Laterally averaged interaction potentials for $^1H_2$ and $^2H_2$ on the (0001) graphite surface. Surf Sci 1982;123:296.
[106] Stan G, Cole MW. Hydrogen Adsorption in Nanotubes. J Low Temp Phys 1998;110:539.
[107] Arellano JS, Molina LM, Rubio A, Alonso JA. Density functional study of adsorption of molecular hydrogen on graphene layers. J Chem Phys 2000;112:8114.
[108] Okamoto Y, Miyamoto Y. Ab initio investigation of physisorption of molecular hydrogen on planar and curved graphenes. J Phys Chem B 2001;105:3470.
[109] Ferre-Vilaplana A. Numerical treatment discussion and ab initio computational reinvestigation of physisorption of molecular hydrogen on graphene. J Chem Phys 2005;122:104709.
[110] Tada K, Furuya S, Watanabe K. Ab initio study of hydrogen adsorption to single-walled carbon nanotubes. Phys Rev B 2001;63:155405.
[111] Girifalco LA, Hodak M. Van der Waals binding energies in graphitic structures. Phys Rev B 2002;65:125404.
[112] Alonso JA, Mañanes A. Long-range van der Waals interactions in density functional theory. Theor Chem Acc 2007;117:467.
[113] Alonso JA, Arellano JS, Molina LM, Rubio A, López MJ. Interaction of molecular and atomic hydrogen with single-wall carbon nanotubes. IEEE Trans Nanotech 2004;3:304.
[114] Arellano JS, Molina LM, Rubio A, López MJ, Alonso JA. Interaction of molecular and atomic hydrogen with (5,5) and (6,6) single-wall carbon nanotubes. J Chem Phys 2002;117:2281.
[115] Ferre-Vilaplana A. Ab Initio Computational Investigation of Physisorption of Molecular Hydrogen on Achiral Single-Walled Carbon Nanotubes. J Chem Phys 2005;122:214724.

[116] Cabria I, López MJ, Alonso JA. Adsorption of Hydrogen on Normal and Pentaheptite Single Wall Carbon Nanotubes. Eur Phys J D 2005;34:279.
[117] Cabria I, López MJ, Alonso JA. Density Functional Study of Molecular Hydrogen Coverage on Carbon Nanotubes. Comput Mater Sci 2006;35:238.
[118] Duplock EJ, Scheffler M, Lindan JD. Hallmark of Perfect Graphene. Phys Rev Lett 2004;92:225502.
[119] Crespi VH, Benedict LX, Cohen ML, Louie SG. Prediction of a pure-carbon planar covalent metal. Phys Rev B 1996;53:R13303.
[120] Boustani I, Quandt A. Nanotubules of bare boron clusters: Ab initio and density functional study. Europhys Lett 1997;39:527.
[121] Boustani I, Quandt A. Boron in ab initio calculations. Comput Mater Sci 1998;11:132.
[122] Ciuparu D, Klie RF, Zhu Y, Pfefferle L. Synthesis of Pure Boron Single-Wall Nanotubes. J Phys Chem B 2004;108:3967.
[123] Gindulyte A, Lipscomb WN, Massa L. Proposed Boron Nanotubes. Inorg Chem 1998;137:6544.
[124] Cabria I, López MJ, Alonso JA. Density Functional Calculations of Hydrogen Adsorption on Boron Nanotubes and Boron Sheets. Nanotechnology 2006;17:778.
[125] Froudakis GE. Why alkali-metal-doped carbon nanotubes possess high hydrogen uptake. Nano Lett 2001;1:531.
[126] Zhao J, Buldum A, Han J, Lu JP. Gas molecule adsorption in carbon nanotubes and nanotube bundles. Nanotechnology 2002;13:195.
[127] Jacobson N, Tegner B, Schroder E, Hyldgaard P, Lundqvist BI. Hydrogen dynamics in magnesium and graphite. Comput Mater Sci 2003;24:273.
[128] Wang P, Orimo S, Tanabe K, Fujii H. Hydrogen in mechanically milled amorphous boron. J Alloys Compd 2003;350:218.
[129] Cabria I, López MJ, Alonso JA. The optimum average nanopore size for Hydrogen storage in Carbon nanoporous materials. Carbon 2007;45:2649.
[130] Patchkovskii S, Tse JS, Yurchenko SN, Zhechkov L, Hrine T, Seifert G. Graphene nanostructures as tunable storage media for molecular Hydrogen. Proc Natl Acad Sci USA 2005;102:10439.
[131] Chen P, Wu X, Lin J, Tan KL. High H_2 uptake by alkali-doped Carbon nanotubes under ambient pressure and moderate temperatures. Science 1999;285:91.
[132] Yang RT. Hydrogen storage by alkali-doped carbon nanotubes-revisited. Carbon 2000;38:623.
[133] Pinkerton FE, Wicke BG, Olk CH, Tibbetts GG, Meisner GP, Meyer MS, et al. Thermogravimetric measurement of Hydrogen absorption in alkali-modified Carbon materials. J Phys Chem B 2000;104:9460.
[134] Dubot P, Cenedese P. Modeling of molecular hydrogen and lithium adsorption on single-wall carbon nanotubes. Phys Rev B 2001;63:241402.
[135] Froudakis GE. Hydrogen interaction with carbon nanotubes: a review of ab initio studies. J Phys Condens Matter 2002;14:R453.
[136] Lee EC, Kim YS, Jin YG, Chang KJ. First-principles study of hydrogen adsorption on carbon nanotubes surfaces. Phys Rev B 2002;66:073415.
[137] Zhu ZH, Lu GQ, Smith SC. Comparative study of hydrogen storage in Li- and K-doped carbon materials–theoretically revisited. Carbon 2004;42:2509.
[138] Maresca O, Pellenq RJM, Marinelli F, Conard J. A search for a strong physisorption site for H_2 in Li-doped porous carbons. J Chem Phys 2004;121:12548.

[139] Deng WQ, Xu X, Goddard WA. New alkali doped pillared Carbon materials designed to achieve practical reversible Hydrogen storage for transportation. Phys Rev Lett 2004;92:166103.

[140] Cabria I, López MJ, Alonso JA. Enhancement of Hydrogen physisorption on graphene and Carbon nanotubes by Li doping. J Chem Phys 2005;123:204721.

[141] Cho JH, Park CR. Hydrogen storage on Li-doped single-walled carbon nanotubes: Computer simulation using the density functional theory. Catal Today 2007;120:407.

[142] Zhu ZH, Lu GQ. Comparative study of Li, Na, and K adsorptions on graphite by using ab initio method. Langmuir 2004;20:10751.

[143] Khantha M, Cordero NA, Molina LM, Alonso JA, Girifalco LA. Interaction of lithium with graphene: An ab initio study. Phys Rev B 2004;70:125422.

[144] Furche F, Ahlrichs R, Weis P, Jacob C, Gilb S, Bierweiler T, et al. The structures of small gold cluster anions as determined by a combination of ion mobility measurements and density functional calculations. J Chem Phys 2002;117:6982.

[145] Gilb S, Weis P, Furche F, Ahlrichs R, Kappes MM. Structures of small gold cluster cations (Au_n^+, $n<14$): Ion mobility measurements versus density functional calculations. J Chem Phys 2002;116:4094.

[146] Hakkinen H, Moseler M, Landman U. Bonding in Cu, Ag, and Au clusters: Relativistic effects, trends, and surprises. Phys Rev Lett 2002;89:033401.

[147] Vajda S, Winans RE, Elam JW, Lee B, Pellin MJ, Seifert S, et al. Supported gold clusters and cluster-based nanomaterials: characterization, stability and growth studies by in situ GISAXS under vacuum conditions and in the presence of hydrogen. Top Catal 2006;39:161.

[148] Mañanes A, Duque F, Méndez F, López MJ, Alonso JA. Analysis of the bonding and reactivity of H and the Al_{13} cluster using density functional concepts. J Chem Phys 2003;119:5128.

[149] Buckart S, Ganteför G, Kim YD, Jena P. Anomalous behavior of atomic Hydrogen interacting with Gold clusters. J Am Chem Soc 2003;125:14205.

[150] Fernández EM, Torres MB, Balbás LC. Planar and cagelike structures of gold clusters: Density-functional pseudopotential calculations. Phys Rev B 2006;73:235433.

[151] Johansson MP, Sundholm D, Vaara J. Au_{32}: A 24-Carat Golden Fullerene. Angew Chem Int Ed 2004;43:2678.

Chapter 8

Laser Induced Crystallization

Andrew Fischer*, R. M. Pagni*, R. N. Compton*,† and D. Kondepudi‡
*Department of Chemistry, University of Tennessee, Knoxville, Tennessee, USA
†Department of Physics, University of Tennessee, Knoxville, Tennessee, USA
‡Department of Chemistry, Wake Forrest University,Winston-Salem, North Carolina, USA

Chapter Outline Head

I. INTRODUCTION

This book is devoted to the physics and chemistry of clusters and their assemblies. Our contribution considers the formation of small "clusters" which may ultimately lead to the formation of crystals. We combine conventional studies of crystal growth occurring from saturated solutions with the formation of crystals induced by pulsed laser light. We will also introduce new studies on the formation of crystals in saturated solutions as a result of intense sound, or compression waves, generated in the solution by intense pulsed lasers. Many tiny crystals are created *throughout* the solution by the laser generated sound waves. The compression waves can be created either by focusing into the liquid cr onto the walls of the container.

Nanoclusters. DOI: 10.1016/S1875-4023(10)01008-9

This new method allows for the instantaneous formation of many "seed" crystals which are then available for further impurity-free crystal growth. More importantly, the tiny "baby" crystals can be harvested before complete growth into an "adult" crystal can occur. This method has the possibility of producing enough "baby" crystals to provide a glimpse into the initial stages of crystal growth using modern microscopy techniques (STM, AFM, TM, etc.). This method of crystal growth may also allow for the generation of crystals which have previously not been possible or are otherwise difficult to produce.

Khanna and Jena [1] have discussed the synthesis of solids from the assembly of atomic clusters. Atomic clusters could be "designed" to mimic a type of Periodic Table which is constructed of atomic clusters or "super atoms." These clusters may then be assembled into new solids and crystals with unique optical, magnetic, electronic, and thermodynamic properties. The simplest example of this model is one of the three known allotropes of carbon: graphite, diamond, and fulleride. Diamond and graphite consist of infinite arrays of three dimensional sp^3 and two-dimensional sp^2 hybridized carbon atoms, respectively. On the other hand, the C_{60} fulleride cluster forms a van der Waal's fcc crystal [2]. Information on the initial stages (i.e., clusters of clusters) of bulk fulleride crystal formation comes from gas phase molecular beam studies. Nozzle jet expansion [3] of C_{60} fullerenes is known to produce clusters of C_{60} having enhanced ion signals in a mass spectrometer corresponding to the well known Mackay icosohedra [4], that is, $(C_{60})_{13}$ and $(C_{60})_{55}$. The first shell of a cubic close packed grouping of equal spheres consists of a central sphere surrounded by twelve other spheres at the vertices of a cubo-octahedron. Succeeding completed n^{th} shells will contain $(10n^2 + 2)$ C_{60} molecules; thus, the first two "super atoms" of fullerenes made from C_{60} would be $(C_{60})_{13}$ and $(C_{60})_{55}$. It is also known that C_{60} molecules pack into a van der Waals solid in which the molecules are free to rotate even at liquid nitrogen temperature.

Crystallization is the result of the ordered growth of a species into an extended structure. In solution, a high concentration of solute by itself is insufficient to cause crystallization. In order for a crystal to grow, a cluster must form and continue to grow in an ordered fashion to result in a crystal. Early studies [5,6] showed that mechanical perturbations, such as agitation, mechanical shock, and pressure gradients, can cause supersaturated solutions (where the concentration in solution is greater than the saturation point) to crystallize. Khamiskii [7] has reviewed crystallization due to external sources, such as electromagnetic effects. Crystallization is normally thought to proceed via two interconnected mechanisms: primary and secondary nucleation. In the case of primary nucleation, nucleation phenomena may be divided into either spontaneous (homogenous) or induced (heterogeneous) nucleation.

A. Primary Nucleation: Thermodynamics and Kinetics

The basic principle underlying crystallization is a thermodynamic imbalance between the liquid phase and a solid, cluster-like phase. The thermodynamic quantity that drives the crystallization process is the chemical potential, μ, for each phase. In all phase transitions, the thermodynamic force driving the transition comes from the difference in μ between the two phases, which is affinity of the transformation A, that is,

$$A = \mu_1 - \mu_2, \tag{1}$$

where the chemical potential is defined as

$$\mu = \mu_0 + RT\ \ln(a) \tag{2}$$

in which μ_0 is the standard potential, R is the gas constant, T is the temperature, and a is the activity.

The driving force A must overcome an energy barrier that stems from the need to form and separate the interface, that is, the new crystal, from the solution [8,9]. As will be shown later, a more descriptive theory for crystallization requires assumptions based upon the mechanism and shape formation of the crystal.

If the homogeneous formation of a crystal were the result of multiple bimolecular collisions, the following scheme would adequately describe the successive addition of molecules to the cluster:

$$\begin{aligned} M + M &\rightleftharpoons M_2 \\ M_2 + M &\rightleftharpoons M_3 \quad , \\ M_{c-1} + M &\rightleftharpoons M_c \end{aligned} \tag{3}$$

where, for each addition, it is as likely to proceed in the reverse direction, that is, lose the recently added molecule, as it is to undergo another successful addition. In the above scheme, M_c represents the critically sized cluster, which is defined as the cluster that, upon further additions of M, proceeds to crystal nucleation. Smaller nuclei can easily form by the above scheme, but, because of the possibility that a molecule of M is lost from the smaller clusters, many precritical nuclei may form before the successful formation of the critically sized cluster.

The exact shape of the cluster is unknown because of its small size. Several theories predict that the cluster exhibits the same morphology as the crystal it will grow into, while other theories predict that the cluster will rearrange into the morphology that minimizes the energy of the small cluster in solution. None of these theories has been proven. Hoare and McInnes [10] have reviewed previous work on the structure and morphology of small molecular clusters.

The chemical potential of cluster depends on its size. For a spherical cluster of radius r, the chemical potential is

$$\mu(r) = \mu_2 + (2\gamma/r)V_{\mathrm{m}}, \tag{4}$$

where μ_2 is the chemical potential of phase 2 in its bulk ($r \to \infty$), σ is the interfacial tension and V_m is the molar volume. Hence the thermodynamic force or affinity for the phase transformation from the solution phase to the solid phase is

$$A = \mu_1 - (\mu_2 + (2\gamma/r)V_m). \tag{5}$$

The transformation from phase 1 to phase 2 will take place only when $A > 0$. Equation (5) shows that for a given $\Delta\mu = (\mu_1 - \mu_2)$, the affinity A is positive only for particles of sufficiently large size. The critical radius above which the affinity is positive is

$$r^* = (2\gamma/\Delta\mu)V_m. \tag{6}$$

The classical theory of nucleation stems from the research of many scientists [11–13] on the condensation of vapor to a liquid. This formalism was adopted to describe the transition from an aqueous cluster to crystals. If one assumes that the cluster is growing into a sphere with radius "r," the change in free energy, ΔG, between the cluster and the solute in solution is the sum of the free energy change of the surface, ΔG_s, and the free energy change of the volume of the cluster, ΔG_V. The excess free energy is given by

$$\begin{aligned} \Delta G &= \Delta G_s + \Delta G_V = 4\pi r^2\gamma - (4\pi r^3/3V_m)(\mu_2 - \mu_1), \\ &= 4\pi r^2\gamma - (4\pi r^3/3V_m)\Delta\mu, \end{aligned} \tag{7}$$

where we have used the fact that $\Delta\mu$ is the free energy change per mole between the bulk states of the two phases. The free energy change for critical-size clusters is

$$\Delta G^* = (16\pi/3)\left(\gamma^3 V_m^2/\Delta\mu^2\right). \tag{8}$$

Only clusters of radius r^* or larger can grow. The formation of critical clusters is a thermally activated process. The rates of thermally activated processes are normally described by the Arrhenius equation:

$$J = A\cdot\exp(-\Delta G/k_B T) \tag{9}$$

where A is the pre-exponential factor, ΔG is the free energy of activation of the process under discussion, k_B is Boltzmann constant, and T the absolute temperature. For nucleation, the free energy of activation is ΔG^*. Using the above expression for G^*, we can write the rate of nucleation:

$$J = A\cdot\exp\left[-(16\pi/3)\left(\gamma^3 V_m^2/\Delta\mu^2 k_B T\right)\right] \tag{10}$$

This expression relates the difference in chemical potential between the two phases $\Delta\mu$ to the nucleation rate. Increase in $\Delta\mu$ results in an increase in the nucleation rate. The supersaturation parameter S is defined as

$$S = c/c_0, \tag{11}$$

where c is the concentration of the solute and c_0 is the saturation concentration. The supersaturation of a solution can be more accurately described by the ratio of the activities of the solute, but the ratio of the concentrations is often used as a valid approximation. In terms of S, the chemical potential difference, the driving force for nucleation can be written as

$$\Delta\mu = RT \ln S. \tag{12}$$

Using this approximation one obtains the nucleation rate as

$$J = A \exp\left[-(16\pi/3)\left(\gamma^3 V^2 / k_B^3 T^3 \; \{\text{In} S\}^2\right]\right. \tag{13}$$

in which $V = V_m/N_A$ is the molecular volume (N_A is the Avogadro number).

Equation (13) is the expression commonly used to describe the rate of primary homogenous nucleation. From this equation, one sees that there are three main variables to describe the rate of nucleation: the temperature, the degree of supersaturation, and the interfacial tension. It was assumed that the nucleus is spherical in all of the above considerations. If that assumption were invalid, a different geometrical factor would have to be used, resulting in a different formulation of Equation (13).

Nielson [14] developed an empirical approach to describe the nucleation process using an induction time, t_{ind}, which is expressed as

$$t_{ind} = k \times c^{1-p}, \tag{14}$$

where k is a constant, c is the concentration of the supersaturated solution, and p is the number of molecules making up a critical nucleus. Equation (10) represents a simplified expression for the complex process of crystallization. The secondary nucleation empirical relationship, together with classical nucleation theories, provides a mechanism for the clustering of molecules, but none of the proposed theories agrees with the relationship between supersaturation and the size of the critical nucleus. Experimental evidence for the size dependence of the critically sized nucleus is necessary to refine further the nucleation theories. Several reviews of nucleation mechanisms have been published [14–16], with the recent review by Kashchiev [17] relating thermodynamics and kinetics to homogenous and heterogenous nucleation.

B. Heterogeneous Nucleation

Heterogeneous nucleation is that induced by noncrystalline matter. Impurities can affect the crystallization process by either promoting or inhibiting the formation of a crystal. In many experiments, spontaneous nucleation is found to be induced by trace amounts of impurities in solution. Generally, aqueous solutions contain many particles that are greater than 1 μm in size; great care can be taken to reduce the number density of the impurities, but total elimination of impurities is virtually impossible. Impurities can also be found frequently trapped within cavities or on sides of the vessels that hold the crystallizing solution, thereby making completely spontaneous nucleation a less likely event.

In order for an impurity to promote crystallization, the free energy change for the formation of the critical nucleus, $\Delta G'_{crit}$, under heterogeneous conditions must be lower than the free energy change, ΔG_{crit}, for homogeneous conditions. The effect of an impurity on the free energy change can be expressed by

$$\Delta G'_{crit} = \Theta \Delta G_{crit}, \tag{15}$$

where Θ is a factor less than unity in the case that the impurity is a promoter. As indicated previously with reference to Equation (13), the interfacial tension, γ, is seen to contribute to the rate of formation of a crystal. Figure 1 shows a phase diagram relating the contact of the crystalline phase, the impurity phase, and the solution phase. The interfacial tensions are denoted by γ_{cl} (interface between crystalline and liquid phase), γ_{sl} (interface between impurity and liquid phase), and γ_{cs} (interface between crystalline and impurity). Combining these forces in a horizontal direction gives

$$\gamma_{sl} = \gamma_{cs} + \gamma_{cl} \cos \Theta. \tag{16}$$

Three scenarios exist for Equation (16); Θ can either equal 0°, Θ can vary between 0° and 180°, or Θ can be 180°. For each of those scenarios, the respective free energies would be

$$\Delta G'_{crit} = 0, \tag{17}$$

$$\Delta G'_{crit} < \Delta G_{crit}, \tag{18}$$

$$\Delta G'_{crit} = \Delta G_{crit}. \tag{19}$$

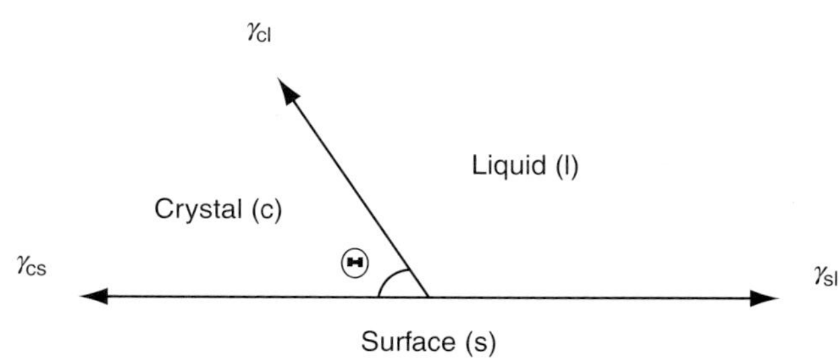

FIGURE 1 Crystallization angle, Θ, relating contact of crystalline phase, impurity phase, and solutions phase where γ_{cl}, γ_{sl}, and γ_{cs} are the crystalline-liquid phase, impurity-liquid phase, and crystalline-impurity interfacial tensions, respectively.

The scenario expressed in Equation (17) corresponds to the impurity being a seed crystal of the material, and represents an example of secondary nucleation. For Equation (18), this corresponds to impurities that are able to promote the crystallization of the material. Equation (19) describes an example in which the material does not change the free energy required for nucleation.

C. Secondary Nucleation

A supersaturated solution will crystallize more quickly if crystals are already present in the solution. In this case, secondary nucleation represents the dominant mode of formation of crystals as a result of the presence of previously formed crystals. Strickland-Constable [15] and Botsaris et al [18]. have previously described various mechanisms by which secondary nucleation may occur. Strickland-Constable proposed four different mechanisms leading to secondary nucleation: initial breeding, needle breeding, polycrystalline breeding, and collisional breeding. Initial breeding is the formation of secondary nuclei as a result of dust swept off of the surface of the seed crystal when it is introduced to the system. Needle and polycrystalline breeding are similar in that each represents the detachment of part of the crystalline structure from the seed crystal (the detachment can occur through physical stress or strain to the system). Collisional breeding is the formation of secondary nuclei through a process that involves the collision of multiple clusters that have formed as a result of the seed crystal.

Qian and Botsaris [19] postulated a theory for the rate of secondary nucleation in which the attraction between clusters and surfaces of the "mother crystal" leads to higher nucleation rates. An empirical rate law for secondary nucleation in stirring experiments is of the form

$$R_s = k(c - c_0)^{\alpha}, \tag{20}$$

where k is a factor that depends upon the rate of stirring, c is the concentration, and c_0 is the concentration at saturation. The exponent α has been suggested to be greater than one [20,21]. This empirical rate law is similar in form to the empirical rate law for primary nucleation. A more fundamental rate law model for secondary nucleation has not been proposed to date.

The principal investigations of secondary crystallization from chiral seeds have focused primarily on three objectives [22]. The first is to identify the origins of secondary nuclei: Are secondary nuclei fragments of a seed crystal or a result of the concentration gradient surrounding the seed? A study by Denk and Botsaris [23] found that, primarily, a crystal formed in the presence of a chiral seed crystal retained the chirality of the seed, though under specific conditions, the chirality could be reversed. Denk and Botsaris concluded that secondary nuclei are generated through three mechanisms: growth and then detachment of irregular surfaces into solution, an impurity concentration gradient at the crystal interface, and an ordering of water molecules near the surface of the seed (though this mechanism seems unlikely).

The second objective regarding the generation of secondary nuclei is to understand whether an enantiomeric seed crystal will preferentially result in additional crystals of the same chirality when the solution is initially racemic or achiral. Such results may indicate the propagation of chirality through nature. Interesting results have been found that address this question. Yokota and Toyokura [24] found that, when an L crystal of (*S*)-carboxymethyl-D-cysteine (SCMC) was immersed in a solution of LSCMC, pits developed on the crystal surface. Davey et al [25]. found similar results from a crystal of triazolylketone immersed in solution. Both groups attributed their findings to the inclusion of the opposite enantiomer into the crystal structure as a contaminant.

The third objective of the study of secondary nucleation is to get an insight into the phenomenon leading to the dominance of certain biochemical enantiomers in nature. This particular phenomenon is termed chiral symmetry breaking, which was realized by Kondepudi [26] with the crystallization of sodium chlorate which will be described in more detail later in the chapter.

In an attempt to describe secondary nucleation, Qian and Botsaris [22] developed a model they entitled Embryos Coagulation Secondary Nucleation (ECSN). The ECSN model combines three facets: classical nucleation theory, attractive van der Waals forces between clusters in solution, and coagulation of colloids in solution. Qian and Botsaris [19] have published quantitative results that reinforce the ECSN model. The ECSN model predicts that a higher concentration of embryos exist surrounding a seed crystal, and the embryos will coagulate to form a cluster of critical size. This process would be an example of the seed crystal aiding the production of other nuclei, but as seen from this postulate, there is no transfer of chirality from the seed to the other nuclei. This particular mechanism of formation would then be in competition with conventional secondary nucleation mechanism (SCN) [27]. Qian and Botsaris show through a variety of experiments that the competition between the ECSN and SCN mechanisms is temperature dependent.

The ECSN model predicts that the precrystal embryos are amorphous and only acquire chirality before the critically sized cluster is formed; this would indicate that the chirality of the crystal is attributed to random factors influencing its formation. The ECSN model is also able to explain the findings of Yokota and Toyokura [24] and Davey et al. [25] as an embryo is attracted toward the surface of the seed crystal through the van der Waals attraction, it is incorporated into the crystal structure, thus affecting the chiral nature of the crystal.

In early crystallization studies, it was noted that many inorganic salts crystallized into less stable polymorphs when their solutions were cooled quickly. Ostwald [28,29] took the results from the salt crystallizations and formulated a general theory; he stated that an unstable system would not necessarily convert to the most stable form of the system, but would rather progress in stages through forms that most closely resembled itself. This theory has been explored theoretically, but thermodynamics has been unable to prove this hypothesis [30]. When combined with theoretical kinetics [31], some facets of Ostwald's rule of stages have been proven, though no complete proof yet exists.

D. Crystallization of Conglomerates and Polymorphs

An interesting and important way in which to study primary and secondary nucleation is to examine the crystallization of compounds that form polymorphs [32] or conglomerates [33]. The amino acid glycine crystallizes in three polymorphic forms, for example: α an achiral crystal, β an unstable crystalline form, and γ a chiral crystal. It is possible to distinguish these polymorphs by solid state ^{13}C NMR or X-ray powder diffraction. The ionic compounds, sodium chlorate and bromate, crystallize from water in the chiral space group $P2_13$ and thus form conglomerate crystals. The chirality of these crystals can be determined by polarimetry on the individual crystals or traditional polarimetry on a slurry of the crystalline powder in a solvent mixture whose refractive index matches that of the powder [34]. The seminal experiment to be performed then is to crystallize compounds of these types, with and without perturbations, and determine the distribution of polymporhs or conglomerates from which one may infer something of the nature of their primary and/or secondary nucleation.

(i) The Behavior of Sodium Chlorate and Bromate [33]

Over 100 years ago, Kipping and Pope examined the frequency of enantiomorphs, that is, crystals rotating polarized light either clockwise (+) or counterclockwise (−), when $NaClO_3$ was crystallized from water [35]. They speculated correctly that there was no reason to believe that one of the chiral forms would crystallize preferentially over the other form. After they collected and examined the handedness of $NaClO_3$ crystals by polarimetry from a single crystallization experiment and then repeating the crystallization experiment a large number of times, the researchers concluded that the crystallization of sodium chlorate is random and thus gives a random distribution of + and − crystals. Primary nucleation randomly generates a + or − crystallite which then serves as a seed to generate more crystals of the same handedness through secondary nucleation.

Another study [23] looked at the effect of secondary nucleation on the distribution of + and − sodium chlorate crystals under a range of conditions (supersaturation, impurity concentration, supercooling, etc.). Some specific conditions would produce only the enantiomorph corresponding to the seed crystal used, while other conditions existed whereby both enantiomorphs would crystallize from the seeded solution. These results led the authors to conclude that multiple mechanisms exist for secondary nucleation.

A startling observation [26] was reported by Kondepudi and coworkers in 1990 that demonstrated spontaneous chiral autocatalytic resolution with crystallizing sodium chlorate crystals. Kondepudi and his students first repeated the experiments of Kipping and Pope and obtained the same results. The researchers then made a simple adjustment to the experimental procedure that produced spectacular results; instead of letting the sodium chlorate solutions evaporate to induce crystallization, Kondepudi and coworkers stirred the evaporating solutions. This one simple adjustment led to each batch of crystals being nearly entirely + or − in every experiment, but with no overall predisposition towards one form over the other.

Kondepudi et al [26]. concluded that secondary nucleation is responsible for their unusual observation. Stirring, it should be noted, had been previously shown to promote secondary nucleation [20]. In another study, Kondepudi et al [36]. used computer simulations to investigate the kinetics of the secondary nucleation. The simulations assumed that the crystallizing solutions were homogeneous in temperature, concentration, etc., which though perhaps not entirely accurate, simplified the problem at hand. The results from their study were that a minimum-sized crystal exists for the secondary nucleation mechanism to spontaneously break symmetry macroscopically. Microscopically, this indicates that the first crystal formed within the solution must grow to a "critical" size before fragmenting into a pair of crystals of the same handedness of the original crystal.

A study by Szurgot and Szurgot [37] brought the results of the stirring experiments of Kodepudi and his group into question. In the new experiments, sodium chlorate crystals were grown in the bottom of a crystallizer at differing temperatures; the first and subsequent crystals were analyzed for their handedness. Szurgot and Szurgot concluded that there was a trimodal distribution of + and – crystals when the number of crystals from the batch was less than 50. A critique of Kipping's and Pope's work [36] was that in the previous experiments there were not enough crystals to guarantee that symmetry breaking was present. That may be the case within this work because the trimodal distribution is only present with the smaller crystal population.

In 2004, Viedma [38] demonstrated that secondary nucleation may not be the only mechanism of autocatalytic chiral resolution in sodium chlorate stirred crystallization. The supersaturated sodium chlorate solutions were prepared and not allowed to spontaneously crystallize; the solutions were then stirred vigorously (approx. 1000 rpm) which resulted in an almost instantaneous appearance of microcrystals. The microcrystals were estimated to be approximately 20 μm in diameter. An analysis of the handedness of the microcrystals revealed that they were almost completely of the same morphology. The rate at which the crystals appeared would seem to indicate that the mechanism of secondary nucleation is not plausible in this instance, yet the handedness of all of the crystals remained the same. It should be noted that these results do not preclude Kondepudi's assertion from being true. The research of Viedma indicates that differing scenarios and mechanisms may be responsible for the spontaneous chiral resolution of sodium chlorate crystals. A more recent study by Viedma [39], using similar experimental conditions to his initial work, shows that any slight abundance of one enantiomer can give rise to complete purity. Viedma concludes that this observation is caused by the combination of nonlinear autocatalytic dynamic of secondary nucleation and the recycling of crystallites when they reach the achiral molecular level during formation.

Mahurin et al. discovered another way in which to perturb the distribution of + and – sodium chlorate crystals formed from evaporating aqueous sodium chlorate [40]. The solutions were exposed to either very energetic beta particles from the nuclear decay of strontium-90 or positrons from the decay of sodium-22. Because of the innate asymmetry of nuclei, the beta particles are predominantly left handed, while the positrons are predominantly right handed. The beta particles afforded an enantiomeric excess of up to 42%, with the + crystals predominating, while the positron afforded an enantiomeric excess of 55%, with the – crystals predominating. There is clearly a strong correlation between the helicity of the perturbing energetic particles and the chirality of the resulting sodium chlorate crystals. Because of the interaction of the electrons and positrons with the aqueous solutions, yielding circularly polarized secondary electrons and gamma rays and a complex chemistry, the mechanism for this asymmetric induction is unknown.

Sodium chlorate and bromate are isomorphous because they are structurally related compounds and crystallize in the same space group, but crystals of the two compounds with the same sign of optical rotation have opposite absolute configurations [33]. These facts have interesting consequences when a crystal of one compound is used to seed the crystallization of the other compound from solution [41]. When aqueous sodium chlorate is seeded with + sodium bromate, – sodium chlorate crystals result. On the other hand, seeding with – sodium bromate crystals yields + sodium chlorate crystals. Likewise, when aqueous sodium bromate is seeded with + or – sodium chlorate crystals, – or + sodium bromate crystals are generated, respectively.

Approximately a decade ago, a very unusual observation was made when sodium bromate was crystallized from water in the absence of any perturbation. As shown in Table 1, which summarizes a series of seventy crystallizations carried out at Wake Forest University (WFU) and The University of Tennessee (UT), an overwhelming number of + sodium bromate crystals were obtained. Similar and equally dramatic results were obtained at Georgia College and State University (Dr. McGinnis and his students) as well as by Dr. Claire Vallance (now at Cambridge University) at The University of Christchurch (New Zealand). In 30 crystallization experiments, 5659 crystals were obtained of which 5598 were +. The results were independent

TABLE 1 The Anomalous Crystallization of Sodium Bromate

WFU				UT			
Run#	(−) $NaBrO_3$	(+) $NaBrO_3$	%(+) $NaBrO_3$	Run#	(−) $NaBrO_3$	(+) $NaBrO_3$	%(+) $NaBrO_3$
1	0	165	100.0	34	1	53	98.1
2	3	182	98.4	35	1	49	98.0
3	0	121	100.0	36	0	52	100.0
4	3	91	96.8	37	0	34	100.0
5	1	54	98.2	38	0	66	100.0
6	56	30	34.9	39	2	23	92.0
7	0	124	100.0	40	0	41	100.0
8	2	36	94.7	41	0	13	100.0
9	0	70	100.0	42	4	31	88.6
10	3	83	96.5	43	3	17	85.0
11	63	111	63.8	44	0	34	100.0
12	0	226	100.0	45	22	23	51.1
13	0	36	100.0	46	4	18	81.8
14	67	24	26.4	47	0	17	100.0
15	0	334	100.0	48	3	42	93.3
16	0	111	100.0	49	8	15	65.2
17	1	44	97.8	50	37	34	47.9
18	2	27	93.1	51	27	18	40.0
19	0	22	100.0	52	14	13	48.1
20	0	49	100.0	53	19	5	20.8
21	8	17	68.0	54	13	13	50.0
22	35	262	88.2	55	15	11	42.3
23	40	115	74.2	56	18	22	55.0
24	35	123	77.8	57	15	10	40.0
25	36	159	81.5	58	16	23	59.0
26	8	90	91.8	59	21	21	50.0
27	44	82	65.1	60	29	14	32.6

TABLE 1 The Anomalous Crystallization of Sodium Bromate—Cont'd

WFU				UT			
Run#	(−) $NaBrO_3$	(+) $NaBrO_3$	%(+) $NaBrO_3$	Run#	(−) $NaBrO_3$	(+) $NaBrO_3$	%(+) $NaBrO_3$
28	40	33	45.2	61	43	21	32.8
29	84	21	20.0	62	17	34	66.7
30	84	161	65.7	63	15	54	78.3
31	9	83	90.2	64	34	6	15.0
32	16	221	93.2	65	42	12	22.2
33	20	78	79.6	66	39	34	46.6
				67	16	16	50.0
				68	23	8	25.8
				69	15	34	69.4
				70	13	12	48.0

of the salt's source and gave the same results when carried out in a glove bag under nitrogen. A variety of spectroscopic experiments demonstrated that there were no measurable concentrations of biological contaminants in the evaporating solutions. Whatever the source of the asymmetric crystallization of sodium bromate, it did not affect the crystallization of sodium chlorate which behaved normally. Viedma has recently reported a cryptochiral environmental effect in the crystallization of both sodium chlorate and sodium bromate [42]. Quite unexpectedly, the unusual behavior has disappeared in recent years at both WFU and UT. It certainly would be nice to know how such anomalous behavior, which occasionally has been reported in the literature for other compounds such as barium nitrite, occurs because this may provide important clues on the nature of primary nucleation.

II. THE EFFECT OF INTENSE LASER RADIATION ON PRIMARY NUCLEATION

In 1996, Garetz et al [43]. published a paper on the photochemically induced nucleation of supersaturated urea solutions. Subjecting the solutions to 1064-nm pulses from a Nd:YAG laser resulted in the spontaneous nucleation of urea and, interestingly, the orientation of the linearly polarized light from the laser dictated the plane of growth for the initially formed crystals. The interpretation of these remarkable results was that the electric field of the light aligned the molecules in its path (vertical or horizontal depending upon the polarization) similar to the optical Kerr effect [44]. Briefly, the optical Kerr effect occurs when an applied electric field induces a dipole moment in the liquid. The applied electric field then simultaneously interacts with the induced dipole moment applying a torque that causes the molecule to align its most polarizable axis parallel to the electric field. If the effect occurs in an amorphous aggregate of sufficient size, primary nucleation will occur. Garetz et al. termed this phenomenon for urea nonphotochemical laser induced nucleation (NPLIN).

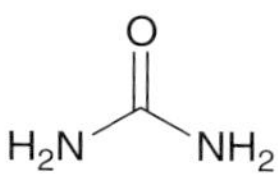

Urea

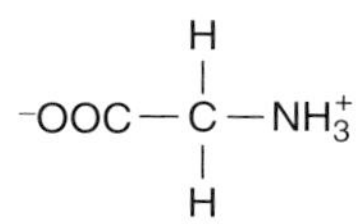

Glycine

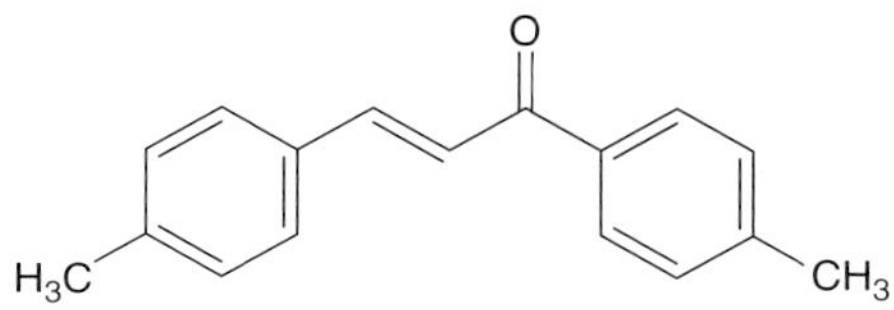

4,4′-Dimethychalcone

Garetz extended the NPLIN work from urea to aqueous solutions of glycine [45]. Before discussing Garetz's results, it is worth spending a moment on the nature of crystalline glycine. Glycine, the simplest amino acid, is achiral and crystallizes into three different polymorphic forms: α, β, and γ, which is chiral. The crystal structure of α-glycine was first established by Albrecht and Corey [46]. The crystal structure reveals double layers of hydrogen bonded molecules packed via van der Waals forces. The β form of glycine is the least stable of the three polymorphs and is generally not formed on crystallization. From the perspective of chiral discrimination, γ glycine is the most interesting of the three polymorphs. Gamma glycine is a strong piezoelectric crystal and crystallizes with trigonal hemihedral symmetry, which was first reported by Iitaka [47]. Interestingly, several publications [48–50] have described different methods of growing γ-glycine crystals. Iitaka [48] reported the growth of γ-glycine by slowly cooling aqueous solutions of glycine and acetic acid or ammonium hydroxide. Bhat and Dharmaprakash [49] reported the formation of γ-glycine from the crystallization of aqueous solutions containing minute quantities of sodium chloride, and Yu and Ng [50] discovered the importance of pH in the formation of γ-glycine.

It is well known that molecules with ionizable groups (NH_3^+ and COOH among others) can crystallize into either neutral species or salts with counter ions. Glycine, for example, with both ammonium and carboxyl groups will exist in solution as an admixture of positively charged, neutral, and negatively charged species, whose composition is pH dependent. The particular form of glycine in solution is obviously paramount to the preferred polymorph formed on crystallization. In α-glycine, cyclic dimers pack together to form a double layer that is hydrogen bonded together, and α-glycine is typically formed unless the solution is acidic or basic [48]. The elemental growth of α-glycine has been previously explained [51] as being pictured as sheets of glycine molecules positioned perpendicular to the b-axis of the molecule. Each sheet of hydrogen bonded molecules forms a chain and is hydrogen bonded to an opposing sheet to form a bilayer, and the bilayers pack together via van der Waals forces. Conversely, γ-glycine crystallizes into helical chains with a threefold symmetry that are packed together hexagonally through lateral hydrogen bonds [48]. Because of its helical nature, γ-glycine possesses chirality that is established through the solid state instead of a traditional chiral carbon center.

Garetz et al. [43] prepared supersaturated solutions of glycine in water, with concentrations ranging from 3.7 to 3.9 M, sealed them in screw-cap vials, and then "aged" for an appropriate time before subjecting them to laser light. It was suggested [51] that "aging" allowed larger glycine clusters to form, thus increasing the probability of nucleation. The light consisted of 1064-nm radiation from a Quanta-Ray DCR Nd:YAG laser, with peak intensities estimated to be 0.7 (±10%) GW/cm^2. It was assumed that the circular vials provided a slight focusing of the radiation. Samples were exposed to radiation for one minute before resealing and allowing the glycine to nucleate. Crystals were typically observed 30 min after exposure of the solutions to the laser, though solutions did not nucleate for every exposure. The crystals that were formed were analyzed with X-ray diffraction (XRD). Surprisingly, γ-glycine crystals were formed during this light-induced process, whereas α-glycine typically crystallized in the absence of laser light [52].

Garetz et al. [53] later investigated the effects of polarization of the incident laser upon the crystallizing glycine solutions. The experimental details for these sets of experiments were consistent with the previous study except that a $\lambda/4$ waveplate was used to generate circularly polarized light. The results of these experiments were that linearly polarized light efficiently produced γ glycine crystals via NPLIN and that α-glycine crystallized when circularly polarized light was used. These results appeared to be consistent with the earlier proposed optical Kerr effect. Briefly, the polarizability of the γ glycine helix most closely resembles that of a rod and is most efficiently aligned via linearly polarized light. Conversely, the bilayers of α-glycine are similar to disks and are most easily aligned by circularly polarized light. Thus, the polarization of the laser dictated the manner in which the glycine clusters are aligned.

4,4′-Dimethylchalcone is a polarizable organic molecule with a long axis which crystallizes in a chiral space group [54]. Kondepudi and his students discovered that crystallization of the molecule from ethyl acetate yielded a random distribution of left-handed and right-handed crystals [55]. When the evaporating solutions were stirred, a bifurcated distribution of left-handed and right-handed crystals were obtained, a result reminiscent of what occurred when sodium chlorate was crystallized from water. Primary nucleation randomly yields a left-handed or right-handed crystal which is amplified by stirring through secondary nucleation. Murphy et al. have extended the crystallization experiments of the chalcone [56]. The unperturbed crystallizations yielded a random distribution of enantiomeric crystals, while stirring yielded a bifurcated distribution, both experiments being in agreement with the results of Kondepudi. In addition, the evaporating ethyl acetate solutions were subjected to beta radiation from a Sr-90 source and linearly, left-handed, and right-handed circularly polarized light with a wavelength of 1064 nm, all three forms of light generated from a pulsed Nd:YAG laser. The beta radiation and the two forms of circularly polarized light yielded random distributions of left-handed and right-handed crystals. The results of the beta radiation are quite different in character from what was observed for sodium chlorate in water. The linearly polarized light afforded a bifurcated distribution of enantiomeric crystals even larger than was obtained by stirring. This is a consequence of the optical Kerr effect. The chalcone exists in columns in the crystal, the molecules in a column being coiled either left handed or right handed depending on which enantiomeric crystal is examined. When a sufficiently large amorphous aggregate is exposed to the linearly polarized light, the chalcone molecules are aligned in the same direction found in the crystal. This facilitates primary nucleation which still generates a left-handed or right-handed crystal randomly. Secondary nucleation then facilitates the formation of crystals of the same asymmetry as that initially formed. It is currently unclear if the light facilitates secondary nucleation. Circularly polarized light had no effect on the crystallization because the chalcone does not exist as discs within the crystal structure.

III. SOUND-INDUCED CRYSTALLIZATION

Following the reports of Garetz et al.[43] on non-photochemical light-induced crystallization, we attempted to reproduce some of their observations. The success and failures of some of these experiments may be found in the Ph.D. dissertation of Fischer.[57] While our initial intention was to verify the results of Garetz on the laser-induced crystallization of glycine from water and then apply the methodology to the crystallization of sodium bromate from water, some very interesting effects were discovered. It was easily observed that the 1064-nm pulses from the Nd:YAG laser also generated sound within the solution which resulted in compression pressure waves within the solution. These waves induced crystal growth throughout the beaker. In order to separate the effect of light from that of sound on the supersaturated solutions, experiments were devised to look at the effect of sound on the crystallization exclusively. This entailed placing a metal boat on the surface of the liquid and then impinging the light onto the surface of the boat. This generated sound at the surface of the liquid which then passed through it. A pictorial representation of the setup for the light-induced and sound-induced crystallization is shown in Figure 1.

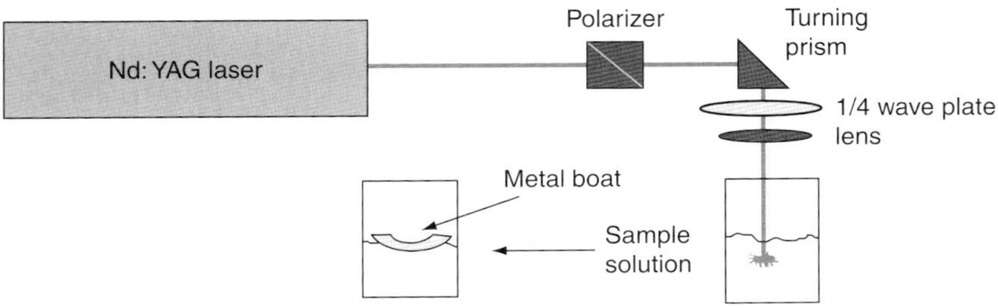

Identification of α- and γ-glycine was carried out through a variety of means, X-ray powder diffraction, solid state ^{13}C NMR, and Raman spectroscopy.[58,59] In the solid state ^{13}C spectrum, the C=O carbon of α-glycine occurs at $\delta = 176.50$ ppm, while the C=O carbon of γ-glycine occurs at $\delta = 174.60$ ppm. The reported peaks are specific to the carboxyl carbon and appear at different chemical shifts due to the slight differences in the hydrogen bonding in the solid state. The Raman spectrum shows that γ-glycine possesses two vibrations around 1340 cm^{-1}, while α-glycine has vibrations at 1320 and 1410 cm^{-1}.

The sodium bromate crystals that were produced during the experiments were often too small to analyze individually for the specific optical activity of each crystal. The methodology developed by Bartus and Vogl [34] was utilized to obtain a quantitative measure of the enantiomeric excess by placing the crystals in a matching index of refraction liquid (carbon tetrachloride plus carbon disulfide). We define the measured rotation per gram of crystals produced as a measure of the enantiomeric excess.

Eight sound experiments were carried out on the supersaturated solutions of glycine, all of which resulted in crystallization. In four experiments, α-glycine predominated, while in the other four γ glycine predominated, as one would expect when the crystallization is induced by achiral sound waves. During the experiments, "micro"-crystals of glycine could be seen "falling" from the bottom of the metal boat. The sound wave experiments resulted in the production of many more crystals of a much smaller size than was typically observed in the light-induced experiments.

Twenty experiments were carried out on the supersaturated sodium bromate solutions, 16 of which resulting in crystallization. The overall average rotation was found to be $-3.91 \pm 2.62°/g$ of sample crystallized, a value deviating somewhat from 0°. Figure 2 shows the results of the sound and control (samples allowed to spontaneously form crystals over week long periods) samples plotted according to their enantiomeric excess, EE, which is given by

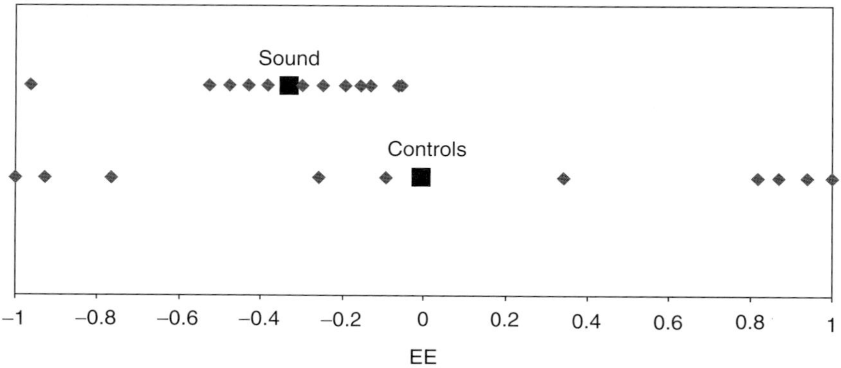

FIGURE 2 Comparison of enantiomeric excess, EE, for sound-induced and control experiments. An EE of 1 or − 1 would refer to a pure (+) or (−) crystal respectively. The square for each series represents the average for the data set.

$$EE = \frac{[\alpha]_0^{\mathrm{D}}}{[\alpha]^{\mathrm{D}}}, \tag{21}$$

where $[\alpha]_0^{\mathrm{D}}$ is the observed optical rotation at the sodium D line and $[\alpha]^D$ is the optical rotation at the sodium D line for a pure (+) crystal. To determine $[\alpha]^{\mathrm{D}}$, single crystals known to be pure (+) form were ground into powder to be measured via Vogl's method [34]. As the measured rotation is dependent upon the amount of sample present, the measurements were standardized to the mass of sample utilized. If an equal number of (+) and (−) crystals were present in solution, the measured rotation would be exactly 0° as the rotational effects would cancel each other, and an EE of 0 would be observed, as is essentially the case for the average of the control samples. As the sound-induced experiments indicate from Figure 2, the average value is offset from 0. It should be noted that as was the case with glycine crystallizations, sodium bromate crystals are seen coming down from the "boat" almost immediately after exposure to the sound wave, and all of them are less than one millimeter in size. Figure 3 shows two of many crystallizations for sodium bromate crystals formed under natural crystal growth conditions (left) and those formed from laser sound-induced crystallization (right).

The hundreds of sound-induced crystals were allowed to grow for a few hours to allow for easier visualization. Experiments are underway to harvest the sound-induced crystals immediately after application of the laser in order to allow observation in a Transmission Electron Microscope. The fact that there is a small excess of (−) crystals produced under laser sound conditions is very similar to the observations noted previously for crystallization of sodium bromate (see above). It is possible that we have not examined enough crystallizations or that there is a chiral impurity in the solution giving rise to

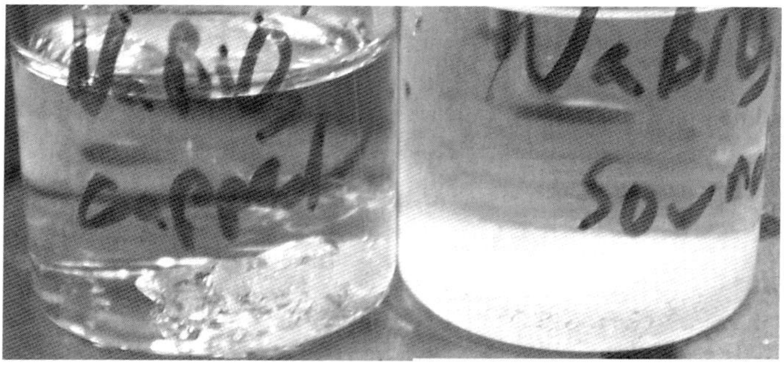

FIGURE 3 Photograph of sodium bromate crystals grown over many days from conventional crystal growth procedure from supersaturated solutions (left photo) as compared with the hundreds of sodium bromate crystals formed from laser sound-induced crystallization (right photo). The tiny laser sound-induced crystals were allowed to grow for a few hours to allow for better visualization.

an excess. One has to include in these possibilities the very unlikely effect of the chiral weak interaction (see discussions in Refs. [33] and [60]).

The increase in nucleation rate due to external factors such as electromagnetic fields (which includes lasers) or sound waves can be explained using the rate expression given in Equation (10). The change in any of the terms, T, $\Delta\mu$, or others, results in a change in the nucleation rate. For sound-induced nucleation, the changes in $\Delta\mu = (\mu_1 - \mu_2)$, the chemical potential difference between the solution phase and the solid phase, and T could be significant as an intense sound wave propagates through the solution. In the case of sound, the dominant changes in the physical properties would be pressure and temperature. Let us assume p and T change from an initial value p_0 and T_0 to $p_0+\delta p$ and $T_0+\delta T$. The corresponding change in $\Delta\mu$ can be written as:

$$\Delta\mu(p_0+\delta p, T_0+\delta T) = \Delta\mu(p_0, T_0) + \int_{p_0}^{p_0+\delta p}\left(\frac{\partial\Delta\mu}{\partial p}\right)_T \mathrm{d}p + \int_{T_0}^{T_0+\delta T}\left(\frac{\partial\Delta\mu}{\partial T}\right)_p \mathrm{d}T. \quad (22)$$

Noting that the derivative $(\Delta\mu/\partial p)_T = V_\mathrm{m}$, the molar volume, and that $(\Delta\mu/\partial T)_p = -S_\mathrm{m}$, the molar entropy, we see that Equation (22) can be written in terms of the changes in molar volume ΔV_m and molar entropy ΔS_m associated with nucleation. For crystallization from solution, in the first approximation, we may assume that ΔV_m and ΔS_m are nearly constant and write the Equation (22) as

$$\Delta\mu(p_0+\delta p, T_0+\delta T) = \Delta\mu(p_0, T_0) + \Delta V_\mathrm{m}\delta p - \Delta S_\mathrm{m}\delta T. \quad (23)$$

During crystallization from solution, the molar volume of the solute in solution phase is larger than that of the solid phase, that is, $\Delta V_m > 0$. Similarly, the molar entropy change ΔS_m is also positive. This implies that increases in p and T have opposite effects on $\Delta\mu$; one increases $\Delta\mu$ while the other causes it to decrease. The overall effect of these changes in the nucleation rate can be calculated by substituting these terms in the expression for the nucleation rate, Equation (10):

$$J = A\exp\left\{-\frac{16\pi}{3}\frac{\gamma^3 V_m^2}{[\Delta\mu(p_0, T_0) + \Delta V_m\delta p - \Delta S_m\delta T]^2 k_B(T_0 + \delta T)}\right\}. \qquad (24)$$

As we have already noted, one could use the approximation $\Delta\mu(p_0,T_0) = RT\ln(S)$, in which S is the supersaturation parameter. For the effects of sound, as changes in pressure and T are periodic functions we may use the root-mean-square values of δp (which in turn is related to the sound intensity) and δT. Using the subscripts "rms" for these quantities, we obtain the nucleation rate for sound-induced crystallization:

$$J = A\exp\left\{-\frac{16\pi}{3}\frac{\gamma^3 V_m^2}{[RT\ln(S) + \Delta V_m\delta p_{rms} - \Delta S_m\delta T_{rms}]^2 k_B(T_0 + \delta T_{rms})}\right\}. \qquad (25)$$

From this expression, it is clear that an increase in nucleation rate due to sound waves will result only when $(\Delta V_m\delta p_{rms} - \Delta S_m\delta T_{rms}) > 0$. By considering the change in the electrochemical potential in the presence of electromagnetic radiation, similar expression for the change in the nucleation rates due to lasers could be obtained.

Acknowledgments

This research was supported by the National Science Foundation (NSF CHE-0650524). We are indebted to Dr M. McGinnis (Georgia College and State University) and Dr Claire Vallance (Oxford University) for sharing their results on crystallization from sodium bromate and sodium chlorate solutions.

REFERENCES

[1] Khanna SN, Jena P. Assembling crystals from clusters. Phys Rev Lett 1992;69:1664.

[2] (a) Krachmer W, Lamb LD, Fostiropoulis K, Huffman DR. Solid c_{60}: A new form of carbon. Nature (London) 1990;347:354. (b) Kroto HW, Heath JR, O'Brien SC, Curl RF, Smalley RE. C_{60}: buckminsterfullerene. Nature (London) 1985;318:162.

[3] Martin TP, Näher U, Schaber H, Zimmermann U. Clusters of fullerene molecules. Phys Rev Lett 1993;70:3079.

[4] Mackay AL. A dense non-crystallographic packing of equal spheres. Acta Crystallogr 1962;15:916.

[5] Young S. Mechanical stimulus to crystallization in super-cooled liquids. J Am Chem Soc 1911;33:148.
[6] Berkeley E. Philos Mag 1912;24:254.
[7] Khamskii E. Crystallization from solutions. New York: Consultants Bureau; 1969.
[8] Christian J. The theory of the transformations of metals and alloys. Oxford: Pergamon Press; 1965.
[9] Kelton K. In: Ehrenreich H, Turnbull D, editors. Solid state physics, vol. 45. New York: Academic Press; 1991.
[10] Hoare M, McInnes J. Statistical mechanics and morphology of very small atomic clusters. Disc Faraday Soc 1976;61:24.
[11] Gibbs J. Collected works, vol. 1, thermodynamics. New Haven: Yale University Press; 1948.
[12] Volmer M. Kinetic der phasenbildung. Leipzig: Steinkopff; 1939.
[13] Becker R, Doring W. Kinetische behandlung der keimbildung in übersättigten dämpfern. Ann Phys 1935;24:719.
[14] Nielsen A. Kinetics of precipitation. Oxford: Pergamon Press; 1964.
[15] Strickland-Constable R. Kinetics and mechanism of crystallization. New York: Academic Press; 1968.
[16] Sohnel O, Garside J. Precipitation: basic principles and industrial applications. Oxford: Butterworth-Heinemann; 1992.
[17] Kaschiev D. Nucleation. Oxford: Butterworth-Heinemann; 2000.
[18] Botsaris G, Denk E, Ersan G. Crystallization-part1. annual review- transport phenomena of nucleation and crystal growth. Ind Eng Chem 1969;61:86.
[19] Qian R, Botsaris G. A new mechanism for nuclei formation in suspension crystallizers: the role of interparticle forces. Chem Eng Sci 1997;52:3429.
[20] Botsaris G, Denk E. Annu Rev Ind Eng Chem 1972;1970:493.
[21] Randolph A, Larson M. Theory of particulate processes. New York: Academic Press; 1971.
[22] Qian R, Botsaris D. Nuclei breeding from a chiral crystal seed of naclo$_3$. Chem Eng Sci 1998;53:1745.
[23] Denk E, Botsaris G. Fundamental studies in secondary nucleation from solution. J. Crystal Growth 1972;13/14:493.
[24] Yokota M, Oguchi T, Arai K, Toyokura K, Inoue C, Naijyo H. In: Myerson A, Toyokura K, editors. Crystallization as a separations process. ACS Symposium No. 438, 1992. p. 271.
[25] Davey R, Black S, William L, McEwan D, Sadler D. The chiral purity of a triazolylketone crystallised from racemic solutions. J Crystal Growth 1990;102:97.
[26] Kondepudi D, Kaufmann R, Singh N. Chiral symmetry breaking in sodium chlorate crystallizaton. Science 1990;250:975.
[27] Saratovkin D. Dendritic crystallization. New York: Consultants Bureau; 1959.
[28] Ostwald W. Lehrbuch der algemeinen chemie. Leipzig: Englemann; 1896.
[29] Ostwad WZ. Studien uber die bildung und umwandlung fester korper. Phys Chem 1897;22:289.
[30] Dufor L, Defay R. Thermodynamics of clouds. New York: Academic Press; 1963.
[31] Cardew P, Davey R. Symposium on the tailoring of crystal growth. London: Institution of Chemical Engineers; 1982 pp. 1.1–1.9.
[32] Bernstein J. Polymorphism in molecular crystals. Oxford: Oxford Clarendon Press; 2002.
[33] Pagni RM, Compton RN. Asymmetric synthesis of optically active sodium chlorate and bromate crystals. Cryst Growth Des. 2002;2:249.
[34] Bartus J, Vogl O. Measurement of optical activity of isotropic compounds in suspension. Monatsh Chem 1993;124:217.
[35] Kipping F, Pope W. Enantiomorphism. J Chem Soc 1898;12:606.

[36] Kondepudi D, Bullock K, Digits J, Hall J, Miller J. Kinetics of chiral symmetry breaking in crystallization. J Am Chem Soc 1993;115:10211.
[37] Szurgot M, Szurgot J. Chiral symmetry breaking in sodium chlorate crystallization from unstirred solution. Cryst Res Technol 1995;30:949.
[38] Viedma C. Experimental evidence of chiral symmetry breaking in crystallization from primary nucleation. J Cryst Growth 2004;261:11.
[39] Viedma C. Chiral symmetry breaking during crystallization: complete chiral purity induced by nonlinear autocatalysis and recycling. Phys Rev Lett 2005;94:065504.
[40] Mahurin S, McGinnis M, Board JS, Hulett LD, Pagni RM, Compton RN. Effect of beta radiation on the crystallization of sodium chlorate from water: a new type of asymmetric synthesis. Chirality 2001;13:636.
[41] Vogl O, Qin M, Bartus J, Jaycox GD. Chiral nucleation. Monatsh Chem 1995;126:67.
[42] Viedma C. Selective chiral symmetry breaking during crystallization: parity violation or cryptochiral environment in control? Cryst Growth Des 2007;7:553.
[43] Garetz BA, Aber J, Goddard NL, Young RG, Myerson AS. Nonphotochemical, polarization-dependent, laser-induced nucleation in supersaturated aqueous urea solutions. Phys Rev Lett 1996;77:3475.
[44] Reinthjes JF. Nonlinear optical parametric process in liquids and gases. Berlin: Springer; 1991.
[45] Zaccaro J, Matic J, Myerson A, Garetz B. Nonphotochemical, laser-induced nucleation of supersaturated aqueous glycine produces unexpected γ-polymorph. Cryst Growth Des 2001;1:5.
[46] Albrecht G, Corey RB. The crystal structure of glycine. J Am Chem Soc 1939;61:1087.
[47] Iitaka Y. The crystal structure of γ-glycine. Acta Crystallogr 1958;11:225.
[48] Iitaka Y. The crystal structure of γ-glycine. Acta Crystallogr 1961;14:1.
[49] Bhat MN, Dharmaprakash SM. Growth of nonlinear optical γ-glycine crystals. J Cryst Growth 2002;236:376.
[50] Yu L, Ng KJ. Glycine crystallization during spray drying: the ph effect on salt and polymorphic forms. Pharm Sci 2002;91:2367.
[51] Gidalevitz D, Feidenhnas'l R, Matlis S, Smilgies D, Christensen MJ, Leiserowitz L. " monitoring in situ growth and dissolution of molecular crystals: towards determination of the growth units. Angew Chem Int Ed 1997;36:95.
[52] Myerson AS, Lo PY. Diffusion and cluster formation in supersaturated solutions. J Cryst Growth 1990;99:1048.
[53] Garetz BA, Matic J, Myerson A. Polarization switching of crystal structure in the nonphotochemical light-induced nucleation of supersaturated aqueous glycine solutions. Phys Rev Lett 2002;89:175501-1–4.
[54] Green B, Rabinovich D, Shakked Z. Acta Crystallogr 1981;B37:1376.
[55] Durand DJ, Kondepudi DK, Moreira P. Generation of molecular chiral asymmetry through stirred crystallization. Chirality 2002;14:284.
[56] Murphy NC, Compton RN, Pagni RM. Effect of chiral and achiral perturbations on the crystallization of 4,4'-dimethylchalcone from ethyl acetate. Cryst Growth Des 2007;7:449.
[57] Fischer A. F. The University of Tennessee: Ph.D. dissertation; 2006.
[58] Krishnan RS, Balsubramanian K. Proc Ind Acad Sci A 1958;48A:55.
[59] Balasubramanian K, Krishnan RS, Iitaka Y. Raman spectrum of γ-glycine. Bull Chem Soc Jpn 1962;35:1303.
[60] Compton RN, Pagni RM. The Chirality of Biomolecules. Adv At Mol Opt Phys 2002;48:219.

Chapter 9

Superatoms: From Motifs to Materials

Arthur C. Reber*, Shiv N. Khanna* and A. W. Castleman. Jr[†]
*Department of Physics, Virginia Commonwealth University, Richmond, Virginia, USA
[†]Departments of Chemistry and Physics, Pennsylvania State University, University Park, Pennsylvania, USA

Chapter Outline Head

I. INTRODUCTION

Cluster assemblies provide an avenue toward building materials with a wide variety of tunable properties. The properties of clusters can be altered by changing the number of atoms, by replacing an atom with a different element, and by assembling them into different architectures [1]. The reactivity [2,3], magnetic properties [4,5], electronic structure, and optical properties of clusters can change strikingly by adding or removing even a single atom. The characterization and cataloging of all of the potential properties is a daunting task, and the organization of the enormous numbers and varieties of clusters into a useful framework would clarify and instruct potential cluster

Nanoclusters. DOI: 10.1016/S1875-4023(10)01009-0

assemblies and properties. A simple and elegant framework in which clusters may be organized is by characterizing those whose properties and reactivity may be approximated by those of an atom and which behave like a superatom [6–12]. This terminology is not meant to minimize the unique character and properties of clusters, but rather to manage them in a reasonable frame in which the general properties of the clusters may be predicted.

The ultimate test of whether the term superatom is merely a provincial organizing principle or a useful scheme for material synthesis is the construction of cluster assembled materials. The idea that clusters might provide building blocks of solids with unique electronic, optical, magnetic, and thermodynamic properties was suggested by Khanna and Jena [12]. More recently, there has been intense interest in synthesizing materials from clusters that *mimic elements* and are termed superatoms [7,8]. For example, a unique 102 atom gold cluster assembled material has in fact been synthesized using thiol ligands [13]. The clusters stability may be reconciled using a simple electron counting rule considering that the cluster is stabilized by 44 sulfur atoms which are electron withdrawing. This results in a 58 electron jellium shell closing, and demonstrates that the material stability and the stability of the pure clusters are connected. Additional gold-based cluster assembled materials have been recently synthesized [14,15]. Numerous Zintl phases also have been synthesized, which consist of polyatomic post transition metal clusters usually mixed with alkali metals as countercations [11,16–18]. Zintl ions have a strong propensity toward a particular oxidation state much like an atom, and in those cases can function as superatoms. While the superatom concept is useful for understanding the chemistry and physics of gas phase clusters, it is the ability to synthesize materials out of these clusters which is most exciting. Indeed, the fact that the superatom element mimics would acquire different periodicity in solids and have different intercluster and intracluster length scales makes the prospects for obtaining materials with unique properties intriguing.

This review focuses on recent progress in the clusters and cluster assembly of aluminum-based superatoms. We begin with a brief introduction of the jellium concept and its importance in identifying electron shell closure in clusters, particularly aluminum ones in the present case. Next, we review the properties of Al_{13} [3,6–8], Al_{14} [8], and Al_7 as they pertain to the superatom concept. This includes the multiple valence observed in Al_7 based clusters [9]. We also discuss the potential for the assembly of aluminum clusters using the superalkali K_3O, and see how the assembly effects the electronic structure [3]. To further illustrate the potential for cluster assembled materials, we discuss some recent work on the reactivity of aluminum-based clusters with O_2 which indicates how tuning the electronic structure can result in highly different reactivity [2]. Lastly, we discuss the future directions and possibilities in how the superatom concept may guide the synthesis of intriguing new materials.

II. THE JELLIUM MODEL

In 1984, Walter Knight and coworkers [19] took a mass spectrum of sodium clusters generated by vaporizing the sodium and passing the vapor through a supersonic nozzle. The resulting mass spectra of the generated clusters showed that clusters containing 2, 8, 18, 20, 40, 58, and 92 atoms were more prominent than other sizes. These numbers were called “magic numbers” and experiments on other alkali metals revealed similar magic species indicating that they were not unique to sodium. Since alkali atoms have one valence electron, the prominence in the mass spectra could either be due to the electronic counts or to other features such as special geometries at these sizes. That the observed phenomenon was primarily electronic was indicated by experiments that showed a drop in ionization potential at sizes corresponding to an electronic count-one above the magic species [19], and low values of electronic affinities at the magic sizes. Further experiments on the dissociation of clusters showed that the magic sizes were the favored fragmentation products [20,21] and that it required more energy to fragment magic species. All these observations revealed that the clusters with these magic numbers are stable species relative to other sizes, and that their stability is probably rooted in their electronic structure rather than their geometric structure, although there may arise a synergistic effect in cases where they both are operative in a particular cluster. Knight and coworkers proposed that the stability of magic clusters could be understood within a simple jellium picture. The concept of superatoms in aluminum clusters relies on the electronic structure of clusters being analogous to the electronic structure in atoms, an analogy which we consider in further detail.

The electronic structure of metals may be described within a simple picture where the valence electrons form a nearly free electron gas weakly perturbed by the potential generated by the ionic cores [22,23]. In fact, the ionic potential is usually replaced by a pseudo-potential that is much weaker and whose lowest eigenstates correspond to the valence states. Therefore, a simpler approximation along these lines is the jellium model where one ignores the lattice structure of the ionic cores altogether and replaces them by a uniform positive background. Ekardt [24] and Knight and coworkers proposed a similar model for clusters where the cluster was replaced by a spherically symmetric potential well, with the positive charge density being uniform inside the sphere and zero outside. The results of two simple jellium models are shown in Figure 1. The electronic levels in such a potential well are characterized by a radial and angular momentum quantum numbers, nl, and results in electronic structures strikingly similar to atoms. Because the radial potential in clusters is different than in atoms, most clusters may adopt the 1S, 1P, and 1D levels before adding radial terms to the wave function to ensure orthogonality. In some smaller clusters and fullerenes, situations may arise in which a radial quantum number no longer applies as the electronic

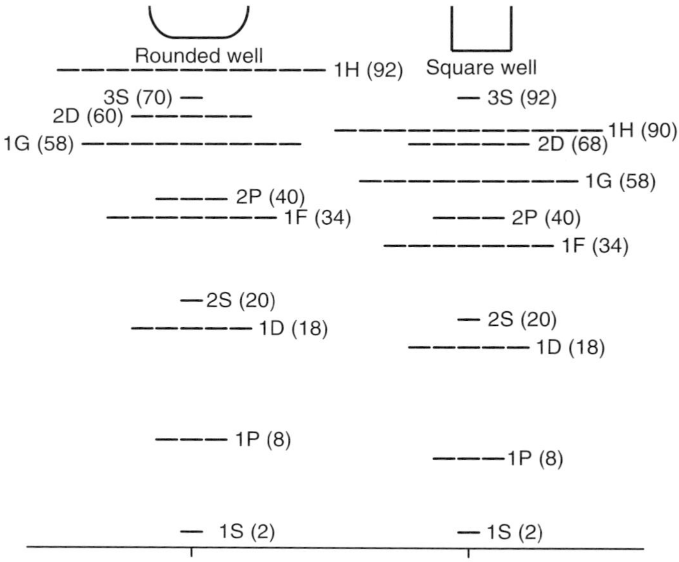

FIGURE 1 Jellium levels for a rounded-bottom and a square well system.

structure is confined to the surface of a sphere, and in other clusters, the geometric shell closings also play a role [25]. Theoretical studies show that the electronic shell structure is maintained, even though the spacing between the shells can depend on the shape of the potential. The key message is that the electronic shell structure is a general property of the confined nearly free electron gas.

It is known that atomic and molecular systems with filled electronic shells and large gaps in the excitation spectrum exhibit enhanced stability. It was, therefore, natural to examine if the origin of magic numbers is rooted in the electronic shell structure. It was pointed out that the stable sizes at 2, 8, 18, 20, 34, 40, and 58 indeed correspond to the electron counts that lead to filled electronic shells, and thus the stability of clusters could be linked to the filling of the electronic shells.

The existence of the electronic shells in a nearly free electron gas and their role in governing the stability raised the question whether stable clusters exhibit electronic features much in the same way as atoms. Consider the case of Al that is trivalent. An Al_{13} cluster has 39 valence electrons and would correspond to the level structure $1S^2$ $1P^6$ $1D^{10}$ $2S^2$ $1F^{14}$ $2P^5$ with one hole in the valence 2P shell. The shell filling is at 40 electrons, so to demonstrate the analogy, the electronic structure of Al_{13}^- is calculated using density functional theory as shown in Figure 2. The anions levels are consistent with the jellium model, and the electron affinity and highest occupied molecular orbital–lowest unoccupied molecular orbital (HOMO–LUMO) gap are quite large exactly as predicted. Khanna and Jena calculated the electronic structure

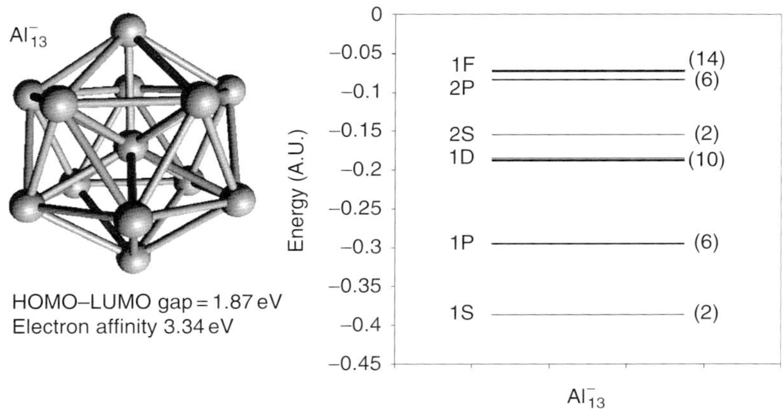

FIGURE 2 The geometry and electronic structure of Al_{13}^-.

of Al_{13}^- and Al_{13} and showed that the Al_{13} had a high electron affinity of 3.34 eV that was comparable to halogen atoms [12]. They also showed that Al_{13}^- had a closed electronic shell with a HOMO–LUMO gap of 1.87 eV. These findings explained the earlier result by Castleman and coworkers [26] who investigated the reactivity of Al_n^- with oxygen in a flow reactor to examine the behavior of Al_{13}^- and related species. Their studies showed that while most clusters are etched away by oxygen, species such as Al_{13}^-, Al_{23}^-, and Al_{37}^- remained intact. These results were intriguing in view of the fact that bulk aluminum surfaces are highly susceptible to etching by oxygen. This again showed that in addition to the jellium model predicting magic numbers, it also predicted the reduced chemical reactivity of the cluster. The inert behavior shows that this cluster effectively corresponds to that of a noble gas. This possibility of adding an electron to greatly enhance the stability revealed the possibility of building a periodic table of clusters [7].

III. Al_{13} AND Al_{14}-BASED SUPERHALOGEN AND SUPERALKALI EARTH CLUSTERS

Superatom chemistry was truly brought into focus in the reactivity studies between aluminum clusters and iodine [7,8]. An Al_{13} has an electronic configuration of $1S^2\ 1P^6\ 1D^{10}\ 2S^2\ 1F^{14}\ 2P^5$ with one hole in the 2P valence shell, although filling the $2P^6$ shell causes it to drop slightly below the $1F^{14}$ levels. This is similar to the case of halogen atoms that also have five electrons in their valence P-shell. It was, therefore, tempting to probe if the properties of Al_{13} were similar to those of a halogen atom. The cluster has a high electron affinity of 3.57 eV [27,28] that is comparable to the known value of 3.36 eV for Br and less than 3.62 eV of Cl. A more stringent test of its halogen properties is whether it bonds with other halogens to form cluster analogies of I_2^- and I_3^-.

The superhalogen nature of Al_{13} became particularly enlightening during our studies on its reactions with HI. In particular, a very prominent $Al_{13}I^-$ species was readily formed [7]. Theoretical studies revealed that the Al_{13} core is almost intact as in Al_{13}^-, and showed that a covalent bond was formed between the aluminum cluster and the iodine atom, as shown in Figure 3. This suggested an analogy to stable halogen dimer anions. In view of its electron affinity, Al_{13} is somewhere between a bromine and chlorine, so the resulting cluster is similar to the well-known ions, BrI^- and ClI^-. Hence, the aluminum cluster is indicated to be behaving as a superhalogen.

Further studies with I_2 in place of HI led to even more exciting results. $Al_{13}I_2^-$ was found to be a prominent species, again with an analogy to well-known halogens, namely BrI_2^- or ClI_2^- [8]. The theoretically deduced charge distribution is shown in Figure 4. Again note that the icosahedral aluminum retains its integrity, and functions as a superhalogen.

The biggest surprise came in studies made with high iodine concentrations. A multiple iodine sequence was first observed which, when etched with oxygen, revealed a stable distribution of iodized aluminum species, reminiscent of the polyhalide series well known in the condensed phase [29]. The dramatic odd–even intensity distributions seen in the figure are characteristics found in the condensed phase, and similar to expectations for XI_n^- anions that are very stable for n = even, where X is also a halogen. One major difference from the condensed phase is the fact that theoretical findings confirm that the halogen atoms are individually bonded to the aluminum cluster halogen mimic, and not as molecular iodine chains.

Another surprising finding was that of a second cluster series, this one involving Al_{14} as a core, with bound iodine atoms in an even–odd series with an exactly opposite odd–even trend to that seen for $Al_{13}I_n^-$. These species are also evident in the mass spectra in Figure 5. A significant aspect is that the

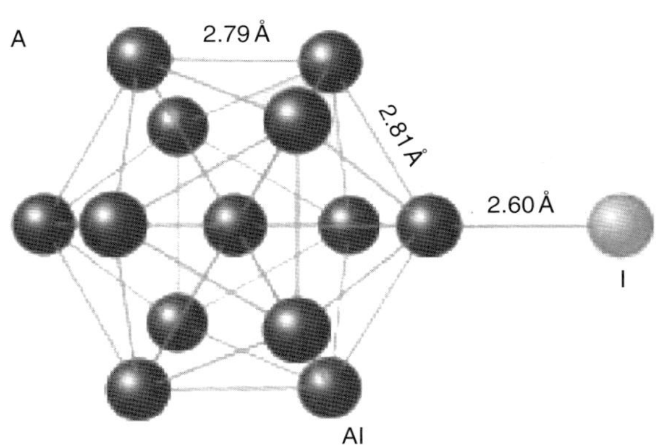

FIGURE 3 Geometry of $Al_{13}I^-$. Adopted from Ref. [7].

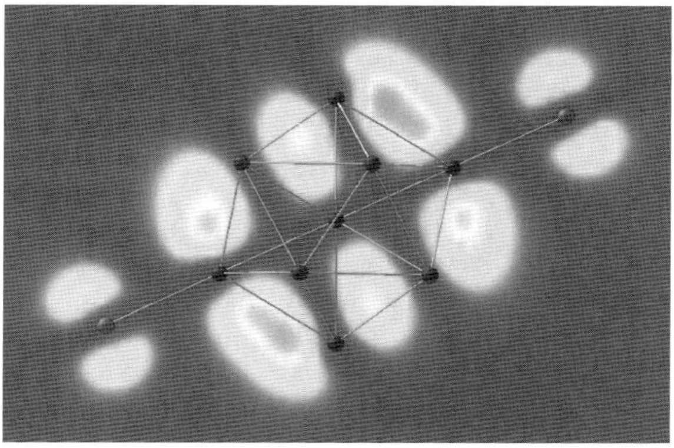

FIGURE 4 Charge density of $Al_{13}I_2^-$. Adopted from Ref. [8].

series commences with $Al_{14}I_m^-$, $m = 3$. This shows that the stable core is striving toward an Al_{14}^{+2} core stabilized by three covalent bonds with iodine. Thus, Al_{14} can be thought of as cluster mimic of an alkaline earth metal. With the calculated charge distribution, the remaining aluminum core effectively takes on a 40-electron closed shell-like character (Figure 6).

As we further considered the implications of the concept of a jellium-based superatom, another interesting feature became evident. The presence of a single iodine bound to the Al_{13}^- cluster results in a half-filled active site on the opposite side of the icosahedron. The origin of this active site is an unexpected consequence of superatom clusters, as the 2P jellium level is perturbed by the presence of the iodine atom as it pushes that level to become the HOMO. This contributes to the high reactivity of the $Al_{13}I_n^-$ clusters with odd number of iodine atoms, although the reactivity with molecular oxygen can primarily be attributed to the cluster having an odd number of electrons (Figure 7).

In order to obtain further evidence of this effect, experiments were conducted in a flow tube reactor to investigate the reactions of aluminum clusters with methyl iodide. The findings revealed that for nascent aluminum clusters, there was only a limited, very slow reaction unless very high concentrations of methyliodide were employed. Significantly, as is evident from Figure 8, upon iodizing aluminum clusters, reactions with CH_3I became much more facile [30].

This work brings out the fact that a variety of clusters may be stabilized by opening a gap whose origins are found in the 40 electron jellium shell closing. This raises the question whether other shell closings may be used to form different superatoms that display desired properties as element mimics. To further investigate this, we next consider the paired 18 electron and 20 electron shell closings in aluminum clusters.

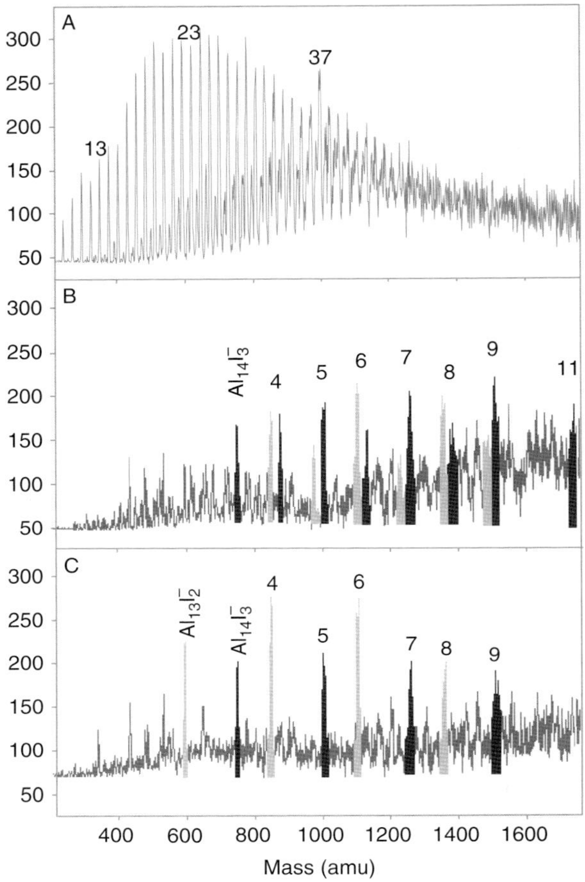

FIGURE 5 Mass spectra of (A) Al_n^- clusters, (B) Al with I_2, and (C) Al with I_2 and etched by O_2. Adopted from Ref. [8].

IV. MULTIPLE VALENCE SUPERATOMS: Al_7 MOTIFS

While the Al_{13} and Al_{14} based superatoms were the first to bring out the analogy between atoms and superatoms, an extension of the concept to clusters which mimic the ability of many atoms which exhibit multiple valences was needed. For example, carbon exhibits both divalent and tetravalent characteristics and, strongly binds with O or Si atoms forming CO and SiC, both of which are stable molecules. A search for such species revealed a new class of superatoms, namely Al_7 based clusters, which exhibit multiple valences, like some of the elements in the periodic table. They have the potential to form stable compounds when combined with multiple atoms. We first demonstrated [9] the exceptional stability of Al_7C^- through the production of Al_nC^- clusters in reactions of

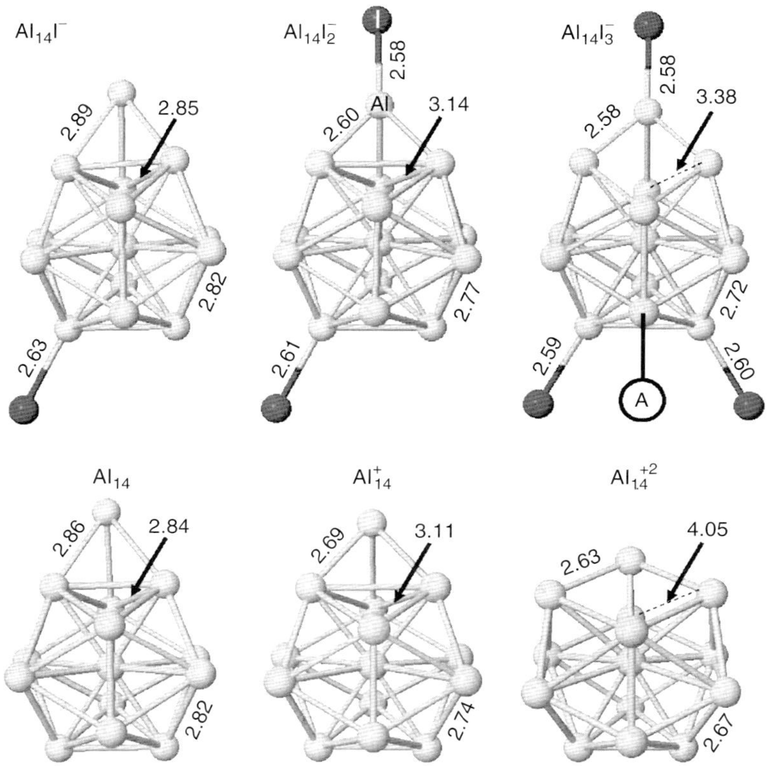

FIGURE 6 The geometry of $Al_{14}I^-$, $Al_{14}I_2^-$, $Al_{14}I_3^-$, and Al_{14}, Al_{14}^+, and Al_{14}^{+2}. Adopted from Ref. [8].

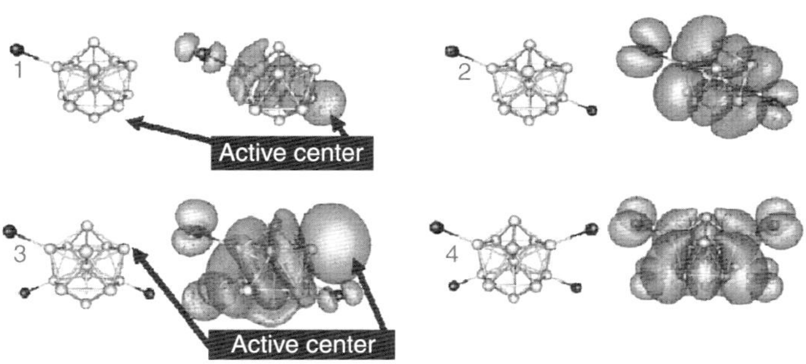

FIGURE 7 The structures and HOMO charge density in $Al_{13}I_n^-$ where $n = 1$–4. Adopted from Ref. [8].

aluminum clusters with benzene, and subsequent etching reactions of the clusters with oxygen to identify the stable species. The mass spectrum (Figure 9) shows the Al_7C^- peak to be even more pronounced than Al_{13}^-. Theoretical

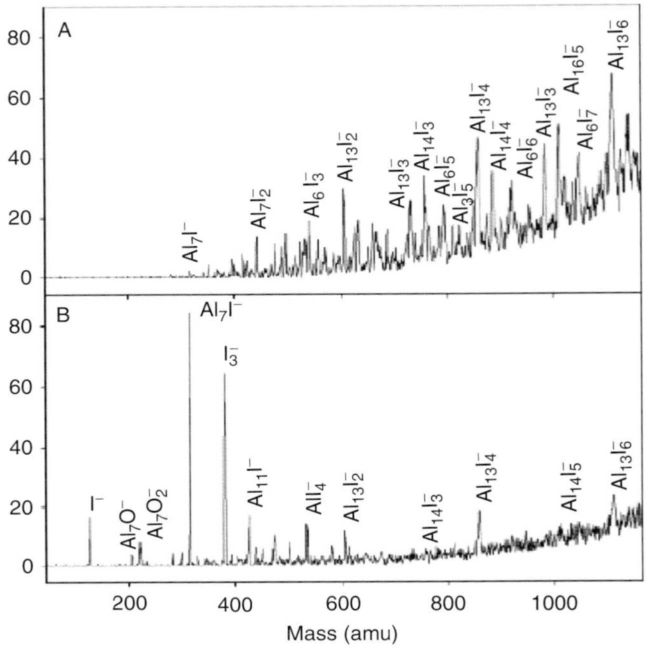

FIGURE 8 Reactions of nascent and iodized aluminum clusters with CH_3I. Adopted from Ref. [30].

considerations indicate that the superatom concept enables one to understand the electronic origin of the exceptional stability often observed for these species. Since aluminum has three valence electrons, an Al_7^- has 22 valence electrons and can form stable compounds by combining with atoms that need four or two electrons to fill their shells. Such considerations also predict stability of other species such as Al_7O^-, Al_7S^-, $Al_7I_2^-$, and Al_7^+ all of which demonstrate that an Al_7^- behaves like a superatom with a valence of 2. This is also seen from Figure 10 that shows the binding energy of Al_7^- to atoms in the second, third, and fourth row of the periodic table. Note that atoms which are tetravalent and divalent bind more strongly than others. Experimental mass spectra show the enhanced stability of Al_7^+ in oxygen etching experiments. These investigations provide further extensive support to the general nature of the superatom concept and show Al_7^- to be a member of the multivalent superatom family.

The origin of stability of Al_7C^- lies in the multiplet nature of Al_7^-. Note that the cluster features an endohedral C atom as shown in Figure 11. The atomic carbon has four unfilled p states, and Al_7^- has 22 valence electrons. The p states of the carbon form bonding states with the aluminum cluster which can be thought of as removing the $2F^2$ and $2S^2$ jellium levels and making them into hybridized bonding orbitals. This is also significant

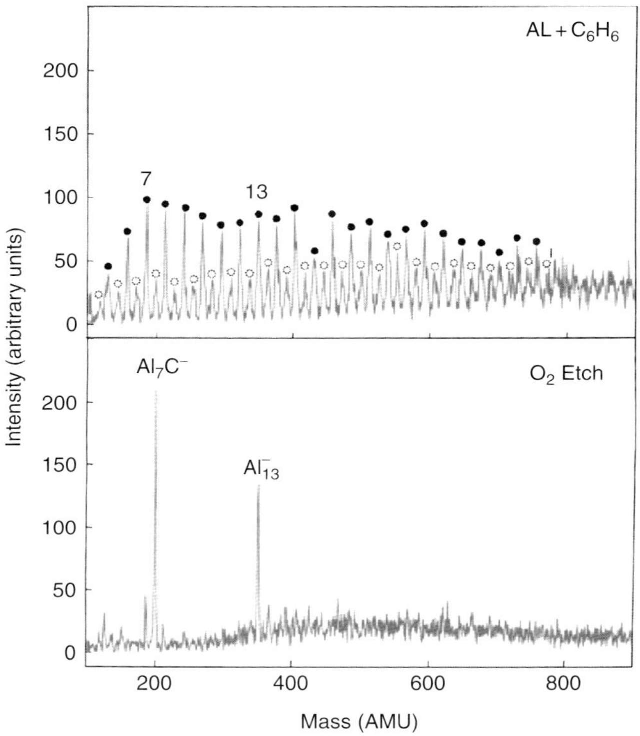

FIGURE 9 $Al_nC_m^-$ mass spectra before and after etching with O_2. Adopted from Ref. [9].

as 18 electrons is a magic number in both the jellium framework and in spherical aromaticity, so there is no problem with the carbon atom being endohedral [31,32]. The Al_7C^- is confirmed to be a closed shell cluster as the HOMO–LUMO gap is found to be 1.69 eV, and the cluster is highly resistant to oxygen etching.

The finding of multiple valences in Al_7-based clusters support the contention that there should be no limitation in finding clusters which mimic virtually all members of the periodic table.

V. ASSEMBLIES OF Al_{13} USING SUPERALKALI COUNTERCATIONS

Al_{13} clusters have a large electron affinity which is comparable to that of a halogen atom, and hence, it should be possible to form ionic compounds by combining Al_{13} with alkali atoms.

Previous studies on $Al_{13}K$ showed it to be an ionically bonded molecule composed of Al_{13}^- and K^+ units much like the ionic salt KCl [33–36]. However, a potassium atom is much smaller than an Al_{13} cluster, and attempts at

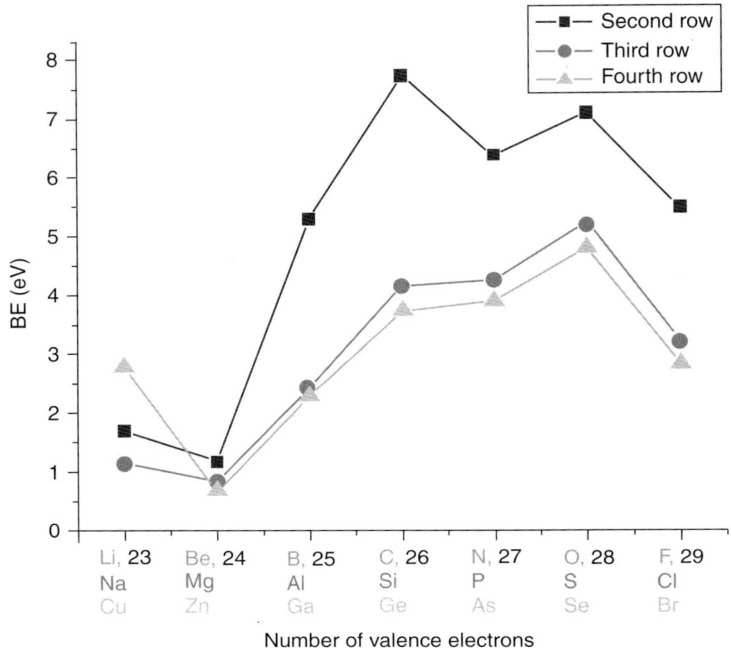

FIGURE 10 The binding energy of Al_7^- with the second, third and fourth row of the periodic table. Adopted from Ref. [9].

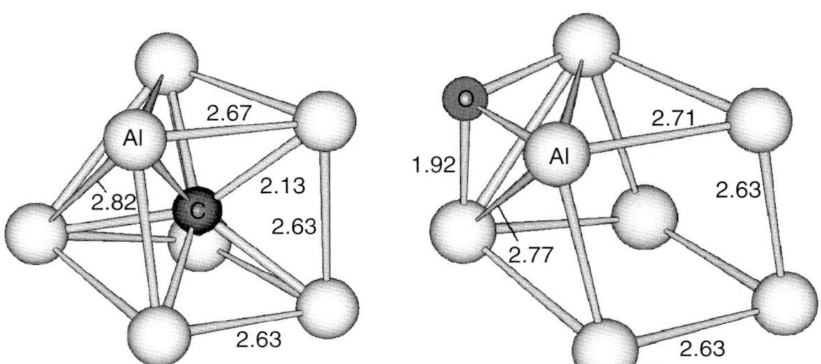

FIGURE 11 The structure of Al_7C^- and the Al_7O^-. Adopted from Ref. [9].

forming materials by assembling $Al_{13}K$ units resulted in strong interactions between the Al_{13} motifs and eventually led to coalescence of the clusters. One way to overcome this obstacle is to use larger molecular units with lower ionization potentials as counter cations. We recently tried K_3O, a molecule that is known to exhibit superalkali behaviors including an ultra low ionization potential of 3.17 eV [37]. Studies of $(Al_{13})_m(K_3O)_n$ clusters were undertaken to demonstrate that superatoms exhibit novel chemical features

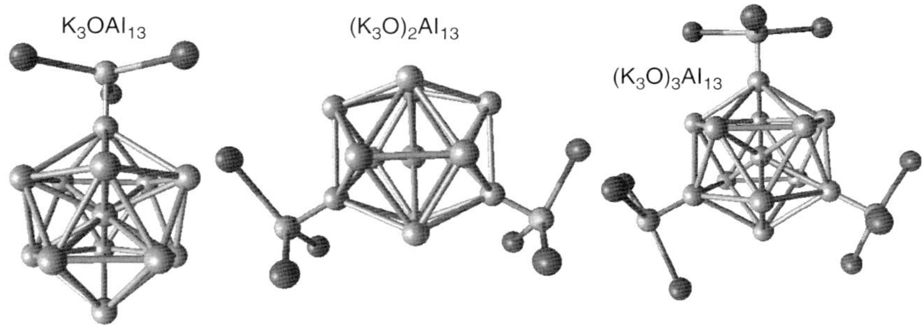

FIGURE 12 The structure of $Al_{13}(K_3O)_n$ where $n = 1$–3. Adopted from Ref. [10].

and to explore the possibility of obtaining cluster assemblies in which the individual units retain their integrity (Figure 12).

We determined that the first ionization potential of $Al_{13}(K_3O)$ is 4.59 eV, and that adding another superalkali to form $Al_{13}(K_3O)_2$ lowers the first ionization potential to 2.88 eV while the second ionization potential remains high at 6.47 eV. Adding a third superalkali to form $Al_{13}(K_3O)_3$ further lowers the first two ionization potentials to 2.27 and 4.51 eV while the third ionization potential remains high at 8.14 eV. The first two ionization potentials are so low that the unit can be thought of as an ultra alkali motif. Further, the HOMO–LUMO gap of neutral $Al_{13}K_3O$, cationic $Al_{13}(K_3O)_2^{+}$, and $Al_{13}(K_3O)_3^{+2}$ are 1.24, 1.27, and 1.22 eV, respectively.

K_3OAl_{13} is bonded by a charge transfer from K_3O to Al_{13}. It is, therefore, an analog of alkali halide, for example, CsCl molecules. Previous studies on alkali halide clusters have revealed that $(CsCl)_n$ units form planar rings at small sizes. To look for similar architectures in K_3OAl_{13}, we investigated the geometries of the $(K_3OAl_{13})_n$ containing up to six units (Figure 13).

Investigations show that the units assemble with Al_{13} and K_3O motifs almost intact. A study of the energetics showed that $(K_3OAl_{13})_4$ is particularly stable. More importantly, the superatom architectures were found to offer novel features. For example, our calculations of the vibrational spectrum of the $(K_3OAl_{13})_4$ cluster showed that it exhibits three low frequency vibrational modes at 37, 41, and 47 cm^{-1} that correspond to intercluster vibrations not exhibited by individual clusters. These findings reconfirm that the world of superatoms does hold novelty along with the possibility of surprising behavior.

VI. SPIN ACCOMMODATION AND REACTIVITY OF ALUMINUM CLUSTERS

Since aluminum particles are used in fuels and since the particles are generally covered with an oxide overcoat, finding new ways to control the reactivity of aluminum clusters with oxygen is an important objective. We have

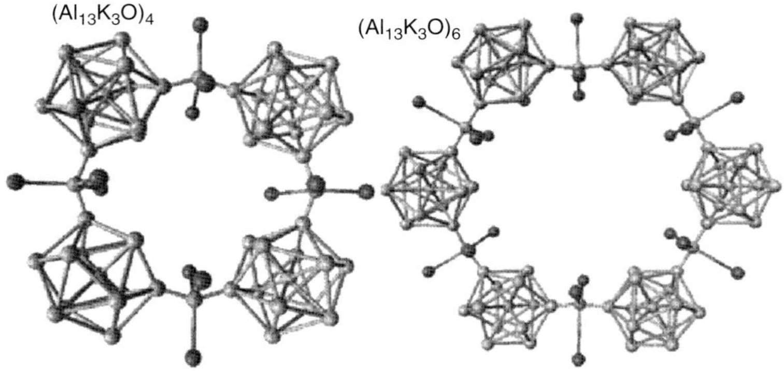

FIGURE 13 Geometry of a $(Al_{13}(K_3O))_4$, $(Al_{13}(K_3O))_6$, and the evolution of the electronic structure with assembly. Adopted from Ref. [10].

recently demonstrated how the reactivity of aluminum clusters can be altered by changing their spin state. Focusing on anion species, small aluminum cluster anions containing up to 12 aluminum atoms are all reactive toward oxygen. We recently studied the addition of hydrogen atoms to aluminum clusters to generate $Al_nH_m^-$ $(1 \leq n \leq 7)$ in an attempt to understand the formation mechanisms of the alane clusters [38–40] and investigate the evolution of electronic structure one electron at a time. We focus on our results in the $Al_4H_n^-$ series to illustrate the new finding. The original series contained all sizes with higher relative abundances at Al_4^- and $Al_4H_7^-$. The clusters were treated with oxygen to eliminate the reactive species. The mass spectra of the reacted species (Figure 14A,B) showed that all the clusters containing even numbers of hydrogen atoms were etched away while those containing odd numbers of hydrogen atoms survived. Further, while the intensity of $Al_4H_3^-$ and $Al_4H_5^-$ decrease very slightly, that of Al_4H^- and $Al_4H_7^-$ grows with the addition of oxygen. Since pure aluminum anions containing up to 12 Al atoms and in particular, Al_4^- is highly reactive to oxygen, the reduced reactivity of Al_4H^- was quite surprising. Our theoretical investigations revealed that the reduction in reactivity was related to the spin. Since molecular oxygen has a triplet spin multiplicity, the reactivity with clusters having even number of valence electrons requires a spin excitation of the metal. For cases where the spin excitation energy of the metal counterpart is low, the O_2 binds more strongly, the O–O bond can break, and the clusters are reactive. On the other hand, when the spin excitation energy is high, the reactivity is reduced. This is evident from Figure 14 which shows the relation between vertical and adiabatic spin excitation energy and reactivity. These findings have important consequences. For example, clusters that are nonreactive can be made reactive by either changing the spin configuration of the cluster (through, e.g., addition of a hydrogen atom) or exciting oxygen to a spin singlet [41]. On the other hand, even reactive aluminum clusters can be made nonreactive through

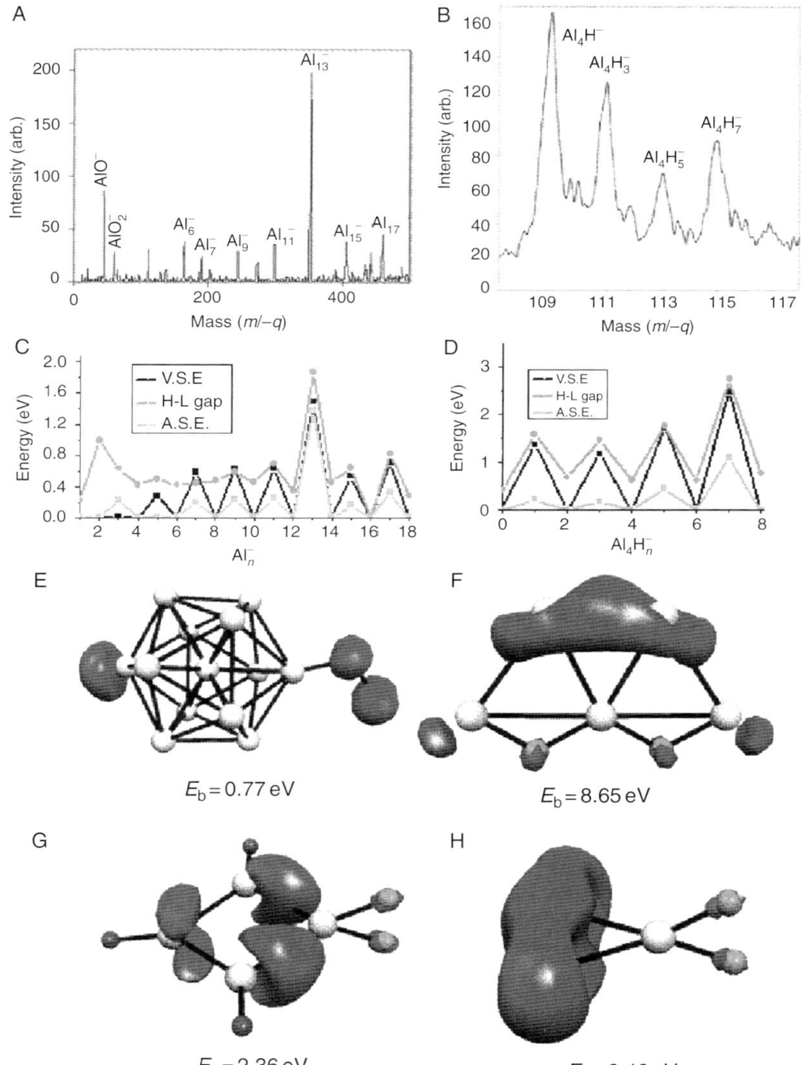

FIGURE 14 Etching spectra of (A) Al_n^-, (B) $Al_4H_m^-$, (C) bandgap and spin excitation energies of Al_n^-, (D) and $Al_4H_m^-$. Spin density and geometry of O_2 with (E) Al_{13}^-, (F) Al_5^-, (G) $Al_4H_3^-$, and (H) Al_3^-. Adopted from Ref. [3].

controlling the spin excitation energy. This can be accomplished either by simply adding hydrogen or by adding transition elements. We also note that oxidation of aluminum clusters with water is strikingly different from that of molecular O_2 and depends on the presence of complementary active sites

[42,43]. This is an exciting development and will play a major role in designing aluminum cluster based materials that are stable under exposure to oxygen and in general for controlling reactivity through spin.

VII. FUTURE DIRECTIONS IN THE CLUSTER PERIODIC TABLE

Efforts to identify new cluster assembled materials and to find new properties which clusters exhibit both in the gas phase and in the solid state continue to produce remarkable results. Recently, a cluster assembled material with a triplet ground state made of main group elements was synthesized [16]. Further, new reactivities of aluminum clusters have been found showing how their reactivity with oxygen may be tuned. The concept of the superatom has been successful in identifying and explaining likely motifs for cluster assembly. Further development and understanding of the superatom concept lie at the core of developing new cluster assembled materials.

Acknowledgments

A. W. C. acknowledges support by the Air Force of Scientific Research Grant No. FA9550-07-1-0151, and Department of Energy Grant No. DE-FG02-02ER46009. A. W. C., A. C. R., and S. N. K. acknowledge support from US Department of Army (MURI) Grant No. W911NF-06-1-0280. A. C. R. and S. N. K. acknowledge support by the Air Force Office of Scientific Research Grant No. FA9550-09-1-0371.

REFERENCES

[1] Castleman Jr AW, Jena P. Proc Natl Acad Sci USA 2006;103:10552.
[2] Wallace WT, Wyrwas RB, Whetten RL, Mitric R, Bonacic-Koutecky V. J Am Chem Soc 2003;125:8408.
[3] Reber AC, Khanna SN, Roach PJ, Woodward WH, Castleman AW. J Am Chem Soc 2007;129:16098.
[4] Reddy BV, Khanna SN, Dunlap BI. Phys Rev Lett 1993;70:3323.
[5] Cox AJ, Louderback JG, Bloomfield LA. Phys Rev Lett 1994;71:923.
[6] Khanna SN, Jena P. Chem Phys Lett 1994;219:479.
[7] Bergeron DE, Castleman Jr AW, Morisato T, Khanna SN. Science 2004;304:84.
[8] Bergeron DE, Roach PJ, Castleman Jr AW, Jones NO, Khanna SN. Science 2005;307:231.
[9] Reveles JU, Khanna SN, Roach PJ, Castleman Jr AW. Proc Natl Acad Sci USA 2006;103:18405.
[10] Reber AC, Khanna SN, Castleman Jr AW. J Am Chem Soc 2007;129:10189.
[11] Castleman Jr AW, Khanna SN, Sen A, Reber AC, Qian M, Davis KM, et al. Nano Lett 2007;7:2734.
[12] Khanna SN, Jena P. Phys Rev B 1995;51:705.
[13] Jadzinsky PD, Calero G, Ackerson CJ, Bushnell DA, Kornberg RD. Science 2007;318:430.
[14] Heaven MW, Dass A, White PS, Holt KM, Murray RW. J Am Chem Soc 2008;130:3754.

[15] Schultz-Dobrick M, Jansens M. Angew Chem 2008;47:2256.
[16] Ugrinov A, Sen A, Reber AC, Qian M, Khanna SN. J Am Chem Soc 2008;130:782.
[17] Goicoechea JM, Sevov SC. J Am Chem Soc 2005;127:7676.
[18] Bag S, Trikalities PN, Chupas PJ, Armatas GS, Kanatzidis MG. Science 2007;317:490.
[19] Knight WD, Clememger K, de Heer WA, Saunders WA, Chou MY, Cohen ML. Phys Rev Lett 1984;52:2141.
[20] Nonose S, Tanaka H, Mizuno T, Kim NJ, Someda K, Kondow T. J Chem Phys 1996;105:9167.
[21] Brechignac C, Cahuzac Ph, Carlier F, de Frutos M, Barnett RN, Landman U. Phys Rev Lett 1994;72:1636.
[22] Brack M. Rev Mod Phys 1985;65:677.
[23] de Heer WA. Rev Mod Phys 1993;65:611.
[24] Ekardt W. Phys Rev B 1984;29:1558.
[25] Martin TP, Bergmann T, Gohlich H, Lange T. Chem Phys Lett 1990;172:209.
[26] Leuchtner RE, Harms AC, Castleman Jr AW. J Chem Phys 1989;91:2753.
[27] Li X, Wang LS. Phys Rev B 2002;65:153404.
[28] Li X, Wu H, Wang XB, Wang LS. Phys Rev Lett 1998;81:1909.
[29] Mizuno M, Tanaka J, Harada I. J Phys Chem 1981;85:1789.
[30] Bergeron DE, Roach PJ, Castleman Jr AW, Jones NO, Reveles JU, Khanna SN. J Am Chem Soc 2005;127:16048.
[31] Chattaraj PK, Giri S. J Phys Chem A 2007;111:11116.
[32] Chen Z, Neukermans S, Wang X, Jannsens E, Zhou Z, Silverans RE, et al. J Am Chem Soc 2006;128:12829.
[33] Liu F, Mostoller M, Kaplan T, Khanna SN, Jena P. Chem Phys Lett 1996;248:213.
[34] Ashman C, Khanna SN, Liu F, Jena P, Kaplan T, Mostoller M. Phys Rev B 1997;55:15868.
[35] Rao BK, Khanna SN, Jena P. Phys Rev B 2000;62:4666.
[36] Ashman C, Khanna SN, Pedersen MR, Kortus J. Phys Rev B 2001;62:16956.
[37] Dao PD, Peterson KI, Castleman Jr AW. J Chem Phys 1984;80:563.
[38] Roach PJ, Reber AC, Woodward WH, Khanna SN, Castleman Jr AW. Proc Natl Acad Sci USA 2007;104:14565.
[39] Li X, Grubisic A, Stokes ST, Cordes J, Gantefor GF, Bowen KH, et al. Science 2007;315:5810.
[40] Grubisic A, Li X, Stokes ST, Cordes J, Gantefor GF, Bowen KH, et al. J Am Chem Soc 2007;129:5969.
[41] Burgert R, Schockel H, Grubisic A, Li X, Stokes ST, Bowen KH, et al. Science 2008;319:438.
[42] Roach PJ, Woodward WH, Castleman Jr AW, Reber AC, Khanna SN. Science 2009;323:492.
[43] Reber AC, Khanna SN, Roach PJ, Woodward WH, Castleman Jr AW. J Phys Chem A 2010;114:6071.

Chapter 10

Silica as an Exceptionally Versatile Nanoscale Building Material: $(SiO_2)_N$ Clusters to Bulk

Stefan T. Bromley*,†

*Institució Catalana de Recerca i Estudis Avançats (ICREA), Barcelona, Spain

†Departament de Química Física & Institut de Química Teòrica i Computacional, Universitat de Barcelona, Barcelona, Spain

Chapter Outline Head

Nanoclusters. DOI: 10.1016/S1875-4023(10)01010-7

I. INTRODUCTION

The technological importance of silica (SiO_2), tailored to possess nanoscale dimensions, can hardly be overstated considering its established utilization in microelectronics [1], adsorption and catalysis [2], and nanocomposite materials [3]. Although other oxides have been tailored at the nanoscale, silica is particularly interesting as a molecular building material due to its bulk stability and the ease with which its inherent structural richness can be exploited via a range of chemical and physical techniques. Silica nanostructures with dimensions of the order of a single nanometer can now be controllably fabricated in a wide variety of distinct topological forms (e.g., nanopores [4], nanospheres [5], nanotubes [6]) leading to promising new applications in photonics [7] and biotechnology [8]. The incredible diversity of silica nano-architecure is also exhibited in some microorganisms [9], exhibiting distinct intricate shell details down to the 10 nm scale. Furthermore, SiO_2 can act as a prototype nano-oxide, with the recent demonstration that silica nanostructures can be isostructurally transformed into other technologically important oxides (e.g., TiO_2, ZrO_2) via shape-preserving displacement reactions [10]. Although the range of synthetic and natural silica nanostructures is impressive, almost all are based upon the condensation and/or manipulation of amorphous silica constituted from essentially random molecular networks of $Si(O_{1/2})_4$ tetrahedral units. In contrast, silica in bulk form exhibits an exotic variety of crystalline polymorphs ranging from fibrous one-dimensional materials, through high-pressure-induced dense phases to very low-density nanoporous frameworks. This spectrum of synthesized polymorphs not only span a wide range of topologies but also occupy a broad spectrum of energetic metastability above the ambient ground state phase of α-quartz. This capacity of silica to maintain a diversity of metastable ordered atomic topologies all of which are highly resistant to structural collapse bodes well for extending its application as versatile nanoscale building material beyond the possibilities offered solely by using the disordered amorphous state. In this chapter, we explore the structures and energies of small silica clusters with a view to their potential as nanoscale building blocks for new materials and nanostructures.

II. EXPERIMENTAL STUDIES OF SILICA CLUSTERS

The control and understanding of nucleation of silica cluster species remain a challenge to experimentalists and theorists alike. Much effort in this area has concentrated on the processes involved in the hydrothermal synthesis of crystalline silicate-based nanoporous materials (e.g., zeolites) involving intricate molecular scale interactions between water and silica, where the silica cluster precursors are hydroxylated species in solution. The complex interactions between silica nanoclusters and water-containing environments are difficult to probe experimentally and have only recently begun to be addressed by accurate theoretical methods [11]. Thus, although the water–silica system is very interesting and of great importance in understanding the nucleation, functionalization, and bio-relevance of silica, this topic is beyond the scope of this chapter and we, instead, concentrate on anhydrous pure $(SiO_2)_N$ clusters and, further, how they may be potentially utilized as nanoscale precursors for new bulk silica materials.

A potentially simpler alternative to solution-based synthesis of silica materials via cluster aggregation is via nucleation of pure SiO_2-species in free space. Vapor phase manipulation of silica and its controlled deposition are already a well-established technology [12] with the production of vapors of SiO_2-based species quite routinely performed through the use of silicon-containing molecular intermediates and highly oxidative conditions. Vapor phase oxidation of silane (SiH_4) [13,14], silicon tetrachloride [15], and tetra-ethyl-orthosililcate (TEOS) [16] molecules, for example, have been successfully employed to produce sources of SiO_2 which can be used to produce very pure-silica thin films and nanoparticles. Unlike with hydrothermal-based silica nucleation, the vapor phase nucleation and oxidation of molecular silicon sources in much faster and less amenable to atomic scale control and involves numerous complex reaction/decomposition routes [13], resulting in amorphous silica products with typical dimensions >5 nm. Although amorphous silica nanoparticles of such dimensions are of interest to the experimental [17] and theoretical community [18], our focus is on small silica clusters (<1 nm diameter) with well-defined atomic structures, which may subsequently be utilized for similarly ordered extended phases.

From the point of view of producing and characterizing small well-defined silica clusters, one would prefer a nucleation process employing precursors containing only silicon and oxygen in order to avoid complex intermediate products. Furthermore, in an ideal experimental set-up, one would like to be able to isolate the clusters before aggregation in order to be able to probe their properties (e.g., chemical composition, structure). Unfortunately, the production of pure SiO_2 molecules and clusters in vapor phase from bulk silica is well known to be complicated by various factors not present in hydrothermal silica syntheses. Unlike in alkali solution, for example, solid silica does not readily decompose in vacuum into a source of monomeric SiO_2-based species. In fact, the vaporization of bulk SiO_2 occurs appreciably only at very high temperatures (>1500 K) and the equilibrium vapor phase above solid SiO_2 primarily leads to the production of SiO molecules rather than SiO_2 species due to the relative high thermodynamic stability of the former [19]. Furthermore, unlike in solution where organic molecules are often used to help direct the silica nucleation process, under high temperature gas phase conditions many potential templating molecules would simply decompose. Rather than creating vapors from heating bulk silica samples, direct local disruption of the surfaces of bulk silica (e.g., via intense laser pulses [20] or heavy ions [21]) has also been attempted as a means of producing pure $(SiO_2)_N$ species but tends to produce transient populations of ionic H-containing silica clusters of rapidly diminishing size with increasing cluster size ($N \approx 1$–25) rather than homogeneous stable sources of SiO_2 species necessary for materials synthesis. Instead of directly using silica, some groups have employed vapors of elemental silicon which upon oxidation produce very small ionic $(SiO_2)_N$ $N=1$–4 clusters [22,23]. Comparison with theoretical calculations has established that these clusters most likely have a chain-like form constituted from Si_2O_2-rings (hereafter referred to as "two-rings") and are terminated with silanone (Si=O) groups (see Figure 1). Unfortunately for the production of larger $(SiO_2)_N$ clusters, silicon vapor itself naturally aggregates into clusters, which above a small size (Si_N $N>4$), appear to be resistant to total oxidation and instead are merely etched away by oxygen in a process producing SiO molecules [24].

Of all the sources of pure vapor phase SiO_2 cluster species for materials synthesis, the most promising appears to be simply via the oxidation of vapors of SiO. Over half a century ago, a synthesis based on the nucleation of SiO_2 in vacuum was performed giving rise to a new crystalline silica material known as silica-W [25] based upon discrete chains of (Si_2O_2) two-rings (see Figure 1). In the experiment, gaseous SiO was produced from solid silicon monoxide in the presence of oxygen at 1550 K to produce a source of SiO_2 species in vacuum. Nucleation of these species was induced only at one point via the introduction of a cold tip into the system, giving rise to the unidirectional fibrous polymorph. More recently, this simple gas phase silica synthesis route has also been used in the study of circumstellar dust whereby quartz and cristobalite nanoparticles containing thousands of atoms were formed from nucleation of oxidized SiO vapors [26]. These studies explicitly demonstrate that gas phase clustering of pure SiO_2 molecular species is a viable means to produce both new silica materials and silica nanostructures. In the remainder of the chapter, we use various computational modeling methods to explore two possible routes to new cluster-based silica materials/nanostructures

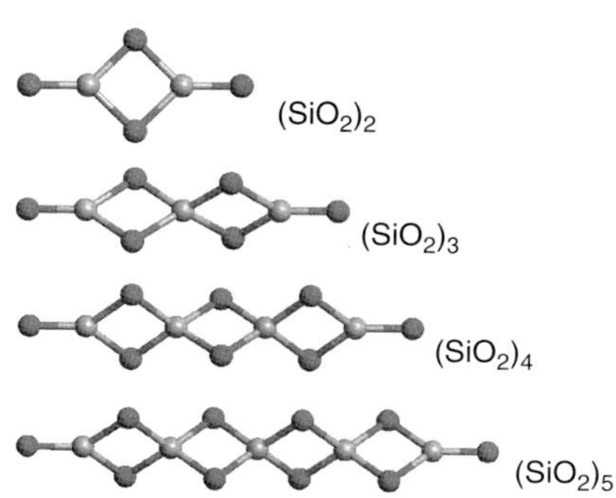

FIGURE 1 Silianone-terminated two-ring chain ground state cluster isomers (D_{2h}) for $(SiO_2)_N$ $N=2$–5.

by examining the likely cluster products of SiO_2 nucleation with particular emphasis on clusters with particularly high stability as potential building blocks.

III. THEORETICAL STUDIES OF LOW-ENERGY $(SIO_2)_N$ CLUSTERS

A. Exploring the Low-Energy Spectrum of $(SiO_2)_N$ Clusters

Although much experimental and theoretical work has focused upon the stabilities and structures of the many silica bulk polymorphs and their surfaces, the low-energy structures and potential energy surface (PES) of nanoscale $(SiO_2)_N$ clusters is relatively unknown. A number of studies in the literature have explored the properties of isomers of small $(SiO_2)_N$ clusters using various combinations of structure derivation and energy minimization: (i) extensive manual construction of cluster isomers based on chemical intuition with results energy-minimized using density functional theory (DFT) [27–30], or bulk-parameterized interatomic potentials [31], (ii) deliberate manual design using repeated stable $(SiO_2)_N$ ($N<5$) subunits with structures energy-minimized using DFT [32–35], (iii) use of classical [36] and *ab initio* [37] molecular dynamics (MD) to explore the low-energy landscape of cluster isomers followed by DFT-based energy minimization. From these, and similar studies, although many small silica nanocluster energies and structures have been evaluated, until relatively recently only the ground state structures of only the five smallest nanoclusters, that is, $(SiO_2)_N$ $N=2$–6, were well established, all having a similar two-ring-based linear chain form with silanone-terminated ends (see Figure 1). Generally, the problem is that with increasing cluster size the number of possible atomic configurations increases exponentially and very quickly manual searches of cluster structures become untenable as a means to exhaustively search for the lowest energy cluster structure. In addition to the formal configurational problem, manual searches also rely on chemical intuition in order to construct low-energy cluster isomers. As we shall see below such trust is largely misplaced and many low-energy silica clusters have very unintuitive structures. Although the use of MD is a more objective way to explore the PES, it relies on the real dynamics of the system and requires that transitional barriers between isomers to be small enough to be easily accessible with respect to the temperatures and timescale of the simulation. In this ideal scenario, an MD simulation then in principle should provide an equilibrium population of isomers at a particular temperature. In reality, for moderately sized clusters, the necessarily finite timescale of a simulation generally does not allow for all possible minima at a particular temperature to be sampled and, moreover, is not a method directed toward finding low-energy clusters. The problem of finite timescale is particularly acute for *ab initio* MD simulations, which, are currently limited to calculations lasting only tens of picoseconds and are thus likely to only sample a selection of low energetically low-laying minima. Even the PES is approximated by the use of potential energy functions to evaluate the relevant atomic interactions and the system is gradually cooled from high temperatures to allow access to the low-energy cluster regime, the increasing complexity of the PES with increasing clusters size is such that searching for low-energy isomers via the inherent dynamics of the system rapidly becomes untenable as a reliable search method for the lowest energy isomers [38].

In order to find low-energy cluster isomers while avoiding explicitly traversing the PES through the vast number of energetic minima, a number of algorithms have been developed which allow the PES of cluster structures to be explored with relatively high efficiency. These methods generically come under the name of global optimization [39,40] and greatly assist in locating energetically low-laying cluster isomers and ultimately the ground state. Although an essential tool for probing low-energy cluster isomers, global optimization relies, for its efficiency, on both and rapid and accurate representation of the energy landscape of cluster configurations. Relatively small (X_M, $M \leq 20$) homogeneous clusters of low atomic weight atoms are now within range of fully quantum mechanical global optimization studies [41] where the PES is calculated at a first principles level of theory. However, for similar sized and larger heterogeneous (i.e., containing two or more atom types) clusters the PES becomes very complex, and thus, a less computationally demanding approach is essential, in part at least, to ensure some degree of tractability. Typically, rather than a fully electronic description of the interatomic interactions within a cluster, analytical interatomic potential energy functions are employed as a coarser grained representation of the true PES. Even at the level of interatomic potentials, one can appreciate the increased complexity of the $(SiO_2)_N$ system with respect to typical homogeneous cluster studies: (i) $(SiO_2)_N$ clusters possess two atom types, {Si,O}, and three two-body interactions, {Si–O, O–O, Si–Si}, meaning that the combinatorial complexity of the system is greatly increased, and (ii) due to the different partial charges on the two atom types long-range electrostatic interactions are important meaning that all atoms interact significantly with all other atoms rather than in homogeneous clusters where all atoms carry approximately the same charge. Such factors, especially when considering a real material, make it particularly crucial that a well-parameterized potential set be employed to accurately represent the complex PES. For nanoscale silica, it is found that commonly used potentials for calculations of bulk and surface silica systems [42,43] are inaccurate for small cluster systems [44]. The reason for this nanoscale breakdown is mainly due to the occurrence of defect states in nanocale SiO_2 not commonly found in bulk silica (e.g., Si_2O_2 two-rings, Si=O terminations). In order to more accurately represent the PES of nanosilica, a potential set has recently been specifically parameterized for nanoscale SiO_2 [44]. This potential, having a two-body Buckingham form complemented with electrostatics, as successfully used in bulk-parameterized SiO_2 potentials [42,43], has been found to be superior to these bulk-potentials and semiempirical methods with respect to *ab initio* nanocluster energies and structures. The explicit form of the short-range part of the potential is described by the equation below:

$$V_{ab}^{\mathrm{Buck}} = A_{ab} \exp\left(\frac{-r}{B_{ab}}\right) - \frac{C_{ab}}{r^6},$$

where r is the distance between two atoms of type a and b ($a, b \in \{\mathrm{Si, O}\}$) and A, B, and C are constants which were determined by fitting them into a set of small optimized and deformed $(SiO_2)_N$ clusters with $N<5$. The somewhat arbitrary partial charges of $2.4e$ for the silicon atoms and $-1.2e$ for the oxygen atoms (as successfully used in other similar potentials) are additionally employed to provide long-range electrostatic interactions between all atoms.

A number of recent studies have employed this potential together with the basin hopping (BH) global optimization approach [45] and subsequent energetic/structural refinement using DFT in order to obtain ground state candidates for $(SiO_2)_N$ $N=6$–27 [46–49]. The BH algorithm uses a combination of Monte Carlo sampling and energy minimization to efficiently sample the immense space of cluster configurations by effectively removing barriers between neighboring minima. Although the BH algorithm is one of the least hindered global optimization methods with respect to the specific topology of the PES [50], as with all global optimization methods one can become trapped in local regions of the PES and thus be isolated from the true ground state. To avoid such effects one can seed the algorithm with a diverse range of initial cluster structures (e.g., random structures, energetically low laying clusters from previous BH runs) to help ensure an even and extensive sampling [45] of the PES. Even with careful extensive use of the BH algorithm one must always remember that the resulting low-energy clusters are all with respect to an approximation to the true PES. Although the potential introduced in reference [44] appears to be relatively accurate for generating many low-energy SiO_2 structures, the direct results of the potential-based BH runs were treated as only a relatively good quality guide to the true isomer energy spectrum. In order to obtain a refined description of the low-energy isomer spectrum, a low-energy subset of candidate $(SiO_2)_N$ structures from each BH run for each N were fully energy-minimized using DFT. For the DFT calculations, the B3LYP hybrid exchange-correlation functional [51] with a 6-31G(d) basis set was used, employing no symmetry constraints. This level of theory has been shown in numerous previous studies to be suitable for calculating accurate structures and energies of SiO_2 nanoclusters [27,28]. The resulting DFT-refined low-energy $(SiO_2)_N$ clusters from this procedure for $N=1$–27 are described below.

B. Low-Energy $(SiO_2)_N$ Clusters, $N=2$–5

For $(SiO_2)_N$ clusters for small values of N the low-energy isomer spectrum contains relatively few isomers and can be explored to a relatively high degree of thoroughness by manual construction of isomers meaning that global optimization is largely unnecessary. A number of studies following this approach chains [27–29] (along with global optimization [46]) that the ground states for $(SiO_2)_N$ clusters $N=2$–5 are silanone-terminated two-ring chains (see Figure 1, note the relaxed SiO_2 monomer is linear which is not shown). This theoretical prediction is also confirmed by cluster beam experiments on $(SiO_2)_N$ clusters containing up to $N=4$ monomers for which the chain-like isomers are also obtained [22,23].

C. Low-Energy $(SiO_2)_6$ Cluster Isomers

For $(SiO_2)_6$ the ground state is still marginally the two-ring chain but other nonlinear isomers incorporating (Si_3O_3) three-rings become energetically competing structures (essentially the chain and the next highest energy isomer are degenerate). This cluster size marks a crossover point in the evolution of silica cluster structure with increasing size where we pass from a preference for one-dimensional growth to more complex growth trends for $(SiO_2)_N$ $N>6$. It is noted that this increase in structural complexity is mirrored in the increased difficultly in mapping the low-energy $(SiO_2)_6$ energy landscape, whereby two of the low-energy isomers in Figure 2 (e.g., 6e, 6f) have been obtained only via use of global optimization [46,49]. In the following sections for $(SiO_2)_N$ $N>6$ clusters all clusters reported are results of global optimization most of which are reported in references [44,46–49].

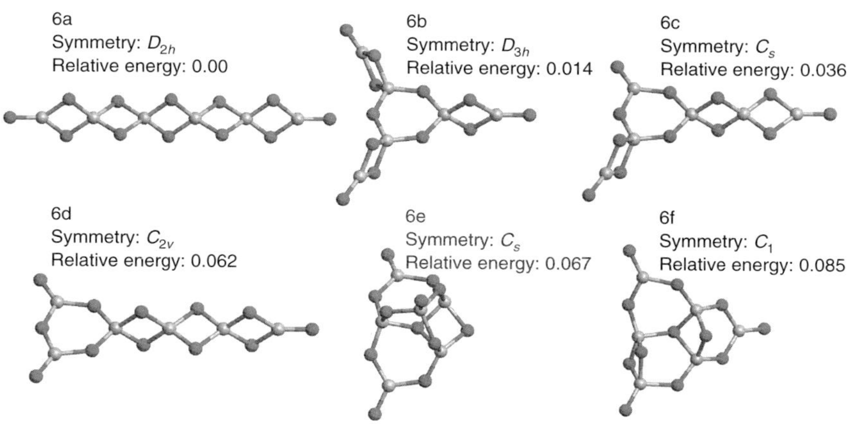

FIGURE 2 *The six lowest energy isomers for $(SiO_2)_6$.* Energies are relative total energies (eV/SiO_2) calculated at a 6-311+G(d)/B3LYP level of DFT theory.

D. Low-Energy $(SiO_2)_7$ Cluster Isomers

For the low-energy isomers of $(SiO_2)_N$ shown in Figure 3 only the higher energy isomers (7e and 7f) were first proposed by manual cluster construction studies with the four lowest energy isomers only first reported by means of DFT-refined global optimization. It is noted that the two-ring chain isomer is now significantly higher in energy than the C_{3v} ground state (7a) and the energetically degenerate C_s isomer (7b) with most cluster isomers now exhibiting less strained (Si_3O_3) three-rings. The 7a ground state is of particular note as it contains an "over-coordinated" three-coordinated oxygen center and corresponding "under-coordinated" oxygen (≡Si–O) termination. Such "compensating-pair" defects appear quite often in small silica clusters and appears to be a reasonably energy efficient manner in which silica reconstructs itself at small length scales. The defect involves a single electron charge transfer from a three-coordinated oxygen center to a singly coordinated oxygen center (i.e., $\equiv O^{+}\cdots$-O^{-}). This type of defect was first proposed to be present in bulk amorphous silica by Greaves [52] and Luckovsky [53] in the 1970s based on the valence alternation pair (VAP) model first applied to chalcogenide glasses by Mott [54]. In low-energy silica nanoclusters, both the singly terminated O^{-} species and the donating $\equiv O^{+}$ center are present exclusively as a surface species [55]. Such surface-terminating VAP defects (or compensated non-bridging oxygen (CNBO) defects [55]) have also been observed in modeling studies of metastable surfaces of α-quartz [56,57].

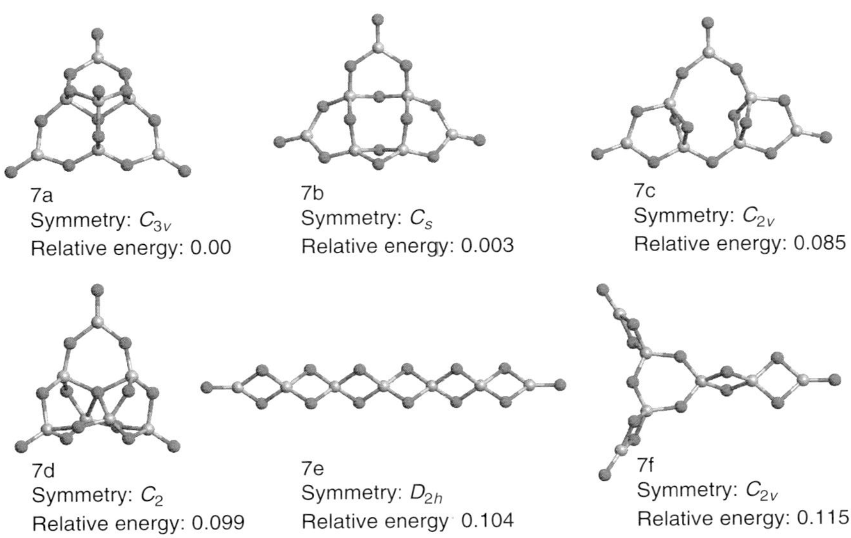

FIGURE 3 *The six lowest energy isomers for* $(SiO_2)_7$. Energies are relative total energies (eV/SiO_2) calculated at a 6-311+G(d)/B3LYP level of DFT theory.

Furthermore, it is notable that when the $\equiv O^+$ center is in a subsurface site, the remaining terminating O^- species is unreactive with water [55] and, much like the related but more reactive terminating silanolate group Si-O^-, may even have the propensity to order water molecules [58] or act as a catalyst [59].

E. Low-Energy $(SiO_2)_8$ Cluster Isomers

The lowest energy $(SiO_2)_8$ cluster is a D_{2d} symmetric cross-like structure containing four type Si=O silanone terminations and no other "defect" centers. Although this cluster has been obtained from global optimization searches, the same cluster has been proposed in the context of designing clusters via Si_3O_3-ring assembly [32]. The highly symmetric form of cluster 8a, seems to be strongly energetically favored with respect to other $(SiO_2)_8$ cluster isomers, with the next lowest energy $(SiO_2)_8$ isomer being 0.128 eV/SiO_2 higher in energy. All four $(SiO_2)_8$ clusters in Figure 4 can be simply regarded as being constructed via the addition of a SiO_2 monomer to either one of the degenerate low-energy $(SiO_2)_7$ isomers 7a an 7b. All $(SiO_2)_8$ isomers contain numerous (Si_3O_3) three-rings but very few strained two-rings.

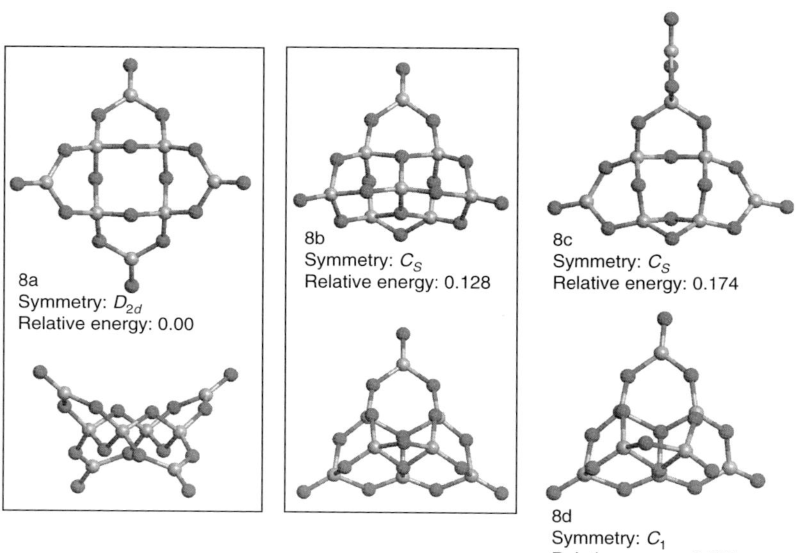

FIGURE 4 *The four lowest energy isomers for $(SiO_2)_8$.* Energies are relative total energies (eV/SiO_2) calculated at a 6-311+G(d)/B3LYP level of DFT theory. The isomers within a box are different views of the same cluster.

F. Low-Energy $(SiO_2)_9$ Cluster Isomers

Progressing to $(SiO_2)_9$ clusters (see Figure 5), the two lowest energy structures found take advantage of the low-energy symmetric core of the $(SiO_2)_8$ ground state cluster. The lowest energy cluster (9a) adds one SiO_2 unit to one of the terminating silanone centers of cluster 9a forming a pendant two-ring and lowering the symmetry to C_s. The next lowest energy cluster 9b results from adding a SiO_2 unit onto the center 8a, creating two compensating-pair defects and a structure having C_{2v} symmetry. Both 9a and 9b structures are essentially degenerate, being exceedingly close in energy. The third lowest energy $(SiO_2)_9$ structure, deviates from being an alteration to the 8a cluster and is found to have a near-planar triangular C_{3v} symmetric form, consisting of alternating two-rings and Si=O-terminated three-rings.

G. Low-Energy $(SiO_2)_{10}$ Cluster Isomers

For the two lowest energy $(SiO_2)_{10}$ clusters there appears to be a tendency to move away from cluster structures built upon smaller low-energy forms, with 10a and 10b, being distinct complex three-dimensional structures predominately formed from interlocking three-membered (Si_3O_3) silica rings (see Figure 6). Both these clusters display four Si=O centers, and have C_2 and C_s symmetries, respectively. The next four energetically lowest lying clusters, 10c–f, however, again follow the patterns of SiO_2-addition to the $(SiO_2)_8$ cross-like cluster found for the two lowest energy $(SiO_2)_9$ clusters (9a and 9b). For clusters 10c and 10d this results in simply the $(SiO_2)_8$ cross cluster with two pendant two-rings, whereas for cluster 10d a novel highly symmetric (T_d) cluster is formed displaying a "nano-diamond" structure with four compensating-pair centers (cluster 10f is a hybrid structure containing both these growth mechanisms). As with clusters 9a and 9b, clusters 10c–f are nearly degenerate in energy. It should be noted that, although one can continue

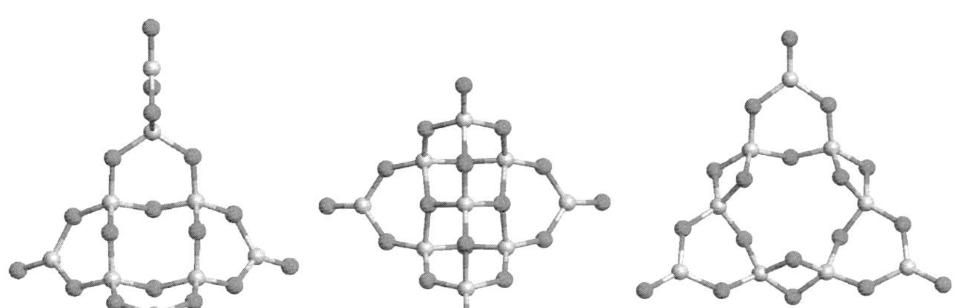

9a
Symmetry: C_s
Relative energy: 0.00

9b
Symmetry: C_{2v}
Relative energy: 0.004

9c
Symmetry: C_{3v}
Relative energy: 0.058

FIGURE 5 *The three lowest energy isomers for $(SiO_2)_9$.* Energies are relative total energies (eV/SiO_2) calculated at a 6-311+G(d)/B3LYP level of DFT theory.

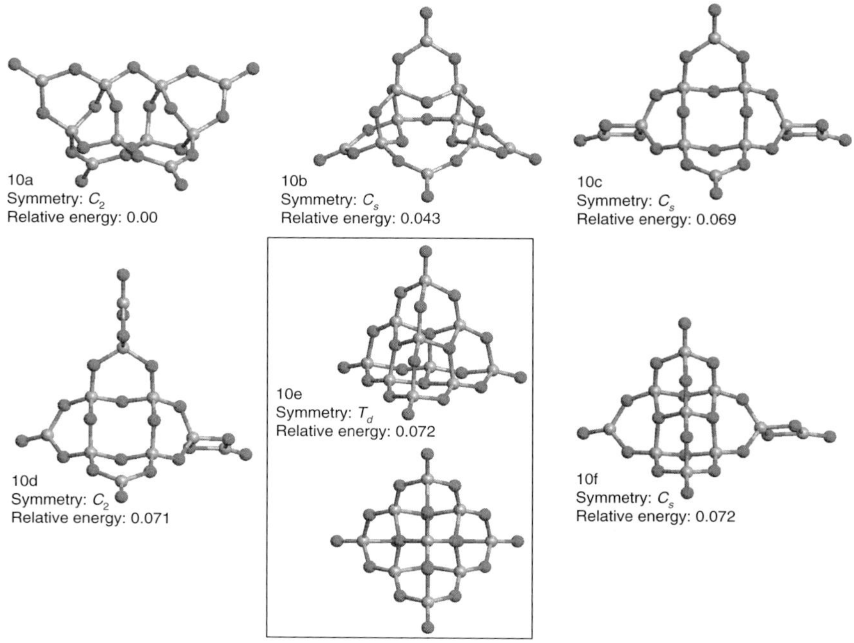

FIGURE 6 *The six lowest energy isomers for $(SiO_2)_{10}$.* Energies are relative total energies (eV/SiO_2) calculated at a 6-311+G(d)/B3LYP level of DFT theory.

adding a pair of two-rings to clusters 10a and 10b (although resulting in clusters that are not particularly low energy), one cannot simply continue the SiO_2 addition pattern 8a→9b→10e to form larger and larger SiO_2 nano-diamonds due to the eventual necessity of creating unstable four-coordinated oxygen atoms. It is of interest that although the emergence of the silica nano-diamond (10e) is the first low-energy indication of a bulk crystalline motif, it is inherently only stable at the nanoscale. This cluster although first reported as the result of global optimization studies [46] was also subsequently reported as the most stable cluster found by a manual search [60]. Although not the most stable isomer, laying considerably higher than the minima obtained employing global optimization, it is indeed a very interesting cluster due to its combination of high symmetry and it only exhibiting compensating-pair terminations, perhaps providing it with unusual optical/electronic properties.

H. Low-Energy $(SiO_2)_{11}$ Cluster Isomers

The two lowest energy $(SiO_2)_{11}$ clusters, 11a and 11b, have a similar three-dimensional triangular structural form and are essentially degenerate in energy, differing by only 0.047 eV (see Figure 7). Isomer 11c seems to be

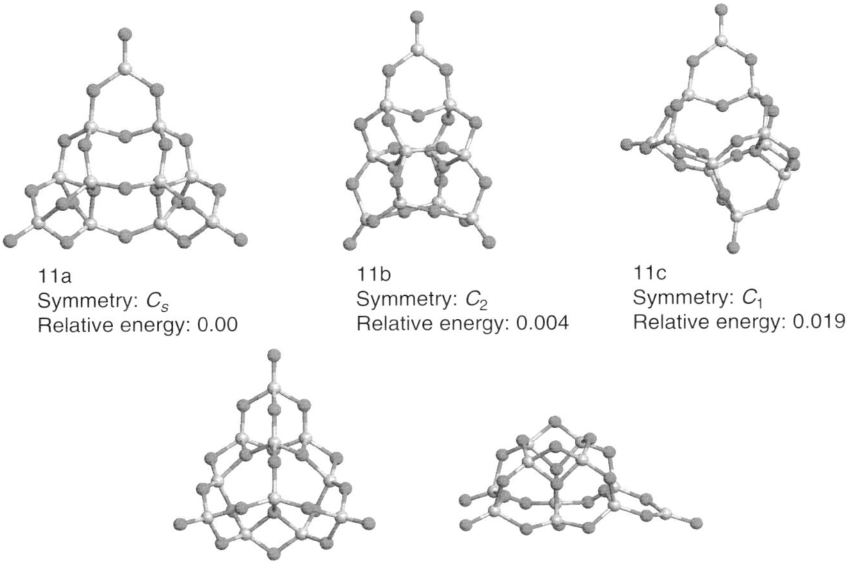

11a
Symmetry: C_s
Relative energy: 0.00

11b
Symmetry: C_2
Relative energy: 0.004

11c
Symmetry: C_1
Relative energy: 0.019

11d
Symmetry: C_s
Relative energy: 0.031

11e
Symmetry: C_s
Relative energy: 0.033

FIGURE 7 *The five lowest energy isomers for $(SiO_2)_{11}$.* Energies are relative total energies (eV/SiO_2) calculated at a 6-31G(d)/B3LYP level of DFT theory.

an irregular asymmetric version of the structure type exhibited in 11a/11b and has a correspondingly higher energy. All three clusters, 11a–c, however, have three similar terminations (one silanone, and two compensating-pair defects). Cluster 11d is, although triangular in form, distinct from 11a,b as it possesses three compensating-pair defects and can be constructed based upon the addition of two SiO_2 monomers to the $(SiO_2)_9$ isomer 9c.

I. Low-Energy $(SiO_2)_{12}$ Cluster Isomers

All three $(SiO_2)_{12}$ lowest energy isomers shown in Figure 8 have a compact elongated form containing numerous four-rings. Cluster isomers 12a and 12c contain a compensating-pair termination at each end, joined together at a single four-ring. These two $(SiO_2)_{12}$ clusters represent two different ways of orientating the two four-ring-joined ends. Cluster 12a has its cylinder ends rotated 180° with respect to each other, about long axis of the cluster, thus causing the two terminations to be diagonally opposed. On the other hand, cluster 12c has its ends rotated by 90° with respect to each other causing a closer alignment of the two terminations and a corresponding slight increase in energy. Cluster 12b, in contrast to 12a and 12c, has three oxygen terminations: two silanones and one compensating-pair defect. This cluster is

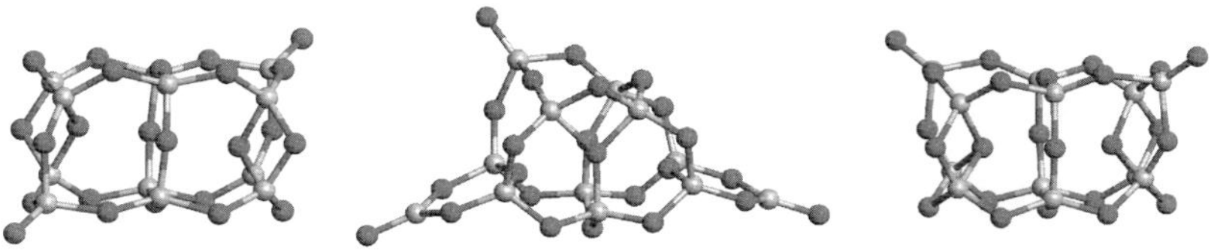

12a
Symmetry: C_{2h}
Relative energy: 0.000

12b
Symmetry: C_1
Relative energy: 0.010

12c
Symmetry: C_2
Relative energy: 0.010

FIGURE 8 *The three lowest energy isomers for* $(SiO_2)_{12}$. Energies are relative total energies (eV/SiO_2) calculated at a 6-31G(d)/B3LYP level of DFT theory.

energetically degenerate with 12c but may be considered as a natural extension of cluster 11e through the addition of a single SiO_2 monomer.

J. Low-Energy $(SiO_2)_{13}$ Cluster Isomers

With regard to $(SiO_2)_N$ cluster structure, the size $N=13$, as for $N=6$, marks a transitional size for cluster structural preference. For $(SiO_2)_N$ $N=6$–13, the clusters seem to have no well-defined structural type but can be roughly characterized as possessing a large proportion of three-rings, typically three or four defective oxygen terminations, and often having pyramidal/trigonal-like structures (see Figure 9). Only for $(SiO_2)_{12}$ does this trend appear to be opposed with the three lowest energy cluster isomers having more compact elongated forms with two or three terminations and more four-rings. For $(SiO_2)_{13}$, we see structures of both types in the three lowest energy isomers: 13a and 13c of the trigonal/three-ring-based type and 13b of the compact elongated form with two defective terminations and a greater proportion of four-rings. For $(SiO_2)_N$ cluster $N>13$ the evolution of cluster structure becomes more regular and below we will continue tracking the structures of low-energy silica clusters with respect the likely ground states (as obtained from DFT-refined global optimizations) and not consider energetically higher laying isomers.

K. $(SiO_2)_N$ Cluster Ground States $N=14$–27

The structural growth trend for $(SiO_2)_N$ $N=14$–22 clusters is found to be remarkably simple and proceeds via the addition of Si_4O_4 four-rings to make progressively longer compact elongated clusters (hereafter referred to as columnar clusters) of a similar form to that of 12a,b and 13b. This four-ring addition growth pattern can be seen through $N=14$, 18, 22 and $N=15$, 19 in Figure 10. For nanoclusters between these values of N, the number of SiO_2 units is not sufficient to make a column of complete four-rings and instead a Si_2O_2 two-ring is inserted into the side of the cluster (e.g., $N=16$, 17, 20, 21). For all columnar clusters the dominant defect termination is via

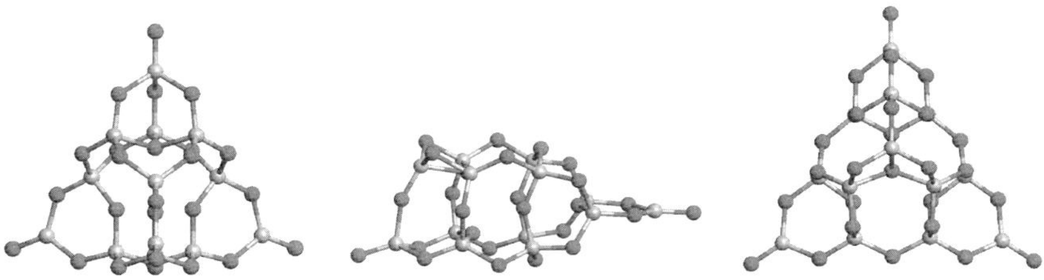

13a
Symmetry: C_{3v}
Relative energy: 0.000

13b
Symmetry: C_s
Relative energy: 0.009

13c
Symmetry: C_s
Relative energy: 0.043

FIGURE 9 *The three lowest energy isomers for* $(SiO_2)_{13}$. Energies are relative total energies (eV/SiO_2) calculated at a 6-31G(d)/B3LYP level of DFT theory.

Si=O, either at both ends of the column for even N, or only at one end for odd N. For clusters $N = 15$–21, the odd number of SiO_2 units does not allow for a twofold symmetric termination with only Si=O groups, and the odd-N columnar clusters are instead terminated by one Si=O defect and a less energetically favorable compensating-pair defect [46]. In all these cases, however, the connected four-ring columnar skeleton is maintained, which still appears to be the energetically favored structural route to obtain the lowest energy for this size range.

Although persistent, the energetic preference for one-dimensional columnar growth is overcome at a transitional $(SiO_2)_N$ cluster size of $N = 23$ by the emergence of a more compact two-dimensional disk-like structure based on the centrally symmetric sharing of three double (Si_5O_5) five-ring cages. For the N-odd $(SiO_2)_N$ clusters for $N \geq 23$, it appears that the having one Si=O termination and one less energetically favored compensating-pair termination in a columnar cluster is outweighed by having three Si=O terminations and a more compact structure. For even N for $N > 23$, the energetic benefit of having a disk-like form as opposed to a columnar form appears to solely be structural with both forms having the same double Si=O defect termination.

L. $(SiO_2)_N$ Cluster Structure $N > 27$

For $(SiO_2)_N$ clusters with $N > 27$ the BH global optimization strategy finds it increasingly more difficult to find low-energy ground state minima due, mainly, to the exponentially increasing complexity of the energy landscape with increasing cluster size. In an attempt to predict the likely structure of larger silica clusters, we have followed the seed of an idea planted in a study of the size-dependent transition from two Si=O defects to a two-ring in a model nanochain-to-nanoring system [61]. In this study, fully coordinated (FC) rings of two-rings were found to be energetically favored over defective chains after a certain size was reached. Recent DFT calculations of larger non-globally optimized compressed nanoslabs of 36–48 SiO_2 units have also

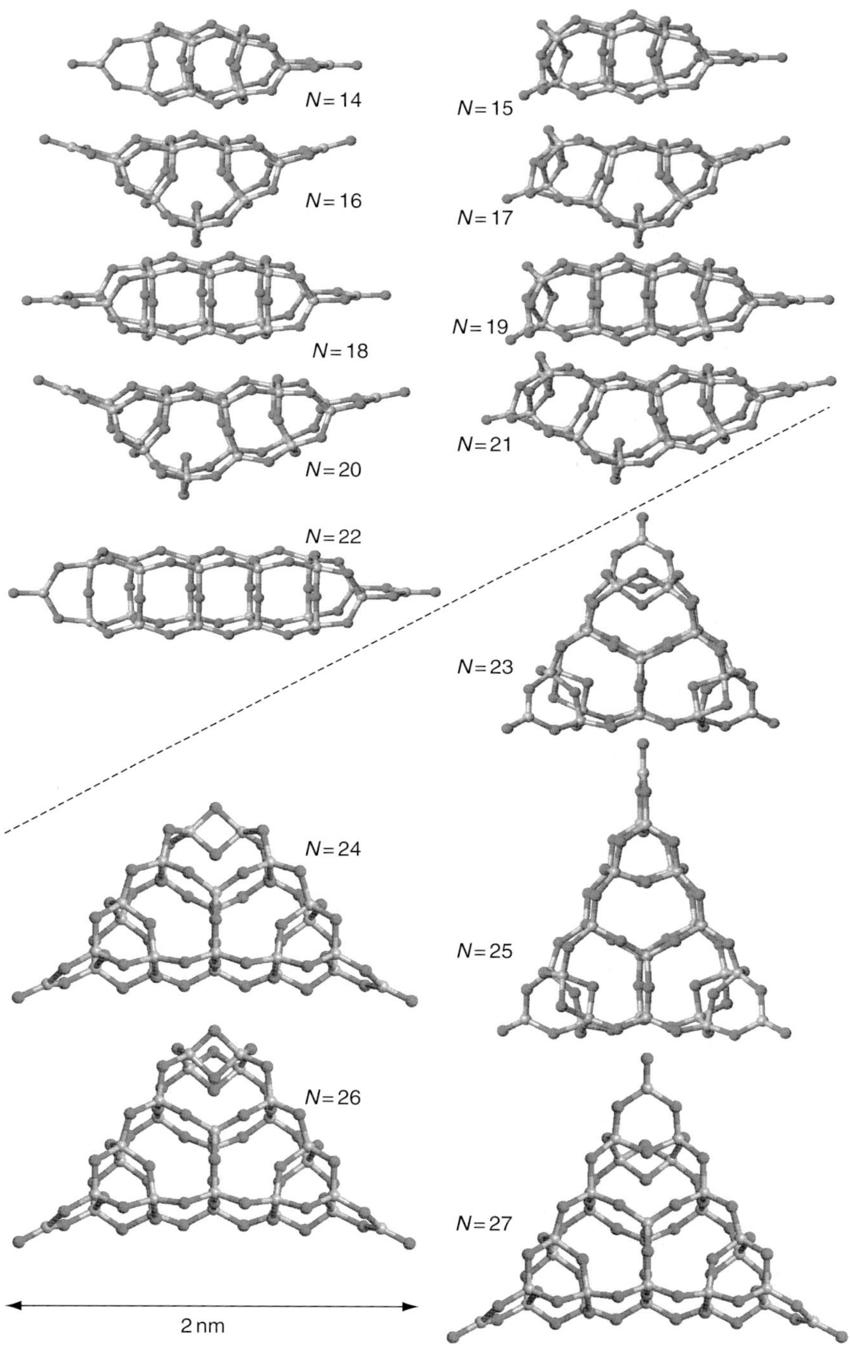

FIGURE 10 *Structures of lowest energy $(SiO_2)_N$ clusters for $N = 14$–27.* The dashed line separates the columnar and disk-like clusters.

observed a competition between silanones and two-rings as surface-energy-reducing mechanisms, noting also that their formation is largely determined by the local bonding environment [62]. As the reaction between two separated silanones to make a FC two-ring is an energetically favorable and barrier-less process [63], depending on the constraints of the bonding topology of the cluster we would expect all silanone terminations to eventually lose out to FC two-rings. Although one can propose a number of plausible FC cluster structures [64], a number of questions remain with respect to the details of the FC silica energy landscape: (i) the topology of the lowest energy FC nanoclusters, (ii) the existence of a defective-to-FC nanoscale crossover, and (iii) at what system size any crossover may occur. Using a new general algorithm for finding the low-energy fully reconstructed forms of directionally bonded discrete nanosystems, it has recently been shown that as silica nanoclusters grow in size they increasingly energetically prefer to adopt complex non-cage-like defect-free FC structures [65]. Using this approach FC clusters have been predicted to become the most stable form of nanosilica beyond a system size of approximately 100 atoms (i.e., $(SiO_2)_N$ of $\sim N > 27$) and before the eventual emergence of bulk crystalline structures. An example of a selection of FC $(SiO_2)_{24}$ nanoclusters, with respect to the ground state is shown in Figure 11A. Plotting the total energy per SiO_2 unit of each of the lowest energy FC isomers and of the correspondingly sized terminated ground state (see Figure 11B) we see that the two lines quickly converge with increasing cluster size. Taking the total energy difference between the two cluster classes and fitting with a power law provides a rough prediction as to the relative stability of each cluster type with further increase in size. This approximate extrapolation, indicates that the lowest energy FC clusters will become the energetic global ground states for cluster $(SiO_2)_N$ sizes $N \geq 27$. With increasing size it is expected that the internal bonded strain will gradually reduce further energetically favoring FC geometries. In Figure 11B, the average O–Si–O angle (a measure of tetrahedral strain) is also plotted (with respect to the optimal unstrained value of 109.47°) for the lowest energy FC clusters showing also the internal bonding strain reduces with increasing cluster size.

IV. FROM $(SIO_2)_N$ CLUSTERS TO BULK MATERIALS

As clusters of any material increase in size, their properties approach those of the respective bulk phase. The rate of approaching the bulk is, however, not the same for all properties and it is quite usual that some properties saturate to a limiting value well before others. Although beyond a certain size the approach the bulk is quite smooth with respect to the variation of all cluster measures up until this point for a range of small clusters the variation in cluster properties can be quite irregular. Within this small cluster size regime, finite size effects dominate often leading to large changes in cluster structure and stability with only small changes in cluster size. Often in this small size

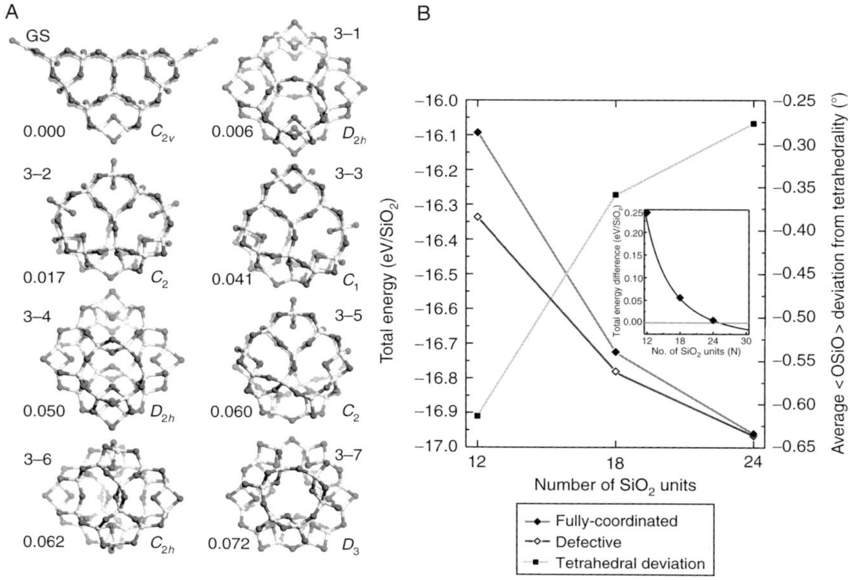

FIGURE 11 (A) Structures, symmetries and total energies (eV/SiO_2) of lowest energy $(SiO_2)_{24}$ FC nanoclusters (3-1 to 3-7) relative to the ground state (GS). (B) Total energies and average deviation from tetrahedrality of GS and lowest energy FC $(SiO_2)_N$ nanoclusters for $N = 12$, 18, 24 (inset: power law fit to the total energy difference between correspondingly sized GS and lowest energy FC clusters).

range particular cluster sizes stand out from the rest respect to their enhanced energetic stability. Clusters possessing these preferred cluster sizes are often referred to as “magic” cluster and have been observed in many systems with conspicuously higher than average abundances in cluster beam experiments [66]. Below we examine the energetic and structural evolution of $(SiO_2)_N$ clusters as they approach to the bulk in order to highlight any indications of “magicness” in the small cluster regime.

A. Evolution of $(SiO_2)_N$ Energetic Stability

It is commonly found that many cluster properties (after the small size regime) scale as $E(N) = E(\text{bulk}) + aN^{-1/3}$ where N is the number of atoms/units in a cluster. Although the assumptions in deriving this relation are based on mono-elemental clusters with spherical form (where $N^{-1/3}$ is proportional to the fraction of surface atoms) a plot of $(SiO_2)_N$ provides us with a rough estimate of the onset of the regime of stable cluster growth where energetic stability converges smoothly to the bulk energy. In Figure 12, we plot the energy of $(SiO_2)_N$ clusters per SiO_2 unit with respect to $N^{-1/3}$ together with a linear best fit passing through zero. We can see that after a certain size

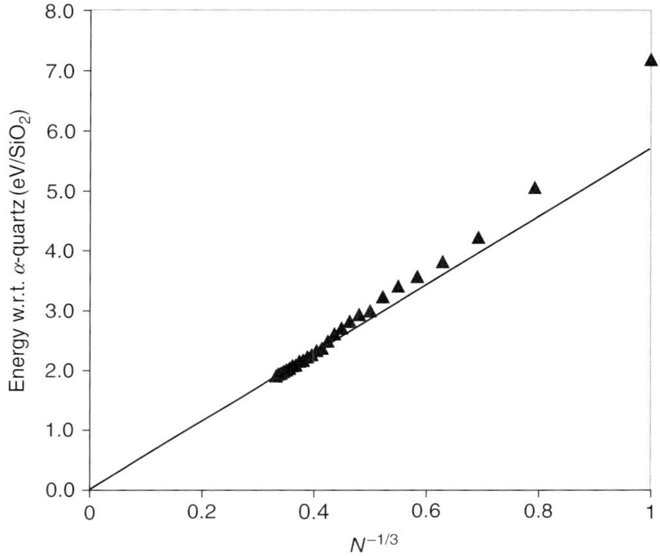

FIGURE 12 Fit of the form $E(N) = E(\text{bulk}) + aN^{-1/3}$ to the DFT calculated energies of $(SiO_2)_N$ clusters from $N = 1$ to 27. The best fit line has $a = 5.7$ and $E(\text{bulk}) = 19$.

the data points converge to the line indicating the onset of smooth convergence to bulk stability. The line of best fit requires a E(bulk) shift of 19.0 eV/SiO_2 to the cluster DFT calculated energies, which can be interpreted as a measure of the energetic stability of the clusters with respect to the thermodynamic limit of bulk SiO_2. The predicted limiting energy of 19.0 eV/SiO_2 is also in good correspondence with the experimentally measured energy of α-quartz (19.2 eV/SiO_2 [67]), giving confidence in the extrapolation. Using the fitted $E(N)$ function, for approximately $N > 7000$ the energy with respect to quartz becomes less than 0.3 eV/SiO_2. Within this energy range, most currently known thermodynamically metastable bulk polymorphs of silica are known to exist giving a rough measure of the onset of bulk stability. Assuming a suitable silica density and taking clusters at this value of N to be spherical it can also be estimated that the SiO_2 nanoscale-to-bulk transition occurs for a particle diameter of approximately 7 nm. The best fit line starts to differ from the cluster energy data at approximately $N = 13$ which also marks the structural transition from irregular three-ring-based clusters to more compact columnar clusters. The transitions between different cluster structure preferences and their respective energies with respect to bulk α-quartz are summarized in Figure 12. From the total energy data in Figures 12 and 13, it is difficult to discern whether any particular clusters are particularly energetically favored over others.

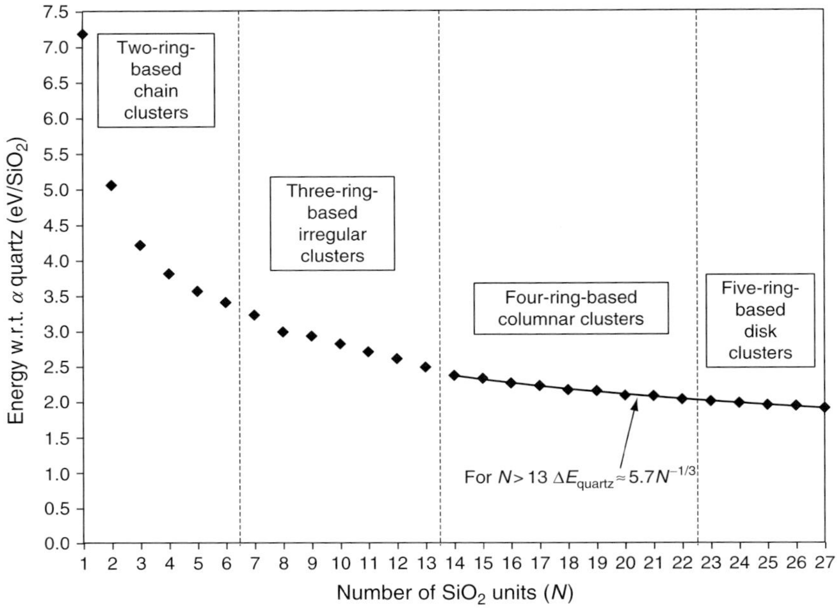

FIGURE 13 DFT calculated energies of $(SiO_2)_N$ clusters from $N=1$ to 27 indicating the corresponding structural changes. The fit of the form $E(N)=E(\text{bulk})+aN^{-1/3}$ is shown for $N>13$ where the data have been shifted such that $E(\text{bulk})=0$.

Using energy differences between neighboring clusters, however, can be used as a more sensitive measure to highlight the energetic stability increase or drop from going from a $(SiO_2)_N$ cluster of size N to one of $N+1$. This energy, $E_{\text{nuc}}=E[N]-E[N-1]-E[SiO_2]$, is termed the nucleation energy and can be used to identify magic clusters with respect to cluster growth. Peaks for $(SiO_2)_x$ in E_{nuc} indicate that nucleation from $(SiO_2)_{x-1}$ clusters to $(SiO_2)_x$ is particularly energetically favorable and further that upon reaching cluster $(SiO_2)_x$ further cluster growth is strongly energetically hindered. The nucleation energy for $(SiO_2)_N$ clusters from $N=1$ to 27 is shown in Figure 14 with respect to the nucleation of two SiO_2 monomers. The E_{nuc} magnitude fluctuates greatly with $(SiO_2)_N$ cluster size but on average increases from $N=2$ to 13 whereupon from $N=14$ to 27 it fluctuates about a mean value of approximately 1.6 eV. Although in this latter stable growth regime there are peaks (e.g., $N=$ 16, 18, and 20) indicating that the preferred columnar clusters have an even number of SiO_2 units these clusters are unlikely to be suitable extended materials building blocks as: (i) they have only two reactive terminations likely to lead to linear one-dimension cluster-based products rather than three-dimensionally extended materials and (ii) they have relatively high atomic weights meaning that they would be relatively difficult to isolate in large quantities with respect to more abundant smaller clusters

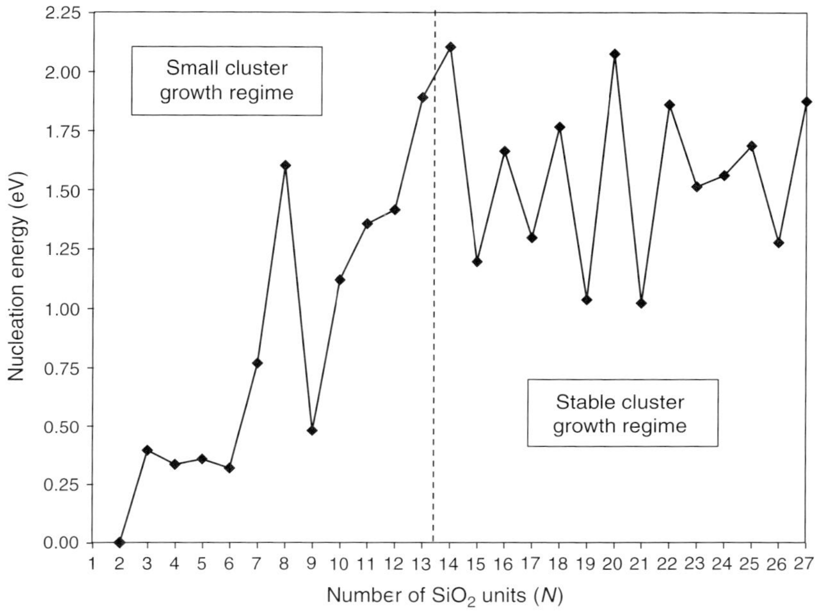

FIGURE 14 The nucleation energy of $(SiO_2)_N$ cluster with sizes $N=2$–27. The scale is set such that the nucleation energy of two SiO_2 monomers is zero.

(e.g., in cluster beams). Although not perhaps ideal for new extended cluster-based materials, the high stability of nanoscale columnar $(SiO_2)_N$ clusters possessing reactive defective ends naturally provides a route to linear coalescence, perhaps linked to the observed high temperature synthesis of silica nanowires [68].

In the small $(SiO_2)_N$ cluster regime there is one significant E_{nuc} peak for $(SiO_2)_8$ only. This large peak corresponds to the highly stable cluster 8a, which has a highly symmetrical D_{2d} structure with four Si=O reactive terminations pointing in mutually opposing direction (see Figure 4). The topology of this isomer evidently provides more scope for extended bulk phases that a two-dimensional column. By further taking the second-order energy difference ($E_{2nd}(N) = -2E[N] + E[N-1] + E[N+1]$), which provides a more sensitive stability measure of a cluster with respect to both its $N-1$ and $N+1$ neighbors, we can see in Figure 15 that the cluster is extremely stable with respect to other nearby cluster sizes. In addition, the $(SiO_2)_8$ ground state cluster is very isolated with respect to other $(SiO_2)_8$ isomers (see Figures 4 and 15; Ref [69]). This strongly indicates that only this particular $(SiO_2)_8$ isomer would be highly abundant in a population of small $(SiO_2)_N$ clusters. In the next section, we explore the potential of the magic $(SiO_2)_8$ ground state for forming extended cluster-based materials.

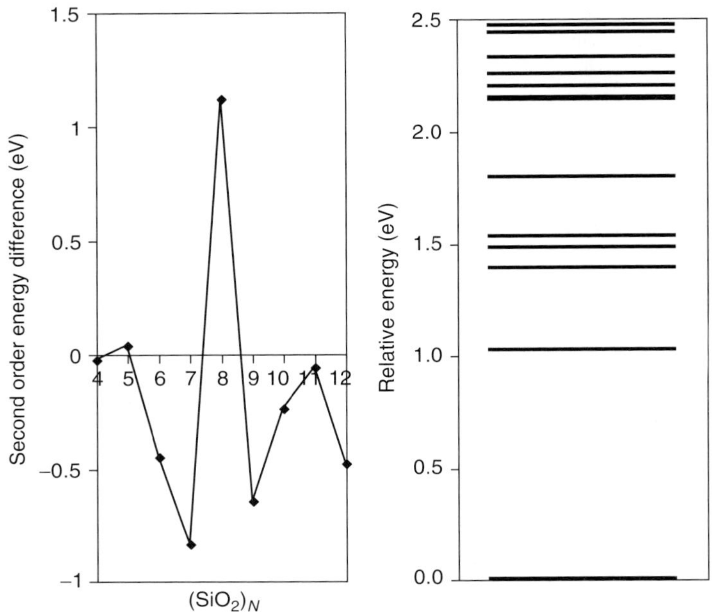

FIGURE 15 Left: the second-order energy difference for clusters $(SiO_2)_N$ $N=4$–12, $E_{2nd}(N)=-2E(N)+E(N-1)+E(N+1)$, derived from ground state cluster energies $E(N)$ calculated at a B3LYP/6-31G(d) level, Right: the low-energy spectrum of the bottom thirteen $(SiO_2)_8$ cluster isomers (B3LYP/6-311+G(d)) showing a large (>1 eV) gap between the magic ground state cluster and the next highest energy isomer.

B. New Bulk Materials Based on $(SiO_2)_8$ Magic Clusters

Recently, the assembly of well-defined rigid molecular building units has been put forward as a new experimental means for the bottom-up synthesis of new materials from solution [70] or co-condensation [71]. For silicate frameworks, through consideration of probable species present in the traditional hydrothermal synthesis mixture, some theoretical efforts have investigated ways in which cubic Si_8O_{12} double four-ring units may be combined into periodic structures [72]. An alternative rationale for choosing such nanoscale building units is to consider small well-defined clusters, which intrinsically display unusually high stability in vacuum, that is, magic clusters. Bottom-up synthesis based upon gas phase clusters requires precise control of (i) the cluster deposition/assembly process and (ii) the cluster species. Numerous experiments have already demonstrated the controlled deposition of silica clusters, usually produced via vapor phase oxidation of silicon or silicon-containing species, to form silicate films [73,74]. In such experiments, there is a wide distribution of silica cluster sizes/types and the resulting deposited phase is amorphous. In order to exert some control over the synthesized material, self-elected magic clusters are particularly good candidates for nanoscale building blocks due to the relative ease with which they can, in principle, be selectively produced in large quantities. Examples of magic cluster building blocks experimentally realized in this manner are C_{60} fullerenes [75], metallocarbohedrene clusters [76], and $Al_{13}I_x^-$ clusters [77]. For silica, we have shown above that the $(SiO_2)_8$ ground state cluster is markedly conspicuous with respect to two defining measures of "magicness": (i) stability with respect to cluster isomers of the same composition, (ii) stability with respect to $N+1$ and $N-1$ ground state isomers (the second-order energy difference; see Figure 15 and Ref. [69]). It is further encouraging that hydroxyl-terminated $(SiO_2)_8$ magic clusters have been shown to exist in laser ablation experiments [20], albeit with different structure to the $(SiO_2)_8$ magic cluster employed herein [78]. In the remainder we investigate magic cluster-based bulk silica phases by exploring numerous ways in which our $(SiO_2)_8$ building blocks may be assembled into ordered materials.

The $(SiO_2)_8$ magic cluster has two clear assets when viewed as a building unit, both with respect to its terminated structure and its overall topology. The cluster possesses four silanone (Si=O) terminations, which, are known to be mutually reactive centers that undergo barrier-less coalescence to form siloxane (Si–O–Si) bridges [79,80], thus providing a natural means to link clusters together to form an extended structure. In a fully connected material possessing no defects, each Si=O termination in the original clusters should thus combine to form two Si–O–Si linkages between clusters. By simply noting the typical energies of a Si–O bond (452 kJ/mol) and a Si=O bond (590 kJ/mol), and the energy difference per cluster between a collection of free clusters and a Si–O–Si-bonded cluster-based material ($4 \times 590 - 8 \times 452 = -1256$ kJ/mol) we can immediately see that, in this ideal scenario, the reaction of the clusters to form an extended material is strongly thermodynamically favored. As this value gives the energy of the connected material per cluster, from a calculation of the energy of a single magic $(SiO_2)_8$ cluster we can estimate the maximum energy gain of a magic cluster-formed material per basic SiO_2 unit. Subtracting from this value the calculated energy of α-quartz gives a measure of the likely minimal energy excess of a magic $(SiO_2)_8$ cluster-based material above the ground state, thus providing a gauge of the viability of such materials. In this manner at an approximate benchmark limit is calculated to be approximately 19 kJ/mol/SiO_2 above α-quartz for an ideal magic cluster-based material.

Due to the fact that the four Si=O terminations of the magic cluster are distributed in a highly symmetric manner with each pointing in a mutually opposing direction the resulting cluster topology can thus be thought of as a super-analogue of the basic SiO_4 tetrahedral silicate unit (see Figure 16).

Extending this concept, the realization of materials employing this topological analogy is highlighted in Figure 17 showing how both corner-sharing and edge-sharing of SiO_4 tetrahedra can be mimicked in a meta-fashion using

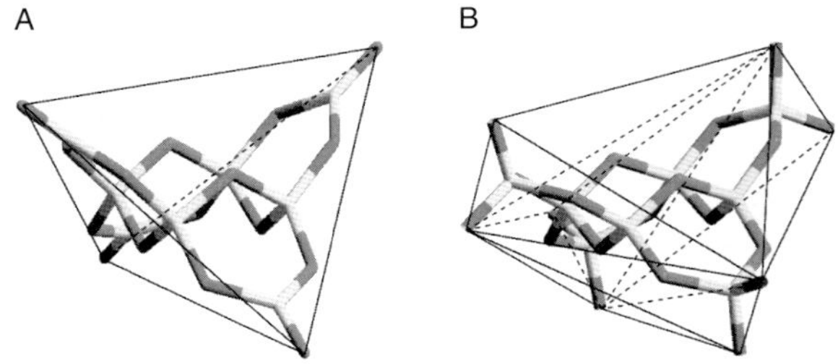

FIGURE 16 (A) The $(SiO_2)_8$ magic cluster schematically encased by a tetrahedron whereby the vertices show the fourfold tetrahedral bonding potential. (B) The $(SiO_2)_8$ magic cluster encased by a tristetrahedron with its vertices indicating the eightfold bonding propensity.

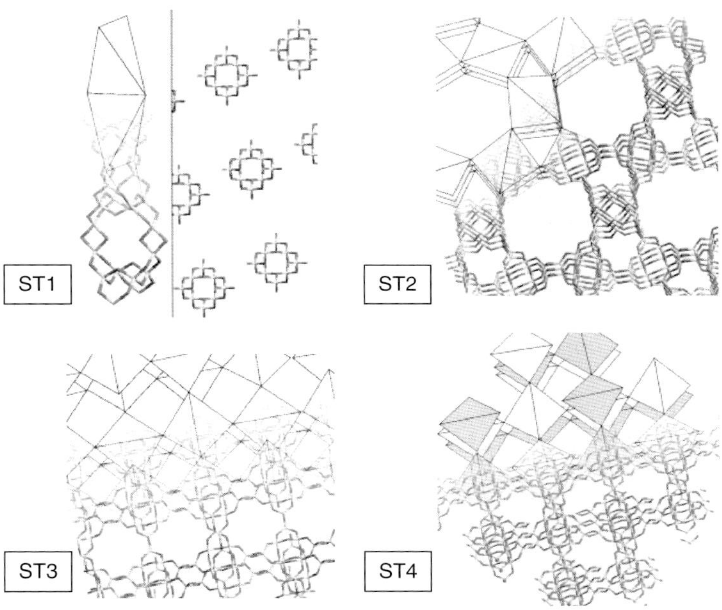

FIGURE 17 *Magic cluster-based ST frameworks showing an atomic representation and a schematic ST representation in case.* ST1: the edge-sharing ST version of the discrete-chain-based polymorph silica-W (left shows an individual super-chain and right shows a view down the *c*-axis of the material), ST2: corner-sharing ST version of quartz, ST3: a corner-sharing ST version of cristobalite, ST4: a ST material formed from two interleaved cristobalite-like frameworks.

the magic clusters to form "super-tetrahedral" (ST) silicate structures. In Figure 17, four such ST silicates (ST1-ST4) are shown: (ST1) a ST version of the edge-sharing one-dimensional chain-like silica polymorph silica-W [25], and ST versions the corner-sharing silicates quartz (ST2) and cristobalite (ST3). Additionally, the ST4 structure shows, how two cristobalite-like frameworks can be interleaved with each other, forming substantially denser material.

In each case the transformation of the usual silica polymorph to its ST form entails an inevitable increase in the unit cell size and also leads to relatively lower density silica materials. In particular, in their ST versions, the two-ring chains of silica-W become more akin to nanowires and the dense silica phases of cristobalite and quartz become open framework materials. A comparison of the experimental crystal parameters and framework densities of the SiO_4-tetrahedral-based silicate materials (where known) and the calculated values for our corresponding ST versions are given in Table 1. Details of the periodic DFT methodology used for all calculating the properties of all magic cluster materials reported in this chapter can be found in Ref. [69].

In order to construct these ST silicates, one must connect the Si=O terminations of distinct clusters in a one-to-one fashion whereby the resulting pairs of siloxane bridges lead to the formation of strained Si_2O_2 two-rings. Examining the energies of the four ST silicates with respect to α-quartz (Table 1), we can see that the coalescence of clusters via two-ring bridges appears to lead to relatively highly energetic materials. Compared to our estimated optimal benchmark energy above α-quartz of approximately 20 kJ/mol for magic cluster-based materials, we see that these structures are a further approximately 36–53 kJ/mol higher in energy and are thus probably not easily viable synthesis targets.

An alternative to joining the Si=O terminations of each cluster in a one-to-one manner is to consider a one-to-two type of connection. In this way, each

TABLE 1 Cell Parameters, Framework Density (FD) and Energies with Respect to α-Quartz, of the Four ST Framework Materials ST1–ST4

	a	*b*	*c*	*α*	*β*	*γ*	FD (Si/1000 Å^3)	$\Delta E_{\alpha\text{-quartz}}$ (kJ/mol/SiO_2)
ST1	12.96 (8.37)	10.12 (5.16)	13.12 (4.76)	95.7 (90)	86.3 (90)	88.4 (90)	9.4 (19.5)	72.9
ST2	17.39 (4.91)	17.39 (4.91)	13.04 (5.41)	90 (90)	90 (90)	120 (120)	7.0 (26.5)	59.2
ST3	17.81 (4.97)	17.67 (4.97)	17.83 (6.92)	90 (90)	90 (90)	90 (90)	5.7 (23.5)	55.4
ST4	13.50	13.50	5.43	95.7	84.3	91.6	15.2	57.0

Comparison of the experimental cell parameters and densities (where available) of the corresponding SiO_4-tetrahedral-based silicates (i.e., ST1: silica-W [25], ST2: quartz [81], ST3: cristobalite [82]) are given in parentheses.

single Si=O termination opens to form two Si–O–Si links with two other magic clusters. With respect to our tetrahedral representation of the magic cluster we can regard this extended bonding mode as an addition of a single connection point (vertex) at each of the four faces of the tetrahedron. Topologically, this leads to the formation of a distorted Triakis tetrahedron (tristetrahedron) [83] geometry, which can be also viewed as a basic building unit (see Figure 2B).

Using this mode of assembly leads to framework materials having clusters that are connected to between two and eight neighbors (super-tristetrahedral [STT] materials), rather than between two and four as in the ST materials. This cluster bonding mode has the advantage that energetically unfavored two-ring linkages are naturally avoided thus allowing greater potential for lower energy materials. In figure 18, we show six STT framework materials (STT1–STT6) in order of decreasing energetic stability, constructed by the connection of magic clusters in a one-to-two fashion. The cell parameters, framework densities, and energies with respect to α-quartz are also given in Table 2.

In order to evaluate the synthetic viability of the STT frameworks, we have investigated both their topology and energetics. An indication of the strain inherent within each STT framework can be assessed by calculating the tetrahedral mismatch (Δ_{TM}) of each framework [84]. Such a measure has been found to be instructive in suggesting limits on the viability of

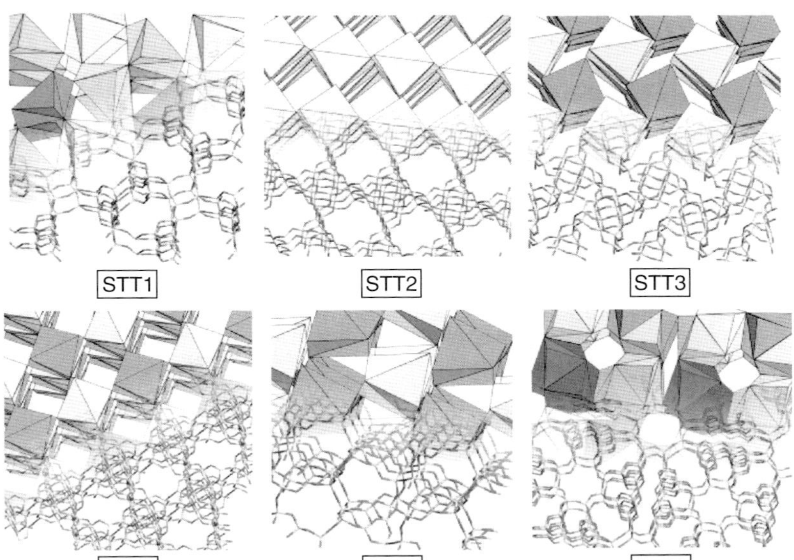

FIGURE 18 Magic cluster-based STT frameworks showing an atomic representation and a schematic super-tristetrahedral representation in case.

TABLE 2 Comparison of Calculated Cell Parameters, Framework Density (FD), Energy with Respect to α-Quartz ($\Delta E_{\alpha\text{-quartz}}$), Tetrahedral Mismatch Δ_{TM}, and Space Group, and Largest Pore Size of the Magic Cluster-Based STT Framework Materials

	a	*b*	*c*	α	β	γ	FD (Si/1000 Å^3)	$\Delta E_{\alpha\text{-quartz}}$ (kJ/mol/SiO_2)	Δ_{TM} (10^{-2})	Space group	Largest pore (number of Si atoms)
STT1	15.81	15.81	9.96	90	90	90	12.9	21.3	2.04	*P4/mbm*	12
STT2	18.31	18.30	5.11	90	90	90	18.7	23.6	2.12	*Amm2*	12
STT3	11.62	11.62	5.20	90	90	90	22.8	25.3	2.39	*P21212*	12
STT4	13.68	13.68	10.49	90	90	90	16.3	25.6	2.23	*I 4/m*	12
STT5	14.17	12.73	10.31	90	90	90	17.2	26.2	3.16	*P222*	8
STT6	20.57	9.74	20.46	90	90	90	15.6	34.6	4.38	*Pmmm*	12

hypothetical framework to be synthesized through hydrothermal means ($\Delta_{TM} < 2.5 \times 10^{-2}$ Å^2) [85]. The tetrahedral mismatch for all six STT materials is given in Table 2. Four of the STT frameworks (STT1–STT4) have Δ_{TM} values that lay within the range shared by known synthesized frameworks suggesting that their bonding topology is not a barrier to their eventual synthesis. Although we do not expect that our STT frameworks to be viable synthesis targets through hydrothermal means, the reasonable range of Δ_{TM} values for STT1–STT4 give us extra confidence that there should exist alternative route to their formation based on magic clusters as proposed herein.

The energies of the STT frameworks STT1–STT5 all lay in a narrow range between 21 and 26 kJ/mol above α-quartz with STT6 being a further 8 kJ/mol higher in energy (see Table 2). The relatively high energy and Δ_{TM} of STT6 is probably due to it exhibiting "super-edge-sharing" whereby some clusters are connected to only six out of a possible eight neighbors. Although all the STT energies are quite high compared to hydrothermal synthesized all-silica zeolites (7–14 kJ/mol [86]), in comparison with the range of experimentally determined enthalpies of formation for mesoporous silica frameworks (19–32 kJ/mol [87]) it is evident that our STT frameworks lay in a thermodynamically accessible window. The higher energies of mesoporous silicas relative to the all-silica zeolites have been attributed to the presence of Si_3O_3 three-membered rings [88]. The incorporation of small rings into zeolitic structures has also been topologically linked to the possibility of creating materials with the highly desirable property of possessing extra-large pores [88,89]. Although, thus far, hydrothermal syntheses have not managed to produce a three-ring-containing pure-silica zeolite, their presence in mesoporous silica frameworks [88] and in biosilicas [90] indicates that this is not a fundamental constraint on crystalline pure-silica. All our cluster-based frameworks incorporate three-rings from their intrinsic presence in the magic cluster (see Figure 1). For small silica nanoclusters, three-rings appear to be a relatively energetically favored motif (see $(SiO)_N$ $N=6$–13 clusters above in Section III and Refs. [46,49]), and thus, the formation of frameworks, perhaps possessing very large pores and/or channels, from such building blocks indicates a novel route to naturally incorporate Si_3O_3 three-membered rings into all-silica materials.

Except for the presence of three-rings, the six STT frameworks do not appear to display physical characteristics that would be deemed particularly atypical of known silicate frameworks. Topologically, however, all STT frameworks (with exception of STT5) have at least one 12-ring pore and are thus so-called large pore frameworks. Most also have 8- or 10-ring pores running in perpendicular directions making such materials potential attractive targets for catalytic and membrane applications. Furthermore, the utilization of a single type of cluster building block does not appear to significantly restrict the variability in framework type. The framework density of the magic cluster-based materials goes from very open (STT1 with a framework density comparable to that of Faujasite) to very dense (STT3 with a framework density comparable to that of α-cristobalite). Similarly, the frameworks range from cage-type frameworks (STT5) via frameworks with pores in only one direction (STT2) to frameworks with pores extending throughout the structure in all three spatial directions (STT1 and STT4). Interestingly, the dense framework STT3 has a slit like 12-ring pore (11.8×5.4 Å) running through it in one direction, in contrast with known frameworks of a similar density (e.g., α-cristobalite, tridymite), which are all nonporous. We note further that, although all our range of frameworks were constructed via bottom-up design, we subsequently discovered that three of them (STT2, STT3, and STT4) also correspond to hypothetical frameworks independently generated through top-down methods based on graph theory [91]. This link between two different predictive approaches provides an important bridge between materials discovery and identifying potential building blocks for their fabrication, suggesting that there may be considerable benefits in combining both approaches.

Taking the energy with respect to α-quartz of our lowest energy framework (21 kJ/mol for STT1—which as yet has not been found to correspond with a previously known hypothetical framework), we can see that the one-to-two mode of construction for STT frameworks can lead to an almost optimal assembly of our magic clusters into a material (with respect to our ideal estimate of $\sim$20 kJ/mol). Although this indicates that further searches into other modes of assembly of this particular magic cluster are unlikely to yield significantly more stable framework materials, it is hoped that this overview of the possibilities offered by considering silica clusters will motivate experimental cluster-based approaches to inorganic materials synthesis (as a gas phase extension to solution-based reticular synthesis [70]) and further stimulate theoretical searches for new viable materials employing other types of cluster building blocks.

REFERENCES

[1] Muller DA, Sorsch T, Moccio S, Baumann FH, Lutterodt K, Timp G. Nature 1999;399:758.

[2] Corma A. J Catal 2003;305:480.

[3] Ke YC, Stroeve P. Polymer-layered silicate and silica nanocomposites. 1st ed. Amsterdam: Elsevier; 2005.

[4] Storm AJ, Chen JH, Ling XS, Zandbergen HW, Dekker C. Nat Mater 2003;2:537.

[5] Hentze H-P, Raghaven SR, McKelvey CA, Kaler EW. Langmuir 2003;19:1069.

[6] Adachi M. Colloid Polym Sci 2003;281:370.

[7] Russell PStJ. Science 2003;299:358.

[8] Luckarift HR, Spain JC, Naik RR, Stone MO. Nat Biotechnol 2004;22:211.

[9] Mann S, Ozin G. Nature 2003;328:313.

[10] Sandhage KH, Dickerson MB, Huseman PM, Caranna MA, Clifton JD, Bull TA, et al. Adv Mater 2002;14:429.

[11] Mora-Fonz MJ, Catlow CRA, Lewis DW. Angew Chem Int Ed 2005;44:3082.

[12] Campbell SA. The science and engineering of microelectronic fabrication. 1st ed. New York: Oxford University Press; 1996.
[13] Suh S-M, Girshick SL, Kortshagen UR, Zachariah MR. J Vac Sci Technol A 2003;21:251.
[14] Cook G, Timms PL, Göltner-Spikermann C. Angew Chem Int Ed 2003;42:557.
[15] Giesenberg T, Hein S, Binnewies M, Kikelbick G. Angew Chem Int Ed 2004;43:5697.
[16] Fujimoto T, Okuyama K, Yamada S, Adachi M. J Appl Phys 1999;85:4196.
[17] Uchino T, Aboshi A, Kohara S, Ohishi Y, Sakashita M, Aoki K. Phys Rev B 2004;69:155409.
[18] Schweigert IV, Lehtinen KE, Carrier MJ, Zachariah MR. Phys Rev B 2002;65:235410.
[19] Brewer L, Edwards RK. J Phys Chem 1954;58:351.
[20] Xu C, Wang W, Zhang W, Zhuang J, Liu L, Kong Q, et al. J Phys Chem A 2000;104:9518.
[21] Schenkel T, Schathölter T, Newman MW, Machicoane GA, McDonald JW, Hamza AV. J Chem Phys 2000;113:2419.
[22] Wang LS, Nicholas JB, Dupuis M, Wu H, Colson SD. Phys Rev Lett 1997;78:4450.
[23] Wang L-S, Desai SR, Wu H, Nichloas JB. Z Phys D 1997;40:36.
[24] Bergeron DE, Castleman AW. J Chem Phys 2002;117:3219.
[25] Weiss A, Weiss A. Z Anorg Allg Chem 1954;276:95.
[26] Kamitsuji K, Suzuki H, Kimura Y, Sato T, Saito Y, Kaito C. Astron Astrophys 2005;429:205.
[27] Lu WC, Wang CZ, Nguyen V, Schmidt MW, Gordon MS, Ho KM. J Phys Chem A 2003;107:6936.
[28] Chu TS, Zhang RQ, Cheung HF. J Phys Chem B 2001;105:1705.
[29] Nayak SK, Rao BK, Khanna SN, Jena PJ. J Phys Chem B 1998;109:1245.
[30] Zang QJ, Su ZM, Lu WC, Wang CZ, Ho KM. J Phys Chem A 2006;110:8151.
[31] Harkless JA, Stillinger DK, Stillinger FH. J Phys Chem 1996;100:1098.
[32] Lu WC, Wang CZ, Ho KM. Chem Phys Lett 2003;378:225.
[33] Zhang D, Zhang RQ. J Phys Chem B 2006;110:15269.
[34] Zhao MW, Zhang RQ, Lee ST. Phys Rev B 2004;69:153403.
[35] Zhang D, Zhang RQ, Han Z, Liu C. J Phys Chem B 2006;110:8992.
[36] Song J, Choi M. Phys Rev B 2002;65:241302R.
[37] Liu S. J Phys Chem A 2007;111:3132.
[38] Wales DJ, Miller MA, Walsh T. Nature 1998;394:758.
[39] Wales D, Scheraga HA. Science 1999;285:1368.
[40] Hartke B. Angew Chem Int Ed 2002;41:1468.
[41] Kiran B, Bulusu S, Zhai H-J, Yoo S, Zeng XC, Wang L-S. Proc Natl Acad Sci USA 2005;102:961.
[42] Tsuneyuki S, Tsukada M, Aoki H, Matsui Y. Phys Rev Lett 1988;61:869.
[43] van Beest BWH, Kramer GK, van Santen RA. Phys Rev Lett 1990;64:1955.
[44] Flikkema E, Bromley ST. Chem Phys Lett 2003;378:622.
[45] Wales DJ. J Phys Chem 1997;101:5111.
[46] Flikkema E, Bromley ST. J Phys Chem B 2004;108:9638.
[47] Bromley ST, Flikkema E. Phys Rev Lett 2005;95:185505.
[48] Bromley ST. Phys Status Solidi A 2006;203:1319.
[49] Bromley ST, Illas F. Phys Chem Chem Phys 2007;9:1078.
[50] Doye JPK, Wales DJ. Phys Rev Lett 1998;80:1357.
[51] Becke ADJ. J Phys Chem 1993;98:5648.
[52] Greaves GN. Philos Mag B 1978;37:447.
[53] Lucovsky G. Philos Mag B 1979;39:513.

[54] Street RA, Mott NF. Phys Rev Lett 1975;35:1293.
[55] Hamad S, Bromley ST. Chem Commun 2008;4156.
[56] Riganese G-M, De Vita A, Charlier J-C, Gonze X, Car R. Phys Rev B 2000;61:13250.
[57] Muurashov VV. J Phys Chem B 2005;109:4144.
[58] Du Q, Freysz E, Shen YR. Phys Rev Lett 1994;72:238.
[59] Goumans TPM, Catlow CRA, Brown W. J Phys Chem C 2008:112;15419.
[60] Zhao MW, Zhang RQ, Lee ST. Phys Rev B 2004;70:205404.
[61] Bromley ST, Zwijnenburg MA, Maschmeyer Th. Phys Rev Lett 2003;90:035502.
[62] Kuo C-L, Lee S, Hwang GS. Phys Rev Lett 2008;100:076104.
[63] Avramov PV, Adomovic I, Ho KM, Wang CZ, Lu WC, Gordon MS. J Phys Chem A 2005;109:6294.
[64] Bromley ST. Nano Lett 2004;4:1427.
[65] Flikkema E, Bromley ST. Phys Rev B 2009;80:035402.
[66] Alivisatos AP, Barbara PF, Castleman AW, Chang J, Dixon DA, Klein ML, et al. Adv Mater 1998;10:1297.
[67] Lide DR. In: Lide DR, editor. CRC handbook of chemistry and physics. 85th ed. Boca Raton: CRC Press; 2004.
[68] Yu DP, Hang QL, Ding Y, Zhang HZ, Bai ZG, Wang JJ, et al. Appl Phys Lett 1998;73:3076.
[69] Wojdel JC, Zwijnenburg MA, Bromley ST. Chem Mater 2006;18:1464.
[70] Yaghi OM, O'Keefe M, Ockwig NW, Chae HK, Eddaoudi M, Kim J. Nature 2003;423:705.
[71] Goldberg N, Hoffmann R. J Am Chem Soc 1996;118:3315.
[72] Mellot-Draznieks C, Girard S, Férey G. J Am Chem Soc 2002;124:15326.
[73] Cook G, Timms PL, Göltner-Spickermann C. Angew Chem Int Ed 2003;42:557.
[74] Chen XY, Lu YF, Tang TJ, Wu YH, Cho BJ, Xu XJ, et al. Appl Phys 2005;97:014913.
[75] Kratschmer W, Lamb L, Fostiropoulos K, Huffman D. Nature 1990;347:354.
[76] Guo BC, Kerns KP, Castleman AW. Science 1992;255:1411.
[77] Bergeron DE, Castleman AW, Morisato T, Khanna SN. Science 2004;304:84.
[78] Bromley ST, Flikkema E. J Chem Phys 2005;122:114303.
[79] Kudo T, Nagase S. J Am Chem Soc 1985;107:2589.
[80] Zachariah MR, Tang W. J Phys Chem 1999;99:5308.
[81] Wright AF, Lehmann MS. J Solid State Chem 1981;36:371.
[82] Downs RT, Palmer DC. Am Mineral 1994;79:9.
[83] Berry LG. Mineralogy: concepts, descriptions, determinations. W.H. Freeman and Co. San Francisco, CA; 1959.
[84] Wells SA, Dove MT, Tucker MG. J Appl Crystallogr 2004;37:536.
[85] Zwijnenburg MA, Simperler A, Wells SA, Bell RG. J Phys Chem B 2005;109:14785.
[86] Piccione PM, Laberty C, Yang S, Camblor MA, Navrotsky A, Davis M. J Phys Chem B 2000;104:10002.
[87] Trofymluk O, Levchenko AA, Tolbert SH, Navrotsky A. Chem Mater 2005;17:3772.
[88] Brunner GO, Meier WM. Nature 1989;337:146.
[89] Zwijnenburg MA, Bromley ST, Jansen JC, Maschmeyer T. Chem Mater 2004;16:12.
[90] Yuan P, He HP, Wu DQ, Wang DQ, Chen LJ. Spectrochim Acta A 2004;60:2941.
[91] Foster MD, Treacy MMM. http://www.hypotheticalzeolites.net/.

Chapter 11

Uncovering New Magnetic Phenomena in Metal Clusters

Mark B. Knickelbein
Chemistry Division and Center for Nanoscale Materials, Argonne National Laboratory, Argonne, Illinois, USA

Chapter Outline Head

I. INTRODUCTION

Although magnetism is normally thought of as a bulk phenomenon, its origins lie in quantum mechanical phenomena at atomic scale. The magnetism displayed by bulk metals is largely a result of spin imbalance due to electrons in incomplete angular momentum shells. In the case iron, cobalt and nickel, the spin-carrying electrons belong to the atomic 3d shells, whereas for the open-shell lanthanide metals, 4f electrons contribute both spin and orbital components to their magnetism. Small clusters containing transition or lanthanide metal atoms (or ions) often exhibit interesting and unexpected magnetic behavior that would not be anticipated on the basis of the magnetic properties

Nanoclusters. DOI: 10.1016/S1875-4023(10)01011-9

of the corresponding bulk solids. Stern–Gerlach molecular beam deflection techniques have proven to be useful for uncovering such behavior, as chemically pure metal clusters can be prepared and studied in isolated form, thus removing the perturbing effects of substrates, matrices, and stabilizing ligands. In this contribution, the results of magnetic deflection studies of selected transition metal cluster systems will be presented.

II. EXPERIMENTAL METHODS

Magnetic moments of metal clusters are measured using a variant of the classic Stern–Gerlach experiment. The experimental methods have been described in detail previously [1,2]. The apparatus consists of a four-stage, differentially-pumped molecular beam apparatus, shown schematically in Figure 1. Cylindrical metal targets are housed in laser vaporization type cluster source through which helium flows continuously. Bare metal clusters are produced via condensation of metal atoms produced by vaporization using the second harmonic (532 nm) of a pulsed Nd:YAG laser. Ligand-covered clusters are produced by adding helium containing a small concentration of reactants (e.g., benzene, CO, or H_2) to the helium carrier gas downstream of the laser vaporization region. The cluster source is coupled to a high aspect ratio flow tube whose temperature can be controlled between 55 and 300 K. The residence time of the clusters within the flow tube (~4 ms) is sufficient to ensure that they were equilibrated to the flow tube temperature prior to expansion into vacuum though a 1.2 mm dia. orifice at the end of the flow tube. At the 10 ± 5 Torr helium backing pressure within the source, the cooling of the clusters in the supersonic

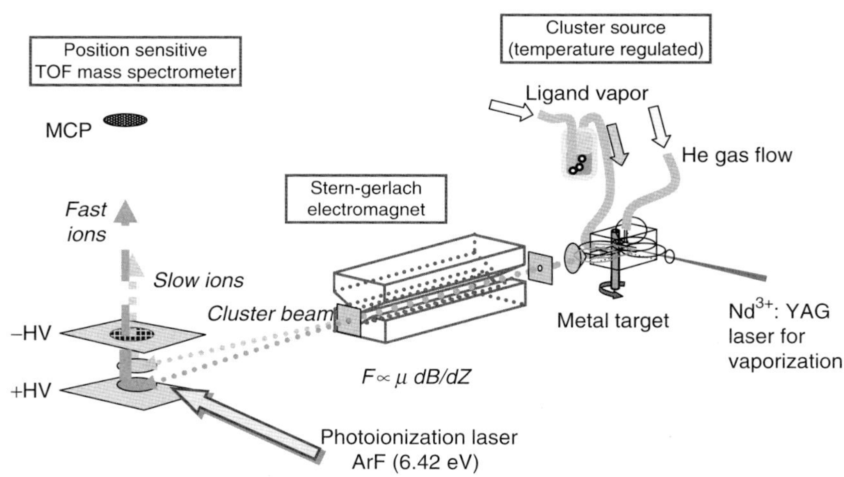

FIGURE 1 Schematic of the experiment. PSTOF mass spectra are recorded sequentially, first with the Stern–Gerlach magnet on and then again with the magnet off.

expansion is minimal, so that the post-expansion cluster temperature is close (within ~5 K) to that of the flow tube. The expanding jet is skimmed into a molecular beam, which passed through a gradient dipole magnet capable of producing B fields of up to ~1.2 T and gradients ($\partial B/\partial z$) up to ~210 T m^{-1} in the center of the gap. The clusters are then ionized with a spatially-expanded ArF excimer laser ($\lambda = 193$ nm), and the resulting singly-ionized clusters are detected via position-sensitive time-of-flight (PSTOF) mass spectrometry, a technique which allows field-induced deflections or broadening of cluster beams to be determined with a spatial sensitivity of about 20 μm. This technique allows the spatial distributions of clusters in a molecular beam to be mapped onto the time-domain and thus recorded using a digital oscilloscope. The spatial deflections or broadening of each cluster size in the beam is independently measured by quantitatively comparing the field–on versus field–off PSTOF peak profiles. Magnetic moments are determined via analysis of the beam broadening and deflection profiles observed in the PSTOF mass spectra, as described in detail below.

III. RESULTS AND DISCUSSION

A. Bare Transition Metal Clusters

Systematic studies of the magnetic properties of isolated metal clusters began in earnest in the early 1990s, when groups led by deHeer and Bloomfield independently showed that beams clusters composed of the ferromagnetic transition metals, Fe_n, Co_n, and Ni_n, are strongly deflected toward high field as they pass through a Stern–Gerlach magnet [3–6]. This behavior is shown in Figure 2 for Ni_8 and Ni_{11}. Khanna and Linderoth pointed out that this high-field-seeking behavior is consistent with superparamagnetism (SP), a phenomenon whereby the spins s of the n individual atoms couple together ferromagnetically to form a giant spin $S \approx ns$ that fluctuates randomly in space, driven by thermal excitations [7]. It had long been held that the occurrence of high-field-seeking behavior in isolated clusters in this type of experiment was an indication that intramolecular spin relaxation [8] occurs on a time scale shorter than the flight time of the clusters through the magnetic field (~0.4 ms), which in turn implies the clusters must serve as their own heat baths. Recently, Xu et al. have argued that high-field-seeking behavior mimicking SP both qualitatively and quantitatively (*vide infra*) can also occur through an adiabatic magnetization process [9].

Experimentally, the induced magnetization $\langle M_z \rangle$ of ensembles of clusters displaying high field-seeking behavior is calculated from the magnitude of the beam deflection Δz as computed from the change in PSTOF mass peak first moments recording with the magnetic field on versus off: [1,10,11]

$$\langle M_z \rangle = C \Delta z m v^2 \left(\frac{\partial B}{\partial z} \right)^{-1} \tag{1}$$

FIGURE 2 Magnified portions PSTOF mass spectra showing spatial profiles for Ni_8/Ni_8O and $Ni_{11}/Ni_{11}O$. Solid traces: B, $\partial B/\partial z=0$; dotted traces: $B=1.23$ T, $\partial B/\partial z=215$ T m^{-1}.

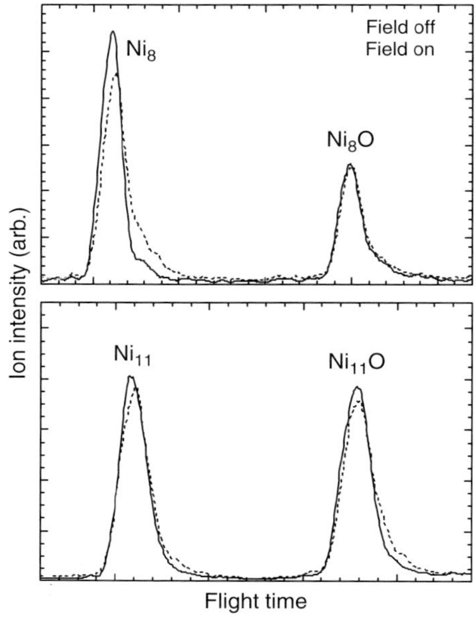

where m is the cluster mass, v is the molecular beam speed, $\partial B/\partial z$ is the field gradient, and C is an apparatus constant. For a magnetic system in thermal equilibrium, magnetization $\langle M_z \rangle$ is related to the intrinsic moment μ exactly via the Brillouin function $B(T)$: [12]

$$\langle M_z \rangle = \mu B(T) \tag{2}$$

At temperatures T for which $\mu B \ll kT$ holds, and when the cluster electronic angular momenta $J \gg 1$, the predictions of the Brillouin function can be approximated by the Curie Law: [7,8,10,11,13]

$$\langle M_z \rangle = \frac{\mu^2 B}{3kT} \tag{3}$$

It is readily shown [8,12] that the *apparent* magnetic moments for a system of angular momentum J derived from $\langle M_z \rangle$ using the Curie law are upper bounds to the true moments, exceeding the true moments by the factor $\sqrt{J(J+1)}/J$ and approach the actual values of μ (as given by Eq. (2)) asymptotically as $J \rightarrow \infty$.

Superparamagnetic behavior was originally observed in clusters of the ferromagnetic transition metals containing a few tens to a few hundreds of atoms: Fe_n [3,10], Co_n [3,14], and Ni_n [5,10]. We have extended the measurements of the magnetic moments of iron, cobalt, and nickel clusters to clusters containing only a few atoms—a size range accessible to accurate first-principles theoretical methods. In most cases, it has been found that the smaller clusters exhibit high-field seeking behavior similar to that displayed by larger clusters. The mean per-atom magnetic moments $\bar{\mu} = \mu/n$, determined using the Curie Law expression for Ni_n, Co_n, and Fe_n are shown in Figure 3. The trend of increasing per-atom moments with decreasing cluster size, first identified in the pioneering studies of deHeer and Bloomfield [3,6], continues in the smaller clusters, albeit non-monotonically.

That bare ferromagnetic clusters display average atomic moments that are significantly larger than the bulk can be traced to the combined effects of enhanced spin polarization and unquenched orbital contributions. For example, the per-atom moments of Fe_{10-12} and Fe_{14} are substantially higher than the fully spin-polarized value of $\sim 3\ \mu_b$, indicating that orbital angular momentum is not completely quenched in these cases; Fe_{12} displays a moment of $5.4 \pm 0.4\ \mu_b$ per atom, very close to the upper limit of 6 μ_b achieved when each Fe atom within the cluster contributes the maximum possible orbital angular momentum ($3\mu_b$), assuming a $3d^7 4s$ valence electron configuration as in the bulk [15]. Nickel and cobalt clusters with $n < 20$ also display per-atom moments far in excess of the corresponding bulk solids, with some sizes exceeding the maximum spin-only values of $\sim 1\mu_b$ (Ni) and $\sim 2\mu_b$ (Co).

B. Clusters Containing Adsorbates

While much progress has been made in understanding the genesis and evolution of magnetism via the study of bare ferromagnetic clusters, there have been fewer systematic investigations of the effects of chemisorption on the magnetic properties of clusters in this same size range. Because a large fraction of the atoms in small magnetic clusters lies at the surface, and because these surface atoms can possess enhanced moments, large changes in the overall cluster magnetic moments induced by chemisorbed species can be anticipated. Moreover, transition metal clusters made up of a few to a few tens of atoms display large, size-dependent variations in their chemisorption behavior, for example, rates of reactions, and binding energies due to the rapid and discontinuous evolution in geometrical structures, suggesting that adsorbate-induced changes in magnetic moments will also be size/structure dependent. Systematic investigation of the size dependence of such adsorbate- induced changes in cluster magnetism thus offers a method to probe the size evolution of cluster geometric/electronic properties and the nature of cluster–adsorbate bonding. The effects of chemisorption on the magnetic properties of larger metal catalyst particles have been investigated by measuring the changes in magnetization that occur with varying adsorbate coverage [16]. Chemical interactions are expected to play an increasingly important role in determining the characteristics of magnetic devices and materials as their dimensions are reduced into the nanometer size range, by inducing changes in the intrinsic magnetic moments of atoms at the interface or via

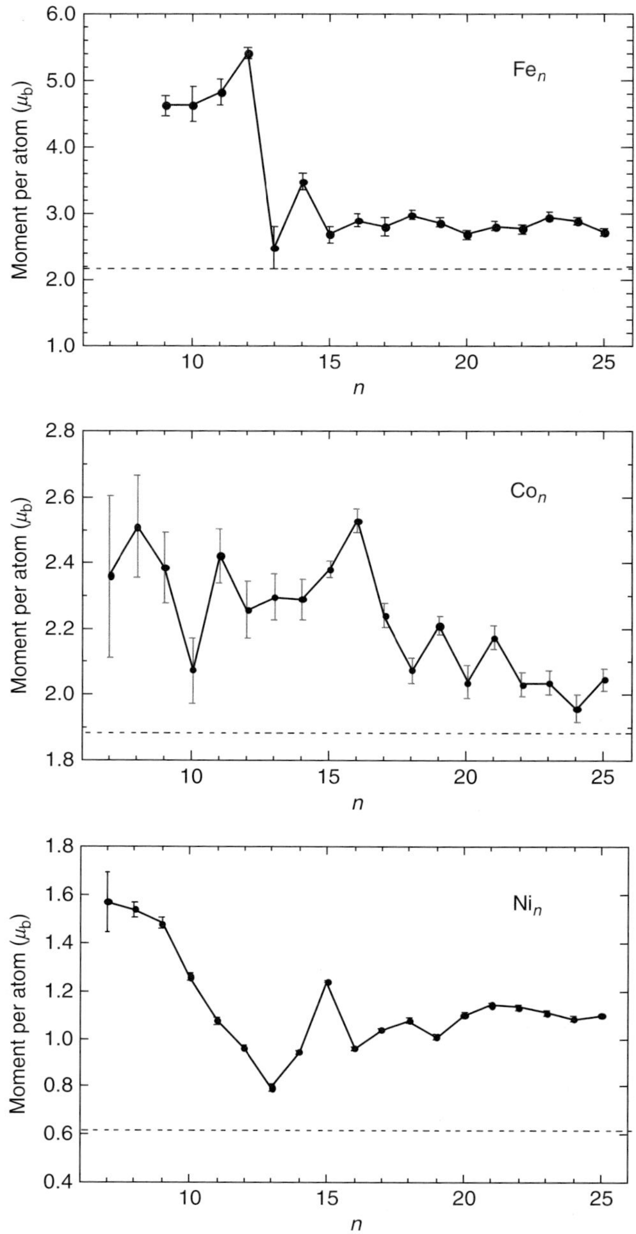

FIGURE 3 Magnetic moments per atom, $\bar{\mu}$, for Fe_n, Co_n, and Ni_n.

modification of surface anisotropy. Cluster studies are expected to contribute to the understanding of microscopic details underlying such interactions.

In the case of nickel, it is nearly always observed that chemisorption is accompanied by a decrease in magnetization because of "spin-quenching" of the surface atoms [16]. One of the first conceptual models devised to explain such phenomona from a microscopic viewpoint was the *bond number* (BN) model, which parameterizes chemisorption-induced changes in magnetization in terms of adsorbate-coordinated surface atoms [16]. This is essentially a "local" model, which makes the simplifying assumption that an adsorbed atom or molecule (of any identity) bound to N surface metal atoms will quench the magnetic moments of those N atoms entirely, while leaving the moments of neighboring metal atoms unaffected. Whereas the BN model assumes that an adsorbed atom or molecule affects only the magnetic moments of the metal atoms to which it is bound, the semiempirical *bond order-rigid band* (BO-RB) model developed by Fournier and Salahub [17] assumes that adsorbates affect the magnetic properties of the metal as a whole, via the changes in the relative populations of the minority and majority d bands. The BO-RB model, basically an electron bookkeeping scheme developed for describing the effects of chemisorption on the magnetic properties of extended surfaces of ferromagnetic metals, assumes as follows:

1. The adsorbed atom or molecule does not significantly perturb the relative energies of the molecular orbitals of the underlying cluster—that is, rigid bands are assumed. This assumption is rigorously valid only in the limit of low adsorbate coverage.
2. Surface-adsorbate bonding can be described by a simple bond order parameter p determined from experimental or theoretical surface-adsorbate bonding studies.
3. Any donation of electrons into the majority and minority spin manifolds resulting from formation of a surface-adsorbate bond occurs in proportion to the respective density of states $\rho^{\uparrow}$ and $\rho^{\downarrow}$ at the Fermi level E_f.
4. The resulting spin polarization is located entirely on the metal cluster, with little or none on the adsorbate(s).

Quantitatively, the BO-RB model predicts that the adsorbate-induced change in magnetic moment, ε, varies with bond order p, the number of occupied surface-adsorbate orbitals m having energies $E < E_f$, the number of electrons k in those m orbitals, and the number of electrons y occupying bonds formed from empty adsorbate orbitals according to the following equation:

$$\varepsilon = P(E_f)[2(p-m)+(k-y)] - 2\frac{m}{N_v}(p-m) \tag{4}$$

In Eq. (4), $P(E_f)$ is the polarization of states at the Fermi level $[\rho^{\uparrow}(E_f)-\rho^{\downarrow}(E_f)]/[\rho^{\uparrow}(E_f)+\rho^{\downarrow}(E_f)]$, μ is the (mean) value of the atomic magnetic moment, and N_v is the number of valence electrons per metal atom. The parameters $P(E_f)$ and N_v are well known for bulk nickel and iron, and their transferability to clusters will be assumed; for μ, the experimental values of μ_{Ni} and μ_{Fe} obtained for bare clusters will be used. The values of p, m, k, and y for each M_n-adsorbate system discussed below are taken from the tabulation of Fournier and Salahub [17], which summarizes the results of detailed electronic structure calculations on model metal-adsorbate systems.

C. Ni_nO

As shown in Figure 2, effect of chemisorbed oxygen on nickel cluster moments depends strongly on cluster size, in some cases resulting in a decreased magnetic moment, in other cases an increased moment or no change [1]. For Ni_{8-10}, adsorbed oxygen atoms decrease the moments, significantly so for Ni_8 and Ni_{10}. For Ni_{11} and Ni_{12}, the trend reverses and for Ni_{13}, no oxygen-induced change in moment is observed. For clusters larger than Ni_{13}, the effects are generally smaller, but several distinct cases stand out: the magnetic moments of Ni_{14}, Ni_{16-17}, and Ni_{22-23} display clear oxygen-induced enhancement, while that of Ni_{15} is reduced.

Application of the BO-RB model to the Ni–O system ($m=3$, $k=4$, $y=0.0$, and $p=1.5$) [17] predicts values of ε ranging from $-0.33\mu_b$ to $-0.56\mu_b$ per adsorbed O atom, with $|\varepsilon|$ varying inversely with μ_{Ni}. The BO-RB prediction is in reasonable accord with studies of chemisorption of oxygen atoms onto nickel catalyst particles, in which decreases in magnetization of $\sim 0.7\ \mu_b$ per adsorbed O atom were observed [18], but is signficantly lower than the measured change of $-2.7\ \mu_b$ per O atom adsorbed on Ni(111) at low coverage [19]. The predictions of the BO-RB model and of the BN model ($N=2$ and 4) are shown in Figure 4 together with the experimental results.

The large magnitude of the oxygen-induced changes in Ni_n magnetic moments, in particular the observation that for some Ni_n adsorbed atomic oxygen increases rather than decreases the moments, indicates a serious failure of the quenching models, even at the qualitative level. In applying the BO-RB and BN models, we have assumed that the adsorbate does not significantly perturb the electronic or geometric structures of the clusters; however, this may not be a valid assumption if there exist structural isomers lying only slightly higher in energy than the ground state. In fact, the existence of multiple low-lying isomers of nickel clusters has been demonstrated by detailed calculations [20–22]. It is thus conceivable that a strongly bound adsorbate such as oxygen ($E_{ads} \approx 5-6$ eV on bulk nickel [23–26]) may induce complete reconstruction, leading to structures having magnetic moments different from those of the corresponding bare clusters. The observation of adsorbate-induced changes in magnetic moment outside the range expected by simple quenching models may reflect such structure changes.

Salahub and coworkers examined the effects of adsorbed O atoms on the magnetic structure of small surface-model clusters [25,27]. Fournier and

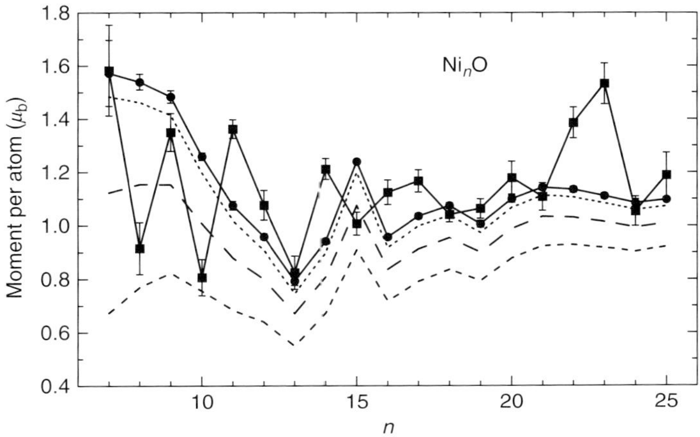

FIGURE 4 Experimentally determined magnetic moments per atom of Ni_n (●) and Ni_nO (■). The predictions of the BN model for $N=2$ and 4 are shown by the coarse and intermediate dashed line; the predictions of the RO-RB model is indicted by the fine dashed curve.

Salahub used two density functional theory-based approaches, scattered wave-local spin density (SW-LSD) and linear combination of Gaussian type orbitals-local spin density (LCGTO-LSD) to examine oxygen atom binding on bridging sites of fcc model clusters (Ni_4 and Ni_{10}) in a simulation of the O/Ni(110) system. For the Ni_4+O system, it was found that the local moments of the Ni atoms directly bonded to the oxygen atom were reduced as compared to bare Ni_4, with the overall magnetic moments either remaining unchanged (LCGTO-LSD) or decreasing by $2\mu_b$ (SW-LSD). For Ni_{10}+O system, no change in overall magnetic moment was found for a single bridge-bound oxygen atom. In an expanded study of the same system, Goursot et al. [25] used the LCGTO-LSD approach to examine the effects of oxygen atoms bound in fourfold hollow site of fcc-like Ni_9 and found similar results: local reductions in moment in the four atoms directly bounded to the oxygen, but no reduction in magnetic moment overall.

D. Ni_nCO

As shown in Figure 5, CO reduces the magnetic moment of most Ni_n species in the $n=8$–18 size range [1,28]. The quenching effect is particularly pronounced for Ni_8, Ni_9, Ni_{15}, and Ni_{18}. The most dramatic effect is observed for Ni_8, the magnetic moment of which is reduced by $\sim 6\ \mu_b$ ($0.78 \pm 0.20\ \mu_b$ per Ni atom) by a single chemisorbed CO molecule, corresponding (in the spin-only picture) to the majority-to-minority band "spin-flipping" of three electrons. For other clusters in this size range, the reductions in moments are smaller. These observations are analogous to the observed quenching

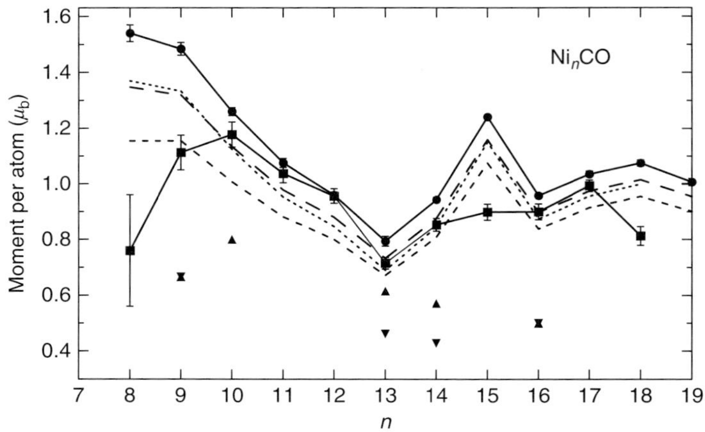

FIGURE 5 Experimentally determined magnetic moments per atom of Ni_n (●) and Ni_nCO (■). The moments per atom calculated by Raatz and Salahub for fcc model Ni_n (⧗) and Ni_nCO (▼) are also shown (see the text). The predictions of the BN model for $N=1$ and 2 are shown by the coarse and intermediate dashed line; the predictions of the RO-RB model for the Ni–O system is indicated by the fine dashed curve.

effects of chemisorbed CO on nickel nanoparticles: At 298 K, chemisorption of CO onto nickel catalyst particles (2–12 nm dia.) was found to be accompanied by decreases in magnetization of 1.1 μ_b per adsorbed CO, independent of particle size, corresponding (in the local picture) to the magnetic quenching of ~2 surface atoms [29].

Carbon monoxide binds to nickel surfaces [30], within polynuclear nickel cluster carbonyl complexes [31,32] and to surface-supported nickel clusters [33] via atop or bridging sites, corresponding to bond numbers of $N=1$ and 2, respectively. As shown in Figure 5, the observed reductions in magnetic moments induced by CO can in most cases be reasonably parameterized using the BN model for $N=1$ or 2. However, as noted above, the reductions in magnetic moments observed for Ni_8, Ni_9, Ni_{15}, and Ni_{18} are larger than can be rationalized by the BN model, demonstrating that the changes in electronic structures induced by CO chemisorption are extended throughout those clusters, rather than local in nature as assumed in the BN model. Application of the BO-RB model to the Ni–CO system ($m=1$, $k=2$, $y=0.3$, and $p=1.0$) [17] predicts $\varepsilon=-1.36$ μ_b per adsorbed CO molecule, independent of the value of μ_{Ni}. As shown in Figure 5, the predictions of the BO-RB model is similar to that of the BN model for $N=1$. The measured CO-induced reductions in μ_{Ni} observed for Ni_8, Ni_{15}, and Ni_{18} are considerably greater than those predicted by either the BN or the BO-RB model, indicating significant nonlocal electronic perturbations or adsorbate-induced structural rearrangements of the underlying clusters.

Raatz and Salahub used a finite cluster-model approach to extended surfaces by performing detailed electronic structure calculations (self-consistent field, local spin density, scattered wave) for selected nickel clusters of Ni_n and Ni_nCO ($n=9$, 13, 14, and 16) that were assumed to have bulk-like (fcc) structures and lattice constants [34,35]. Adsorption of CO via atop (Ni_9 and Ni_{14}), bridging (Ni_{16}), and fourfold (Ni_5 and Ni_{13}) sites was found to lead to a reduction in the overall magnetic magnetic moment of $2\mu_b$, with the exception of Ni_9 and Ni_{16} in which no change in overall moment was observed. Detailed orbital analysis revealed that formation of the Ni_n–CO bond is accompanied by destabilization of a doubly-occupied nickel-CO antibonding orbital to energies above the Fermi level. It is the subsequent transfer of the two antibonding electrons into empty minority spin d orbitals that leads to the $2\mu_b$ reduction in magnetic moment. Although the structures of these fcc model clusters are different from those experimentally produced, their finite nature allows a comparison of their total magnetic moments and CO-induced reductions in moments to experimental data. The calculated magnetic moments for Ni_n and Ni_nCO results are shown in Figure 5 along with the experimental data. Although the magnitudes of the calculated Ni_n magnetic moments are 40–60% lower than those measured in the present study, the calculations do predict a net spin-quenching effect by a single CO molecule for Ni_{13} and Ni_{14}, in qualitative agreement with the present study. The calculations found no changes in overall magnetic moment induced by CO for Ni_9 or Ni_{16}, even though both clusters exhibited significant local decreases in the magnetic moments of the Ni atoms to which CO was bound. That CO quenching is theoretically predicted to occur for some cluster sizes and not for others is in general agreement with the experimental observations.

E. $Ni_nH_m(H_2)_p$

At low temperatures (<100 K), nickel clusters react with hydrogen to form complexes containing both chemisorbed atomic hydrogen and a more weakly bound layer of physisorbed molecular hydrogen [36]. As shown in Figure 6, the magnetic moments of most nickel clusters are quenched by the overlayer of chemisorbed atomic hydrogen [1]. Application of the BO-RB model to the Ni–H system ($m=1$, $k=1$, $y=0.0$, and $p=1.0$) predicts $\varepsilon=-0.80$ μ_b per adsorbed H atom, independent of μ_{Ni}. It is assumed that the physisorbed H_2 molecules are non-interacting "spectators" and do not affect the cluster magnetic moments of the underlying Ni_nH_m complexes [16]. The total H-induced decrease in Ni_n moments, calculated from this value of ε and the observed saturation values of m [36] are shown in Figure 6 along with the experimental data.

Detailed electronic structure calculations of the effect of hydrogen chemisorption on the magnetic moments of fcc model clusters have been performed by Salahub and coworkers [37–39]. The model Ni_n systems ($n=4$, 5, 7, 10,

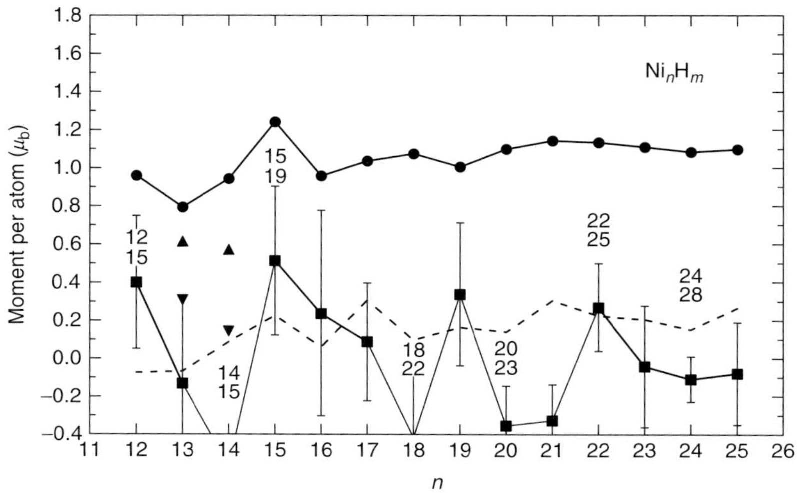

FIGURE 6 Experimentally determined magnetic moments per atom of Ni_n (●) and Ni_nH_m (■). The saturation values of m are indicated for selected n [36] as $_m{}^n$. The moments per atom calculated by Raatz and Salahub for Ni_n (▼) and Ni_nH_2 (▲) ($n=12$, 14) are also shown (see the text). The predictions of the RO-RB model for the Ni–H system is indicated by the dashed curve (see the text). As with the RO-RB model, the BN model (not shown) also predicts essentially complete magnetic quenching of the underlying nickel clusters.

13, and 14) were constructed so as to display the atop, threefold, and fourfold sites presented by Ni(100) and Ni(111) surfaces. For a single H atom adsorbed to either a threefold (Ni_4, Ni_7, and Ni_{10}) or fourfold (Ni_5) site, an overall change in moment of –1.0 μ_b was calculated, with the Ni atoms bound to the hydrogen displaying the largest decrease in moment [37,39]. A first-order molecular orbital description of the quenching in this case involved simple donation of the "extra" electron of the H atom into the minority *d* band of the cluster, a result consistent with the predictions of the BO-RB model. However, in cases of adsorption of two H atoms, overall changes in magnetic moment of –4 μ_b (Ni_{13}, fourfold sites) and –6 μ_b (Ni_{14}, atop sites) were calculated, values significantly larger than those predicted by the BO-RB model. In these cases, adsorption of two H atoms induced a significant nonlocal change in electronic structure, namely a reduction in the exchange splitting, the result of which was the transfer of one (Ni_{13}) or two (Ni_{14}) additional electrons from the majority spin manifold to the minority spin manifold—essentially a partial collapse of the magnetic structure of the cluster.

From surface studies [40] and calculations [24,41,42], it is known that atomic hydrogen is strongly bound to nickel surfaces and clusters, ~2.3–3.0 eV per H atom, depending on the type adsorption site occupied. It can be anticipated that chemisorption of hydrogen atoms on nickel clusters may reconstruct the underlying nickel clusters in some cases; however, the large number of adsorbed H atoms in the Ni_nH_m species is expected to dominate any such structure-change-related changes in μ_{Ni}, leading instead to complete quenching of cluster magnetism, as is observed experimentally.

F. $Fe_nH_m(H_2)_p$

Like nickel clusters, iron clusters react readily with hydrogen to form fully saturated complexes containing dissociatively chemisorbed hydrogen [43–45]. As shown in Figure 7, the magnetic moments of most Fe_nH_m complexes in the $n = 10$–25 size range are comparable to or greater than those of the corresponding Fe_n species [46]. Hydrogen-induced enhancements are particularly striking for $n = 13$–18. This finding is in contrast to analogous studies of nickel clusters, in which hydrogenation completely quenches magnetic moments. Studies of supported iron catalyst nanoparticles [16] have shown that dissociatively chemisorbed hydrogen has little or no measurable effect on magnetization, aside from those small effects due to perturbations of the particles' internal magnetic anistropies—assumed to be a negligible effect in small iron clusters at the temperature employed in the experiment (110 K).

Unlike the situation for nickel, the density of majority spin states at the Fermi level for iron exceeds the density of minority spin states, leading to the prediction that bonding interactions involving any adsorbate will lead to a net increase in magnetic moment. When quantitatively applied to the iron + hydrogen system, the BO-RB model predicts a net increase in magnetic moment of 0.4 μ_b per adsorbed H atom. The magnetic moments (per Fe atom) of Fe_nH_m predicted by the BO-RB model assuming that $\mu_{at}(Fe_nH_m) = [\mu_{cluster}(Fe_n) + 0.4m]/n$ are plotted in Figure 7 along with the experimentally determined values of $\mu(Fe_nH_m)$. The measured increases in

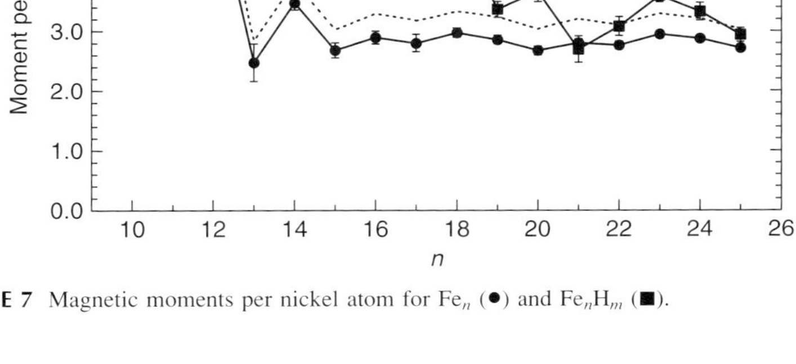

FIGURE 7 Magnetic moments per nickel atom for Fe_n (●) and Fe_nH_m (■).

μ_{at} are of the same magnitude as those predicted by the BO-RB model for $n=11$, 19, and 22–25, while for $n=13$–18 and $n=20$, they are significantly larger than those predicted. That the BO-RB model fails to quantitatively account for the observed changes in magnetic moment for many clusters in this size range is not surprising given that the high coverage conditions of the experiment do not comply with assumption (1) of the BO-RB model. Furthermore, this simple band structure-based model is grounded in a bulk-like description of the electronic structure of iron and thus does not account for size-dependent variations. Quantitative prediction of cluster magnetic moments and their variation with adsorbate coverage will require detailed electronic calculations in which cluster structure is explicitly considered.

IV. MAGNETIC ORDERING IN CLUSTERS OF NONFERROMAGNETIC TRANSITION METALS

Why is it that among the transition metals, only iron, cobalt, and nickel exhibit spontaneous ferromagnetism? [47] All of the transition metals other than Fe, Co and Ni are paramagnetic with the exception of Mn and Cr, which are antiferromagnetically ordered below 95 and 312 K, respectively. "Anomalous" magnetic ordering is sometimes observed in some of the nonmagnetic transition metals when they are produced in spatially-confined forms such as thin films, an effect attributed to the enhancement of densities of d states at the Fermi level because of reduced atomic coordination [48]. The effects of spatial confinement on the magnetic properties of transition metals can be put on a quantitative framework using the Stoner model of itinerant ferromagnetism. This model, despite its simplicity, is a remarkably useful predictor of spontaneous magnetization in bulk transition metals, successfully predicting ferromagnetic ordering in iron, cobalt, and nickel (and only those transition metals) [12,49,50]. According to this model, the paramagnetic susceptibility χ is determined by the density of d states at the Fermi level $N(E_f)$ and the exchange function J,

$$\chi = \frac{\mu_0 \mu_b^2 N(E_f)}{1 - JN(E_f)}$$

where μ_0 is the permeability of free space and μ_b is the Bohr magneton. The *Stoner enhancement* term given by $[1-JN(E_f)]^{-1}$ increases rapidly as $JN(E_f) \rightarrow 1$, and for $JN(E_f) \geq 1$, χ (formally) becomes negative, signifying a ferromagnetic instability and thus spontaneous magnetic ordering [50]. The Stoner model thus provides a simple framework with which to predict the emergence of "unexpected" magnetic ordering in the transition elements as spatial dimensions are reduced. The d-band narrowing that typically accompanies low atomic coordination [48] produces an enhancement in $N(E_f)$ which, if sufficiently large, can lead to ferromagnetic instability: spontaneous magnetic ordering occurs.

There are now several examples of transition metal cluster systems that show unexpected magnetic ordering. The Virginia group showed that bare rhodium clusters (Rh_{9-60}) display magnetic moments [51,52] as high as 0.8 μ_b per atom (Rh_{9-11}) indicative of ferromagnetic or ferrimagnetic ordering, even though bulk rhodium is a Pauli paramagnet at all temperatures [53]. More recently, the same group has measured magnetic moments for Cr_{8-156} that ranged from ~0.5 to ~1.0 μ_b per atom, clearly indicating that these clusters do not display antiferromagnetic ordering as in bulk chromium, but rather ferromagnetic or ferrimagnetic ordering [54].

A. Manganese Clusters

Below its Néel temperature of 95 K, α-manganese is antiferromagnetically ordered, but is paramagnetic above 95 K [53,55,56]. The large paramagnetic susceptibility displayed by bulk Mn [53] suggests that it is on the verge of ferromagnetic instability, as predicted by the Stoner model [12,49]. Experiments at Argonne have shown that manganese clusters in fact do display magnetic moments far in excess of what they would if they were simply tiny bits of the bulk, indicating that they are magnetically ordered in some way [2,57]. At nonzero fields, two types of deflection behavior, exemplified in Figure 8, are observed. The PSTOF mass peaks of Mn_5 and Mn_6 broaden symmetrically about the undeflected center line of the beam ($z=0$), with little net deflection in the z direction. By contrast, the PSTOF peaks for Mn_7 and larger clusters shift uniformly to later arrival times upon application of the gradient field, corresponding to a net spatial shift in the $+z$ direction toward higher fields, as is observed for clusters of the ferromagnetic metals (*vide supra*). This high field-seeking behavior for Mn_{7-99} is consistent with superparamagnetic behavior, whereas symmetric broadening of the PSTOF peaks observed for Mn_5 and Mn_6 indicates that their magnetic moments are strongly coupled to their atomic frameworks (via magnetocrystalline or other types of anisotropy) and thus rotate with the clusters as a whole. Under conditions in which $\mu B \ll kT_{rot}$ and in the limit of high rotational quantum numbers, the distribution function of moments $R(\mu_z)$ takes on a simple analytical form that is independent of T_{rot}: [58]

$$R(\mu_z) = \frac{1}{2\mu} \ln\left(\frac{\mu}{|\mu_z|}\right) \tag{5}$$

In practice, μ is determined by convoluting the unbroadened beam profile with the distribution function R, and least-squares fitting the resulting convolution to the field-broadened beam profile [2]. The mean per-atom magnetic moments $\bar{\mu}$, determined for Mn_5 and Mn_6 obtained via the adiabatic rotor model are shown in Figure 8 together with the values obtained above Mn_{7-99} using the Curie's Law expression given above. Generally, the

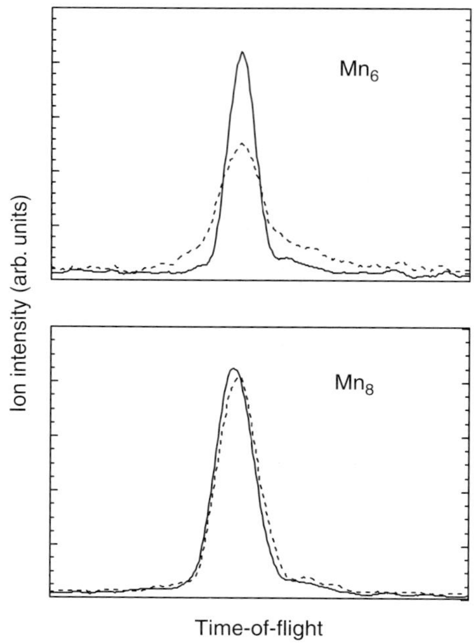

FIGURE 8 A magnified view of the manganese cluster time-of-flight spectrum showing the PSTOF profiles for Mn_6 and Mn_8. The PSTOF mass peaks shown here were numerically smoothed for illustrative purposes. Solid traces: B, $\delta B/\delta z=0$; dashed traces: $B=0.97$ T, $\delta B/\delta z=192$ T m^{-1}.

per-atom moments of Mn_n are of the order of 1 μ_b per atom or smaller, with a few smaller clusters (e.g., Mn_{12}) displaying somewhat larger moments (Figure 9). At first glance, this would appear to be a surprising result, as the ground state Mn atom possesses a half-filled 3d shell and thus a spin moment of $5\mu_b$. Detailed calculations [59–73] show, however, that the ground states of Mn clusters tend to adopt a ferrimagnetic arrangement of spins, with small but nonzero moments resulting from the imbalance of up versus down spins. This situation is illustrated in Figure 10 which shows the structures of ferrimagnetic isomers of Mn_5, Mn_6, and Mn_7. These ferrimagnetic isomers possess moments that are in good agreement with the experimental values, whereas energetically nearby ferromagnetic isomers have moments of 3–4 μ_b per atom. The measurement of magnetic moments thus provides a method to distinguish among cluster isomers in cases where their magnetic moments differ substantially.

V. RARE EARTH CLUSTERS

The group 3 (rare earth) metals, Sc, Y, and La, which are Pauli paramagnets as bulk solids, also display magnetism when produced as small clusters. At 60 ± 5K, locked moment behavior is displayed by Y_5, Y_6, La_5, and all Sc_n clusters except Sc_{10} and Sc_{20}. By contrast the beam profiles of $Sc_{10–20}$, $Y_{7–10,13–18}$, and $La_{6–9,11,13–20}$ shift uniformly to later arrival times upon

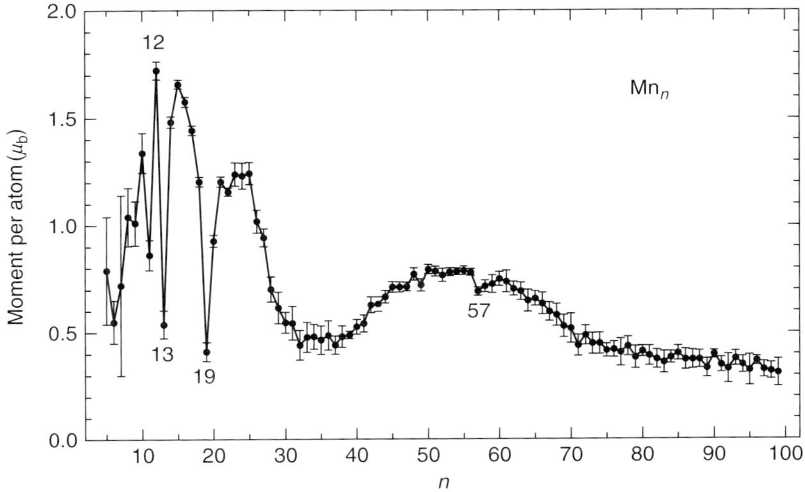

FIGURE 9 Magnetic moments per atom for Mn$_n$ produced at 68 K.

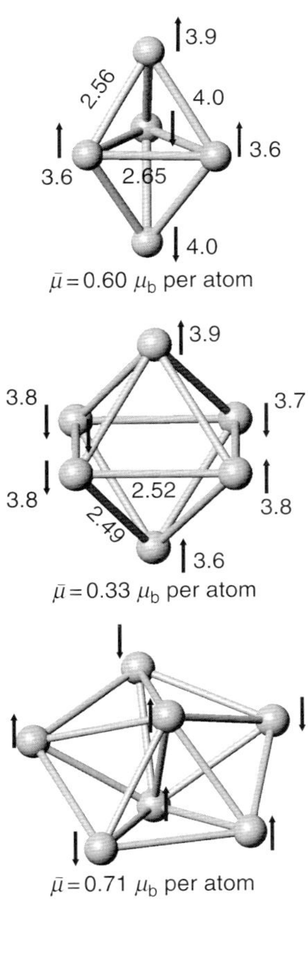

FIGURE 10 Equilibrium structures of low-lying ferrimagnetic isomers of Mn_5, Mn_6, and Mn_7 as reported in Refs. [65,67]. In the case of Mn_5 and Mn_7, the structures shown are the minimum energy isomers. The structure shown for Mn_6 lies 0.03 eV above the higher-moment (4.3 per atom) ground state also having an octahedral structure.

application of the gradient field, consistent with superparamagnetic behavior. The magnitudes of the broadening and shifting are significantly larger than would be expected based on the susceptibilities of the corresponding bulk elements, implying that these clusters are not simply small fragments of the corresponding paramagnetic solids, but rather are molecular magnets.

The per-atom moments $\bar{\mu}$ derived using the adiabatic rotor analysis for Sc_{5-9}, Sc_{11-19}, Y_{5-6}, and La_5 are shown in Figure 11, along with the Curie Law values obtained for the superparamagnetic clusters. The beam profiles for $Y_{11,12,19,20}$ and $La_{12,20}$ display no measurable deflection or broadening (beyond statistical uncertainty) even at the highest fields; accordingly their moments are taken to be $<0.1\mu_b$ per atom. The moments for most Sc_n, Y_n, and La_n clusters are quite small ($<0.3\ \mu_b$ per atom). Notable exceptions are Sc_{13}, Y_8, Y_{13}, and La_6, with total moments of 6.0, 5.5, 8.8, and $4.8\mu_b$, respectively, thus putting these clusters in the class of true high-spin molecular magnets. The size variations of the per-atom moments of scandium clusters and yttrium clusters display many of the same features (e.g., local maxima at $n=6$, 8, 13, and 18), with the moments of yttrium cluster tending to be slightly larger. This similarity may imply a common structural motif for these two series of clusters (bulk Sc and Y both adopt hcp packing). By contrast, the moments of lanthanum clusters evolve with a different pattern than those displayed by scandium and yttrium clusters, suggesting that they adopt a different structural motif (unlike Sc and Y, bulk La adopts dhcp packing). Recent theoretical calculations of the structures and magnetic properties of scandium

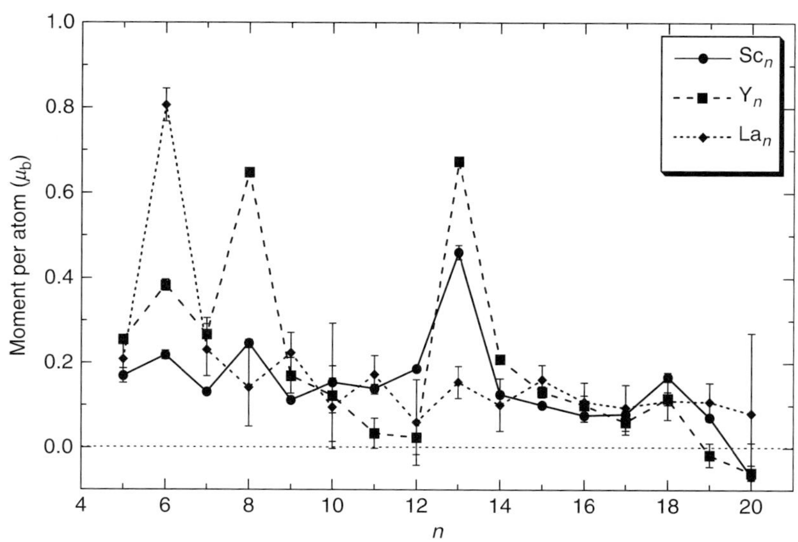

FIGURE 11 Magnetic moments per atom measured for Sc_n, Y_n, and La_n.

[74] and lanthanum [75] clusters confirm the existence many low-lying isomers of competing cubic and icosahedral growth motifs. The ground states, the structures of which bear little resemblance to the corresponding bulk solids, possess small spontaneous moments that are in reasonably good agreement with the measured values.

VI. SUMMARY

Small transition metal cluster display a variety of surprising magnetic phenomena that could not be anticipated on the basis of the behavior of the corresponding bulks solids and extended surfaces. The present results were obtained for isolated clusters and may be quite different for clusters supported on surfaces or within matrices. As the characteristic length scales of magnetic materials are pushed into the nanometer regime, it can be anticipated that normal magnetic behavior (e.g., as exhibited by conventional ferromagnetic particles) may give way to unusual, unanticipated, confinement-induced magnetic behavior, including anomalous magnetic ordering, in normally non-magnetic metals and chemisorption-induced magnetic enhancement.

Acknowledgments

The author thanks Ken Miyajima for his expert preparation of the experimental schematic shown in Figure 1. This research was supported by the U. S. Department of Energy, Office of Science, Office of Basic Energy Sciences, under Contract No. DE-AC02-06CH11357 and by the CREST (Core Research for Evolutional Science and Technology) program of the Japan Science and Technology Agency (JST).

REFERENCES

[1] Knickelbein MB. J Chem Phys. 2002;116:9703.
[2] Knickelbein MB. Phys Rev B. 2004;70:014424.
[3] Billas IML, Châtelain A, de Heer WA. Science. 1994;265:1682.
[4] Billas IML, Châtelain A, de Heer WA. J Magn Magn Mater. 1997;168:64.
[5] Apsel SE, Emmert JW, Deng J, Bloomfield LA. Phys Rev Lett. 1996;76:1441.
[6] Bloomfield LA, Bucher JP, Douglass DC. On clusters and clustering. In: Reynolds PJ, editor. Atoms to fractals. Amsterdam: Elsevier; 1993.
[7] Khanna SN, Linderoth S. Phys Rev Lett. 1991;67:742.
[8] Knickelbein MB. J Chem Phys. 2004;121:5281.
[9] Xu X, Yin S, Moro R, de Heer W. Phys Rev Lett. 2005;95:237209.
[10] Becker JA, de Heer WA. Ber Bun Phys Chem. 1992;96:1237.
[11] Bucher JP, Bloomfield LA. Int J Mod Phys B. 1993;7:1079.
[12] Blundell S. Magnetism in condensed matter. Oxford: Oxford University Press; 2001.
[13] Châtelain A. Philos Mag B. 1999;79:1367.
[14] Douglass DC, Cox AJ, Bucher JP, Bloomfield LA. Phys Rev B. 1993;47:12874.

[15] Castro M, Jamorsky C, Salahub DR. Chem Phys Lett. 1997;271:133.
[16] Selwood PW. Chemisorption and magnetization. New York: Academic Press; 1975.
[17] Fournier R, Salahub DR. Surf Sci. 1990;238:330.
[18] Martin GA, Primet M, Dalmon JA. J Catal. 1978;53:321.
[19] Elmers HJ, Gradmann U. Surf Sci. 1988;193:94.
[20] Jellinek J, Güvenç ZB. In: Farrugia LJ, editor. The synergy between dynamics and reactivity at clusters and surfaces. Dordrecht: Kluwer; 1995. p. 217.
[21] Stave MS, DePristo AE. J Chem Phys. 1992;97:3386.
[22] Curotto E, Matro A, Freeman DL, Doll JD. J Chem Phys. 1998;108:729.
[23] Egelhoff Jr. WF. Phys Rev B. 1984;29:3681.
[24] Au C-T, Liao M-S, Ng C-F. J Phys Chem A. 1998;102:3959.
[25] Goursot A, Mele F, Russo N, Salahub DR, Toscano M. Int J Quant Chem. 1993;48:277.
[26] Panas I, Siegbahn P, Wahlgren U. J Chem Phys. 1989;90:6791.
[27] Fournier R, Salahub DR. Surf Sci. 1991;245:263.
[28] Knickelbein MB. J Chem Phys. 2001;115:1983.
[29] Primet M, Dalmon JA, Martin GA. J Catal. 1977;46:25.
[30] Hoffmann FM. Surf Sci Rep. 1983;3:107.
[31] Longoni G, Chini P, Cavalieri A. Inorg Chem. 1976;15:3025.
[32] Longoni G, Chini P. Inorg Chem. 1976;15:3029.
[33] Vanolli F, Heiz U, Schneider W-D. Chem Phys Lett. 1997;277:527.
[34] Raatz F, Salahub DR. Surf Sci. 1984;146:L609.
[35] Raatz F, Salahub DR. Surf Sci. 1986;176:219.
[36] Zhu L, Ho J, Parks EK, Riley SJ. Z Phys D. 1993;26:313.
[37] Fournier R, Salahub DR. Int J Quant Chem. 1986;29:1077.
[38] Russier V, Salahub DR, Mijoule C. Phys Rev B. 1990;42:5046.
[39] Mlynarski P, Salahub DR. J Chem Phys. 1991;95:6050.
[40] Christmann K. Surf Sci Rep. 1988;9:1.
[41] Taft CA, Guimarães TC, Pavão AC, Lester WA. Int Rev Phys Chem. 1999;18:163.
[42] Whitten JL, Yang H. Surf Sci Rep. 1996;218:55.
[43] Parks EK, Liu K, Richtsmeier SC, Pobo LG, Riley SJ. J Chem Phys. 1985;82:5470.
[44] Parks EK, Nieman GC, Pobo LG, Riley SJ. J Phys Chem. 1987;91:2671.
[45] Knickelbein MB, Koretsky GM, Jackson KA, Pederson MR, Hajnal Z. J Chem Phys. 1998;109:10692.
[46] Knickelbein MB. Chem Phys Lett. 2002;353:221.
[47] Ashcroft NW, Mermin MD. Solid State Physics. Philadelphia: Sounders; 1976.
[48] Himpsel FJ, Ortega JE, Mankey GJ, Willis RF. Adv Phys. 1998;47:511.
[49] Wohlfarth EP. In: Wohlfarth EP, editor. Ferromagnetic materials, vol. 1. Amsterdam: North-Holland; 1980. p. 1.
[50] Janak JF. Phys Rev B. 1977;16:255.
[51] Cox AJ, Louderback JG, Bloomfield LA. Phys Rev Lett. 1993;71:923.
[52] Cox AJ, Louderback JG, Apsel SE, Bloomfield LA. Phys Rev B. 1994;49:12295.
[53] Adachi K, Bonnenberg D, Franse JJM, Gersdorf R, Hempel KA, Kanematsu K, et al. In: Wijn HPJ, editor. Landolt-Börnstein, vol. 19. Berlin: Springer-Verlag; 1986.
[54] Bloomfield LA, Deng J, Zhang H, Emmert JW. In: Jena P, Khanna SN, Rao BK, editors. Proceedings of the International Symposium on Cluster and Nanostructure Interfaces. Singapore: World Publishers; 2000. p. 131.
[55] Hobbs D, Hafner J, Spisak D. Phys Rev B. 2003;68:014407.
[56] Hobbs D, Hafner J, Spisak D. Phys Rev B. 2003;68:014408.

[57] Knickelbein MB. Phys Rev Lett. 2001;86:5255.
[58] Bertsch GF, Yabana K. Phys Rev A. 1994;49:1920.
[59] Pederson MR, Reuse F, Khanna SN. Phys Rev B. 1998;58:5632.
[60] Nayak SK, Jena P. Chem Phys Lett. 1998;289:473.
[61] Nayak SK, Rao BK, Jena P. J Phys Condens Matter. 1998;10:10863.
[62] Nayak SK, Nooijen M, Jena P. J Phys Chem A. 1999;103:9853.
[63] Briere TM, Sluiter HF, Kumar V, Kawazoe Y. Phys Rev B. 2002;66:064412.
[64] Bobadova-Parvanova P, Jackson K, Srinivas S, Horoi M. Phys Rev A. 2003;67:061202(R).
[65] Khanna SN, Rao BK, Jena P, Knickelbein MB. Chem Phys Lett. 2003;378:374.
[66] López-Urías F, Díaz-Ortiz . Phys Rev B. 2003;68:180406(R).
[67] Jones NO, Khanna SN, Baruah T, Pederson MR. Phys Rev B. 2004;70:045416.
[68] Bobadova-Parvanova P, Jackson KA, Srinivas S, Horoi M. J Chem Phys. 2005;122:014310.
[69] Longo RC, Noya EG, Gallego LJ. J Chem Phys. 2005;122:226102.
[70] Morisato T, Khanna SN, Kawazoe Y. Phys Rev B. 2005;72:014435.
[71] Gutsev GL, Mochena MD, Bauschlicher Jr. CW. J Phys Chem A. 2006;110:9758.
[72] Kabir M, Mookerjee A, Kanhere DG. Phys Rev B. 2006;73:224439.
[73] Jellinek J, Acioli PH, García-Rodeja J, Zheng W, Thomas OC, Bowen Jr. KH. Phys Rev. 2006;A74:153401.
[74] Yuan HK, Chen H, Ahmed AS, Zhang JF. Phys Rev B. 2006;74:144434.
[75] Lyalin A, Solov'yov AV, Greiner W. Phys Rev A. 2006;74:043201.

Chapter 12

Metal Clusters, Quantum Dots, and Trapped Atoms: From Single-Particle Models to Correlation

M. Manninen* and S. M. Reimann†

*Department of Physics, Nanoscience Center, FI-40014 University of Jyväskylä, Jyväskylä, Finland

†Department of Mathematical Physics, Lund Institute of Technology, Lund, Sweden

Chapter Outline Head

Nanoclusters. DOI: 10.1016/S1875-4023(10)01012-0

I. INTRODUCTION

One common feature of most small quantum-mechanical systems is the discreteness of the quantum states. In systems with high symmetry, the single-particle energy levels are degenerate, which may lead to shell structure. This is known to happen not only in free atoms but also in nuclei [1]. Spherical metal clusters [2], where the particles move in a spherically symmetric mean field, provide another example. In semiconductor quantum dots with circular symmetry, shell structure was observed by conductance spectroscopy by Tarucha et al. [3] (for a review see Ref. [4]). The "universality" of shell structure bridges these fields of physics. However, there are also fundamental differences: in atoms and in quantum dots, the fixed external potential dominates, leading to Hund's rule with maximum spin at mid-shell to resolve the degeneracy of the spherical confinement. The valence electrons in metal clusters, or the neutrons and protons of a nucleus, however, move in a mean-field potential determined solely by the particle dynamics. To resolve degeneracies for non-closed shells, metal clusters and nuclei exhibit spontaneous shape deformation, while atoms and quantum dots do not. Consequently, the often used name "artificial atom" is well suited for semiconductor quantum dots, but would be misleading for free metal clusters.

Many properties of metal clusters can be calculated by using so-called *ab initio* electronic structure calculations and molecular dynamics. These computational results often are in very good agreement with experimental data—as, for example, in photoemission spectroscopy. They can pin-point the detailed ground-state geometries of particular cluster sizes [5]. However, many overall features can even be understood using simple models [2,6]. This also holds for semiconductor quantum dots, where often, simple single-particle models have been very successful [4].

The purpose of this (brief, and by no means complete) review is to summarize the simple models, their advantages, and limitations in describing overall properties of metal clusters and quantum dots, and to draw analogies between these finite quantal systems to the more recently emerging field of cold (bosonic or fermionic) atom gases in traps.

Let us begin by looking at the simple, but relevant many-particle physics of the harmonic oscillator. These results are then applied to understand the jellium model for metal clusters and electronic states in quantum dots. The universality of deformation is shortly described using simple models. We finally turn to a comparison of quantum dots at strong magnetic fields, and weakly interacting bosonic systems that are set rotating.

II. MANY-PARTICLE PHYSICS IN HARMONIC OSCILLATOR

The harmonic oscillator confining a single-particle is solved in about all text books of quantum mechanics. However, adding more particles immediately makes it more challenging to describe the system theoretically, and new interesting phenomena appear. The many-body Hamiltonian is then written as:

$$H = \sum_i \left(-\frac{\hbar^2}{2m} \nabla_i^2 + \frac{1}{2} m\omega_0^2 r_i^2 \right) + \sum_{i \neq j} v(|\mathbf{r}_i - \mathbf{r}_j|), \tag{1}$$

where ω_0 is the frequency of the confining harmonic oscillator and $v(r)$ the interparticle two-body interaction. The position vector r_i and the Laplace operator ∇_i^2 may be three- (3D), two- (2D), or one-dimensional (1D) depending on the system in question. Sometimes, we want to use the occupation number representation and write

$$H = \sum_{i,\sigma} \varepsilon_i c_{i,\sigma}^+ c_{i,\sigma} + \sum_{\{i,\sigma\}} v_{i_1,i_2,i_3,i_4} c_{i_1,\sigma_1}^+ c_{i_2,\sigma_2}^+ c_{i_4,\sigma_4} c_{i_3,\sigma_3}, \tag{2}$$

where c^+ and c are the normal creation and annihilation operators (as here, for fermions), and ε_i is the single-particle energy of the form $\varepsilon_i = \hbar\omega_0(n_i + d/2)$, d being the dimension of the system. Most conveniently, one uses the single-particle states of the confining harmonic oscillator as a basis. It is important to note that even if the spin index appears in this formulation of the Hamiltonian (as a summation index), here we consider only spinless interactions, that is, the Hamiltonian is, as obvious from Eq. (1), independent of the spin.

The perhaps most important feature of a harmonic confinement is, that the center-of-mass motion separates from the internal motion, regardless of the interaction between the particles. This can easily be shown for both classical and quantum systems [7]. As a consequence, the selection rule for the dipole oscillations only allows the center-of-mass excitation. In the case of simple metal clusters and quantum dots, this is the plasmon resonance, with energy $\hbar\omega_0$, where ω_0 is the frequency of the harmonic confinement [6]. In connection with 2D quantum dots [8–12], the effect of the separation of the center-of-mass motion was earlier often referred to as “Kohn's theorem” [13]. In the case of atomic nuclei, the related excitation is called the “giant resonance,” where the proton and neutron distributions oscillate with respect to each other [1].

Another important property of the harmonic confinement is the separation of the spatial coordinates from the center-of-mass motion (or single-particle motion). This means that the level structure in the most general case (with different oscillation frequencies ω_i along different directions) is simply

$$\varepsilon_{n_1,n_2,n_3} = \hbar\omega_1 \left(n_1 + \frac{1}{2} \right) + \hbar\omega_2 \left(n_2 + \frac{1}{2} \right) + \hbar\omega_3 \left(n_3 + \frac{1}{2} \right). \tag{3}$$

For spherically symmetric potentials, labeling the energy levels by their radial and angular momentum indices, one obtains the harmonic energy shells given in Table 1. Including the spin degree of freedom with a factor of two, the

TABLE 1 Shell Structure of Three- and Two-Dimensional (3D and 2D) Harmonic Oscillators

Shell	Levels	3D			2D		
		g	N	$2N$	g	N	$2N$
1	1s	1	1	2	1	1	2
2	1p	3	4	8	2	3	6
3	2s1d	6	10	20	3	6	12
4	2p1f	10	20	40	4	10	20
5	3s2d1g	15	35	70	5	15	30
6	3p2f1h	21	56	112	6	21	42

g is the degeneracy of the shell and N the cumulatice number of states without spin-degeneracy.

"magic numbers" of the harmonic oscillator in three dimensions occur at particle numbers 2, 8, 20, 40, 70, ..., and at 2, 6, 12, 20, 30, 42, ... in two dimensions. In a non-harmonic potential, the additional degeneracy of different radial states disappears, and other degeneracies occur. A famous example is the Wood–Saxon potential,

$$V_{\mathrm{WS}}(r) = -\frac{V_0}{1 + \mathrm{e}^{(R-r)/a}}, \tag{4}$$

frequently used in nuclear physics (where the spin–orbit interaction of the nucleons further splits the shells [1]). Single-particle states in this potential are filled following the sequence 1s, 1p, 1d, 2s, 2p, 1f, and so on. The Woods–Saxon potential Eq. (4) is a good approximation for the mean-field potential in metal clusters [14], with energetically dominant magic numbers at 2, 8, 20, 40, 58, 92, and so on. Note, however, that the first few magic numbers, as here 2, 8, and 20, are mainly determined by the angular momentum degeneracy and are nearly independent of the radial shape of the spherical potential.

In the 2D harmonic confinement, the noninteracting electron states can be solved analytically also in the presence of a magnetic field [4], the resulting single-particle states being the so-called Fock–Darwin states [15–17]. Consequently, the harmonic confinement has been widely utilized when studying the quantum Hall effect in finite systems [18].

III. JELLIUM MODEL OF METAL CLUSTERS

Many properties of simple metals, such as alkalis, alkali earths, and even aluminum, can be explained as properties of the interacting, homogeneous electron gas. The role of the ions is then merely to keep the electron gas together at its equilibrium density. In the jellium model [19], inside the metal the charge of the ions is smoothened out and replaced by a homogeneous background charge with the same density as the electron gas. At the surface, the background charge goes abruptly to zero. The electron density is usually described by the density parameter r_s defined as the radius of a sphere (in units of Bohr radius) containing one electron: $4\pi r_s^3/3 = 1/n_0$, with n_0 being the number density of the electrons. The density functional method in the local density approximation (LDA) is ideally suited for the jellium model which naturally has a smooth and slowly varying electron density.

This approach was first used to study metal surfaces [20], lattice defects [21], and impurities in metals [22]. The first application to metal clusters was made by Martins et al. [23], who studied the size variation of the ionization energy. Similar work had been successful for calculating the work function of planar surfaces of alkali metals [24].

In the density functional Kohn–Sham method, the electrons move in an effective mean-field potential:

$$V_{\mathrm{eff}}^{\sigma}(\mathbf{r}) = -e\phi(\mathbf{r}) + V_{\mathbf{xc}}^{\sigma}(\mathbf{r}), \tag{5}$$

where ϕ is the total electrostatic potential of the background charge and electron-density distribution, and V_{xc} is the exchange-correlation potential which depends locally on the electron density and spin polarization [25]. In the case of a spherical jellium cluster the effective potential resembles a finite potential well with rounded edge. It can be well approximated by the above mentioned Woods–Saxon potential (Eq. 4). The spherical jellium model suggests that, like in free atoms, the ionization potential is largest for the magic clusters, and at a minimum when only one electron occupies an open shell. The experimental results, however, show a much richer structure as a function of the cluster size, which in alkali metals is dominated by a marked odd–even staggering [2]. The reason behind is shape deformations, as described in the next section.

In the spherical jellium model for metal clusters, the the background charge is a homogeneously charged sphere of radius $R = \sqrt[3]{N} r_s$. The (external) potential caused by this sphere is

$$V_{\mathrm{sphere}}(r) = \begin{cases} -\frac{Ne^2}{8\pi\varepsilon_0 R^3}(3R^2 - r^2), & \text{if } r \le R, \\ -\frac{Ne^2}{4\pi\varepsilon_0 r}, & \text{if } r > R. \end{cases}$$

Note that the potential inside the sphere is harmonic and can be written as:

$$V_{\mathrm{sp}}(r) = \frac{1}{2} m\omega_{\mathrm{sp}}^2 r^2, \tag{6}$$

where

$$\omega_{\mathrm{sphere}}^2 = \frac{e^2}{4\pi\varepsilon_0 r_s^3 m} = \frac{n_0 e^2}{3\varepsilon_0 m} = \frac{\omega_p^2}{3} \tag{7}$$

is the square of the plasmon frequency of a metallic sphere. As the electrons only slightly spill out from the region of the harmonic potential, the plasmon is the dominating dipole absorption mechanism for spherical jellium clusters [6]. Ekardt [26] used the spherical jellium model in connection with the time-dependent density functional theory to study optical absorption. He found that the anharmonicity of the background potential caused fragmentation of the single plasmon peak to a distribution of close-lying absorption peaks. Similar work was subsequently done by several groups, using the RPA method [6,27].

Koskinen et al. [28] used shell-model methods from nuclear physics to try to solve the electronic structure and photo-absorption of the jellium clusters beyond the mean-field approach. For up to eight electrons, they could diagonalize the many-electron Hamiltonian nearly exactly. Already for 20 electrons, however, the configuration interaction (CI) method showed a much too slow convergence as a function of the size of the basis set (in fact, it was shown that the error in the correlation energy was $\propto E_{\text{cutoff}}^{-3/2}$). For eight electrons, their result agreed with those of the RPA calculations. For positively charged jellium spheres, the fragmentation of the plasmon peak disappears as the confinement of the electrons becomes harmonic [28,29].

Historically, it is interesting to note that Martins et al. [23] corrected the pure jellium results by including ion pseudo-potentials via first-order perturbation theory, in a similar fashion that Lang and Kohn [20] had done for metal surfaces. While for planar surfaces the correction had only aminor effect (in alkali metals), it became dominating for large metal clusters, and completely diminished the effects of the electronic shell structure of the pure jellium sphere. A similarly large effect of the pseudopotential correction was observed for large voids in metals, shown to be due to the low-index surfaces present in spherical systems cut from an ideal lattice [30]. The notion of the possible importance of the lattice potential made the theoreticians cautious in making too strong predictions of the applicability of the jellium model to real metal clusters [31,32], until the magic numbers of alkali metal clusters were observed [33].

The degeneracy of the open shell clusters should lead to Hund's rules like in the case of free atoms. In the spherical jellium models, the clusters with open shells should have a large total spin and magnetic moment [31]. This was predicted prior to the success of the jellium model by Geguzin [34], who studied highly symmetric cub-octahedral Na_{13} clusters. For free clusters, however, deformation wins over Hund's rule and removes both the degeneracy and magnetism [35].

We conclude that the simple spherical shell structure explains well the magic numbers in the experimental mass spectra of sodium clusters. It lies behind the so-called super shells [36,37] observed in alkali metal clusters [38], as well as the importance of the collective plasmon resonance.

The simple jellium model also accounts for some properties of noble metals. Recently, it has been observed that even in gold clusters some features can be explained most easily with arguments based on the jellium model [39].

IV. DEFORMED JELLIUM

The similarity of small nuclei and simple metal clusters is not limited to magic numbers and to the existence of the plasmon-type giant resonance, but extends even to the internal deformation of the system. It is clear that the smallest clusters can be viewed as well-defined molecules with a geometry determined by the atomic configurations. Quantum chemistry can be used to characterize the ground state and spectroscopic properties of clusters with only a few atoms. For larger clusters ($N > 10$), the early theories assumed spheres cut from a metal lattice [40], or faceted structures with shapes determined by the Wulff polyhedra [41]. In reality, however, the clusters exhibit geometries very different from these ideal structures. Many metals form icosahedral clusters [42,43]. Jahn–Teller deformations are important even in quite large clusters, as manifested, for example, by the odd–even staggering of the ionization potential [2].

In the early cluster beam experiments, the temperature of the clusters was lowered only by evaporative cooling. The resulting cluster temperatures were so high that the clusters were most likely liquid [44]. The clusters showed the electronic shell structure as well as deformation, as determined by the splitting of the plasmon peak [45]. In fact, the super-shell structure could only be seen in liquid sodium clusters. Solid clusters formed icosahedral structures which governed the abundance and ionization potential spectra [46].

To model cluster deformations, Clemenger [47] was the first to apply the Nilsson model familiar from nuclear physics [1]. He was able to explain qualitative features of the abundance spectrum of sodium clusters, including the observed odd–even staggering. A more general model, based on the Strutinsky–model of nuclei [48], was developed by Reimann et al. [49], and applied to triaxial geometries by Yannouleas and Landmann [50] as well as Reimann et al. [51,52]. It could explain nearly quantitatively the stabilities and deformation of small sodium clusters.

A. Ultimate Jellium Model

The simplest way to include deformation in the jellium model is to assume the uniform background charge density to be a spheroid, or an ellipsoid [53]. The model explains qualitatively the splitting of the plasmon peak and the size dependence of the ionization potential of alkali metal clusters. However, the optimal deformation shape determined by the electronic structure is not an ellipsoid, but a more generally shaped jellium background [54]. In the ultimate limit, the energy is minimized, when the background density equals the electron density—as suggested by Manninen already in 1986 [55]. In this so-called "ultimate jellium model" (UJM) [56], the density of the background is not fixed, but in a large cluster adjusts itself to correspond to $r_2 \approx 4.2a_0$, a value close to the equilibrium electron density in sodium. (Here, a_0 is the Bohr radius). The ground-state densities for clusters with $N = 2$–22 electrons are shown as constant density surfaces in Figure 1. Clearly, the magic numbers at small N, here for 2, 8, and 20 electrons, correspond to spherical symmetry of the freely deformable "ultimate jellium" droplet. Off-shell, however, the shapes of the clusters exhibit breaking of axial and inversion symmetries. In general, the resulting ground-state geometries are far from ellipsoidal. Clusters which lack inversion symmetry are very soft against odd-multipole deformations [56].

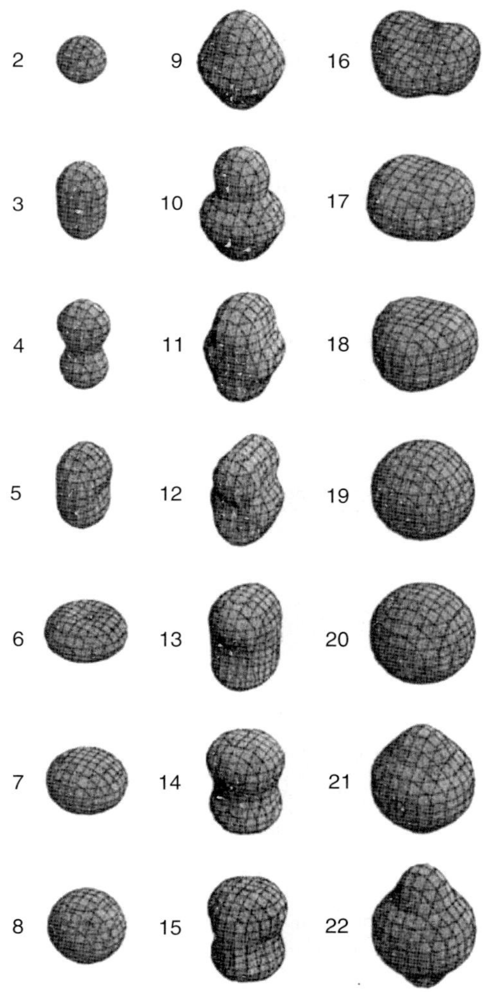

FIGURE 1 Constant-density surfaces of "ultimate jellium" clusters with up to 22 electrons. After Koskinen et al., see Ref. [56] for details and scales).

Remarkably, the results obtained from the UJM for deformations are very close to those of *ab initio* calculations for sodium [57], as shown in Figures 2 and 3.

Koskinen et al. [59] applied the UJM to determine the shape deformations of small nuclei. Their method gave rather good agreement with experimental results, and surprisingly, nearly exactly the same geometries as for the electron-gas jellium. Häkkinen et al. [58] studied further the idea of this "universal deformation" and found that in the LDA, density functional theory predicts similar deformations for all small fermion clusters.

This shape universality can be easily understood in systems where the particles move in a mean field caused by the particles themselves. When the number of the particles is small, there are only a small number of single-particle states which determine the shape. For example, for four particles, only the 1s and, say, $1p_x$ states are filled. Consequently, the shape is prolate along the x-direction. This corresponds to the basis of the Nilsson model [1].

The robustness of the shape on the specific model was further studied by Manninen et al. [60], who showed that deformations of the UJM are in very good agreement with results of the "ultimate" tight-binding model: the Hückel model for clusters [61].

For nuclei, the simple universal model only needs two parameters, the bulk modulus and the average binding energy per nucleon (the first term in the so-called mass formula [1]), to give good quantitative approximations to the deformation parameters and even excitation energies of shape isomers, as shown in Figure 4.

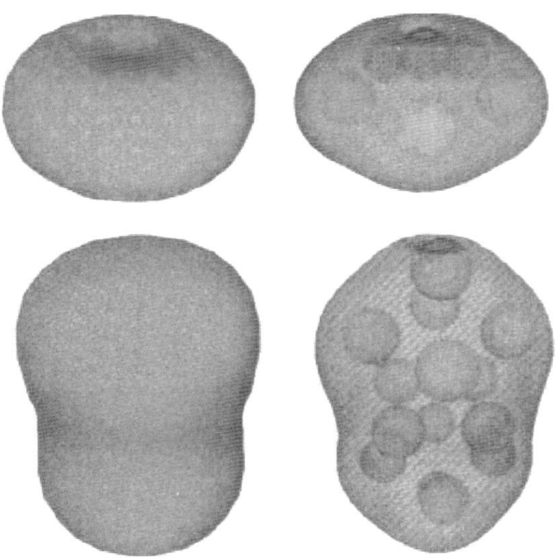

FIGURE 2 Comparison of shapes of UJM clusters (left) to those of DFT-LDA molecular-dynamics methods (right), for Na-6 (upper panel) and Na-14 (lower panel). In all cases, the outer surface shown corresponds to the same particle density. Blue spheres represent the ions. From Ref. [58].

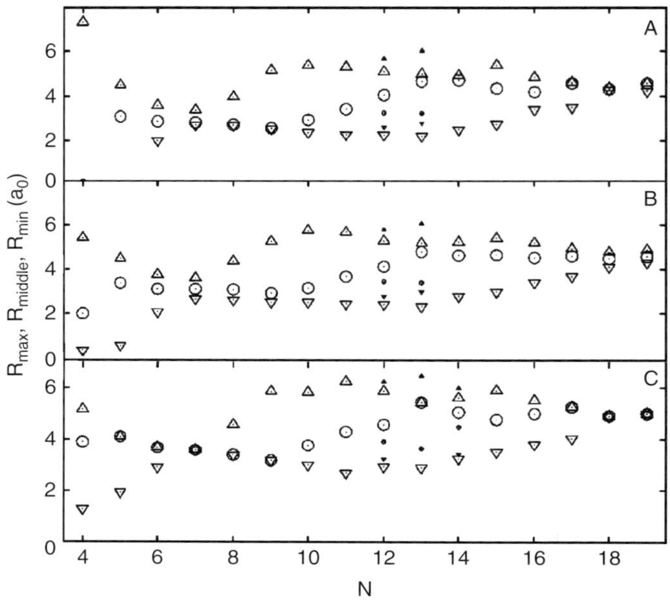

FIGURE 3 The three radii of the anionic sodium clusters along the principal axis, plotted versus the number of atoms in the cluster. Down-triangles, circles, and up-triangles correspond to R_{min}, R_{middle}, and R_{max}, respectively. (A) Radii corresponding to the ground-state geometry of *ab initio* calculations, (B) thermally averaged radii from room-temperature simulations, and (C) radii calculated within the UJM. From Ref. [57].

B. Triangles and Tetrahedra

The jellium model has also been applied to quasi 2D clusters, as, for example, in the early studies by Kohl et al. [62,63]. A physical realization of 2D clusters could be sodium clusters on an inert surface, or even 2D electron–hole liquids in semiconductors. Reimann et al. [64] analyzed systematically the UJM ground-state shapes for quasi 2D sodium clusters. Contours of the self-consistent ground-state densities of these 2D fermion droplets are shown in Figure 5, calculated for a 2D layer thickness of $3.9a_0$. The shape systematics reveals that for electron numbers 6, 12, 20, and 30 the 2D clusters have triangular shape. Initially, this result appeared puzzling, as these shell closures correspond to those of the circular 2D harmonic oscillator, and one should thus expect azimuthal symmetries of the ground-state densities. The explanation was, however, that in 2D, a triangular cavity has precisely these magic numbers [65], and only in the large-N limit, the increased surface tension at the corners makes the oscillator shells more stable. In 2D, the shell closings are rather weak, with favorable energy minima (gaps at the fermi level) appearing mainly in the small-N limit. Given the freedom of unrestricted shape deformations, a pronounced odd–even staggering appears in the ground-state energies, as seen in Figure 6. Incidentally,

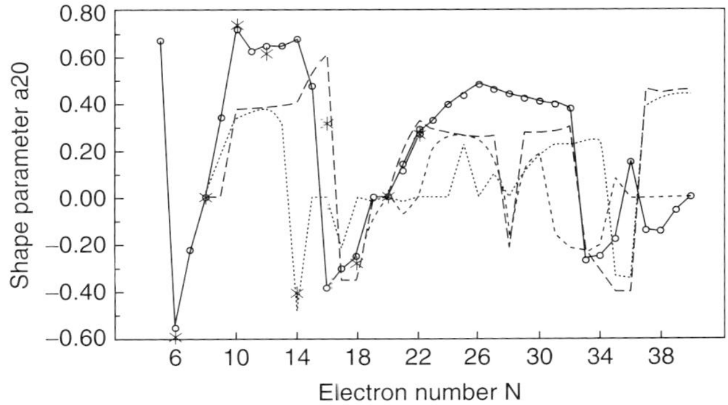

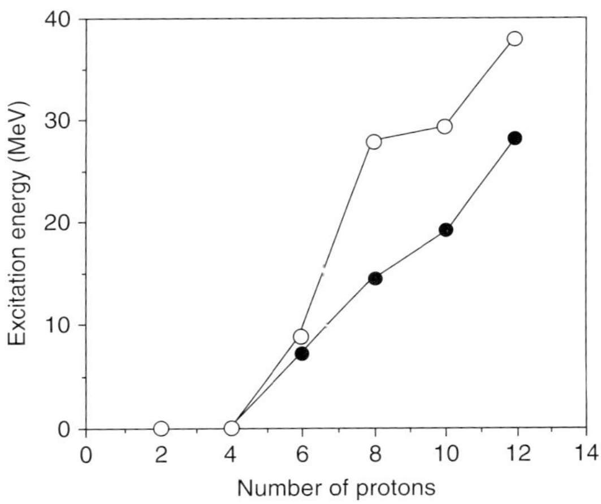

FIGURE 4 Left: Shape parameter a_{20} for fermion clusters from 5 to 40 particles calculated with the UJM (black dots connected with solid line) compared to the experimental results for even–even nuclei (stars). Right: Excitation energies of linear isomers calculated with the UJM for nuclei (open circles) and compared to the experimental results (black dots). From Ref. [60].

these shell fillings for the triangular geometries (without spin-degeneracy) equal precisely the number of atoms forming a close-packed triangle. In fact, the same holds in three dimensions: at small N, tetrahedral shell structure is preferred [66,67], with magic numbers at $N=2$, 8, 20, 40, 70, and 112. These numbers correspond precisely to twice the numbers of atoms in a close-packed thetrahedral cluster geometry (see Figure 7). One should expect that the compact tetrahedral geometry at an electronic magic number stabilizes theses clusters. However, first principles calculations have shown that this is not generally the case. Mg_{10} has an overall tetrahedral shape, but is not a perfect tetrahedron [67]. Na_{20} [68] and Mg_{20} [67] are not tetrahedra, but Au_{20} seems to be [69]. The experimental abundance spectrum of Mg shows a maximum at Mg_{35} [70]—but so far, there is no evidence that its geometry is a tetrahedron like

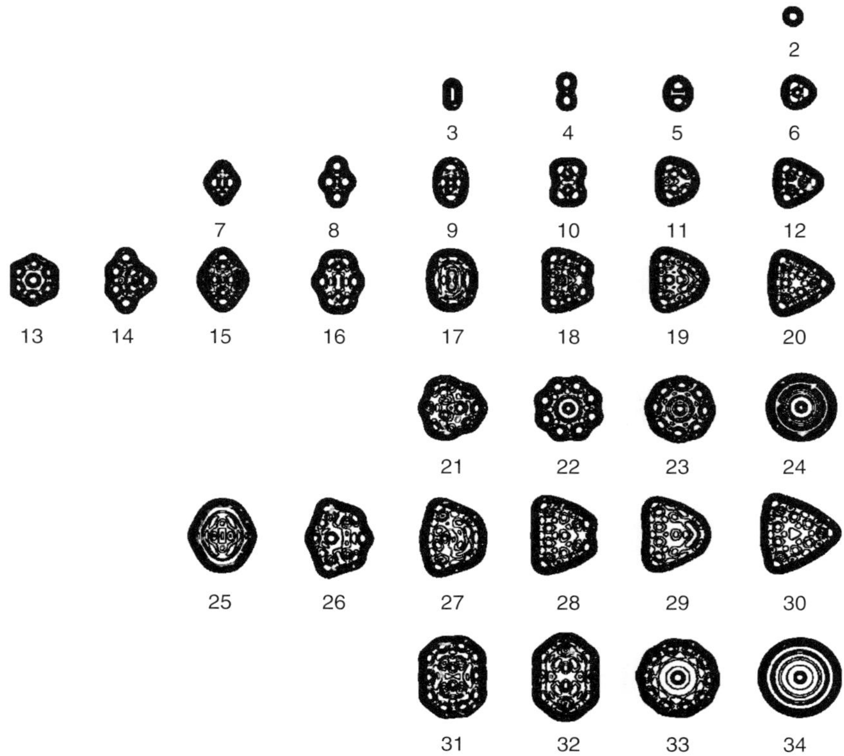

FIGURE 5 Contours of the self-consistent ground-state densities of two-dimensional UJM clusters for electron numbers $N \leq 2 \leq 34$, calculated for a 2D layer thickness of $3.9a_0$. From Ref. [64] (see this Ref. [64] for details).

the one shown in Figure 7. The above results suggest that trivalent metals on an oxide or graphite surface could favor triangular shapes.

In fact, advances in the experimental realization of surface-supported planar clusters have been recently reported by Chiu et al. [71]. They found magic numbers in quasi 2D Ag clusters grown on Pb islands, and studied the transition from electronic to geometric shell structure.

We finally mention that high stability of tetrahedral shapes has also been discussed in nuclear physics [72,73], predicting tetrahedral ground states for some exotic nuclei around ^{110}Zr (see Schunck et al. [74]).

V. SEMICONDUCTOR QUANTUM DOTS

Generally speaking, a quantum dot is a system where a small number of electrons are confined in small volume in all three spatial directions. It can be,

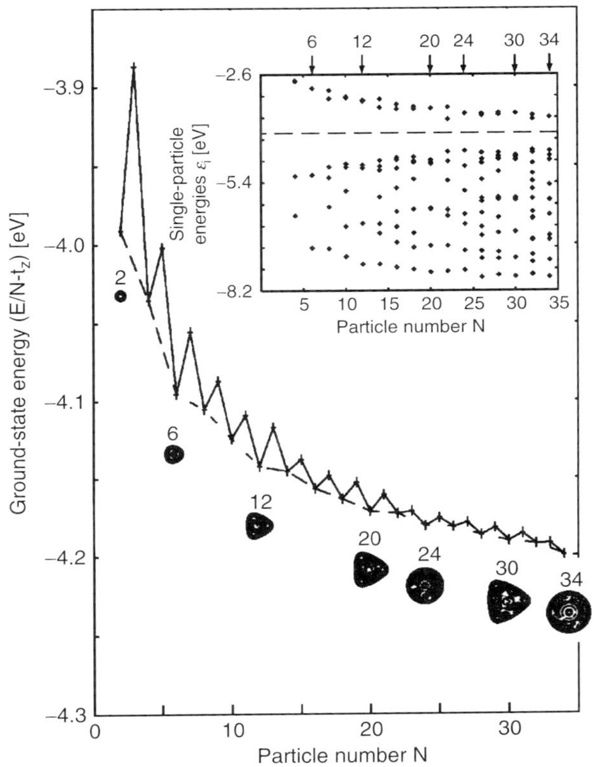

FIGURE 6 Ground-state energies per electron of two-dimensional clusters, as a function of cluster size N. (The kinetic energy contribution in z-direction, t_z, was subtracted). The inset shows the self-consistent single-particle Kohn–Sham energies for even particle numbers. From Ref. [64].

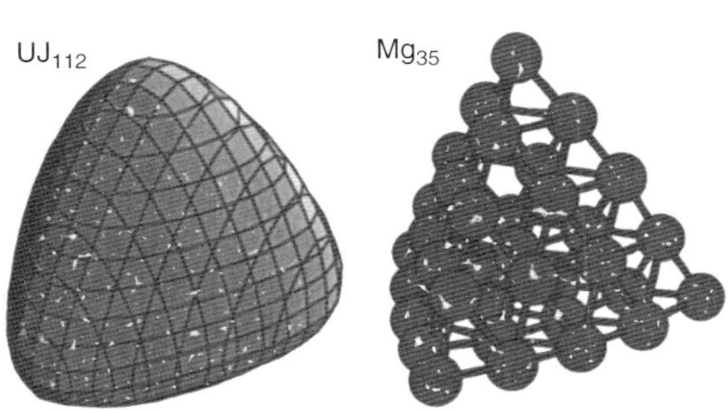

FIGURE 7 Left: Ground-state shape of the UJM for 112 electrons. Right: A possible structure of an Mg_{35} cluster with 70 valence electrons. From Ref. [64].

for example, a 3D atomic cluster or a 2D island of electrons formed by external gates in a semiconductor heterostructure [4,75]. In this chapter, we shall only consider 2D semiconductor quantum dots. Most often, they are formed from AlGaAs–GaAs layered structures, where a low-density 2D conduction electron gas is formed in the AlGaAs layer. The quantum dot is formed by removing the electrons outside the dot region with external gates (lateral dot), or by etching out the material outside the dot region (vertical dot). In both cases, the resulting confining potential is, to a good approximation, harmonic. The underlying lattice of the semiconductor material can be taken into account by using an effective mass for the conduction electrons, and a static dielectric constant, reducing the Coulomb repulsion.

The resulting generic model for a semiconductor quantum dot is a 2D harmonic oscillator with interacting electrons. This in fact is like a 2D jellium model, with the simplification that now the harmonic confinement has infinite range and the center-of-mass motion separates out exactly (Kohn's theorem [13]). This means that in the ideal case (in zero magnetic field) there is only one dipole absorption peak, as seen in experiments [76].

Conductance spectroscopy can be used on one single dot. The dot is weakly connected to leads and the current is measured as a function of the gate voltage which determines the chemical potential and thus the number of electrons in the dot [4]. When the electron number in the dot is large, the energy of an additional electron can be estimated from the capacitance C of the dot, as $\Delta E = e^2/C$. The resulting conductance then shows equidistant peaks as a function of the gate voltage.

When the number of electrons is small, the individual single electron levels in the dot become important and their shell structure can be seen in the conductance spectrum. Tarucha et al. [3] were the first to successfully determine the shell structure of circular quantum dots. Their result is shown in the lower panel of Figure 8, where the second derivative of the total energy of the dot is plotted as a function of the number of electrons, N. For comparison, the corresponding result of the LSDA calculation for electrons in a harmonic oscillator is included, too.

The density functional Kohn–Sham method for semiconductor quantum dots usually assumes that (i) the system is 2D, (ii) only the conduction electrons are considered, with an effective mass m^* and their Coulomb interaction screened by the static dielectric function ε of the material in question, and (iii) they move in a harmonic confinement $m\omega_0^2 r^2$. A local approximation is used for the spin-dependent exchange-correlation energy, derived from the functionals for the 2D electron gas [77]. For details see Ref. [4].

Shell structure with main shell fillings (magic numbers) at $N = 2$, 6, 12, and 20 appears very clearly in the addition energy differences $\Delta_2(N)$ (see Figure 8). Furthermore, like in free atoms, because of Hund's rule at mid-shell the total spin is maximal. This means that (just like for the spherical jellium model discussed above), any half-filled shell shows as a weak "magic" number, with increased stability. This is clearly seen in Figure 8 where the second derivative of the total energy shows maxima at $N = 4$ and 9 in addition to the clear peaks at the filled shells, $N = 2$, 6, and 12. Figure 8 also shows the calculated total spin as a function of the number of electrons in the dot.

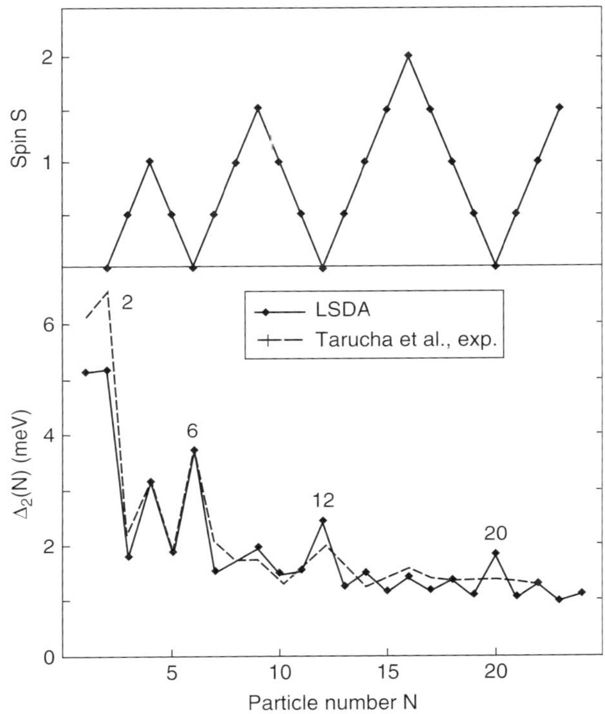

FIGURE 8 Second derivative of the total energy of electrons in a quantum dot as a function of the number of electrons. The magic numbers are shown. The experimental result is from Tarucha et al. [3]. The upper panel shows the calculated total spin.

The self-consistent data appear to agree very nicely with the experimental data. However, we notice that this agreement becomes worse with increasing N, showing very clear deviations between theory and experiment after the third shell, that is, around $N=20$. Another series of experimental data was later published by the same group in 2001. In Ref. [78], addition energies for 14 different quantum dot structures, all similar to the device used in the earlier work by Tarucha et al. [3], were analyzed. Strong variations in the spectra were reported, very clearly differing from device to device and seemingly indicating that each of these vertical quantum dots indeed has its own properties: a comparison to the theoretically expected shell structures needs to be made with care. Progress with vertical quantum dots was achieved more recently, where few-electron phenomena could be studied by tunneling spectroscopy through quantum dots in nanowires [79,80].

The self-consistent electronic structure calculations for quantum dots for some electron numbers showed internal symmetry-breaking of the spin-density [81], leading to a static "spin-density wave" (SDW). Figure 9 shows, as an

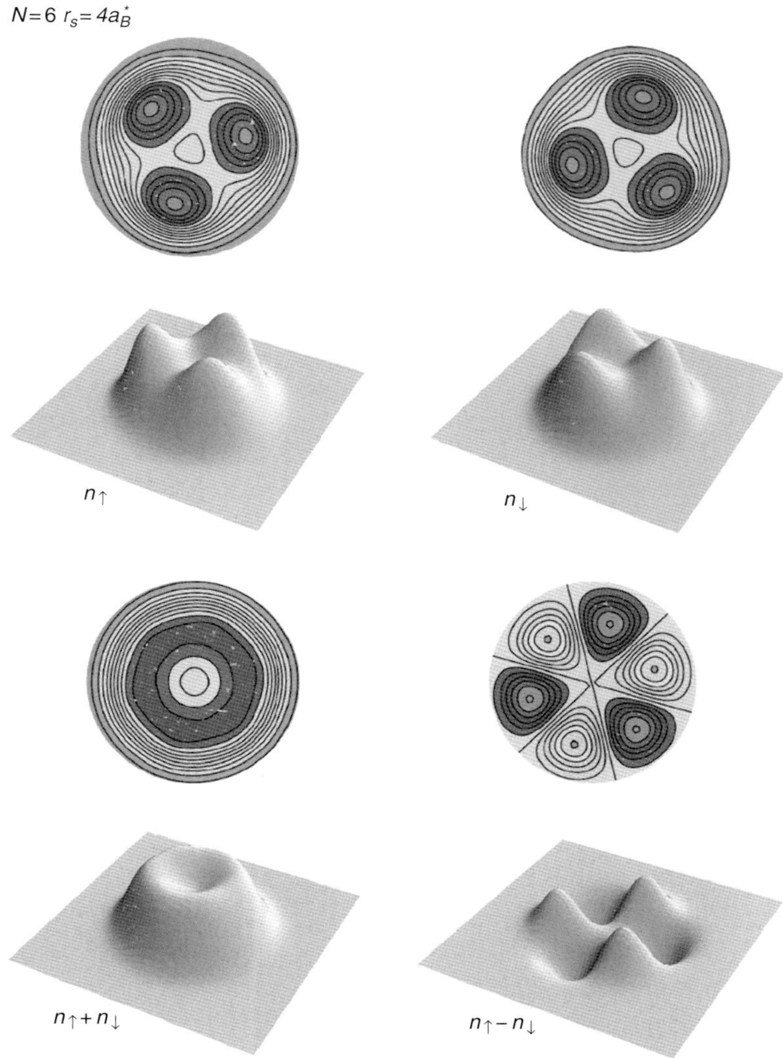

FIGURE 9 DFT spin densities $n^\uparrow$ and $n^\downarrow$ (upper panel) and total density $(n^\uparrow+n^\downarrow)$ as well as (un-normalized) spin polarization $(n^\uparrow+n^\downarrow)$ (lower panel) for a six-electron quantum dot at $r_s=4a_0{}^*$, shown as 3D plots and their contours. From Ref. [82].

example, the intriguing ground-state spin polarization for a quantum dot with six electrons. For not too small densities of the electron gas, that is, $r_s \leq 6a_0{}^*$, this quantum dot still has a closed-shell configuration, with $S=S_z=0$. This result is obtained from spin-density functional theory (SDFT). The total density obtained by the SDFT method is circularly symmetric, with zero net polarization ($S=0$). However, the spin polarization (which equals the difference between the spin densities $n^\uparrow + n^\downarrow$ normalized by the total density, $n^\uparrow + n^\downarrow$), in standard SDFT breaks the azimuthal symmetry of the confinement, showing a regular spin structure. Figure 9 shows this very clearly for the example of a six-electron quantum dot at $r_s = 4a_0$. Both spin-up and spin-down densities exhibit three clear bumps, which are twisted against one another by an angle of $\pi/3$. This resembles very much an antiferromagnet-like structure, with alternating up- and down spins, on a ring. Such states were obtained both with the Tanatar–Ceperley [77] and the more recent Attaccalite–Moroni–Gori-Giorgi–Bachelet (AMGB) [83] functionals for exchange-correlation. As the AMGB functional depends explicitly on the spin polarization, there is left no doubt that the SDW states are *not* simply an artifact of the *ad hoc* approximation to the correlation energy, which is usually interpolated following the polarization-dependence of the exchange energy [81]. In quasi-1D quantum rings (see Section VII.B), these SDW states become more distinctive, as shown in Figure 10.

The existence of the nonspherical spin-densities in quantum dots was disputed in the literature, as the spherical symmetry of the Hamiltonian dictates spherical symmetry [85,86]. However, as well known from nuclear physics, a meal field theory (like KS-LSDA) can lead to *internal* symmetry-breaking. In some cases [82], it reveals the internal structure which, in fact, can be very difficult to extract from the exact wave function. We will repeatedly meet this

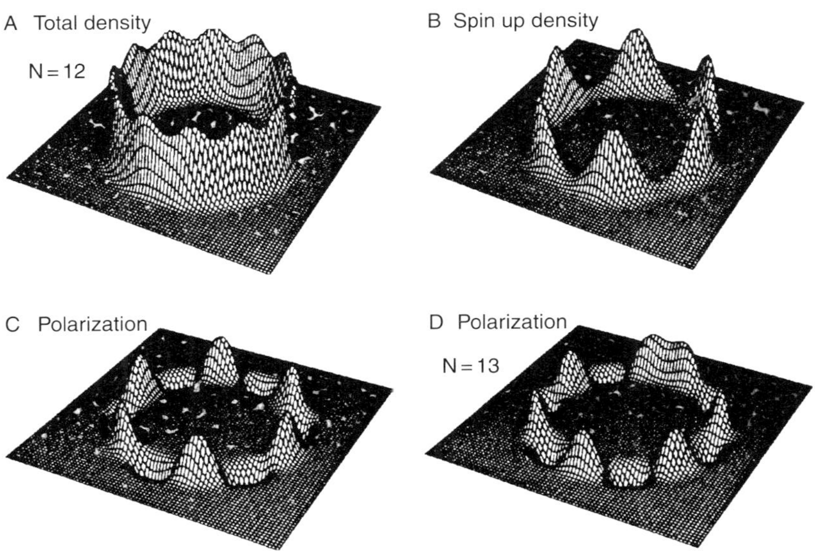

FIGURE 10 Quantum rings with $N=12$ and 13 electrons, showing antiferromagnetic spin-ordering along the ring. The maximum electron density in the 12-electron ring is at $n_{\max} = 0.157 a_0^{*-2}$. From Ref. [84].

problem, for example, when studying vortices and localization in rotating quantum systems; see Sections **VI** and VI.D below.(for further reading on the internal symmetry-breaking, we refer to the recent review articles [4,87,148]).

The full quantum-mechanical problem of a few electrons in a 2D harmonic oscillator can be solved using the so-called CI technique, numerically diagonalizing a large Hamiltonian matrix. This method is often called "exact diagonalization," although there is always an approximation due to the necessary restrictions in the basis set or the number of configurations included. Nevertheless, up to say 6 or even 10 particles (depending on the confinement strength), the results can be viewed as practically exact.

In general, the results of the exact diagonalization agree well with those obtained within the LSDA. The same spins dictated by the Hund's rule are obtained and the total energies agree with good accuracy. Also, the electron-density and spin-density profiles are in excellent agreement. However, the exact diagonalization can reveal the existence of the internal symmetry-breaking only via the pair correlation:

$$g_{\sigma\sigma'}(\mathbf{r}, \mathbf{r}') = \langle \Psi | \hat{n}_{\sigma}(\mathbf{r}) \hat{n}_{\sigma'}(\mathbf{r}') | \Psi \rangle, \tag{8}$$

where Ψ is the many-particle quantum state and $\hat{n}_{\sigma}$ is the spin-density operator. This pair correlation function is also called as "conditional probability," as it gives the probability of finding an electron with spin σ' at r' when an electron with spin σ is located at $\mathbf{r}$. As an example, we show in Figure 11 the pair correlations for the above discussed six-electron quantum dot, here at $r_s = 3.8a_0{}^*$. The top panel shows the (up, down)-correlations and the bottom panel the (up–up) correlations. Clearly, the internal structure of the exact ground state resembles the SDFT result described above: the probability maxima appear on a ring, with six alternating maxima of the up- and down correlations.

For a more detailed discussion, we refer to the recent work by Borgh et al. [82] on broken-spin-symmetry in SDFT ground states and the reliability of SDFT. Here, we only note that SDFT and CI results generally agree very well in the case of *singlet* states, as it was examplified above by the six-electron quantum dot. However, if the true ground state is a spin-multiplet, SDFT introduces an artificial splitting of multiplet states which may be misleading, and even become a real pitfall when determining ground state energies and symmetries where CI or other exact results to compare with, are not at hand.

A. Wigner Molecules

The exact diagonalization method has also been used to study electron localization in low-density quantum dots. At extremely low densities, the homogeneous electron gas forms a Wigner crystal [88] also in the bulk. This happens in 3D, 2D, and 1D, although in 1D the true long-range order is fading with distance.

In three dimensions, the critical value at which crystallization occurs was determined to be $r_s = 100a_0$ by Ceperley and Alder [89]. In two dimensions,

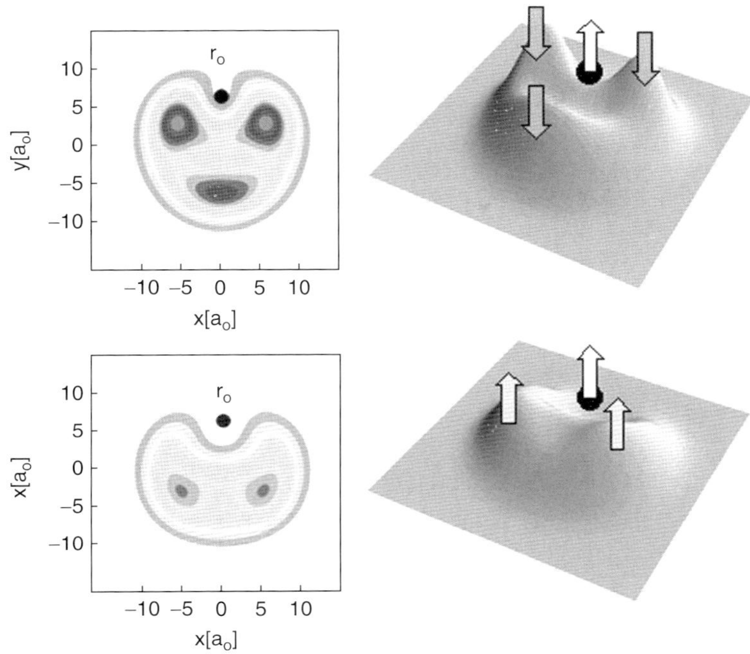

FIGURE 11 Pair correlations for a quantum dot with $N=6$ electrons, here at $r_s=3.8a_0^*$, obtained with the CI method. The top panel shows the (up–down)-correlations, the bottom panel the (up–up) correlations. The black dot marks the reference point, with given spin. From Ref. [82].

the transition occurs at smaller r_s values, according to Tanatar and Ceperley [77] at $r_s > 37a_0$*. Breaking of translational invariance in 2D lowers this value to $r_s \approx 7.5a_0$*. Thus, in finite systems, localization may happen at even smaller values, as discussed, for example, by Creffield et al. [90], Egger et al. [91], or Yannouleas and Landmann [92].

For finite number of electrons in a quantum dot, the localized state is often called a "Wigner molecule" [91]. The LDA cannot produce properly the localized states because of the lack of exact cancelation of the direct and exchange Coulomb interactions. The most direct notion of electron localization can be found by using the unrestricted Hartree–Fock approximation. The (complicated) mean-field character of the approach can lead to broken-symmetry solutions, showing the electron localization directly in the electron density [92,93]. This method also indicates a clear "phase transition" point (N-dependent r_s) where the crystallization occurs. However, going beyond Hartree–Fock to the exact diagonalization results makes the situation more complicated: there is no clear phase transition in these finite systems, but the localization gradually becomes stronger when r_s increases [94].

The most studied system in this context is the two-electron quantum dot, the so-called "quantum dot helium" [95], which is in some cases exactly solvable [96–100]. Zhub et al. [100] have shown that at low densities (weak confinement, small ω_0) the many-particle excitation spectrum can be described with the rotation–vibration spectrum of two localized electrons. We will return to this method in the case of rotating systems in the next section.

As mentioned earlier, the LDA does not support electron localization because of the incomplete cancelation of the direct and exchange Coulomb interaction. The introduction of spin-dependence (LSDA) increases slightly the tendency of localization. If more internal degrees of freedom are included, as it could be done in the case of multi-valley semiconductors, the localization of electrons is expected to happen also in the local approximation. This possibility was studied by Kärkkäinen et al. [101]. Figure 12 shows the localization of eight electrons in a multicomponent electron system when the confinement becomes weaker.

VI. ROTATING SYSTEMS IN 2D HARMONIC OSCILLATOR

A semiconductor quantum dot in the presence of a perpendicular magnetic field is a finite-size realization of the quantum Hall liquid (QHL), which has been an exciting system of study for both experimentalists and theorists. In fact, when Laughlin [18] suggested his celebrated wave function for the fractional quantum Hall state, he used exact diagonalization calculations for a finite quantum dot to test his ingenious Ansatz. Since then, it has been one of the systems used to mimic also the infinite systems in many-particle physics of QHL.

The conductance measurements through a quantum dot show that as a function of the magnetic field B, the conductance has rather complicated oscillations at small B-values [102]. These are caused by successive changes in the quantum states of the electrons, characterized by changes in the angular momentum and spin quantum numbers. However, at a certain field range, the oscillations disappear and it is believed that the electrons form an integer QHL (with filling factor one). At this state, the electron system is fully polarized.

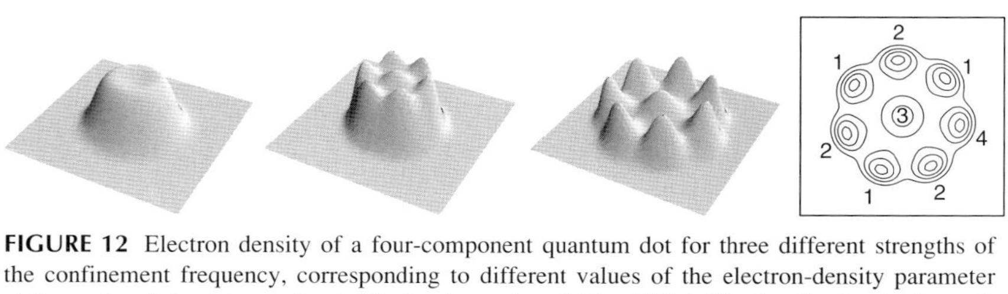

FIGURE 12 Electron density of a four-component quantum dot for three different strengths of the confinement frequency, corresponding to different values of the electron-density parameter at the center of the dot: $r_s = 2a_0$ (left), $6a_0$ (center), and $14a_0$ (right). The localization in the multicomponent LDA is made possible by the fact that the neighboring localized electrons belong to different components as indicated by the numbers in the contour plot.

The density functional theory has been extended to treat 2D electrons in the presence of magnetic fields [103]. In the so-called current-density functional method, the exchange-correlation energy of the electrons depends on the *local current density* of the electrons, in addition to its dependence on the electron density and polarization. Although the method can be disputed in being not uniquely defined in all cases and its functionals are not well established [104,105], it has been useful in understanding the general "phase diagram" of the conductance, and has been successful to suggest new kinds of symmetry-broken ground states, with localized edge states [84,106] and vortices [107] (see below) as prominent examples.

The perpendicular magnetic field in the 2D harmonic confinement has two effects: it interacts directly with the magnetic moments of the electrons causing a Zeeman term $g\mu_B S_Z B$, and changes the single electron kinetic energy from $p^2/2m$ to $(\mathbf{p}-e\mathbf{A})^2/2m$. Using the symmetric gauge for the vector potential, $\mathbf{A}=(1/2)B(-y,\ x,\ 0)$, and the definition of the cyclotron frequency $\omega_c=eB/m$ the single-particle Hamiltonian becomes

$$h=\frac{\hbar^2}{2m}\nabla^2+\frac{1}{2}m\left(\omega_0^2+\frac{1}{4}\omega_c^2\right)r^2+\frac{1}{2}\omega_c\hat{l}, \tag{9}$$

where $\hat{l}$ is the z-component of the orbital angular momentum operator. Note that in 2D systems this is the only component, and thus, the same as the total angular momentum. We denote the single-particle angular momentum by m and the many-particle angular momentum by L. Clearly, the single-particle problem is exactly solvable, as discussed already by Fock [15], Darwin [16], and Landau [17]. The single-particle energies can be written as:

$$\varepsilon_{nm}=\hbar\omega_h(2n+|m|+1)+\frac{1}{2}\hbar\omega_c l, \tag{10}$$

where $\omega_h=\sqrt{\omega_0^2+\omega_c^2/4}$ and the radial quantum number is $n=0,1,\ldots$ with the angular momentum m being an integer. Figure 13 shows the single-particle states as a function of the cyclotron frequency (magnetic field). Only the levels with $|l|\leq 7$ are shown to illustrate clearly the separation of the levels to different Landau bands at large values of the magnetic field (large ω_c). The lowest of these consists of states with $n=0$ and $m=0,-1,-2,\ldots$ in increasing order of energy.

Normally, the Landé factor g in the Zeeman energy is nonzero. Consequently, in a strong magnetic field, the electron system will polarize. In this case, the noninteracting electrons fill the N lowest energy states of the lowest Landau level (LLL). The single-particle states of the LLL are simply

$$\psi_m(r,\phi)=C_m r^l \mathrm{e}^{-r^2/2\ell_h^2}\mathrm{e}^{il\phi}, \tag{11}$$

where C_m is the normalization constant and $\ell_h=\sqrt{\hbar/m\omega_h}$ is the effective oscillator length. In the theory of QHL, it is customary to describe the electron

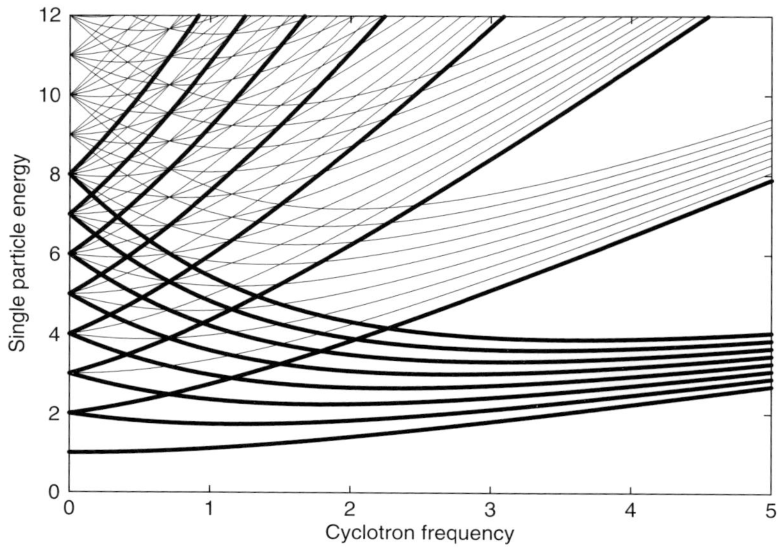

FIGURE 13 Single electron states in 2D harmonic oscillator in a perpendicular magnetic field. The levels are plotted as a function of the cyclotron frequency ω_c. The levels with $n=0$ in Eq. (10) are shown as thick lines.

coordinates by a complex number $z=x+iy$, where $x=r\cos\phi$ and $y=r\sin\phi$. The ground state of polarized noninteracting electrons is a Slater determinant formed from the N lowest single-particle states. Conveniently, it can be written (in the complex plane) as:

$$\Psi_{\mathrm{MDD}}(z_1, z_2, \ldots, z_N) = \prod_{i<j}^{N} (z_i - z_j)\mathrm{e}^{-\Sigma|z_k|^2/2\ell_h^2}, \tag{12}$$

where the normalization is omitted. This state is called the maximum density droplet (MDD) and is the finite-size analog of an infinite integer QHL. Note that the state is antisymmetric and has the total angular momentum $L_{\mathrm{MDD}}=N(N-1)/2$.

The electron density of the MDD is constant inside the dot, as illustrated in Figure 14, (note that the density of a single Slater determinant is simply $n(r)=\Sigma|\psi_m|^2$). In the case of noninteracting, polarized electrons, the increase of the magnetic field does not change the structure of the system, but the MDD becomes smaller and smaller as ℓ_h decreases when B (or ω_c) increases. Before we turn to the much more interesting case of interacting electrons, let us note a few facts about the excited states of noninteracting fermions in the LLL.

The only way to excite electrons (for a fixed B or ω_c) in the LLL is to increase the single-particle angular momenta m such that the total angular momentum increases by ΔL. This gives an excitation energy of $\Delta E=\hbar\omega_h L$. However, the degeneracy of the state is in general large, as there are many

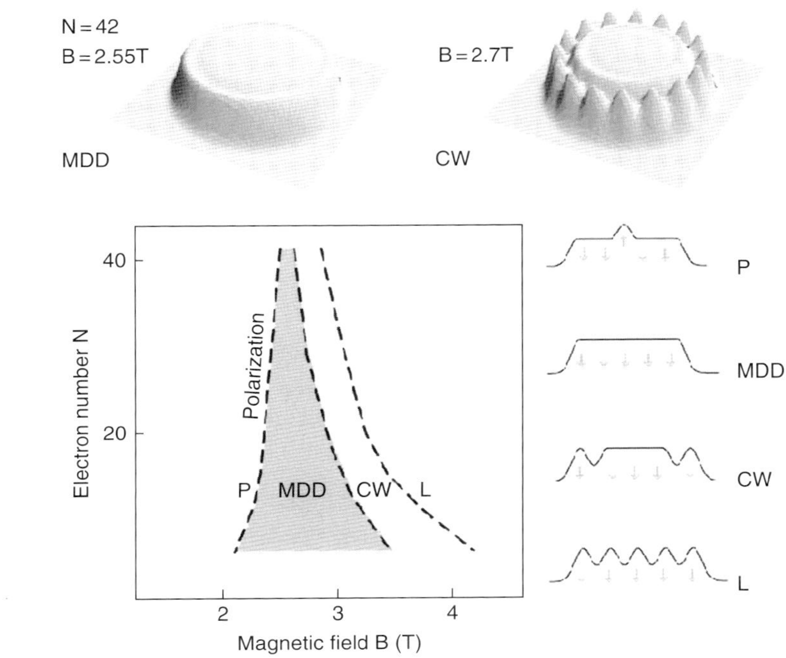

FIGURE 14 "Phase diagram" for electrons in a harmonic confinement in the presence of a magnetic field: *P* denotes the region of where the polarization happens, MDD is the maximum density droplet, CW is the region of the edge reconstruction, and *L* denotes the high-field region where electron localization sets in. Schematic densities and spin configurations of the different regions are shown at the right. The two figures on top show calculated electron densities for 42 electrons in the region of the MDD (left) and CW (right). The confinement strength was set to $\hbar\omega = 4.192N^{-1/4}$ meV, corresponding to typical GaAs values.

ways to distribute the single-particle states in the LLL so that the total angular momentum is $L_{\mathrm{MDD}} + \Delta L$. The wave function can be written in the complex coordinates as:

$$\Psi_{\delta L} = P(z_1, z_2, \ldots, z_N) \prod_{i<j}^{N} (z_i - z_j) \mathrm{e}^{-\Sigma |z_k|^2 / 2\ell_h^2}, \tag{13}$$

where P is any homogeneous *symmetric* polynomial of order ΔL. The proper antisymmetry is provided by the determinant $\prod(z_i - z_j)$.

A. Interacting Electrons in the LLL

The electron–electrons interactions can be included at different levels of approximations. The current-density functional theory in the LSDA takes into account the interactions on a mean-field level and allows including the magnetic field as described above. Using the material parameters (m^* and ε) of

GaAs, Reimann et al. [84] showed, in agreement with the experiments, that for each electron number there exists a region where the ground state is the MDD. This droplet slightly shrinks with increasing magnetic field. The "phase diagram" shown in Figure 14 demonstrates how this region of MDD ground states becomes narrower when the number of electrons in the dot increases (Here, the average electron density in the dot was chosen to be approximately constant, setting the confinement strength to $\hbar\omega = 4.192N^{-1/4}$meV, which corresponds to a typical value for GaAs.). When the magnetic field becomes too large, the MDD breaks down. At large electron numbers, this begins from the surface of the droplet. A ring of electrons separates from the inner, still compact, droplet. The results of current-density functional calculations suggest that in this split-off ring, the electrons are localized [84], as shown in Figure 14. Again, this broken internal symmetry was disputed in the literature. However, calculations based on other many-particle methods have shown similar localization tendency of this so-called Chamon–Wen edge in the correlation functions [108,109].

Figure 15 compares the correlation functions obtained from the CI calculations, to the corresponding result in mean-field current spin-density functional theory (CSDFT). As an example, we here chose the 20-electron quantum dot at high rotation, or equivalently, strong magnetic fields. The broken-symmetry along the so-called Chamon–Wen edge is reproduced

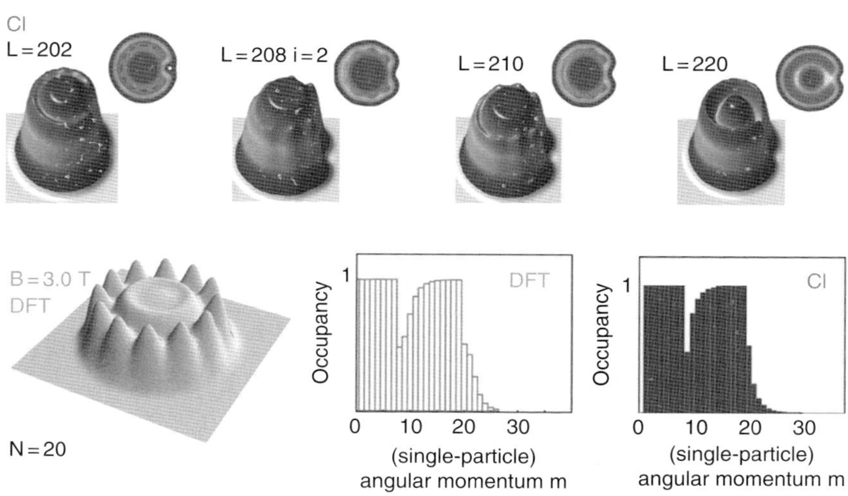

FIGURE 15 Comparison of CI correlation functions (upper panel) and mean-field densities, for a 20-electron quantum dot at high rotation, or equivalently, strong magnetic fields. The CI results were obtained for rotation in the lowest Landau level (LLL) only, for fixed angular momentum as specified. The mean-field result (lower panel, left) was calculated in CSDFT, at an effective magnetic field of $B = 3.0$ T ($r_s = 2a_0^*$). The two plots at the right-hand side of the lower panel compare the occupancies of the single-particle levels in the LLL, characterized by their single-particle angular momentum m.

in the CI correlations. The occupancies of the single-particle levels in the LLL, characterized by their single-particle angular momentum, m, agree remarkably well, demonstrating the success of CSDFT in describing the correlated electronic structure at strong magnetic fields.

B. Rotation Versus Magnetic Field

A magnetic field applied to the 2D harmonic oscillator leads to the simple Hamiltonian Eq. (9). For a fixed angular momentum l, the last term of the Hamiltonian is a constant and the solutions are the harmonic oscillator energies and wave functions for the effective confinement $\omega_{\mathrm{h}} = \sqrt{\omega_0 + \omega_{\mathrm{c}}/4}$. This is an important notion: we can equivalently study the rotational spectrum of the harmonic oscillator. For simplicity, we will now neglect the Zeeman effect, that is, the direct interaction between the electron spins and the magnetic field, $g\mu_B S_z B$ (In fact, in semiconductors, the effective Landé factor g can be reduced to zero.).

Similarly, for the many-particle system, even when the interactions are included, the effect of the magnetic field for a fixed L is to increase the strength of the confinement. Clearly, the Hamiltonian can be written as:

$$H = \sum_i \left(-\frac{\hbar^2}{2m} \nabla_i^2 + \frac{1}{2} m\omega_{\mathrm{h}}^2 r_i^2 \right) + \sum_{i \neq j} v(|\mathbf{r}_i - \mathbf{r}_j|) + \frac{1}{2}\omega_{\mathrm{c}} \hat{L}, \qquad (14)$$

where now $\hat{L}$ is the total angular momentum. Again, if the total angular momentum is fixed, the last term reduces to a constant: in the case of a 2D harmonic confinement, the effect of the magnetic field is only to put the system in rotation, and to increase the strength of the confinement. In Figure 16, the results of exact diagonalization for six electrons are shown for three different strengths of the field. Clearly, the relative structure of the spectra is very similar, and the effect of the field is only to tilt the spectrum toward higher angular momenta and to determine the energy scale via ω_h. The rotational spectrum alone reveals all the effects the magnetic field can have (apart from the simple Zeeman term), making the direct comparison to other rotating systems (like, e.g., cold, atomic quantum gases) meaningful.

C. Localization of Particles at High Angular Momenta

We will now study the interacting system in a rotational state with a very high angular momentum. First, let us consider fermions. For small particle numbers, the exact diagonalization technique can be used with the harmonic oscillator states as the single-particle basis. When the angular momentum is large, all the low-energy states are in the LLL as shown in Figure 13, and the basis set can thus be restricted to include only the LLL. With this restriction, the matrix size will be finite (for a fixed L) and for a small particle number no other approximations are needed.

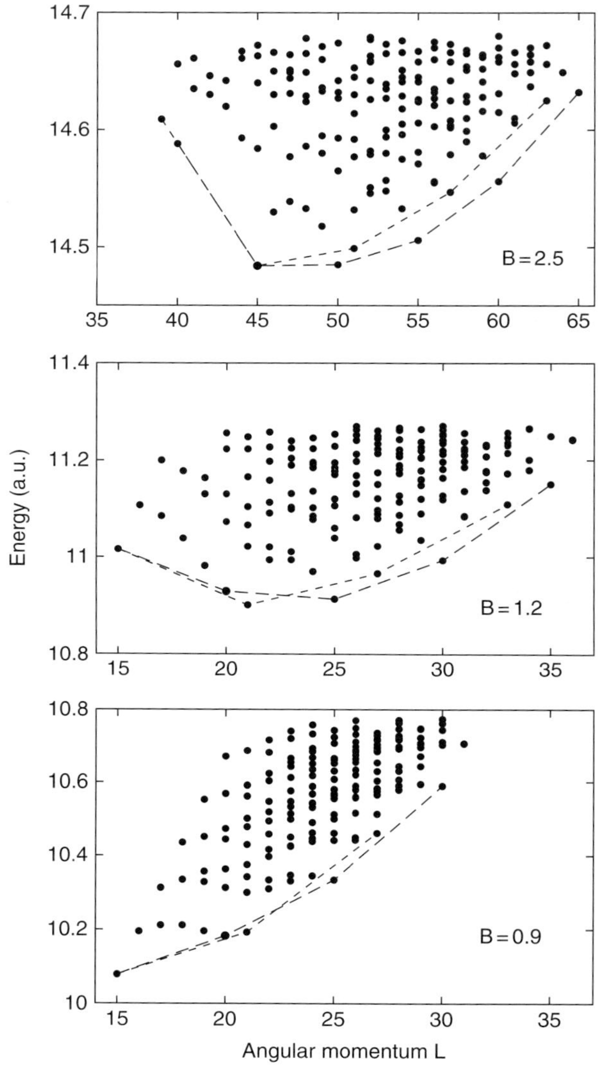

FIGURE 16 Many-particle energy spectrum as a function of the total angular momentum for three different values of the magnetic field (given in atomic units). From Ref. [110].

Using the formalism of second quantization, the Hamiltonian for the polarized electrons (we drop the spin index) is

$$H = \sum_{l} \hbar\omega_0 l c_l^+ c_l + \sum_{\{i\}} v_{i_1,i_2,i_3,i_4} c_{i_1}^+ c_{i_2}^+ c_{i_3} c_{i_4}. \tag{15}$$

For a fixed angular momentum L, the diagonal term of the Hamiltonian gives the energy $\hbar\omega_0 L$ for all configurations, thus, just adding a constant. The diagonalization of the Hamiltonian is thus reduced to the non-diagonal interaction term. The effect of the confinement frequency ω_0 (or ω_h) is to provide the single-particle basis and to determine the energy scale through the interaction matrix elements v_{i_1,i_2,i_3,i_4}. The many-particle states are completely independent of the confinement strength, when only the LLL is included in the basis. When studying the rotational energy spectrum, it is thus customary to plot the interaction energy, instead of the total energy. When the angular momentum of the system increases, the systems expands and the interparticle interactions decrease. The interaction energy then decreases with increasing angular momentum, as seen in the figures below.

Figure 17 shows the energy spectrum calculated for four electrons as a function of the total angular momentum. Two features are distinct. First, each appearing new energy is repeated for all higher angular momentum values. This is due to the center-of-mass excitations. As discussed in Section **II**, the center-of-mass motion separates from the internal motion, and its excitation energy is $\hbar\omega_0 n$. In the LLL, each center-of-mass excitation increases the angular momentum by $\Delta L = 1$, but since this does not change the interaction energy, it remains constant.

The second important feature of Figure 17 is the periodic oscillation of the lowest energy state as a function of the angular momentum. These "yrast" states, that is, the states with highest possible angular momentum at a fixed energy, are connected with a continuous line in the figure. Actually, the name "yrast" comes from Swedish language for "the most dizzy," and originates

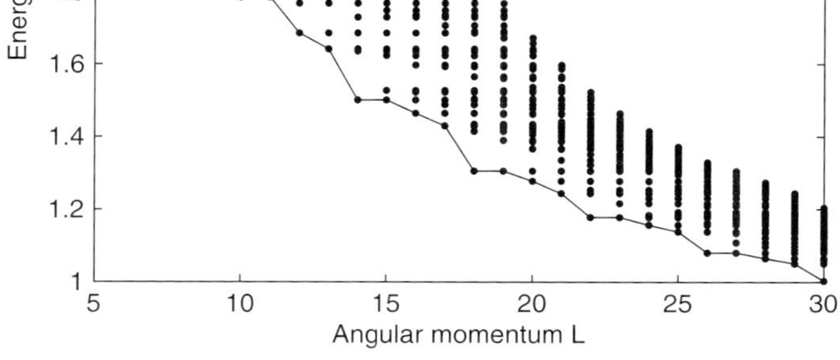

FIGURE 17 Many-particle energy spectrum (the interaction energy) for four electrons in a harmonic confinement as a function of the angular momentum. The lowest energy states are connected with a line to illustrate the period of four.

from nuclear physics [1,111]. The periodic oscillation, which becomes more distinct when the angular momentum increases, is a caused by localization of the electrons [112,113]. Assuming that the electrons are localized in a Wigner molecule, which in the case of four electrons has the geometry of a square, the rigid rotation of this molecule can be quantized. The symmetry requirements of the total wave function allow only every fourth angular momentum for a rigid rotation [113]. These L-values correspond precisely to the low-energy cusps of the yrast line. The points in between cannot be pure rigid rotations and must be other internal excitations. One possibility is, for example, the vibrational modes of the Wigner molecule.

To understand the rotation–vibration spectrum of the Wigner molecules, one can use methods familiar from molecular physics. The corresponding energy levels are

$$E_{\rm cl} = E_{\rm cl}^0 + \frac{L^2}{2I} + \sum_k \hbar\omega_k \left(n_k + \frac{1}{2} \right) + \hbar\omega_0(n_{\rm o} + 1), \tag{16}$$

where $I = \sum mr_i$ is the moment of inertia of the molecule, ω_k the vibration frequencies, and the last term gives the energy of the center-of-mass motion. The difference between the Wigner molecule and a normal molecule is that in the former case the Coriolis force is essential for determining the vibrational frequencies. In practice, they have to be determined in a rotational frame [112,114]. Another important difference is the drastic expansion of the Wigner molecule as a function of the angular momentum. This causes not only the decrease of the vibration frequency but also the increase of the moment of inertia.

For four electrons, the vibrational modes can be solved analytically [114] and the resulting energy spectrum can be constructed by considering which combinations of vibrational modes and rotational states can be used to construct an antisymmetric state. This can be done with the help of group theory [112,115]. Figure 17 shows part of the rotational spectrum, computed with exact diagonalization of the quantum-mechanical system. It is compared to the spectrum obtained from Eq. (16), that is, using classical mechanics and group theory. The figure shows an excellent agreement between the spectra. This demonstrates clearly that at such high angular momenta, the four-electron system is just a vibrating Wigner molecule of localized electrons.

The pair correlation functions shown in the lower panel of Figure 18 further support this conclusion. For the cusp states (as here at $L=50$, 54, 58 for $N=4$), the pair correlations show clearly the localization of the electrons in a square geometry, while the points in between these angular momenta reflect the properties of the two different vibrational states.

Finally, let us discuss how this relates to the fractional quantum Hall effect. Laughlin [18] showed already in his pioneering work that the maximum amplitude of the many-electron state in the fractional QHL

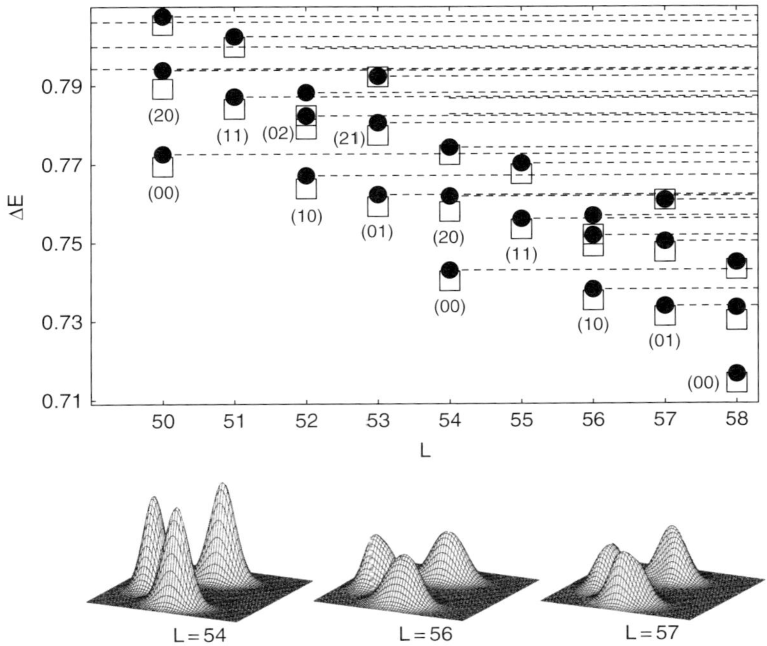

FIGURE 18 Many-particle energy spectrum (the interaction energy) for four electrons in a harmonic confinement as a function of the angular momentum. Solid points, exact diagonalization; squares, model Hamiltonian (the numbers indicate the vibrational state in question). The lower panel shows the pair correlation functions for some yrast states.

$$\Psi_q(z_1, z_2, \ldots, z_N) = \prod_{i<j}^{N} \left(z_i - z_j\right)^q e^{-\Sigma |z_k|^2/2l_h^2}, \tag{17}$$

is obtained when the electrons are localized at their classical equilibrium positions. In the wave function above, q is an odd integer (the filling fraction of the LLL is $\nu = 1/q$). The localization becomes more pronounced when q increases [110]. In the region where the true Wigner crystal is formed, the above wave function is not any more accurate. In small systems, however, already in the region of $\nu = 1/3$ ($q = 3$), the exact energy spectrum shows the periodic oscillation of the yrast spectrum caused by the electron localization (see Figure 17). The spectrum in Figure 18 is from the region $\nu \sim 1/9$.

The classical geometry of the localized electrons depends on their number [116]. Generally, the electrons tend to form concentric rings. Up to five electrons, they form a single ring, but for six electrons the ground state is a fivefold ring with one electron at the center, as schematically shown in Figure 19. This figure displays the classical equilibrium positions, for the example of 6 (upper panel) and 10 electrons (lower panel), respectively (after

FIGURE 19 Classical electron positions for small particle numbers, $N \leq 10$, in a parabolic well. After Bolton and Rössler [117].

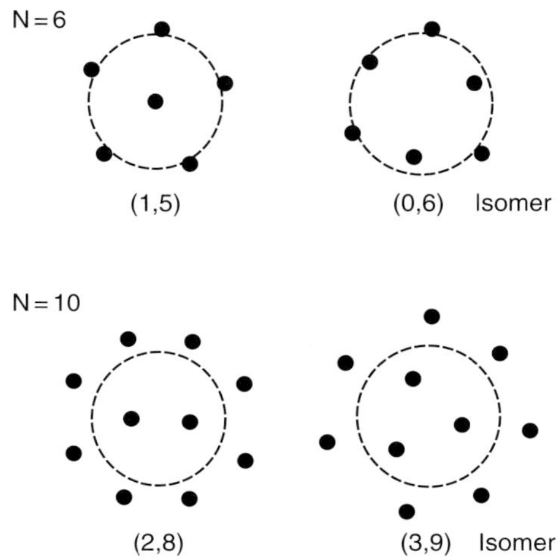

Bolton and Rössler [117]; see also the discussion in Ref. [4]). The localization caused by the highly rotational state is not limited to electrons in a harmonic confinement, but is a more general phenomenon to occur for all particles with long-range interactions. The reason behind is simple: at large angular momenta, the system can be described by the classical rotations and vibrations of Eq. (16). The different symmetry requirements can, however, select different allowed vibration modes for fermions and bosons (at a given angular momentum).

Figure 20 shows as an example, the interaction energy as a function of the angular momentum, for six bosons interacting by Coulomb repulsion. Subtracting a smooth function of angular momentum (third-order polynomial) from the yrast line, pronounced and regular oscillations of period $\Delta L = 5$ are visible in the large-L limit, originating from the localization into a fivefold ring with one boson at the center, just as for the fermion case discussed above. This localization is confirmed by the pair correlations, shown as an inset in the same figure. Reimann et al. [118] have furthermore shown that the energy spectra of small numbers of bosons and fermions are nearly identical at high angular momenta.

The Laughlin wave function, Eq. (17), is also applicable for bosons. In this case, naturally the exponent q needs to be even. This suggests that there is a relation between the boson and fermion wave functions. As the boson wave function is symmetric, a proper fermion wave function can be constructed by multiplying the boson wave function with the determinant $\prod(z_i - z_j)$. Indeed, for the wave functions at large angular momenta, this construction gives excellent approximations for the fermion wave functions. The overlap between this construction and the exact fermion wave function for four electrons at high angular momenta is typically 99 % [119]. Note, that this is not only true for the rigidly rotating states but also for states with internal vibrations.

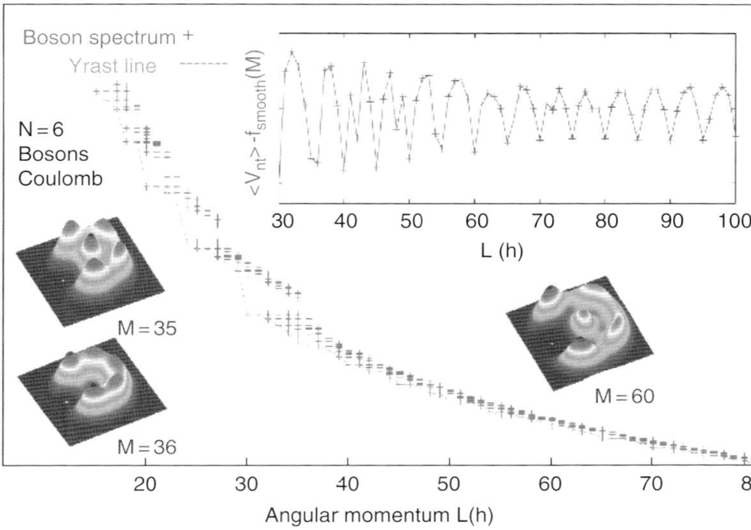

FIGURE 20 Interaction energy of $N=6$ bosons as a function of the angular momentum. The inset shows the yrast line with a smooth function of angular momentum (third-order polynomial) subtracted from the energies, in order to make the oscillations more visible. The large-L limit is dominated by a regular oscillation with $\Delta L=5$. The pair correlation functions to the left clearly demonstrate localization in Wigner molecule geometries at high angular momenta. While at smaller L-values, the (1,5) and the (0,6) configurations compete, at extreme angular momenta fivefold symmetry dominates.

Why does the rotational motion localize the particles in a harmonic confinement? In the case of electrons with long-range Coulomb interactions one could think that this is caused by Wigner crystallization. When the angular momentum increases, the electron cloud expands because of the centrifugal force and eventually, Wigner crystallization sets in. However, we have already seen that the external magnetic field increases the strength of the confinement. In fact, the average electron density remains essentially constant when localization occurs.

D. Vortices in Polarized Fermion Systems

Vortex formation in type-II superconductors is a well-known phenomenon [120]. When the magnetic field increases, at the first critical field strength at a given temperature, vortices penetrate the superconductor forming a regular triangular lattice. Similarly, in rotating ^{3}He, vortex formation has been observed by optical measurements [121]. Vortex formation in rotating systems has been considered as a definite signature of superfluidity.

In the case of semiconductor quantum dots, vortex formation was discussed theoretically by Saarikoski et al. [107] using the current-density functional formalism. Later, this was confirmed by exact diagonalization calculations [122,123]. The vortices appeared when the magnetic field was increased beyond formation of the MDD, but at field strengths below those where the fractional QHL occurs.

Just as for localization of the electrons, as discussed above, in mean-field theory the vortices are visible directly as distinct minima in the total electron density, with the electron current showing circulation of the vortex core, as illustrated in Figure 21. The vortices seem to localize a in a regular "molecule," with geometries resembling those observed for the finite-size Wigner crystallites discussed above.

To analyze the vortex solutions gained by the exact diagonalization method is not an easy task [124]. Naturally, for the exact solution of the many-body Hamiltonian, the total density is circularly symmetric and one has to study correlation functions—just as explained above for the case of Wigner localization. Figure 22 shows the electron–electron pair correlation for 36 electrons at a highly rotational state. Clearly, in addition to the exchange-correlation hole around the reference electron, there are four distinct minima in the pair correlation. These are four localized vortices—the reference electron pins their position, making them visible.

Other ways to observe the internal symmetry-breaking in the exact diagonalization study are to break the circular symmetry, for example by an ellipsoidal confinement [125,126], or by using perturbation theory [122]. The electron density at the vortex core is zero, and the phase of the wave function changes by 2π when a coordinate is rotated around the vortex core. In the case of the many-particle wave functions, these characteristics are difficult to use. It was suggested by Saarikoski et al. [107] to determine the phase change of the many-particle wave function by by fixing the positions of $N-1$ coordinates when the Nth coordinate is rotated around the vortices (fixing the other coordinates fixes also the positions of the vortices). The phase maps created in this way [107,122] show that in addition to the "free" vortices there

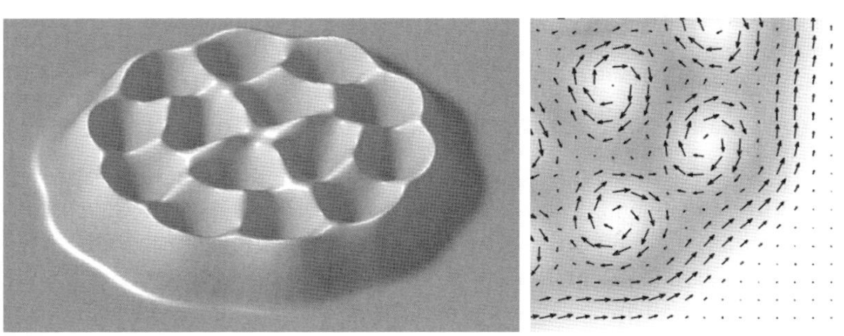

FIGURE 21 Electron density of a 24-electron quantum dot showing 14 vortices (left) and the corresponding currents (right). Results from a current-spin-density functional calculation by Saarikoski et al. From Ref. [107].

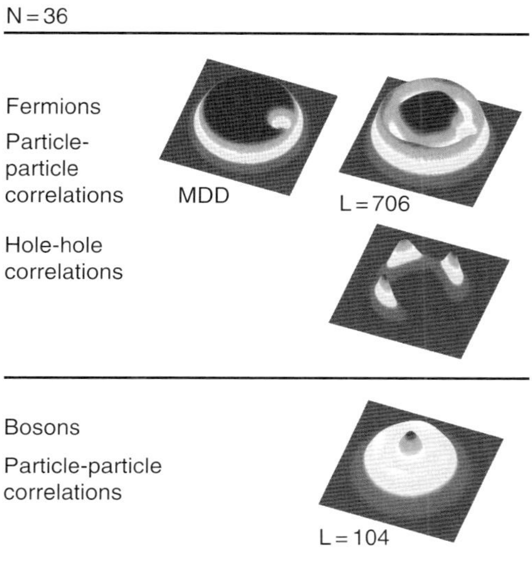

FIGURE 22 Pair correlation functions calculated for 36 electrons. The *upper panel* shows the electron–electron correlations for the MMD, $L=630$ (*left*), for particles at $L=706$ showing four vortices (*right*), and for holes at the same angular momentum (*lower right*). The *lower panel* shows the corresponding correlation function for a bosonic four-vortex state at angular momentum $L=104$ (note the absence of the exchange hole in the bosonic case.) From Ref. [124].

is one vortex attached to each electron. In the language of the QHL, each electron carries a flux quantum (in the case of fractional QHL with filling factor $\nu=1/3$, each electron carries three flux quanta). Electrons with attached flux quanta (or vortices) are also called composite fermions [127].

In the polarized case, there is a simple way to understand the occurrence of free vortices. They are *holes* in the otherwise filled Fermi sea, that is, holes in the MDD, where all states up to the single-particle angular momentum $L_{\text{MDD}}=N(N-1)/2$ are filled. When the angular momentum is increased, we create holes (missing electrons) corresponding to small angular momenta relative to L_{MDD}. Formally, we can define the creation (annihilation) operator of a hole as $d^+ = c$ $(d = c^+)$ and write the Hamiltonian Eq. (15) in terms of these,

$$\begin{aligned} H = {} & \sum_i m_i \hbar\omega_0 \left(1 - d_i^+ d_i\right) + 2\sum_{i,j,k} \left(V_{ijkj} - V_{ijjk}\right) d_k^+ d_i \\ & + \sum_{i,j,k,l} V_{ijkl} d_l^+ d_k^+ d_j d_i + \text{constant}. \end{aligned} \tag{18}$$

Note that the interactions between the holes are the same as those between the particles, but the second term means that the particles do not any longer move in a strictly harmonic confinement. Naturally, the solution of this Hamiltonian leads to an equivalent result as that of the original Hamiltonian, requiring the same computational effort.

We will now show that—even within a limited single-particle space—the holes localize to a Wigner molecule. Let us consider n holes in a system with N electrons and restrict the single-particle basis to its minimum possible value in the LLL; that is, the maximum single-particle angular momentum is $l_{\max} = N + n - 1$. When the angular momentum of the electron system is $L_c = L_{MDD} + \Delta L$, the angular momentum of the system of holes is $L_h = (N + n)(N + n - 1) - L_e = 2nN + n(n - 1) - \Delta N$. For example, for the $N = 4$ particle system (cf. Figure 22), $L_h = 66$ corresponds to such high angular momentum that the quasi-particles (now holes) are well localized.

This localization suggests that the excitation spectrum can be determined from the "classical" rotations and vibrations, resulting in similar periodic oscillations as found above for localized electrons at high angular momenta. This indeed is the case, as shown by Manninen et al. [123,124]. Figure 23 shows the energy spectrum for 20 electrons as a function of angular momentum. The yrast line shows oscillations, in the beginning with period 2, followed by oscillations of period 3 and then period 4 (in units of angular momentum). These regions correspond to the formation of two, three, and four vortices, respectively. This means that the main features of the many-particle spectrum at these angular momenta are determined by the rotation–vibration spectrum of localized vortices. In Section VI.E, we will see that similar oscillations reveal the existence of vortices in rotating boson systems.

Like for fermions, also in the bosonic case, the structure of the bosonic wave function can be understood in terms of Laughlin-type wave functions [122,124]. The simplest Ansatz for the single vortex at the center is the Bertsch–Papenbrock [128] wave function

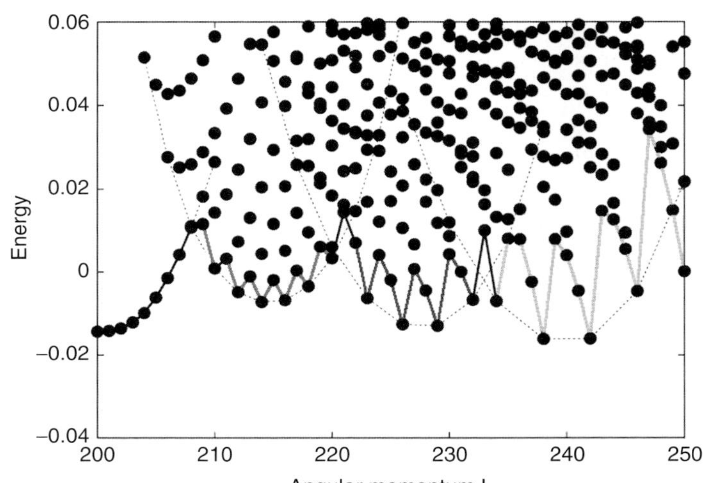

FIGURE 23 Energy spectrum as function of the total angular momentum for 20 electrons. A smooth function is subtracted from the total energy to show the oscillations of the yrast line (thick line). The thin lines show the lowest energy states with 1, 2, 3, and 4 vortices.

$$\Psi_{1v} = \prod_i (z_i - z_0)\Psi_{\mathrm{MDD}}, \tag{19}$$

where $z_0 = \sum z_i/N$ is the center-of-mass coordinate. This wave function is a good approximation for the wave function calculated using only the LLL. For example, for 10 electrons the overlap between these two states is $|\langle \Psi_{1v}|\Psi_{\mathrm{exact}}\rangle|^2 = 0.90$. In a large quantum dot, the center-of-mass can be approximated as fixed at the origin. Similarly, having n vortices in a ring, we can approximate the wave function as:

$$\begin{aligned}\Psi_{kV} &= \prod_{j_1}^{N}\left(z_{j_1} - ae^{i\alpha_1}\right) \times \cdots \times \prod_{j_k}^{N}\left(z_{j_k} - ae^{i\alpha_k}\right)\Psi_{\mathrm{MDD}} \\ &= \prod_{j}^{N}\left(z_j^k - a^n\right)\Psi_{\mathrm{MDD}},\end{aligned} \tag{20}$$

where k is the number of vortices, a is the distance of the vortices from the origin and $\alpha_j = 2\pi j/k$. Clearly, the above wave function does not have a good angular momentum. Projecting to good angular momentum means collecting out states with a given power of a. We obtain a state

$$\Psi_{kV} = a^{k(N-K)}\mathcal{S}\left(\prod_j^K z_j^k\right)\Psi_{\mathrm{MDD}} \tag{21}$$

which now corresponds to a good angular momentum $M = M_{\mathrm{MDD}} + kK$ (here, $\mathcal{S}$ symmetrizes the polynomial). The above wave function corresponds to the most important configuration of the exact wave function: the n holes are next to each other in consecutive angular momenta. Toreblad et al. [122] called this state a "vortex-generating configuration." (However, the wave function (21) does not localize the vortices but rather keeps them delocalized at a distance a from the origin).

E. Vortices in Rotating Bose Systems

The observation of Bose–Einstein condensation in atomic traps once again increased the interest in the many-particle physics of the harmonic potential (for a review, see Ref. [129]). The experimental observation of vortex lattices in rotating systems was a further milestone. By external fields, the trap can be made to be 3D or quasi-2D. In a highly rotational state, the cloud of atoms forms a (quasi-2D) disc, with the effective confining potential in the rotating frame being

$$V_{\mathrm{ext}}(r,z) = \frac{1}{2}m\omega_{\mathrm{eff}}^2 r^2 + \frac{1}{2}m\omega_0^2 z = \frac{1}{2}m\left(\omega_0^2 - \omega_{\mathrm{r}}^2\right)r^2 + \frac{1}{2}m\omega_0^2 z, \tag{22}$$

where ω_{r} is the angular velocity of the rotation. For large enough ω_{r}, only the lowest energy state along the z-direction is occupied (Note that the rotation velocity cannot exceed ω_0.). We can then approximate the rotating Bose system as particles confined in a 2D harmonic trap, and directly compare to fermions confined in a quantum dot.

The interaction between the atoms in the dilute condensate consists of individual scattering events which are described by the scattering length. The *contact* interaction, being a standard model interaction for cold atom gases, is written as $v(\mathbf{r}_i - \mathbf{r}_j) = g\delta(\mathbf{r}_i - \mathbf{r}_j)$, where $g = 4\pi a_s \hbar^2/m$, a_s being the scattering length (for s-wave scattering). In the dilute gas, the total energy per particle is proportional to the density. Consequently, in the LDA, the effective potential will also be proportional to the density. In a Bose system at zero temperature, all particles are in the same quantum state, and the density is simply $\rho(\mathbf{r} = |\psi(\mathbf{r})|^2$, where the single-particle wave function ψ is the solution of the so-called Gross–Pitaevskii equation

$$-\frac{\hbar^2}{m}\nabla^2\psi(\mathbf{r}) + \frac{1}{2}m\omega_0^2 r^2 + g\left|\psi\left(\mathbf{r}\right)\right|^2\psi(\mathbf{r}) = \varepsilon\psi(\mathbf{r}), \tag{23}$$

which is a mean-field equation in close correspondence to the Kohn–Sham LDA equations for the electron system [130]. The nonlinearity of the equation makes symmetry-breaking possible. Indeed, for a rotating Bose gas, the equation has solutions showing vortex patterns very similar to the ones discussed above for the fermion case [131,132]. Figure 24 shows the boson spectra, with oscillations in the yrast line resembling to vortex structures as discussed above in the fermion case.

For small numbers of bosons in a harmonic potential, the problem can be solved exactly. In the case of weak interactions between the bosons, the basis set can be restricted to the LLL. The only difference to the fermion system discussed above is that now the wave function has to be symmetric. For contact interactions, it has been shown that the Bertsch–Papenbrock Ansatz

$$\Psi_{\mathrm{BP}} = \prod_i (z_i - z_0)\mathrm{e}^{-\Sigma|z_k|^2/2l_0^2} = \prod_i (z_i - z_0)\Psi_{\mathrm{BEC}} \tag{24}$$

is *exact* for the state with a single vortex at the center (ℓ_0 is now the oscillator length of the pure confinement). This state has total angular momentum $L = N$. Increasing the angular momentum creates more vortices. A second vortex appears at $L = 1.7N$, the third at $L = 2.1N$, and the fourth at $L = 2.8N$ [132]. An approximation for the n vortices in a ring is again the vortex-generating state, Eq. (20), where now the fermion MDD is replaced by the ground state of the BEC. But again, as for fermions, the exact solution is more complicated and supports vortex localization in a much more effective way.

For bosons, the localization of vortices is not as easily seen in the pair correlations as for fermions—mainly, because the bosonic occupancies in the Fock states make it difficult to interpret the vortices directly as unoccupied states or holes in the MDD (as in the fermion case), as the

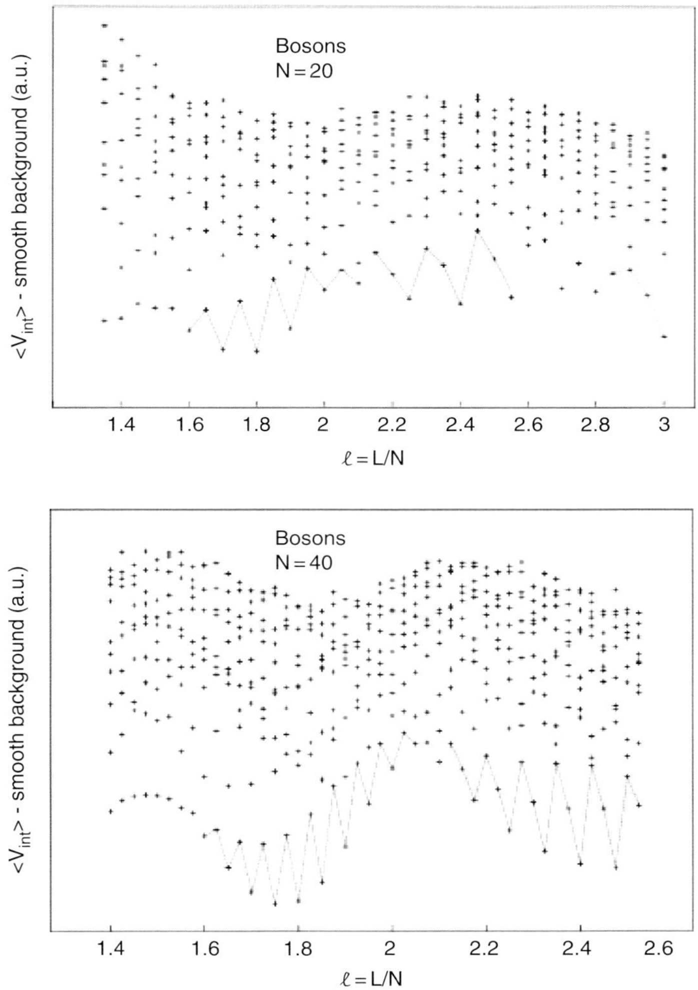

FIGURE 24 Energy spectrum as function of the total angular momentum for $N=20$ bosons (upper panel) and $N=40$ bosons (lower panel), with Coulomb interactions (L is the angular momentum and N is the number of bosons). A smooth function is subtracted from the total energy to show the oscillations of the yrast line (thick line).

occupation number is not limited. Nevertheless, for small particle numbers, we can transform the boson wave function to its fermionic equivalent, by multiplying it with the determinant $\prod(z_i - z_j)$. Figure 25 shows the vortex–vortex pair correlations determined in this way, for $N = 20$ bosons with three vortices ($L=48$). For comparison, the corresponding fermion state

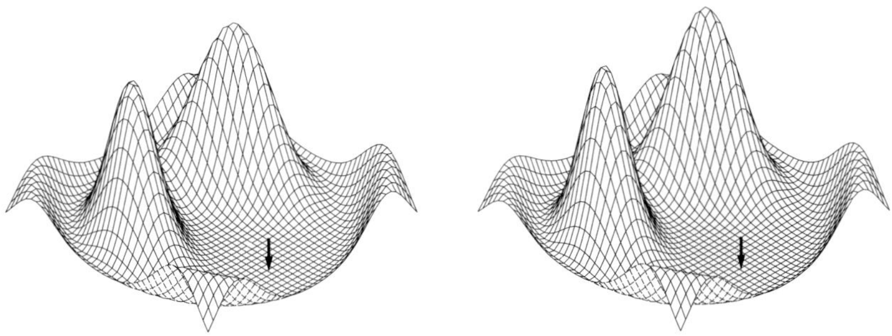

FIGURE 25 Vortex–vortex correlation functions for three vortices in a fermion and boson system with $N=20$ particles (the difference in total angular momentum is due to the MDD in the fermion case, with $L_{\mathrm{MDD}}=N(N-1)/2$). The boson wave function was first transformed to fermion Fock states, as described in the text. The arrows show the site of the reference vortex.

(which in this case is not the ground state) is shown as well ($L=328$). The correlation functions appear surprisingly similar.

The above analysis showed very clearly that vortex localization occurs both in the bosonic and the fermionic case, and is mapped out very directly by studying the corresponding correlation functions. The rotational spectra confirmed-this observation.

Figure 24 shows the energetically low-lying many-body energies for $N=20$ and 40 bosons, respectively. As for fermions, we observe the oscillatory behavior of the yrast line. We saw above that the period of the oscillations corresponds to the number of the first few vortices that localize on a ring. Note that, on the horizontal axis, we have now given the ratio L/N in order to demonstrate that for bosons, the regions for different vortex numbers depend only on L/N. This is not the case in fermion systems [124], where for larger systems with $N \geq 14$, the vortices appear closer to the surface of the MDD, leaving its center unaffected.

Finally, we mention the possibility of vortex formation in boson and fermion systems where the particles have an internal degree of freedom, such as spin or pseudospin. In this case, the single-component vortex patterns are still observed; however, they are not any longer lowest-energy excitations. This holds for fermions [133] as well as for bosons.

Concluding our discussion of vortices in harmonically confined quantum systems that are set rotating, we should emphasize that the vortex formation gives characteristic oscillations in the yrast spectrum [124]. The low-energy states of the rotational spectrum are determined by the rigid rotation and vibrational states of Wigner molecules of vortices [123]. The vortex formation is similar for bosons and fermions and it is nearly independent of the form of the repulsive interparticle interaction [122,134].

VII. 1D SYSTEMS

A. 1D Harmonic Oscillator

Let us finally discuss interacting electrons confined by a 1D harmonic oscillator, as well as a quasi-1D quantum ring. In an anisotropic oscillator, $V_{\text{ext}} = (1/2)m(\omega_x^2x^2 + \omega_y^2y^2 + \omega_z^2z^2)$, choosing the frequencies ω_y and ω_z of two spatial directions is so large that the particles occupy only the lowest state in the perpendicular direction, and the system becomes effectively 1D.

The 1D system of fermions is very different from the 2D and 3D cases. The exchange interaction, or the Pauli exclusion principle, becomes dominating. As two electrons with the same spin cannot be in the same place, in 1D this means that electrons with the same spin cannot pass each other. This enhances drastically the tendency to form a spin-density wave. In fact, an infinite 1D electron gas is unstable against the so-called spin-Peierls transition: a static spin density makes a spin-dependent mean-field potential (e.g., LSDA) with a wave length of $\pi/2k_{\text{F}}$ and consequently opens a gap at the Fermi level (remember that the Fermi surface consists of only two points in 1D). Figure 26 shows the result of an LSDA calculation for 12 electrons in a quasi-1D harmonic potential, showing very clearly the resulting SDW. The total density shows 12 maxima corresponding to

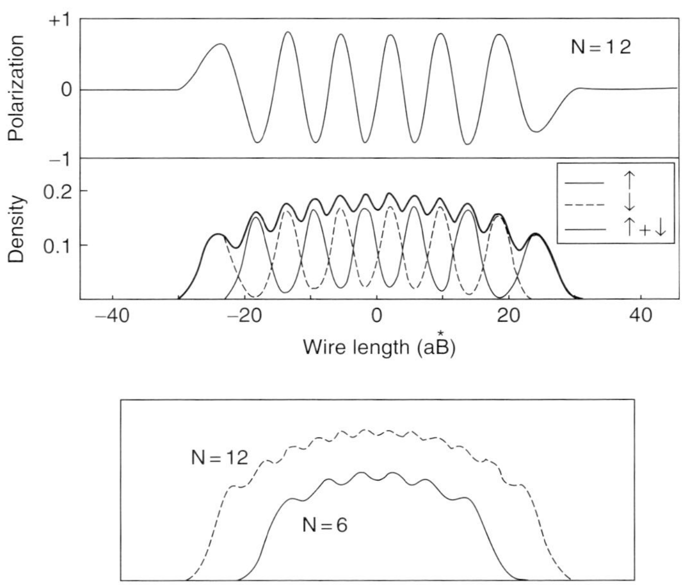

FIGURE 26 Upper, spin polarization, as well as spin- and total-electron densities at the center of a quasi-1D wire with 12 electrons, calculated with the LSDA; lower, electron densities for $N = 6$ and $N = 12$ noninteracting spinless fermions in a 1D harmonic oscillator.

"localized" electrons forming an antiferromagnetic chain. Even in 1D, the LSDA cannot properly localize the electrons.

It is interesting to compare the self-consistent electron density to that of noninteracting electrons, shown also in Figure 26. The density for 12 *spinless* electrons is quite similar to the LSDA result, while the density of 12 electrons with spin has only six maxima as each single-particle state now occupies two electrons. However, the similarity of the LSDA density to that of the noninteracting spinless electrons does not reach to the individual single-particle wave functions. Figure 27 shows the densities of the single-particle wave functions of the LSDA calculation. Interestingly, the last occupied state $i=6$ is localized at the end of the electron cloud. This *end state* is related to the surface states in a metal surface. The existence of the periodic potential which ends at the surface makes localized states possible [41]. In our 1D case, the periodic potential is provided by the spin-Peierls transition and the static SDW.

B. Quantum Rings

The observations of Aharonov–Bohm oscillations [135] and persistent currents [136] have made quasi-1D quantum rings a playground for simple theories. Indeed, the one-dimensionality as such gives a multitude of interesting properties [87,137]. Here, we will only study the spectral properties of

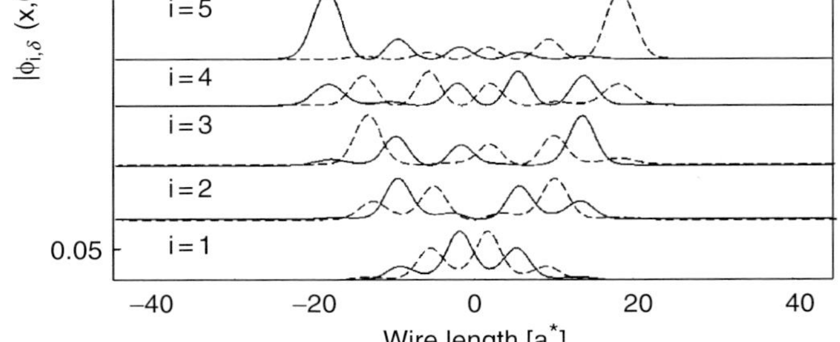

FIGURE 27 Single-particle (Kohn–Sham) densities and energy eigenvalues (inset) for a linear finite wire with 12 electrons. Note that the last occupied state ($i=6$) is localized at either end of the wire. The energy gap between occupied and unoccupied states is denoted by Δ in the inset. From Ref. [64].

finite rings as they are directly related to what we discussed earlier in connection to rotational states in a 2D harmonic confinement.

In the strictly 1D case, the single-particle eigenvalues are $\varepsilon_l = \hbar^2 l^2/2mR^2$, where R is the radius of the ring and l the angular momentum eigenvalue. The corresponding single-particle states are $\psi(\phi) = \exp(il\phi)$. The total angular momentum and energy for noninteracting particles are as follows:

$$L = \sum_i^N l_i, \quad E = \sum_i^N \varepsilon_{l_i}. \tag{25}$$

Let us first consider noninteracting polarized (spinless) fermions. It is easy to determine their energy as a function of the total angular momentum using Eq. (25). The results are shown for $N = 8$ fermions in Figure 28 as black dots. The yrast line shows a period of eight, suggesting that the electrons are localized in an octagon, the downward cusps corresponding to purely rotational states of the octagon. The black dots correspond to internal vibrations of the Wigner molecule. The fact that *noninteracting polarized electrons form a Wigner molecule* is a special property of 1D systems. It can be shown that particles interacting with $1/r^2$ interaction in a 1D ring have the same energy spectrum as noninteracting particles (or particles interacting with an infinitely strong interaction of delta-function type) [87]. Figure 28 shows (as open circles) also the classically determined energies

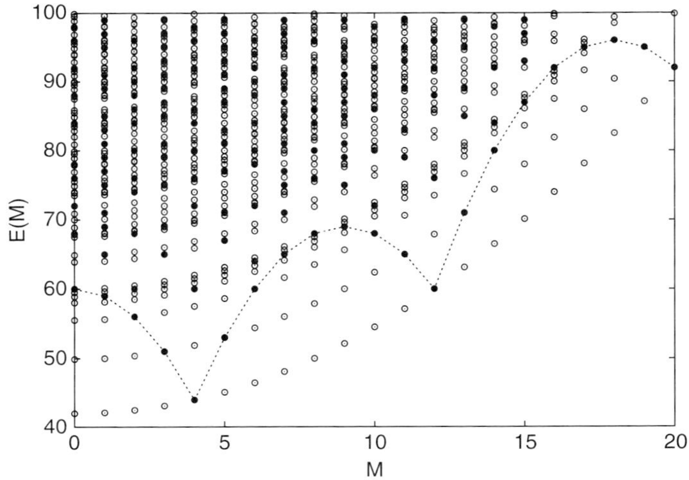

FIGURE 28 Many-particle energy spectrum of eight noninteracting polarized electrons in a strictly one-dimensional quantum ring (black dots) compared to the rotation–vibration spectrum of classical particles interacting with $1/r^2$ interaction (open circles). The dotted curve shows the yrast line of the polarized electrons. From Ref. [87].

$$E = E_{\rm rot} + E_{\rm vib} = \frac{\hbar^2 L^2}{2NmR^2} + \sum_{\nu} n_\nu \hbar\omega_\nu \tag{26}$$

for the $1/r^2$ interaction. Each vibrational level forms a rotational band. We can see that the spectrum of noninteracting polarized fermions (black dots) consists only of points at the classical energies. In fact, for electron with spin and infinite strong delta-function interaction ($v(r) = A\delta(r)$, where $A \to \infty$) one obtains all the classical points (open circles).

The reason why noninteracting spinless electrons localize and have vibrational modes simply follows from the fact that the electrons cannot pass each other. If the electron–electron distance is d, each electron is then localized between its neighbors in a region $2d$. Its kinetic energy will then be proportional to $1/d^2$. This effectively leads to a $1/r^2$ interaction between the electrons.

Interacting electrons in 1D systems have been extensively studied using the Hubbard model (for reviews see Refs. [87,137]). The energy spectrum can be solved exactly using the Bethe Ansatz [138]. There, several analytic results exist. For a half-filled Hubbard band (with one electron per site), it is rather easy to show that with the large U-limit the Hubbard model becomes an antiferromagnetic Heisenberg model. However, the Heisenberg model seems to be a good approximation also for small filling [87,139]. This is important, as the low-filling limit of the Hubbard model approaches to free electrons with delta interaction (this is the same as the tight-binding model approaching the free electron model at the bottom of the band [140]). Thus, free electrons with spin also localize in antiferromagnetic order as long as they have strong enough repulsive interactions between them.

Koskinen et al. [141] performed exact diagonalization calculations for electrons confined in a quasi-1D ring described with the external 2D potential $V_{\rm ext} = m\omega_0^2(r - r_0)^2/2$, where $r = (x, y)$. The rotational spectrum for six particles is shown in Figure 29 for two different values of the narrowness of the ring. The upper panel corresponds to a very narrow ring. In this case, the different vibrational bands are clearly separated and correspond quantitatively to the energies determined by solving the vibrational frequencies of the classical linear chain of electrons on the ring. The lower panel shows the result obtained for a wider, less 1D ring. In this case, only the vibrational ground state is clearly separated, with the different spin states separating in energy. With high accuracy, these different spin states correspond to those of an antiferromagnetic Heisenberg model for six electrons on a ring [141].

In narrow quantum rings, the rotational spectrum is very robust. It is insensitive to the interparticle interaction or the specific model for the confinement. Even the discrete Hubbard model gives similar results as the continuum approaches [87]. However, this demonstrates once more that the clearest indication of Wigner molecules in the ground states of high-symmetry systems can be obtained by analyzing the rotation–vibration spectrum.

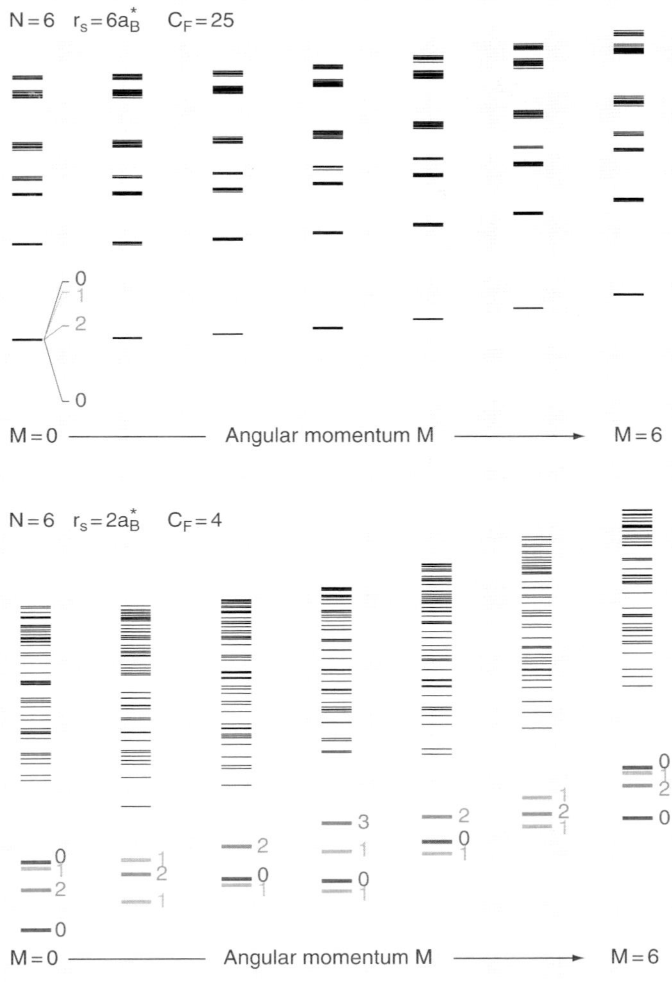

FIGURE 29 Energy spectra for two quasi-one-dimensional continuum rings with six electrons (in zero magnetic field). The upper panel is for a narrow ring and it shows several vibrational bands. The lower panel is for a wider ring which shows stronger separation of energy levels corresponding to different spin states (shown as numbers next to the energy levels). Note that also the narrow ring has the same spin-ordering of the nearly degenerate state as expanded for the lowest $L=0$ state.

VIII. CONCLUDING REMARKS

In this short review, we summarized some characteristic aspects of finite quantal systems that have their origin in the quantized level structure in—and beyond—the mesoscopic regime. We discussed shell structure and deformation, as well as the occurrence of Hund's rule in finite fermion systems, conjointly for metallic clusters, quantum dots in semiconductor heterostructures, or cold atoms in traps—seemingly different, but nevertheless in many aspects rather similar quantum systems. Like for atoms, shell structure does not only determine the stability and chemical inertness of metallic clusters, but also determines the conductance of a small quantum dot—both close to, and far away from equilibrium.

The experimental realization of Bose–Einstein condensation in an atomic gas [142–145] opened up a whole new research field on ultra-cold atoms and coherent matter. In a cloud of bosonic atoms that is set rotating, vortices may form. We discussed the fact that this vortex formation is not unique for bosonic systems, but may occur in a very similar way for (nonpaired) fermions under rotation, showing many analogies to the physics of the quantum Hall effect. Extreme rotation causes strong correlations, and the system is formally equivalent to charged particles in a strong magnetic field. We finally gave a short summary of the physics of a finite fermionic system in quasi-one dimension.

As a final remark, we wish to emphasize that the many analogies existing between nanostructures such as quantum dots and quantum wires, and cold atom gases will become more important in the future—last but not least because of the fact that these systems can be built much more "clean," and thus more coherent, than their semiconductor counterparts. An example for the cross-fertilization between these different sub-fields of physics is the recently discussed possibility of *van der Waals blockade* [146], which is expected to play a key role in transport experiments on confined cold atoms, and in *atomtronic* devices [147].

Acknowledgments

This work was financially supported by the Swedish Research Council and the Swedish Foundation for Strategic Research, as well as the European Community Project ULTRA-1D (NMP4-CT-2003-505457). We thank J. Akola, M. Borgh, P. Singha Deo, H. Häkkinen, G. Kavoulakis, M. Koskinen, P. Lipas, B. Mottelson, P. Nikkarila, M. Toreblad, S. Viefers, and Y. Yu for their collaboration on the subjects discussed in this review.

REFERENCES

[1] Bohr Å, Mottelson BR. Nuclear structure. London: W.A. Benjamin; 1975.
[2] de Heer W. Rev Mod Phys 1993;65:611.
[3] Tarucha S, Austing DG, Honda T, van der Haage RJ, Kouwenhoven L. Phys Rev Lett 1996;77:3613.
[4] Reimann SM, Manninen M. Rev Mod Phys 2002;74:1283.

[5] Akola J, Manninen M, Häkkinen H, Landman U, Li X, Wang L-S. Phys Rev B 1999;60: R11297.
[6] Brack M. Rev Mod Phys 1993;65:677.
[7] Talmi I. Simple models for complex nuclei. Chur: Harwood; 1993.
[8] Pfannkuche D, Gerhardts RR. Phys Rev B 1991;44:13132.
[9] Wixforth A, Kaloudis M, Rocke C, Ensslin K, Sundaram M, English JH, et al. Semicond Sci Technol 1994;9:215.
[10] Hawrylak P, Schulz PA, Palacios JJ. Solid State Commun 1995;93:909.
[11] Darnhofer T, Rössler U, Broido DA. Phys Rev B 1995;52:14379.
[12] Heitmann D, Bollweg K, Gudmundsson V, Kurth T, Riege SP. Phys E 1997;1:204.
[13] Kohn W. Phys Rep 1961;123:1242.
[14] Frauendorf S, Pashkevich VV. Z Phys D 1993;26:98.
[15] Fock V. Z Phys 1928;47:446.
[16] Darwin CG. Proc Cambridge Philos Soc 1930;27:86.
[17] Landau L. Z Phys 1930;64:629.
[18] Laughlin RB. Phys Rev B 1983;27:3383.
[19] Lang ND. Solid State Phys 1973;28:225.
[20] Lang ND, Kohn W. Phys Rev B 1970;1:4555.
[21] Manninen M, Nieminen R, Hautojärvi P, Arponen A. Phys Rev B 1975;12:4012.
[22] Puska MJ, Nieminen RM, Manninen M. Phys Rev B 1981;24:3037.
[23] Martins JL, Car R, Buttet J. Surf Sci 1981;106:265.
[24] Monnier R, Perdew JP. Phys Rev B 1978;17:2595.
[25] Parr RG, Yang W. Density-functional theory of atoms and molecules. Oxford: Oxford University Press; 1989.
[26] Ekardt W. Phys Rev Lett 1984;52:1925.
[27] Yannouleas C. Phys Rev B 1998;58:6748.
[28] Koskinen M, Lipas PO, Manninen M. Phys Rev B 1994;49:8418.
[29] Yannouleas C, Broglia RA, Brack M, Bortignon PF. Phys Rev Lett 1989;63:255.
[30] Manninen M, Nieminen R. J Phys F 1978;8:2243.
[31] Hintermann A, Manninen M. Phys Rev B 1983;27:7262.
[32] Ekardt W. Phys Rev B 1984;29:1558.
[33] Knight WD, Clemenger K, de Heer WA, Saunders WA, Chou MY, Cohen ML. Phys Rev Lett 1984;52:2141.
[34] Geguzin II. Sov Phys Solid State 1982;24:248.
[35] Mottelson B. Nucl Phys A 1994;574:365c.
[36] Balian R, Bloch C. Ann Phys 1970;60:401.
[37] Nishioka H, Hansen K, Mottelson B. Phys Rev B 1990;42:9377.
[38] Pedersen J, Bjørnholm S, Borgreen J, Hansen K, Martin TP, Rasmunssen HD. Nature 1991;353:733.
[39] Walter M, Häkkinen H. Phys Chem Chem Phys 2006;8:5407.
[40] Martins J, Car R, Buttet J. Surf Sci 1981;106:261.
[41] Zangwill A. Physics at surfaces. Cambridge: Cambridge University Press; 1988.
[42] Martin TP. Phys Rep 1996;273:201.
[43] Baletto F, Ferrando R. Rev Mod Phys 2005;77:371.
[44] Bjørnholm S, Borggreen J. Philos Mag B 1999;79:1321.
[45] Chandezon F, Bjørnholm S, Borggreen J, Hansen K. Phys Rev B 1997;55:5485.
[46] Martin TP, Näher U, Schaber H, Zimmermann U. J Chem Phys 1994;100:2322.
[47] Clemenger K. Phys Rev B 1985;32:1359.

[48] Brack M, Damgaard J, Jensen AS, Pauli HC, Strutinsky VM, Wong CY. Rev Mod Phys 1972;44:320.
[49] Reimann SM, Brack M, Hansen K. Z Phys D 1993;28:235.
[50] Yannouleas C, Landman U. Phys Rev B 1993;48:8376.
[51] Reimann SM, Frauendorf S, Brack M. Z Phys D 1995;34:125.
[52] Reimann SM, Frauendorf S. Surf Rev Lett 1996;3:25.
[53] Penzar Z, Ekardt W. Z Phys D 1990;17:69.
[54] Hirschmann T, Brack M, Meyer J. Ann Phys 1994;3:336.
[55] Manninen M. Phys Rev B 1986;34:6886.
[56] Koskinen M, Lipas PO, Manninen M. Z Phys D 1995;35:285.
[57] Moseler M, Huber B, Hakkinen H, Landman U, Wrigge G, Hoffmann MA, et al. Phys Rev B 2003;68:165413.
[58] Häkkinen H, Kolehmainen J, Koskinen M, Lipas PO, Manninen M. Phys Rev Lett 1997;78:1034.
[59] Koskinen M, Lipas PO, Manninen M. Nucl Phys A 1995;591:421.
[60] Manninen M, Kolehmainen J, Koskinen M, Reimann SM. Phys Scr 2000;T88:90.
[61] Yoshida A, Dossing T, Manninen M. J Chem Phys 1994;101:3041.
[62] Kohl C, Montag B, Reinhard P-G. Z Phys D 1996;38:81.
[63] Kohl C, Reinhard P-G. Z Phys D 1997;39:225.
[64] Reimann SM, Koskinen M, Helgesson J, Lindelof PE, Manninen M. Phys Rev B 1998;58:8111.
[65] Reimann SM, Koskinen M, Häkkinen H, Lindelof PE, Manninen M. Phys Rev B 1997;56:12147.
[66] Hamamoto I, Mottelson BR, Xie H, Zhang XZ. Z Phys D 1991;21:163.
[67] Akola J, Rytkönen K, Manninen M. Eur Phys J D 2001;16:21.
[68] Röthlisberger U, Andreoni W. J Chem Phys 1991;94:8129.
[69] Li J, Li X, Zhai HJ, Wang LS. Science 2003;299:864.
[70] Diederich Th, Döppner T, Fennel Th, Tiggesbumker J, Meiwes-Broer K-H. Phys Rev A 2005;72:023203.
[71] Chiu Y-P, Huang L-W, Wei C-M, Chang C-S, Tsong T-T. Phys Rev Lett 2006;97:165504.
[72] Dudek J, Góźdź A, Schunck N, Miśkiewicz M. Phys Rev Lett 2002;88:252502.
[73] Dudek J, Curien C, Dubray N, Dobaczewski J, Pangon V, Olbratowski P, et al. Phys Rev Lett 2006;97:072501.
[74] Schunck N, Dudek J, Góźdź A, Regan PH. Phys Rev C 2004;69:061305(R).
[75] Chakraborty T. Quantum dots: a survey of the properties of artificial atoms. Amsterdam: North-Holland; 1999.
[76] Demel T, Heitmann D, Grambow P, Ploog K. Phys Rev Lett 1990;64:788.
[77] Tanatar B, Ceperley DM. Phys Rev B 1989;39:5005.
[78] Matagne P, Leburton JP, Austing DG, Tarucha S. Phys Rev B 2001;65:85325.
[79] Thelander C, Mårtensson T, Björk MT, Ohlsson BJ, Larsson MW, Wallenberg LR, et al. Appl Phys Lett 2003;83:2052.
[80] Fuhrer A, Fröberg LE, Pedersen JN, Larsson MW, Wacker A, Pistol M-E, et al. Nano Lett 2007;7:243.
[81] Koskinen M, Manninen M, Reiman SM. Phys Rev B 1997;56:12147.
[82] Borgh M, Toreblad M, Koskinen M, Manninen M, Åberg S, Reimann SM. Int J Quant Chem 2005;105:817.
[83] Attaccalite C, Moroni S, Gori-Giorgi P, Bachelet GB. Phys Rev Lett 2002;88:256601 and Erratum, Phys. Rev. Lett. 91, 109902 (2003).

[84] Reimann SM, Koskinen M, Manninen M, Mottelson B. Phys Rev Lett 1999;83:3270.
[85] Hirose K, Wingreen NS. Phys Rev B 1999;59:4604.
[86] Harju A, Räsänen E, Saarikoski H, Puska MJ, Nieminen RM, Niemälä K. Phys Rev 2004; B69:153101.
[87] Viefers S, Koskinen P, Deo PS, Manninen M. Phys E 2004;21:1.
[88] Wigner EP. Phys Rev 1934;46:1002.
[89] Ceperley DM, Alder BJ. Phys Rev Lett 1980;45:566.
[90] Creffield CE, Häusler W, Jefferson JH, Sarkar S. Phys Rev B 1999;59:10719.
[91] Egger R, Häusler W, Mak CH, Grabert H. Phys Rev Lett 1999;82:3320.
[92] Yannouleas C, Landman U. Phys Rev Lett 1999;82:5325.
[93] Müller HM, Koonin SE. Phys Rev B 1996;54:14532.
[94] Reimann SM, Koskinen M, Manninen M. Phys Rev B 2000;62:8108.
[95] Pfannkuche D, Gerhardts RR, Maksym PA, Gudmundsson V. Phys B 1993;189:6.
[96] Taut M. J Phys A 1994;27:1045.
[97] El-Said M. Phys Status Solidi B 1996;193:105.
[98] González A, Quiroga L, Rodriguez BA. Few-Body Syst 1996;21:47.
[99] Dineykhan M, Nazmitdinov RG. Phys Rev B 1997;55:13707.
[100] Zhub J-L, Zhub S, Zhuc Z, Kawazoec Y, Yaoc T. Phys Lett A 1998;246:157.
[101] Kärkkäinen K, Koskinen M, Reimann SM, Manninen M. Phys Rev B 2004;70:195310.
[102] Oosterkamp TH, Janssen JW, Kouwenhoven LP, Austing DG, Honda T, Tarucha S. Phys Rev Lett 1999;82:2931.
[103] Vignale G, Rasolt M. Phys Rev Lett 1987;59:2360.
[104] Capelle K, Vignale G. Phys Rev B 2002;65:113106.
[105] D'Agosta R, Vignale G. Phys Rev B 2005;71:245103.
[106] Koskinen M, Kolehmainen J, Reimann SM, Toivanen J, Manninen M. Eur J Phys D 1999;9:487.
[107] Saarikoski H, Harju A, Puska MJ, Nieminen RM. Phys Rev Lett 2004;93:116802.
[108] Goldmann E, Renn SR. Phys Rev B 1999;60:16611.
[109] Toreblad M, Yu Y, Reimann SM, Koskinen M, Manninen M. J Phys B Atom Mol Opt 2006;39:2721.
[110] Manninen M, Viefers S, Koskinen M, Reimann SM. Phys Rev B 2001;64:245322.
[111] Grover JR. Phys Rev 1967;157:832.
[112] Maksym PA. Phys Rev B 1996;53:10871.
[113] Maksym PA, Imamura H, Mallon GP, Aoki A. J Phys Condens Matter 2000;12:R299.
[114] Nikkarila J-P, Manninen M. Solid State Commun 2007;141:209.
[115] Tinkham M. Group theory and quantum mechanics. New York: McGraw-Hill; 1964.
[116] Bedanov VM, Peeters FM. Phys Rev B 1994;49:2667.
[117] Bolton F, Rößler U. Superlatt Microstr 1993;13:139.
[118] Reimann SM, Koskinen M, Yu Y, Manninen M. New J Phys 2006;8:59.
[119] Borgh M, Koskinen M, Christensson J, Manninen M, Reimann SM. Phys Rev A 2008;77:033615.
[120] Tinkham M. Introduction to superconductivity. Krieger Publishing Company; 1975.
[121] Manninen AJ, Pekola JP, Kira GM, Ruutu JP, Babkin AV, Alles H, et al. Phys Rev Lett 1992;69:2392.
[122] Toreblad M, Borgh M, Manninen M, Koskinen M, Reimann SM. Phys Rev Lett 2004;93:090407.
[123] Manninen M, Reimann SM, Koskinen M, Yu Y, Toreblad M. Phys Rev Lett 2005;94:106405.

[124] Reimann SM, Koskinen M, Yu Y, Manninen M. Phys Rev A 2006;74:043603.
[125] Manninen M, Koskinen M, Reimann SM, Mottelson B. Eur Phys J D 2001;16:381.
[126] Saarikoski H, Reimann SM, Räsänen E, Harju A, Puska MJ. Phys Rev B 2005;71:035421.
[127] Jain JK, Kamilla RK. In: Heinonen O, editor. Composite fermions: a unified view of the quantum Hall effect. River Edge, NJ: World Scientific; 1998. p. 1.
[128] Bertsch GF, Papenbrock T. Phys Rev Lett 1999;83:5412.
[129] Pethick CJ, Smith H. Bose–Einstein condensation in dilute gases. Cambridge: Cambridge University Press; 2002.
[130] Capelle K, Borgh M, Kärkkäinen K, Reimann SM. Phys Rev Lett 2007;99:010402.
[131] Butts DA, Rokshar DS. Nature (London) 1999;397:327.
[132] Kavoulakis GM, Mottelson B, Pethick CJ. Phys Rev A 2000;62:063605.
[133] Koskinen M, Reimann SM, Nikkarila J-P, Manninen M. J Phys Condens Matter 2007;19:076211.
[134] Vorov OK, Hussein MS, van Isacker P. Phys Rev Lett 2003;90:200402.
[135] Aharonov Y, Bohm D. Phys Rev 1959;115:485.
[136] Webb RA, Washburn S, Umbach CP, Laibowitz RB. Phys Rev Lett 1985;54:2696.
[137] Kolomeisky EB, Straley JP. Rev Mod Phys 1996;68:175.
[138] Lieb EH, Wu FY. Phys Rev Lett 1968;20:1445.
[139] Yu N, Fowler M. Phys Rev B 1992;45:11795.
[140] Manninen M, Mansikka-aho J, Hammarén E. Europhys Lett 1991;15:423.
[141] Koskinen M, Manninen M, Mottelson B, Reimann SM. Phys Rev B 2001;63:205323.
[142] Anderson MH, Ensher JR, Matthews MR, Wieman CE, Cornell EA. Science 1995;269:189.
[143] Davis KB, Mewes MO, Andrews MR, van Druten NJ, Drufee DS, Kurn DM, et al. Phys Rev Lett 1995;75:3969.
[144] Ketterle W. Rev Mod Phys 2002;74:1131.
[145] Cornell EA, Wieman CE. Rev Mod Phys 2002;74:875.
[146] Capelle K, Borgh M, Kärkkäinen K, Reimann SM. Phys Rev Lett 2007;99:010402.
[147] Seaman BT, Krämer M, Anderson DZ, Holland MJ. Phys Rev A 2006;75:023615.
[148] Yannouleas C, Landman U. Proc Natl Acad Sci USA 2006;103:10600.

Chapter 13

Tailoring Functionality of Clusters and Their Complexes with Biomolecules by Size, Structures, and Lasers

Vlasta Bonačić-Koutecký, Roland Mitrić, Christian Bürgel and Jens Petersen
Institut für Chemie, Humboldt-Universität zu Berlin, Berlin, Germany

Chapter Outline Head

I. INTRODUCTION

Functionality of clusters and their complexes is closely related to their size and structure selective ground state properties, which is particularly attractive in the context of nanostructured materials. Moreover, important aspects of

Nanoclusters. DOI: 10.1016/S1875-4023(10)01013-2

functionality involve size and structure dependent absorption properties of clusters in different environments as well as reaction dynamics over excited and ground electronic states which can be often influenced or even controlled by tailored laser fields. In this contribution, we wish to address structural, electronic, and dynamical aspects.

We first present optical properties of silver clusters at support (e.g., MgO) and optical properties of hybrid silver-biomolecule (e.g., tryptophan) systems. The aim is to unravel the mechanism responsible for interactions among excitations between different subunits and to use the fundamentals gained from this study to propose size and structure selective emissive centers which might be useful in the context of optical storage and for biosensing, respectively.

Furthermore, we propose to introduce intrinsic dynamical properties as a new criterion for promoting reactivity of small size noble metal reactive centers relevant for heterogenous catalysis. We show that the different nature of internal vibrational redistribution (IVR) is responsible for significantly different sticking probability of O_2 to gold and to silver clusters for the given cluster size, allowing the reaction with CO to occur only in the case of gold and not in the case of silver.

All these novel properties arise in the nonscalable size regime in which each atom counts, which means that not only structural but also dynamical properties drastically change on adding or removing one single atom. This will be illustrated on strongly size selective photoinduced processes in small anionic gold clusters.

Finally, we show that tailored laser pulses can be used to drive isomerization processes in complex systems. This has been demonstrated on two prototype examples representing rigid molecules (Na_3F cluster) and floppy molecules (glycine). From the shapes of the optimized pulses which are very different for these two classes of systems, the underlying dynamical processes can be revealed, allowing us to use control as a tool for analysis.

For all above topics, it will be demonstrated that the role of theory is not only to provide interpretation of the experimental results but also to establish conditions under which experimental observation of selected processes can be realized, thus to stimulate new experiments.

II. OPTICAL PROPERTIES OF SUPPORTED SMALL SILVER CLUSTERS

After UV or visible light excitation, small silver clusters in rare gas matrices or in helium droplets exhibit light emission which was also theoretically confirmed [1–6]. Moreover, the observed photoactivated fluorescence upon illumination of silver oxide films was assigned to the formation of small silver clusters [7]. It was proposed that these findings might be useful for optical storage with high data capacities because of efficient multicolor writing which is followed by readout via fluorescence excitation in the visible region [8–10]. However, the mechanism responsible for such emissive properties of supported clusters is still unknown. In order to be able to distinguish contributions from the cluster and the support which might lead to absorption and emission in the visible regime, we have chosen ionic MgO support which is relatively weakly interacting with silver clusters. We wish to show the influence of the cluster size and the site of support on absorption and emission of supported clusters. Therefore, we present here the results for two selected cluster sizes Ag_4 and Ag_8 at two different sites of MgO support, stoichiometric and F_s-center defect, on the basis of the time-dependent density functional method and the embedded cluster approach [11].

In our model, the full quantum mechanical treatment is restricted to $Mg_{13}O_{13}$ and $Mg_{13}O_{12}$ clusters for the stoichiometric MgO (100) surface and the F_s-center, respectively, and the distant part of the electrostatic field of the support is represented by an array of positive point charges (PC's) ($13\times13\times10$). In order to compensate for strong polarization by positive PC's, 16 Mg^{2+} cations represented by empty pseudopotentials without basis functions have been introduced at the boundary of the model.

The structural and binding properties of Ag_n at MgO have been determined using the DFT method with gradient corrected Perdew–Burke–Ernzerhof (PBE) functional [12] and triple zeta plus polarization atomic basis set (TZVP) together with 19-electron relativistic core potential from the Stuttgart group with corresponding AO basis set [13]. The optimization of supported cluster structures has been carried out either keeping the Mg and O coordinates fixed, or allowing for relaxation of neighboring atoms for the F_s-center. For the calculation of absorption spectra and optimization of excited state geometries of supported clusters, we employ for silver atoms an 11-electron relativistic effective core potential (11e-RECP) with corresponding basis sets which is designed for accurate description of excited states of silver clusters [14]. Our embedded cluster approach is suitable for qualitative analysis of interactions between excitations within the silver cluster and the support. The spectra of clusters obtained by time dependent density functional theory (TDDFT) compare well with those obtained with the more accurate method EOM-CCSD [14,15].

Calculated absorption patterns for supported Ag_4 and Ag_8 clusters are shown in Figure 1, where the comparison with free clusters is also given. The qualitative analysis of excitations between Kohn-Sham orbitals shown in Figure 2 allows identifying the nature of leading configurations contributing to dominant transitions.

The silver tetramer is perpendicularly bound to the stoichiometric surface and to the F_s-center of MgO as shown in Figure 1A. The absorption pattern of Ag_4 at MgO (100) is characterized by interaction of the orbitals which are polarized parallel to the surface. Consequently, the lowest energy intense transition at 2.67 eV (cf. Figure 1A) which is dominated by $P_x \rightarrow D_{x^2}$ excitation

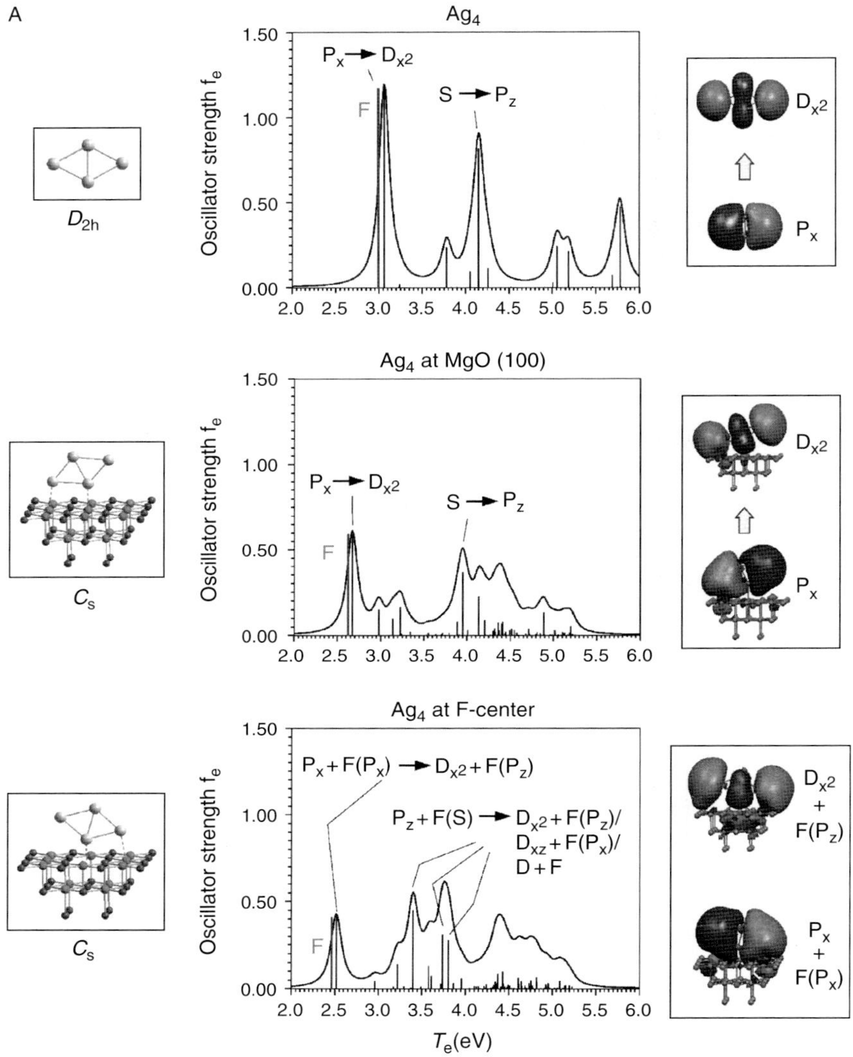
A
Ag4
D2h
Oscillator strength fe
Px → Dx2
S → Pz
F
Dx2
Px
Ag4 at MgO (100)
Cs
Px → Dx2
S → Pz
F
Dx2
Px
Ag4 at F-center
Cs
Px + F(Px) → Dx2 + F(Pz)
Pz + F(S) → Dx2 + F(Pz)/
Dxz + F(Px)/
D + F
F
Dx2
+
F(Pz)
Px
+
F(Px)
Te(eV)

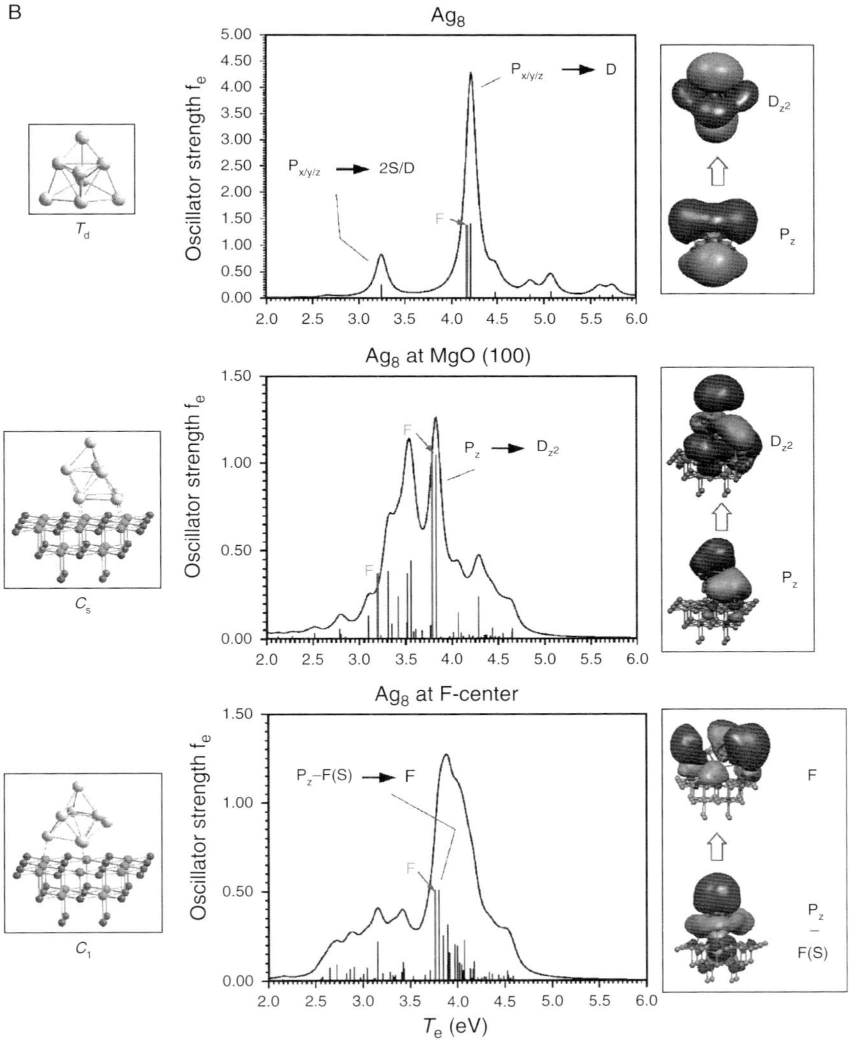

FIGURE 1 (A) Comparison of absorption spectra for free Ag_4, Ag_4 at stoichiometric MgO (100), and Ag_4 at F_S-center defect obtained for optimized structures. The types of characteristic excitations involved in intense transitions are assigned and the corresponding orbitals are shown. The position of the fluorescence line (red line) obtained by optimizing the geometry of the corresponding excited state is labeled by F. (B) Comparison of absorption spectra for free Ag_8, Ag_8 at stoichiometric MgO (100), and Ag_8 at F_S-center defect obtained for optimized structures. Reprinted from Ref. [11], with kind permission of the European Physical Journal (EPJ). (For interpretation of the references to color in this figure legend, the reader is referred to the Web version of this chapter.)

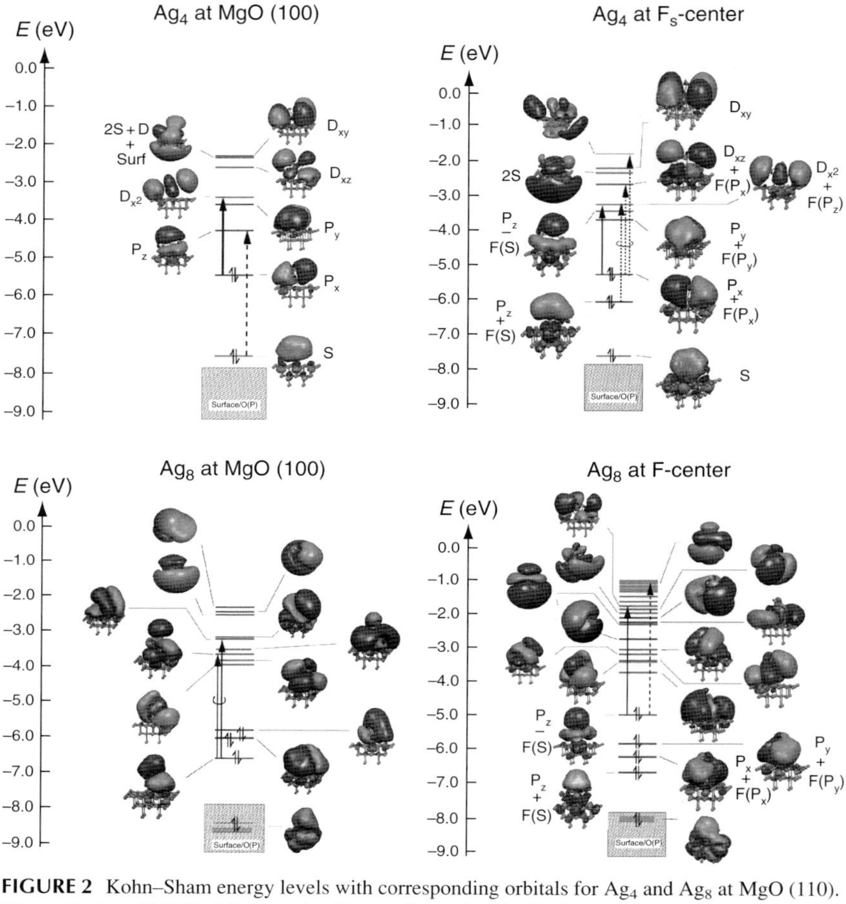

FIGURE 2 Kohn–Sham energy levels with corresponding orbitals for Ag_4 and Ag_8 at MgO (110). The excitations (full line) indicate leading configurations in the lowest energy intense transition. The excitations (dashed lines) correspond to leading configurations of intense transitions with higher energies in correspondence with labeled transitions in Figure 1.

(HOMO→LUMO+2 cf. Figure 2) is red shifted with respect to the one calculated for the free cluster because of polarization in x-direction parallel to surface. In contrast, the second intense transition located at 3.95 eV dominated by S → P_z excitation (HOMO-1→LUMO cf. Figure 2) remains almost unchanged with respect to the one of the free clusters. As the relaxation of excited state geometry of the lowest energy transition leads to a fluorescence line located close to the vertical transition, Ag_4 at Mg (100) is a good candidate for an emissive center in the visible regime, provided that the cluster is sufficiently stabilized by the support (e.g., the binding energy is 1.17 eV).

In the case of Ag_4 at F_s-center, the $P_x + F(P_x)$ orbital becomes occupied (cf. Figure 2). Therefore the lowest energy transition at ~2.5 eV is characterized by leading excitation $P_x + F(P_x) \rightarrow D_{x^2} + F(P_x)$ which is of P → D type as shown also in Figure 1A. This transition is red shifted with respect to the analogous transition in the free cluster (cf. Figure 1A) because of the contribution from the $F(P_x)$ orbital. As this transition is well isolated from the others and the binding energy is also convenient, Ag_4 at F_s-center is particularly a good candidate for emissive center. The group of intense transitions located at 3.40, 3.74, and 3.80 eV corresponds to the excitations from the $P_x + F(P_x)$ and $P_z + F(S)$ orbitals to higher D-type orbitals (cf. Figure 2).

As the next example, we have chosen to present the supported Ag_8 cluster because of completely different spectroscopic features due to the three-dimensional structure of the cluster. The filled S- and P-type valence orbitals give rise to absorption patterns with dominant transitions distributed in a large energy interval as shown in Figure 1B. In the case of Ag_8 at stoichiometric surface, the excitations giving rise to intense transitions are of P → D type. The splitting of intensities is due to the lowering of symmetry with respect to the gas phase Ag_8 with the T_d structure. The polarization by the surface causes again the red shift of the spectroscopic pattern of the supported cluster. The locations of the fluorescence bands are close to the intense transitions for the selected excited states which indicate that emission might be observed in a broad energy range. The increased density of low lying excited states can strongly influence emissive properties, and therefore, the determination of lifetimes is mandatory for predicting the efficiency of emissive centers. For Ag_8 at F_s-center, the $P_z - F(s)$ orbital which is of D-type becomes HOMO (cf. Figure 2). Therefore, a large number of D → F type excitations contribute to the intense transitions which are spread in the energy interval between 2.5 and 3.5 eV and above 3.75 eV as shown in Figure 1B. The distinct feature of absorption pattern of Ag_8 at the F_s-center is that although there is a red shift for the dominant transition with respect to the free cluster, no separated intense transitions in the visible regime are present as it is the case for smaller supported clusters (e.g., Ag_4 at F_s center). Therefore, it is to be expected that different emissive properties will arise. It is interesting to notice that inspite of discrete energy levels the spectroscopic pattern of supported Ag_8 already resembles those expected for larger nanoparticles. The above described results show that absorption and emission properties of small silver clusters at the MgO support can be tuned by the cluster size, cluster structure, and the site of the support. The inherent characteristic features of the free clusters remain partly preserved while the support causes red shift of intense transitions. The location of intense absorption and fluorescence bands in the visible regime which is experimentally easy accessible is present only for supported Ag_4. Fluorescence in the higher energy intervals for supported Ag_8 cannot be excluded but determination of the life times is necessary for predicting efficiency of emissive centers. The future investigation should involve the following: (i) The influence of further increase of the cluster size on characteristics of absorption patterns has to be determined; the connection between the nature of the filled cluster-support-site orbitals from which excitations could take place and absorption patterns should be established in order to develop functionalized supported clusters. (ii) Absorption properties of silver clusters at other defect sites of MgO should be investigated. (iii) The interplay of the life times of excited states with the size and the structure of supported clusters has to be explored in order to provide information about experimental realization of the emissive centers.

III. PHOTOABSORPTION AND PHOTOFRAGMENTATION OF ISOLATED CATIONIC SILVER CLUSTER-TRYPTOPHAN HYBRID SYSTEMS

Large nanoparticles have been used in the biomolecule detection for a long time in the frame of the surface enhanced Raman spectroscopy (SERS) [16–18]. Recently, it has been proposed that small silver clusters are suitable probes in biomolecular sensing because of their biocompatibility in contrast to large plasmon-supporting nanoparticles [19]. In fact, silver clusters with only few atoms give rise to scaffold-specific single molecule Raman signals making them ideal optical probes for the biomolecular environment [19]. Therefore, recently we proposed that these fascinating optical properties of small silver clusters interacting with an adequate environment can provide fundamentals for developing biosensoring materials. The interaction between excitations within silver clusters and within the biomolecules might give rise to novel optical properties. We have chosen the tryptophan molecule because of its known fluorescence properties which occur in the similar energy interval as for small silver clusters. In such hybrid cluster-biomolecule systems, excitations within both interacting subunits are of molecular nature [20–22] in contrast to plasmonic excitations in large nanoparticles. Therefore, these systems are extremely suitable for tuning functionality by varying the size of involved subunits because of the structure dependent nature of interactions. The gas phase study is valuable for gaining basic knowledge about the mechanism of interactions in such hybrid systems before addressing more complex systems, including different environments.

Therefore, we present here results on photoabsorption of Trp-Ag_3^+ and Trp-Ag_9^+ [21,22]. The choice of these two systems has been made to show that different types of structures (charge solvated (CS) and zwitterionic (ZW)) are responsible for different types of charge transfer (CT) in excited states. This has been evidenced by photofragmentation dissociative channels and direct MD simulations [22]. The structural properties and stationary absorption spectra were determined using DFT with hybrid B3LYP functional [23,24]. For Trp, the 6–311 G** AO basis set augmented by diffuse functions for appropriate description of excited states was used. The same RECP with corresponding AO basis for Ag atoms was employed as for supported silver clusters [11]. The simulation of the ground and excited state dynamics "on the fly" has been performed in the framework of the semiempirical CI method on the basis of the AM1 parameterization in order to simulate main fragmentation channels observed experimentally. The nonadiabatic dynamics in exited electronic states has been carried out employing the Tully's surfaces hopping algorithm [25] and the nonadiabatic coupling was calculated with the semiempirical CI method [26]. We wish to show that the absorption patterns and fragmentation channels of hybrid silver tryptophan can be classified according to the cluster size and the structure type of the complex (CS vs. ZW). In the CS structures, the positively charged cluster is coordinated to the free electron pairs on the oxygen and nitrogen atoms of the side chain as well as to the π-system of the indole ring. In the ZW structure, the charged COO^- and NH_3^+ groups are present, and the positively charged cluster is bound to the COO^- group. Comparison of absorption spectra calculated for pure Ag_3^+ and Ag_9^+ and of their complexes with Trp for CS and ZW structures are shown in Figure 3. The calculated spectrum of Trp-Ag_3^+ for the most stable structure belonging to the CS class exhibits resonant transitions with the absorption of Ag_3^+ in the energy interval around 300 and 200 nm. The analysis of the leading excitations contributing to the transition located at $\sim$300 nm shows that CT between both subunits and excitation within silver occur (cf. Figure 4A). This type of CT is due to the interaction between the silver cluster and the indole ring in the CS structure. The intensity of the transition at 300 nm in Trp-Ag_3^+ is of an order of magnitude larger than in Trp-H^+ and Trp-Ag^+ measured and calculated.

The CT nature of the optical transition around 300 nm and the assignment to the CS structure have been confirmed by wavelength dependent fragmentation channels exhibiting the loss of a single silver atom and the nonadiabatic MD simulations starting in the corresponding excited states [22]. The loss of a single Ag atom occurs between 1 and 2 ps as shown in Figure 5 corresponding to the main observed fragmentation channel which is a consequence of the CT excitation due to an electron transfer to the antibonding orbital of Ag_3^+, thus weakening the bonding within the Ag_3 subunit. In contrast, the MD in the electronic ground state shows loss of the intact Ag_3^+ subunit (cf. also Figure 5) in agreement with the collision-induced dissociation (CID) spectrum.

In addition, there is clear evidence that the CS structure can be assigned to the observed cross-section since the ZW structure which lies 0.24 eV higher in energy gives rise to an intense transition in the UV regime above 350 nm which has not been observed.

The situation changes drastically for larger silver cluster-tryptophan complexes such as Trp-Ag_9^+ as the 3D structures of the cluster give rise to dominant intense transitions in the narrow energy interval $\sim$300 nm (cf. Figure 3). The reminiscence of the spectroscopic pattern of the cluster remains reflected in the absorption spectrum of the complex. The ZW structure is lower in energy than the CS structure. Although spectroscopic patterns

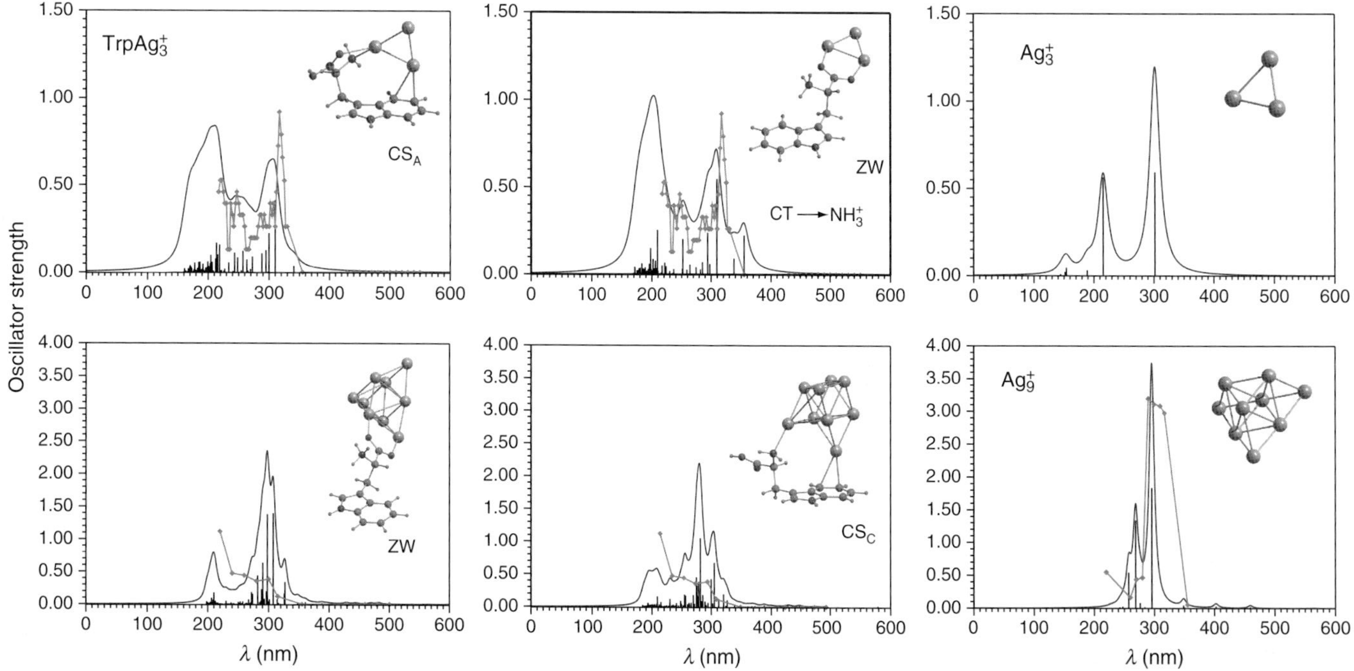

FIGURE 3 Calculated absorption spectra for two types of structures CS and ZW for Trp-Ag_3^+ and Trp-Ag_9^+ and for the free clusters Ag_3^+ and Ag_9^+. The broadening of the lines is simulated by a Lorentzian function with a width of 20 nm (blue curves). In the case of CS structures, we label different subclasses according to their bonding to different atoms of the side chain and/or to the π-systems in the following way: In the CS_A type of structures, the silver cluster is bound to the N and O atoms of the side chain, and to the π-system; in the CS_C structure, the Ag_n^+ is bound to the π-system and to the N-atom. The measured photoabsorption cross sections are shown in red. For Trp-Ag_3^+, there is good agreement with the calculated spectrum for the CS_A structure. (For interpretation of the references to color in this figure legend, the reader is referred to the Web version of this chapter.)

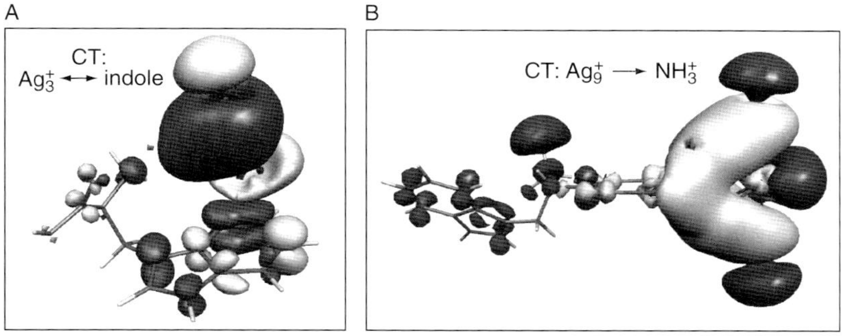

FIGURE 4 (A) Charge density difference between the ground and excited state of Trp-Ag_3^+ (CS_A) located at 310 nm. (B) Charge density difference between the ground and excited state of Trp-Ag_9^+ (ZW) located at 308 nm. Blue regions correspond to accumulation and gray regions to depletion of the electron density. (For interpretation of the references to color in this figure legend, the reader is referred to the Web version of this chapter.)

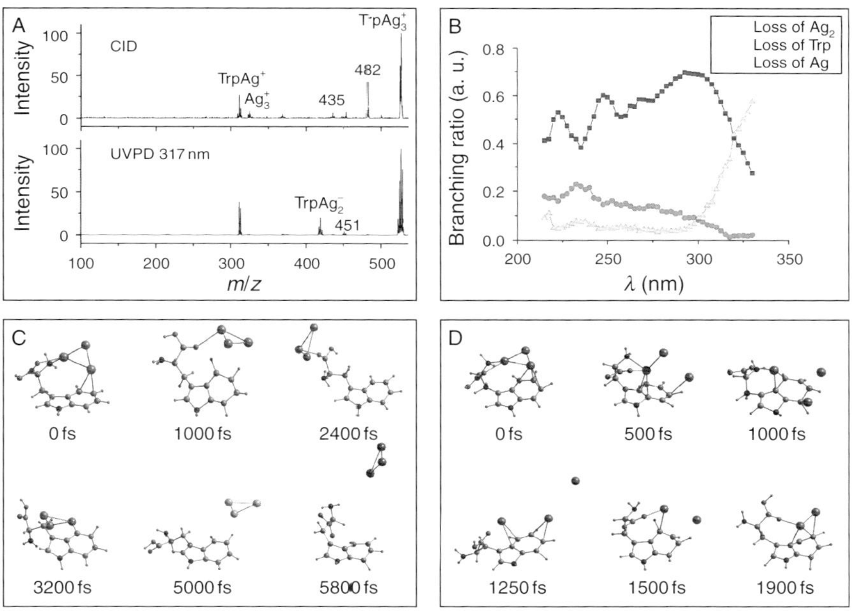

FIGURE 5 (A) Measured CID spectrum of Trp-Ag_3^+ and UV photodissociation at 317 nm. (B) Branching ratio for the different fragmentation channels of Trp-Ag_3^+ as function of the wavelength. (C) Snapshots of the MD "on the fly" in the electronic ground state from a selected trajectory at high temperature simulating the fragmentation process. (D) Snapshots of the MD in the electronic excited state located at 310 nm from a selected trajectory at 300 K. Reprinted with permission from Ref. [22]. Copyright 2007, American Institute of Physics.

calculated for ZW and CS structures exhibit intense transitions around 300 nm, the nature of these intense transitions is completely different and it is typical for these two classes of structures. In the case of the ZW structure, the transitions located at 299 and 308 nm are characterized by excitation within the silver cluster and the CT to the NH_3^+ group (cf. Figure 4B) [22]. In contrast, in the case of the CS structure, the excitations dominantly responsible for intense transitions are due to CT between both subunits and due to excitations within the metallic subunit as it is the case for the CS structure of Trp-Ag_3^+.

The comparison with experimentally measured cross sections (cf. Figure 6) shows that the locations of intense transitions are in agreement with theoretical results. However, the fragmentation yield of Trp-Ag_9^+ is considerably smaller than in the case of Trp-Ag_3^+, in spite of the resonances between excitations within the two subunits [22].

This remarkable finding is expected to be due to energy dissipation by fluorescence. The fragmentation yield for pure Ag_9^+ clusters which is six times larger than in the complex is in agreement with previous experimental findings as well as with theoretical results as shown in Figure 3. This large difference between the fragmentation yields in Trp-Ag_3^+ and Trp-Ag_9^+ might be due to efficient and fast IVR in the excited state of Trp-Ag_9^+ with a larger number of degrees of freedom. Such efficient IVR may be followed by fluorescence which is then responsible for the lower experimental fragmentation

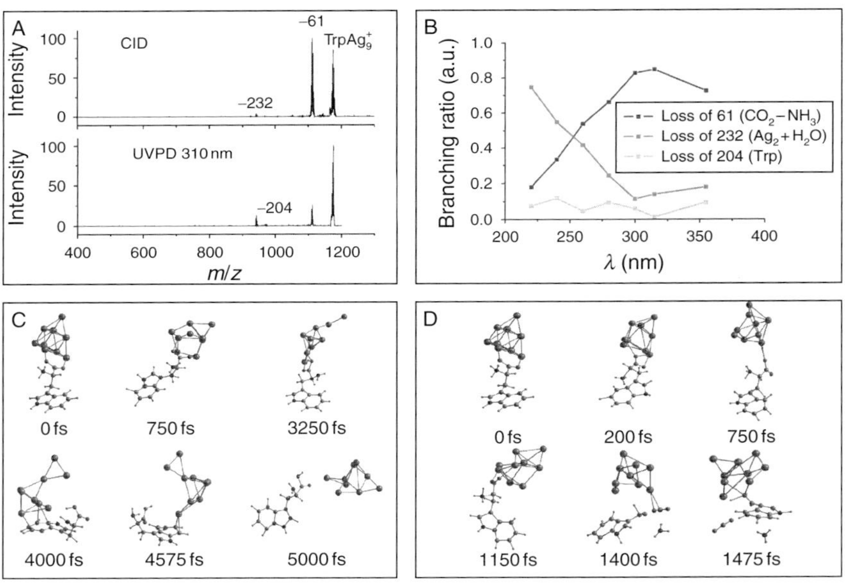

FIGURE 6 (A) Measured CID spectrum of Trp-Ag_9^+ and UV photodissociation at 310 nm. (B) Branching ratio for the different fragmentation channels of Trp-Ag_9^+ as function of the wavelength. (C) and (D) Snapshots of the MD "on the fly" in the electronic ground state from selected trajectories at high temperature illustrating the minor fragmentation channel leading to loss of the tryptophan unit (C) and the dominant one leading to the loss of CO_2 and NH_3 (D). Reprinted with permission from Ref. [22]. Copyright 2007, American Institute of Physics.

yield. This interesting aspect which involves drastic changes in IVR between the two subunits by increasing size of metallic subunits by only few atoms still needs further investigation. The photofragmentation channel which is characteristic for the ZW structure and involves the loss of CO_2+NH_3 has been confirmed by MD simulations (cf. Figure 6). It is initiated by a CT from the Ag_9^+ cluster to the NH_3^+ group which results in successive evaporation of NH_3 and CO_2 and with formation of a bond between silver and indole subunits as shown in the snapshots at 1475 fs in Figure 6. In summary, we provided an insight into the nature of excitations in interacting nanoparticle-biomolecule subunits and identified characteristic spectral features for two classes of structures: CS and ZW. Moreover we assigned two types of CT: from π-system of Trp to silver cluster for CS structures and from silver cluster to NH_3^+ for the ZW class of structures. This classification of CT has been confirmed by experimental fragmentation channels and MD simulations. Our findings that Trp-Ag_3^+ does not fluoresce and Trp-Ag_9^+ might fluoresce are significant because they show that interactions between excitations from different subunits give rise to very specific optical properties of hybrid systems, which seem to be different from those found for supported clusters. In the case of pure silver clusters, at surfaces the influence of the support has to be minimized in order to preserve optical and emissive properties of pure clusters. In contrast, in the case of hybrid cluster-biomolecule systems both subunits play equally important roles. However, differences in size selectivity of emissive centers in hybrid systems and in supported clusters require further investigation. Furthermore, in the future two directions will be pursued: First, nonradiative lifetimes have to be determined theoretically and experimentally in order to establish the dependence of emissive properties on the size of the cluster. Second, Trp-Ag_n^+ hybrid systems stabilized by support such as MgO or/and TiO_2 will be investigated in order to provide the basis for biosensing materials.

IV. NEW REACTIVITY CRITERION BASED ON INTERNAL VIBRATIONAL ENERGY REDISTRIBUTION

A prominent example that small metallic clusters can exhibit new chemical properties because of size and structure selective confinement is the unexpected finding that small nanosized gold particles can catalyze oxidation reactions occurring in, for example, combustion [27–31]. Also, silver is widely used as catalyst in important industrial processes such as epoxidation [31]. However, mechanisms responsible for oxidation reactions still remain to be unraveled. The knowledge about mechanisms is a required prerequisite for rational design of efficient nanocatalysts. Until now, the investigation of size dependent reactivity of small gold and silver clusters in the gas phase or at the support has been focused on structure-reactivity relationship based on important energetic aspects [29,30,32–39].

Dynamical processes such as IVR between interacting subunits during the reaction have not been considered yet. Therefore, we wish to introduce the dynamical aspect as a new reactivity promoting criterion [40,41]. We will show that the resonant IVR between the reactants promotes reactivity toward adsorbates in contrast to the dissipative IVR which prevents reactions. This novel concept will be illustrated on the examples of anionic gold versus silver clusters interacting with molecular oxygen. We address this issue first in the gas phase, which allows to introduce and understand clearly dynamical aspects before considering the important role of the interface.

Before presenting the results on dynamics, we briefly summarize previous findings: Even/odd alternating binding energies for adsorption of O_2 on Au_n^- and Ag_n^- have been found [35,37,39,42], as clusters with an odd number of electrons strongly bind O_2 which is an electron acceptor. Moreover, the binding and electron affinity patterns are related to each other [35,37]. The clusters with even number of electrons have higher electron affinities and they are inert toward O_2. Studies of oxidation of CO by anionic gold and silver clusters [29,30,32–34] provide the finding that the cluster alone cannot break the O-O bond without cooperative effects [33,43]. However, a difference between reactivity of gold and silver clusters has been observed. $Au_2O_2^-$ [33] and $Au_6O_2^-$ [34] oxidize CO but $Ag_2O_2^-$ and $Ag_6O_2^-$ do not [44], in spite of the fact that these gold and silver oxides have similar structural and energetic properties. The only difference is in relativistic effects which are substantially more pronounced in the case of gold than of silver clusters because of a smaller s–d energy gap in the former case. Therefore, the contribution of d-electrons in Au clusters is responsible for more directional bonding than in the case of Ag clusters [45]. Consequently, gold clusters remain planar for larger sizes than the silver clusters do [46–49]. In order to answer the question why gold promotes reactivity of O_2 and silver does not, we investigated reaction dynamics and IVR during the collisions of Au_6^- und Ag_6^- with O_2. For this purpose, we performed MD simulations with DFT using the PBE functional [12] and 19e-RECP for gold and silver combined with triple-zeta quality of AO basis sets [13]. The lowest energy structures of Ag_6^- and Au_6^- assume planar trapezoidal structures as shown in Figure 7. However, the other isomers are closer lying in energy for Ag_6^- than for Au_6^- in which case all three isomers assume planar structures. Collisions with O_2 have been carried out starting with the most stable isomers and impact orientations have been sampled over a sphere around the cluster with 400 initial conditions for each cluster. An impact kinetic energy of 0.25 eV has been chosen corresponding to $T \approx 300$ K. The snapshots of MD simulations are presented in Figure 8 together with the sticking probability of O_2. Figure 8 shows that the gold cluster remains rigid during longer times than the silver cluster and that sticking of O_2 to Au_6^- is considerably lower than in the case of Ag_6^- (cf. Ref. [40]).

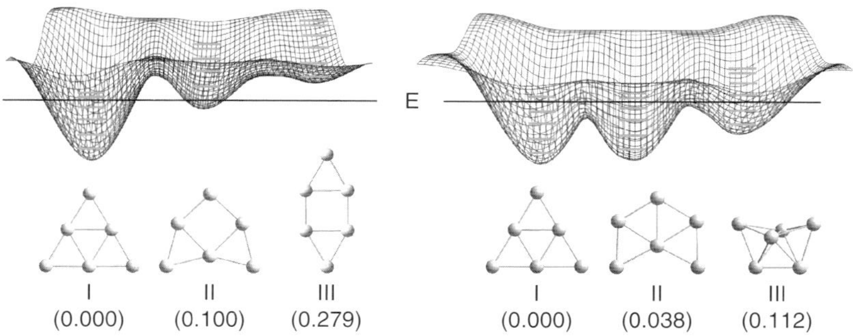

FIGURE 7 The lowest energy structures for Au_6^- (left) and Ag_6^- (right). Schematically, the PES with vibrational levels (red) is shown to demonstrate the higher vibrational density of states for Ag_6^- for some given energy E (black horizontal line). Reprinted from Ref. [40]. (For interpretation of the references to color in this figure legend, the reader is referred to the Web version of this chapter.)

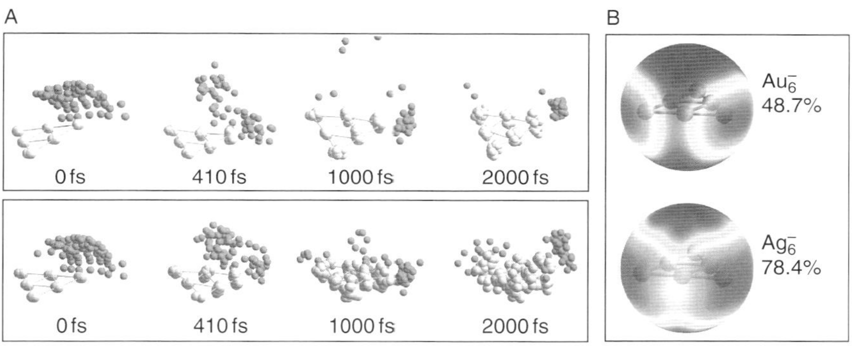

FIGURE 8 (A) Snapshots of MD simulations for collision of Au_6^- (top, yellow colored) and Ag_6^- (bottom, gray) with molecular O_2 (red). Initial conditions ($t = 0$) sample the whole space due to D_{3h} symmetry of the noble metal hexamer. (B) Sticking probability for O_2 projected onto a sphere around the metal cluster. Red color reflects impact angles for successful collision with O_2. Reprinted from Ref. [40]. (For interpretation of the references to color in this figure legend, the reader is referred to the Web version of this chapter.)

These differences are however not due to the different masses of gold and silver but are caused by inherently different dynamical processes which are due to more rigid Au_6^- originating from relativistic effects versus floppy behavior of Ag_6^- due to s-like metallic character. This has been verified by a series of MD simulations with varying masses of metal atoms under the same initial conditions. In Figure 9A, we show that for collisions of O_2 and Au_6^-, the transferred kinetic energy increases with decreasing mass, but O_2 does not stick to Au_6^- independently from the mass. In contrast, in the case of Ag_6^-, O_2 sticks to the cluster independently whether the mass is increased to that of gold or not. Higher rigidity of Au_6^- is also demonstrated by temperature dependent Lindemann indices which are used as a signature of phase transition in

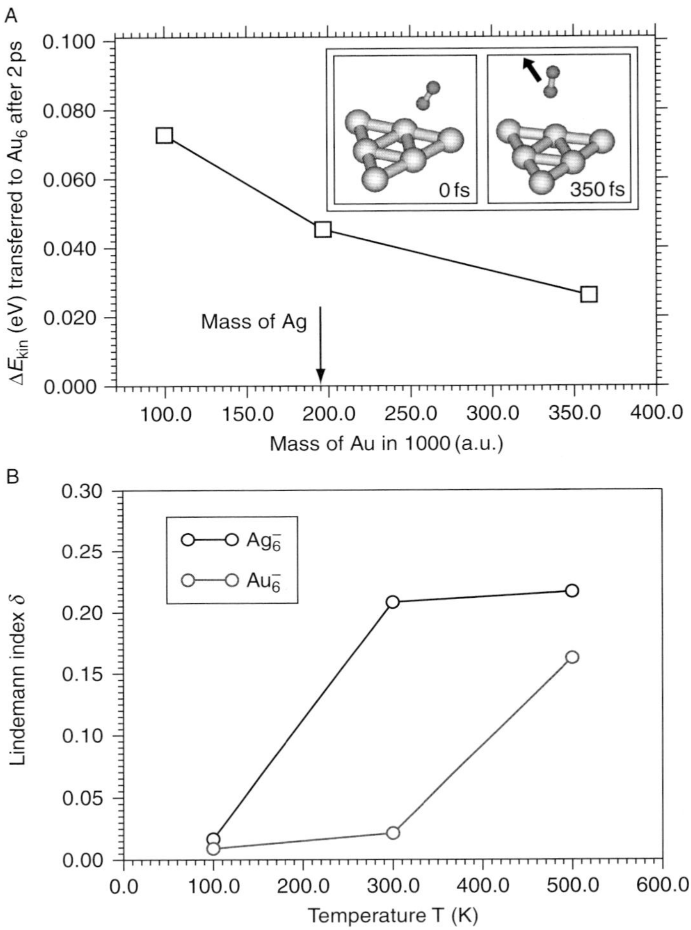

FIGURE 9 (A) Kinetic energy transferred from O_2 to the Au_6^- upon collision is shown for different masses of gold. Notice that ΔE_{kin} is showing a mass dependence, but in contrast to silver, O_2 does not bind in any of the MD simulations for gold with the given initial conditions. (B) Lindemann indices for Ag_6^- and Au_6^-. The curve for gold is shifted to higher temperatures indicating a higher rigidity of the cluster. Reprinted from Ref. [40].

infinite systems. They show melting-like behavior for Au_6^- occurring at much higher temperature ($T \approx 450$ K) than for Ag_6^- ($T \approx 200$ K) (cf. Figure 9B).

Finally in Figure 10, we present results from which it is possible to identify the different natures of IVR for collisions of Au_6^- and Ag_6^- with O_2. In the case of efficient dissipative IVR, accumulation of kinetic energy in the Ag_6^- after $\sim$1.4 ps takes place, and therefore, the O–O bond remains only weakly vibrationally excited as shown in Figure 10. In contrast, resonant IVR has been identified for Au_6^- colliding with O_2 exhibiting oscillatory

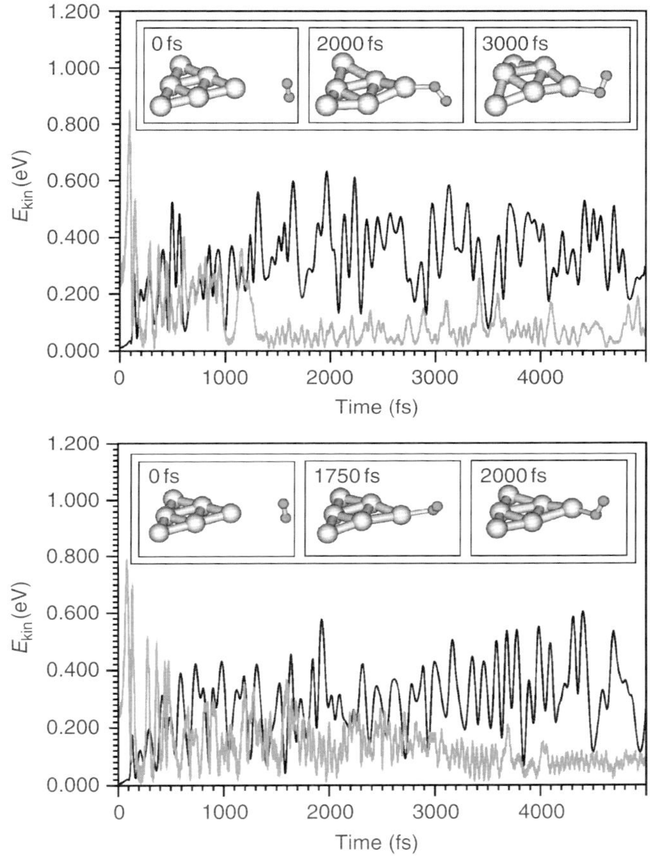

FIGURE 10 Kinetic energy partition between Ag_6 (top), Au_6 (bottom), labeled by black curves, and O_2 (red curves) for two reactive collisions demonstrating a dissipative IVR for Ag_6^- and resonant IVR for Au_6^-. For illustration, a single representative trajectory has been chosen. Reprinted from Ref. [40]. (For interpretation of the references to color in this figure legend, the reader is referred to the Web version of this chapter.)

exchange of kinetic energy between Au_6^- and O_2 subunits. Thus, the kinetic energy remains localized at the O_2 adsorption site leading to a vibrationally highly excited O–O bond (cf. Figure 10). In addition to the chemical activation of the strong O–O bond by CT, the vibrational excitation of the O–O bond induced by Au_6^- is responsible for higher oxidation reactivity confirming experimental observations. The role of IVR emphasizes the necessity to consider all degrees of freedom in reaction dynamics.

On the basis of our described findings, we propose the nature of IVR as new criterion for promoting the reactivity. Resonant IVR between the cluster and adsorbate is responsible for an activation mechanism which facilitates overcoming the barriers and therefore enhances oxidation reactivity (cf. Ref. [40]). This means that the mechanism responsible for oxidation reactions of coinage metal particles relevant for heterogenous catalysis is based on intrinsic dynamical properties which are closely connected with the nature of chemical bonding of these species (e.g., directional vs. delocalized bonding).

V. SIZE DEPENDENT DYNAMICS AND EXCITED STATES OF ANIONIC GOLD CLUSTERS: FROM OSCILLATORY MOTION TO PHOTOINDUCED MELTING

Femtosecond-time resolved photoelectron spectroscopy (TRPES) using the pump-probe technique is a powerful tool for the real time investigation of electronic and nuclear dynamics (see, e.g., Refs. [50,51]). It is based on the preparation of a nonequilibrium nuclear configuration by optical excitation of a stable species in the pump step and exploration of its time evolution by the probe step. The binding energy of the emitted electron corresponds to the energy difference between the excited intermediate and final state. Thus, TRPES offers the unique chance to follow structural changes with high precision in real time, for example, wavepacket motion [52], isomerization [53], and desorption [54]. So far, photoinduced geometry changes have been difficult to observe in metal clusters owing to a comparatively high electronic density of states in the valence region [55–57]. In contrast, gold behaves differently since it has nominally a closed d-shell. Although due to relativistic effects, the d-contribution to the valence orbital space in gold clusters is important [45,46], it is not as prominent as in transition metals and, therefore, a relatively large HOMO—LUMO band gap and discrete level density exists in small Au clusters [58]. In particular, the dipole allowed excited state near 1.5 eV is a well isolated electronic state [59]. Therefore, small gold clusters are excellent candidates to study nuclear dynamics using TRPES.

We present here photoinduced nuclear dynamics of Au_5^- and Au_7^- in real time. Depending on the cluster size, we observe oscillatory motion in a bound excited state (Au_5^-) and photoinduced melting (Au_7^-). By a joint experimental and theoretical study, we demonstrate for the first time that a melting-like behavior of a finite system takes place on a time scale of $\approx$1 ps. This occurs because of a large excess of energy after return from the excited state to the ground state after internal conversion (IC) [60].

The experimental setup is described in detail in Ref. [61]. In short, gold cluster anions below room temperature [58,62], that is, in the ground state structure, are produced in a pulsed laser vaporization source. Before irradiation with two subsequent fs-laser pulses, the cluster anions are mass selected and decelerated. The kinetic energy of the detached electrons is analyzed by a magnetic bottle time-of-flight photoelectron spectrometer. For photodetachment, an amplified Ti:Sa femtosecond laser system (pulse duration

40 fs full width at half-maximum (FWHM), energy per pulse < 1mJ, spectral bandwidth 40 meV FWHM, and laser intensity $> 10^{11}$ W/cm^2) was employed. We use the fundamental (1.56 eV) as the pump and the 2nd harmonic (3.12 eV) as the probe pulse. Absorption of a pump photon leads to the population of an excited electronic state. The ensuing dynamics of the excited intermediate state is mapped by photodetachment with the probe pulse. The photon energy of the pump pulse is distinctly lower than that of the photodetachment and/or dissociation energy of the clusters [63], which assures the observation of bound state dynamics.

TRPE spectra were simulated in the framework of our Wigner distribution approach [64,65] based on the propagation of an ensemble of classical trajectories "on the fly." This approach allows accurate simulation of ultrafast processes and femtosecond signals in complex systems, involving both adiabatic and nonadiabatic dynamics [65]. As here the dynamics is required in excited electronic states of anionic gold clusters, we employ the TDDFT in its linear response formulation for calculation of forces needed to carry out dynamics in excited states. We use the gradient corrected PBE functional [12] together with Stuttgart group 19e-RECP with corresponding gaussian AO basis set which has been reoptimized for better description of optical properties of gold clusters [66]. The simulation of the TRPES involves three steps: (i) The ensemble of initial conditions has been generated by sampling the Wigner distribution function corresponding to the canonical ensemble at $T=300$ K. (ii) In the second step, the trajectories are propagated on the excited electronic state allowing nonadiabatic transition to the ground state when the conical intersection is reached. Parallel to the propagation of the ensemble, the time dependent energy gaps to the neutral state are calculated for each trajectory. (iii) The TRPES is calculated by averaging over the ensemble of trajectories employing the analytical expression derived in the framework of the Wigner distribution approach [65].

The time-dependent spectra of Au_5^- are shown in Figure 11A. Up to ≈ 3 ps, the experimental photodetachment peak oscillates between 1.5 and 1.8 eV. An oscillation period of 315 fs is deduced from a sine fit to the data. The theoretically derived oscillation period amounts to 420 fs and is in qualitative agreement with the experiment as shown in Figure 11B. This oscillatory behavior is due to vibrational wavepacket motion which can also be seen from the snapshots of the 30 calculated trajectories shown in Figure 11C. Periodic motions corresponding to breathing-like modes can be recognized from the snapshots. With increasing time delay the wavepacket or ensemble of trajectories broadens due to population of different vibrational modes which might be an explanation for the fading of photoelectron intensity observed at 10 ps and later.

The TRPE spectrum of Au_7^- shown in Figure 12A exhibits strikingly different features. At $t=0.1$ ps, a peak is seen at a binding energy of 2 eV which corresponds to the transition from the first electronically excited state A^-, to the neutral ground state X. The peak shifts to higher binding energy until it stabilizes at BE$=2.6$ eV after around 10 ps. The simulated TRPE-spectra of

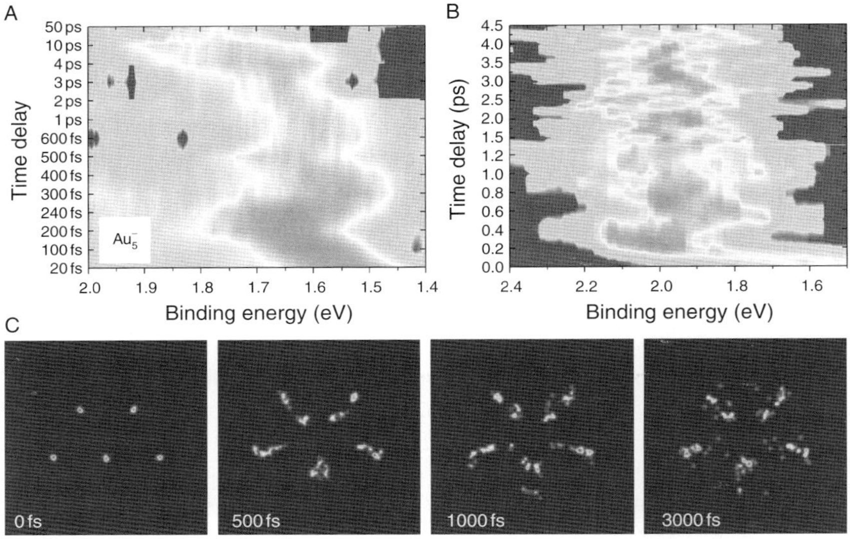

FIGURE 11 Au_5^-. Experimental (A) and simulated (B) TRPE-spectra are shown as a contour plot. (C) Geometry snapshots calculated for 30 trajectories. At $t = 3000$ fs the atoms are much more localized than in Au_7^- (Figure 12B).

Au_7^- are presented in Figure 12C. The experimentally observed binding energy shift as well as the timescales are in agreement with the theoretical spectrum (cf. Figure 12A). The TDDFT calculations reveal that the energy shift can be explained by IC from the initially populated A^- state to the anionic ground state X^-. IC takes place on the same time scale as the experimentally observed energy shift. In Figure 12B, the kinetic energy of the peak maximum ($E_{el} = h\nu_{probe} - BE$) is plotted as a function of time and fitted to an exponential function with $\tau = 1.8$ ps. This time constant is in agreement with the IC time obtained from the calculated population dynamics ($\tau_{calc} = 1.7$ ps, Figure 12D). The reason for the relation between energy shift and IC becomes evident from Figure 13A where the time evolution of the potential energy of A^-, X^-, and X is shown for a selected trajectory. The binding energy of the detached electron corresponds to the calculated energy difference between X and A^-. At $t = 0$ fs, the dynamics is initiated on A^-. After 340 fs, A^- crosses with X^- and the kinetic energy of the nuclear system (KE) rises rapidly. This is associated with an increase in the binding energy of the detached electron. Note that our calculations reveal an energy flow into the nuclear system of 0.69 eV in average, which corresponds to a vibrational temperature of ≈ 1070 K, leading to strong structural fluctuations. This loss of well defined structures corresponds to a melting-like transition in finite systems. According to the Lindemann criterion [67], a melting transition in Au_7^- occurs at a vibrational temperature between 800 and 900 K. Thus, the energy

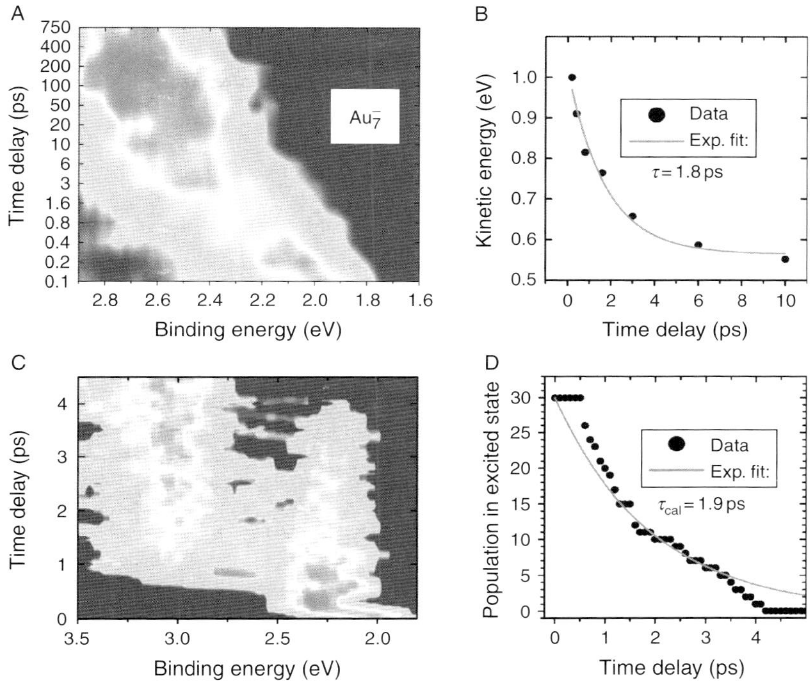

FIGURE 12 Au_7^-. Experimental (A) and simulated (C) TRPE-spectra as a contour plot. (B) The kinetic energy of the experimental peak maximum is plotted as a function of time. The data are fitted to an exponential function with a decay time of $\tau = 1.8$ ps. (D) Calculated population dynamics of A^- as a function of time obtained from 30 trajectories. After $\tau_{calc} = 1.7$ ps, the population of A^- has declined by the factor 1/e (red line). (For interpretation of the references to color in this figure legend, the reader is referred to the Web version of this chapter.)

flow into the vibrational system is sufficient for a melting-like transition of Au_7^- as also evidenced by the geometry snapshots presented in Figure 13B.

In summary, this joint experimental and theoretical study shows the extremely size-dependent nuclear dynamics of Au_5^- and Au_7^- clusters. The oscillating wavepacket movement in Au_5^- demonstrates the long lived bound state character of the lowest excited state. In contrast, in Au_7^-, fast IC into the electronic ground state leads to an energy flow into the nuclear system which initiates a melting-like transition after ≈ 1.8 ps.

VI. OPTIMAL CONTROL OF MODE SELECTIVE FEMTOCHEMISTRY IN MULTIDIMENSIONAL SYSTEMS

The optimal control of molecular processes by shaped laser fields which has been first proposed theoretically [68–72] has flourished after numerous experimental realizations [73,74] of the closed-loop learning loop optimal control

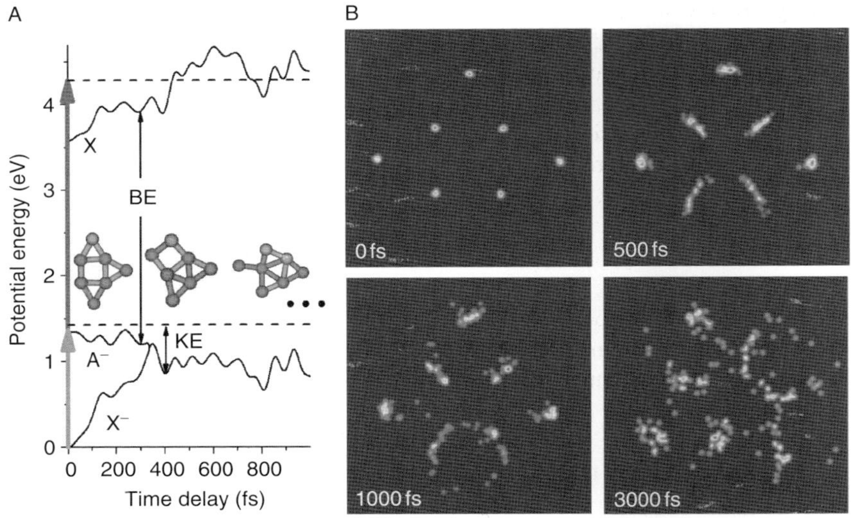

FIGURE 13 Au_7^-. (A) Temporal evolution of the potential energy of the X^-, A^- and X electronic states. (B) Calculated geometry snapshots for the 30 trajectories. Each point corresponds to the position of a single atom (30 × 7 atoms in total). At $t=0$fs, the initial conditions involve the ground state structure at room temperature. At $t=3000$ fs, the atomic positions are blurred because of melting-like behavior. Reprinted with permission from Ref. [60]. Copyright 2007, American Institute of Physics.

scheme (CLL) [75]. The idea of this approach is to use a learning algorithm coupled to a laser system with a pulse shaper in order to produce laser pulses which are able to control the outcome and yield of photochemical processes. The role of the evolutionary algorithm is to iteratively optimize the electric field until the yield of the desired product or process is maximized. As the control of MD requires in principle the knowledge of the global potential energy surface (PES) which is not available for complex systems, the aim of the CLL was to avoid this difficulty [75]. However, in spite of tremendous successes in various fields of applications, the disadvantage of this technique is the lack of knowledge about the processes and the mechanism underlying the control. In contrast, by analyzing the shaped pulses obtained from the theoretical optimal control (OC) and comparing with pulses produced by the experimental CLL technique [76,77] this information can be gained. In order to achieve this for complex systems, we have proposed a strategy for optimal control including all degrees of freedom in multidimensional systems without the need for precalculation of the PES [78]. Our approach has united the direct quantum chemical MD "on the fly" [79,80] with the optimal control theory formulated in the frame of Wigner representation of quantum mechanics [81,82] and it opens the perspective to control the dynamics in systems with increasing complexity. The scope of our approach has been successfully demonstrated on the optimal control of pump-dump processes using the strategy of the intermediate target [36,83]. One of the ultimate goals of the femtochemistry is to design ultrashort pulses in the infrared spectral region which drive ground state isomerization processes and chemical reactions [84]. This opens the perspective to control the photoinduced functionalism of biomolecules [85] and optical switches [86], offering, in combination with Wigner dynamics optimal control, also the additional information about underlying processes. For this purpose, ultrashort IR pulses on a subpicosecond time scale are mandatory in order to "beat" the IVR processes which prevent the efficient localization of energy in selected vibrational modes. Such ultrashort pulses in the mid-IR region have been experimentally realized and successfully used to induce molecular dissociation on a sub-statistical time scale [87]. So far, the IR control of the ground state dynamics based on full quantum treatment has been limited to low dimensional models [88,89].

Here, we introduce Wigner dynamics optimal control of the isomerization processes in electronic ground state on the basis of the MD "on the fly" [41]. This approach allows taking into account all degrees of freedom and is not limited to model potentials. We demonstrate the scope of this new strategy and the broadness of its potential applications on two prototype examples: the Na_3F cluster [90–92] and the glycine amino-acid, which are representatives of highly symmetrical rigid molecules and complex floppy molecules with low frequency modes, respectively. We show that selective isomerization driven by ultrashort tailored laser fields can be achieved and point out that prototype specific processes can be identified in optimized pulse shapes. Thus, the Wigner dynamics optimal control provides an intuitive picture of the MD allowing to identify the underlying processes in the shaped pulses. Therefore, it can be used as a tool for the analysis of dynamical processes, thus achieving the complete goal of the optimal control. Our theoretical approach for the ground state optimal control is based on the classical limit of the quantum Liouville-von Neumann equation for the time evolution of the density operator $i\hbar\hat{\rho} = \left[\hat{H}_0 - \overrightarrow{\mu}\cdot\overrightarrow{E}(t), \hat{\rho}\right]$, where $\hat{H}_0$ is the field free Hamiltonian of the molecular system, and the interaction with the laser field is described within the dipole approximation. The semiclassical limit of a quantum mechanical operator arises most naturally if the Wigner phase space representation is employed:

$$A(\mathbf{q}, \mathbf{p}, t) = \frac{1}{(2\pi\hbar)^N}\int c\mathbf{s}\, e^{-i\mathbf{p}\mathbf{s}/\hbar}\left\langle \mathbf{q} - \frac{\mathbf{s}}{2}|\hat{A}|\mathbf{q} + \frac{\mathbf{s}}{2}\right\rangle, \tag{1}$$

where $(\mathbf{p}, \mathbf{q})$ represent the classical phase space variables. The time evolution of the Wigner phase space density is given by

$$\frac{\partial\varrho(\mathbf{q}, \mathbf{p}, t)}{\partial t} = -\frac{2}{\hbar}H(\mathbf{q}, \mathbf{p})\sin\left(\frac{\hbar}{2}\Lambda\right)\varrho(\mathbf{q}, \mathbf{p}, t), \tag{2}$$

where Λ is a Poisson bracket operator defined as

$$\Lambda = \frac{\overleftarrow{\partial}}{\partial \mathbf{q}} \frac{\overrightarrow{\partial}}{\partial \mathbf{p}} - \frac{\overleftarrow{\partial}}{\partial \mathbf{p}} \frac{\overrightarrow{\partial}}{\partial \mathbf{q}}. \tag{3}$$

The expansion of the operator $\sin(\frac{\hbar}{2}\Lambda)$ in Eq. 2 in a series allows transforming Eq. 2 into

$$\frac{\partial \varrho}{\partial t} = \frac{\partial H}{\partial \mathbf{q}} \frac{\partial \varrho}{\partial \mathbf{p}} - \frac{\partial \varrho}{\partial \mathbf{q}} \frac{\partial H}{\partial \mathbf{p}} - \frac{\hbar^2}{24} \frac{\partial^3 H}{\partial \mathbf{q}^3} \frac{\partial^3 \varrho}{\partial \mathbf{p}^3}. \tag{4}$$

If terms in the lowest order in $\hbar$ only are retained in Eq. 4, a classical Liouville equation of motion under the influence of the laser field $E(t)$ is obtained:

$$\frac{\partial \varrho}{\partial t} = \frac{\partial H_0}{\partial \mathbf{q}} \frac{\partial \varrho}{\partial \mathbf{p}} - \frac{\partial \varrho}{\partial \mathbf{q}} \frac{\partial H_0}{\partial \mathbf{p}} - E(t) \frac{\partial \mu}{\partial \mathbf{q}} \frac{\partial \varrho}{\partial \mathbf{p}}. \tag{5}$$

In general, Eq. 5 is solved by propagating a swarm of classical trajectories with quantum initial conditions sampled from the canonical Wigner distribution function which includes the temperature of the system [36]. For the propagation of trajectories, quantum chemical *ab initio* or semiempirical MD "on the fly" is used, in which both forces and dipole derivatives are calculated analytically. For the treatment of complex systems such as biomolecules surrounded by the solvent, even more approximate methods such as force fields can be implemented allowing also to treat solvent effects.

Notice that Eq. 5 is exact and includes all quantum effects if the Hamiltonian is at most quadratic (harmonic). In the anharmonic case, the inclusion of quantum effects will lead to increased spreading of the wavepacket, but the center of the wavepacket will still evolve according to the classical equations (cf. Ehrenfest theorem). Thus, while our approach does not account for all quantum effects (e.g., tunneling), it describes adequately the main features of the laser driven MD on the considered timescales. This has been explicitly demonstrated on two prototype examples by simulating pump-probe spectra of Ag_3 [93] and Na_3F [91]. In both cases the comparison with full quantum dynamical treatment provided a good agreement.

The laser field which drives the MD and can be represented by any analytic form in the time- or frequency domain. Here we employ

$$E(t) = \sum_{i=1}^{n} E_i \exp\left(-\frac{(t-\alpha_i)^2}{2\beta_i^2}\right) \cos\left(\gamma_i t + \delta_i t^2 + \varepsilon_i\right), \tag{6}$$

representing Gaussian modulated linearly chirped cosine terms and using 12 parameters for $n=2$. The aim of the optimal control is to maximize the expectation value of the quantum mechanical operator $\hat{A}$ in the Wigner representation at the given time t_f:

$$J(t_{\mathrm{f}}) = \int\int \mathrm{d}\mathbf{q}\mathrm{d}\mathbf{p}A(\mathbf{q},\mathbf{p})\varrho(\mathbf{q},\mathbf{p},t), \tag{7}$$

where A(**p, q**) is the target operator. Here, as the population of the desired isomer is maximized, the target operator is defined as a function representing the deviation from the target structure:

$$J(t_{\mathrm{f}}) = \frac{1}{N_{\mathrm{traj}}}\sum_{i}^{N_{\mathrm{traj}}} ||\mathbf{D}_i(t_{\mathrm{f}}) - \mathbf{D}_{\mathrm{isomerII}}||, \tag{8}$$

where $\mathbf{D}_i(t_f)$ and $\mathbf{D}_{\mathrm{isomerII}}$ are distance matrices containing all pairs of interatomic distances for the product ensemble and the desired isomer II, respectively. All pulse parameters are iteratively optimized within the specified range using the standard genetic algorithm with binary coding of parameters and the usual selection, crossover, and mutation operations [94] allowing also to restrict the total energy of the pulse. Alternatively, the pulse optimization in the framework of the optimal control theory requires forward and backward propagation of the phase space ensemble and the calculation of the overlap between them. Therefore, an enormous number of trajectories would be required in order to obtain a smooth representation of the ensemble [95]. Thus, direct optimization with genetic algorithm is more practicable for control of isomerization in complex systems.

In the following, we illustrate our approach on controlling the isomerization processes in the Na_3F cluster and the glycine amino-acid. In both cases, the goal is to transfer the population to the second energetically higher isomer (isomer II) that is not populated initially. Notice that competing processes such as ionization or fragmentation which might occur in particular in strong laser fields are not treated here. The schemes for the infrared control of isomerization in Na_3F and the optimal pulse are shown in Figure 14A and D, respectively. The two isomers of Na_3F differ in energy by 0.09 eV and are separated by the isomerization barrier of 0.12 eV [96–98]. The initial canonical ensemble corresponding to $T = 50$ K consists of 32 trajectories obtained using *ab initio* MD "on the fly" [96]. The optimal pulse presented in Figure 14D drives the isomerization process from the initial to the final ensemble at $t_f = 400$ fs, which are both localized and are well represented by a discrete ensemble consisting of 32 trajectories. The optimal pulse has almost a single cycle [99] sinusoidal shape with a period of 250 fs. This period corresponds to the vibrational normal mode involving the motion of the fluorine atom along the C_2 axis (cf. Figure 14C). The snapshots of the ensemble dynamics are presented in Figure 14C. The selective population transfer from isomer I to isomer II in this symmetric and rigid system (due to highly polar Na–F bonds) is achieved by an ultrafast excitation of a single normal mode (cf. Figure 14B). After 25 generations, an almost uniform pulse generation consisting only of the replicas of the optimal pulse has been obtained.

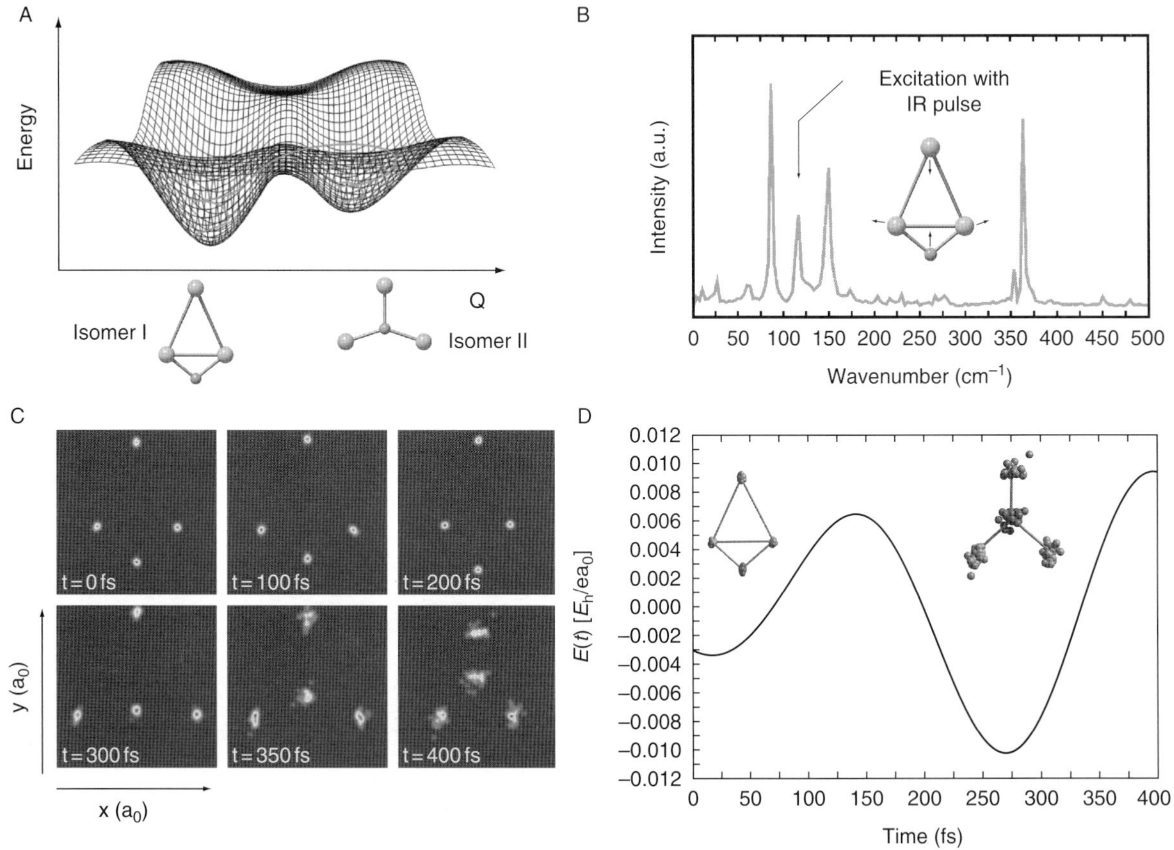

FIGURE 14 (A) Schematic representation of the PES for the isomerization process of the Na_3F cluster from isomer I to isomer II symbolized by the red trajectory. (B) Simulated IR spectrum of the Na_3F cluster obtained from the MD "on the fly" by taking the Fourier transform of the dipole autocorrelation function $I(\omega) = \frac{1}{2\pi}\int_{-\infty}^{+\infty} e^{i\omega t}\langle\mu(t)\mu(0)\rangle dt$. (C) Snapshots of the ensemble dynamics driven by the optimized laser pulse projected onto the (x,y) molecular plane. In order to obtain a smooth phase space distribution, the individual trajectories have been folded by a Gaussian functions with the width $\sigma = 0.05$. (D) Optimal laser pulse driving the isomerization of Na_3F from the initial to the final ensemble which are both shown in as insets. Reprinted from Ref. [78].

In order to demonstrate the scope of our method with possible future applications in biomolecular systems with low frequency modes, we choose the glycine aminoacid as the second prototype. The two gas phase isomers of glycine differ in energy by 0.33 eV. Again, an ensemble of 32 trajectories has been propagated using the semiempirical AM1 method [100]. As the conformational change from the isomer I to the isomer II in glycine involves a low frequency torsional mode, a longer time interval of $t_f = 1500$ fs had to be selected. The optimal pulse driving the isomerization processes and the snapshots of the selected trajectory representing the laser driven dynamics are shown in Figure 15. The optimal pulse has two temporal components: one weak slowly varying subpulse with a half-period of 1500 fs and a second subpulse which is a strong Gaussian centered at 850 fs with a width of 200 fs. The shape of the optimal pulse is very different from the one obtained for Na_3F. This can be attributed to the larger number of degrees of freedom for glycine and to the substantially larger energy separation of both isomers. Therefore, because of both factors, the IVR process is expected to be much faster for glycine. As a consequence, the simple conformational change by

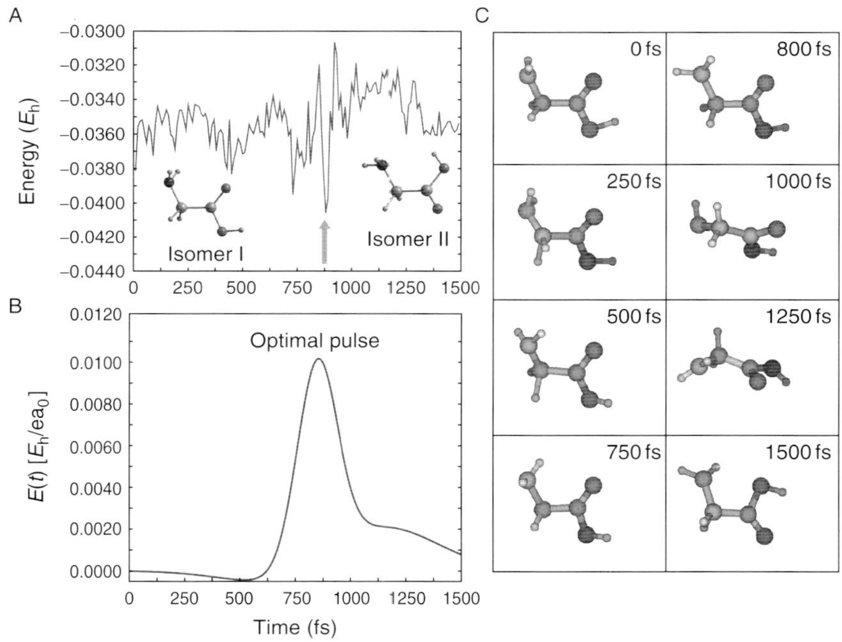

FIGURE 15 (A) Total energy of glycine during isomerization from isomer I to isomer II under the influence of the laser field. The red arrow corresponds to impulsive excitation represented by the maximum of the pulse, (B) Optimal pulse driving the isomerization and (C) Snapshots of the laser driven dynamics. Reprinted from Ref. [78]. (For interpretation of the references to color in this figure legend, the reader is referred to the Web version of this chapter.)

periodic excitation of the torsional normal mode is prevented by fast dissipation of energy which would lead to heating of the molecule. Evidently, the optimal pulse avoids such situation. Instead, the conformational change is achieved by an ultrashort impulsive Gaussian subpulse which induces the isomerization process by pumping $\sim$0.30 eV of energy into the system at $t=850$ fs (cf. Figure 15A). The slowly varying subpulse subsequently cools the molecule so that effectively only 0.07 eV of energy is accumulated. These results represent prototype information, valuable for control of larger biomolecules in which a similar situation is expected.

In summary, we have presented Wigner dynamics optimal control of the ground state isomerization in complex systems by shaped infrared fields allowing to include all degrees of freedom and without the need for using precalculated PESs. We have demonstrated that the mechanism underlying optimal control can be significantly different in symmetric rigid systems and in molecules with floppy low frequency modes. Our approach provides a basis to control the conformational changes in functional biomolecules by shaped laser pulses and also provides information about underlying processes. In particular, our combination of the optimal control with MD "on the fly" using the whole spectrum of methods, from ab initio quantum chemical methods to parameterized force fields, offers a powerful tool for infrared driven control of a complex molecular system in the gas phase as well as in different environments.

VII. CONCLUSIONS

We have presented the study of the optical and emissive properties of silver clusters at surfaces and their interaction with tryptophan. We showed that the role of support is mainly to stabilize the clusters. The influence of different support defects on the absorption and emission properties of silver clusters needs still to be further investigated. We have characterized the absorption properties of cationic silver clusters interacting with tryptophan and shown that size and structural classes of the complexes are responsible for different types of CT in excited states which can be identified by photofragmentation channels. The role of excitations within both subunits of these hybrid systems is equally important and can lead to enhancement of either absorption or emission. The results on both topics serve as a starting point toward understanding of processes which might open interesting applicational aspects such as optical storage and biosensing.

Our dynamical studies presented here serve two purposes. First, we introduce the nature of IVR as new criterion for promoting reactivity of noble metal clusters toward adsorbates. Resonant IVR between the cluster and O_2 is responsible for an activation mechanism which enhances the oxidation reactivity. This occurs in the case of Au_6^--O_2 collisions. In contrast, dissipative IVR is unfavorable, because it increases the sticking probability of O_2 as in the case of Ag_6^--O_2 collisions. The reason for the different behavior of gold and silver clusters is due to pronounced relativistic effects in gold which cause rigidity in the structures of gold clusters preventing dissipative IVR. The later is more favorable for floppy systems such as silver clusters because of s-like metallic character. Second, the study of dynamical properties of excited states of anionic gold clusters Au_5^- versus Au_7^- demonstrates that by changing the cluster size by two atoms the bound excited states in Au_5^- become repulsive, causing IC for Au_7^-. This photo-induces a melting type of behavior in the ground state because of the large excess of energy. These processes can be revealed from simulated and measured TRPE spectra. Finally, in order to demonstrate how tailored laser pulses can be used to drive isomerization over a barrier, we proposed a new strategy for IR control on the basis of the Wigner representation of quantum mechanics and MD "on the fly" which is applicable to complex systems considering all degrees of freedom. This has been illustrated again on two prototype systems, a rigid one such as Na_3F and a floppy one such as the glycine molecule. The resulting optimal pulses reflect the underlying dynamical processes: A periodic pulse is responsible for selective mode excitation to drive the isomerization process in the rigid Na_3F system. Participation of a considerably higher number of normal modes in the case of isomerization control in the glycine molecule gives rise, in addition to the impulsive component, to longer components of the pulse, which have a specific role in the control of isomerization processes in floppy molecules. Therefore, we could show that dynamical processes can be assigned directly to pulse shapes providing information about the mechanism driving one process such as isomerization and suppressing another one such as IVR. This clearly illustrates that complex systems are also controllable and therefore encourages new experiments.

Acknowledgments

We wish to thank to A. Kulesza for his contribution on silver clusterbiomolecule hybrid systems as well as to our experimental partners Prof. M. Broyer, Dr. P. Dugourd, Dr. R. Antoine, Prof. W. Eberhardt, and Dr. M. Neeb for fruitful cooperation. We are grateful to the Deutsche Forschungsgemeinschaft, SFB 450, for financial support.

REFERENCES

[1] König L, Rabin I, Schulze W, Ertl G. Science 1996;274:1353.
[2] Rabin I, Schulze W, Ertl G. J Chem Phys 1998;108:5137.
[3] Rabin I, Schulze W, Ertl G, Félix C, Sieber C, Harbich W, Buttet J. Chem Phys Lett 2000;320:59.
[4] Félix C, Sieber C, Harbich W, Buttet J, Rabin I, Schulze W, Ertl G. Phys Rev Lett 2001;86:2992.

[5] Sieber C, Buttet J, Harbich W, Félix C, Mitrić R, Bonačić-Koutecký V. Phys Rev A 2004;70:041201.
[6] Diederich T, Tiggesbäumker J, Meiwes-Broer KH. J Chem Phys 2002;116:3263.
[7] Peyser LA, Vinson AE, Bartko AP, Dickson RM. Science 2001;291:103.
[8] Gleitsmann T, Stegemann B, Bernhardt TM. Appl Phys Lett 2004;84:4050.
[9] Lee T-H, González JI, Zheng J, Dickson RM. Acc Chem Res 2005;38:534.
[10] Gleitsmann T, Bernhardt TM, Wöste L. Appl Phys A 2006;82:125.
[11] Bonačić-Koutecký V, Bürgel C, Kronik L, Kuznetsov AE, Mitrić R. Eur J Phys D 2007;45:471.
[12] Perdew JP, Burke K, Ernzerhof M. Phys Rev Lett 1996;77:3865.
[13] Andrae D, Haeussermann U, Dolg M, Stoll H, Preuss H. Theor Chim Acta 1990;77:123.
[14] Bonačić-Koutecký V, Pittner J, Boiron M, Fantucci P. J Chem Phys 1999;110:3876.
[15] Bonačić-Koutecký V, Veyret V, Mitrić R. J Chem Phys 2001;115:10450.
[16] Moskovits M. Rev Mod Phys 1985;57:783–826.
[17] Katz E, Willner I. Angew Chem Int Ed 2004;43:6042.
[18] Zhao LL, Jensen L, Schatz GC. Nano Lett 2006;6:1229.
[19] Peyser-Capadona L, Zheng J, González JI, Lee T-H, Patel SA, Dickson RM. Phys Rev Lett 2005;94:058301.
[20] Antoine R, Tabarin T, Broyer M, Dugourd P, Mitrić R, Bonačić-Koutecký V. Chemphyschem 2006;7:524.
[21] Compagnon I, Tabarin T, Antoine R, Broyer M, Dugourd P, Mitrić R, Petersen J, Bonačić-Koutecký V. J Chem Phys 2006;125:164326.
[22] Mitrić R, Petersen J, Kulesza A, Bonačić-Koutecký V, Tabarin T, Compagnon I, Antoine R, Broyer M, Dugourd P. J Chem Phys 2007;127:134301.
[23] Becke AD. J Chem Phys 1993;98:5648.
[24] Lee C, Yang W, Parr RG. Phys Rev B 1988;37:785.
[25] Tully JC. J Chem Phys 1990;93(2):1061.
[26] Granucci G, Persico M, Toniolo A. J Chem Phys 2001;114:10608.
[27] Haruta M. Catal Today 1997;36:153.
[28] Valden M, Lai X, Goodman DW. Science 1998;281:1647.
[29] Sanchez A, Abbet S, Heiz U, Schneider WD, Häkkinen H, Barnett RN, Landman U. J Phys Chem A 1999;103:9573.
[30] Yoon B, Häkkinen H, Landman U, Wörz AS, Antonietti J-M, Abbet S, Judai K, Heiz U. Science 2005;307:403.
[31] Gates BC, editor. Catalytic chemistry. New York, Singapore: Wiley; 1992.
[32] Kimble ML, Castleman Jr. AW, Mitrić R, Bürgel C, Bonačić-Koutecký V. J Am Chem Soc 2004;126:2526.
[33] Socaciu LD, Hagen J, Bernhardt TM, Wöste L, Heiz U, Häkkinen H, Landman U. J Am Chem Soc 2003;125:10437.
[34] Wallace WT, Whetten RL. J Am Chem Soc 2002;124:7499.
[35] Lee TH, Ervin KM. J Phys Chem 1994;98:10023.
[36] Bonačić-Koutecký V, Mitrić R. Chem Rev 2005;105:11.
[37] Salisbury BE, Wallace WT, Whetten RL. Chem Phys 2000;262:131.
[38] Ho J, Ervin KM, Lineberger WC. J Chem Phys 1990;93:6987.
[39] Bernhardt TM. Int J Mass Spectrom 2005;243:1.
[40] Mitrić R, Bürgel C, Bonačić-Koutecký V. Proc Natl Acad Sci USA 2007;104:10314.

[41] Mitrić R, Bonačić-Koutecký V. In: Kühn O, Wöste L, editors. Analysis and control of ultrafast photoinduced reactions. Springer series in chemical physics, vol. 87. Berlin/Heidelberg: Springer-Verlag; 2007.
[42] Cox DM, Brickman RO, Creegan K, Kaldor A. Z Phys D 1991;19:353.
[43] Kim YD, Ganteför G. Chem Phys Lett 2004;383:80.
[44] Socaciu LD, Hagen J, Le Roux J, Popolan D, Bernhardt TM, Wöste L, Vajda S. Chem Phys 2004;120:2078.
[45] Bonačić-Koutecký V, Burda J, Mitrić R, Ge M, Zampella G, Fantucci R. J Chem Phys 2002;117:3120.
[46] Hakkinen H, Moseler M, Landman U. Phys Rev Lett 2002;89(3):033401.
[47] Furche F, Ahlrichs R, Weis P, Jacob C, Gilb S, Bierweiler MM, Kappes T. J Chem Phys 2002;117:6982.
[48] Bonačić-Koutecký V, Fuchs C, Gaus J, Pittner J, Koutecký . Z Phys D 1993;26:192.
[49] Bonačić-Koutecký V, Češpiva L, Fantucci P, Koutecký J. J Chem Phys 1994;100:490.
[50] Lochbrunner S, Larsen JJ, Shaffer JP, Schmitt M, Schultz T, Underwood JG, Stolow A. J Electron Spectros Relat Phenomena 2000;112(1–3):183.
[51] Stolow A. Annu Rev Phys Chem 2003;54:89.
[52] Assion A, Geisler M, Helbing J, Seyfried V, Baumert T. Phys Rev A 1996;54(6):R4605.
[53] Blanchet V, Zgierski MZ, Seideman T, Stolow A. Nature 1999;401(6748):52.
[54] Luttgens G, Pontius N, Bechthold PS, Neeb M, Eberhardt W. Phys Rev Lett 2002;88 (7):076102.
[55] Pontius N, Bechthold PS, Neeb M, Eberhardt W. Phys Rev Lett 2000;84(6):1132.
[56] Verlet JRR, Bragg AE, Kammrath A, Cheshnovsky O, Neumark DM. J Chem Phys 2004;121(20):10015.
[57] Bragg AE, Verlet JRR, Kammrath A, Cheshnovsky O, Neumark DM. J Chem Phys 2005;122:054314.
[58] Hakkinen H, Yoon B, Landman U, Li X, Zhai HJ, Wang LS. J Phys Chem A 2003;107 (32):6168.
[59] Gilb S, Jacobsen K, Schooss D, Furche F, Ahlrichs R, Kappes MM. J Chem Phys 2004;121(10):4619.
[60] Stanzel J, Burmeister F, Neeb M, Eberhardt W, Mitrić R, Bürgel C, Bonačić-Koutecký V. J Chem Phys 2007;127:164312.
[61] Pontius N, Bechthold PS, Neeb M, Eberhardt W. Appl Phys B 2000;71(3):351.
[62] Luttgens G, Pontius N, Friedrich C, Klingeler R, Bechthold PS, Neeb M, Eberhardt W. J Chem Phys 2001;114(19):8414.
[63] Spasov VA, Shi Y, Ervin KM. Chem Phys 2000;262(1):75.
[64] Hartmann M, Pittner J, Bonačić-Koutecký V, Heidenreich A, Jortner J. J Chem Phys 1998;108:3096.
[65] Bonačić-Koutecký V, Mitrić R. Chem Rev 2005;105:11.
[66] Gilb S, Weis P, Furche F, Ahlrichs R, Kappes MM. J Chem Phys 2002;116(10):4094.
[67] Schmidt M, Kusche R, Kronmuller W, vonIssendorff B, Haberland H. Phys Rev Lett 1997;79(1):99.
[68] Tannor DJ, Rice SA. J Chem Phys 1985;83:5013.
[69] Shapiro M, Brumer P. J Chem Phys 1986;84:4103.
[70] Peirce AP, Dahleh MA, Rabitz H. Phys Rev A 1988;37:4950.
[71] Rice SA, Zhao M. Optical control of molecular dynamics. New York: Wiley; 2000.
[72] Brumer PW, Shapiro M. Principles of the quantum control of molecular processes. Berlin: Wiley-VCH; 2003.

[73] Assion A, Baumert T, Bergt M, Brixner T, Kiefer B, Seyfried V, Strehle M, Gerber G. In: Elsasser T, Fujimoto JG, Wiersma DA, Zinth W, editors. Ultrafast phenomena XI. Springer series in chemical physics, vol. 63, Berlin: Springer-Verlag;1998.
[74] Brixner T, Gerber G. Chemphyschem 2003;4:418.
[75] Judson RS, Rabitz H. Phys Rev Lett 1992;62:1500.
[76] Daniel C, Full J, Gonzalez L, Lupulescu C, Manz J, Merli A, Vajda S, Wöste L. Science 2003;299:536.
[77] Schäfer-Bung B, Mitrić R, Bonačić-Koutecký V, Bartelt A, Lupulescu C, Lindinger A, Vajda SM, Weber SM, Wöste L. J Phys Chem A 2004;108:4175.
[78] Mitrić R, Bonačić-Koutecký V. Phys Rev A 2007;76:031405(R).
[79] The basic idea of the MD "on the fly" is to compute forces from the electronic structure calculations "on the fly". This method, introduced by Car and Parinello [80] based on DFT was originally used for the ground state dynamics and was later extended for different quantum chemical methods applicable to ground and excited electronic states [36].
[80] Car R, Parrinello M. Phys Rev Lett 1985;55:2471.
[81] Wigner E. Phys Rev 1932;40:749.
[82] Hillary M, O'Connel RF, Scully MO, Wigner EP. Phys Rep 1984;106:1984.
[83] Mitrić R, Hartmann M, Pittner J, Bonačić-Koutecký V. J Phys Chem A 2002;106:10477.
[84] Zare RN. Science 1875;1998:279.
[85] Vogt G, Krampert G, Niklaus P, Nuerenberger P, Gerber G. Phys Rev Lett 2005;94:068305.
[86] Vogt G, Nuerenberger P, Brixner T, Gerber G. Chem Phys Lett 2006;433:211.
[87] Windhorn L, Yeston JS, Witte T, Fuß W, Motzkus M, Proch D, Kompa KL, Moore CB. J Chem Phys 2003;119:641.
[88] Jakubetz W, Manz J, Schreier HJ. Chem Phys Lett 1990;165:100.
[89] Gollub C, Korff BMR, Kompa KL, de Vivie-Riedle R. Phys Chem Chem Phys 2007;9:369.
[90] Durand G, Heitz M-C, Spiegelman F, Meier C, Mitrić R, Bonačić-Koutecký V, Pittner J. J Chem Phys 2004;121:9898.
[91] Heitz M-C, Durand G, Spiegelman F, Meier C, Mitrić R, Bonačić-Koutecký V. J Chem Phys 2004;121:9906.
[92] Mitrić R, Bonačić-Koutecký V, Pittner J, Lischka H. J Chem Phys 2006;125:024303.
[93] Andrianov I, Bonačić-Koutecký V, Hartmann M, Manz J, Pittner J, Sundermann K. Chem Phys Lett 2000;318:256.
[94] Goldberg DE. Genetic algorithms in search, optimization and machine learning. Addison-Wesley:Boston; 1989.
[95] Donoso A, Martens CC. Phys Rev Lett 2001;87:223202.
[96] Both Hartree-Fock and DFT with the hybrid B3LYP functional have been used for determination of structures and MD "on the fly."
[97] Lee C, Yang W, Parr RG. Phys Rev B 1985;37:785.
[98] Becke AD. Phys Rev A 1988;98:3098.
[99] Uiberacker C, Jakubetz W. J Chem Phys 2004;120:11532.
[100] Dewar MJS, Zoebisch EG, Healy EF, Stewart JJP. J Am Chem Soc 1985;107:3902.

Chapter 14

Interfacing Cluster Physics with Biology at the Nanoscale

Carl Leung*,† and Richard E. Palmer*
*Nanoscale Physics Research Laboratory, School of Physics and Astronomy, The University of Birmingham, Edgbaston, Birmingham, United Kingdom
†London Centre for Nanotechnology, London, United Kingdom

Chapter Outline Head

Nanoclusters. DOI: 10.1016/S1875-4023(10)01014-4

I. INTRODUCTION

In his visionary lecture in 1959, Richard Feynman spoke of physical tools capable of making imprints at the nanoscale, moving atoms to create new structures, adapting biological systems to store information, and deploying them for useful purposes. In the years to follow, many of these predictions have become a reality in the form of microcontact printing [1], atomic manipulation [2], and biological sensors, respectively [3]. In the same decade, John Kendrew and his coworkers were studying the atomic structure of proteins by X-ray crystallographic methods, which culminated in the three-dimensional model of myoglobin at 6 Å resolution in 1957 [4]. Kendrew subsequently received the Nobel Prize in 1962 for laying the foundations of a new era of protein science. In half a century since the pioneering work of Kendrew and the vision of Feynman, mainstream biology and physics have been converging rapidly, with the emergence of newly coined disciplines such as "bio-nanotechnology."

A recent contribution in the field has been the creation of nanostructured surfaces, with lateral feature sizes between 2 and 10 nm, by depositing clusters onto selected surfaces at specific energies [5,6]. The size of these deposited clusters corresponds to the length scale of individual protein molecules, which creates the exciting possibility of studying the interface between inorganic and biological systems on the nanometer-scale by surface science techniques. The approach has also allowed the demonstration of protein biochips with enhanced sensitivity.

II. PRODUCTION AND DEPOSITION OF SIZE-SELECTED CLUSTERS

Figure 1 shows the cluster beam source used to generate the size-selected clusters in this work. The source consists of three differentially pumped sections: sputtering and cluster formation, beam focusing and acceleration, and, finally, mass selection.

The cluster production takes place in the inner liquid-nitrogen-cooled chamber (see schematic inset of Figure 1). A magnetron is mounted on a long axial support, enabling the distance between the front of the magnetron and the end of the chamber to be varied. The sputter gas (Ar) is injected from small orifices at the front of the magnetron, and the Ar plasma is ignited by applying a radio frequency high voltage. The sputtering process creates a dense vapor of atoms, ions, and small clusters in front of the target, and cluster condensation from this vapor is induced by helium He gas injected from the back of the chamber. The cluster and gas mixture leaves the inner

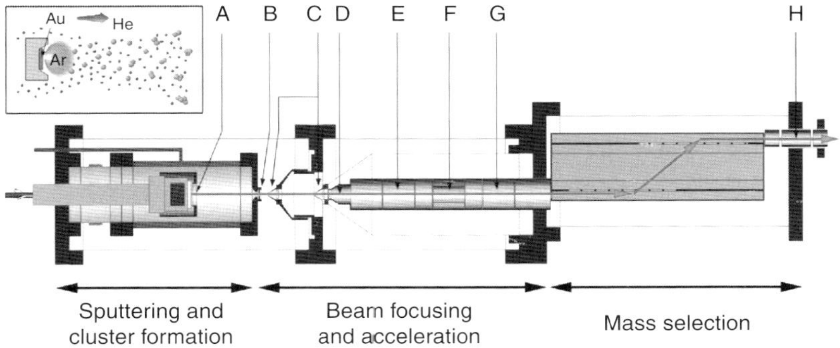

FIGURE 1 Setup of the magnetron gas condensation cluster source with a lateral time-of-flight mass filter for deposition and pinning of mass-selected nanoclusters. (A) The magnetron gun with a gold target; (B) adjustable-diameter nozzle; (C) electrostatic skimmers; (D) high-voltage lens; (E) Einzel lens for controlling the kinetic energy of the clusters; (F) *X–Y* deflector plates for spatial control of the cluster beam; and (G) and (H) Einzel lenses. The entire setup is differentially pumped, and the mass filter operates at pressures of 10^{-7} mbar. Inset: Formation of ionized (Au) nanoclusters in the argon plasma by gas aggregation. Arrows indicate the trajectory of the clusters in the instrument. Only clusters of the selected size will be transmitted and deposited at the exit of the cluster source. The deposition stage is omitted in this diagram. Figure adapted from Ref. [7].

chamber via an adjustable nozzle, allowing independent control of the gas flow rate and gas pressure. Two conical skimmers are placed after the nozzle. The gas jet undergoes supersonic expansion through the nozzle before entering the first skimmer. The adjustable mounting of this skimmer allows its distance from the nozzle to be varied, allowing appropriate selection of the central portion of the supersonically expanding gas jet.

Since a significant proportion of the clusters are ionized in the plasma, positive cluster ions can be accelerated and focused into an ionized cluster beam by the negative potentials applied to a set of ion optics. Small negative voltages are also applied to both the nozzle and skimmers, and these three apertures practically serve as the first lens elements for forming the cluster beam. The ion optics are used to generate a well-focused ion beam that is required for the mass selection process by the transverse time-of-flight (TOF) mass filter which operates by the lateral deflection of the cluster beam.

The lateral TOF mass selection principle is illustrated in Figure 2. As the cluster beam enters the mass filter, the TOF elements are all at the same negative voltage corresponding to the beam energy. The top and bottom plates are connected to high-voltage switches. A section of the incoming beam experiences the first perpendicular acceleration pulse provided by the bottom plate, which raises its potential (to zero). After this first pulse, all the plates

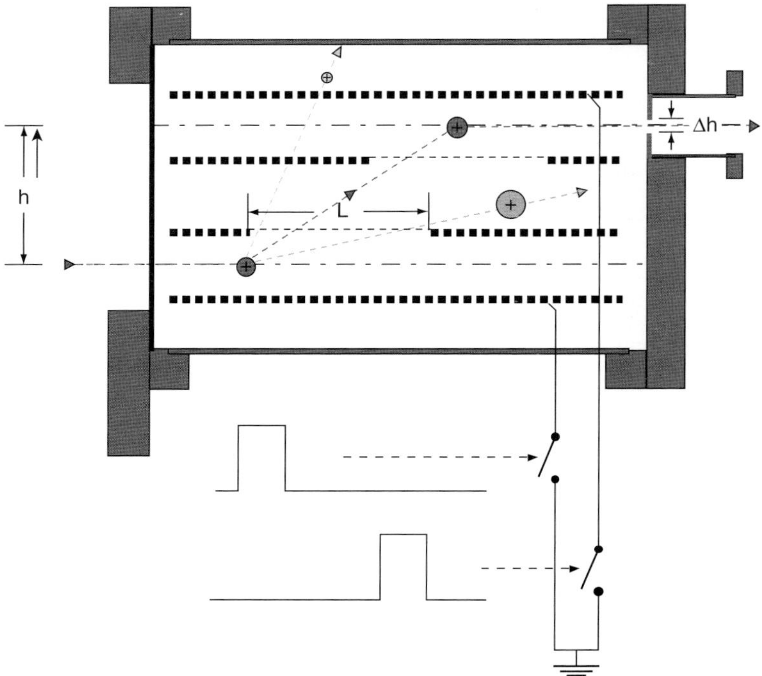

FIGURE 2 A cross-sectional schematic of the time-of-flight mass filter for cluster deposition experiments. The top and bottom plates of the TOF are pulsed to displace the incoming cluster beam laterally and only ions of the selected mass (colored in red in the schematic) can reach the exit aperture. L is the useful part of the cluster beam for each pulsing cycle, h the lateral displacement of the cluster beam, and Δh is the width of the exit aperture as described in the main text. (For interpretation of the references to color in this figure legend, the reader is referred to the Web version of this chapter.)

are biased back to the beam potential, while the accelerated section of the beam drifts upward in the central field-free region. The drift velocities are dependent on the ion masses. A second pulse identical to the first is then applied to the top plate to cancel the initial acceleration. The particles with different mass-to-charge (m/e) ratios are vertically separated. Only a certain portion can pass through the exit aperture. With a typical beam energy of less than 1 keV, it can be shown that the central selected mass m is given by

$$m = \frac{eU_a\alpha}{d_1 x f^2}$$

where α is the delay between opposite pulses, U_a the voltage applied to the acceleration/deceleration plates, x the total displacement of the transmitted ions, d_1 the distance between the acceleration and deceleration plates, e the charge of the ions, and f the frequency of the square wave pulses. Since the incoming beam has no initial perpendicular velocity, and hence no vertical velocity distribution, the mass resolution is independent of mass. It is, in fact, equal to the ratio $h/\Delta h$, where Δh is the width of the exit aperture. In the current setup, the theoretical mass resolution is $m/\Delta m \sim 25$. In practice, the absolute value of the mass has to be calibrated from a peak of "known" mass. The cluster source can be tuned to produce only Ar^+ ions, which are therefore used for the calibration of other mass peaks.

In conventional (linear) TOF mass spectrometry, the separation of different masses occurs in the same direction as the beam propagation. Accordingly, only a small section of the beam can be selected. With this perpendicular TOF mass filter, the transmission of the beam has been measured experimentally to be more than 50%. A Faraday cup is installed at the end of the lower section of the TOF for measuring the current of the "white" (non-mass-selected) beam. The transmission efficiency was measured by comparing the white beam current with the size-selected beam current for an incoming Ar ion beam. The theoretical value depends on the geometry of the TOF mass selector. The lower plate of the drift section has an opening of length L for the beam to pass through. This determines the portion of the incoming beam that can be transmitted during one cycle of mass selection. The mass range of this TOF mass filter is not limited by theoretical factors. In practice, the maximum frequency limit (600 kHz) of the high-voltage switches constrains the lowest detectable mass to be around that of He^+. The upper limit of the mass filter has not been determined; however, Au clusters of 20 nm in diameter have been successfully mass-selected with this system.

After the TOF, another ion optics set focuses the size-selected cluster beam into a high-vacuum chamber for cluster deposition. The cluster beam current is detected by a simple Faraday cup setup, consisting of an aperture in front of a metal plate, and is measured by an electrometer at a noise level of approximately 1 pA. The Faraday cup is attached to a vertical linear manipulator which also contains a sample holder accommodating 16 samples. The sample holder can be biased with a high voltage, which defines the deposition energy of the clusters on the surface.

III. CREATION OF CLUSTER-DECORATED SURFACES

The energetic beam of ionized clusters can be deposited on a substrate at various kinetic energies. Depending on the deposition energy employed, the nanoclusters can, in the case of the model graphite surface, (i) diffuse across the surface (low impact energy, <10 eV per atom [8]), (ii) implant into the surface (high impact energy, >50 eV per atom [9]), or (iii) (of most interest for the present work) attach to their point of impact on the surface at intermediate deposition energies (between 10 and ~ 50 eV/atom [5]). The underlying mechanism for this "pinning" process is the displacement of a surface carbon atom to create a reactive binding site which prevents the characteristic diffusion and aggregation of clusters observed at lower incident energies. The exact impact energy required for pinning or implantation varies with the cluster material as well as the size of the cluster selected.

For fundamental studies, the size-selected clusters are typically deposited in a circular section of about 3 mm in diameter onto, for example, a rectangular graphite substrate (10×5 mm). A typical deposition time for a sample is 20 min at a current of 2 pA, which results in approximately 1.5×10^9 clusters on the graphite surface. Following the deposition and pinning of the clusters, the samples are transferred to an inert argon atmosphere. The cluster films are stable not only at room temperature but also at temperatures above 200 °C, depending on the impact energy of the clusters [10]. They are also stable if placed in an autoclave (130 °C for 2 h in high-pressure steam) to sterilize the surface. Many other physical properties can be expected to be sensitive to the cluster morphology, and varying the cluster shape via the impact energy is a potential method of "tuning" the film behavior. The surfaces in our experiments can thus be rather uniformly structured at the nanoscale level with lateral features typically of the order of a few nanometers in size.

Figure 3A shows a scanning tunneling microscope (STM) image of a gold-nanocluster film (Au_{55}) on graphite, prepared using the magnetron sputtering cluster source. The total number of nanoclusters per square micrometer is roughly 1100, which corresponds to about a billion nanoclusters in a typical 3-mm diameter circle on graphite. The distribution of the pinned nanoclusters across the surface of the graphite is random. Figure 3B shows a close-up view of a single Au_{55} nanocluster from Figure 3A. Each of these nanoclusters consists of 55 ± 1 gold atoms (in accordance with the resolution of the mass filter). The average cluster height is 0.5 nm, and the average cluster–cluster distance, which is controlled by the cluster beam deposition time, is 25 nm. The selection of gold nanoclusters containing <100 atoms, with a diameter typically <3 nm, is made with the aim of ensuring that only one protein can interact with the surface of a single nanocluster. The deposition of size-selected clusters using this mass filter is applicable for different types of surfaces, for example, amorphous carbon for transmission electron microscopy (TEM) studies (Figure 3C) and insulating surfaces (Figure 3D), while also providing the option to use different cluster sizes and target materials (Figure 3E and F).

IV. PHYSICAL MEASUREMENTS OF PROTEIN MOLECULES

Proteins are the working molecules of the cell, and carry out the program of activities encoded by the genes [15]. They fulfill a number of activities which include catalyzing a range of chemical reactions, providing structural rigidity to the cell, controlling the flow of material through membranes, acting as sensors and switches, causing motion, and controlling gene function, among

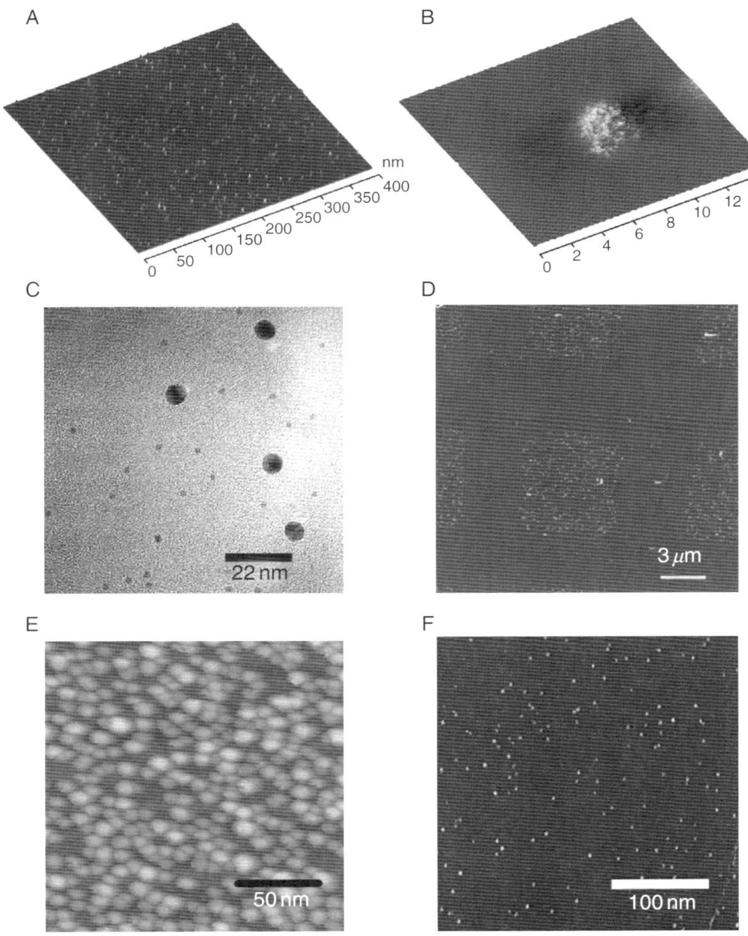

FIGURE 3 Examples of size-selected clusters deposited on surfaces. (A) 400 × 400 nm STM scan of Au_{55} clusters pinned onto a graphite substrate. (B) A high-resolution image of a single Au_{55} cluster from (A) with the graphite lattice visible in the background. (C) TEM image of size-selected $Au_{10,000\pm500}$ and $Au_{330\pm17}$ clusters, co-deposited on an amorphous carbon film [11]. (D) Tapping-mode AFM of size-selected $Au_{20,000}$ clusters deposited on a PMMA thin film coated quartz substrate through a copper mesh [12]. (E) STM image of Ag_{2700} clusters deposited on pre-sputtered graphite at a bias voltage of 1.0 V and a tunneling current of 0.1 nA [13]. (F) STM image of Pd_{150} clusters pinned on graphite at 1.8 keV, obtained with a surface bias voltage of 0.3 V and tunneling current of 0.1 nA [14].

others. Many proteins of interest are in the size range of 2–10 nm and thus they are truly nanoscale machines designed for specific tasks in the body. However, it is only when a protein adopts its correct three-dimensional structure or conformation that it is able to function efficiently, failure of which may lead to diseases such as Alzheimer's disease, Parkinson's disease, Creutzfeldt–Jakob disease, and so on [16]. The spatial organization of proteins is a key to understanding how they work and is a primary focus in the emerging field of proteomics. Proteomics, the large-scale study of protein structure and function, can be broken down into several activities which include (a) identifying all the proteins made in a given cell, tissue, or organism; (b) determining how those proteins form networks akin to electrical circuits; and (c) probing the precise three-dimensional structures of the proteins in an effort to find where drugs might turn their activity off or on. It is precisely in these latter activities that nanoscale science and technology can be advantageously exploited to provide new insights on the working of proteins.

V. ANALYTICAL TOOLS FOR PROTEIN STRUCTURE DETERMINATION

In the last decade, there has been growing interest in coupling proteomics with nanotechnology, as new tools are being developed which can probe protein structures at the nanometer or even subnanometer level. Traditionally, the primary tools for structural examination of biological molecules have been X-ray diffraction (XRD) and nuclear magnetic resonance (NMR). Both of these techniques are capable of yielding angstrom resolution of protein structures. However, new microscopy techniques, especially TEM and atomic force microscopy (AFM), have emerged, which can overcome the limitations associated with XRD and NMR (see Table 1 for a qualitative comparison of the microscopy techniques). AFM is particularly useful in probing the tertiary and quaternary levels of the structure of proteins and their complexes, where its nanometer (vs. angstrom) scale resolution is still valuable, and especially where surface interactions are the focus.

VI. AFM FOR IMAGING PROTEINS ON SURFACES

The AFM was introduced in 1986 by Binnig, Quate, and Gerber as an alternative scanning probe microscopy technique [17]. In its simplest form, it consists of a cantilever with a sharp tip at its end (see Figure 4). The tip is brought in close proximity to a sample surface, and the force between the tip and the sample leads to a deflection of the cantilever. Typically, the deflection is measured using a laser spot reflected from the top of the cantilever into a position-sensitive photodetector. The laser signal measured at the photodetector as the tip is scanned across the surface is then used to determine the vertical displacement of the tip due to surface features. A feedback mechanism is usually employed to adjust the tip–sample distance to keep the force on the sample constant. The movement of the tip is controlled by a tube scanner made from piezoelectric ceramics allowing sub-angstrom positioning over

TABLE 1 A Comparison of Microscopy Techniques for Characterizing Biological Molecules

	TEM	STM	AFM
Spatial resolution	Limited by lens aberration, usually a few angstroms	Approximately 1 Å	Usually >10 nm, determined by tip size
Sample preparation	Freeze drying, replication, staining	Direct deposition onto substrate—requires drying, shadowing	Direct deposition onto substrate—no drying necessary
Area under examination	From hundreds of micrometers to nanometers	Few micrometers to nanometer scan sizes	Tens of micrometers to approximately 100 nm scan sizes
Imaging mechanism	Mass thickness contrast	Electron tunneling, ionic currents?	Force interaction between tip and sample
Raw data	Two-dimensional image	Topographical, conductivity	Topographical information
Operating environment	High vacuum	Usually in air or vacuum	Ambient air, vacuum, or liquid
Sample requirement	Electron transparent material	Smooth, electronically homogeneous sample preferred	Smooth samples preferred
Substrate requirement	Thin films, for example, amorphous carbon	Conductive substrate and immobilization of sample	Immobilization of sample onto substrate

AFM, atomic force microscopy; STM, scanning tunneling microscope; TEM, transmission electron microscopy.

the surface. Aimed at overcoming the requirement for electrically conducting samples imposed by STM, the AFM has proven to be a valuable tool for characterizing biological molecules, with one of the main attractions for biologists being its capability of imaging even in aqueous environments, which is a prerequisite for observing proteins at work.

Membrane proteins were among the first group of proteins to be studied by the AFM [19,20]. In these studies, the proteins were generally found to self-assemble into regular arrays with long-range order. Even at this early stage, structural features in the arrays could be resolved. Soon afterward, the usefulness of AFM was demonstrated in the imaging of a variety of biological materials including live cells [21], gap junctions [22], and DNA molecules [23],

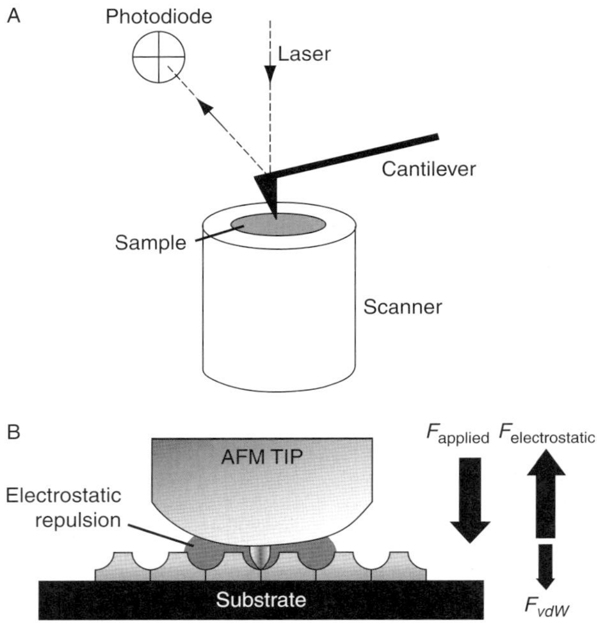

FIGURE 4 Atomic force microscopy (AFM) for protein imaging. (A) The basic operation of the AFM. The AFM produces a topographic image of a sample by scanning the surface with a sharp, nanometer-scale probe (tip) attached to a flexible cantilever. A laser beam is made to fall on the cantilever, and deflections in the laser beam as the cantilever scans the surface are sensed by a photodiode and used to reconstruct the image. (B) Forces interacting between the AFM tip and sample in electrolyte solution. While the electrostatic double layer interacts via long-range forces with a larger area of the macromolecular assembly, the short-range van der Waals attraction and Pauli repulsion interact with individual microscopic protrusions. The force effectively interacting at the tip apex is a composite of all interacting forces. If the force due to the electrostatic double layer is negligible or eliminated, the effective force is equal to the sum of the applied force and the attractive van der Waals force. However, a sufficiently high electrostatic double layer force will partially compensate for the applied force and the van der Waals attraction. Thus, under these conditions, the effective force is smaller than the applied force. Adapted from Ref. [18].

to name but a few. The number of biological systems studied using AFM is too large to detail here, and comprehensive reviews include Refs. [24,25].

In the above-mentioned pioneering work, the spatial resolution achievable by AFM was a few nanometers at best. It was generally thought that the ultimate spatial resolution of the AFM would be limited by the size of the microscope tip. However, progress in improving the spatial resolution of protein topographs has been made in several laboratories by optimizing parameters such as sample preparation and image acquisition methods as well as by continuous developments of the instrumentation. In contrast to some microscopy techniques, the high signal-to-noise ratio of the AFM topograph allows not only the imaging of single molecules but also the observation of their structural details. The breakthrough in attaining subnanometer spatial resolution was achieved by Engel and coworkers, who demonstrated that by careful control of the electrolyte concentration, the protein surface can be imaged under electrostatically balanced conditions, resulting in a local reduction of the vertical forces [18]. In such conditions, the deformation of the native protein is minimized, and the sample surface can be reproducibly contoured at a lateral resolution of 0.6 nm or better (see Figure 4B).

The AFM can also acquire data in real time, which can be advantageously exploited to detect (slow) molecular motions. For example, in the work by Viani et al. [26], a prototype small AFM cantilever was used to observe the association and dissociation of individual GroEL–GroES complexes (in the presence of ATP) in real time. When the tip is scanned in one dimension (enabling higher temporal resolution—up to about 15 Hz), height variations corresponding to time-dependent changes in the protein structure were observed. Such valuable information is, of course, much more difficult or impossible to obtain using conventional techniques such as TEM or XRD.

AFM studies can be categorized as topographical or non-topographical applications. The first group includes all the studies already discussed, where the main objective consists of obtaining an image of the sample surface for structural or sometimes dynamic characterization. Non-topographical applications represent a promising and interesting area, as the interaction between tip and sample is exploited to yield new data, such as the study of inter and intramolecular forces in proteins. In these examples, the tip of the microscope can be used not only as a probe but also as a tool for sample manipulation, allowing it to drag [27], dissect [28], or remove individual proteins in biological assemblies [29]. In some cases, imaging at subnanometer resolution is a prerequisite to observe single biological macromolecules before using the tip to induce controlled conformational changes in the structure of proteins.

The non-topographical applications of the AFM can provide quantitative data through the measurement of surface forces, elasticity, adhesion, rigidity, friction, viscosity, and so on. These properties are usually measured or derived from force against distance plots or force against time plots [30]. In non-topographical applications, the tip of the AFM (which may or may not be functionalized) is typically lowered over the sample and, as the tip and the sample interact with each other, the cantilever deflection is measured as a function of the relative tip displacement. The resulting data are also known as a force curve. The advent of force against distance plots, in particular, has led to the creation of a new experimental possibility, called single-molecule force spectroscopy. This mode of operation of AFM allows for what was previously considered impossible: the measurement of the mechanical properties of single molecules, with piconewton sensitivity [31]. The protein that has attracted the most interest in the area of force spectroscopy is the multidomain, giant, striated muscle protein, titin. Some related publications of note include contributions by Gaub et al. [32] and by Marszalek et al. [33]. These works have essentially demonstrated that by repeated stretching of titin by the AFM tip, single-molecule force spectroscopy can be applied to investigate the magnitude of the forces involved in the unfolding of these proteins. It is of particular interest to note that the strength of covalent bonds can also be estimated using such a technique. For instance, when Grandbois et al. [34] mechanically pulled single polysaccharide molecules covalently anchored between a surface and an AFM tip, they measured a Au–S bond strength of several nanonewtons.

VII. THE CHAPERONIN GroEL: A MODEL PROTEIN ON EXTENDED AND NANOSTRUCTURED SURFACES

Chaperonins are proteins that guide other proteins along the proper pathways for folding. The chaperonin GroEL is a multimeric protein, that is, it contains 14 identical GroEL molecules forming two sevenfold rings packed back to back to each other [35]. *In vivo*, the functional chaperonin complex consists of two GroEL rings capped by a GroES ring. The complex has a cavity open at one end to receive proteins to be folded, and the two GroEL rings communicate with each other via structural changes which are important to the mechanism of the chaperonin. From the top view, a GroEL ring is 15 nm wide with a protein-sized cavity of about 5 nm in diameter in the center (see Figure 5). Furthermore, the GroEL ring is highly flexible and is capable of undergoing

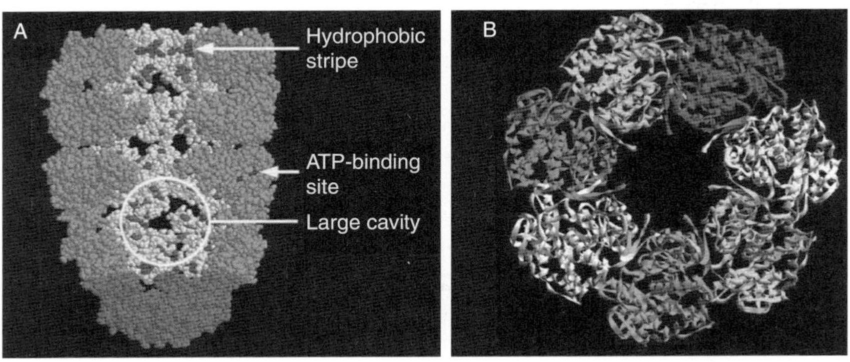

FIGURE 5 The chaperonin system. (A) Ball and stick representation of the chaperonin complex with three of the subunits in each GroEL ring removed to show the interior, leaving four subunits in each ring. Carbon-rich amino acids are colored blue. The bottom cavity is capped by the GroES (colored pink). The large cavity allows a protein chain trapped inside (not shown) to fold on its own. Adapted from the Protein Data Bank [36]. (B) A ribbon schematic of the unbound GroEL ring with each constituent subunit colored differently. The molecule is about 15 nm across, with the central cavity being approximately 5 nm. (For interpretation of the references to color in this figure legend, the reader is referred to the Web version of this chapter.)

very large conformational changes [35]. The ring shape and function of the GroEL therefore makes it an attractive protein to be probed by the AFM.

VIII. GroEL ON EXTENDED SURFACES

The negatively charged surface of mica in aqueous solution implies that many biological molecules can be readily attracted by electrostatic interactions and immobilized for probing by AFM. In some cases, mica can be modified by treatment with a solution of divalent ions such as Mg^{2+} or Ni^{2+} ions which replace the K^{+} in the surface [37]. In the literature, AFM imaging of GroEL usually involves the use of chemical fixatives to immobilize the GroEL ring for high-resolution imaging [38]. However, at present the AFM resolution of GroEL is low in comparison to that achieved on other regular biological samples, such as reconstituted membrane proteins. The major limitation for high-resolution imaging of GroEL rings is the problem of bisection of the protein by the scanning tip, probably due to the weak interring binding [27].

When a sub-monolayer concentration of the GroEL is deposited on the mica surface, the contact mode AFM scans are largely unstable due to the lateral movement of the protein as the AFM tip sweeps along, producing typical "scan stripes." This situation is therefore not conducive to high-resolution imaging of the GroEL ring, but it does suggest that the interactions between protein and substrate are relatively weak. The most common method of stabilizing the GroEL ring on the surface is to increase the concentration of the protein so that at least a monolayer or indeed a multilayer film is formed on the surface. In these conditions, the protein–protein lateral interactions are often sufficient for the AFM scan to yield a stable image, such as that in Figure 6A. In this image, the bright spots are 15–18 nm in size, close to the expected diameter of the native GroEL ring. This area can be scanned repeatedly without significant changes to the protein film. If the scan is zoomed into a selected area, the 5-nm internal cavity of the GroEL ring where protein folding takes place can be clearly distinguished (Figure 6A, inset). This clearly implies that the spatial resolution in this AFM image is better than 5 nm, which is generally associated with small asperities (typically about 2 nm on oxide-sharpened silicon nitride tips) at the end of the AFM tip. This effect is commonly exploited for high-resolution imaging of membrane proteins [18].

When GroEL is deposited on the bare, highly oriented pyrolytic graphite (HOPG) surface, the AFM scans in Figure 6B reveal (a) discrete circular spots between 20 and 25 nm in diameter distributed uniformly over the surface and (b) larger spots between 60 and 100 nm in diameter located at the step edges of the graphite. The latter behavior is attributable to the smooth nature of the graphite surface, which enables ready lateral diffusion of the protein. The GroEL rings are thus likely to agglomerate at the more reactive (hydrophilic) binding sites provided by steps or grain boundaries, or the occasional surface point defect.

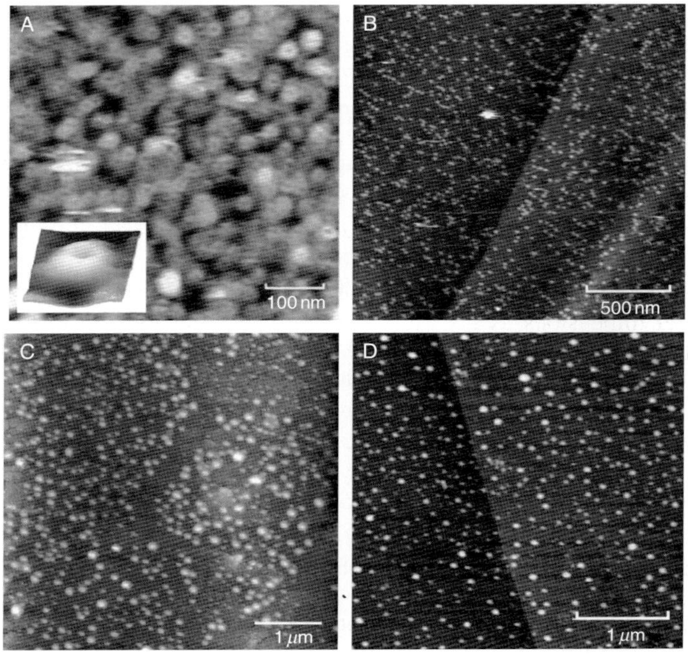

FIGURE 6 GroEL on extended and nanostructured surfaces. (A) GroEL rings imaged by contact mode AFM on mica clearly reveal the 5-nm internal cavity where protein folding takes place. (B) On bare graphite, the GroEL rings can still be observed but the internal cavity is now absent. (C) On an extended film of gold, in addition to GroEL rings, large islands of the protein can be observed, and finally on a film of size-selected nanoclusters (D), protein islands with diameters from 15 to 130 nm are formed. All images were collected in buffer solution in contact mode AFM after incubation at room temperature for at least 30 min. The imaging buffer was 50 mM HEPES, 50 mM KCl, and 10 mM $MgCl_2$ at a pH of 7.5 with ultrapure water added to compensate for evaporation during imaging.

The essentially monomodal 22 ± 3 nm discrete circular spots are dispersed uniformly over the graphite surface, with a nearest neighbor spacing of about 40 nm. Although the size of these spots is relatively close to the expected 15 nm diameter of an individual GroEL ring (as determined on the mica substrate), the measured height of the features is only about 3 nm. This discrepancy from the expected height of a GroEL ring (14 nm) lying flat on the surface, even when compression of the protein by the AFM tip is taken into account, can be explained if the GroEL double rings are bisected into single rings by the scanning AFM tip. The resultant height of the GroEL (single) ring is effectively halved to about 7 nm [27]. Crucially, the internal cavity of the GroEL ring was not visible on the graphite surface, in contrast to the mica surface, suggesting that conformational changes also occur within the protein.

The possibility of proteins being adsorbed differently in the presence of the hydrophobic graphite substrate as compared to the hydrophilic mica is well documented in the literature [27]. Moreover, other studies have also shown that proteins may adopt different conformations when exposed to the graphite surface (e.g., fibrinogen [39], antifreeze glycoproteins [40]). For these proteins, hydrophobic residues which are normally buried within their structure can be exposed to the graphite surface with the molecule lying flat down on the surface and distorted (GroEL contains extensive hydrophobic regions). When scanned by the AFM, these conformational changes result in a drastic decrease in the measured height, accompanied by an increase in the lateral dimensions as the protein spreads over the hydrophobic surface. When the surface in Figure 6B is scanned repeatedly by the AFM tip, the GroEL rings are gradually displaced but, although the GroEL rings can be moved by the scanning AFM tip in contact mode, they are generally much more stable on the graphite surface as compared to the mica surface at sub-monolayer concentrations (where the stabilizing lateral interactions are absent). However, we note that the increased interaction between the GroEL ring and the surface can also lead to the collapse of the GroEL ring.

As compared to the previous extended surfaces, a clean gold film is typically less hydrophilic than mica but not as hydrophobic as HOPG. Figure 6C shows the AFM image when the GroEL is deposited at roughly monolayer concentration on the extended gold film. There is clearly a uniform distribution of bright spots of about 20 nm in diameter and 6 nm in height populating the surface of the gold film, which can be attributed to GroEL rings. Along with the GroEL rings are much larger spots which can measure above 60 nm in diameter and exceed 16 nm in height. These larger spots, GroEL "islands," were also apparent on the graphite surface even at sub-monolayer concentrations of the protein, suggesting that protein island formation is not concentration dependent. When the tip is scanned repeatedly over the same area, the GroEL rings and GroEL islands remain stable, as compared to the previous cases (on HOPG or mica) whereby the molecules were swept aside by the AFM tip. The internal cavity of the GroEL is not visible in any of the images scanned, which suggests the likelihood that the protein undergoes conformational changes when exposed to the gold surface.

In contrast to graphite, which is relatively inert, a gold surface can form covalent bonds with sufur-containing cysteine residues in a protein. The covalent binding of proteins to gold surfaces is well documented in the literature. In some instances, proteins were found to retain their biological activity while immobilized on the surface [41]. As each GroEL molecule contains three cysteines, a total of 21 cysteines in a single GroEL ring can potentially bind in a covalent manner to the gold film. The strength of the Au–S bond (about 2 eV [34]) suggests that covalent bonds can provide stability to the immobilized GroEL molecules so that they are not displaced by the scanning AFM tip. The stability of the AFM images is a strong indication that the cysteine to gold covalent bonds have an effect in immobilizing the GroEL molecules—a clear difference from GroEL on mica (immobilization via hydrophilic interactions) or on graphite (immobilization via hydrophobic interactions).

IX. GroEL ON SIZE-SELECTED GOLD CLUSTERS

The decoration of the graphite surface with an array of pinned gold-nanoclusters alters the behavior of the surface with respect to the adsorption of proteins. Contact-mode AFM imaging of the gold cluster film in buffer solution generates (Figure 6D) a dispersed array of protein islands with diameters from 15 to 130 nm, which can be assigned to the nucleation of protein islands on the gold clusters dispersed over the surface. However, the 5-nm cavity in the middle of the chaperonin ring was again not evident from the smallest features (15–20 nm in diameter and 3–4 nm in height) attributed to individual complexes. These features sizes are comparable with the GroEL rings deposited onto an extended gold film (Figure 6C). Since the chaperonin ring contains 21 cysteine residues (but no disulphide bridges), we can speculate that on both the extended gold film and the gold clusters, the protein is chemisorbed via one or more surface thiolate bonds, thereby conferring the observed stability of the protein molecules under repeated contact mode AFM imaging. The presence of protein islands larger than the GroEL ring therefore suggests the formation of protein aggregates around GroEL monomers anchored by the gold clusters which act as the nuclei for island growth.

X. HRP ON GRAPHITE AND Au CLUSTERS

In direct contrast to mica, graphite is a very hydrophobic substrate. Proteins in their native state normally consist of a hydrophilic exterior shell and an interior hydrophobic core. However, on hydrophobic surfaces, the inner hydrophobic core may become exposed to the hydrophobic surface via conformational changes. One factor that confers stability on proteins is the formation of disulfide bridges between adjacent cysteine residues. The role of cysteines in the protein structure is very dependent on the cellular location of the protein in which they are contained. Within extracellular proteins, cysteines are frequently involved in disulfide bonds, where pairs of cysteines are oxidized to form a covalent bond. These bonds serve mostly to stabilize the protein structure, and the structure of many extracellular proteins is almost entirely determined by the topology of multiple disulfide bonds. In this section, we consider the horseradish peroxidase (HRP) molecule which contains four such disulfide bridges on the hydrophobic graphite surface. *In vivo*, HRP acts as an enzyme to catalyze the reduction of carcinogenic hydrogen peroxide to water. HRP is a single-chain polypeptide and has a molecular weight of approximately 34,000 amu. HRP normally exists in the form of a hexameric complex which measures about 16 × 16 × 11 nm, whereas a single subunit of HRP is estimated to be approximately 5 nm in size as shown in Figure 7. Each HRP molecule also contains four disulfide bridges, which are highlighted.

To obtain a thick film of HRP on bare graphite, the protein is deposited at a concentration of 0.1 mg/mL, which is necessary to ensure dense packing and the lateral constraints that prevent the sweeping effect by the AFM tip in contact mode. The HRP can be observed to self-organize into densely packed arrays with local hexagonal packing as illustrated in Figure 8. Based on X-ray crystallographic data, the bright spots are consistent with the dimension of an HRP hexamer (16 × 16 × 11 nm) or possibly larger multimeric forms of the protein, rather than single HRP monomers (5 × 4 × 3 nm). The size of the HRP molecules we imaged in aqueous solution by AFM is also in concordance with the findings of previous workers in the field, for example, Hobara et al. [42]. The self-organization of the HRP molecules is presumably due to protein–protein interactions, but in contrast to membrane proteins, the hexagonal lattice ordering of the HRP molecules does not extend beyond five or six molecules. This is probably attributed to the nonspherical shape of the HRP molecule. The same features can be resolved even after repeated scans over the same area, demonstrating the stability of the thick close-packed layers.

The reduced vertical height, that is, apparently flattened morphology of the proteins, as shown in the inset to Figure 8, can be explained by two factors, namely, the tip locus effect, and the force applied by the AFM cantilever during imaging. Regarding the latter, it can be expected that the imaging

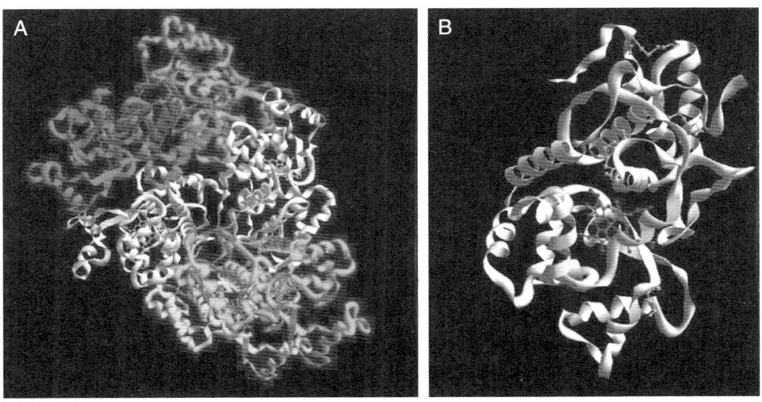

FIGURE 7 The structure of horseradish peroxidase (HRP). (A) Ribbon representation of the hexameric form of HRP. Each subunit of the complex is assigned a different color. (B) The monomeric form of the same protein, with the four disulfide bridges highlighted in yellow and the heme ring colored red. (For interpretation of the references to color in this figure legend, the reader is referred to the Web version of this chapter.)

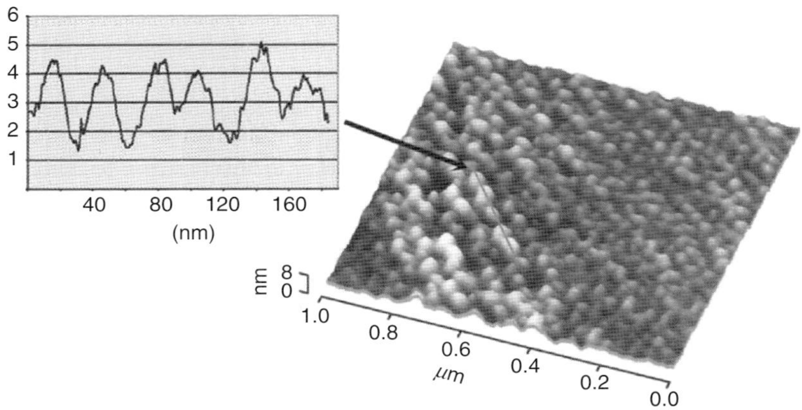

FIGURE 8 A thick HRP film deposited onto bare graphite imaged by AFM in liquid. The scan area is 1 μm^2. Inset: a line profile as traced by the red line in the main figure. From the line profile, the HRP molecules are on average about 20 nm in diameter and 3 nm in height. This image was collected in contact mode at pH 4 and a protein concentration of 0.1 mg/mL. (For interpretation of the references to color in this figure legend, the reader is referred to the Web version of this chapter.)

process will induce some degree of structural deformation to the HRP molecules during the scanning of the AFM tip.

Figure 9A shows a large-area scan where the highlighted area has been repeatedly rastered by the AFM tip, revealing the underlying HRP molecules immobilized by Au_{17} nanoclusters pinned to the graphite surface. The outskirts of this area are still covered by a loosely bound film of molecules. This thick film appears similar to what has been observed when HRP is deposited on bare graphite (Figure 8), indicating that the topmost layer of molecules is largely unaffected by the underlying gold cluster film. The contact mode AFM image of Figure 9B was obtained by zooming into the highlighted area in Figure 9A. The bright features in this AFM image can be attributed to the formation of protein islands that are attached to the Au_{17} clusters.

The stability of the protein islands immobilized by the gold clusters is demonstrated in Figure 10A and B where multiple scans (in contact mode) over the same area were performed over a 30-min interval. Although the effect of piezo scanner drift is apparent between the images, the protein islands, both small and large, are still present as highlighted in the images. This stability of the protein islands is not observed when bare graphite or mica is utilized as substrates.

A plot comparing the diameter of the HRP molecules on the gold-nanocluster film and the HRP islands that make up the thick protein film on bare graphite is shown in Figure 11. Inspection of the diameter distribution shows that the HRP islands on the gold-nanocluster film generally show a

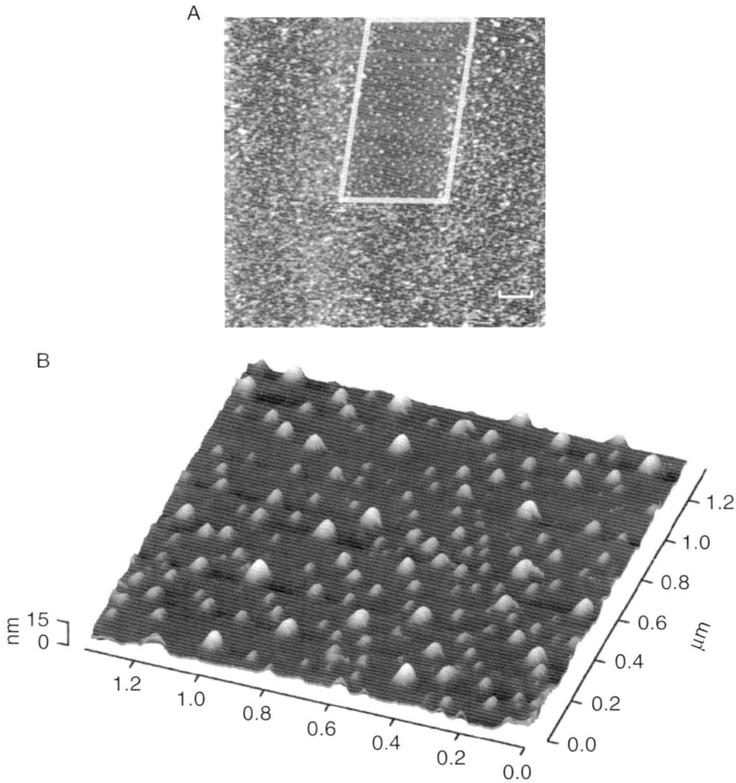

FIGURE 9 HRP on size-selected gold clusters. (A) A large 5 × 5 μm scan showing an area that has been repeatedly rastered by the AFM tip to remove the loosely bound proteins (highlighted box) and the surrounding thick protein film. Zooming into the highlighted area yields the AFM image in (B), whereby the underlying HRP is immobilized onto a film of Au_{17} gold clusters. The images were collected in contact mode at 0.5 nN applied force, pH 4, and a protein concentration of 0.1 mg/mL.

much broader size distribution, with some islands as small as 20 nm and other islands exceeding 100 nm in diameter. Even if possible tip locus effects are taken into account, the size of the largest islands in Figure 9B does not correspond to HRP hexamers, but instead suggests the formation of assemblies of multiple HRP molecules.

Presumably, the larger protein islands in Figure 9B, with diameters up to about 130 nm, are formed by the binding of HRP molecules to a nucleus defined by the first chemisorbed protein. The occasional scan streaks evident in Figure 10 are probably due to the displacement of the more loosely bound proteins on top of the HRP assembly. The main features of interest, however, are the smaller features in Figure 9B. The minimum lateral feature size of

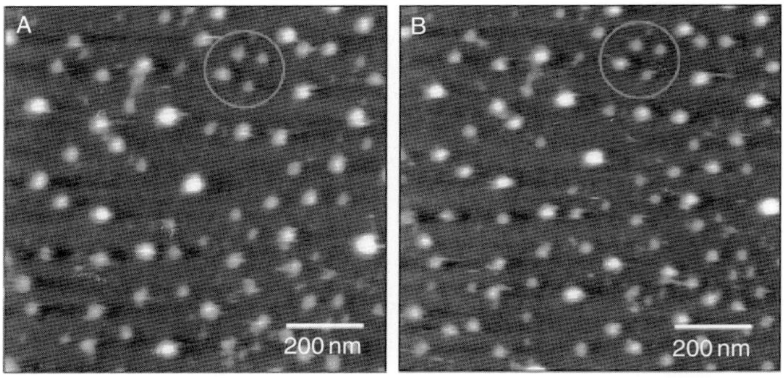

FIGURE 10 Stability of the HRP islands. Panels (A) and (B) are repeated scans over the same region of the substrate, demonstrating the stability of the islands highlighted by the circled region. The islands even appear to be resistant to small changes in the applied cantilever force. Piezo scanner drift is also apparent in the images after this extended scanning period.

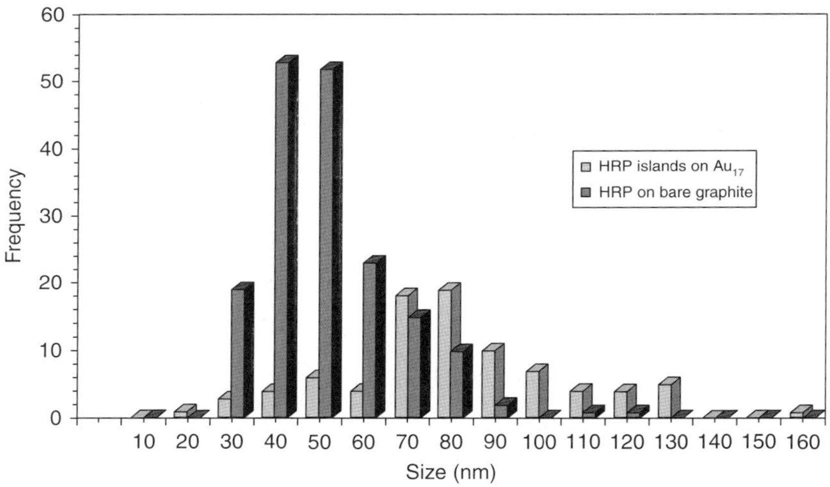

FIGURE 11 A comparison of size distributions of (i) HRP islands induced by the presence of gold clusters and (ii) HRP molecules deposited onto bare graphite. Size distribution data were collected and analyzed by Image SXM [43].

protein complexes on the gold cluster film lies in the range 15–25 nm. This is obviously larger than the size of a single HRP molecule (about $5 \times 4 \times 3$ nm) but consistent with the size of the hexameric HRP units (about $16 \times 16 \times 11$ nm) [44]. These results from the AFM images strongly indicate that the presence of the gold clusters onto the graphite substrate affects the behavior of the proteins on the nonhomogeneous nanostructured surface.

The immobilization of HRP on gold nanoclusters has also been demonstrated in a more technologically relevant context, for example, protein biochip application [45]. For instance, a transparent glass surface can be coated with a layer of the protein repellent molecule, polyethylene glycol (PEG) and then decorated with Au nanoclusters such that the deposited proteins diffuse and bind to the nanoclusters as opposed to the surrounding surface. Figure 12A shows a tapping-mode AFM image from the edge of the patterned size-selected Au_{6000} clusters on PEG-coated glass prior to protein deposition. After incubation (immersion) of this surface in the HRP solution, the sample was rinsed in deionized water and was imaged again in air immediately afterward. The AFM image of Figure 12B reveals that the height of a proportion of the clusters increases by 3 nm, which is consistent with the size of the HRP

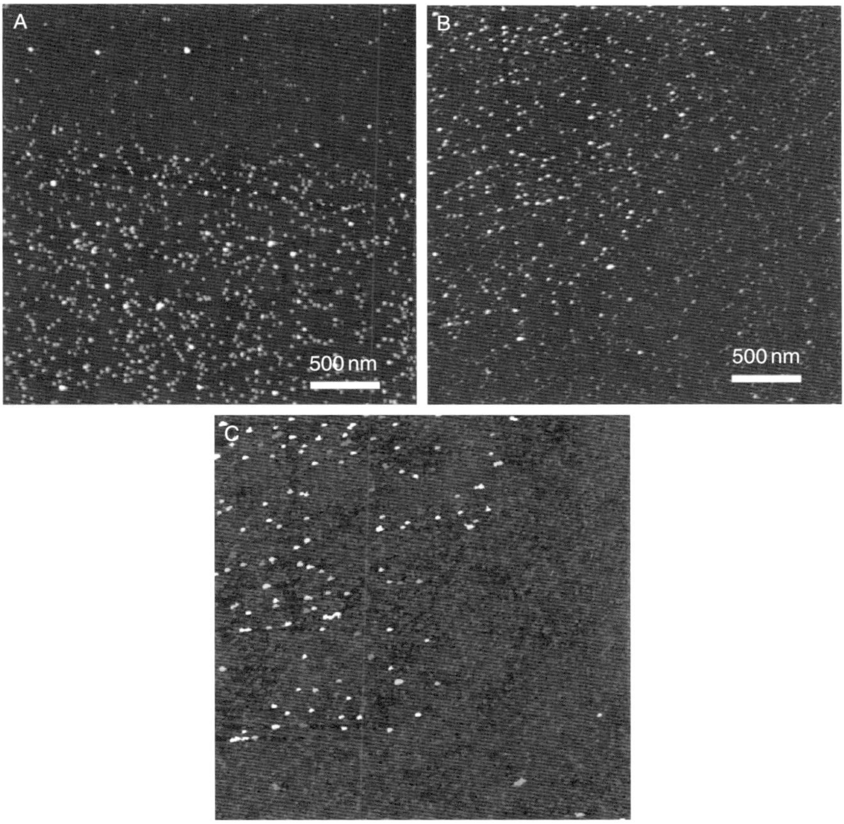

FIGURE 12 HRP for protein biochip applications [45]. (A) Tapping-mode AFM image in air of a patterned area of size-selected Au_{6000} clusters on PEG-coated glass before protein deposition. (B) Tapping-mode AFM image in air of a patterned area of size-selected Au_{6000} clusters on PEG-coated glass after protein immobilization with corresponding height profile. (C) Processed image, based on (B), in which the lowest features have been filtered. It is evident that HRP islands were largely confined to the cluster area.

monomer. Figure 12C shows a processed image, in which features below 2 nm in height have been filtered out, and the surviving features can be clearly observed within the cluster-decorated area but not outside.

XI. ROLE OF CYSTEINES IN PROTEIN BINDING TO SIZE-SELECTED CLUSTERS

We have already proposed that the GroEL and HRP molecules can be immobilized via covalent bonds between cysteine thiol groups and the gold clusters. When the GroEL or HRP is presented to the hydrophobic HOPG decorated with size-selected clusters, dispersed protein islands ranging between 15 and 130 nm in diameter can be observed on the surface. These are the key experimental observations that need to be addressed to establish a mechanism for protein adsorption onto the gold nanoclusters. Comparison with the behavior of other proteins, notably oncostatin M (OSM), green fluorescent protein (GFP), luciferase, and lysozyme represents an important test of our thinking.

XII. GFP AND OSM INTERACTION WITH GOLD CLUSTERS

So far, both proteins investigated (GroEL and HRP) have shown the inclination to bind to the size-selected gold clusters, forming stable protein islands of various sizes in the process. The stability of the protein islands has been attributed to the formation of covalent bonds between their cysteine residues and the underlying gold nanocluster. An important consideration is the location of the cysteine residues within the protein structure. In the simplest case, a cysteine residue buried inside a protein is not accessible to the solvent, and, by extension, is inaccessible to the size-selected gold cluster, barring any conformational changes in the protein structure. Likewise, the degree of accessibility of a cysteine residue near the protein surface can also vary, which will be discussed in the following sections. To gain a better understanding of the protein binding mechanism and its implications, a comparative study involving two distinctively different proteins, GFP and human OSM, has been performed by Prisco et al. [46]. These two proteins are structurally well characterized and present different arrangements of cysteine residues in their tertiary structures. In this section, we will demonstrate that OSM can bind to gold clusters, whereas GFP does not, as its cysteine residues are not available.

Figure 13A shows the crystal structure of GFP obtained by XRD [47]. This protein is intrinsically fluorescent and can be isolated from the jellyfish, *Aequorea victoria*. It absorbs light at 470 nm, and the fluorescence emission spectrum peaks at 509 nm. As such, GFP offers considerable scope as a non-invasive marker in living cells, and is used for numerous applications such as cell lineage tracing, reporting of gene expression, and measurement of protein–protein interactions [48]. GFP consists of a sequence of 238 amino

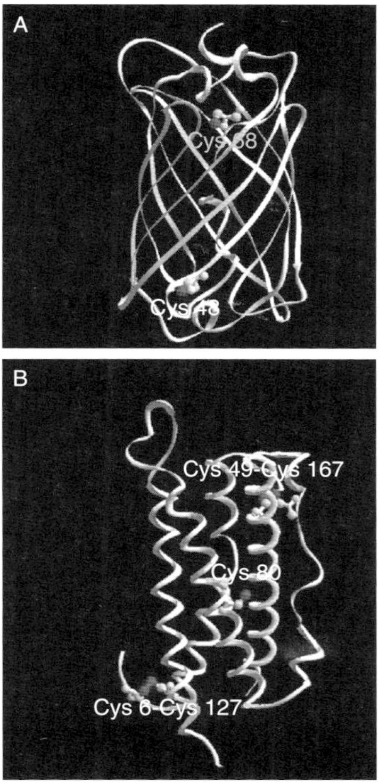

FIGURE 13 Ribbon structures of GFP and OSM, with cysteine residues highlighted. (A) The β-can structure of GFP: the molecule is about $45 \times 35 \times 35$ Å in size. (B) Structure of OSM: the molecule is about $60 \times 24 \times 20$ Å. The cysteine residues are represented by balls and sticks, and the sulfur atoms are colored red. (For interpretation of the references to color in this figure legend, the reader is referred to the Web version of this chapter.)

acids, and it has two cysteine residues along the backbone, Cys 48 and Cys 68. The cysteine residues are buried within the molecule and should play no role in binding to the gold clusters. GFP has the shape of a cylinder, comprising 11 strands of β sheet on the outside with an α-helix on the inside, forming what is called a "β-can." In the crystal structure, two protomers pack closely together to form a dimer, with the fluorophores protected inside the cylinders. It is believed that GFP is monomeric in dilute solution (below 2 mg/mL) and dimeric in concentrated solution (above 5 mg/mL). Therefore, in the present work, we expect that the protein exists in its monomeric form.

OSM is a growth-regulating cytokine that affects a number of normal and tumor cells. It exerts inhibitory effects on the growth of melanoma, lung carcinomas, and other cancer cells [49]. The structure of OSM is shown in Figure 13B. It is a monomeric protein composed of 186 amino acids, with a molecular weight of 21,000 amu [50]. OSM contains five cysteine residues. Two pairs of these residues form disulfide bonds, Cys 6–Cys 127 and Cys 49–Cys 167, with one additional free cysteine residue, Cys 80. Thus, the protein is a good candidate for binding to gold nanoclusters, either through the free cysteine or via the disulfide bonds.

Contact-mode AFM imaging of GFP molecules on gold nanoclusters in a variety of buffers and different pH values clearly indicate that GFP does not bind covalently to the gold nanoclusters. At sub-monolayer concentrations of the protein, neither stable, single proteins nor protein agglomerates can be repeatedly imaged in the cluster-decorated area of the graphite substrate. The behavior of the protein is the same outside the cluster region, that is, on the bare graphite surface. After the first scan, even with minimum contact force, the proteins are displaced by the action of the tip, and sometimes caused to agglomerate at the HOPG steps. This situation is typical of other proteins on bare HOPG [39]. A typical result of a scan in the cluster area of a graphite surface decorated with gold nanoclusters (Au_{70}) is shown in the summary series of scans in Figure 14. Figure 14A was collected after about five consecutive contact mode scans, and Figure 14B was obtained by zooming out from the previous area. The square region wiped clear of proteins in the prior scans is visible and demonstrates that GFP is not immobilized by the size-selected gold nanoclusters. Attempts to immobilize GFP onto different size-selected clusters (Au_{26}, Au_{55}, and Au_{70}) were also not successful, suggesting that the cluster size is not a critical parameter for protein immobilization.

The results of depositing OSM molecules onto Au_{40} clusters on graphite are shown in Figure 14C. In contrast to GFP (Figure 14A), stable globular features similar to the cases of GroEL and HRP are visible on the surface, which are attributed to the protein molecules immobilized on the clusters. The observed protein features were unmoved by repeated scans, and even when an increased value of the interaction force between tip and surface was employed. When the applied force is increased in this way, the imaged protein features can be stretched laterally by the tip, but immediately after reducing the applied force to about 0.5 nN, they regain their original globular shape. The immobilized proteins were stable when imaged again after several days. The experiment was repeated with Au_{55} clusters yielding similar results. No immobilized proteins were observed outside the cluster area as illustrated in Figure 14D.

XIII. MOLECULAR SURFACES OF PROTEINS

It can reasonably be assumed that the sulfur-containing cysteine residues are important in whether a protein chemisorbs to the gold nanoclusters. In general, only residues at the surface of the protein can participate in the interaction of the protein, whereas residues buried within the structure (interior residues) do not. For simplicity, it is helpful to consider the protein as a solid object with a surface that can interact with its surroundings. On this view, the larger the exposed molecular surface associated with the cysteine residue (or the sulfur atom therein), the more accessible it is to an external "probe," such as the gold nanoclusters in these experiments. To determine whether the cysteine residues are present on the surface, we have therefore applied the concept of protein molecular surfaces.

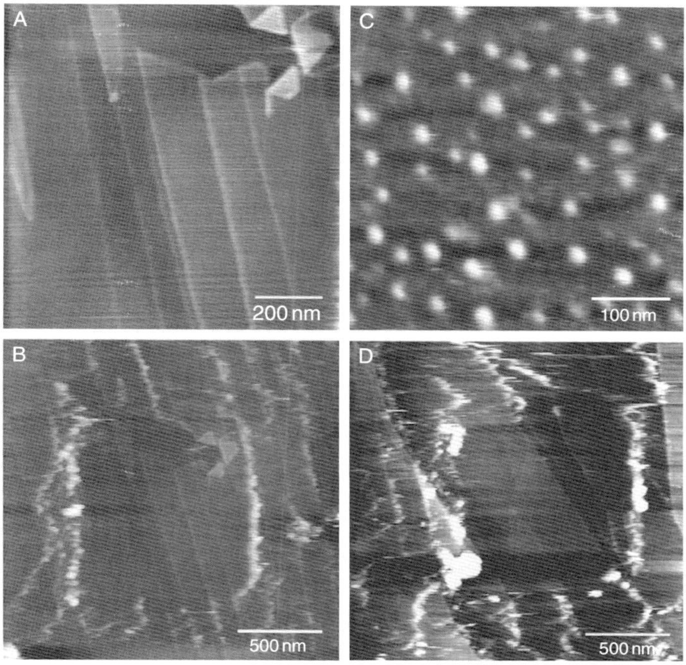

FIGURE 14 Interaction of GFP and OSM with a film of size-selected gold nanoclusters. (A) AFM image after five scans in buffer solution of a gold nanocluster area on graphite after GFP deposition. Zooming out from (A) yields (B), showing the area that has been swept clear by the tip and protein aggregation at the edges; the imaging buffer was 55 mM MES at pH 6.4. The protein was imaged in native buffer solution at a concentration of 0.22 mg/mL. (C) AFM image in buffer solution of the Au_{40} cluster-decorated area on graphite after OSM deposition. Globular features that are stable under repeated scanning by the AFM tip can be readily observed. (D) AFM image of OSM outside the nanocluster area displaying the sweeping away of the OSM molecules on the bare graphite surface. The protein concentration used in this experiment was 0.15 mg/mL and the imaging buffer was 172 mM PBS at a pH of 7.3. For all AFM experiments in aqueous environments, imaging parameters were chosen carefully so that the interaction force between the AFM tip and the protein was repulsive, thereby increasing the measurement sensitivity. Figure adapted from Ref. [46].

For small molecules, the van der Waals surface usually gives a good representation of the outer surface and overall shape; however, for larger molecules, most of the van der Waals surface is buried in the interior, so clearly some other method is required to define the outer surface of a macromolecule. Molecular surface area (MSA) calculations consist of two parts: the contact surface and the reentrant surface. The contact surface is that part of the van der Waals surface of the atoms that is accessible to a probe sphere representing a solvent molecule (usually a water molecule) whereas the reentrant surface originates from the inward-facing surface of the probe sphere when it is simultaneously in contact with more than one atom. Recent examples of biophysical applications of MSA calculations include protein folding [51], macromolecular docking [52], and calculation of solvation energies [53], among others.

XIV. MOLECULAR SURFACES OF GFP AND OSM

The molecular surface calculations for the cysteine residues of both the GFP and the OSM molecules are summarized in Table 2. In these calculations, a probe radius of 1.4 Å was used, as it matches closely with the radius of a water molecule [54]. Incidentally, this value is also very close to the atomic radius of the gold atom (1.4–1.7 Å), which suggests that the results may also apply to gold atom within a cluster which effectively acts as a probe on the protein surface.

Referring to Table 2, the MSA calculations reveal that the entirety of Cys 68 in the GFP molecule is well buried within the can-like structure of the protein, as postulated above, and therefore is inaccessible to the probe sphere. It is thus unlikely that Cys 68 can participate in the binding to gold clusters. An initial inspection shows that the Cys 48 residue in the GFP is only marginally exposed, with an MSA of about 10 Å^2. However, this MSA contribution is due to the backbone oxygen (with a concave surface) of the cysteine residue and not to the sulfur atom in the side chain. Furthermore, since the Cys 48 residue lies within a rigid beta sheet, and the sulfur atom is directed toward the inside of the GFP molecule, as illustrated in Figure 13A, it is even less likely to be available for binding to the gold nanoclusters.

The five cysteine residues of the OSM molecule present a range of MSAs. Cys 49 in OSM is a similar case to Cys 48 in the GFP, in that its contribution toward the MSA is entirely from the amino acid backbone and not from the sulfur atom in the side chain. The curvature values also suggest that the surface is highly concave, which reduces the probability of Cys 49 binding to gold nanoclusters even further. Cys 49 is also covalently linked to Cys 167, and with the latter residue completely buried inside the protein, the resulting disulfide bridge is thus likely to be unavailable on the protein surface. Therefore, from MSA calculations it is possible to eliminate one of the potential binding sites for the gold nanoclusters. By contrast, Cys 6 has an MSA of 48 Å^2, of which about 15 Å^2 originates from the sulfur atom in the side chain. This makes the Cys 6 a possible binding site for the gold nanoclusters. Cys 6 and Cys 127 together form another disulfide bridge; both sulfur atoms contribute approximately 15 Å^2 to the MSAs of their respective residues. It can be noted that the sulfur in the side chain of Cys 127 is the only portion of the residue that is on the molecular surface, and presents a significantly more concave molecular surface than the sulfur in Cys 6. We therefore envisage that this particular disulfide bridge (especially the Cys 6) is a binding site to the gold nanoclusters. The lone Cys 80 in OSM presents the most accessible molecular surface listed in Table 2, with the sulfur atom contributing

TABLE 2 Molecular Surface Area (MSA), Accessible Surface Area (ASA), and Curvature Results for the Cys Residues in Green Fluorescent Protein and Oncostatin

Protein	Cys residue	Atom	ASA	MSA	Curvature
Green fluorescent protein	48	N	0	0	0
		Cα	0	0	0
		C	0	0	0
		O	3.14	10.23	0.26
		Cβ	0	0	0
		Sγ	0	0	0
	68	N	0	0	0
		Cα	0	0	0
		C	0	0	0
		O	0	0	0
		Cβ	0	0	0
		Sγ	0	0	0
Oncostatin	6	N	3.64	5.07	0.07
		Cα	0.45	2.76	0.36
		C	0.02	0.96	0.51
		O	18.91	11.49	–0.17
		Cβ	14.21	12.66	–0.06
		Sγ	8.86	15.23	0.07
	49	N	0	0	0
		Cα	0	0	0
		C	1.13	3.96	0.31
		O	0.07	0.68	0.39
		Cβ	2.23	8.80	0.28
		Sγ	0	0	0
	80	N	1.82	2.36	0.11
		Cα	6.69	9.36	0.09
		C	0	0	0
		O	0	0	0

Continued

TABLE 2 Molecular Surface Area (MSA), Accessible Surface Area (ASA), and Curvature Results for the Cys Residues in Green Fluorescent Protein and Oncostatin—Cont'd

Protein	Cys residue	Atom	ASA	MSA	Curvature
		Cβ	18.96	14.72	−0.12
		Sγ	33.31	32.97	0.08
	127	N	0	0	0
		Cα	0	0	0
		C	0	0	0
		O	0	0	0
		Cβ	0	0	0
		Sγ	1.36	14.34	0.47
	167	N	0	0	0
		Cα	0	0	0
		C	0	0	0
		O	0	0	0
		Cβ	0	0	0
		Sγ	0	0	0

The constituent atoms of the Cys residues are labeled in sequence: Cα, C, O, Cβ, Sγ. The corresponding sulfur atom in each Cys residue (Sγ) is highlighted. A negative curvature indicates an exposed convex surface. Values quoted are in angstrom squared ($Å^2$) and calculations performed in SurfRace using a 1.4-Å probe radius [54].

approximately 33 $Å^2$, that is, more than half of the MSA of the entire residue. In addition, the molecular surface of the sulfur atom in Cys 80 is comparably concave with respect to Cys 6, and far less concave than Cys 127. Figure 15A illustrates the molecular surface of OSM, and in summary it is clear from the picture that the Cys 80 and the disulfide bridge Cys 6–Cys 127 are the most likely candidate residues to bind to the gold nanoclusters.

If OSM molecules can be anchored to the clusters via either of possible binding sites, then this is expected to result in two different orientations of the protein on the surface, as illustrated in Figure 16. A direct measurement of the orientation of the protein on the gold nanoclusters is not possible, but it is clear that these two possible orientations will each result in a different height, which, at least in principle, is observable in the AFM data. In the literature, a variety of protein orientations on extended gold films have already

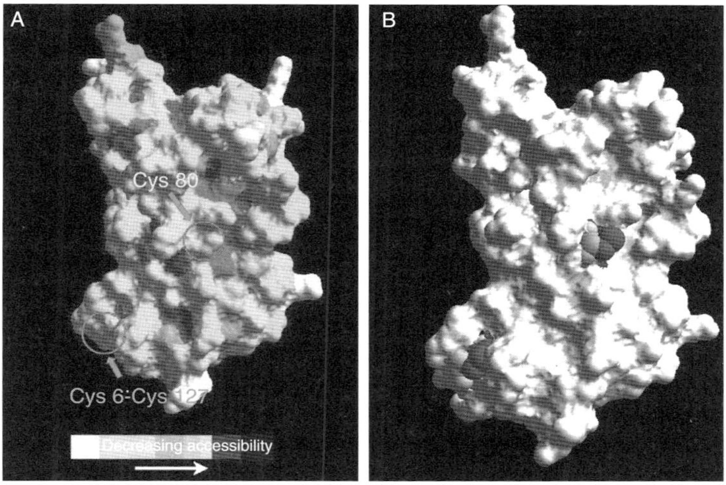

FIGURE 15 (A) The molecular surface of OSM with color-coded accessibility; orange/yellow regions are the most accessible, while dark blue regions are the least accessible. Cys 80 and the disulfide bridge Cys 6–Cys 147 both present large molecular areas for chemisorption onto gold nanoclusters, and both candidate binding sites for gold nanoclusters are moderately accessible as compared to the rest of the residues in the protein. The Cys 49–Cys 167 disulfide bridge is hidden from view as it is on the reverse face of the protein. (B) A composite molecular surface/van der Waals' image of OSM with the Cys side chains in red and the backbones in yellow. The sulfur atoms for these residues (Cys 80 and the disulphide bridge Cys 6–Cys 147) are clearly on the surface and make the protein conducive to binding with the size-selected gold nanoclusters. The images are generated using the PDB ID "1EVS" from the Protein Data Bank. (For interpretation of the references to color in this figure legend, the reader is referred to the Web version of this chapter.)

been theoretically modeled for other proteins [41,55]. If Cys 80 is the binding site for OSM to the gold nanoclusters, the molecule is then lying flat on the surface (as in Figure 16B) and the resulting height is about 20 Å. If the disulfide bridge Cys 6–Cys 127 is the binding site, the OSM molecule is instead expected to stand up with the long axis at an angle to the surface. This configuration may be less stable than lying flat on the surface with respect to AFM imaging, and the OSM may swing about the disulfide bridge binding site while being scanned by the AFM tip. In this situation, the measured height of the protein may therefore take a spread of values between 24 and 45 Å (based on the Protein Data Bank crystallographic data).

A statistical analysis of the height of the OSM array on the size-selected gold nanoclusters on graphite is shown in Figure 16C. The height distribution shown can be interpreted in terms of two main peaks: a first strong maximum at 30–40 Å (corresponding to the binding of the OSM at the Cys 80 site) and a second weaker maximum at 50 Å (corresponding to the Cys 6–Cys 127 disulfide bridge anchor point). The bimodal height distribution is thus consistent with the two predicted orientations of the protein on the surface. We note that

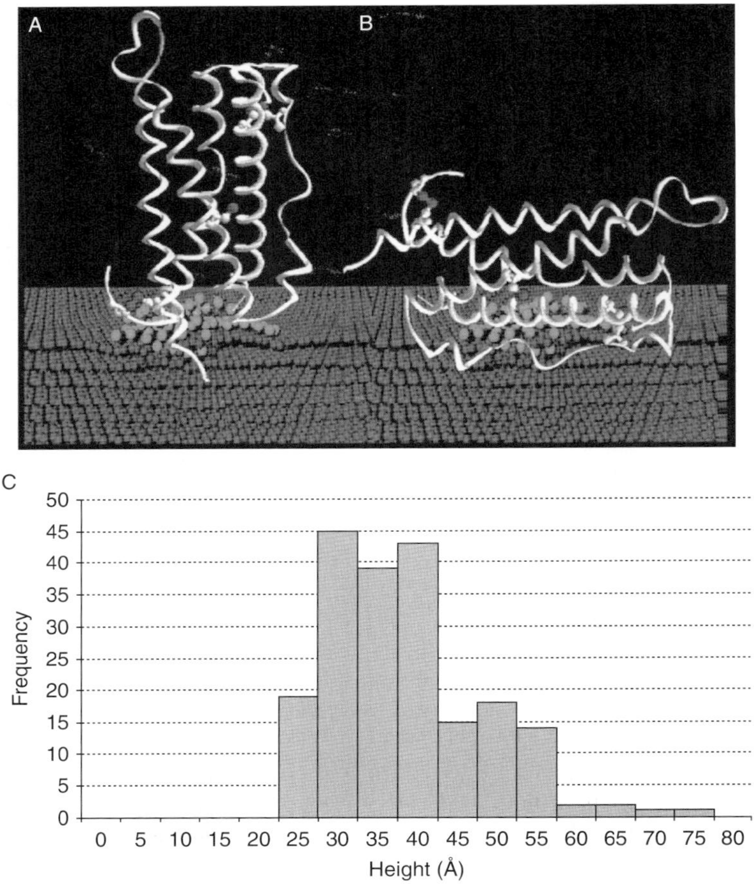

FIGURE 16 Hypothetical orientations of the OSM on gold nanoclusters. (A) Anchoring of the OSM via the Cys 6–Cys 127 disulfide bridge and (B) via the Cys 80. An estimated height difference of up to 20 Å is apparent due to the two possible orientations of the molecule. (C) Histogram of the (maximum) height distribution of 200 OSM molecules immobilized on Au_{40} nanoclusters on graphite. Adapted from Ref. [46].

deformation (compression) of a protein is likely even at minimal applied force in AFM, which may account for the slightly lower than expected heights.

We note that oriented immobilization of proteins of the type discussed provides better spatial accessibilities for species attraction to active binding sites than non-oriented immobilization. It has been shown, for instance, that randomly attached Fab fragments of an antibody cannot be immobilized at as high a density as oriented fragments. The randomly oriented antibodies show lower specific activity, suggesting that oriented immobilization gives better sensitivity and specificity for biosensing applications [56].

XV. MSA CALCULATIONS: PREDICTING PROTEIN IMMOBILIZATION ON NANOCLUSTERS

In addition to OSM, several other proteins have been successfully immobilized on the size-selected gold nanoclusters and probed by the AFM, while other proteins behave like GFP and cannot be immobilized covalently. In this section, we shall review some of the experimental AFM results and examine whether the binding of the proteins to gold nanoclusters can be predicted by MSA calculations and/or molecular visualization. Table 3 shows the results of MSA calculations performed on a series of proteins. Some of the results have already been reported, such as GroEL and HRP, while two more proteins, lysozyme [57] (from hen egg-white) and luciferase [58] (from the firefly *Photinus pyralis*), are added to this list to widen the study. All the proteins listed in the table have been tested experimentally for binding with the gold nanoclusters.

HRP, which exists either in monomeric or hexameric form, has been shown to bind to gold nanoclusters in earlier sections. The MSA results for the cysteines are identical in both the hexamer and monomer of HRP, that is, only Cys 91 and Cys 177 have their sulfur atoms exposed. Cys 91 (with exposed sulfur MSA of 15 Å^2) and Cys 177 (4 Å^2) are thus both candidate binding sites. However, the highly concave surface of the Cys 177 is expected to make this site far less conducive than the Cys 91 residue as the binding site for gold nanoclusters. This is further supported in Figure 17A where Cys 91 appears to be lodged in a deep cavity.

The MSA calculations for GroEL in heptamer form (single ring) shows that no sulfur atoms are exposed at any of the cysteine residues, although parts of the cysteine backbones are on the molecular surface. However, if the seven constituent subunits of the GroEL are investigated separately (e.g., by examining chain A in isolation so that the subunits are no longer considered to be interfacing with each other), then the sulfur atom of Cys 519 becomes readily available. The exposed (convex) molecular surface of the sulfur atom measures about 15 Å^2 and is shown graphically in Figure 17B. Although under normal circumstances the GroEL ring is expected to remain intact, it is feasible that GroEL may undergo conformational changes leading to the collapse of the internal cavity on certain surfaces, for example, on a uniform gold film as demonstrated by AFM studies previously. It is therefore plausible that at least part of the GroEL ring may distort, freeing up the cysteine to bind to the gold nanoclusters.

By applying a similar logic, it can be deduced that lysozyme should bind readily to gold nanoclusters via at least one (Cys 6) of its eight cysteine residues, whereas in luciferase, the limited MSA from Cys 81 and Cys 258 (as well as the high concavity at these surfaces) would not suggest binding to the gold nanoclusters. These predictions have indeed been verified experimentally by depositing these proteins on size-selected gold clusters and performing the AFM scans in buffer solution.

TABLE 3 A Series of Proteins: GroEL (Heptamer and One Subunit in Isolation), HRP (Hexamer and One Subunit in Isolation), Luciferase, and Lysozyme, Studied by MSA Calculations

Protein	PDB ID	Cys residue	Total MSA	Sulfur MSA	Sulfur curvature	Will bind to Au clusters
Peroxidase (hexamer)	1ATJ	11	1.40	0.00	0.000	Yes
		44	3.52	0.00	0.000	
		49	0.00	0.00	0.000	
		91	23.02	15.11	0.134	
		97	0.00	0.00	0.000	
		177	4.86	4.86	0.673	
		209	0.00	0.00	0.000	
		301	6.30	0.00	0.000	
Peroxidase (monomer) [11–91], [44–49] [97–301], [177–209]	1ATJ (chain A)	11	1.40	0.00	0.000	Yes
		44	3.52	0.00	0.000	
		49	0.00	0.00	0.000	
		91	23.02	15.11	0.134	
		97	0.00	0.00	0.000	
		177	4.86	4.86	0.673	
		209	0.00	0.00	0.000	
		301	6.30	0.00	0.000	
GroEL (heptamer)	1OEL	138	18.33	0.00	0.000	No
		458	11.72	0.00	0.000	
		519	0.00	0.00	0.000	
GroEL (one subunit)	1OEL (chain A)	138	20.90	0.00	0.000	Yes
		458	16.67	0.00	0.000	
		519	48.48	15.16	-0.050	
Luciferase	1LCI	81	9.69	5.32	0.546	No
		216	0.00	0.00	0.000	
		258	13.74	5.57	0.571	
		391	0.00	0.00	0.000	

Continued

TABLE 3 A Series of Proteins: GroEL (Heptamer and One Subunit in Isolation), HRP (Hexamer and One Subunit in Isolation), Luciferase, and Lysozyme, Studied by MSA Calculations—Cont'd

Protein	PDB ID	Cys residue	Total MSA	Sulfur MSA	Sulfur curvature	Will bind to Au clusters
Lysozyme [6–127], [30–115] [64–80], [76–94]	1DPX	6	49.00	15.93	0.001	Yes
		30	5.51	5.51	0.395	
		64	2.77	0.00	0.000	
		76	22.26	0.00	0.000	
		80	8.36	0.00	0.000	
		94	9.61	0.00	0.000	
		115	2.95	0.00	0.000	
		127	24.06	0.00	0.000	

HRP, horseradish peroxidase; MSA, molecular surface area.
The numbers in square brackets denote residues which form disulfide bridges within the protein.
Values quoted are in ($Å^2$) and calculations performed in SurfRace using a 1.4-Å probe radius.

XVI. EVIDENCE FOR WEAK PROTEIN–AU NANOCLUSTER INTERACTIONS

In AFM contact mode imaging, any noncovalent interaction between protein and cluster is likely to be too weak to withstand the lateral forces imposed by the scanning AFM tip, even at the lowest applied force. However, in tapping mode these lateral forces are minimized and, for the case of GFP deposited onto gold nanoclusters, GFP protein islands can be observed decorating the surface. This can be explained by the fact that tapping-mode imaging is less disruptive to the molecule-surface binding, whereas in contact mode imaging the AFM tip can sweep away any material interacting weakly with the surface. The weak interaction between protein and cluster identified from the tapping-mode AFM images is analogous to the "physisorption" regime interaction of small molecules with sulfur. Such a state may function as a "precursor state" to subsequent chemisorption in the case where free cysteines are available.

Figure 18A shows a tapping-mode AFM image of GFP deposited at approximately monolayer concentration onto a graphite surface decorated with Au_{147} clusters. This height image clearly shows the presence of features ranging from about 4 nm in height (single GFP molecules) to 20 nm (multilayer GFP protein islands). In the same image, the presence of a roughly 5–10-nm high protein film surrounding the protein islands is also clearly

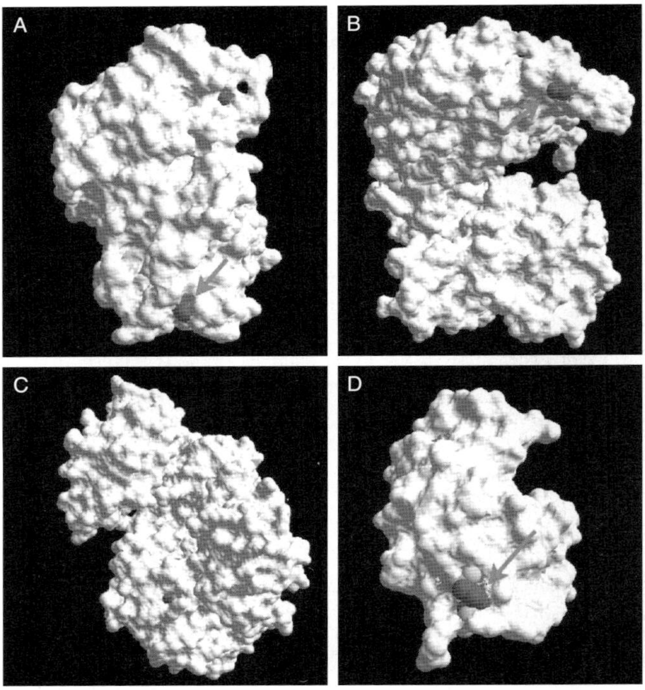

FIGURE 17 Composite molecular surface–van der Waals' surface of the proteins (not to scale) listed in Table 3. (A) Chain A of HRP with Cys 91(at the bottom) and Cys 177 lodged in a crevice. (B) Chain A of a GroEL ring showing Cys 519. (C) Luciferase with Cys 81 displayed. (D) Lysozyme with Cys 6 displayed. The side chains of the proteins are colored red, while the protein backbone is in yellow. Green arrows point to the most likely binding site for gold nanoclusters based on the MSA analysis. (For interpretation of the references to color in this figure legend, the reader is referred to the Web version of this chapter.)

visible. We also found that in phase imaging (Figure 18B) of the same area, the continuous GFP film exhibits an enhanced contrast when compared to the GFP islands. This GFP film is less stable than the GFP attached to the Au_{147} clusters, which suggests that the lateral interaction between GFP molecules is weaker than the GFP–Au nanocluster interaction. This is illustrated in Figure 18C, which was collected after about 1 h of continuous scanning. It is apparent that the protein film has been broken up by the action of the AFM tip scanning in tapping mode. GFP islands are now observed where the film was present, because GFP molecules from the film have relocated to vacant Au_{147} cluster sites. The corresponding phase image (Figure 18D) confirms that the scan area is essentially depleted of the loose GFP film. Furthermore, as illustrated in Figure 18E and F, on a bare graphite surface the formation and growth of GFP islands cannot be observed and thus can only be attributed to the presence of gold clusters.

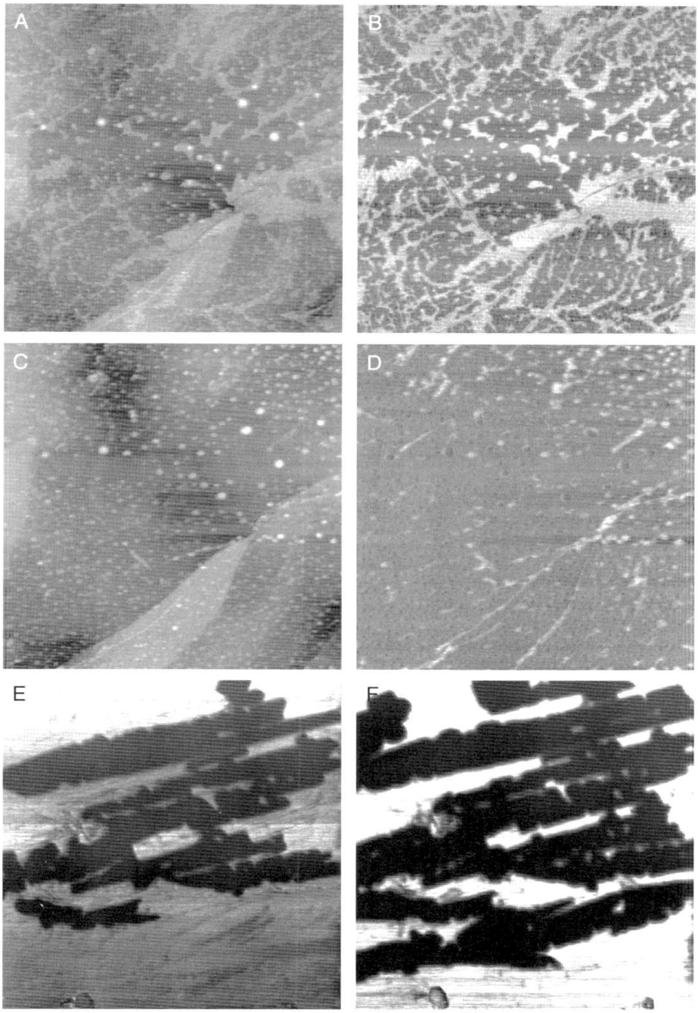

FIGURE 18 Protein film reorganization in the presence of Au clusters. (A) 8×8 μm tapping-mode imaging of GFP molecules on gold-nanocluster-decorated graphite surface AFM at the initial stages of protein island formation. (B) Phase image of the area in (A), with the lighter areas clearly showing the presence of the loose protein film over the surface. (C) The same region as in (A) after approximately 1 h of continuous scanning. The protein islands are much more prominent, while the amount of loose GFP film is far less than in (A). (D) The corresponding phase image from (C) displays a much lower overall density of the loose protein film (i.e., the lighter areas in the phase image) which appear to have been replaced by dark features corresponding to the GFP protein islands, as inferred from the height image. (E) Tapping-mode image of GFP molecules deposited on a bare graphite surface. The film is clearly visible on the HOPG, but continuous scanning in the same area after half an hour yields (F) where the protein film has clearly broken up, but no protein islands are formed in the absence of gold nanoclusters.

XVII. SUMMARY AND CONCLUSION

The immobilization of proteins by size-selected gold clusters has been investigated with six different proteins: the chaperonin GroEL, HRP, GFP, OSM, lysozyme, and firefly luciferase. The experimental studies have been performed by *in situ* imaging of the proteins using AFM and thereafter complemented by modeling of the protein structures to explain their behavior in the presence of the gold nanoclusters.

Some general rules can be drawn to predict whether a protein will bind to gold nanoclusters: (a) the protein needs to have at least one surface accessible cysteine residue and in particular a sulfur atom; (b) the sulfur atom should not point inward (as is the case with GFP); (c) the sulfur atom should have at least 15 $Å^2$ of exposed area and preferably lie on a convex surface; and (d) for multimeric proteins which may undergo large structural conformation changes (e.g., GroEL), the subunits also need to be analyzed separately. With regard to (c), the exact value of the molecular surface of the sulfur atom that needs to be exposed in the Cys residue before it can then interact with the gold nanoclusters is debatable. However, the experiments conducted to date suggest that those proteins that can bind to the clusters exhibit an MSA of 15 $Å^2$ or more.

The MSA approach is not perfect, however. A protein molecule in solution is continuously undergoing substantial fluctuations in the relative positions of its constituent atoms. Thus, the instantaneous accessible areas of individual atoms and residues will vary with time. These fluctuations may be crucial to biological and chemical reactivity and, consequently, this must be kept in mind when analyzing molecular structures obtained from XRD measurements in this way. Likewise, the possibility of protein denaturing cannot, in general, be ignored. While the evidence suggests that proteins such as GFP or HRP are very likely to retain their conformations on the cluster/graphite surface in the solution environment presented here, and in particular the general agreement between MSA prediction and the experimental measurements of immobilization, it is possible that other structurally "weaker" proteins may not.

In many state-of-the-art single-molecule measurements, an important requirement is to immobilize the molecule of interest to an appropriate surface. To avoid intermolecular interactions, and to achieve single-molecule resolution in the case of optical measurements, a dilute array of binding sites, each designed to immobilize no more than one protein molecule, would be ideal. Moreover, strong bonds (e.g., covalent bonds) will help immobilize the proteins on the surface and make them amenable to probing by a scanning AFM tip, even in contact mode. Thus, the surfaces discussed here consisting of size-selected gold clusters pinned onto graphite, enable proteins containing free cysteine residues, or disulfide bridges, to chemisorb directly to the bare gold cluster surface. It is to be noted that selection of gold clusters in the size range 1–100 atoms (i.e., <3 nm in diameter even for two-dimensional monolayer platelets) should ensure that typically only one protein can bind directly to the cluster surface. Size-selected clusters of the same or smaller size than the protein also present minimal contact between the gold and the protein, thereby minimizing the risk of protein denaturation as compared to extended gold films. The creation of tailored binding sites for protein chemisorption, based on cluster deposition on inert surfaces, therefore appears to represent an attractive route toward single-molecule science studies of protein morphology and interactions. Moreover, immobilization of proteins on solid surfaces plays an important role in several other fields of modern biology and biotechnology. In the development of biosensors, for instance, the oriented immobilization of biomolecules at interfaces can play a crucial role. To obtain highly sensitive sensing or array surfaces, it is necessary to present a receptor molecule so that the corresponding ligand can be bound without steric restrictions. In this respect, the cluster immobilization approach providing the promise of oriented, covalent immobilization of proteins on surfaces, exemplifies the multidisciplinary strategy of interfacing biology and physics at the nanoscale, both for basic studies in molecular biology and for applications such as protein microarrays.

REFERENCES

[1] Feynman RP. There's plenty of room at the bottom. Eng Sci 1960;23(5):22–36.
[2] Mrksich M, Whitesides GM. Patterning self-assembled monolayers using microcontact printing—a new technology for biosensors. Trends Biotechnol 1995;13(6):228–35.
[3] Eigler DM, Schweizer EK. Positioning single atoms with a scanning tunneling microscope. Nature 1990;344(6266):524–6.
[4] Kendrew JC, Bodo G, Dintzis HM, Parrish RG, Wyckoff HW, Philips DC. A three-dimensional model of the myoglobin molecule obtained by X-ray analysis. Nature 1958;181 (4610):662–6.
[5] Carroll SJ, Pratontep S, Streun M, Palmer RE, Hobday S, Smith R. Pinning of size-selected Ag clusters on graphite surfaces. J Chem Phys 2000;113(18):7723–7.
[6] Pratontep S, Carroll SJ, Xirouchaki C, Streun C, Palmer RE. Size-selected cluster beam source based on radio frequency magnetron plasma sputtering and gas condensation. Rev Sci Instrum 2005;76(4):045103.
[7] Palmer RE, Leung C. Immobilisation of proteins by atomic clusters on surfaces. Trends Biotechnol 2007;25(2):48–55.
[8] Carroll SJ, Seeger K, Palmer RE. Trapping of size-selected ag clusters at surface steps. Appl Phys Lett 1998;72(3):305–7.
[9] Pratontep S, Preece P, Xirouchaki C, Palmer RE, Sanz-Navarro CF, Kenny SD, et al. Scaling relations for implantation of size-selected Au, Ag, and Si clusters into graphite. Phys Rev Lett 2003;90(5):055503.
[10] Yin F, Xirouchaki C, Guo QM, Palmer RE. High-temperature stability of size-selected gold nanoclusters pinned on graphite. Adv Mater 2005;17(6):731–4.
[11] Di Vece M, Young NP, Li ZY, Chen Y, Palmer RE. Co-deposition of atomic clusters of different size and composition. Small 2006;2(11):1270–2.

[12] Palomba S, Palmer RE. Patterned films of size-selected Au clusters on optical substrates. J Appl Phys 2007;101(4):044304.

[13] Claeyssens F, Pratontep S, Xirouchaki C, Palmer RE. Immobilization of large size-selected silver clusters on graphite. Nanotechnology 2006;17(3):805–7.

[14] Gibilisco S, Di Vece M, Palomba S, Faraci G, Palmer RE. Pinning of size-selected Pd nanoclusters on graphite. J Chem Phys 2006;125(8):084704.

[15] Lodish H, Berk A, Zipursky SL, Matsudaira P, Baltimore D, Darnell J. In: Freeman, WH editor. Molecul cell biol. 4th ed. New York; 2000.

[16] Dobson CM. Protein folding and misfolding. Nature 2003;426(6968):884–90.

[17] Binnig G, Quate CF, Gerber C. Atomic force microscope. Phys Rev Lett 1986;56(9):930–3.

[18] Muller DJ, Fotiadis D, Scheuring S, Muller SA, Engel A. Electrostatically balanced subnanometer imaging of biological specimens by atomic force microscope. Biophys J 1999;76(2):1101–11.

[19] Butt HJ, Downing KH, Hansma PK. Imaging the membrane-protein bacteriorhodopsin with the atomic force microscope. Biophys J 1990;58(6):1473–80.

[20] Egger M, Ohnesorge F, Weisenhorn AL, Heyn SP, Drake B, Prater CB, et al. Wet lipid protein membranes imaged at submolecular resolution by atomic force microscopy. J Struct Biol 1990;103(1):89–94.

[21] Henderson E, Haydon PG, Sakaguchi DS. Actin filament dynamics in living glial-cells imaged by atomic force microscopy. Science 1992;257(5078):1944–6.

[22] Hoh JH, Lal R, John SA, Revel JP, Arnsdorf MF. Atomic force microscopy and dissection of gap-junctions. Science 1991;253(5026):1405–8.

[23] Weisenhorn AL, Gaub HE, Hansma HG, Sinsheimer RL, Kelderman GL, Hansma PK. Imaging single-stranded-DNA, antigen-antibody reaction and polymerized langmuir-blodgett-films with an atomic force microscope. Scan Microsc 1990;4(3):511–6.

[24] Santos NC, Castanho MARB. An overview of the biophysical applications of atomic force microscopy. Biophys Chem 2004;107(2):133–49.

[25] El Kirat K, Burton I, Dupres V, Dufrene YF. Sample preparation procedures for biological atomic force microscopy. J Microsc (Oxford) 2005;218:199–207.

[26] Viani MB, Pietrasanta LI, Thompson JB, Chand A, Gebeshuber IC, Kindt JH, et al. Probing protein-protein interactions in real time. Nat Struct Biol 2000;7(8):644–7.

[27] Schiener J, Witt S, Hayer-Hartl M, Guckenberger R. How to orient the functional groel-sri mutant for atomic force microscopy investigations. Biochem Biophys Res Commun 2005;328(2):477–83.

[28] Henderson E. Imaging and nanodissection of individual supercoiled plasmids by atomic force microscopy. Nucleic Acids Res 1992;20(3):445–7.

[29] Muller DJ, Baumeister W, Engel A. Controlled unzipping of a bacterial surface layer with atomic force microscopy. Proc Natl Acad Sci USA 1999;96(23):13170–4.

[30] Sagvolden G. Protein adhesion force dynamics and single adhesion events. Biophys J 1999;77(1):526–32.

[31] Gimzewski JK, Joachim C. Nanoscale science of single molecules using local probes. Science 1999;283(5408):1683–8.

[32] Rief M, Gautel M, Oesterhelt F, Fernandez JM, Gaub HE. Reversible unfolding of individual titin immunoglobulin domains by AFM. Science 1997;276(5315):1109–12.

[33] Marszalek PE, Lu H, Li HB, Carrion-Vazquez M, Oberhauser AF, Schulten K, et al. Mechanical unfolding intermediates in titin modules. Nature 1999;402(6757):100–3.

[34] Grandbois M, Beyer M, Rief M, Clausen-Schaumann H, Gaub HE. How strong is a covalent bond? Science 1999;283(5408):1727–30.

[35] Xu ZH, Horwich AL, Sigler PB. The crystal structure of the asymmetric groel-groes-(adp)(7) chaperonin complex. Nature 1997;388(6644):741–50.

[36] Berman HM, Westbrook J, Feng Z, Gilliland G, Bhat TN, Weissig H, et al. The protein data bank. Nucleic Acids Res 2000;28(1):235–42.

[37] Brayshaw DJ, Berry M, McMaster TJ. Optimisation of sample preparation methods for air imaging of ocular mucins by AFM. Ultramicroscopy 2003;97(1–4):289–96.

[38] Mou JX, Sheng ST, Ho RY, Shao ZF. Chaperonins groel and groes: views from atomic force microscopy. Biophys J 1996;71(4):2213–21.

[39] Marchin KL, Berrie CL. Conformational changes in the plasma protein fibrinogen upon adsorption to graphite and mica investigated by atomic force microscopy. Langmuir 2003;19(23):9883–8.

[40] Sarno DM, Murphy AV, DiVirgilio ES, Jones WE, Ben RN. Direct observation of antifreeze glycoprotein-fraction 8 on hydrophobic and hydrophilic interfaces using atomic force microscopy. Langmuir 2003;19(11):4740–4.

[41] Bizzarri AR, Bonanni B, Costantini G, Cannistraro S. A combined atomic force microscopy and molecular dynamics simulation study on a plastocyanin mutant chemisorbed on a gold surface. Chemphyschem 2003;4(11):1189–95.

[42] Hobara D, Uno Y, Kakiuchi T. Immobilization of horseradish peroxidase on nanometre-scale domains of binary self-assembled monolayers formed from dithiobis-n-succinimidyl propionate and 1-tetradecanethiol on Au(111). Phys Chem Chem Phys 2001;3(16):3437–41.

[43] Image SXM, University of Liverpool. Available at http://www.liv.ac.uk/~sdb/ImageSXM/. Accessed October 20, 2010.

[44] Gajhede M, Schuller DJ, Henriksen A, Smith AT, Poulos TL. Crystal structure of horseradish peroxidase c at 2.15 angstrom resolution. Nat Struct Biol 1997;4(12):1032–8.

[45] Palomba S, Beck M, Palmer RE.; unpublished data.

[46] Prisco U, Leung C, Xirouchaki C, Jones CH, Heath JK, Palmer RE. Residue-specific immobilisation of protein molecules by size-selected clusters. J R Soc Interface 2005;1(3):169–75.

[47] Ormo M, Cubitt AB, Kallio K, Gross LA, Tsien RY, Remington SJ. Crystal structure of the *Aequorea victoria* green fluorescent protein. Science 1996;273(5280):1392–5.

[48] Zimmer M. Green fluorescent protein (gfp): applications, structure, and related photophysical behavior. Chem Rev 2002;102(3):759–81.

[49] Economides AN, Ravetch JV, Yancopoulos GD, Stahl N. Designer cytokines—targeting actions to cells of choice. Science 1995;270(5240):1351–3.

[50] Robinson RC, Grey LM, Staunton D, Vankelecom H, Vernallis AB, Moreau JF, et al. The crystal-structure and biological function of leukemia inhibitory factor—implications for receptor-binding. Cell 1994;77(7):1101–16.

[51] Livingstone JR, Spolar RS, Record MT. Contribution to the thermodynamics of protein folding from the reduction in water-accessible nonpolar surface-area. Biochemistry 1991;30 (17):4237–44.

[52] Jackson RM, Sternberg MJE. A continuum model for protein-protein interactions—application to the docking problem. J Mol Biol 1995;250(2):258–75.

[53] Raschke TM, Tsai J, Levitt M. Quantification of the hydrophobic interaction by simulations of the aggregation of small hydrophobic solutes in water. Proc Natl Acad Sci USA 2001;98 (11):5965–9.

[54] Tsodikov OV, Record MT, Sergeev YV. Novel computer program for fast exact calculation of accessible and molecular surface areas and average surface curvature. J Comput Chem 2002;23(6):600–9.

[55] Andolfi L, Cannistraro S, Canters GW, Facci P, Ficca AG, Van Amsterdam IMC, et al. A poplar plastocyanin mutant suitable for adsorption onto gold surface via disulfide bridge. Arch Biochem Biophys 2002;399(1):81–8.

[56] Peluso P, Wilson DS, Do D, Tran H, Venkatasubbaiah M, Quincy D, et al. Optimizing antibody immobilization strategies for the construction of protein microarrays. Anal Biochem 2003;312(2):113–24.

[57] Weiss MS, Palm GJ, Hilgenfeld R. Crystallization, structure solution and refinement of hen egg-white lysozyme at pH 8.0 in the presence of MPD. Acta Crystallogr D Biol Crystallogr 2000;56:952–8.

[58] Conti E, Franks NP, Brick P. Crystal structure of firefly luciferase throws light on a superfamily of adenylate-forming enzymes. Structure 1996;4(3):287–98.

Chapter 15

Dynamics of Biomolecules From First Principles

Ivan M. Degtyarenko and Risto M. Nieminen
Laboratory of Physics, Helsinki University of Technology, P.O.B. 1100, Finland

Chapter Outline Head

I. INTRODUCTION

A. Molecular Modeling

Discovery of the deoxyribonucleic acid (DNA) in the early 1950s [1,2] has established the primary importance of molecular structure for the

Nanoclusters. DOI: 10.1016/S1875-4023(10)01015-6

understanding of the function of molecular systems and led to the revealing of the structures of proteins and enzymes. From that time, the structures of a very large number of proteins and other biological molecules have been determined experimentally [3,4]. The realization that the protein must be flexible in order to perform its biological functions, for example that of hemoglobin [5], indicated the need to investigate the dynamical properties of the proteins.

A number of experimental techniques have been developed. However, the information content from these studies of dynamics is generally averaging in nature, affording little insight into the atomic details of these fluctuations [6]. Atomic-resolution information of the dynamics of biomolecules [7], in their natural environment rather than in crystalline form, as well as the relationship of the dynamics to molecular functions and properties, is the field where computational studies can extend our knowledge beyond what is accessible to experimentalists. Molecular modeling is not only the way to mimic the behavior of molecular systems, but also the way to understand experiments which are difficult to interpret. Nowadays, molecular modeling is associated mostly with computer simulations. However, computers simply allow the higher level complexity problems to be solved.

B. Biological System Specificity

Biological systems are characterized by specificity and complexity, both in their structure and in chemical reactivity [8]. The origins of complexity are numerous: the size of the systems, the relatively long time of biological reactions and processes, the lack of systematic periodicity, the variety of intermolecular interactions which must be considered, and the need for careful treatment of the effects of the environment. The length scale varies from a few angströms of chemical bond length to the size of the earth's ecosystem. The time scale varies from femtoseconds of quantum level effects to the 10^9 years of evolution time.

Myoglobin is a representative example of a molecular structure which biophysics deals with every day (Figure 1). Myoglobin is a medium-sized hemoprotein consisting of a polypeptide chain of 151 amino acids, associated with a single heme group, expressed solely in cardiac myocytes and oxidative skeletal muscle fibres [9]. Its crystalline structure, investigated originally by John Kendrew et al. [10], contains in all 1260 atoms excluding hydrogen; in addition there are some 400 atoms of liquid and salt solution, a number of which are bound to fixed sites on the surface of the molecule. Myoglobin was called the hydrogen atom of biology due to its molecular properties [11].

This example shows the level of complexity of the biological problems. Experimental techniques allow these problems to be investigated to some degree. However, nowadays simulation tools are the key to understanding biology at the molecular scale and interpreting experimental results.

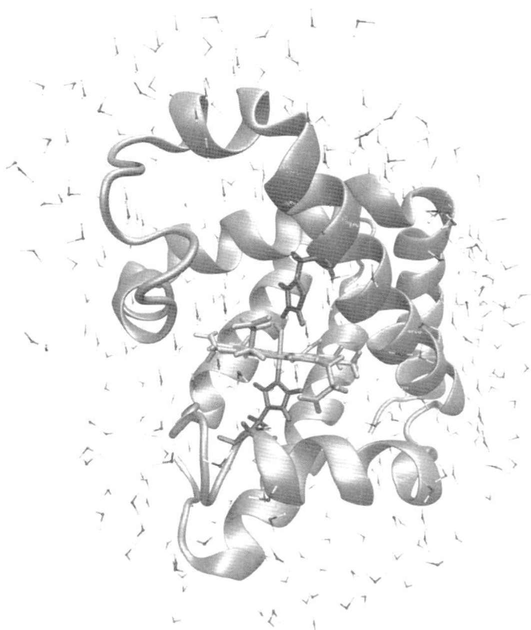

FIGURE 1 Myoglobin in a water environment. The active site, heme, the proximal and distal residues are highlighted. The backbone is shown as a tube, and the water molecules are in ball and stick representation.

C. Computational Methods in Molecular Modeling

The computational techniques demands are mainly dictated by two of the already mentioned factors of biomolecular systems complexity: the large molecular sizes and the long times of biological reactions and processes. To model the dynamics of such complex systems and thus overcome the length and time scales, one must employ sufficient computational techniques [12]. Nowadays, the force-field methods and density functional theory (DFT) are the main workhorses in biosimulations at the molecular mechanics (MM) and quantum mechanics (QM) [13] levels, respectively.

The parameters that enter the MM (or force-field) potential energy function are fitted to experimental or higher level computational data. The molecules are modeled as atoms held together by bonds and are basically described by a "ball and spring" model, and the quantum aspects are neglected. The methods are computationally "cheap" (compared to QM methods) and used for molecular dynamics (MD) simulations of large systems ($\sim 10^6$ atoms) [14] at long time scales (~ 500 μs) [15]. However, the methods suffer from several limitations, and the worst is neglecting quantum level effects. Thus the methods cannot be used to simulate reactions where covalent bonds are broken/formed. The force-field methods also require extensive parameterization and thus depend on the available experimental and QM calculation data. The energy estimates obtained are not accurate and the approach is limited in their abilities for providing accurate details regarding the chemical environment.

The step beyond the atomistic simulations is the coarse-grained methods. Here, rougher approximations are used. Instead of explicitly representing every atom of the system, one uses pseudo atoms to represent groups of atoms [16].

First-principles calculations (or *ab initio*, from Latin "from the beginning") are methods aimed at solving the electronic Schrödinger equation without reference to experimental data. *Ab initio* methods do not contain empirical parameters and are mathematically rigorous and thus accurate, but computationally expensive. DFT is the first-principles method [17] which deals with the electronic total density as opposed to electronic wavefunctions in the Hartree–Fock approach. [18] However, in case of DFT, the level of theory depends on the number of approximations used, and mainly on the exchange-correlation functional (*xc*-functional) choice [19]. The key size range at which the realistic dynamical properties of the molecular system can be studied is up to hundreds of atoms and time scale is up to tens of picoseconds. An example of such calculations is demonstrated in this work. DFT carefully addresses the electron correlation effect, which is critical for most biologically relevant problems. The method normally scales as N^3 and potentially allows static calculations of systems with thousands of atoms. The density functional method has become the method of choice when high accuracy in the treatment of conformational energetics and nonbonded interactions is required [20]. The approach is unique due to its appealing combination of computational efficiency and accuracy. While it does not yet uniformly achieve chemical accuracy (1 kcal/mol maximum error for heats of atomization), the performance of the best functionals with hybrid methods achieves an average error of 3.1 kcal/mol for several hundred heats of atomization, ionization potentials, electron affinities, and proton affinities over a wide range of small molecules [21].

Despite its obvious advantages, it is evident that a price has to be paid for putting MD on mechanical QM level: the simulation times and system sizes that are accessible are much smaller than what is affordable via MM methods. However, the fundamental advantages of the first-principles computational methods are the accuracy and the truly predictive power. Empirical methods allow seeing only what was foreseen, while *ab initio* methods will show what was not foreseen before starting the simulations.

It is a fact that the size range of hundreds and even thousands of atoms achievable by DFT is not enough for simulations of macromolecules. The chosen method in computational studies is always a compromise between efficiency and accuracy of simulations. To overcome the system size limitations, one should improve the underlying theory [22] or employ a number of approximations.

The way to improve the theory while still keeping it at the *ab initio* level is the use of linear-scaling techniques, or the so-called *O(N)* methods [23]. The new algorithms enable the calculations to be less CPU time and memory size consuming, and thus are essential tools for large system calculations. Another way is by using semiempirical methods. Semiempirical approaches are set up with the same general structure as a Hartree–Fock scheme but certain pieces of the Hamiltonian are approximated or simply omitted and some parameterization is used in order to avoid errors [24]. Semiempirical methods allow modeling of systems with tens of thousands of atoms.

Irrespective of the computational technique, the simplest approximation often used is making the studied system smaller by isolating those parts of the system that are believed to be important to the process under investigation. Such a mimetic model can be treated fully quantum mechanically. However, one must ensure that, when enlarging the system, the results of the calculation do not change significantly. A better approximation is to embed the fragment treated quantum mechanically in a larger system which is described by a classical MM approach. These QM/MM approaches allow the simulation of entire proteins including solvating water molecules [25,26]. The difficulty lies in achieving a good description of the interface between two parts, where covalent bonds may be cut. In solutions, the effect of solvation can be approximated by a dielectric continuum model, using self-consistent reaction field (SCRF) methods [27].

II. STRUCTURE OF L-ALANINE AMINO ACID

A. L-Alanine Amino Acid in Different Environments

As a study case, we demonstrate the simulations of a small biomolecule, the amino acid L-alanine. Amino acids are the building blocks of the proteins. L-alanine (LA) is the smallest naturally occurring chiral amino acid with a nonreactive hydrophobic methyl group ($-CH_3$) as a side chain (Figure 2). LA has the zwitterionic form ($^{+}NH_3-C_2H_4-COO^{-}$) both in crystal [28] and in aqueous solution over a large range of pH values [29]. Zwitterion is the most biologically relevant form of amino acids as it is their most abundant form in living organisms: they are synthesized in the zwitterionic form and are essentially always in this form at neutral pH. In contrast, in the gas phase, where interactions with environment are not present, amino acids are mostly in their neutral nonionic form [30].

B. L-Alanine Ionic Form Transformation

An isolated LA molecule is one of the representative examples of simulations in which the molecular system cannot be treated classically due to proton transfer. We performed a number of simulations, starting from different initial molecular conformations, but the results for all of them were the same. The conformations

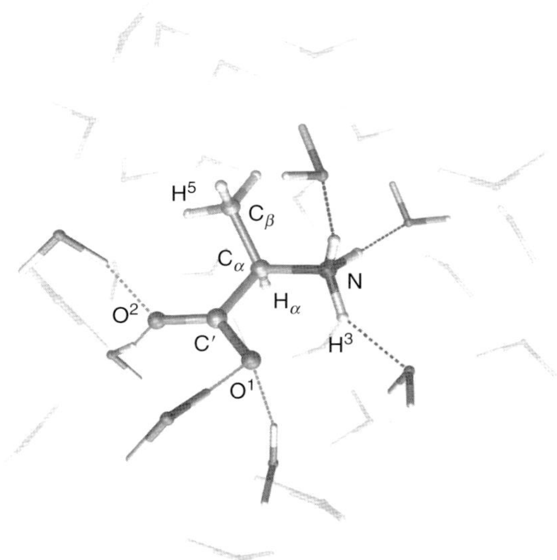

FIGURE 2 Ball and stick representation of L-alanine amino acid in water. Water molecules within the first hydration shell are shown in solid. Dotted lines indicate hydrogen bonding.

of the isolated LA zwitterion, in the absence of water molecules or neighbor amino acids as in the crystal, ended up being neutral nonionic. For all starting geometries, the LA amino acid adopted conformations in which the acidic carboxylate group received a proton from the basic ammonium group and the molecule became nonionic ($NH_3^+ \cdots COO^- \rightarrow NH_2 \cdots COOH$; Figure 3).

Additional simulations of nonionic fully hydrated LA in water have shown the reverse effect. The hydrated nonionic molecule takes one of the ionic forms depending on the environment pH.

Our simulations demonstrate the case where a H-atom transfer reaction effect cannot be omitted. The fact that amino acids are not zwitterions in the gas phase, but they are in water, implies that interaction with the water environment is a key determinant of the stable zwitterionic structure.

C. Experimental Studies on L-Alanine Zwitterion

The crystalline state is well defined structurally and can be successfully used for a detailed examination of a broad range of molecular properties. The low-temperature data were used for a detailed analysis of the electrostatic properties of the LA molecule [31,32]. X-ray diffraction [33], infrared spectroscopy [34], Raman scattering [35], and coherent inelastic neutron scattering [36] were employed to study the vibrational dynamics. The crystalline L-and D-alanine enantiomers were also used to investigate parity violation [37,38].

In contrast to the success in studying LA in the solid state, experimental studies by infrared, Raman and neutron diffraction spectroscopic techniques

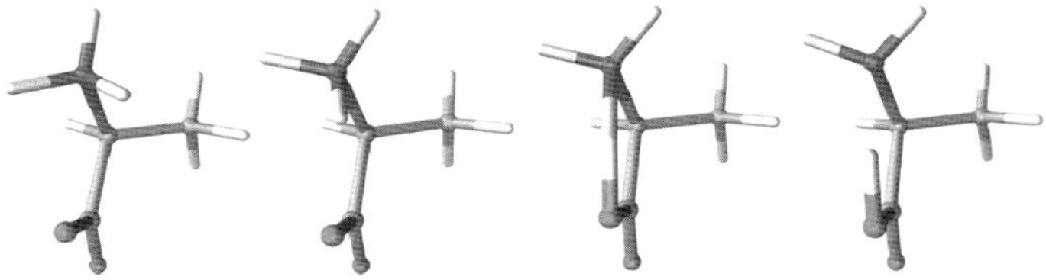

FIGURE 3 Evolution of the L-alanine zwitterion to the nonionic form in the gas-phase simulations.

have not yielded any conformational information about α-alanine in aqueous solution, having mainly focused on information concerning the structure of the hydration shell and the interactions between water molecules and amino acid residues [39–44]. The structural parameters of alanine molecules in water come predominantly from computational studies [45–47].

As a result of this lack of experimental information regarding the LA's structure in water, the experimental zwitterionic structure of the molecule, which is derived from solid-state crystallographic data, is often considered to be also valid for the molecule in aqueous media. However, the origin for the LA's zwitterionic form is different in crystal and in water. In crystals, all three available protons of the ammonium group (NH_3^+) are used to form single N–H···O hydrogen bonds with oxygen atoms of three carboxylate groups (COO^-) of nearest amino acid molecules, thus linking the molecules together to form a three-dimensional crystal structure. In contrast, the key determinant of the stable zwitterionic structure in aqueous solution is hydrogen-bonding interactions with surrounding water molecules (N–H···O_w and O···H_wO_w type hydrogen bonds).

Another reason that makes the LA study interesting is that above-mentioned experimental techniques bring ambiguous results. Hetch et al. observed correlations between solute-induced perturbations of the solvent structure and amino acid's hydrophobicity. Whereas later Ide et al. concluded, also on the basis of Raman spectroscopy, that the structure of water in solution of various amino acids at neutral pH does not depend on the nature of side chains. Recently, Kameda et al. obtained data on the number of water molecules inside the hydration shell and on the nearest neighbor distances by means of neutron diffraction with the isotopic substitution technique. They determined that the hydration number for the ammonium group differs for glycine and alanine, which are 3.0 and 2.4, respectively [48]. This means that the hydration structure of the amino acid depends on the type of molecule.

Molecular modeling, which would fully cover the aspects of the dynamics of hydrated alanine amino acids, can be used here in order to understand and explain the results of the experiments. Although first-principles MD simulations are computationally costly, they have several advantages over empirical-potential methods. The major advantage is that, by treating the system fully quantum mechanically, we are able to observe a proton transfer that can occur as has been demonstrated above and by other studies [49].

III. DYNAMICS OF THE L-ALANINE AMINO ACID

A. Initial Structures and Computational Methods

We have performed Born–Oppenheimer MD simulations of a LA in its zwitterionic form in an aqueous environment. To stabilize the structure of the LA zwitterion, the overall number of water molecules was chosen to be large enough to accommodate two solvation shells of the LA molecule. Since *ab initio* modeling of an amino acid surrounded by a significant amount of water molecules is computationally very expensive, it is crucial to have a good initial guess regarding the number of water molecules as well as their positions and orientations around the amino acid. Therefore, we first performed classical MD simulations of an LA molecule solvated in a box with large number of water molecules. Then, we extracted a well-equilibrated structure of LA with a limited number of water molecules nearest to the amino acid; all other water molecules were removed. The resulting structure was then used for DFT Born–Oppenheimer MD simulations. Overall, our simulations were performed in three steps.

(i) Classical atomic-scale MD simulations of LA in a box of water were first carried out with the use of an empirical force field. Force-field parameters for LA were taken from the full-atom Gromacs force field supplied with the GROMACS package (ffgmx2 set) [50]. A molecule of LA was solvated by around 500 water molecules; the simple point charge (SPC) model [51] was used to represent water. The MD simulations were performed in the N*p*T ensemble with temperature and pressure kept constant, using the Berendsen scheme [52]. The temperature was set to 300 K and the pressure was set to 1 bar. The Lennard–Jones interactions were cut off at 1 nm. For the long-range electrostatic interactions, we used the particle-mesh Ewald (PME) method [53]. The time step used was 1 fs and the total simulation time was 100 ps. The classical MD simulations were performed using the GROMACS suite.

(ii) The final structure of step (i) was used for preparing the initial structure for subsequent *ab initio* simulations. This was accomplished by removing the water molecules located farther than 6 Å from any atom of LA. Using this criterion only 50 water molecules were eventually left around the LA zwitterion. The resulting structure was then fully relaxed.

(iii) Finally, first-principles simulations were performed using a numerical atomic orbitals DFT approach as implemented in the SIESTA code [54]. The *ab initio* calculations were carried out within the generalized-gradient approximation, in particular with the Perdew–Burke–Ernzerhof (PBE) exchange-correlation functional [55], and a basis set of numerical atomic orbitals at the double-ζ polarized level [56]. The choice of the exchange-correlation PBE functional used in our study is based on its reliability in describing strong and moderate hydrogen bonds [57]. It is known to give accurate molecular bond lengths with a mean absolute error 0.012–0.014 Å [58,59]. Core electrons are replaced with norm-conserving pseudopotentials in their fully nonlocal representation [60]. The integrals of the self-consistent terms of the Kohn–Sham Hamiltonian were obtained using a regular real space grid to which the electron density was projected. A kinetic energy cutoff of 150 Ry was used for the MD run, which gave a spacing between the grid points of $\sim$0.13 Å. A cutoff of 300 Ry was used for structure optimization. The initial equilibration of the system was done at 300K; the thermostat was then switched off during production, so that the microcanonical ensemble was probed and the "flying ice cube" effect was avoided [61]. The DFT MD simulations were run for 40 ps with a time step of 1 fs; only the last 38 ps were used for the subsequent analysis. For visualization and trajectory analysis, the Visual Molecular Dynamics (VMD) package was employed [62]. The same methodology has successfully been used in recent studies of problems of biological relevance, including modeling of proteins, DNA, and liquid water [63].

B. Stable Zwitterion

The major result of the first-principles MD simulations of the fully hydrated LA is that the molecule is stable in its zwitterionic form. In the gas phase (isolated molecule) the protonation reaction takes place between carboxylate and ammonium group within 1 ps. The protonation reaction time window is 2–3 ps. In contrast, the molecular transformation from the zwitterion to the neutral form was not observed for the entire 38 ps of the production run of our simulations of LA in the water environment. Essentially, polar water molecules around the charged groups, $NH_3^+ \cdots COO^-$, stabilize the LA zwitterion, preventing a proton of the ammonium group from transferring to the carboxylate group. The fact that an alanine zwitterion is stable in the presence of water implies that interaction with solvent is the key determinant of its stability. Nevertheless, this trivial and always-assumed fact has never been shown by the first-principles MD studies.

C. Molecule Dynamics and Trajectory

The total trajectory time (38 ps) has been divided into three periods. The first 14-ps period is the time when the molecule remains fully hydrated, and all the sites, carboxylate (COO^-), ammonium (NH_3^+), and methyl (CH_3) stay inside the water droplet. The last 22-ps period is the time when the LA molecule is close to the water droplet surface. The carboxylate and ammonium ends remain surrounded by water molecules, while the methyl group and the α-hydrogen are exposed to the droplet surface. The 2-ps time period in between the fully hydrated and surface phases is a transient phase.

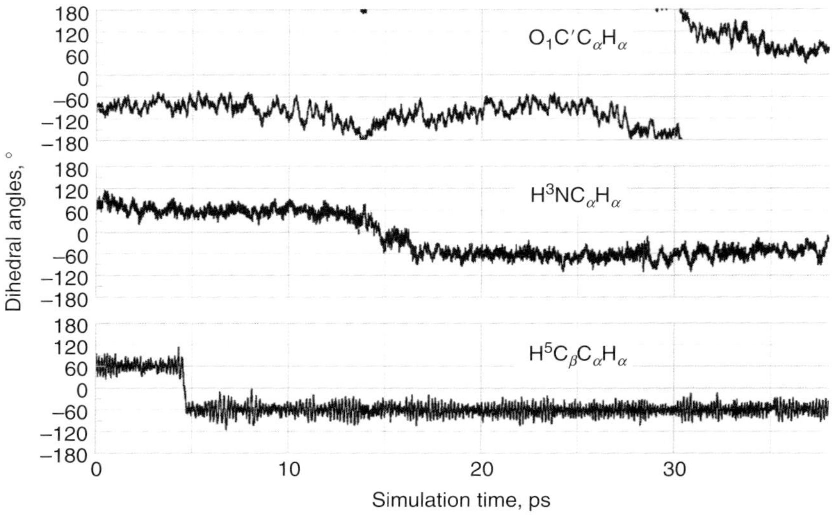

FIGURE 4 Evolution of the LA active sites related dihedral angles. $O^1C'C_{\square}H_{\square}$ attributes to the carboxylate group, $H^3NC\alpha H_{\alpha}$ to the ammonium group, and $H^5C_{\beta}C_{\alpha}H_{\alpha}$ to the methyl group. See labeling in Figure 2.

The course of the MD simulations has demonstrated that the hydrated LA zwitterion is a flexible molecule. We have observed the fluctuations of LA inside the water droplet as well as rapid rotational motions of the active sites, (COO^-, NH_3^+, and CH_3 groups) of the alanine (Figure 4).

Our simulations suggest that the LA zwitterion does not have any preferred conformation in aqueous solution at room temperature. The trajectory analysis shows that jump-like 120° rotational motions of the ammonium and the methyl group takes place during the first phase of the simulations when the amino acid molecule is fully hydrated. Such rotational motions were not observed during the second phase of the simulations when the molecule is only partially in water and methyl group and its hydrophobic part are exposed out of water. In contrast to the ammonium and methyl sties, the carboxylate group performs gradual rotations with 60° jumps occurring only once. The gradual rotational motions occur both in clockwise and counterclockwise directions. However, the ammonium and methyl groups prefer to stay in an almost staggered conformation with respect to H_{α} and in unstaggered (eclipsed) conformation with respect to each other.

The results of the simulations suggest that the structural properties of the LA molecule in aqueous solution can differ from those in the crystalline phase where the molecule can be considered as static. Overall, our findings strongly suggest that the generally accepted approach of extending the structural information acquired from crystallographic data to the LA molecule in aqueous solution should be used with caution.

The explanation to the high mobility of the LA molecule in water is that the whole system adjusts itself in order to minimize the free energy. LA is one of the aliphatic amino acids that has the important property of not interacting favorably with water. The aliphatic amino acids (alanine, valine, leucine and isoleucine) have methylene and methyl groups on their side chains. They interact more favorably with each other and with other nonpolar atoms than with the polar water molecules. It is one of the main factors in stabilizing the folded proteins. It is also the factor that defines the behavior of the molecules in an aqueous solution. Initially, the fully hydrated LA amino acid moves toward the water droplet surface. The methyl group's free solvation energy is positive. The entropy is large and negative since water is more ordered around the methyl group, while the enthalpy is small and also negative due to formation of additional hydrogen bonds between water molecules surrounding the methyl group.

Therefore, the system will try to minimize its free energy by pushing the hydrophobic side chain out of the water phase; this hydrophobic effect gives rise to the net force that defines the molecule's behavior inside the water droplet.

The Kohn–Sham energy plot in Figure 5 and dihedral angles plots in Figure 4 demonstrate that the molecule in water is flexible and adjusts its conformation in order to minimize the free energy. The Kohn-Sham energy plot justifies the existence of the above two parts of the simulation trajectory. The initially fully hydrated amino acid rotates and moves in order to minimize contacts of its hydrophobic methyl group and α-hydrogen with water. The energy of the first part steadily decreases as the molecule searches for the most favorable energetic state. The α-hydrogen is released first from the water droplet, with a corresponding energy drop at 14 ps. Then, shortly after the α-hydrogen release, the methyl group gets fully released at 16 ps. The energy peak at 17 ps corresponds to a partial α-hydrogen dipping back to the water shell, which is energetically costly. Eventually, the molecule adjusts itself in such a way that only the hydrophobic part is mainly out of the water droplet, while its carboxylate and ammonium charged groups remain fully in water.

The averaged system energy values (Figure 5) for the initial fully hydrated stage and after the hydrophobic part has been released differ by about 15 kcal/mol. Additional structure optimization calculations of two energy minimum states extracted from the MD trajectory show that the energy of the first minimum (when the alanine molecule is fully covered by water) can be up to 5 kcal/mol

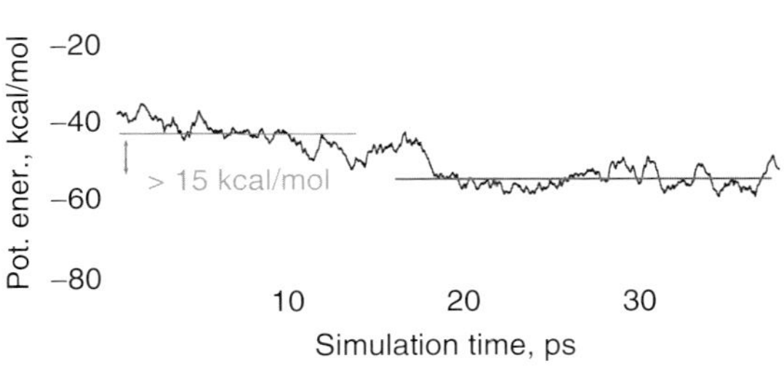

FIGURE 5 Kohn–Sham energy evolution during the course of the simulation. The running average is taken every 500 fs. The red lines indicate the average values for the "fully hydrated" and "at surface" phases.

higher than the second minimum (when the methyl group and α-hydrogen are exposed at the droplet's surface). Thus, the system state with hydrophobic groups exposed out of water is found to be the most energetically favorable.

D. The Hydration Shell Structure

The radial distribution functions (RDFs) for atoms of the ammonium and carboxylate groups and atoms of water molecules are shown in Figures 6 and 7. We have calculated the RDFs, for both simulation periods, for a fully hydrated molecule (first 14 ps) and for the second phase (last 22 ps) when the alanine hydrophobic part is exposed at the surface of a droplet (see the green and brown lines in Figures 6 and 7). In both cases, the high first peaks and low minima of the RDFs suggest a highly structured first solvation shell around the charged groups. The RDFs peaks indicate the average nearest neighbor and donor–acceptor distances, Figures 6 and 7, respectively. Water molecules around both charged sites take preferred orientations in which an oxygen atom of each water molecule near the ammonium group faces the ammonium-hydrogen atom(s) and each hydrogen atom of water molecules near the carboxylate group faces the carboxylate-oxygen atom(s).

The green and brown curves' first peaks and minima practically coincide. This means, that the LA amino acid, being a zwitterion, always has a highly ordered hydration shell around its charged ionic ends, whether or not its side chain is surrounded by water molecules.

Our results support the Raman spectroscopy study performed by Ide et al. on aqueous solutions of various amino acids (including LA), which suggested that perturbations of the structure of water due to interactions between amino

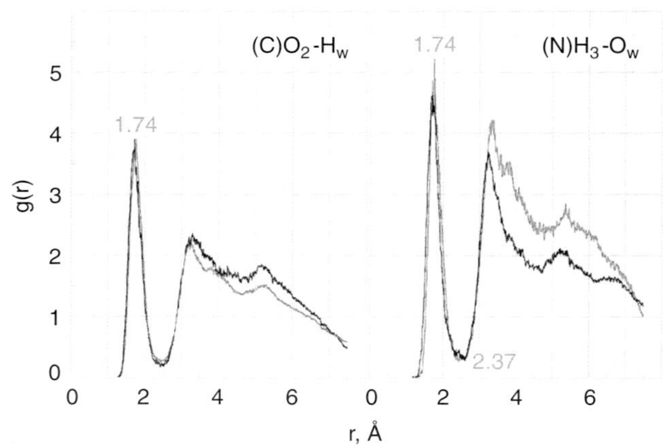

FIGURE 6 The nearest neighbor atom radial distribution functions. The hydrogen atoms of waters (H_w) relatively the oxygen atoms of CO_2^- group ($(C)O_2$-H_w)on the left and the oxygen waters relative to the hydrogen atoms of the NH_3^+ group ($(N)H_3$-O_w) on the right. The green line corresponds to the "fully hydrated" simulation period, and the brown line corresponds to the phase when the hydrophobic part is exposed to the surface.

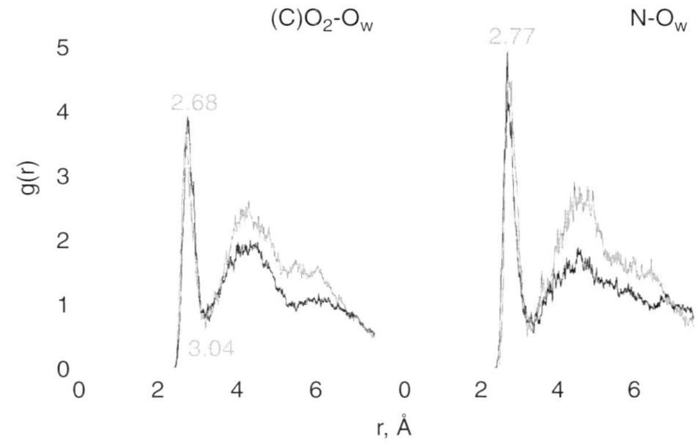

FIGURE 7 The donor–acceptor radial distribution functions. The hydrogen atoms of waters (H_w) relatively the oxygen atoms of CO_2^- group ((C)O_2-H_w) on the left and the oxygen waters relative to the hydrogen atoms of the NH_3^+ group ((N)H_3-O_w) on the right. Two periods of the simulation have shown, which are same as in Figure 6.

acids and water molecules were localized around the ammonium and carboxylate functional groups and that the structure of water was only slightly affected by a side chain.

E. Properties of the First Hydration Shell

We define the first hydration shell of LA as a set of water molecules explicitly interacting with the carboxylate and ammonium groups of the amino acid. A water molecule is defined to be inside of the first hydration shell if the distance between its oxygen atom and ammonium-hydrogen/carboxylate-oxygen atoms falls within the first peaks of the corresponding (N)H_3–O_w and (C) O_2–O_w RDFs (Figures 6 and 7). The first minima of these distributions are considered as the hydration radii, 2.37 and 3.04 Å for the ammonium and carboxylate groups respectively. Integration over the first peaks, up to hydration radius, yields the hydration numbers, that is, numbers of water molecules inside the hydration shells: 2.76 for the NH_3^+ group and 4.06 for the CO_2^- group. Our results are in good agreement with the recent neutron diffraction experimental data.

Thus, in total, the average number of water molecules within the first hydration shell of the LA zwitterion is close to 7.

To characterize the dynamic properties of water molecules we calculated mean-square displacements over the full trajectory for water molecules within the first hydration shell and in bulk. The diffusion coefficient of the water molecules in the first hydration shell is found to be about 2.2 times smaller than that of water molecules in bulk. These results suggest that water inside the first hydration shell is less mobile than water in bulk and is relatively rigid.

By employing a geometrical definition of the hydrogen bond, the donor–acceptor distance no larger than the corresponding hydration radius, and the angle donor–H–acceptor in the range 120–180°, we found that a water molecule inside the first hydration shells of the corresponding charged groups is hydrogen-bonded on average for 96.5% and 98.2% of full simulation time for NH_3^+ and CO_2^-, respectively.

IV. SUMMARY AND CONCLUSIONS

Nowadays, depending on the accuracy required and the origin of the problem, one can obtain precise atomization energy for a methane molecule or perform long-timescale MD simulations for systems containing millions of atoms. Undoubtedly, computer simulation methods have the predictive power and nowadays play a key role in understanding biological processes. Although many of these processes involve long time and length scales and therefore can be handled only by simulations at the empirical-potential level, first-principles simulations are very much needed in this area. Despite its obvious advantages, it is evident that a price has to be paid for putting MD on a QM level: the simulation times and system sizes that are accessible are much smaller than what is affordable via MM methods. However the fundamental advantages of the first-principle computational methods are their accuracy and the truly predictive power. Empirical methods allow us to see only what was foreseen, while *ab initio* methods will show what was not foreseen before starting the simulations.

As a study case, we demonstrated the simulations of a small biomolecule LA. The crystalline structure of the LA zwitterion has been known since the early works by Bernal [64] and Levy and Corey [65]. In great contrast, as far as the zwitterionic form of LA in aqueous solution is concerned, its conformational properties have remained poorly understood for almost 70 years. As a consequence, it was generally accepted that the structure of the LA zwitterion obtained from the crystalline phase could also be applied to the molecule in aqueous solution.

In this study we have employed first-principles computer simulations to predict the structure of the LA amino acid in water at room temperature. We used a density-functional linear scaling approach that allowed us to study relatively large molecular systems at the *ab initio* level of theory. Using this approach, we were able to treat fully quantum mechanically the structure of crystalline LA as well as LA solvated in 50 water molecules.

The overall goal of the work was to answer open questions not answered or emerging form experimental data and to provide information for a better understanding of the selected system. In order to apply *ab initio* calculations to large systems, such as biological systems, it is necessary to strike a balance between accuracy and computational efficiency. The work presented here is an example of how QM techniques can be successfully applied to biologically relevant problems in rather large and complex systems.

Acknowledgments

This work was supported by the Academy of Finland (Center of Excellence Grant 2006-2011). The computer resources were provided by the Laboratory of Physics in Helsinki University of Technology (M-grid project).

REFERENCES

[1] Olby R. The path to the double helix: discovery of DNA. London: MacMillan; 1974.
[2] Campbell N, Reece J. Biology. 6th ed. San Francisco: Benjamin Cummings; 2002.
[3] Creighton TE. Proteins: structure and molecular properties. 2nd ed. New York: Freeman; 1993.
[4] There are currently over 40,000 biological macromolecule structures available, including proteins, nucleic acids, and protein-nucleic acid complexes (http://www.pdb.org/). Berman HM, Westbrook HM, Feng Z, Gilliland G, Bhat TN, Weissig H, et al. Nucleic Acid Res 2000;28(1):235–42.
[5] Perutz M, Fermi G, Luisi B, Shaanan B, Liddington RC. Acc Chem Res 1987;20(9):309–21.
[6] Serdyuk IN, Zaccai NR, Zaccai J. Methods in molecular biophysics. New York: Cambridge; 2006.
[7] Karplus M, Petsko GA. Nature 1990;347(18):631–9.
[8] Frauenfelder H, Wolynes PG, Austin RH. Rev Mod Phys 1999;71(2):S419–30.
[9] Ordway GA, Garry DJ. J Exp Biol 2004;207:3441–6.
[10] Kendrew JC, Bodo G, Dunitz HM, Parrish RG, Wyckoff H, Phillips DC. Nature 1958;181:662–6.
[11] Kendrew JC. Nobel lectures, Chemistry 1942–1962. Amsterdam: Elsevier; 1964.
[12] Program code parallelization and growing computer power also make larger simulations possible.
[13] Segall MD. J Phys: Condens Matter 2002;14:2957–73.
[14] (a) Freddolino PL, Arkhipov AS, Larson SB, McPherson A, Schulten K. Structure 2006;14:437–49. (b) Sanbonmatsu KY, Joseph S, Tung C-S. Proc Natl Acad Sci USA 2005;102:15854–9.
[15] Jayachandran G, Vishal V, Garcia AE, Pande VS. J Struct Biol 2007;157(3):491–9.
[16] Girard S, Müller-Plathe F. In: Karttunen M, Vattulainen I, Lukkarinen A, editors. Novel methods in soft matter simulations. Berlin: Springer-Verlag; 2004. p. 327–54.
[17] Kohn W. Rev Mod Phys 1999;71:1253–66 and references therein.
[18] Szabo A, Ostlund NS. Modern quantum chemistry. Introduction to advanced electronic structure theory. New York: Dover; 1996.
[19] Koch W, Holthausen MC. A chemist's guide to density functional theory. 2nd ed. Weinheim: Wiley; 2000.
[20] However, one notable exception being the dispersion forces. Dobson JF, McLennan K, Rubio J, Wang J, Gould T, Le HM, Dinte BP. Aust J Chem 2001;54:513–27.
[21] Curtis L, Redfern PC, Raghavachari K. J Chem Phys 2005;123:124107–19 and references therein.
[22] Friesner RA. Proc Natl Acad Sci USA 2005;102(19):6648–53.
[23] Goedecker S. Rev Mod Phys 1999;71(4):1085–123.
[24] Jensen F. Introduction to computational chemistry. Chichester: Wiley; 2001. Ch 3, p. 81–97.

[25] Lyne PD, Walsh OA. In: Becker OM, MacKerell Jr. AD, Roux B, Watanabe M, editors. Computational biochemistry and biophysics. New York: Dekker; 2001. p. 221–36.

[26] Norberg J, Nilsson L. Q Rev Biophys 2003;36(3):257–306.

[27] Besides including the solvent molecules explicitly, one can also employ a reaction field continuum model, or use both approaches and model the solute molecule in its first hydration shell surrounding this "supermolecule" with continuum solvent. Tomasi J, Mennucci B, Cammi R. Chem Rev 2005;105:2999–3093 and references therein.

[28] (a) The crystal structure of alanine was studied by both X-ray (at T=298 K and T=23 K) (a, b) and neutron diffraction (at T=298 K) (c) Simpson HJ, Marsh RE. Acta Cryst 1966;20:550. (b) Destro R, Marsh RE, Bianchi R. J Phys Chem 1988;92:966. (c) Lehmann TF, Koetzle TF, Hamilton WC. J Am Chem Soc 1972;94(8):2657.

[29] The actual pH range is 3.5–8.0. Above 3.5 the carboxylic acid groups of 19 amino acids and 1 imino acid of proteins are almost entirely in their carboxylate forms and below 8.0 the amino groups are in their ammonium ion forms.

[30] (a) Blanco S, Lessari A, Lopez JC, Alonso JL. J Am Chem Soc 2004;126:11675–83. (b) Császár AG. J Phys Chem 1996;100:3541–51 and references therein.

[31] Destro R, Bianchi R, Gatti C, Merati F. Chem Phys Lett 1991;186(1):47.

[32] Destro R, Bianchi R, Morosi G. J Phys Chem 1989;93:447.

[33] Barthes M, Bordallo HN, Denoyer F, Lorenzo J-E, Zaccaro J, Robert A, Zontone F. Eur Phys J B 2004;37:375.

[34] Bandekar J, Genzel L, Kremer F, Santo L. Spectrochim Acta A 1983;39(4):357.

[35] Migliori A, Maxton PM, Clogston AM, Zirngiebl E, Lowe M. Phys Rev B 1988;38(18):13464.

[36] Micu AM, Durand D, Quilichini M, Field MJ, Smith JC. J Phys Chem 1995;99:5645.

[37] (a) Wang W, Yi F, Ni Y, Zhao Z, Jin X, Tang Y. J Biol Phys 2000;26:51. (b) Wang W-Q, Min Z, Liang Z, Wang LY, Chen L, Deng F. Biophys Chem 2003;103:289.

[38] Wilson CC, Myles D, Ghosh M, Johnson LN, Wang W. New J Chem 2005;29:1318.

[39] Ide M, Maeda Y, Kitano H. J Phys Chem B 1997;101:7022.

[40] Kameda Y, Sugawara K, Usuki T, Uemura O. Bull Chem Soc Jpn 2003;76:935.

[41] Kameda Y, Sasaki M, Yaegashi M, Tsuji K, Oomori S, Hino S, Usuki T. J Solution Chem 2004;33(6–7):733.

[42] Ellzy MW, Jensen JO, Hameka HF, Kay JG. Spectrochim Acta A 2003;59(11):2619.

[43] Hetch D, Tadesse L, Walters L. J Am Chem Soc 1993;115:3336.

[44] Fischer WB, Eysel H-H. J Mol Struct 1997;415:249.

[45] Jalkanen KJ, Degtyarenko IM, Nieminen RM, Nafie LA, Zhu F, Barron LD. Theor Chem Acc 2008;119(1–3):191–210.

[46] Tajkhorshid E, Jalkanen KJ, Suhai S. J Phys Chem B 1998;102:5899.

[47] Frimand K, Bohr H, Jalkanen KJ, Suhai S. Chem Phys 2000;255:165.

[48] Kameda Y, Ebata H, Usuki T, Uemura O, Misawa M. Bull Chem Soc Jpn 1994;67:3159–64.

[49] Park S-W, Ahn D-S, Lee S. Chem Phys Lett 2003;371:74–9.

[50] Lindahl E, Hess B, van der Spoel DJ. Mol Model 2001;7:306.

[51] Berendsen HJC, Postma JP, van Gunsteren WF, Hermans J. In: Pullman B, editor. Intermolecular forces. Dordrecht: Reidel; 1981.

[52] Berendsen HJC, Postma JP, van Gunsteren WF, DiNola A, Haak JR. J Chem Phys 1984;81:3684.

[53] (a) Darden T, York D, Pedersen L. J Chem Phys 1993;98:10089. (b) Essman U, Perera L, Berkowitz T, Darden T, Lee H, Pedersen LG. J Chem Phys 1995;103:8577.

[54] (a) Soler JM, Artacho E, Gale JD, Gárcia A, Junquera J, Ordejón P, Sánchez-Portal D. J Phys Condens Matter 2002;14:2745. (b) Ordejón P, Artacho E, Soler JM. Phys Rev B 1996;53:10441.
[55] Perdew JP, Burke K, Ernzerhof M. Phys Rev Lett 1996;77:3865.
[56] Junquera J, Paz O, Sánchez-Portal D, Artacho E. Phys Rev B 2001;64:235111.
[57] Ireta J, Neugebauer J, Schefler M. J Phys Chem A 2004;108:5692.
[58] Ernzerhof M, Scuseria G. J Chem Phys 1999;110(11):5029.
[59] Adamo C, Barone V. J Chem Phys 1999;110(13):6158.
[60] (a) Troullier N, Martins JL. Phys Rev B 1991;43:1993. (b) Kleinman L, Bylander DM. Phys Rev Lett 1982;48:1425.
[61] Harvey SC, Tan RK-Z, Cheatham III TE. J Comp Chem 1998;19(7):726.
[62] Humphrey W, Dalke A, Schulten K. J Mol Graphics 1996;14:33.
[63] (a) Sánchez-Portal D, Ordejón P, Canadell E. Struct Bond 2004;113:103 and references therein. (b) Degtyarenko I, Nieminen RM, Rovira C. Biophys J 2006;91(6):2024.
[64] Bernal JD. Z Kristallogr 1931;78:363.
[65] Levy HA, Corey RB. J Am Chem Soc 1941;63:2095.

Index